TRENDS IN MATHEMATICS

Trends in Mathematics is a series devoted to the publication of volumes arising from conferences and lecture series focusing on a particular topic from any area of mathematics. Its aim is to make current developments available to the community as rapidly as possible without compromise to quality and to archive these for reference.

Proposals for volumes can be sent to the Mathematics Editor at either

Birkhäuser Verlag
P.O. Box 133
CH-4010 Basel
Switzerland

or

Birkhäuser Boston Inc.
675 Massachusetts Avenue
Cambridge, MA 02139
USA

Material submitted for publication must be screened and prepared as follows:

All contributions should undergo a reviewing process similar to that carried out by journals and be checked for correct use of language which, as a rule, is English. Articles without proofs, or which do not contain any significantly new results, should be rejected. High quality survey papers, however, are welcome.

We expect the organizers to deliver manuscripts in a form that is essentially ready for direct reproduction. Any version of $\TeX$ is acceptable, but the entire collection of files must be in one particular dialect of $\TeX$ and unified according to simple instructions available from Birkhäuser.

Furthermore, in order to guarantee the timely appearance of the proceedings it is essential that the final version of the entire material be submitted no later than one year after the conference. The total number of pages should not exceed 350.

Number Theory

R. P. Bambah
V. C. Dumir
R. J. Hans-Gill
Editors

Birkhäuser Verlag
Basel · Boston · Berlin

Editors' address:

Centre for Advanced Study in Mathematics
Panjab University
Chandigarh 160 012
India

1991 Mathematical Subject Classification 11-02

A CIP catalogue record for this book is available from the
Library of Congress, Washington D.C., USA

Deutsche Bibliothek Cataloging-in-Publication Data

Number theory / R. P. Bambah ... ed. - Basel ; Boston ; Berlin : Birkhäuser, 2000
(Trends in mathematics)
ISBN 978-3-0348-7025-2 ISBN 978-3-0348-7023-8 (eBook)
DOI 10.1007/978-3-0348-7023-8

ISBN 978-3-0348-7025-2

Softcover reprint of the hardcover 1st edition 2000
Authorized edition by Birkhäuser Verlag, P.O. Box 133, CH-4010 Basel, Switzerland,
for exclusive distribution worldwide except India

ISBN 978-3-0348-7025-2

9 8 7 6 5 4 3 2 1

Contents

Preface

The Indian National Science Academy on the occasion of the Golden Jubilee Celebration (Fifty years of India's Independence) decided to publish a number of monographs on the selected fields. The editorial board of INSA invited us to prepare a special monograph in Number Theory. In reponse to this assignment, we invited several eminent Number Theorists to contribute expository/research articles for this monograph on Number Theory. Although some of those invited, due to other preoccupations could not respond positively to our invitation, we did receive fairly encouraging response from many eminent and creative number theorists throughout the world. These articles are presented herewith in a logical order.

We are grateful to all those mathematicians who have sent us their articles. We hope that this monograph will have a significant impact on further development in this subject.

R. P. Bambah
V. C. Dumir
R. J. Hans-Gill

A Centennial History of the Prime Number Theorem

Tom M. Apostol

The Prime Number Theorem

Among the thousands of discoveries made by mathematicians over the centuries, some stand out as significant landmarks. One of these is the **prime number theorem**, which describes the asymptotic distribution of prime numbers. It can be stated in various equivalent forms, two of which are:

$$\pi(x) \sim \frac{x}{\log x} \text{ as } x \longrightarrow \infty, \tag{1}$$

and

$$p_n \sim n \log n \text{ as } n \longrightarrow \infty. \tag{2}$$

In (1), $\pi(x)$ denotes the number of primes $p \leq x$ for any $x > 0$. In (2), p_n is the nth prime number when they are listed in increasing order; thus, $p_1 = 2$, $p_2 = 3$, $p_3 = 5, \ldots$.

Interest in primes goes back to Euclid's *Elements* (Book IX, Proposition 20) where it is shown that there are infinitely many primes. Euclid's theorem implies that $\pi(x) \rightarrow \infty$ as $x \rightarrow \infty$, and that $p_n \rightarrow \infty$ as $n \rightarrow \infty$.

The behaviour of $\pi(x)$ as a function of x, and of p_n as a function of n, have been the object of intense study by many celebrated mathematicians. Around 1791 Gauss (at age 14) conjectured that $\pi(x)$ is approximately equal to the integral $\int_2^x dt/\log t$. This is now called the logarithmic integral and is denoted by Li(x). Except for numerical evidence, Gauss provided no argument to support his conjecture.

The first proof of the prime number theorem came a century later, in 1896, the culmination of efforts by many talented individuals. This paper tells the story of this proof, which has several heros. First, there is **Euclid** himself, who started it all with his theorem on the infinitude of the primes. Then we have **Euler, Gauss, Legendre, Dirichlet, Chebyshev, Riemann, von Mangoldt, Hadamard,** and **de la Vallée Poussin**. As the story unfolds, it reveals how each contributed a significant stepping stone to ultimate success.

Euler's Contribution

In 1737, Euler was the first to introduce methods of analysis to study primes when he proved Euclid's theorem on the infinitude of primes by showing that the infinite series

$$\sum_p \frac{1}{p}$$

(extended over all primes p) is divergent. Every calculus student learns that the harmonic series $\sum_{n=1}^{\infty} \frac{1}{n}$ diverges. If you discard all the composite numbers in the harmonic series there are still enough terms left over to make the series diverge. Euler's proof was based on the formal identity

$$\sum_{n=1}^{\infty} \frac{1}{n^s} = \prod_{p} \frac{1}{1-p^{-s}}, \tag{3}$$

which he established for real $s > 1$. The infinite product on the right is extended over all primes p. It's not hard to see where this identity comes from. The key is our old friend the geometric series:

$$\frac{1}{1-x} = 1 + x + x^2 + x^3 + \cdots,$$

which is also familiar to every student of calculus. The series converges whenever $|x| < 1$. Take $x = \frac{1}{p^s}$ and expand each factor in the infinite product as a geometric series to get

$$\frac{1}{1-p^{-s}} = 1 + \frac{1}{p^s} + \frac{1}{p^{2s}} + \cdots.$$

Note that each denominator on the right is a prime power p^m raised to the power s. When you multiply all these series together and rearrange the terms according to increasing denominators you end up with the series $\sum_{n=1}^{\infty} \frac{1}{n^s}$ because of the fundamental theorem of arithmetic, which states that every integer n greater than 1 can be factored in one and only one way as a product of prime powers, apart from the order of the factors. Because we are multiplying an infinite number of infinite series, there are some delicate questions of convergence involved. Euler ignored these questions but the steps can be justified.

Euler's infinite product identity can be regarded as the analytic equivalent of the fundamental theorem of arithmetic, and it forms the basis for nearly all subsequent work on the distribution of primes.

The Conjectures of Gauss and Legendre

It seems that Gauss[5] and Legendre[8] were the first to consider the distribution of primes. In 1791, at age 14, Gauss examined a table of primes compiled by Lambert. Gauss counted the primes in blocks of 100, 1000, and 10,000 consecutive integers and made a note in his diary that the function $1/\log x$ was a good approximation to the average density of distribution (number of primes per unit interval). A few years later in 1796, when Vega published an extended table of primes up to 400,031, Gauss substantiated his hypothesis further, and he kept returning to this work as new tables of primes appeared. Many years later, in 1849, he communicated his observations in a letter to the astronomer Encke[6], and the results were published posthumously in 1862. (Gauss died in 1855.) Based on tables listing primes up to 3 million, Gauss observed that $\pi(x)$ is closely approximated by the integral of the density function.

Here are some excerpts adapted from his letter to Encke. This table shows $\pi(x)$ and $\mathrm{Li}(x)$ for various x between $\frac{1}{2}$ million and 3 million.

x	$\pi(x)$	LI(x)	% error
500,000	41,556	41,604.4	.12
1,000,000	78,501	78,627.5	.16
1,500,000	114,112	114,263.1	.13
2,000,000	148,883	149,054.8	.11
2,500,000	183,016	183,245.0	.12
3,000,000	216,745	216,970.6	.10

The agreement between $\pi(x)$ and Li(x) is certainly striking. The error in each of these approximations is only about one-tenth of one percent.

Legendre[8] also considered this question in the 2nd edition of his number theory text, published in 1808. A page in this book lists approximations for $\pi(x)$ for various x up to a million. Legendre asserts that $\pi(x)$ is closely approximated by the quotient

$$\frac{x}{\log x - 1.08366}.$$

On a later page he states that

$$\pi(x) \approx \frac{x}{\log x - A(x)},$$

where $A(x)$ is a function of x that approaches 1.08366 as $x \to \infty$.

Neither Gauss nor Legendre revealed how he arrived at the appearance of the natural logarithm in their formulas. Nor did they make any explicit statement about how good these approximations were thought to be outside the range of the tables. It is generally understood that both intended to imply that the ratio of $\pi(x)$ to each approximating formula on the right tends to the limit 1 as x tends to infinity.

Integration by parts shows that the value of Gauss's integral is approximately $x/\log x$, so both Gauss's and Legendre's relations are equivalent to the simpler asymptotic relation

$$\pi(x) \sim \frac{x}{\log x} \text{ as } x \longrightarrow \infty,$$

which means

$$\frac{\pi(x)}{\left(\frac{x}{\log x}\right)} \longrightarrow 1 \text{ as } x \longrightarrow \infty.$$

This result is now known as **the prime number theorem**, one of the most astonishing results in all of mathematics. It describes a simple relation between primes and the natural logarithm function which, at first glance, has nothing to do with prime numbers.

How to Conjecture the Prime Number Theorem by Examining a Table of Primes!

It is natural to ask why Gauss and Legendre used the natural logarithm in their formulas. They did not leave any written clues; they simply recorded their formulas and the supporting data. Let's see how one might be led to conjecture the prime number theorem by examining a table of primes. Here are some entries from a table of values of $\pi(x)$:-

x	10^2	10^4	10^6	10^8	10^{10}	10^{12}	10^{14}
$\pi(x)$	25	1,229	78,498	5,761,455	455,052,512	37,607,912,018	3,204,941,750,802

We have listed $\pi(x)$ for successive powers of 10 increasing by a factor of 10^2. Gauss had access to tables that only went up to 3 million. We have added the last four entries from more modern tables. What can we learn by looking at these numbers? Since we want to find how fast $\pi(x)$ grows with x, it is natural to look at the ratio $x/\pi(x)$ that compares the two quantities. This table shows the corresponding ratios:-

x	10^2	10^4	10^6	10^8	10^{10}	10^{12}	10^{14}
$\pi(x)$	25	1,229	78,498	5,761,455	455,052,512	37,607,912,018	3,204,941,750,802
$\frac{x}{\pi(x)}$	4.000	8.137	12.739	17.357	21.975	26.590	31.202

Notice the gaps between successive entries in this last row of numbers:

gaps: 4.137 4.602 4.618 4.618 4.615 4.612

As each exponent of 10 increases by 2, the ratio $\frac{x}{\pi(x)}$ increases by an almost constant amount, 4.6, which is 2.3 times the change in the exponent of 10. But if x is expressed as a power of 10, then the exponent of x is the logarithm of x to the base 10. So the table indicates that the change in the ratio $\frac{x}{\pi(x)}$ is approximately 2.3 times the change in $\log_{10} x$. What about this strange factor 2.3? A bright fourteen-year old such as Gauss would immediately realize that 2.3 is very nearly the logarithm of 10 to the base e, so

$$2.3 \log_{10} x = (\log_e 10)(\log_{10} x) = \log_e x.$$

This suggests that we compare the ratio $x/\pi(x)$ with $\log x$ (the natural logarithm of x). Our table now looks like this:-

x	10^2	10^4	10^6	10^8	10^{10}	10^{12}	10^{14}
$\pi(x)$	25	1,229	78,498	5,761,455	455,052,512	37,607,912,018	3,204,941,750,802
$\frac{x}{\pi(x)}$	4.000	8.137	12.739	17.357	21.975	26.590	31.202
$\log x$	4.605	9.210	13.816	18.421	23.026	27.361	32.236
$\frac{\log x}{x/\pi(x)}$	1.151	1.132	1.085	1.061	1.048	1.039	1.033

Anyone who looks at this last row of numbers would surely be tempted to conjecture that they approach 1 as $x \to \infty$. Gauss, Legendre, and many other eminent mathematicians of the early 19th century apparently thought so, but they were not able to prove it. As far as we know, neither Gauss nor Legendre made any significant progress toward a proof.

Legendre's Conjecture on Primes in Arithmetical Progressions: Dirichlet's Contribution

In the 1808 edition of his number theory book, Legendre formulated another conjecture on prime numbers that plays a role in this story. It has to do with primes in arithmetic progressions. An arithmetic progression of integers with first term h and common difference k consists of all numbers of the form $kn + h$ as n runs through all the nonnegative integers $0, 1, 2, \ldots$. If h and k have a common prime factor p then each term of the progression is divisible by p and there can be no more than one prime in the progression. Legendre asked whether there must be infinitely many primes in the progression if h and k have no common prime factor.

In a celebrated memoir published in 1837, **Dirichlet**[4] showed that every arithmetic progression $kn + h$, where h and k have no prime factor in common, must contain infinitely many primes, a result now known as **Dirichlet's theorem on primes in arithmetical progressions**. Inspired by Euler's proof of the infinitude of primes, Dirichlet used an ingenious argument to show that the sum of the reciprocals of all the primes in the progression $kn + h$ diverges. He did this by establishing the following asymptotic formula:

$$\sum_{\substack{p \leq x \\ p \equiv h \pmod{k}}} \frac{1}{p} \sim \frac{1}{\varphi(k)} \log(\log x) \text{ as } x \longrightarrow \infty.$$

This states that the partial sums of the series containing the reciprocals of all primes $p \leq x$ in the progression has the asymptotic value $\frac{1}{\varphi(k)} \log(\log x)$, where $\varphi(k)$ is the Euler φ function (the number of integers from 1 to k that have no prime factor in common with k). Since $\log(\log x) \to \infty$ as $x \to \infty$ this formula implies that the infinite series extended over all primes in the progression diverges, so the progression must contain infinitely many primes.

Dirichlet's proof of the infinitude of primes in arithmetical progressions was a landmark achievement. It marked the birth of **analytic number theory**, a new branch of mathematics in which problems concerning the integers are attacked by methods of analysis. This paper undoubtedly had an effect on subsequent work on the prime number theorem. The ideas introduced in Dirichlet's proof laid the basis for areas of mathematical research that have had profound applications to both analytic and algebraic number theory.

Chebyshev's Contributions

After Dirichlet's work on primes in arithmetical progressions in 1837, the first real step towards a proof of the prime number theorem itself was made in 1848 by the Russian mathematician, P.L. Chebyshev[1]. He proved that:

If the ratio $\frac{\pi(x) \log x}{x}$ has a limit as $x \to \infty$, then this limit must equal 1.

But Chebyshev was unable to prove that this ratio actually tends to a limit. In 1850, he[2] proved that $\pi(x)$ lies between two numerical constants times the ratio $x/\log x$, so $x/\log x$ does, indeed, represent the true order of magnitude of $\pi(x)$. The problem is difficult because there is no useful formula that generates the primes, and there is no obvious way to obtain information about their distribution.

Chebyshev introduced two new functions $\vartheta(x)$ and $\psi(x)$ that are somewhat more convenient to deal with than $\pi(x)$. They are defined as follows:-

$$\vartheta(x) = \sum_{p \le x} \log p$$

and

$$\psi(x) = \sum_{m=1}^{\infty} \sum_{p^m \le x} \log\, p = \vartheta(x) + \vartheta(x^{\frac{1}{2}}) + \vartheta(x^{\frac{1}{3}}) + \cdots .$$

In the sum over m the term with $m = 1$ is just $\vartheta(x)$. The term with $m = 2$ is the sum over all primes with $p^2 \le x$, or $p \le \sqrt{x}$, so this is $\vartheta(x^{\frac{1}{2}})$. The term with $m = 3$ is $\vartheta(x^{\frac{1}{3}})$, and so on. The sum on m is a finite sum because it terminates when $p^m > x$, or $m > (\log x)/(\log p)$. The second term is

$$\vartheta(x^{\frac{1}{2}}) = \sum_{p \le \sqrt{x}} \log p \le \sqrt{x} \log \sqrt{x} = \frac{1}{2}\sqrt{x} \log x .$$

The remaining terms are smaller, and there are at most (log x)/log 2 terms altogether, so $\psi(x)$ and $\vartheta(x)$ are related as follows:-

$$0 \le \psi(x) - \vartheta(x) \le \frac{\sqrt{x} \log^2 x}{2 \log 2} .$$

Divide this inequality by x and let $x \to \infty$ to obtain the limit relation

$$\lim_{x \to \infty} \left(\frac{\psi(x)}{x} - \frac{\vartheta(x)}{x} \right) = 0.$$

This shows that both ratios $\psi(x)/x$ and $\vartheta(x)/x$ exhibit the same asymptotic behaviour for large x.

The Connection with $\pi(x)$

As might be expected, the Chebyshev functions are related to $\pi(x)$. For example, since $\pi(x)$ is a step function with a jump of 1 at each prime, we can use Stieltjes integrals to write

$$\vartheta(x) = \sum_{p \le x} \log p = \int_1^x \log t \, d\pi(t).$$

Integration by parts gives us

$$\vartheta(x) = \pi(x) \log x - \int_1^x \frac{\pi(t)}{t} dt,$$

from which we find

$$\frac{\vartheta(x)}{x} = \frac{\pi(x) \log x}{x} - \frac{1}{x} \int_1^x \frac{\pi(t)}{t} dt.$$

From this and the relation between $\vartheta(x)$ and $\psi(x)$ it is easy to show that the following three statements are equivalent formulations of the prime number theorem:

$$\lim_{x\to\infty} \frac{\pi(x)\log x}{x} = 1,$$

$$\lim_{x\to\infty} \frac{\vartheta(x)}{x} = 1,$$

$$\lim_{x\to\infty} \frac{\psi(x)}{x} = 1.$$

Chebyshev proved that if any one of these three limits exists and equals 1, then the others also exist and equal 1. But he was unable to prove that any of these limits exists.

Incidentally, in 1881 Sylvester wrote a paper in the *American Journal of Mathematics* describing Chebyshev's methods. He praised Chebyshev's contributions and concluded:

> "We shall probably have to wait until someone is born into the world as far surpassing Chebyshev in insight and penetration as Chebyshev has proved himself superior in these qualities to the ordinary run of mankind."

Well, when Sylvester wrote these words, three people had already been born with these qualities: Riemann, born in 1826, Hadamard, born in 1865, and de la Vallée Poussin, born in 1866, the year Riemann died.

Riemann's Contributions

The next significant step toward the proof of the prime number theorem was made in 1859 by G.F.B. Riemann[9] in a famous 8-page paper, the only one he wrote on prime numbers.

Riemann attacked the problem with a new method based on Euler's infinite product identity (3). He replaced the real variable s by a complex variable, $s = \sigma + it$, and showed that the distribution of primes is intimately related to properties of the function $\zeta(s)$ obtained by analytic continuation of the series

$$\zeta(s) = \sum_{n=1}^{\infty} \frac{1}{n^s}. \tag{4}$$

The function $\zeta(s)$, now called the **Riemann zeta function**, is initially defined by the series in (4) in its half-plane of convergence $\sigma > 1$. Riemann showed that this function can be continued analytically beyond the line $\sigma = 1$. It exists as a meromorphic function in the entire s-plane, its only singularity being a simple pole at $s = 1$ with residue 1. Moreover, $\zeta(s)$ and $\zeta(1-s)$ are related by the functional equation

$$\zeta(1-s) = 2(2\pi)^{-s}\Gamma(s)\cos\left(\frac{\pi s}{2}\right)\zeta(s).$$

We'll see in a moment that the zeros of the zeta function play an important role in the distribution of primes. Where are the zeros? First we note that $\zeta(s)$ is represented by a convergent infinite product for $\sigma > 1$; convergent products are never zero, so the zeta function has no zeros for $\sigma > 1$. But it does have zeros in the negative half-plane. This can be seen from the functional equation. When s is a negative integer, say $s = -n$, $\zeta(1-s)$ becomes $\zeta(1+n)$ which is finite. The functional equation gives us

$$\zeta(1+n) = 2(2\pi)^n \Gamma(-n)\cos\left(\frac{\pi n}{2}\right)\zeta(-n).$$

But the gamma function has a simple pole at each negative integer so this pole must be canceled by a zero of one of the other factors. The factor $\cos(\frac{\pi n}{2})$ is zero at the negative *odd* integers; but it doesn't vanish when n is even, so when n is even the pole of the gamma function must be canceled by a zero of the zeta function. In other words,

$$\zeta(-2n) = 0 \quad \text{for} \quad n \geq 1.$$

These are called the **trivial zeros** of $\zeta(s)$. There are infinitely many trivial zeros at the points $-2, -4, -6$, etc. The remaining zeros of the zeta function must lie in the strip $0 \leq \sigma \leq 1$, which is called the **critical strip**. Riemann showed that there were infinitely many zeros in the critical strip. The nontrivial zeros are denoted by ρ. They occur in conjugate pairs and the functional equation shows that they are symmetrically located about the midline $\sigma = \frac{1}{2}$, which is called the **critical line**.

To connect the zeta function with the distribution of primes, Riemann took the logarithmic derivative of the Euler product. Taking logarithms in the equation

$$\zeta(s) = \prod_p \frac{1}{1-p^{-s}},$$

we find

$$\log\zeta(s) = -\sum_p \log(1-p^{-s}).$$

Following Riemann, we differentiate this equation, but the details presented here are not the same as those given by Riemann. Differentiate and change sign to get

$$-\frac{\zeta'(s)}{\zeta(s)} = \sum_p \frac{p^{-s}}{1-p^{-s}}\log p = \sum_{m=1}^{\infty}\sum_p \frac{\log p}{p^{ms}}.$$

But $\psi(x)$ is a step function with jump $\log p$ at the prime powers, so the last sum can be written as a Riemann-Stieltjes integral and we get

$$-\frac{\zeta'(s)}{\zeta(s)} = \int_1^{\infty}\frac{d\psi(x)}{x^s} = s\int_1^{\infty}\frac{\psi(x)}{x^{s+1}}dx$$

after integration by parts. This is a Mellin transform of the function $\psi(x)$, and it can be inverted to give

$$\psi(x) = \frac{-1}{2\pi i}\int_{2-i\infty}^{2+i\infty}\frac{x^s}{s}\frac{\zeta'(s)}{\zeta(s)}ds.$$

The path of integration is along the vertical line $\sigma = 2$. Riemann's idea was to move the line of integration as far to the left as possible, all the way to $-\infty$. As we cross the poles, we have to add the residues of the integrand.

The integrand has a pole at $s = 1$ from the factor $\frac{\zeta'(s)}{\zeta(s)}$, which has residue -1, so the pole at $s = 1$ has residue x.

There is a pole at $s = 0$ from the factor $1/s$ with residue $-\frac{\zeta'(0)}{\zeta(0)}$.

And there is a pole at each zero ρ of $\zeta(s)$ with residue $-x^{\rho}/\rho$.

We separate the nontrivial zeros from the trivial zeros. This gives us

$$\begin{array}{ccccccc} \psi(x) = & x & - \dfrac{\zeta'(0)}{\zeta(0)} & - \displaystyle\sum_{\rho} \frac{x^{\rho}}{\rho} & - \displaystyle\sum_{n=1}^{\infty} \frac{x^{-2n}}{-2n}, \\ & (s=1) & (s=0) & (s=\rho) & (s=-2n) \end{array}$$

where the sum over ρ is the contribution from the nontrivial zeros, and the sum on n is the contribution from the trivial zeros. The sum over n is just the power series for $\frac{1}{2}\log(1-x^{-2})$, and this tends to 0 as x tends to ∞, so we have the explicit formula

$$\psi(x) = x - \frac{\zeta'(0)}{\zeta(0)} - \sum_{\rho} \frac{x^{\rho}}{\rho} + o(1) \text{ as } x \longrightarrow \infty.$$

This remarkable formula tells us that $\psi(x)$ is equal to x, plus a constant, plus an error term depending on the nontrivial zeros of the Riemann zeta function. We divide by x, and rewrite this as follows:

$$\frac{\psi(x)}{x} - 1 = -\sum_{\rho} \frac{x^{\rho-1}}{\rho} + o(1) \text{ as } x \longrightarrow \infty.$$

Incidentally, Riemann expressed his explicit formula somewhat differently. There are many technical details that have to be taken care of to justify this process, and Riemann was not able to provide full justification. The version given here was derived by von Mangoldt[10] in 1894, with full justification, and it has superceded Riemann's formula in more modern treatments. This explicit formula tells us that the prime number theorem is equivalent to the statement

$$\lim_{x\to\infty} \sum_{\rho} \frac{x^{\rho-1}}{\rho} = 0.$$

If the limit could be taken term by term we would obtain the prime number theorem if we could show that each term $x^{\rho-1}$ tends to 0, and this would follow if $Re\ \rho < 1$, or, in other words, if the zeta function has no zeros on the line $\sigma = 1$. Riemann was not able to show this. He said it was likely that all the nontrivial zeros of the zeta function had real part $\frac{1}{2}$, but admitted that he could not prove this. This statement has become one of the most famous unsolved problems in mathematics. It is called the **Riemann hypothesis**, and it states that each nontrivial zero of the zeta function has the form $\rho = \frac{1}{2} + it$.

The Riemann hypothesis implies (and is implied by) the relation

$$\psi(x) = x + O(x^{\frac{1}{2}+\epsilon}) \quad \text{(for every } \epsilon > 0\text{)}.$$

Riemann died in 1866 at the age of 40 and, as far as we know, he made no further progress toward a proof of the prime number theorem. But Riemann's work stimulated a great deal of activity on this problem.

In 1885, Stieltjes announced that he had proved the Riemann hypothesis and, therefore, the prime number theorem, but Stieltjes never published a proof. Stieltjes died in 1894 without having either substantiated or retracted his claim of having proved the Riemann hypothesis.

In an effort to justify some of Riemann's formal techniques, Hadamard developed the theory of entire functions of finite order, and in 1893 he provided proofs of some of the results that Riemann had stated with incomplete proofs. Following Hadamard's work on entire functions, von Mangoldt in 1894 gave the first correct proof of the explicit formula for $\psi(x)$ referred to above.

Then in 1896, a little more than a century ago, the prime number theorem was finally proved – independently and almost simultaneously – by two young mathematicians, Jacques Hadamard, age 31, and Charles-Jean de la Vallée Poussin, age 30. We turn now to some remarks about their contributions.

Contributions by Hadamard and de la Vallée Poussin

Following Riemann, both Hadamard and de la Vallée Poussin realized that the prime number theorem follows from the fact that the zeta function has no zeros on the line $\sigma = 1$, and both proved this fact for the first time, but they used different methods.

There is a diplomatic remark in Hadamard's paper[7] that has some historical interest. He says: "Stieltjes proved, in accordance with Riemann's expectations, that all the zeros of $\zeta(s)$ in the critical strip have the form $\frac{1}{2} + it$, where t is real; but his proof has never been published, and it has not even been established that the function ζ has no zeros on the line $\sigma = 1$. It is this last assertion that I propose to prove here." And so he did, thus providing a proof of the prime number theorem.

De la Vallée Poussin's first contributions to prime numbers were published in 1896 in three memoirs[3]; the first contains his proof of the prime number theorem, the second extends his method to obtain a prime number theorem for arithmetic progressions, and the third is on primes representable by quadratic forms.

Of the two proofs that $\zeta(1 + it) \neq 0$, Hadamard's is the simpler, and it is based on a relation between $\zeta(\sigma + it)$ and $\zeta(\sigma + 2it)$. In a two-page note at the end of his third memoir, de la Vallée Poussin acknowledges that Hadamard's method was the simpler of the two, and then shows how Hadamard's method can be simplified even further. He adapts Hadamard's idea to the function $\zeta'(s)/\zeta(s)$ and shows in just a few lines that the nonvanishing of $\zeta(s)$ on the line $\sigma = 1$ follows quite easily from the elementary trigonometric identity for the cosine of a double angle:

$$1 - \cos 2\theta = 2(1 - \cos^2 \theta),$$

together with the inequality $0 \leq 1 - \cos\theta \leq 2$.

The two trigonometric relations imply

$$3 + 4\cos\theta + \cos 2\theta \geq 0,$$

a trigonometric inequality that de la Vallée Poussin used to establish the nonvanishing of $\zeta(1+it)$. He then points out that the use of these trigonometric relations shortens his original proof in the first memoir by 24 pages, and that the same relations can be used to simplify the second and third memoirs as well.

Here's the connection with the zeta function. From the formula for the logarithm of the zeta function,

$$\log \zeta(s) = \sum_{p} \sum_{m=1}^{\infty} \frac{1}{mp^{ms}}$$

obtained earlier, we see that $\zeta(s)$ is the exponential of this double sum, and the absolute value of $\zeta(s)$ is the exponential of its real part. Now $p^{-ms} = \exp\{-ms \log p\} = p^{-m\sigma} \exp\{-imt \log p\}$, so we get

$$|\zeta(\sigma + it)| = \exp \left\{ \sum_{p} \sum_{m=1}^{\infty} \frac{\cos(mt \log p)}{mp^{m\sigma}} \right\}.$$

De la Vallée Poussin then established the inequality

$$\zeta^3(\sigma)|\zeta(\sigma + it)|^4 |\zeta(\sigma + 2it)| \geq 1 \tag{5}$$

as follows. The product on the left is equal to

$$\exp \left\{ \sum_{p} \sum_{m=1}^{\infty} \frac{3 + 4\cos\theta + \cos 2\theta}{mp^{m\sigma}} \right\}, \quad \text{where } \theta = mt \log p.$$

Because of the trigonometric inequality, each term of this series is ≥ 0 so the exponential is ≥ 1, hence we obtain inequality (5). From this inequality, he deduced that $\zeta(1 + it) \neq 0$ for all real t. The argument, by contradiction, is very simple. Suppose that $\zeta(1 + it) = 0$ for some real $t \neq 0$. Inequality (5) can be rewritten as follows by dividing both sides by $\sigma - 1$, where $\sigma > 1$:

$$(\sigma - 1)^3 \zeta^3(\sigma) \left| \frac{\zeta(\sigma + it) - \zeta(1 + it)}{\sigma - 1} \right|^4 |\zeta(\sigma + 2it)| \geq \frac{1}{\sigma - 1}.$$

Now let σ approach 1 from the right. On the left, the product $(\sigma - 1)\zeta(\sigma)$ tends to 1 because $\zeta(s)$ has a pole at 1 with residue 1, the second factor tends to the fourth power of the absolute value of the derivative $\zeta'(1)$, which is finite, the third factor tends to the absolute value of $\zeta(1 + 2it)$, which is finite, so the left member tends to a finite value. But the right member tends to infinity, and we get a contradiction! Therefore, $\zeta(1 + it) \neq 0$ for all real t, and, as noted above, this implies the prime number theorem.

The original memoir of de la Vallée Poussin develops more profound knowledge of the properties of $\zeta(s)$. For example he showed that $\zeta(\sigma + it) \neq 0$ in a region of the form $\sigma \geq 1 - A/\log t$ for sufficiently large t. He later used these properties to extend the prime number theorem with the more general asymptotic formula

$$\pi(x) \sim \frac{x}{\log x} + 1!\frac{x}{(\log x)^2} + \cdots + (n - 1)!\frac{x}{(\log x)^n} \quad \text{as } x \longrightarrow \infty.$$

The case $n = 1$ is the prime number theorem. He also expressed the prime number theorem in the form

$$\pi(x) = \mathrm{Li}(x) + O(xe^{-a\sqrt{\log x}}),$$

where a is a positive constant and $\mathrm{Li}(x)$ is the logarithmic integral introduced by Gauss. Apart from the value of a, the exponential factor in the error term remained unimproved until 1921, when Littlewood showed that the function $\log x$ under the square root sign could be replaced by $\log x \log\log x$. This has only historical interest now because the error was later improved by I.M. Vinogradov and H.M. Korobov, who showed that $\sqrt{\log x}$ could be replaced by $(\log x)^{2/3}$.

There is also a prime number theorem for arithmetic progression first proved by de la Vallée Poussin. It states that the number of primes $\leq x$ in the progression $kn + h$ is asymptotic to $\frac{1}{\varphi(k)}\frac{x}{\log x}$. There are $\varphi(k)$ reduced residue classes modulo k, and de la Vallée Poussin's theorem shows that in the limit the primes are equally distributed among each of these $\varphi(k)$ reduced residue classes. Each residue class gets its fair share of primes.

Elementary Proof of the Prime Number Theorem

The first proof of the prime number theorem, as given by Hadamard and de la Vallée Poussin, was simplified by Landau, and by Hardy & Littlewood, in the early part of the 20th century, and new proofs were later discovered, by Norbert Wiener and others, all using sophisticated methods of real and complex analysis, and the fact that $\zeta(s)$ doesn't vanish on the line $\sigma = 1$. Then in 1949, Paul Erdös and Atle Selberg discovered an ***elementary proof*** that makes no use of the Riemann zeta function or complex function theory. But this so-called elementary proof is more difficult to understand than the analytic proofs.

In 1980 Donald J. Newman gave a concise version of the analytic proof that uses very little complex analysis beyond Cauchy's theorem. Newman's proof also uses the fact that $\zeta(s)$ doesn't vanish on the line $\sigma = 1$, but he has a clever way of deducing this.

Another Form of the Prime Number Theorem

Earlier we stated that the prime number theorem can also be expressed by describing the asymptotic behaviour of the nth prime p_n. We can deduce this behaviour from that of $\pi(x)$. Start with the prime number theorem in the form

$$\lim_{x\to\infty} \frac{\pi(x)\log x}{x} = 1. \tag{1}$$

Take logarithms of both sides to obtain

$$\lim_{x\to\infty} \{\log \pi(x) + \log\log x - \log x\} = 0,$$

or

$$\lim_{x\to\infty} \left\{\log x \left(\frac{\log \pi(x)}{\log x} + \frac{\log\log x}{\log x} - 1\right)\right\} = 0.$$

Since $\log x \to \infty$ the last factor multiplying $\log x$ must tend to 0, so

$$\lim_{x\to\infty}\left(\frac{\log \pi(x)}{\log x}+\frac{\log\log x}{\log x}-1\right)=0.$$

The quotient $\frac{\log\log x}{\log x}$ tends to 0, hence

$$\lim_{x\to\infty}\frac{\log \pi(x)}{\log x}=1.$$

Multiply this relation by (1), cancel $\log x$, and we get

$$\lim_{x\to\infty}\frac{\pi(x)\log \pi(x)}{x}=1.$$

Now let $x = p_n$, so that $\pi(x) = n$. Then the last formula becomes

$$\lim_{n\to\infty}\frac{n\log n}{p^n}=1.$$

This says that the nth prime p_n is asymptotically equal to $n \log n$ as $n \to \infty$, or, in other words:

For large *n* the *n*th prime grows like *n* log *n*.

It can be shown that this also implies the prime number theorem, so this statement is logically equivalent to the prime number theorem.

Concluding Remarks

The prime number theorem is important not only because it makes an elegant and simple statement about primes and has many applications but also because much new mathematics was created in the attempts to find a proof. This is typical in number theory. Some problems, very simple to state, are often extremely difficult to solve, and mathematicians working on these problems often create new areas of mathematics of independent interest.

Another example is the Fermat conjecture, which has received more publicity as an unsolved problem than any other result in mathematics. It is interesting to note that Gauss considered the Fermat conjecture to be of only minor importance, and refused to work on it.

The prime number theorem and Fermat's last theorem are two outstanding examples of problems that attract the intellectual curiosity of many individuals but resist efforts at solution. Repeated failure by eminent mathematicians to settle these problems by known procedures stimulates the invention of new methods, new approaches, and new ideas that, in time, become part of the mainstream of mathematics, and even change the way mathematicians think about their subject. This is certainly true of the prime number theorem. Early attempts to prove the prime number theorem stimulated the development of the theory of functions of a complex variable, a branch of mathematics that is the life blood of mathematical analysis. Efforts to prove Fermat's last theorem led to the development of algebraic number theory, one of the most active areas of modern mathematical research, with ramifications far beyond the Fermat equation. One unexpected application of algebraic number theory is in designing security systems for computers.

In number theory alone there are hundreds of unsolved problems. New problems arise more rapidly that the old ones are solved, and many of the old ones have remained unsolved for centuries. Progress of our knowledge of numbers is advanced not only by what we already know about them, but also by realizing that there is much we do not know about them.

References

1. P. Chebyshev, *Oeuvres* **1** (1899), 27–48.
2. ———, *Oeuvres* **1** (1899), 49–70.
3. C-J. de la Vallée Poussin, *Ann. Soc. Sci. Bruxelles* **20** (1896), 183–256, 281–397.
4. L. Dirichlet, *Mathematische Abhandlungen, Bd.* **1** (1889), 313–342.
5. C.F. Gauss, *Werke* **II** (1872), 436–442.
6. ———, *Werke* **II** (1872), 444–447.
7. J. Hadamard, *Bull. Soc. Math. de France* **24** (1896), 199–220.
8. A.M. Legendre, *Essai sur la Théorie des Nombres*, 2nd ed., 1808, Paris.
9. B. Riemann, *Gesammelte Mathematische Werke*, 2nd Aufl. (1892), 145–155.
10. H. von Mangoldt, *Sitz. König. Preus. Akad. Wiss. zu Berlin* (1894), 337–350, 883–895.

California Institute of Technology
1–70 Caltech
Pasadena, CA 91125
USA
E-Mail: apostol@caltech.edu

Non-homogeneous Problems: Conjectures of Minkowski and Watson

R.P. Bambah, V.C. Dumir and R.J. Hans-Gill

§1 Introduction

Here we shall survey the developments regarding two well known problems in Geometry of Numbers. The first is a conjecture of Minkowski about the product of non-homogeneous real linear forms. The second one is a conjecture of Watson concerning non-homogeneous real indefinite quadratic forms. Whereas the first one is still resisting solution in the general case, the second one has been completely proved.

In order to be able to express the conjectures in a geometrical language also, we give some definitions.

Here $\mathbb{R}^n$ denotes the n-dimensional Euclidean space. For any n linearly independent points $A_1, \ldots, A_n$ in $\mathbb{R}^n$, the set $\wedge = \{u_1A_1 + \cdots + u_nA_n : u_1, \ldots, u_n \text{ are integers}\}$ is called a *lattice* and $A_1, \ldots, A_n$ is called a *basis* of $\wedge$. The determinant of $\wedge$ is defined as $d(\wedge) = |\det(A_1, \ldots, A_n)|$. It equals the volume of the parallelotope generated by the vectors $A_1, \ldots, A_n$ (called a *cell* of $\wedge$). The lattice $\wedge$ can be generated by infinitely many bases; but $d(\wedge)$ is independent of their choice.

For any subset S of $\mathbb{R}^n$, we say that $\wedge$ is a *covering lattice* for S or that $(S, \wedge)$ is a *covering* if

$$\mathbb{R}^n \subset \cup \{S + A : A \in \wedge\},$$

where

$$S + A = \{X + A : X \in S\} = \text{translate of } S \text{ through } A.$$

If $f : \mathbb{R}^n \to \mathbb{R}$ and $S = \{X : |f(X)| \leq 1\}$ then $\wedge$ is a covering lattice for S if and only if for any P in $\mathbb{R}^n$ there is a point A in $\wedge$ such that

$$|f(A + P)| \leq 1.$$

This is equivalent to saying that for any P in $\mathbb{R}^n$ and any basis $A_1, \ldots, A_n$ of $\wedge$, there exist integers $u_1, \ldots, u_n$ such that

$$|f(u_1A_1 + \cdots + u_nA_n + P)| \leq 1.$$

§2 Minkowski's Conjecture

Let $L_i = a_{i1}x_1 + \cdots + a_{in}x_n$, $1 \leq i \leq n$, be n real linear forms in n variables $x_1, \ldots, x_n$ and having determinant $\Delta = \det(a_{ij}) \neq 0$. The following conjecture is attributed to H. Minkowski:

For any given real numbers $c_1, \dots, c_n$ there exist integers $x_1, \dots, x_n$ such that

$$|(L_1 + c_1) \dots (L_n + c_n)| \le |\Delta|/2^n. \qquad (*)$$

Equality is necessary iff after a suitable unimodular transformation the linear forms L_i have the form $2c_i x_i$ for $1 \le i \le n$.

It is obvious that equality occurs in (∗) for the cases mentioned in the conjecture. So this conjecture would be proved if we can show that for all other linear forms, (∗) holds with strict inequality.

In the geometrical language of coverings, Minkowski's conjecture can be stated as follows:

Any lattice $\wedge$ in $\mathbb{R}^n$ is a covering lattice for the set

$$S : |x_1 \dots x_n| \le 2^{-n}\, d(\wedge).$$

Further, $\wedge$ is a covering lattice for the interior of S iff $\wedge$ does not have a basis consisting of the vectors of the type $B_i = (0, \dots, 0, b_i, 0, \dots, 0)$, $1 \le 1 \le n$.

It is clear that it is enough to consider lattices of determinant 1.

§2.1 Known Results: Minkowski's Conjecture

This result is trivial for $n = 1$. For $n = 2$, a proof was first given in 1899 by Minkowski[105] (See [107] pages 42–45 or [106] §4 Chap. XVI). Several mathematicians: Mordell[108,109] (1928, 1941) Landau[88] (1931), Perron[129] (1938), Pall[128] (1943), Macbeath[89,90] (1948, 1961), Sawyer[132] (1948) (see also Mordell[110] (1953)) and Heilbronn (See Hardy and Wright [83] §24.7 and page 413) have obtained a variety of proofs, partly in an effort to find a proof which would generalize to higher dimensions.

The proof by Sawyer follows from an interesting result of Delone[46] (1947), though he used a slightly weaker result. For any point $A \in \mathbb{R}^2$, the set $\Gamma = \wedge + A$ is called a *grid* and for any cell $\mathcal{P}$ of $\wedge$ and point Y in Γ, $\mathcal{P} + Y$ is called a *cell* of Γ. A cell of Γ is called a *divided cell* if it has a vertex in each of the quadrants. Delone proved that every grid in $\mathbb{R}^2$ which has no point on the co-ordinate axes has a divided cell. Delone showed that the analogous result does not hold in $\mathbb{R}^3$.

Minkowski's conjecture has so far been proved for $n \le 5$. For $n \ge 3$, the proofs have been obtained by mathematicians listed below:

$n = 3$	Remak[130] (1923)
	Davenport[38] (1939)
	Birch and Swinnerton-Dyer[20] (1956)
	Narzullaev[117] (1968)
$n = 4$	Dyson[62] (1948)
	Skubenko[135] (1973)
	Bambah and Woods[13] (1974)
$n = 5$	Skubenko[135] (1973)
	Bambah and Woods[15] (1980)

We briefly describe the three methods of approach that have been successful for $n = 3$. Only the first has been successfully extended to give proofs for $n = 4, 5$.

I. Remak-Davenport method
II. Birch and Swinnerton-Dyer method
III. DOTU method

I. Remak-Davenport method suggests that one way to approach the conjecture of Minkowski is to prove the following two results:

(i) For any lattice $\wedge$ in $\mathbb{R}^n$ there is an ellipsoid

$$E : a_1 x_1^2 + \cdots + a_n x_n^2 < 1$$

which contains no point of $\wedge$ other than O but has n linearly independent points of $\wedge$ on its boundary.

(ii) If $\wedge$ is a lattice of determinant 1 and there is a sphere $|X| < R$ which contains no point of $\wedge$ other than O and has n linearly independent points of $\wedge$ on its boundary then $\wedge$ is a covering lattice for $|X| \leq \sqrt{n/4}$.

It can be easily seen that Minkowski's conjecture (except for the determination of critical cases) follows for any n for which both (i) and (ii) hold. Consider any lattice $\wedge$ of determinant 1. On making the linear transformation

$$\sqrt{a_i} x_i = (a_1 \ldots a_n)^{1/2n} y_i,\ 1 \leq i \leq n,$$

which is an automorphism of the form $x_1 \ldots x_n$, and replacing $\wedge$ by the transformed lattice we can suppose that the ellipsoid E found in (i) is a sphere. Then (ii) implies that $\wedge$ is a covering lattice for the sphere $K : |X| \leq \sqrt{(n/4)}$. Using the inequality of arithmetic and geometric means we can easily see that K is contained in the region $|x_1 \ldots x_n| \leq 1/2^n$.

For $n = 3$, Remak's proof was lengthy and complicated. It was considerably simplified by Davenport[38] (1939). Mahler[103] (1940) gave another proof of (ii).

For $n = 4$, (ii) was proved by Hofreiter[84] (1933). Dyson[62] (1948) proved both (i) and (ii) and hence Minkowski's conjecture for $n = 4$. The proof is considered difficult; proof of (i) used strong tools form Algebraic Topology. Cleaver[35] (1965) gave another proof of (ii). A simpler proof was given by Woods[146] (1965). Using some results of Birch and Swinnerton-Dyer[20] (1956) (mentioned below) which allow one to restrict the attention to lattices having no points on co-ordinate planes, Bambah and Woods[13] (1974) gave proof of (i), hence completing an elementary proof of Minkowski's conjecture for $n = 4$. Skubenko[135] (1973) had sketched another elementary proof of (i) for $n = 4$. He also gave a proof of (i) for $n = 5$, in which many details were somewhat obscure. The result (ii) had already been proved for $n = 5$ by Woods[147] (1965). Bambah and Woods[15] (1980) gave clear detailed proof of (i) for $n = 5$ on the lines of Skubenko's proof. Again, they used Birch and Swinnerton-Dyer's result to restrict attention to lattices having no points in co-ordinate planes.

For $n = 6$, the result (ii) has been proved by Woods[150] (1972), while counterpart (i) is still awaiting proof.

II. Birch and Swinnerton-Dyer[20] (1956) gave a different proof of Minkowski's conjecture for $n = 3$. It is based on a homogeneous reduction of linear forms. Though they could not extend the method to obtain a proof of the conjecture for $n = 4$, they succeeded in proving very useful general results stated below:

(A) Suppose that the Minkowski's conjecture has been proved for $n = 1, 2, \ldots, N-1$. Then to prove if for $n = N$, it is enough to consider only those sets of forms $L_1, \ldots, L_n$ for which the homogeneous minimum

$$m_H = \inf\{|L_1 \ldots L_n| : x_1, \ldots, x_n \text{ integers not all zero}\}$$

is attained and non-zero.

(B) Minkowski's conjecture is true for all lattices in a neighbourhood of the conjectured critical lattice (such a neighbourhood is specified explicitly).

The result (A) has been used in the proofs of Minkowski's conjecture for $n = 4, 5$ by Bambah and Woods, and Skubenko to restrict their attention to lattices having no points in the co-ordinate planes.

Macbeath[91] (1961) formed a connection of the Minkowski's conjecture with factorization of matrices in a particular manner. He called a non-singular matrix M of order n *Minkowski* if the lattice $M\mathbb{Z}^n$ is a covering lattice for the region

$$|x_1 \ldots x_n| \le 2^{-n}|\det M|.$$

Thus Minkowski's conjecture can be stated as:

Every non-singular matrix of order *n* is a Minkowski matrix.

Minkowski had himself proved that every rational non-singular matrix is a Minkowski matrix. The result (B) of Birch and Swinnerton-Dyer stated above can be restated as:

For every n, there is a neighbourhood of the identity matrix which consists entirely of Minkowski matrices. Macbeath showed that more generally, every rational non-singular matrix has a neighbourhood consisting of Minkowski matrices. For this he used a very simple method. He called a non-singular matrix M a DOTU-matrix if it can be expressed as a product

$$M = \text{DOTU},$$

where D is diagonal, O is orthogonal, U is unimodular and T is unit triangular i.e., if T$= (t_{ij}), t_{ij} = 1$ for $1 \le i \le n$ and $t_{ij} = 0$ for $i < j$.

Macbeath proved that every DOTU-matrix is a Minkowski matrix. Further, he showed that if $M =$ DOTU and $O = (a_{ij})$ with det $(a_{ij}^2) \neq 0$ then M is an inner point of the family of Minkowski Matrices. Since every rational non-singular matrix can be written as a product DTU, the result applies to these. Macbeath remarked that for $n = 2$, every non-singular matrix is a DOTU-matrix can be seen with little difficulty. This is essentially, the basis of the proof due to Heilbronn given in Hardy and Wright [83]. Narzullaev[117,118,119] (1968, 1969, 1975) established the corresponding result for $n = 3$.

Gruber[71,72] (1970, 1976) and Ahmedov[4,5,6] (1972, 1975, 1977) showed the existence of non-singular matrices which are not DOTU. The proof depends upon the existence of totally real algebraic number fields of degree n whose discriminant is $< n^n$. The existence of such fields is guaranteed by some results of class field theory. Skubenko[139] (1981) gave an example for $n = 2880$.

For results on the measure of the set of lattices for which Minkowski's conjecture holds and for a discussion of related results see Section (xvi) of Gruber and Lekkerkerker [74].

§2.2 *Chebotarev's Theorem and Asymptotic Estimates*

Lot of work has been done for obtaining weaker results which are valid for all n. Let μ_n be the infimum of the numbers μ such that any lattice Λ is a covering lattice for

$$|x_1 \dots x_n| \le \mu\, d(\Lambda).$$

Then Minkowski's conjecture states that $\mu_n = 2^{-n}$.

Chebotarev[32] (1934) obtained the interesting result $\mu_n \le 2^{-n/2}$, for all n. The proof is simple and elegant. It essentially depends on the fact that the union K of two translates $S \pm (1, \dots, 1)$ of the hyperboloid $S : |x_1 \dots x_n| \le 1$ contains the box $\max_{1\le i\le n}|x_i| \le 2$ to which the Minkowski's convex body theorem is applied. Several authors have worked on obtaining better estimates of μ_n by using a convex body of bigger volume contained in K and applying Minkowski's theorem, or applying Blichfeldt's theorem and its refinements.

Writing $\mu_n = 2^{-n/2}\eta_n^{-1}$ and $\eta = \lim_{n\to\infty}\eta_n$, we can summarise these improvements in terms of values of η as follows

$2e - 1$	$\sim$	4.43656	Davenport[40](1946)
$2(2e - 1)$	$\sim$	8.873127	Woods[145](1958)
$3.0001(2e - 1)$	$\sim$	13.310134	Bombieri[25](1963)
$3(2e - 1)$	$\sim$	13.309690	Gruber[68](1967).

Skubenko[137,138] (1977, 1978) obtained $\eta_n = e^{-2}n^{1/3}\log^{-2/3} n$ for large n, which is considerably better than the earlier estimates, since here the limit η equals ∞. For improvements in this direction see Narazullaev and Skubenko[125] (1979), Mukhsinov[113,114] (1981), Andrijasjan, Ilin and Malyshev[7] (1986). Mukhsinov[115] (1981) and Ilin[85] (1991) have obtained improved estimates for small values of n.

§2.3 *Asymmetric Inequalities*

It is interesting though not surprising that some best possible general results which can be regarded as asymmetric versions of Minkowski's conjecture have been obtained with simple, elegant methods. Chalk[29] (1947) showed that every lattice Λ is a covering lattice for the region

$$x_1 \dots x_n \le d(\Lambda), \quad x_j > 0 \quad \text{for} \quad 1 \le j \le n.$$

Considering the integral lattice, we see that this result is best possible.

Cole[37] (1952) showed that every lattice Λ is a covering lattice for the region

$$x_1 \dots x_{n-1}|x_n| \le \frac{1}{2}d(\Lambda), \quad x_j > 0 \quad \text{for} \quad 1 \le j \le n-1.$$

These results suggest a natural CONJECTURE:

Every lattice Λ is a covering lattice for the region

$$x_1 \dots x_{n-r}|x_{n-r+1} \dots x_n| \le 2^{-r}d(\Lambda), \quad x_j > 0 \quad \text{for} \quad 1 \le j \le n-r.$$

(For $r = n$, this is Minkowski's conjecture).

For $n = 2$, there are only two cases, which are taken care of by Chalk's theorem and Cole's Theorem. For $n = 3$, the missing third case was confirmed by Bambah and Woods[14] (1977). The above conjecture is open for $n \geq 4$ and $2 \leq r \leq n$ except for $r = n = 4, 5$, which are the confirmed cases of Minkowski's conjecture.

Davenport[42] (1948), in an effort to establish an analogue of Minkowski's conjecture for (ternary) quadratic forms (which topic is the subject of our exposition in the second part) proved another type of asymmetric inequality. He showed that any plane lattice Λ is a covering lattice for

$$-\sigma d(\Lambda) \leq x_1 x_2 \leq \rho d(\Lambda), \rho > 0, \sigma > 0, \rho\sigma \geq 1/16.$$

Sawyer[133] (1950) gave a simple geometrical proof which is a modification of his proof mentioned earlier for $\rho = \sigma = 1/4$. Woods[148] (1981) proved the corresponding results for the regions

$$-\sigma d(\Lambda) \leq x_1 |x_2| \leq \rho d(\Lambda), \quad \rho\sigma \geq 1/16$$

and

$$-\sigma d(\Lambda) \leq x_1 x_2 |x_3| \leq \rho d(\Lambda), \quad \rho\sigma \geq 1/64.$$

The corresponding best result for the region

$$-\sigma d(\Lambda) \leq x_1 x_2 x_3 \leq \rho d(\Lambda)$$

has still not been proved. Grover[66] (1989) proved it for $\rho\sigma \geq 1/16.81$, instead of the expected $\rho\sigma \geq 1/64$. He could prove it with $\rho\sigma \geq 1/64$, for lattices with homogeneous minimum zero.

§2.4 Other Directions

Several investigations in other directions have been made for the product of homogeneous linear forms. For example, bounds on the non-homogeneous minimum related to the homogeneous minima of product of linear forms have been obtained by Birch and Swinnerton-Dyer[20] (1956), Gruber[70] (1970), Bakiev, Pen and Skubenko[9] (1978), Bakiev[8] (1981), Skubenko and Bakiev[140] (1979). Forms associated with algebraic number fields have been discussed in some detail by Clarke[34] (1951) and Davenport[41] (1952).

For introductory reading one can refer to §§24.7–10 of Hardy and Wright [83] and §11.4 of Cassels [28]. For a detailed description of many results and history, Gruber and Lekkerkerker [74] is excellent.

§3 Non-homogeneous Indefinite Quadratic Forms

Minkowski's theorem for $n = 2$ can be formulated in terms of non-homogeneous minima of indefinite binary quadratic forms. Let $L_1(x, y) = \alpha x + \beta y$, $L_2(x, y) = \gamma x + \delta y$ be two real linear forms of determinant $\Delta = \alpha\delta - \beta\gamma \neq 0$. Let

$$Q(x, y) = L_1(x, y) L_2(x, y) = ax^2 + bxy + cy^2. \tag{3.1}$$

Then $Q(x, y)$ is an indefinite quadratic form of determinant $D = ac - b^2/4 = \Delta^2/4$. Given real numbers c_1, c_2 we can find reals x_0, y_0 such that

$$L_1(x_0, y_0) = c_1, L_2(x_0, y_0) = c_2 \tag{3.2}$$

so that

$$(L_1 + c_1)(L_2 + c_2) = Q(x + x_0, y + y_0). \tag{3.3}$$

Conversely, given an indefinite binary quadratic form $Q(x, y)$ and real numbers x_0, y_0 we can find real linear forms L_1, L_2 and real numbers c_1, c_2 such that (3.1), (3.2), (3.3) hold. An equivalent alternative way of stating Minkowski's result is the following:

Theorem 1 *Let $Q(x, y)$ be an indefinite binary quadratic form of determinant $D \neq 0$. Then given any real numbers x_0, y_0 there exist integers, x, y such that*

$$|Q(x + x_0, y + y_0)| \leq \left(\frac{1}{4}|D|\right)^{1/2}. \tag{3.4}$$

Equality is needed in (3.4) *iff $Q(x, y) \sim \rho xy, \rho \neq 0$ and $(x_0, y_0) \sim (\frac{1}{2}, \frac{1}{2})$* (mod1).

Remark 1 We say that two quadratic forms $f(X) = f(x_1, \ldots, x_n)$ and $g(X) = g(x_1, \ldots, x_n)$ are equivalent, and we denote this by $f \sim g$, if there exists a unimodular transformation T such that $f(TX) = g(X)$. Further we say $X = (x_1, \ldots, x_n) \sim (y_1, \ldots, y_n) = Y (\text{mod } 1)$ if $Y - TX$ is a point with integer co-ordinates. A natural question is to generalize the above result to indefinite quadratic forms in more than two variables. Blaney[21] (1948) proved the following result

Theorem 2 *Let $Q = Q(x_1, \ldots, x_n)$ be an indefinite quadratic form in n variables with determinant $D \neq 0$. For any given γ a number $\Gamma = \Gamma(\gamma, n)$ exists such that for any Q and any real numbers $\alpha_1, \ldots, \alpha_n$, integers $x_1, \ldots, x_n$ exist such that*

$$\gamma|D|^{1/n} < Q(x_1 + \alpha_1, \ldots, x_n + \alpha_n) \leq \Gamma|D|^{1/n}. \tag{3.5}$$

The proof is by induction on n. In particular it implies that there exists a constant $C = C(n)$ such that for real $\alpha_1, \ldots, \alpha_n$ there exist integers $x_1, \ldots, x_n$ such that

$$|Q(x_1 + \alpha_1, \ldots, x_n + \alpha_n)| \leq C(|D|)^{1/n}. \tag{3.6}$$

His proof shows that we can take $C = 2^{n-2}$.

The inequality (3.6) of Blaney was strengthened by D.M.E. Foster[65] (1956) by showing that we can take $\Gamma(\gamma, n) = \gamma + C_n\gamma^{1/2^n} + C_{n'}$, where $C_n, C_{n'}$ are positive numbers depending only on n. C.A. Rogers[131] (1952) using the minima of positive definite quadratic forms showed that the inequality (3.6) is true with $C = \frac{1}{4}n\gamma_n$, where γ_n is the Hermite's constant. Since $\gamma_n \leq \delta_n \sim \frac{n}{\pi e}$ we have $C \leq \beta_n \sim \frac{n^2}{4\pi e}$.

Davenport[42] (1948) extended Minkowski's theorem to indefinite ternary forms by proving

Theorem 3 *Let $Q(x, y, z)$ be an indefinite quadratic form in three variables with determinant $D \neq 0$. Then given any real numbers x_0, y_0, z_0 there exist integers x, y, z such that*

$$|Q(x+x_0, y+y_0, z+z_0)| \leq \left(\frac{27}{100}|D|\right)^{1/3}. \tag{3.7}$$

Equality is needed in (3.7) *iff*

$$Q(x, y, z) \sim \rho(x^2+5y^2-z^2+5yz+zx), \rho \neq 0 \tag{3.8}$$

and

$$(x_0, y_0, z_0) \sim \left(\frac{1}{2}, \frac{1}{2}, \frac{1}{2}\right) \text{ (mod 1)}.$$

For the proof of (3.7) *Davenport*[42] *obtained the following generalization of Minkowski's theorem.*

Theorem 4 *Let L_1, L_2 be linear forms as above. If $\rho > 0, \sigma > 0$ are such that $\rho\sigma \geq \frac{1}{16}$ then given any reals c_1, c_2 there exist integers x, y such that*

$$-\sigma|\Delta| < (L_1+c_1)(L_2+c_2) \leq \rho|\Delta|. \tag{3.9}$$

We shall discuss this result in greater detail in Section 5.

If $Q(x_1, \ldots, x_n) = \sum_{i,j=1}^n a_{ij}x_ix_j$, $a_{ij} = a_{ji}$ is an indefinite quadratic form of determinant $D = \det(a_{ij}) \neq 0$, then Q can be written as

$$Q(x_1, \ldots, x_n) = L_1^2 + \cdots + L_r^2 - L_{r+1}^2 - \cdots - L_{r+s}^2,$$

where $1 \leq r, s < n, r+s = n, L_1, \ldots, L_{r+s} = L_n$ are linear forms in $x_1, \ldots, x_n$ of determinant $\Delta \neq 0$, where $|D| = |\Delta|^2$. The numbers r, s are uniquely determined by Q. We say Q is a quadratic form of type (r, s) and the number $\sigma = r - s$ is called the signature of Q.

Let $C_{r,s}$ denote the infimum of all those constants C such that for any indefinite quadratic form $Q(x_1, \ldots, x_n)$ of type (r, s), determinant $D \neq 0$ and real numbers $c_1, \ldots, c_n$ there exists integers $x_1, \ldots, x_n$ such that

$$|Q(x_1+c_1, \ldots, x_n+c_n)| \leq (C|D|)^{1/n}. \tag{3.10}$$

One of the basic problems in Geometry of Numbers is to determine $C_{r,s}$ and to determine all those quadratic forms Q of type (r, s) for which equality is needed in (3.10) with $C = C_{r,s}$ for some $c_1, \ldots, c_n$. Such forms we shall call *critical forms*. Results of Minkowski and Davenport cited above imply that $C_{1,1} = \frac{1}{4}$, $C_{2,1} = C_{1,2} = \frac{27}{100}$.

Birch[19] (1958) generalized Minkowski's result to all forms of signature 0, by proving that $C_{r,r} = \frac{1}{4}$ for all $r \geq 1$. More explicitly, he proved

Theorem 5 *Let $Q(x_1, \dots, x_{2r})$ be an indefinite quadratic form of signature 0 and determinant $D \neq 0$. Then for any real numbers $c_1, \dots, c_{2r}$ there exist integers $x_1, \dots, x_{2r}$ such that*

$$|Q(x_1 + c_1, \dots, x_{2r} + c_{2r})| \leq \left(\frac{1}{4}|D|\right)^{1/2r}. \tag{3.11}$$

Equality is necessary in (3.11) *iff*

$$Q \sim \rho(x_1x_2 + \cdots + x_{2r-3}x_{2r-2} + 2x_{2r-1}x_{2r}),\ \rho \neq 0$$

and

$$(c_1, \dots, c_{2r}) \sim \left(0, \dots, 0, \frac{1}{2}, \frac{1}{2}\right) (\operatorname{mod} 1).$$

His proof is by induction on r and by dealing with the zero forms and non-zero forms separately. A quadratic form $Q(x_1, \dots, x_n)$ is said to be a zero form if there exist integers $x_1, \dots, x_n$ not all zero such that $Q(x_1, \dots, x_n) = 0$, otherwise it is called a non-zero form.

If $Q(x_1, \dots, x_n) = Q_n$ is a quadratic form in n variables we define the inhomogeneous minimum $M_I(Q_n)$ by

$$M_I(Q) = \sup_{c_1,\dots,c_n} \left\{ \inf_{x_i \equiv c_i \ (\operatorname{mod} 1)} |Q(x_1, \dots, x_n)| \right\}. \tag{3.12}$$

The results proved by Birch are

Theorem 6 *Let Q_n be an indefinite non-singular quadratic form in n variables with incommensurable coefficients. Suppose that either (i) $n \geq 3$ and Q_n represents arbitrarily small non-zero values, or (ii) $n \geq 4$ and Q_n represents zero properly.*

Then $M_I(Q_n) = 0$.

Theorem 7 *Let Q_{2r} be a quadratic form in $2r$ variables of signature zero and determinant $D \neq 0$, which does not represent zero properly. Then*

$$M_I(Q_{2r}) \leq \left|\frac{1}{4}D\right|^{1/2r} \min\left(1, \left(\frac{5}{6}\right)^{(r-4)/3}\right).$$

Theorem 8 *Let Q_{2r} be a rational quadratic form of signature zero and determinant $D \neq 0$ that represents zero properly. Then*

$$M_I(Q_{2r}) \leq \left|\frac{1}{4}D\right|^{1/2r}.$$

Theorem 6 is derived from Blaney's theorem stated above and the following results of Oppenheim[126,127] proved in 1953.

Theorem 9 *If $Q(x_1, \dots, x_n)$ is an indefinite quadratic form such that for every $\varepsilon > 0$, the inequality*

$$0 < Q(x_1, \dots, x_n) < \varepsilon$$

is solvable in integers $x_1, \dots, x_n$, then for $n \geq 3$, the inequality

$$0 < -Q(x_1, \dots, x_n) < \varepsilon$$

is also solvable in integers $x_1, \dots, x_n$ for every $\varepsilon > 0$.

Theorem 10 *If the indefinite, non-singular quadratic form $Q(x_1, \dots, x_n)$ is an incommensurable form, which represents zero properly then the inequality*

$$0 < |Q(x_1, \dots, x_n)| < \varepsilon$$

is solvable in integers.

For the proof of Theorem 7, using results from ternary and quaternary forms, Birch developed a reduction in which Q_{2r} represents a binary quadratic form of small determinant. More precisely he proved.

Theorem 11 *Let Q_{2r} be a quadratic form of signature zero in $2r \geq 4$ variables of determinant $D \neq 0$ such that Q_{2r} does not take small values, then*

$$\begin{aligned} Q_{2r} &\sim \alpha(x_1 + a_2x_2 + \cdots)^2 - \beta(x_2 + b_3x_3 + \cdots)^2 + Q_{2r-2}(x_3, \dots, x_{2r}) \\ &= \psi(x_1 + a_2x_2, \dots, x_2 + b_3x_3 + \cdots) + Q_{2r-2}(x_3, \dots, x_{2r}), \end{aligned}$$

where $\psi(x, y) = \alpha(x + a_2y)^2 - \beta y^2$ is an indefinite binary quadratic form of determinant $-\delta$, with

$$\delta^r \leq \lambda|D|$$

where λ is a suitable constant.

To derive the result by induction he proved the following asymmetric inequality for non-homogeneous indefinite binary quadratic forms.

Lemma 1 *Let $Q(x, y)$ be an indefinite binary quadratic form of determinant $-\delta$. Then given any real numbers x_0, y_0 and μ, we can find $(x, y) \equiv (x_0, y_0) \pmod 1$ such that*

$$|Q(x, y) + \mu| \leq max(2^{-1/2}\delta^{1/2}, \delta^{1/4}|\mu|^{1/2}). \tag{3.13}$$

If $Q \sim P_2 = x^2 + xy - y^2$ we can solve the stronger inequality

$$|P_2(x, y) + \mu| \leq max(1, |\mu|^{2/3}). \tag{3.14}$$

For the zero rational quadratic forms Birch proved the following reduction result:

Theorem 12 *Let Q_{2r} be a rational quadratic form of determinant $D \neq 0$ of signature zero that represents zero properly. Then*

$$\begin{aligned} Q_{2r} \sim\ & H_1(x_1 + a_{12}x_2 + \cdots)x_2 + H_2(x_3 + a_{24}x_4 + \cdots)x_4 \\ & + \cdots + H_m(x_{2m-1} + a_{m,2m}x_{2m} + \cdots)x_{2m} + Q_{2n-2m}, \end{aligned}$$

where $|a_{ij}| \leq \frac{1}{2}$, $H_i > 0$, $\frac{H_{i+1}}{H_i}$ is an integer and either $m = n$ in which case Q_{2n-2m} is omitted or $m = n-1$ or $n-2$ in which case Q_{2n-2m} is a binary or a quaternary non-zero form of signature zero.

From these results one easily deduces the following

Corollary 1 *If Q_{2r} is a rational quadratic form of determinant $D \neq 0$, signature zero and which represents zero, then we have*

$$Q \sim \psi(x_1 + \alpha x_2 + \ldots, x_2 + \beta x_3 + \cdots) + Q_{2r-2}(x_3, \ldots, x_{2r}),$$

where either $\psi = H(x_1 + a_2x_2 + \cdots)x_2$ with $(\frac{1}{4}H)^{2r} \leq \frac{1}{4}|D|$ or ψ is binary form with det d with $|d|^r < \frac{1}{4}|D|$.

Theorems 7 and 8 follow easily by induction on r and using Theorem 11, Corollary 1 and the Lemma 1.

Watson[142] (1960) extended Theorem 6 of Birch by proving

Theorem 13 *Let $f(x_1, \ldots, x_n)$ be an indefinite non-singular quadratic form of determinant $D \neq 0$ in $n \geq 3$ variables which takes arbitrarily small non-zero values for integers $x_1, \ldots, x_n$. Then for any real $\alpha, c_1, \ldots, c_n$ and $\varepsilon > 0$, the inequality*

$$|f(x_1 + c_1, \ldots, x_n + c_n) - \alpha| < \varepsilon$$

has a solution in integers $x_1, \ldots, x_n$.

Theorem 14 *Let $f(x_1, \ldots, x_n)$ be an indefinite non-singular quadratic form with $n \geq 3$ which is a multiple of a rational form and represents zero non-trivially. Let $c_1, \ldots, c_n$ be real numbers which are not all rational. Then for any real α and $\varepsilon > 0$, there exist integers $x_1, \ldots, x_n$ such that*

$$|f(x_1 + c_1, \ldots, x_n + c_n) - \alpha| < \varepsilon.$$

Theorem 13 is proved in the same manner as Theorem 6. Theorem 14 after a suitable reduction is derived from the following result on uniform distribution mod 1.

Lemma 2 *If $P = P(x_1, \ldots, x_n)$ is a polynomial whose coefficients other than the constant term are not all rational, then the fractional part of P for integers x_i takes values which are everywhere dense in the interval $(0, 1)$.*

Next major development in the determination of $C_{r,s}$ was done by Watson[143] in 1962 who determined $C_{r,s}$ for all r, s such that $n = r + s \geq 21$ i.e. he solved the problem completely for forms in $n \geq 21$ variables. The precise result proved by him is

Theorem 15 *Let $Q(x_1, \dots, x_n)$ be an indefinite quadratic form in $n \geq 21$ variables of determinant $D \neq 0$ of type (r, s) and signature $\sigma = r - s$. Then for any real numbers $c_1, \dots, c_n$ there exist integers $x_1, \dots, x_n$ such that*

$$|Q(x_1 + c_1, \dots, x_n + c_n)| \leq (C_{r,s}|D|)^{1/n} \tag{3.15}$$

where

$$C_{r,s} = \begin{cases} 1/4 & \text{if } \sigma \equiv 0 \text{ or } \pm 1 \pmod 8 \\ 1/3 & \text{if } \sigma \equiv \pm 2 \pmod 8 \\ 1/2 & \text{if } \sigma \equiv \pm 3 \pmod 8 \\ 1 & \text{if } \sigma \equiv \pm 4 \pmod 8 \end{cases} \tag{3.16}$$

Equality is needed in (3.15) *for each r, s for suitable Q and $c_1, \dots, c_n$.*

Watson reduces the problem to consideration of values of suitable integral quadratic polynomials in n variables. A well known conjecture of Oppenheim states that if $Q(x_1, \dots, x_n)$ is an incommensurable indefinite quadratic form of determinant $D \neq 0$ in $n \geq 5$ variables then $Q(x_1, \dots, x_n)$ takes arbitrarily small values. This conjecture was proved to be true for forms in $n \geq 21$ variables in a series of papers by Davenport, Birch and Ridout (see [44]) using analytical and other techniques.

Thus the result for incommensurable forms follows from Theorem 6. For commensurable forms, by homogeneity we can assume Q to be a primitive integral form. Since $n \geq 21$, Q is a zero form. If $c_1, \dots, c_n$ are not all rational, then again (3.15) is true with any $\varepsilon > 0$ on the right hand side by Theorem 14. We may therefore assume that c_i are rational and write

$$c_i = \frac{u_i}{q}, i = 1, \dots, n$$

where u_i, q are integers, $q > 0$ and $(u_1, \dots, u_n, q) = 1$. We can therefore write

$$f(x_1, \dots, x_n) = Q(x_1 + c_1, \dots, x_n + c_n) = Q(x_1, \dots, x_n) + K^{-1}(b_1 x_1 + \dots + b_n x_n) + c$$

where c is a rational constant, $b_1, \dots, b_n, K$ are integers with $K > 0$, $(K, b_1, \dots, b_n) = 1$. The values of f for integral x_i are of the form $c + K^{-1}N$. Any such value is assumed by f if and only if the Diophantine equation

$$KQ(x_1, \dots, x_n) + b_1 x_1 + \dots + b_n x_n = N \tag{3.17}$$

is soluble in integers x_i.

If one can show that (3.17) is soluble for at least one of every H consecutive integers, then it will follow that

$$(C_{r,s}|D|)^{1/n} \leq \frac{H}{2K}.$$

It is known that (3.17) has solution iff

$$KQ(x_i, \dots, x_n) + b_1 x_1 + \dots + b_n x_n \equiv N \pmod{p^t}$$

has a solution for all prime powers p^t. Using p-adics it is easy to see that we need to consider only finitely many prime powers which are related to the determinant of Q. A careful analysis of this reduces the problem to one concerning distribution of quadratic residues.

Watson in fact by this method also proves a more general gap theorem, namely he obtains the smallest value of β depending only on σ, n, D such that for any α

$$\alpha < Q(x_1 + c_1, \ldots, x_n + c_n) \leq \alpha + \beta$$

always has a solution for any Q of signature σ. We shall discuss this in greater detail in Section 6.

It may be remarked that the condition $n \geq 21$ is not very essential for this p-adic argument but avoids lot of tedious computation. However, as he himself says, the method will not work for small n.

Watson's Conjecture:

The formulae for $C_{r,s}$ given in (3.16) hold for $n \geq 4$.

Dumir (1967) proved this conjecture for $n = 4$. For $n = 5$, the conjecture was proved by Hans-Gill and Madhu Raka[79,80] (1979, 1980). They also obtained all the critical forms. Their methods for $n = 4, 5$ like that of Davenport, are based on reduction of the problems to forms in fewer variables and obtaining and using suitable asymmetric inequalities for such forms.

It may be remarked that the conjectured values of $C_{r,s}$ for $n \geq 4$ depend only on the class of $\sigma \equiv r - s \pmod 8$.

Next obvious step was to prove Watson's conjecture for quadratic forms in any number of variables having a fixed signature $\sigma = 0, \pm 1, \pm 2, \pm 3$, and 4. For forms of signature 0, it was already done by Birch. Madhu Raka[93,94,95] (1981, 1983 a,b) using the method of Birch proved Watson's conjecture for all forms of signature $\sigma = \pm 1, \pm 2, \pm 3, \pm 4$. Thus together with Birch's result for $\sigma = 0$, Watson's conjecture was known for a complete set of residue system mod 8.

After this, in order to prove Watson's conjecture completely, one just needed to prove that $C_{r,s} = C_{r',s'}$ if $r+s = r'+s' = n$ and $r - s \equiv r' - s' \pmod 8$. This was proved by Dumir, Hans-Gill and Woods[60] (1994). Margulis[104] in 1987 proved Oppenheim's conjecture. He in fact proved

Theorem 16 *If $f(X)$ is an indefinite incommensurable quadratic form of determinant $D \neq 0$ in $n \geq 3$ variables, then for any $\varepsilon > 0$ there exists X in $\mathbb{Z}^n$ such that*

$$0 < |f(X)| < \varepsilon.$$

Thus for incommensurable forms $f(X)$ in $n \geq 3$ variables, Theorem 13 implies that for any real α, $C = (c_1, \ldots, c_n)$ and $\varepsilon > 0$

$$|f(X + C) - \alpha| < \varepsilon \tag{3.18}$$

has a solution in $X \in \mathbb{Z}^n$.

Further if $f(X)$ is commensurable and $n \geq 5$ and C is in $\mathbb{R}^n - Q^n$, then again (3.18) is true by Theorem 14.

Thus to prove that $C_{r,s}$ depends only on signature mod 8, we need to restrict ourselves to only primitive integral forms $Q(X)$ and $C \in Q^n$. Using results on congruentally equivalent integral forms Dumir, Hans-Gill and Woods[60] in 1994 proved the following general theorem on non-homogeneous indefinite quadratic forms with integral coefficients and having the same signature modulo 8.

Theorem 17 *Let $\alpha < \beta$ be real numbers and let $n \geq 6$ be an integer. Let σ, σ_0 be integers such that $|\sigma| < n, |\sigma_0| < n, \sigma \equiv \sigma_0 \pmod 8$ and $n \equiv \sigma \pmod 2$. Suppose that for each integral quadratic form $g(X)$ in n variables of determinant $D \neq 0$ and signature σ_0 and for any $C \in Q^n$, the inequality*

$$\alpha|D|^{1/n} < g(X+C) \leq \beta|D|^{1/n} \tag{3.19}$$

is solvable for X in $\mathbb{Z}^n$.

Then for any integral quadratic form $f(X)$ in n variables of determinant D and signature σ and for any $C \in Q^n$, there exists X in $\mathbb{Z}^n$ such that

$$\alpha|D|^{1/n} < f(X+C) \leq \beta|D|^{1/n}. \tag{3.20}$$

For convenience if we denote by C_{n,σ_0} the constant in Watson's conjecture for quadratic form of signature σ_0 in n variables then we have

$$C_{r,s} = \begin{cases} 1/4 & \text{if } \sigma_0 \equiv 0 \text{ or } \pm 1 \pmod 8 \\ 1/3 & \text{if } \sigma_0 \equiv \pm 2 \pmod 8 \\ 1/2 & \text{if } \sigma_0 \equiv \pm 3 \pmod 8 \\ 1 & \text{if } \sigma_0 \equiv \pm 4 \pmod 8 \end{cases} \tag{3.21}$$

If we take $|\sigma_0| \leq 4, -\alpha = \beta = (C_{n,\sigma_0})^{1/n}$, Watson's conjecture follows from Theorem 17 and the results of Birch and Madhu Raka cited above.

§4 Positive Values of Non-homogeneous Indefinite Quadratic Forms

Since indefinite quadratic forms take both positive and negative values one can consider only one sided inequalities and ask:

Do there exist numbers Γ depending only on r, s such that if $Q(x_1, \ldots, x_n)$ is a real indefinite quadratic form of type (r, s) and determinant $D \neq 0$, then for any given real $c_1, \ldots, c_n$ there exist integers $x_1, \ldots, x_n$ satisfying

$$0 < Q(x_1 + c_1, \ldots, x_n + c_n) \leq (\Gamma|D|)^{1/n}? \tag{4.1}$$

As remarked earlier the existence of such Γ was proved by Blaney[21] (1948).

Let $\Gamma_{r,s}$ denote the infimum of all such constants Γ for which (4.1) has a solution. The problem is to evaluate $\Gamma_{r,s}$ for different r, s and to determine all those forms Q and $c_1, \ldots, c_n$ for which equality is needed in (4.1) with $\Gamma = \Gamma_{r,s}$.

For indefinite binary quadratic forms this result was proved by Davenport and Heilbronn[43] (1947) who showed that

$$0 < Q(x + x_0, y + y_0) \leq (4|D|)^{1/2} \tag{4.2}$$

always has a solution if $Q(x, y)$ is an indefinite binary quadratic form of determinant $D \neq 0$. Clearly equality is needed in (4.2) if

$$Q(x, y) = xy, \quad x_0 = y_0 = 0$$

Thus $\Gamma_{1,1} = 4$.

Blaney[23] (1950) showed that $\Gamma_{2,1} = 4$ and also determined all the critical forms. This result was rediscovered by Barnes[18] in 1961.

$\Gamma_{1,2} = 8$ was proved by Dumir[48] in 1967 thereby completing the result for ternary forms.

Dumir[49,50] (1968; a,b) proved that $\Gamma_{3,1} = 16/3, \Gamma_{2,2} = 16$. Dumir and Hans-Gill[53] (1981) proved that $\Gamma_{1,3} = 16$, thereby completing the results for quaternary forms.

$\Gamma_{3,2} = 16, \Gamma_{4,1} = 8$ were proved by Hans-Gill and Raka[81,82] in 1980 and 1981 respectively. Methods used in the proofs of the above results are purely arithmetical, relying on results for asymmetric, homogeneous and non-homogeneous forms in 2, 3, 4 variables, some of which were known and some were discovered for purpose of applications.

$\Gamma_{2,3} = (7/4)^5$ was proved by Bambah, Dumir and Hans-Gill[12] in 1984 who in fact proved more which we shall discuss later. Bambah, Dumir and Hans-Gill[10,11,12] in a series of papers in 1981, 1983, 1984, using a reduction similar to that of Birch[19] (1958) cited earlier, proved that for $n \geq 6, -1 \leq \sigma \leq 3$, where $\sigma = r - s$ is the signature of the quadratic form $Q(x_1, \ldots, x_n)$ of types (r, s),

$$\Gamma_{r,s} = \frac{2^n}{|\sigma| + 1}. \tag{4.3}$$

For incommensurable forms they use method similar to that of Birch as in Theorem 11 and the proof is completed by induction by using the following analogue of Lemma 1 proved by them.

Lemma 3 *Let $\phi(x, y)$ be an indefinite binary quadratic form of determinant $-\delta$ which does not represent zero trivially. Then given any real numbers x_0, y_0 and $\mu > 0$, we can find $(x, y) \equiv (x_0, y_0)(\text{mod}1)$ such that*

$$0 < \phi(x, y) + \mu \leq max\left(2\sqrt{\delta}, \sqrt{\delta + 4\mu\sqrt{\delta}} - \sqrt{\delta}\right). \tag{4.4}$$

If $Q \sim P_2 = x^2 + xy - y^2$, we can solve the stronger inequality

$$0 < P_2(x, y) + \mu \leq max\,[2.1, (\sqrt{5}\mu^2)^{1/3}]. \tag{4.5}$$

For commensurable forms, which for $n \geq 5$ are necessarily zero forms, though they use the reduction of Birch, but the proof for this case is not done by Birch's method but by a gap theorem of T.H. Jackson[86] (1971) for zero forms. It relies very heavily on a Lemma of Macbeath[92] (1951) on the non-homogeneous minima for a parabolic region. Macbeath's Lemma can be formulated as

Lemma 4 *Let α, β, A be real numbers with $\alpha \neq 0$. Let $2h, K$ be positive integers such that*

$$\left|h - k^2|\alpha|\right| + \frac{1}{2} < A. \tag{4.6}$$

Further suppose that either $|\alpha| \neq \frac{h}{k^2}$ or $\beta \not\equiv \frac{h}{k}(mod\ \frac{1}{k}, 2\alpha)$ i.e. $\beta - \frac{h}{k}$ is not an integral combination of $\frac{1}{k}$ and 2α, then for any real number ν, there exist integers x,y satisfying

$$0 < x + \beta y + \alpha y^2 + \nu < A. \tag{4.7}$$

If $Q(x_1, \ldots, x_n)$ is a zero rational form, by using Birch reduction and homogeneity, we can suppose

$$Q(x_1, \ldots, x_n) = (x_1 + a_2x_2 + \cdots)x_2 + \varphi(x_3, \ldots, x_n).$$

Suppose φ represents a number $-a$, primitively with $a \neq 0$, satisfying a suitable inequality. In this case we can assume

$$\varphi(x_3, \ldots, x_n) = -ax_3^2 + \ldots$$

so that

$$Q(x_1, \ldots, x_n) = (x_1 + a_2x_2 + \ldots)x_2 - ax_3^2 + \ldots$$

Suppose we are interested in solving an inequality of the type

$$0 < Q(x_1 + c_1, \ldots, x_n + c_n) < (\lambda|D|)^{4/n} = d, \text{ say.}$$

We can take $|c_i| \leq \frac{1}{2}$. In case $d > \frac{1}{2}$ and $c_2 \neq 0$, then for $x_1 = x, x_2 = \ldots = x_n = 0$, $Q(x + c_1, c_2, \ldots, c_n) = xc_2 + \mu$, for some μ. Now choose integer x such that

$$0 < Q(x + c_1, c_2, \ldots, c_n) = xc_2 + \mu \leq |c_2| \leq \frac{1}{2} < d.$$

Again if $c_2 = 0$ and $d > 1$ we are through as above if we take $x_2 = 1, x_3 = \ldots = x_n = 0$ and x_1 suitably.

In other cases we take $x_2 = c_2$ if $0 < |c_2| \leq \frac{1}{2}, x_2 = 1$ if $c_2 = 0, x_1 = x, x_3 = y, x_4 = \ldots = x_n = 0$, so that

$$\begin{aligned} Q(x_1 + c_1, x_2 + c_2, \ldots, x_n + c_n) &= [x + c_1 + a_2x_2 + a_3(y + c_3) + a_4x_4 + \ldots]x_2 \\ &\quad -a(y + c_3)^2 + \ldots \\ &= xx_2 + \beta' y - ay^2 + \nu' \end{aligned}$$

for some β', ν'. Therefore,

$$0 < Q(x_1 + c_1, \ldots, x_n + c_n) < d$$

is soluble if there exist integers, x, y such that

$$0 < xx_2 + \beta' y - ay^2 + \nu < d$$

i.e.

$$0 < \pm x + \frac{\beta'}{|x_2|}y - \frac{a}{|x_2|}y^2 + \frac{\nu'}{|x_2|} < \frac{d}{|x_2|}.$$

This is an inequality of the type (4.7) and hence is soluble if we can satisfy the conditions of Lemma 4 for suitable h, k.

It was conjectured by Bambah, Dumir and Hans-Gill[10] that for $n \geq 6$,

$$\Gamma_{r,s} = \begin{cases} \frac{2^n}{|\sigma|+1} & \text{if } |\sigma| \leq 3 \\ \frac{2^n}{\sigma} & \text{if } \sigma = 4 \end{cases} \tag{4.8}$$

and

$$\Gamma_{r,s} = \Gamma_{r',s'}$$

if

$$r + s = r' + s' = n, r - s \equiv r' - s' (\text{mod } 8) \tag{4.9}$$

except perhaps for a finite number of exceptions.

Mary Flahive[64] (1988) proved the conjecture of Bambah, Dumir and Hans-Gill for $n \geq 21$ by the technique similar to that of Watson[143] (1962). Aggarwal and Gupta[1,2,3] (1988, 1991a,b) determined $\Gamma_{r,s}$ for $\sigma = 2$ and $n \geq 8$, $\sigma = -3$ and $n \geq 9$ and $\sigma = 4$ with $n \geq 6$ and confirmed the conjecture of Bambah et al. in these cases.

From the general theorem (Theorem 3.7) of Dumir, Hans-Gill and Woods[60] proved in 1994, the second part of Bambah et al. conjecture follows by taking $\alpha = 0$, $\beta = \Gamma_{r,s}^{1/n}$ from the known results for forms of signature $-3 \leq \sigma \leq 4$.

This argument determines $\Gamma_{r,s}$ except for $\Gamma_{2,5}, \Gamma_{2,4}, \Gamma_{1,4}$.
$\Gamma_{2,5} = 32$ was proved by Dumir and Sehmi[55] (1994) and $\Gamma_{2,4} = 64/3$ was proved by Dumir, Hans-Gill and Sehmi[61] in 1995; confirming the conjecture of Bambah et al.

This leaves open only the determination of $\Gamma_{1,4}$. It is conjectured that $\Gamma_{1,4} = 8$. Dumir and Sehmi[59] (1994) proved that $8 \leq \Gamma_{1,4} \leq 16$. Madhu Raka and Urmila Rani[96] (1997) have improved this to $8 \leq \Gamma_{1,4} \leq 12$. Madhu Raka and Urmila Rani observe that it may be possible to extend this method to show that $\Gamma_{1,4} < 32/3$, but the number of cases to be dealt with is very large. There is difficulty in applying Macbeath's Lemma if one attempts to obtain a smaller upper bound.

It may be noted that for large values of n, the evaluation of $C_{r,s}, \Gamma_{r,s}$ is relatively easy. For small values of n a detailed and careful analysis is needed. However it may be noted that after the proof of Margulis of Oppenheim's conjecture, for $n \geq 5$ we need to deal with rational forms only and these are zero forms by Meyer's theorem; so Birch's reduction for zero rational forms and Macbeath's Lemma can be used.

§5 Asymmetric Inequalities for Non-homogeneous Indefinite Quadratic Forms

As described in §3 and §4 to determine the constants $C_{r,s}$ and $\Gamma_{r,s}$ one needs to prove more general asymmetric inequalities for non-homogeneous indefinite quadratic forms. We briefly describe these in this section.

Let $Q(x_1, \ldots, x_n)$ be an indefinite quadratic form of determinant $D \neq 0$ of type (r, s). Let t be real. One is interested in finding a function $f(t)$ (which will depend on r, s also) such that for any real $c_1, \ldots, c_n$ there exist integers $x_1, \ldots, x_n$ such that

$$-t(f(t)|D|)^{1/n} < Q(x_1 + c_1, \ldots, x_n + c_n) \leq (f(t)|D|)^{1/n}. \tag{5.1}$$

One would be interested in finding the best $f(t)$ for each t which will work for all quadratic form Q of type (r, s). However for the purpose of applications it will be more appropriate to find continuous functions $f(t)$ which will work for t in a given interval but which need not be best possible for all t in that interval.

Determination of such $f(t)$ is related to the non-homogeneous minima of the region

$$-t < x_1^2 + \cdots + x_r^2 - x_{r+1}^2 - \cdots - x_{r+s}^2 < 1.$$

Theorem 4 proved by Davenport[42] in 1948 shows that for indefinite binary quadratic forms Q(x, y) we can solve (5.1) with

$$f(t) = \frac{1}{4\sqrt{t}}.$$

For $t = 1$ it reduced to Minkowski's result.

Blaney[24] (1950) improved upon this result. For binary forms there is no loss of generality in assuming that $t \leq 1$. Blaney reproves Davenport's theorem and shows that the result is best possible for $t = 1$ and $t = \frac{m}{m+2}$, $m = 1, 2, 3, \ldots$ and obtained all the forms for which equality is needed for these t. For $0 \leq t \leq 1/3$, Blaney proved that the inequality (5.1) is soluble with $f(t) = [(t + 1)(1 + 9t)]^{-1/2}$ and the inequalities are best possible for $t = \frac{1}{4m-1}$, $m = 1, 2, \ldots$, and $t = 0$. For $t < 0$, he obtains a suitable function $f(t)$ which is best possible for only finitely many values of t. The first and third results are adaptation of the geometric method of Davenport while the other proofs are arithmetical in nature. These results of Blaney have been extensively used in literature for determining different types of non-homogeneous inequalities for indefinite forms in larger number of variables. Dumir and Grover[52] (1986) improved upon some of the results of Blaney for t lying in different sub-intervals and showed that their results were best possible for infinitely many new values of t. Grover and Raka[67] (1991) obtained further improvements for some range of t.

Dumir[51] (1969) determined such a function $f(t)$ for indefinite ternary forms of type (2,1) for $t \geq 0$. $f(t)$ is specified as a cubic polynomial in six different intervals. For example $f(t) = 8/(1 + t)^3$ for $7 \leq t \leq \infty$. The results is best possible for eight values of t namely $t = 0, 1/7, 3/5, 1, 9/7, 3/7$ and ∞. The earlier best known values were $f(1)$ due to Davenport[42], $f(0)$ due to Blaney[22], Barnes[18], and $f(\infty)$ due to Dumir[48]. Hans-Gill and Raka[76] (1979) were able to sharpen Dumir's result in the interval $1/7 < t < 1/3$ by utilizing known inequalities for the values of homogeneous and inhomogeneous binary quadratic forms. Hans-Gill and Raka[77] (1980) also gave a function $f(t)$ for $0 \leq t < 1$ such that the inequality

$$t(f(t)|D|)^{1/3} < Q(x + x_0, y + y_0, z + z_0) \leq (f(t)|D|)^{1/3}$$

is soluble for quadratic forms Q of type (2,1). The result is best possible only for $t = 0$ and 1/9.

Results of similar nature for other quadratic forms of special type have been determined by Dumir, Hans-Gill and Raka and have been applied for determining $C_{r,s}$ and $\Gamma_{r,s}$.

§6 Gaps between Values of Non-homogeneous Indefinite Quadratic Forms

Watson[143] in his 1962 paper also considered another type of asymmetric inequality for non-homogeneous indefinite quadratic forms. Let $Q(x_1, \ldots, x_n)$ be an indefinite quadratic form of determinant $D \neq 0$, and type (r, s). Watson considered inequalities of the type

$$\alpha < Q(x_1 + c_1, \ldots, x_n + c_n) \leq \alpha + \beta \tag{6.1}$$

and defined $G = G(Q)$ as the least upper bound of the real numbers β for which for suitable real $\alpha = \alpha(Q, \beta)$ the inequality (6.1) is not soluble in integers $x_1, \ldots, x_n$ for some $c_1, \ldots, c_n$. Clearly the inhomogeneous minima M_I of Q satisfies.

$$2M_I \leq G.$$

Watson proved the following results:

Theorem 18 *Let $Q(x_1, \ldots, x_n)$ be a real indefinite non-singular quadratic form in $n \geq 21$ variables, of determinant D, type (r, s) and signature σ. Then*

$$G \leq 2|D|^{1/n}. \tag{6.2}$$

Equality in (6.2) *is needed iff either*

(*i*) $Q \sim x_1^2 + \cdots + x_r^2 - x_{r+1}^2 - \cdots - x_{r+s}^2, (c_1, \ldots, c_n) \sim (1/2, \ldots, 1/2)$

or

(*ii*) *n even,* $8 \not| \sigma$.

$$Q \sim (\mathrm{sgn})\, \sigma \sum_{0<8i\leq|\sigma|} E(x_{8i-7}, \ldots, x_{8i}) + \sum_{|\sigma|<2i\leq n} x_{2i-1}x_{2i},$$
$$(c_1, \ldots, c_n) \sim (0, \ldots, 0)$$

where

$$E = E(x_1, \ldots, x_8) = \frac{1}{4}\sum_{i=1}^{4}(x_i^2 + (2x_{i+4} + x_i - x_1 - x_2 - x_3 - x_4)^2.$$

Watson in fact showed that for $n \geq 21$ if $f(x_1, \ldots, x_n) = Q(x_1 + c_1, \ldots, x_n + c_n)$ is not a multiple of a rational form, then we have

$$G^n < \frac{2^n}{5}|D|. \tag{6.3}$$

T.H. Jackson[86] (1971) improved upon Watson's result for zero forms for all n by proving

Theorem 19 *If $Q(x_1, \ldots, x_n)$ is a non-singular indefinite zero form of determinant D, then*

$$G(Q) \leq 2|D|^{1/n} \tag{6.4}$$

and the result is best possible for all n.

He determined all the critical forms also. This proof is by induction on n. Jackson conjectured the result to be true for $n \geq 4$, without the hypothesis that $Q(x_1, \dots, x_n)$ is a zero form. The result is false for $n = 2, 3$.

In 1983, Bambah, Dumir and Hans-Gill[11] proved the following weaker version of conjecture of Jackson for non-zero indefinite quadratic forms of signature 0, ± 1 or ± 2.

Theorem 20 *Let $Q(x_1, \dots, x_n)$ be a real indefinite quadratic form of determinant $D \neq 0$ of signature 0, ± 1 or ± 2. Let $|\alpha| \leq |D|^{1/n}$. For any real $c_1, \dots, c_n$ there exist integers $x_1, \dots, x_n$ such that*

$$|Q(x_1 + c_1, \dots, x_n + c_n) - \alpha| \leq |D|^{1/n}.$$

It may be observed that in view of the result of Margulis on Oppenheim's conjecture, Jackson's conjecture is true for non-zero forms in $n \geq 4$ variables which are not multiples of rational forms using Watson's Theorem, cited above. Since by Meyer's theorem, rational indefinite quadratic forms in $n \geq 5$ variables represent zero, Jackson's conjecture is true for $n \geq 5$. Thus Jackson's conjecture needs to be verified for non-zero rational forms in four variables only.

§7 Isolated Minima of Non-homogeneous Indefinite Quadratic forms

A natural question is whether the constants $C_{r,s}$, $\Gamma_{r,s}$ are isolated i.e. if we omit the forms equivalent to the critical forms for $C_{r,s}(\Gamma_{r,s})$ can the inequality 3.10 (4.1) be solved for some $C < C_{r,s}(\Gamma_{r,s})$. It is known that $C_{1,1} = 1/4$, $\Gamma_{1,1} = 4$ are not isolated i.e. for any $\varepsilon > 0$ the inequality

$$|Q(x + x_0, y + y_0)| < \left(\left(\frac{1}{4} - \varepsilon\right)|D|\right)^{1/n}$$

is not satisfied for some $Q(x + x_0, y + y_0) \not\sim (x + \frac{1}{2})(y + \frac{1}{2})$.

Similarly, for any $\varepsilon > 0$ the inequality

$$0 < Q(x + x_0, y + y_0) < ((4 - \varepsilon)|D|)^{1/2}$$

is not soluble for some $Q(x + x_0, y + y_0) \not\sim (x + 0)(y + 0)$.

However it is known that for $n \geq 3$, the constants $C_{r,s}$, $\Gamma_{r,s}$ are isolated (see Vulakh[141] (1985)). Davenport[42] (1948) himself had shown that the constant $C_{2,1} = C_{1,2} = \frac{27}{100}$ is isolated. The exact value of the next two successive minima for ternary forms were determined by Barnes[16,17] (1954, 1956). If we consider the complete spectrum of values $m(Q, c_1, \dots, c_n) = \inf \frac{|Q(x_1+c_1,\dots,x_n+c_n)|}{|D|^{1/n}}$ for indefinite quadratic forms $Q(x_1, \dots, x_n)$ of type (r, s) and determinant D, where the infimum is taken over all integers $x_1, \dots, x_n$, then

$$M_I|Q| = \sup_{c_1,\dots,c_n} m(Q, c_1, \dots, c_n)$$

and

$$C_{r,s} = \sup_{Q} (M_I(Q))^n.$$

Vulakh[141] (1985) has shown that for $n \geq 3$ the set of numbers $m(Q, c_1, \ldots, c_n)$ has zero as its only limit point. We arrange the numbers $m^n(Q, c_1, \ldots, c_n)$ in descending order. The first term is $C_{r,s}$.

Let $C_{r,s}^{(k)}$ be the kth term of this sequence. Then $C_{r,s}^{(k)}$ is called the kth successive minima for non-homogeneous quadratic forms of type (r, s). Following results are known

$C_{2,1}^{(2)} = C_{1,2}^{(2)} = \frac{4}{15}$	Barnes[16](1954)
$C_{2,1}^{(3)} = C_{1,2}^{(3)} = \frac{1}{4}$	Barnes[17](1956)
$C_{3,2}^{(2)} = C_{2,3}^{(2)} = \frac{1}{6}$, $C_{3,2}^{(3)} = C_{2,3}^{(3)} = \frac{1}{8}$	Dumir and Sehmi[58](1995)
$C_{4,1}^{(2)} = C_{1,4}^{(2)} = \frac{1}{4}$	Dumir and Sehmi[55](1994)
$C_{3,1}^{(2)} = C_{1,3}^{(2)} = \frac{1}{4}$	Raka and Urmila Rani[98](1997)
$C_{r+2,r}^{(2)} = C_{r,r+2}^{(2)} = \frac{1}{4}$	Raka and Urmila Rani (unpublished).

Similarly let $\Gamma_{r,s}^{(k)}$ denotes the kth successive minima for positive values of non-homogeneous quadartic form of type (r, s). In this case following results are known

$\Gamma_{2,3}^{(2)} = 16$	Bambah, Dumir and Hans-Gill[12](1983)
$\Gamma_{1,2}^{(2)} = \frac{27}{4}$	Dumir and Hans-Gill[54](1997)
$\Gamma_{3,2}^{(2)} = 8$	Dumir and Sehmi[56](1992)
$\Gamma_{r+1,r}^{(2)} = 2^{2r-1}$	Dumir and Sehmi[57](1992)
$\Gamma_{2,1}^{(2)} = \frac{8}{3}$	Raka and Urmila Rani[101](1996)

Rieger gave a wrong proof of $\Gamma_{3,1}^{(2)} = 4$ in 1976. Raka and Urmila Rani[99,100] (1994) gave a correct proof of Rieger's result and used it to prove that

$$\Gamma_{r+2,r}^{(2)} = 2^{2r}.$$

Raka[102] (1993) and Raka and Rani[97] (1998) have also obtained some results valid for zero forms of types (2, 1) and (2, 2).

The questions about the general behaviour of these sequences are still open.

References

[1] S.K. Aggarwal and D.P. Gupta, Positive values of inhomogeneous quadratic forms of signature - 2, *J. Number Theory* **29** (1988), 138–165.

[2] ———, Least positive values of inhomogeneous quadratic forms of signature –3, *J. Number Theory* **37** (1991), 260–278.

[3] S.K. Aggarwal and D.P. Gupta, Positive values of inhomogeneous forms of signature 4, *J. Indian Math. Soc.* **57** (1991), 1–23.

[4] N.S. Ahmedov, Abstract, all-Union Conference on Number Theory (Samarkand 1972), (1972), 42–43.

[5] N.S. Ahmedov, Representations of matrices in the form DOTU, *Trudy Samarkand Gos. Univ.* **286** (1975), 18–27.

[6] ———, Representation of square matrices by the product of diagonal, orthogonal, triangular and integral and unimodular matrices (in connection with the inhomogeneous Minkowski conjecture), *Zap. Naucn. Sem. Leningrad. Otdel. Mat. Inst. Steklov. (LOMI)* **67** (1977), 86–94, 255 = *J. Soviet Math.* **16** (1981), 820–826.

[7] A.K. Andrijasjan, I.V. Il'in and A.V. Malyšev, An application of a computer to the estimate of the Cebotarev type in the non-homogeneous Minkowski conjecture. *Zap. Naucn. Sem. Leningrad. Otdel Math. Inst. Steklov. (LOMI)* **151** (1986), 7–25.

[8] K.B. Bakiev, An upper estimate for the product of linear non-homogeneous forms for lattices with a small homogeneous minimum, *Zap. Naučn. Sem. Leningrad. Otdel. Mat. Inst. Steklov. (LOMI)* **106** (1981), 5–16, 170 = *J. Soviet Math.* **23** (1983), 2107–2114.

[9] K.B. Bakiev, A.S. Pen and B.F. Skubenko, On an upper bound for the product of linear inhomogeneous forms. *Mat. Zametki* **23** (1978), 789–797.

[10] R.P. Bambah, V.C. Dumir and R.J. Hans-Gill, Positive values of non-homogeneous indefinite quadratic forms, *Proceedings Colloqu. Classical Number Theory, Budapest (Hungary)* (1981), 111–170.

[11] ———, On a conjecture of Jackson on non-homogeneous quadratic forms. *J. Number Theory* **16** (1983), 403–419.

[12] ———, Positive values of non-homogeneous indefinite quadratic forms, II, *J. Number Theory* **18** (1984), 313–341.

[13] R.P. Bambah and A.C. Woods, On a theorem of Dyson, *J. Number Theory* **6** (1974), 422–433.

[14] ———, On a product of three inhomogeneous linear forms, in: *Number Theory and Algebra*, 7–18, Academic Press, New York, 1977.

[15] ———, Minkowski's conjecture for n = 5: A theorem of Skubenko, *J. Number Theory* **12** (1980), 27–48.

[16] E.S. Barnes, The inhomogeneous minimum of a ternary quadratic form, *Acta Math.* **92** (1954), 13–33.

[17] ———, The inhomogeneous minimum of a ternary quadratic form II, *Acta Math.* **96** (1956), 67–97.

[18] ———, The positive values of inhomogeneous ternary quadratic forms, *J. Austral. Math. Soc.* **2** (1961), 127–132.

[19] B.J. Birch, The inhomogeneous minimum of quadratic forms of signature zero, *Acta Arith.* **3** (1958), 85–98.

[20] B.J. Birch and H.P.F. Swinnerton-Dyer, On the inhomogeneous minimum of the product of n linear forms, *Mathematika* **3** (1956), 25–39.

[21] H. Blaney, Indefinite quadratic forms in n variables, *J. London Math. Soc.* **23** (1948), 153–160.

[22] ———, Indefinite ternary forms, *Quart. J. Math. Oxford Ser. (2)* **1** (1950), 262–269.

[23] ———, On the Davenport-Heilbronn theorem, *Monat. Math.* **61** (1957), 1–36.

[24] ———, Some asymmetric inequalities. *Proc. Camb. Phil. Soc.* **46** (1950), 359–376.

[25] E. Bombieri, Sul teorema di Tschebotarev, *Acta Arith.* **8** (1963), 273–281.

[26] ———, Alcune osservazioni sul prodotto di n forme linear reali non omogenee, *Annali Mat. Pura Appl. (4)* **61** (1963), 279–285.

[27] ———, Un principio generale della Geometria dei Numeri e sue applicazioni ai problemi non omogenei, *Rend. Accad. Naz. Lincei* (1962), 45–48.

[28] J.W.S. Cassels, An introduction to Geometry of Numbers, *Grundl. Math. Wiss.* **99**. Berlin, Springer, 1971.

[29] J.H.H. Chalk, On the positive values of linear forms, Quarterly *J. Math.* **18** (1947), 215–227.

[30] ———, On the positive values of linear forms II, ibid. **19** (1948), 67–80.

[31] J.H.H. Chalk, On the product of non-homogeneous linear forms, *J. London Math. Soc.* **25** (1950), 46–51.

[32] N. Chebotarev, (= Tschebotaröw = Tschebotareff) Beweis des Minkowski'schen Satzes über lineare inhomogene Formen (Russian) ucen. *Zapiki Kazansk. Gos. Univ.* **94** (1934), 3–16 = *Vjschr. Naturforsch. Ges. Zürich* **85** (1940), 27–30.

[33] L.E. Clarke, On the product of three non-homogeneous linear forms, *Proc. Cambridge Phil Soc.* **47** (1951), 260–265.

[34] ———, Non-homogeneous linear forms associated with algebraic fields, *Quarterly J. Math. (2)* **2** (1951), 308–315.

[35] F.L. Cleaver, On a theorem of Voronoi, *Trans. Amer. Math. Soc.* **120** (1965), 390–400 (1966).

[36] ———, On coverings of four-space by spheres, *Trans. Amer. Math. Soc.* **120** (1965), 402–416 (1966).

[37] A.J. Cole, On the product of n linear forms, *Quarterly J. Math. (2)* **3** (1952), 56–62.

[38] H. Davenport, A simple proof of Remak's theorem on the product of three linear forms, *J. London Math. Soc.* **14** (1939), 47–51.

[39] ———, On the product of three non-homogeneous linear forms, *Proc. Cambridge Phil. Soc.* **43** (1947), 137–152.

[40] ———, On a theorem of Tschebotareff. *J. London Math. Soc.* **21** (1946), 28–34, Corrigendum, *ibid.* **24** (1949), 316.

[41] ———, Linear forms associated with an algebraic number field. *Quarterly J. Math. (2)* **3** (1952), 32–41.

[42] ———, Non-homogeneous ternary quadratic forms, *Acta Math.* **80** (1948), 65–95.

[43] H. Davenport and H. Heilbronn, Asymmetric inequalities for non-homogeneous forms *J. London Math. Soc.* **22** (1947), 53–61.

[44] H. Davenport and D. Ridout, Indefinite Quadratic forms, *Proc. London Math. Soc.* **9** (1959), 544–555.

[45] H. Davenport and H.P.F. Swinnerton-Dyer, Products of inhomogeneous linear forms, *Proc. London Math. Soc. (3)* **5** (1955), 474–499.

[46] B.N. Delone, An algorithm for the 'divided cells' of a lattice (Russian), *Izvestiya Akad. Nauk SSSR. Ser. Mat.* **11** (1947), 505–538.

[47] V.C. Dumir, Inhomogeneous minimum of indefinite quaternary quadratic forms, *Proc. Cambridge Philos Soc.* **63** (1967), 277–290.

[48] ———, Asymmetric inequalities for non-homogeneous ternary quadratic forms, *Proc. Cambridge Philos Soc.* **63** (1967), 291–303.

[49] ———, Positive values of inhomogenous quadratic forms I, *J. Austral. Math. Soc.* **8** (1968), 87–101.

[50] ———, Positive values of inhomogenous quadratic forms II, *J. Austral. Math. Soc.* **8** (1968), 287–303.

[51] ———, Asymmetric inequalities for non-homogeneous ternary quadratic forms. *J. Number Theory* **1** (1969), 326–345.

[52] V.C. Dumir and V.K. Grover, Some asymmetric inequalities for non-homogeneous indefinite binary quadratic forms, *J. Indian Math. Soc.* **50** (1986), 21–28.

[53] V.C. Dumir and R.J. Hans-Gill, On positive values of non-homogeneous quaternary quadratic forms of type (1,3), Indian *J. Pure Appl. Math.* **12** (1981), 814–825.

[54] V.C. Dumir and R.J. Hans-Gill, The second minimum for positive values of non-homogeneous ternary quadratic forms of type (1,2), *Ranchi University Math. Journal* **28** (1997), 65–75.

[55] V.C. Dumir and R. Sehmi, Positive values of non-homogeneous indefinite quadratic forms of type (2,5), *J. Number Theory* **48** (1994), 1–35.

[56] V.C. Dumir and R. Sehmi, Positive values of non-homogeneous indefinite quadratic forms of type (3,2), *Indian J. Pure Appl. Math.* **23** (1992), 812–853.

[57] ———, Positive values of non-homogeneous indefinite quadratic forms of signature 1, *Indian J. Pure Appl. Math.* **23** (1992), 855–864.

[58] ———, Isolated minima of non-homogeneous indefinite quadratic forms of type (3,2) or (2,3), *J. Indian Math. Soc. (N.S.)* **61** (1995), 197–212.

[59] ———, Positive values of non-homogeneous indefinite quadratic forms of type (1,4), *Proc. Indian Acad. Sci. (Math. Sci.)* **104** (1994), 557–579.

[60] V.C. Dumir, R.J. Hans-Gill and A.C. Woods, Values of non-homogeneous indefinite quadratic forms, *J. Number Theory* **47** (1994), 190–197.

[61] V.C. Dumir, R.J. Hans-Gill and R. Sehmi, Positive values of non-homogeneous indefinite quadratic forms of type (2,4), *J. Number Theory* **55** (1995), 261–284.

[62] F.J. Dyson, On the product of four non-homogeneous linear forms, *Annals Math.* **49** (1948), 82–109.

[63] V. Ennola, On the first inhomogeneous minimum of indefinite binary quadratic forms and Euclid's algorithm in real quadratic fields, *Ann. Univ. Turku A* **I 28** (1958), 1–58.

[64] M. Flahive, Indefinite quadratic forms in many variables, *Indian J. Pure Appl. Math.* **19** (10) (1988), 931–959.

[65] D.M.E. Foster, Indefinite quadratic polynomials in n variables, *Mathematika* **3** (1956), 111–116.

[66] V.K. Grover, The asymmetric product of three inhomogeneous linear forms, *Jl. Austral. Math. Soc., Ser. A* **46** (1989), 236–250.

[67] V.K. Grover and Madhu Raka, On inhomogeneous minima of indefinite binary quadratic forms, *Acta Math* **167** (1991), 287–298.

[68] P. Gruber, Eine Erweiterung des Blichfeldtschen Satzes mit einer Anwendung auf inhomogene Linearformen, *Monat. Math.* **71** (1967), 143–147.

[69] ———, Über das Produkt inhomogener Linearformen, *Acta Arth.* **13** (1967/68), 9–27.

[70] ———, Über einige Resultate in der Geometrie der Zahlen, in: Number Theory, *Coll. J. Bolyai Math. Soc. 2* (Debrecen 1968), 105–110, Amsterdam, North-Holland, 1970.

[71] ———, Über einen Satz von Remak in der Geometrie der Zahlen, *J. reine angew. Math.* **245** (1970), 107–118.

[72] ———, Eine Bemerkung uber DOTU-Matrizen, *J. Number Theory* **8** (1976), 350–351.

[73] ———, Geometry of numbers, in Contributions to Geometry, *Proc. Geometry Sympos.* (Siegen 1978), 186–225, Basel, Birkhauser, 1979.

[74] P. Gruber and C.G. Lekkerkerker, Geometry of Numbers, Second edition, North-Holland Mathematical Library, **37** (1987).

[75] P. Gruber, P. Erdös and J. Hammer, Lattice Points, Longman Scientific and Technical (1989).

[76] R.J. Hans-Gill and Madhu Raka, An asymmetric inequality for non-homogeneous ternary quadratic forms. *Monat. fur Math.* **87** (1979), 281–285.

[77] ———, An inequality for indefinite ternary quadratic forms of type (2,1), *India J. Pure & Applied Maths.* **11** (1980), 994–1006.

[78] ———, Some inequalities for non-homogeneous quadratic forms. *Indian J. Pure and Applied Maths.* **11** (1980), 60–74.

[79] ———, Inhomogeneous minimum of indefinite quadratic forms in five variables of type (3,2) or (2,3): A conjecture of Watson, *Monat. Math.* **88** (1979), 305–320.

[80] R.J. Hans-Gill and Madhu Raka, Inhomogeneous minium of indefinite quadratic forms in five variables of type (4,1) or (1,4): A conjecture of Watson, *Indian J. Pure Appl. Math.* **11** (1980), 75–91.

[81] ———, Positive values of inhomogeneous 5-ary quadratic forms of type (3,2), *J. Austral. Math. Soc. Ser. A* **29** (1980), 439–453.

[82] ———, Positive values of inhomogeneous quinary quadratic forms of type (4,1), *J. Austral. Math. Soc. Ser. A* **31** (1981), 175–188.

[83] G.H. Hardy and E.M. Wright, An introduction to the theory of Numbers, Oxford University Press, 5th Edition, 1979.

[84] N. Hofreiter, Über einen Approximationssatz von Minkowski, *Monatsh. Math. Phys.* **40** (1933), 351–392.

[85] I.V. Il'in, Chebotarev estimates in the inhomogeneous Minkowski conjecture for small dimensions, Algebraic systems (Russian), *Ivanov. Gos. Univ. Ivanovo* (1991), 115–125.

[86] T.H. Jackson, Gaps between the values of quadratic polynomials. *J. London Math. Soc.* **3** (1971), 47–58.

[87] Chr. Karanikolov, Sur un cas particulier d'une hypothese de H. Minkowski godisnik. *Viss. Tehn. Ucebn. Zaved, Mat.* **4** (1967), kn. **1** (1971), 27–30.

[88] E. Landau, Neuer Beweis eines Minkowskischen Satzes, *J. reine angew. Math.* **165** (1931), 1–3.

[89] A.M. Macbeath, Non-homogeneous linear forms, *J. London Math. Soc.* **23** (1948), 141–147.

[90] ———, A theorem on non-homogeneous lattices. *Annals Math.* **56** (1952), 269–293.

[91] ———, Factorization of martrices and Minkowski's conjecture, *Proc. Glasgow Math. Ass.* **5** (1961), 86–89.

[92] ———, A new sequence of minima in the geometry of numbers. *Proc. Camb. Phil. Soc.* **47** (1951), 266–273.

[93] Madhu Raka, Inhomogeneous minimum of indefinite quadratic forms of signature ± 1, *Math. Proc. Cambridge Philos. Soc.* **89** (1981), 225–228.

[94] ———, Inhomogeneous minimum of indefinite quadratic forms in six variables: A conjecture of Watson, *Math. Proc. Cambridge Philos. Soc.* **94** (1983), 1–8.

[95] ———, On a conjecture of Watson, *Math. Proc. Cambridge Philos. Soc.* **9** (1983), 9–22.

[96] Madhu Raka and Urmila Rani, Positive values of non-homogeneous indefinite quadratic form of type (1,4), *Proc. Indian Acad. Sci. (Math. Sci.)* **107** (1997), 329–361.

[97] ———, Inhomogeneous minima of a class of quaternary forms of type (2,2), *Contemporary Maths. AMS* **210** (1998), 275–298.

[98] ———, The second minima of inhomogeneous quaternary forms of type (3,1) and (1,3), *Ranchi Univ. Math. Jl.* **28** (1997) 5–34.

[99] ———, Positive values of non-homogeneous indefinite quadratic forms of type (3,1). *Osterreich Akad Wiss Math. Abt. II*, **203** (1994–95), 175–197.

[100] ———, Positive values of non-homogeneous indefinite-quadratic forms of signature 2. *Osterreich Acad. Wiss Abt II*, **203** (1994–95), 198–213.

[101] ———, Positive values of inhomogeneous indefinite quadratic form of type (2,1). *Hokkaido Math. Jl.* **25** (1996), 215–230.

[102] Madhu Raka, Inhomogeneous minima of a class of Ternary Quadratic Forms, *J. Austral. Math. Soc. Ser. A* **55** (1993), 334–354.

[103] K. Mahler, On a property of positive definite ternary quadratic forms, *J. London Math. Soc.* (1940), 305–320.

[104] G.A. Margulis, Indefinite quadratic forms and unipotent flows on homogeneous spaces, *C.R. Acad. Sci Paris* (1987), 249–253.

[105] H. Minkowski, Über die Annherung an eine reelle Grosse durch rationale Zahlen, *Math. Annalen*. **54** (1901), 91–124.

[106] H. Minkowski, Gesammelte Abhandlungen, 1, 2 Leipzig-Berlin, Teubner 1911, reprint, New York, Chelsea, 1967.

[107] ———, Diophantische Approximationen, Leipzig-Berlin 1907 and 1927.

[108] L.J. Mordell, Minkowski's theorem on the product of two linear forms, *J. London Math. Soc.* **3** (1928), 19–22.

[109] ———, On the product of two non-homogeneous linear forms, ibid. **16** (1941), 86–88.

[110] ———, Note on Sawyer's paper "The product of two non-homogeneous linear forms" ibid. **28** (1953), 510–512.

[111] ———, Note on the product of n inhomogeneous linear forms, *J. Number Theory* **4** (1972), 405–407.

[112] K.H. Mukhsinov, An application of the transport theorem to Cebotarev's problem, and the method of convex bodies in Minkowski's problem, *Trudy Samarkand. Gos. Univ. (N.S.)* **181** (1970), 99–108.

[113] ———, A refinement of the estimate for the arithmetical minimum of the product of non-homogeneous linear forms for large dimensions. *Zap. Naucn. Sem. Leningrad. Otdel. Math. Inst. Steklov. (LOMI)* **106**, (1981) 82–103, 171–172 = *J. Soviet Math.* **23** (1983), 2163–2176.

[114] ———, Sharpening of estimates of the arithmetical minimum of the product of inhomogeneous linear forms for small dimensions. *Dokl. Akad. Nauk. Uz. SSr.* (1981), 10–11.

[115] ———, Estimates in Minkowski's non-homogeneous conjecture for small dimensions. *Zap. Naucn. Sem. Leningrad. Otdel. Math. Inst. Steklov. (LOMI)* **106** (1981), 104–133, 172 = *J. Soviet Math.* **23** (1983), 2177–2200.

[116] ———, On the inhomogeneous Minkowski's conjecture (Letter to the editor), *Zap. Naucn. Sem. Leningrad. Otdel. Mat. Inst. Steklov. (LOMI)* **121** (1983), 195–196 = *J. Soviet Math.* **29** (1985), 1370–1371.

[117] H.N. Narzullaev, On Minkowski's problem relative to a system of inhomogeneous linear forms, *Dokl. Akad. Nauk SSSR* **180** (1968), 1298–1299 = *Soviet Math. Dokl.* **9** (1968), 766–767.

[118] ———, Minkowski's conjecture on a system of linear inhomogeneous forms, *Mat. Zametki* **5** (1969)107–116 = *Math. Notes* **5** (1969), 66–72.

[119] ———, Concerning the product of linear non-homogeneous forms, *Trudy Samarkand. Gos. Univ. (N.S.)* **181** (1970), 75–78.

[120] ———, The product of linear inhomogeneous forms, *Mat. Zametki* **16** (1974), 365–374 = *Math. Notes* **16** (1974), 806–812 (1975).

[121] ———, Representation of unimodular matrices in the form DOTU for n = 3, *Mat. Zametki* **18** (1975), 213–221 = *Math. Notes* **18** (1975), 713–719.

[122] ———, On the neighbourhood of the DOTU-matrices, *Zap. Naucn. Sem. Leningrad. Otdel. Mat. Inst. Steklov. (LOMI)* **91** (1979), 171–179 = *J. Soviet Math.* **17** (1981), 2201–2206.

[123] ———, Isolation theorems for DOTU-matrices, *Zap.Naucn. Leningrad. Otdel. Mat. Inst. Steklov. (LOMI)* **112** (1981), 159–166, 201 = *J. Soviet Math.* **25** (1984), 1084–1089.

[124] H.N. Narzullaev and B.F. Skubenko, On the question of the inhomogeneous Minkowski problem, *Trudy Samarkand. Gos. Univ. (N.S.)* **181** (1970), 129–142.

[125] ———, A refinement of the estimate of the arithmetical minimum of the product of inhomogeneous linear forms (on the inhomogeneous Minkowski conjecture). *Zap, Naucn. Sem. Leningrad. Otdel. Mat. Inst. Steklov. (LOMI)* **82** (1979), 88–94, 166 = J. Soviet Math. **18** (1982), 913–918.

[126] A. Oppenheim, The inhomogeneous minimum of quadratic forms I, II. Quart, *J. Math. Oxford Ser.* (2) **4** (1953), 54–59, 60–66.

[127] ———, Values of quadratic forms III. *Monat. Math.* **57** (1953), 97–101.

[128] G. Pall, On the product of linear forms, *Amer. Math. Monthly* **50** (1943), 173–175.

[129] O. Perron, Neuer Beweis eines Satzes von Minkowski, *Math. Annalen* **115** (1938), 656–657.

[130] R. Remak, Verallgemeinerung eines Minkowskischen Satzes, *Math. Zeitschr.* **17** (1923), 1–34, *Math. Annalen* **18** (1923), 173–200.

[131] C.A. Rogers, Indefinite quadratic ternary quadratic forms in n variables. *J. London Math. Soc.* **27** (1952), 314–319.

[132] D.B. Sawyer, The product of two non-homogeneous linear forms, *J. London Math. Soc.* **23** (1948), 250–251.

[133] ———, A note on the product of two non-homogeneous linear forms, ibid. **25** (1950), 239–240.

[134] B.F. Skubenko, On Minkowski's conjecture for n = 5, *Dokl. Akad. Nauk SSSR* **205** (1972), 1304–1305 = *Soviet Math. Dokl.* **13** (1972), 1136–1138.

[135] ———, A proof of Minkowski's conjecture on the product of n linear inhomogeneous forms in n variables for $n \leq 5$, *Zap. Naucn. Sem. Leningrad. Otdel. Mat. Inst. Steklov. (LOMI)* **33** (1973), 6–63.

[136] ———, A new variant of the proof of the inhomogeneous Minkowski's conjecture for n = 5, *Trudy Mat. Inst. Steklov.* **142** (1976), 240–253 = *Proc. Steklov Inst. Math.* **142** (1979), 261–275.

[137] ———, On a theorem of Čebotarev. *Dokl. Akad. Nauk SSSR* **233** (1977), 302–303 = *Soviet Math. Dokl.* **18** (1977), 348–350.

[138] ———, On Minkowski's conjecture for large n. *Trudy Mat. Inst. Steklov.* **148** (1978), 218–224 = *Proc. Steklov Inst. Math.* **148** (1980), 227–233.

[139] ———, There exist real square matrices in any dimension $n \geq 2880$ which are not DOTU-matrices, *Zap. Naucn. Sem. Leningrad. Otdel. Mat. Inst. Steklov. (LOMI)* **106** (1981), 134–136 = *J. Soviet Math.* **23** (1983), 2201–2203.

[140] B.F. Skubenko and K. Bakiev, Translation theorem in the non-homogeneous Minkowski problem, *Zap. Naucn. Sem. Leningrad. Otdel. Mat. Inst. Steklov. (LOMI)* **91** (1979), 119–124, 181 = *J. Soviet Math.* **17** (1981), 2162–2165.

[141] I.Y. Vulakh, On minima of rational indefinite quadratic forms, *J. Number Theory* **21** (1985), 275–285.

[142] G.L. Watson, Indefinite quadratic polynomials, *Mathematika* **7** (1960), 141–144.

[143] ———, Indefinite quadratic forms in many variables, inhomogeneous minimum and a generalization, *Proc. London Math. Soc. (3)* **12** (1962), 564–576.

[144] ———, Asymmetric inequalities for indefinite quadratic forms, *Proc. London Math. Soc.* **18** (1968), 95–113.

[145] A.C. Woods, On a theorem of Tschebotareff, *Duke Math. J.* **25** (1958), 631–638.

[146] ———, The densest double lattice packing of four spheres, *Mathematika* **12** (1965), 138–142.

[147] ———, Lattice coverings of five space by spheres, *Mathematika* **12** (1965), 143–150.

[148] ———, The asymmetric product of three inhomogeneous linear forms, *J. Austral. Math. Soc. A* **31** (1981), 439–455.

[149] ———, The product of inhomogeneous linear forms, *Proc. 1972 Number Th. Conf. Univ. Colorado* (Boulder 1972), 243–246.

[150] ———, Covering six space with spheres, *J. Number Theory* **4** (1972), 157–180.

Centre for Advanced Study in Mathematics
Panjab University
Sector 14
Chandigarh-160014
India

On the Oscillation Theorems of Pringsheim and Landau

Paul T. Bateman and Harold G. Diamond

1 Introduction

Our theme is a relation between the sign of a real function and the analytic behaviour of its associated generating function at a special point on the boundary of convergence. The central idea is the plausible principle: Let f denote a real valued function with support contained in $[0, \infty)$ or the nonnegative integers and let $\hat{f}$ denote the generating function associated with f. If f is ultimately of one sign, then extremal behaviour of $|\hat{f}|$ occurs around the real point with largest real part on the boundary of the region of convergence of the series or integral that defines $\hat{f}$.

The main results treated are (1) that analyticity of the generating function at this point entails an infinitude of sign changes of f and (2) if f is everywhere nonnegative, then we can give upper estimates for $|\hat{f}|$ in terms of its size near the distinguished point. Each of these results has interesting applications in analysis and number theory.

To fix ideas, we state the first theorem on the subject. Let $f(n)$ denote a real valued function defined on the nonnegative integers and let $\hat{f}(z) = \sum_{n=0}^{\infty} f(n)z^n$ denote the associated generating function, a power series converging for $|z| < R$ for some positive R. It is well known that $\hat{f}$ has at least one singular point on its circle of convergence.

Theorem 1 (Pringsheim). *If f and $\hat{f}$ are as above and $\hat{f}$ can be continued as an analytic function at the positive real point $z = R$, then the radius of convergence exceeds R or the original function f has an infinite number of changes of sign. Equivalently, if f is of one sign from some point onward, then the radius of convergence exceeds R or $\hat{f}$ must have a singularity at the point $z = R$.*

The first statement and proof of the theorem were given in 1894 by Pringsheim [11], but the name of Vivanti is sometimes appended to the theorem. A year earlier the latter had noted informally and without proof a form of the theorem with the extra condition that the coefficients of the power series are rapidly decreasing [15]. In his 1901 book on analytic function theory [16], Vivanti himself ascribed Theorem 1 to Pringsheim. This result was given also in Hadamard's 1901 book on analytic functions (p. 21 of [6]), but with no attribution.

Proof: (This is the argument given in [6], [11], and [16].) If we consider the Taylor expansion of the function $\hat{f}$ around the point $z = r$, where r is some fixed positive number less than R, then clearly the radius of convergence ρ of that Taylor series is at least $R - r$. But if the point $z = R$ is not a singular point, then $\rho > R - r$. Now the Taylor series for $\hat{f}$

around the point $z = re^{i\theta}$ for any real θ is majorized by the Taylor series for $\hat{f}$ around the point $z = r$, since the j-th coefficients of the two Taylor series are

$$\sum_{n=j}^{\infty} f(n)\binom{n}{j}(re^{i\theta})^{n-j}, \quad \sum_{n=j}^{\infty} f(n)\binom{n}{j}r^{n-j}$$

respectively. Hence the radius of convergence of the Taylor series for $\hat{f}$ around the point $z = re^{i\theta}$ is at least ρ for every real θ. Thus, if $z = R$ is not a singularity of $\hat{f}$, then the original Taylor series $\sum f(n)x^n$ must have radius of convergence at least $\rho + r > R$. □

Here are a few simple examples, each with f defined on $\mathbf{N}\cup\{0\}$ and radius of convergence $R = 1$.

$$f_1(n) = (-1)^n, \quad \hat{f}_1(z) = 1/(1+z),$$
$$f_2(n) = 1, \quad \hat{f}_2(z) = 1/(1-z),$$
$$f_3(n) = 1, n \text{ odd}, \ = -1/2^n, \ n \text{ even}, \quad \hat{f}_3(z) = \frac{z}{1-z^2} - \frac{1}{1-z^2/4}.$$

The first two examples illustrate the two forms of the theorem, while the third example shows that the theorem does not have a valid converse. For simplicity, the functions given here have simple poles; in fact other types of singularities can occur. For example, the generating function of $f_4(n) = 1$, n a square, $f_4(n) = 0$, otherwise, is

$$\sum_{m=0}^{\infty} z^{m^2},$$

a function which is singular at every point of the unit circle.

It was noted by Landau [9] that an analogue of Pringsheim's Theorem holds for generalized Dirichlet series. This form of the result has turned out to have more applications than the power series version; this is reasonable in view of the non-compact boundary of the region of convergence. Here, we shall treat Dirichlet series $\sum_{n=1}^{\infty} a_n n^{-s}$ with real coefficients or, more generally, Mellin transforms $\hat{F}(s) = \int_{1-}^{\infty} x^{-s}\, dF(x)$, where F is a right continuous real function supported on $[1, \infty)$ which is of bounded variation on each finite interval $[0, X]$. A function F having these properties will be said to be of the class $\mathcal{V}$. A few remarks on various transforms are appropriate here. The choice $F(x) = \sum_{n\le x} a_n$ enables us to express a Dirichlet series as a Mellin transform, and we will use the same notation in both cases. The change of variable $u = \log x$ and setting $G(u) = F(x)$ leads to the unilateral Laplace-Stieltjes transform $\int_0^{\infty} e^{-su}\, dG(u)$.

For any of these transforms the open region of convergence, if such exists, is a half plane $\{s \in \mathbf{C} : \Re s > \alpha\}$ for some real number α, called the abscissa of convergence. The transform may converge at all, some, or no points of the line $\{s : \Re s = \alpha\}$. In contrast to the behaviour of power series on the (compact!) circle of convergence, the analytic function associated with a Dirichlet series or Mellin transform may have no singularities on the line of convergence. For example, we have

$$\sum_{n=1}^{\infty}(-1)^n n^{-s} = (1 - 2^{1-s})\zeta(s), \quad \Re s > 1,$$

where $\zeta(s)$ denotes the Riemann zeta function. It is known that $\zeta(s)$ is analytic on $\mathbf{C} - \{1\}$, and has a simple pole at $s = 1$. Thus $(1 - 2^{1-s})\zeta(s)$ is entire, while the series $\sum_{n=1}^{\infty}(-1)^n n^{-s}$ is divergent at $s = 0$.

The analogue of Pringsheim's theorem for Dirichlet series and Mellin transforms is –

Theorem 2 (Landau's Oscillation Theorem). *If $\sum_1^{\infty} a_n n^{-s}$ is a Dirichlet series (respectively $\int_{1-}^{\infty} x^{-s}\,dF(x)$ a Mellin transform) that converges for $\Re s > \alpha$ and if the associated analytic function is regular at the point $\Re s = \alpha$ then either the series (transform) converges to the left of the point $s = \alpha$ or else the sequence a_n is not ultimately of one sign (F is not ultimately monotonic). Equivalently, if a_n or dF is of one sign from some point onward, then the abscissa of convergence exceeds α or $\hat{F}$ must have a singularity at the point $s = \alpha$.*

2 Proof of Landau's Theorem

We prove the result for Mellin transforms, assuming that α is the abscissa of convergence of $\hat{F}(s) = \int_{1-}^{\infty} x^{-s}\,dF(x)$, F is monotonic from some point onward, and that $\hat{F}$ is analytic at α. We shall show that the integral defining $\hat{F}$ converges to the left of α, contradicting the assumption upon α.

Suppose WLOG that $F \uparrow$ on (x_0, ∞). We expand $\hat{F}$ as a Taylor series about some real point $\beta > \alpha$:

$$\hat{F}(s) = \sum_0^{\infty} \hat{F}^{(j)}(\beta)(s-\beta)^j/j!.$$

For $j = 0, 1, 2, \ldots$ and any $X \geq x_0$, we have

$$(-1)^j \hat{F}^{(j)}(\beta) = \int_{1-}^{\infty} x^{-\beta} \log^j x\,dF(x) \geq \int_{1-}^{X} x^{-\beta} \log^j x\,dF(x).$$

By assumption, $\hat{F}$ is analytic on $\{s : \sigma > \alpha\} \cup \{s : |s - \alpha| < \delta\}$ for some $\delta > 0$. Let $R > \beta - \alpha$ be a number such that the disk of center β and radius R lies within the domain of analyticity of $\hat{F}$. Let s be a real number such that $\beta - R < s < \alpha$.

We have

$$\hat{F}(s) \geq \sum_0^{\infty} \frac{(\beta - s)^j}{j!} \int_{1-}^{X} x^{-\beta} \log^j x\,dF(x) = \int_{1-}^{X} \sum_0^{\infty} \frac{(\beta - s)^j}{j!} x^{-\beta} \log^j x\,dF(x),$$

since the series converges uniformly for $1 \leq x \leq X$ by Weierstrass' M-test. If we sum the series, we obtain

$$\hat{F}(s) \geq \int_{1-}^{X} e^{(\beta - s)\log x} x^{-\beta}\,dF(x) = \int_{1-}^{X} x^{-s}\,dF(x).$$

We see that the last integral is bounded above for all X. Since the integrand is nonnegative and F is monotone on (x_0, ∞), the integral defining $\hat{F}(s)$ converges at s. This contradicts the condition that α is the abscissa of convergence of $\hat{F}$.

3 Some Applications of Landau's Theorem in Number Theory

There are several applications of Landau's Theorem that are well known, one of which we mention here in passing; details can be found in Ch. 5 of [7]. Let $\psi(x) = \sum \log p$, where the sum extends over all prime powers $p^\alpha \le x$. By examining the behaviour of the left hand side of the identity

$$\frac{-\zeta'(s)}{s\zeta(s)} - \frac{1}{s-1} - \frac{C}{s-\alpha} = \int_1^\infty x^{-s-1}(\psi(x) - x - Cx^\alpha)\,dx,$$

we see that $\psi(x) - x - Cx^\alpha$ changes sign infinitely often for any real value of C and any value of $\alpha < \beta$, where β denotes the sup of the real parts of the zeros of the Riemann zeta function. Now we turn to a few applications of Landau's Theorem that may be less familiar.

Nonvanishing of L Functions

One of Landau's first uses of this result was to give (in §2 of [10]) a simple proof that $L(1, \chi) \neq 0$ for χ any real non principal residue character, where

$$\mathrm{L}(s, \chi) = \sum_{n=1}^{\infty} \chi(n) n^{-s} = \prod_p (1 - \chi(p)p^{-s})^{-1},$$

with the sum representation valid for $\Re s > 0$ and the product valid for $\Re s > 1$. We shall not prove just this result, but instead will use Landau's theorem to establish a more general assertion following the lines of [2].

Theorem 3 *If χ is a residue character and α is a real number (with $\alpha \neq 0$ in case χ is the principal character for some modulus), then $L(1 + i\alpha, \chi) \neq 0$.*

Proof: We first remark that if $\alpha = 0$ and χ is the principal character for some modulus k, then $L(s, \chi)$ has a simple pole at $s = 1$. In the other cases, $L(1+i\alpha, \chi)$ is finite. We assume that $L(1 + i\alpha, \chi) = 0$ and apply Landau's theorem to get a contradiction. By complex conjugation $L(1 - i\alpha, \overline{\chi})$ also is zero. We consider the Dirichlet series $\sum a(n)n^{-s}$ of the function f defined for $\Re s > 1$ by

$$f(s) = \zeta(s)^2 L(s + i\alpha, \chi) L(s - i\alpha, \overline{\chi}).$$

It follows from our assumption of the vanishing of $L(1 + i\alpha, \chi)$ and $L(1 - i\alpha, \overline{\chi})$ that f is regular at $s = 1$, since the double pole of ζ^2 is canceled by these zeros. Then f is regular at every point of the positive real axis. We are going to show that $a(n) \ge 0$ for all n and that the series $\sum a(n)n^{-s}$ fails to converge for some positive values of s, a contradiction to Landau's Theorem.

From the Euler products for ζ and L it is seen that a is a multiplicative arithmetic function. To prove that $a(n) \ge 0$ for all n it thus suffices to establish the relation for each prime power value of n. Also, we have from the Euler product

$$1 + \frac{a(p)}{p^s} + \frac{a(p^2)}{p^{2s}} + \ldots = \left(1 - \frac{1}{p^s}\right)^{-2}\left(1 - \frac{\epsilon(p)}{p^s}\right)^{-1}\left(1 - \frac{\overline{\epsilon}(p)}{p^s}\right)^{-1}, \quad (\Re s > 1),$$

where $\epsilon(p) = \chi(p)p^{-i\alpha}$, a function whose modulus is either 0 or 1. Using the representation

$$(1-z)^{-1} = \exp\left(\sum_{n=1}^{\infty} z^k/k\right) \qquad (|z| < 1),$$

we obtain

$$1 + \frac{a(p)}{p^s} + \frac{a(p^2)}{p^{2s}} + \ldots = \exp\left(\sum_{k=1}^{\infty} \frac{2+\epsilon(p)^k + \overline{\epsilon}(p)^k}{kp^{ks}}\right). \qquad (*)$$

Since $2 + \epsilon(p)^k + \overline{\epsilon}(p)^k = 2 + 2\Re\epsilon(p)^k \geq 0$, for each p and k, use of the exponential series shows that a is non negative.

From $(*)$ it follows that $a(p) = 2 + \epsilon(p) + \overline{\epsilon}(p) \geq 0$ and that

$$\begin{aligned} a(p^2) &= (2+\epsilon(p)+\overline{\epsilon}(p))^2/2 + (2+\epsilon(p)^2+\overline{\epsilon}(p)^2)/2 \\ &= 2 - \epsilon(p)\overline{\epsilon}(p) + (1+\epsilon(p)+\overline{\epsilon}(p))^2 \\ &\geq 2 - |\epsilon(p)|^2 \geq 1. \end{aligned}$$

Thus

$$\sum_{n=1}^{\infty} \frac{a(n)}{n^{1/2}} \geq \sum_p \frac{a(p^2)}{p} \geq \sum_p \frac{1}{p}.$$

From the divergence of the last sum, we see that $\sum a(n)/n^{1/2}$ diverges. Thus the abscissa of convergence of $\sum a(n)/n^s$ is positive, contradicting the regularity of f on the positive real axis. Thus the assumption that $L(1+i\alpha, \chi) = 0$ has led to a contradiction. □

Schnirelmann Density of k-th Power-free Integers

Let Q_k denote the set of k-th power-free integers, i.e. the integers not divisible by the k-th power of any prime. The elements of the set $Q_2 = \{1, 2, 3, 5, 6, 7, 10, \ldots\}$ are called square-free integers. Let $Q_k(x)$ denote the number of elements of Q_k not exceeding x. We have

$$Q_k(x) = \sum_{n \leq x} \sum_{d^k | n} \mu(d) = \sum_{d=1}^{\infty} \left[\frac{x}{d^k}\right]\mu(d) = \frac{x}{\zeta(k)} + O(x^{1/k}),$$

where ζ again denotes the Riemann zeta function. Thus $d(Q_k)$, the asymptotic density of Q_k, is $1/\zeta(k)$. The Schnirelmann density of Q_k is, by definition, $\sigma(Q_k) := \inf_{x\geq 1} Q_k(x)/[x]$. It is known, e.g., that $\sigma(Q_2) = Q_2(176)/176 = 0.60227\ldots < 6/\pi^2 = 1/\zeta(2)$. We have the following comparison of $\sigma(Q_k)$ and $d(Q_k)$, a result first proved by Stark in [13].

Theorem 4 *$\sigma(Q_k) < d(Q_k)$ for all $k \geq 2$.*

Proof: For $\Re s > 1$ we have

$$\frac{\zeta(s)}{\zeta(ks)} = \sum_{n \in Q_k} \frac{1}{n^s} = \int_{1-}^{\infty} x^{-s}\, dQ_k(x) = s \int_1^{\infty} \frac{Q_k(x)}{x^{s+1}}\, dx,$$

so that

$$\frac{1}{s}\left\{\frac{\zeta(s)}{\zeta(ks)} - \frac{\zeta(s)}{\zeta(k)}\right\} = \int_1^{\infty} \frac{Q_k(x) - [x]/\zeta(k)}{x^{s+1}}\, dx.$$

Now $\zeta(s)/\zeta(ks) - \zeta(s)/\zeta(k)$ has no singularities on the positive real axis, but it has many singularities with real part $1/(2k)$ (arising from zeros of $\zeta(s)$ with $\Re s = 1/2$). Thus the abscissa of convergence of the Mellin transform is at least $1/(2k)$. It follows from Landau's theorem that $Q_k(x) - [x]/\zeta(k)$ changes sign infinitely often. In particular, $Q_k(x)/[x] < 1/\zeta(k)$ holds for some values of x, and thus we have shown – in one step – that $\sigma(Q_k) < 1/\zeta(k)$ for each integer $k \geq 2$. □

We note that the same argument shows that $Q_k(x) - [x]/\zeta(k) + Cx^{\alpha}$ changes sign infinitely often, for any real value of C, if $k\alpha$ is less than the sup of the real parts of the zeros ρ of ζ for which ρ/k is not a zero of ζ.

A Power Series Involving the Möbius Function

It is well known that the Dirichlet series $\sum_1^{\infty} \mu(n)n^{-s}$ for the Möbius function is the reciprocal of the Riemann zeta function, a representation valid in the half plane $\{s : \Re s > 1\}$, and further that $1/\zeta(s)$ is continuable as a meromorphic function on $\boldsymbol{C}$. Thus, the divergent infinite series $\sum \mu(n)$ may be considered to have some evanescent connection with $1/\zeta(0)$, whose value is known to be -2. The behaviour of the related power series $\sum_1^{\infty} \mu(n)r^n$ as $r \to 1-$ was recently investigated by Delange [3]. Empirical data in [5] strongly suggest that the limit of the power series is -2, but this is completely misleading. Landau's Theorem enables us to show that the series is in fact unbounded from above and below as $r \to 1-$. It is convenient for our calculations to use $e^{-1/x}$, $x > 0$, in place of r.

Theorem 5 *Let* $0 < \theta < 1/2$ *and let* C *be real. Then*

$$\sum_{n=1}^{\infty} \mu(n)e^{-n/x} - \frac{C}{(1-e^{-1/x})^{\theta}}$$

changes sign infinitely often on the interval $0 < x < \infty$.

Proof: For $\Re s > 1$ we have

$$\Gamma(s)n^{-s} = \int_0^{\infty} u^{s-1}e^{-nu}\, du = \int_0^{\infty} x^{-s-1}e^{-n/x}\, dx,$$

and by multiplying by $\mu(n)$, summing, and changing sum and integral we get

$$\Gamma(s)/\zeta(s) = \int_0^{\infty} x^{-s-1} \sum_{n=1}^{\infty} \mu(n)e^{-n/x}\, dx.$$

Similarly,

$$\Gamma(s)\zeta(s+1-\theta) = \int_0^\infty x^{-s-1} \sum_{n=1}^\infty n^{\theta-1} e^{-n/x}\, dx,$$

so that

$$\Gamma(s)\{1/\zeta(s) - c\zeta(s+1-\theta)\} = \int_0^\infty x^{-s-1} \left\{ \sum_{n=1}^\infty \mu(n) e^{-n/x} - c \sum_{n=1}^\infty n^{\theta-1} e^{-n/x} \right\} dx$$

holds for any real c. Now the last integral is not of the Mellin form required for Theorem 2, because the integration range extends to zero. But,

$$\begin{aligned} A(s) :&= \int_0^1 x^{-s-1} \left\{ \sum_{n=1}^\infty \mu(n) e^{-n/x} - c \sum_{n=1}^\infty n^{\theta-1} e^{-n/x} \right\} dx \\ &= \int_1^\infty u^{s-1} \left\{ \sum_{n=1}^\infty \mu(n) e^{-nu} - c \sum_{n=1}^\infty n^{\theta-1} e^{-nu} \right\} du \end{aligned}$$

is an entire function of s, and we consider

$$\begin{aligned} &\Gamma(s)\{1/\zeta(s) - c\zeta(s+1-\theta)\} - A(s) \\ &= \int_1^\infty x^{-s-1} \left\{ \sum_{n=1}^\infty \mu(n) e^{-n/x} - c \sum_{n=1}^\infty n^{\theta-1} e^{-n/x} \right\} dx. \end{aligned}$$

The left-hand side here has complex singularities with real part $1/2$ but no real singularity greater than θ. Thus

$$\sum_{n=1}^\infty \mu(n) e^{-n/x} - c \sum_{n=1}^\infty n^{\theta-1} e^{-n/x}$$

changes sign infinitely often for any real c. Since

$$\sum_{n=1}^\infty n^{\theta-1} e^{-n/x} \sim \Gamma(\theta)/(1-e^{-1/x})^\theta$$

as $x \to \infty$, the theorem follows. □

Remark: The above theorem shows that if $0 < \theta < 1/2$ then

$$(1-r)^\theta \sum_{n=1}^\infty \mu(n) r^n$$

is unbounded from above and below as $r \to 1-$. By a refinement of the above argument, Delange has shown that

$$(1-r)^{1/2} \sum_{n=1}^\infty \mu(n) r^n$$

has a positive limit superior and negative limit inferior as $r \to 1-$.

4 An L^2 Theorem of Wiener and Wintner

An L^2 analogue of Landau's theorem for the transform of a nonnegative function was treated by Wiener and Wintner in [17]. For a nonnegative function f and an arbitrary compact set $E \subset \mathbf{R}$, we can estimate $\int_E |\hat{f}|^2$ in terms of $\int |\hat{f}|^2$ taken over a small neighborhood of the distinguished point. We begin with a power series version of this result which was proved earlier by Erdös and Fuchs in [4].

Theorem 6 *Suppose that $f(z) = \sum_{n=0}^{\infty} a_n z^n$ denotes a power series that converges for $|z| < R$, and suppose that $a_n \geq 0$ for all n. Let $f_r(\theta) := f(re^{i\theta})$, $r < R$, and supppose that $0 < \epsilon < \pi$. Then*

$$\int_{-\pi}^{\pi} |f_r(\theta)|^2 \, d\theta \leq \left[\frac{2\pi}{\epsilon} + 1\right] \int_{-\epsilon}^{\epsilon} |f_r(\theta)|^2 \, d\theta.$$

Proof: The argument depends on the following inequality. Let g, h denote functions on $\mathbf{R}$ of period 2π belonging to $L^2[-\pi, \pi]$; suppose they have Fourier series $\sum \hat{g}(n)e^{in\theta}$ and $\sum \hat{h}(n)e^{in\theta}$ respectively satisfying $\hat{g}(n) \geq 0$, $\hat{h}(n) \geq 0$, $n \in \mathbf{Z}$. Then for all real c,

$$\left| \int_{-\pi}^{\pi} g(\theta)\overline{h}(\theta - c) \, d\theta \right|$$

$$= 2\pi \left| \sum_n \hat{g}(n)\hat{h}(n)e^{inc} \right| \leq 2\pi \sum_n \hat{g}(n)\hat{h}(n) = \int_{-\pi}^{\pi} g(\theta)\overline{h}(\theta) \, d\theta.$$

We choose

$$g(\theta) = |f_r(\theta)|^2 = \sum_{m,n=0}^{\infty} a_m a_n r^{m+n} e^{i(m-n)\theta} = \sum_{k=-\infty}^{\infty} A_{k,r} e^{ik\theta},$$

with

$$A_{k,r} = \sum_{n=0}^{\infty} a_{n+k} a_n r^{2n+k} \geq 0,$$

and

$$h(\theta) = \frac{\epsilon}{2\pi} + \frac{\epsilon}{2\pi} \sum_{n \neq 0} \left(\frac{\sin \epsilon n/2}{\epsilon n/2}\right)^2 e^{in\theta} = (1 - |\theta|/\epsilon)^+, \quad -\pi \leq \theta \leq \pi.$$

It follows that

$$\int_{-\pi}^{\pi} |f_r(\theta)|^2 h(\theta - c) \, d\theta \leq \int_{-\pi}^{\pi} |f_r(\theta)|^2 h(\theta) \, d\theta \leq \int_{-\epsilon}^{\epsilon} |f_r(\theta)|^2 \, d\theta.$$

Now by superposition with $c = 0, \epsilon, \ldots, [2\pi/\epsilon]\epsilon$, we obtain the claimed inequality, since $h(\theta - j\epsilon) + h(\theta - (j+1)\epsilon) = 1$ for $j\epsilon \leq \theta \leq (j+1)\epsilon$, $j = 0, 1, \ldots [2\pi/\epsilon] - 1$, and $h(\theta - [2\pi/\epsilon]\epsilon) + h(\theta) > 1$ for $[2\pi/\epsilon]\epsilon \leq \theta \leq 2\pi$. □

Using the completeness of L^2, an immediate consequence of the theorem is –

Corollary 6.1 *Let f_r and ϵ be as in the theorem, and suppose that the restriction of f_r to a segment $-\epsilon \le \theta \le \epsilon$ has a limit in $L^2[-\epsilon, \epsilon]$ as $r \to R-$. Then f_r has a limit in $L^2[-\pi, \pi]$ as $r \to R-$.*

By partial summation and a simple estimate we can get the following extension of the theorem.

Corollary 6.2 *Let f_r, R, and ϵ be as in the theorem, but in place of the condition that a_n be nonnegative, suppose that a) $0 < R < 1$ and b) all Cesaro sums of a_n of some fixed order are nonnegative. Then*

$$\int_{-\pi}^{\pi} |f_r(\theta)|^2 \, d\theta \ll \int_{-\epsilon}^{\epsilon} |f_r(\theta)|^2 \, d\theta.$$

(The condition $0 < R < 1$ was not stated in [17], but the result is not valid if $R \ge 1$, as the example $a_n = (-1)^n$ shows.)

For the Mellin version of Wiener and Wintner's results, let $\hat{F}(s) = \int_{1-}^{\infty} x^{-s} \, dF(x)$, where $F \in \mathcal{V}$. We write $s = \sigma + it$ with σ and t real and let $\sigma_c(\hat{F})$ denote the abscissa of convergence of $\hat{F}$.

Theorem 7 *Let $F \in \mathcal{V}$ be monotone increasing, $0 < \epsilon \le 1$, $b \ge \epsilon$, d be any real number, and $\sigma > \sigma_c(\hat{F})$. Then*

$$\frac{1}{2b} \int_{d-b}^{d+b} |\hat{F}(\sigma + it)|^2 \, dt \le 2\Big(\frac{1}{2\epsilon} + \frac{1}{2b}\Big) \int_{-\epsilon}^{\epsilon} |\hat{F}(\sigma + it)|^2 \, dt.$$

Proof: This time we consider functions $h(t) := (1 - |t|/\epsilon)^+$, with Fourier transform

$$\hat{h}(x) = \int_{-\infty}^{\infty} h(t) e^{-ixt} \, dt = \epsilon \Big(\frac{\sin \epsilon x/2}{\epsilon x/2}\Big)^2 \ge 0,$$

and

$$\begin{aligned} g(t) := |\hat{F}(\sigma + it)|^2 &= \iint_{x,y \ge 1-} (xy)^{-\sigma} (y/x)^{-it} \, dF(x) dF(y) \\ &= \int_{-\infty}^{\infty} e^{-itu} \, dG(u), \end{aligned}$$

where

$$G(u) := \iint_{T(u)} (xy)^{-\sigma} \, dF(x) dF(y), \quad T(u) := \{(x, y) : x, y \ge 1-, \ y \le e^u x\}.$$

Thus g is the Fourier transform of a positive measure.

For any real c we have

$$\int_{c-\epsilon}^{c+\epsilon} |\hat{F}(\sigma+it)|^2\Big(1-\frac{|t-c|}{\epsilon}\Big)\,dt = \int_{-\infty}^{\infty} g(t)h(t-c)\,dt$$
$$= \int_{-\infty}^{\infty}\int_{-\infty}^{\infty} e^{-itu}\,dG(u)h(t-c)\,dt = \int_{-\infty}^{\infty} \hat{h}(u)e^{-icu}\,dG(u)$$
$$\le \int_{-\infty}^{\infty} \hat{h}(u)\,dG(u) = \int_{-\epsilon}^{\epsilon} |\hat{F}(\sigma+it)|^2\Big(1-\frac{|t|}{\epsilon}\Big)\,dt \le \int_{-\epsilon}^{\epsilon} |\hat{F}(\sigma+it)|^2\,dt.$$

The result follows by taking $c = d - b + j\epsilon$ with $j = 0, 1, \ldots [2b/\epsilon] + 1$ and adding the results. □

As before, we have some corollaries. For the first of them, we define repeated integrals of $F \in \mathcal{V}$ by setting $F_0(x) = F(x)$ and for $n \ge 1$ let $F_n(x) = \int_1^x F_{n-1}(t)\,dt$.

Corollary 7.1 *Let F, b, ϵ be as in the theorem. Suppose that $\sigma_c(\hat{F}) > 0$ and, for some positive integer N, we have $F_N \ge 0$. If*

$$\lim_{\sigma\to\sigma_c+} \hat{F}(\sigma+it) \in L^2[-\epsilon,\epsilon],$$

it follows that

$$\lim_{\sigma\to\sigma_c+} \hat{F}(\sigma+it) \in L^2[-b,b].$$

The following estimate is useful in proving Halasz's theorem on mean values of multiplicative functions.

Corollary 7.2 *Let F be as in the theorem with $\sigma_c = a$ and suppose that $\hat{F}(s) \ll 1/|s-a|$ holds in a region $\{s : \sigma > a,\ |s-a| \le \epsilon\}$. Then*

$$\int_{-\infty}^{\infty} |\hat{F}(\sigma+it)|^2 \frac{dt}{\sigma^2+t^2} \ll \frac{1}{\sigma-a}.$$

5 Quantitative Versions of Landau's Oscillation Theorem

Landau's Oscillation Theorem is a relatively straightforward analytic tool and does not always give the best possible results. The theorem has several quantitative extensions, of which the Wiener-Wintner theorems are examples. Here we discuss a few more results of this general type. The first of these estimates the oscillation of a function in terms of the principal part of the generating function at a pole on the abscissa of convergence. This is an idea which goes back at least as far as Erhard Schmidt's paper [12].

Theorem 8 *Suppose that F is in the function class $\mathcal{V}$ (defined in §1), $\hat{F}$ exists and has a continuation as a meromorphic function to some domain containing $\{s : \Re s \ge \beta\}$ with a* *pole of order $m \geq 1$ at some point $\beta + i\gamma$ with $\beta > 0$ and $\gamma > 0$. Assume that the principal part there is*

$$\sum_{j=-1}^{-m} c_j (s - \beta - i\gamma)^j.$$

Finally, suppose that there exists no singularity of $\hat{F}$ on the real segment $[\beta, \infty)$. Then

$$\limsup_{x\to\infty} \frac{F(x)}{x^\beta (\log x)^{m-1}} \geq \frac{|c_{-m}|}{|\beta + i\gamma|\,\Gamma(m)}, \quad \liminf_{x\to\infty} \frac{F(x)}{x^\beta (\log x)^{m-1}} \leq \frac{-|c_{-m}|}{|\beta + i\gamma|\,\Gamma(m)}.$$

Proof: It suffices to show the second relation; the first follows upon replacing F by $-F$. Suppose that c is a positive number such that $F(x) + cx^\beta (\log x)^{m-1}/\Gamma(m)$ is positive for all sufficiently large values of x. If no such c exists, then $F(x)/(x^\beta (\log x)^{m-1})$ has limit inferior $-\infty$ and there is nothing further to show.

For $\sigma > \beta$ we have

$$\frac{\hat{F}(s)}{s} + \frac{c}{(s-\beta)^m} = \int_1^\infty x^{-s} \left\{ F(x) + \frac{cx^\beta}{\Gamma(m)} (\log x)^{m-1} \right\} \frac{dx}{x}$$

or briefly $g(s) = \int x^{-s} f(x)\, dx$. This representation is valid for all $\sigma > \beta$ by Landau's Theorem. Since f is assumed positive for all $x \geq X$, for some X, we have

$$|g(s)| \leq \int_1^X x^{-\sigma} |f(x)|\, dx + \int_X^\infty x^{-\sigma} f(x)\, dx$$

$$\leq \int_1^X x^{-\sigma} \{|f(x)| - f(x)\}\, dx + g(\sigma) = g(\sigma) + O(1).$$

Now take $s = \sigma + i\gamma$ and let $\sigma \to \beta+$. We have

$$\frac{\{1 + o(1)\}|c_{-m}|}{(\sigma - \beta)^m\, |\beta + i\gamma|} = |g(\sigma + i\gamma)| \leq g(\sigma) + O(1) = \frac{c}{(\sigma - \beta)^m} + O(1).$$

The last inequality implies that $c \geq |c_{-m}|/|\beta + i\gamma|$. Thus, if we choose $c' < |c_{-m}|/|\beta + i\gamma|$, then we must have $F(x) + c'x^\beta (\log x)^{m-1}/\Gamma(m) \leq 0$ for a sequence of arbitrarily large values of x. □

If β denotes the supremum of the real parts of the zeros of the Riemann zeta function and if the supremum is actually attained, then the preceding theorem gives Schmidt's result that

$$\limsup_{x\to\infty} (\psi(x) - x)/x^\beta > 0, \quad \liminf_{x\to\infty} (\psi(x) - x)/x^\beta < 0.$$

Theorem 8 also yields the result of Delange quoted at the end of Section 3.

If one is in possession of further information about the generating function, one can use more sophisticated methods. For example, Ingham established in [8] quantitative oscillation estimates using many singularities of the generating function on the abscissa of convergence.

His method was based on an identity occurring in the most familiar proof of the Wiener-Ikehara tauberian theorem; an excellent discussion of this method is given in [1].

Finally, for F absolutely continuous with a piecewise continuous derivative ω, we mention the problem of estimating the number of sign changes of ω. If the generating function

$$\hat{F}(s) = \int_1^\infty x^{-s}\omega(x)\,dx$$

is regular at the real point on the abscissa of convergence, then Laudau's theorem guarantees that ω changes sign infinitely often. A theorem of Steinig [14] provides a lower bound on the number of changes of sign of ω in any interval $[1, X]$ under the assumption that the generating function $\hat{F}$ has a continuation as a meromorphic function to an open half plane that properly includes the half plane of convergence.

References

1. Robert J. Anderson and Harold M. Stark, *Springer LNM* **899** (1981), 79–106.
2. Paul T. Bateman, *L'Enseignement Math.* **43** (1997), 1–4.
3. Hubert Delange, Lecture at the International Number Theory Conference, Zakopane-Kościelisko, Poland, 7 July 1997.
4. Paul Erdös and Wolfgang H.J. Fuchs, *J. London math. Soc.* **31** (1956), 67–73.
5. Carl-Erik Fröberg, *Nordisk Tidskr Informationsbehandlung (BIT)* **6** (1966), 191–211.
6. Jacques Hadamard, *La Série de Taylor et son Prolongement Analytique*, Gauthier-Villars, Paris (1901).
7. A.E. Ingham, *The Distribution of Prime Numbers*, Cambridge University Press (1932).
8. A.E. Ingham, *Amer. J. Math.* **64** (1942), 313–319.
9. Edmund Landau, *Math. Ann.* **61** (1905), 527–550.
10. Edmund Landau, *Sitz. Preuss. Akad. Wiss. Berlin* (1906), 314–320.
11. Alfred Pringsheim, *Math. Ann.* **44** (1894), 41–56.
12. Erhard Schmidt, *Math. Ann.* **57** (1903), 195–204.
13. Harold M. Stark, *Proc. Amer. math. Soc.* **17** (1966), 1211–1214.
14. John Steinig, *Comment. Math. Helv.* **44** (1969), 385–400.
15. Giulio Vivanti, *Rivista di Matematica* **3** (1893), 111–114.
16. Giulio Vivanti, *Teoria della funzioni analitiche*, Hoepli, Milano (1901).
17. Norbert Wiener and Aurel Wintner, *Rev. Math. Cuyana* **2** (1956), 53–59.

Department of Mathematics
University of Illinois
Urbana, IL
USA
E-Mail: bateman@symcom.math.uiuc.edu

Department of Mathematics
University of Illinois at Urbana-Champaign
Urbana, IL - 61801
USA
E-Mail: diamond@math.uiuc.edu

Modular Equations in Ramanujan's Lost Notebook

Bruce C. Berndt

0 Introduction

Ramanujan recorded several hundred modular equations in his three notebooks [7]; no other mathematician has ever discovered nearly so many. Complete proofs for all the modular equations in Ramanujan's three notebooks can be found in Berndt's books [1]–[3]. In particular, Chapters 19–21 in Ramanujan's second notebook are almost exclusively devoted to modular equations. Ramanujan used modular equations to evaluate class invariants, certain q-continued fractions including the Rogers-Ramanujan continued fraction, theta-functions, and certain other quotients and products of theta-functions and eta-functions [3].

In his lost notebook, and in a few fragments published with the lost notebook [9], Ramanujan organized some of his modular equations by type, rather than by degree as he did in his second notebook. These lists cover the most important kinds of modular equations. Although many of these modular equations are found in his notebooks [7], some are not. The purpose of this paper to is provide a list and discussion of all these modular equations and to give proofs for those not found elsewhere in Ramanujan's notebooks.

Each modular equation is equivalent to a certain theta-function identity, but a theta-function identity may not have an equivalent modular equation. Ramanujan's lost notebook contains many new and beautiful theta-function identities which we will not discuss here. These are being proved by the author and/or his past and current Ph.D. students in a long series of papers currently being written. These students include H.H. Chan, Y.-S. Choi, S.-S. Huang, S.-Y. Kang, W.-C. Liaw, J. Sohn, S.H. Son, and L.-C. Zhang.

In the first section, we examine the modular equations on page 55 of the lost notebook. These have been called P–Q modular equations (Part IV [2, p. 204]), or eta-function identities, or Schläfli-type modular equations [6]. These are among the most elegant and beautiful modular equations found by Ramanujan and they have been most useful in the applications mentioned above.

In Section 2, we examine a fragment on pages 350–352 of [9] containing six groups of modular equations. These include modular equations associated with the names of A.M. Legendre, H. Schröter, and R. Russell. The last of the six sets contains Ramanujan's beautiful formulas for multipliers.

The brief Section 3 is devoted to a fragment found on page 349 of [9].

Before proceeding further, we provide some definitions in preparation for defining a modular equation, as Ramanujan would have understood it.

The complete elliptic integral of the first kind associated with the *modulus* k, $0 < k < 1$, is defined by

$$K := K(k) := \int_0^{\pi/2} \frac{d\theta}{\sqrt{1-k^2\sin^2\theta}}.$$

The complementary modulus k' is defined by $k' = \sqrt{1-k^2}$; set $K' = K(k')$. If $q = \exp(-\pi K'/K)$, then one of the central theorems in the theory of elliptic functions asserts that

$$\varphi^2(q) = \frac{2}{\pi}K(k) = {}_2F_1\left(\tfrac{1}{2}, \tfrac{1}{2}; 1; k^2\right), \tag{0.1}$$

where φ denotes the classical theta-function defined by

$$\varphi(q) = \sum_{k=-\infty}^{\infty} q^{k^2},$$

${}_2F_1(\frac{1}{2}, \frac{1}{2}; 1; k^2)$ denotes the ordinary hypergeometric function, and where the last equality in (0.1) follows from expanding the integrand in a binomial series and integrating termwise. It is (0.1) upon which all of Ramanujan's modular equations ultimately rests.

Let K, K', L, and L' denote complete elliptic integrals of the first kind associated with the moduli k, k', ℓ, and $\ell' := \sqrt{1-\ell^2}$, respectively, where $0 < k, \ell < 1$. Suppose that

$$n\frac{K'}{K} = \frac{L'}{L} \tag{0.2}$$

for some positive integer n. A relation between k and ℓ induced by (0.2) is called a *modular equation of degree n*. In fact, modular equations are always algebraic equations. After Ramanujan, set $\alpha = k^2$ and $\beta = \ell^2$. In the sequel, we shall frequently say that β has degree n over α. Lastly, the multiplier m is defined by

$$m = \frac{K}{L}.$$

At the end of Section 2, we shall state several formulas for multipliers in terms of α and β. These can be regarded as transformations of elliptic integrals, or, by (0.1), transformations for hypergeometric functions.

1 Eta-function Identities

After Ramanujan, define

$$f(-q) := q^{-1/24}\eta(z) := \prod_{n=1}^{\infty}(1-q^n) = \sum_{k=-\infty}^{\infty}(-1)^k q^{k(3k-1)/2}, \quad q = \exp(2\pi i z),$$

where $\eta(z)$ denotes the Dedekind eta-function, $|q| < 1$, and the last equality is Euler's pentagonal number theorem.

There are four sets of modular equations on page 55 of [9]. For the first set, Ramanujan puts

$$u = \frac{f(-q)}{q^{1/6}f(-q^5)} \quad \text{and} \quad v = \frac{f(-q^n)}{q^{n/6}f(-q^{5n})}. \tag{1.1}$$

Five identities comprise the first set.

Entry 1 *For $n = 2$ in (1.1), the functions u and v satisfy the modular equation of degree 5,*

$$uv + \frac{5}{uv} = \left(\frac{u}{v}\right)^3 + \left(\frac{v}{u}\right)^3.$$

Entry 1 is identical to Entry 53 in Chapter 25 of Part IV [2, p. 206].

Entry 2 *If $n = 3$ in (1.1), then the functions u and v satisfy the modular equation of degree 15,*

$$(uv)^3 + \left(\frac{5}{uv}\right)^3 = -\left\{\left(\frac{u}{v}\right)^6 - \left(\frac{v}{u}\right)^6\right\} - 9\left\{\left(\frac{u}{v}\right)^3 + \left(\frac{v}{u}\right)^3\right\}.$$

Entry 2 is the same as Entry 63 of Chapter 25 of Part IV [2, p. 223].

Entry 3 *If $n = 4$ in (1.1), then the functions u and v satisfy the modular equation of degree 5,*

$$\begin{aligned}(uv)^3 + \left(\frac{5}{uv}\right)^3 &= \left(\frac{u}{v}\right)^5 + \left(\frac{v}{u}\right)^5 - 8\left\{\left(\frac{u}{v}\right)^3 + \left(\frac{v}{u}\right)^3\right\} \\ &\quad + 4\left(\frac{u}{v} + \frac{v}{u}\right) + 4\Big/\left(\frac{u}{v} + \frac{v}{u}\right).\end{aligned} \tag{3.1}$$

Proof: Let β have degree 5 over α, and let m denote the multiplier of degree 5. From Entries 12(ii) and 12(iv) in Chapter 17 (Part III [1, p. 124]),

$$u = \sqrt{m}\left(\frac{1-\alpha}{1-\beta}\right)^{1/6}\left(\frac{\alpha}{\beta}\right)^{1/24} \quad \text{and} \quad v = \sqrt{m}\left(\frac{1-\alpha}{1-\beta}\right)^{1/24}\left(\frac{\alpha}{\beta}\right)^{1/6}, \tag{3.2}$$

respectively. Recall the definition (Part III [1, p. 284, eq. (13.3)])

$$\rho = (m^3 - 2m^2 + 5m)^{1/2} \tag{3.3}$$

and the representations (Part III [1, p. 286, eq. (13.12)])

$$\left(\frac{\alpha}{\beta}\right)^{1/4} = \frac{2m+\rho}{m(m-1)} \quad \text{and} \quad \left(\frac{1-\alpha}{1-\beta}\right)^{1/4} = \frac{2m-\rho}{m(m-1)}. \tag{3.4}$$

Thus, from (3.2) and (3.4),

$$\frac{u}{v} = \left(\frac{1-\alpha}{1-\beta}\right)^{1/8}\left(\frac{\alpha}{\beta}\right)^{-1/8} = \sqrt{\frac{2m-\rho}{m(m-1)}}\sqrt{\frac{m(m-1)}{2m+\rho}} = \sqrt{\frac{2m-\rho}{2m+\rho}}. \tag{3.5}$$

Hence, after a modicum of elementary algebra,

$$\frac{u}{v}+\frac{v}{u}=\frac{4m}{\sqrt{4m^2-\rho^2}}. \tag{3.6}$$

Next, by (3.2) and (3.4),

$$(uv)^3=m^3\left(\frac{1-\alpha}{1-\beta}\right)^{5/8}\left(\frac{\alpha}{\beta}\right)^{5/8}=\frac{(4m^2-\rho^2)^{5/2}}{m^2(m-1)^5},$$

and so

$$(uv)^3+\left(\frac{5}{uv}\right)^3=\frac{(4m^2-\rho^2)^{5/2}}{m^2(m-1)^5}+\frac{125m^2(m-1)^5}{(4m^2-\rho^2)^{5/2}}. \tag{3.7}$$

Hence, by (3.6), (3.7), and (3.3),

$$\begin{aligned}\left(\frac{u}{v}+\frac{v}{u}\right)\left((uv)^3+\left(\frac{5}{uv}\right)^3\right) &= 4m\left(\frac{(4m^2-\rho^2)^2}{m^2(m-1)^5}+\frac{125m^2(m-1)^5}{(4m^2-\rho^2)^3}\right)\\ &= 4m\left(\frac{(-m^3+6m^2-5m)^5+125m^4(m-1)^{10}}{m^2(m-1)^5(-m^3+6m^2-5m)^3}\right)\\ &= 4m\left(\frac{(m-1)^5(m-5)^5+125(m-1)^{10}/m}{(m-1)^8(m-5)^3}\right)\\ &= 4m\left(\frac{(m-5)^5-125(m-1)^5/m}{(m-1)^3(m-5)^3}\right). \end{aligned} \tag{3.8}$$

Next, by (3.5) and (3.3),

$$\begin{aligned}\left(\frac{u}{v}\right)^3+\left(\frac{v}{u}\right)^3 &= \left(\frac{2m-\rho}{2m+\rho}\right)^{3/2}+\left(\frac{2m+\rho}{2m-\rho}\right)^{3/2}\\ &= \frac{16m^3+12m\rho^2}{(4m^2-\rho^2)^{3/2}}=\frac{4m(3m^3-2m^2+15m)}{(-m^3+6m^2-5m)^{3/2}}. \end{aligned} \tag{3.9}$$

Hence, combining (3.6) and (3.9), with the use of (3.3), we find that

$$\left(\frac{u}{v}+\frac{v}{u}\right)\left(\left(\frac{u}{v}\right)^3+\left(\frac{v}{u}\right)^3\right)=\frac{16(3m^3-2m^2+15m)}{(-m^2+6m-5)^2}. \tag{3.10}$$

Next, by (3.5) and (3.3),

$$\begin{aligned}\left(\frac{u}{v}\right)^5+\left(\frac{v}{u}\right)^5 &= \frac{64m^5+160m^3\rho^2+20m\rho^4}{(4m^2-\rho^2)^{5/2}}\\ &= \frac{m^3(20m^4+80m^3+24m^2+400m+500)}{(-m^3+6m^2-5m)^{5/2}}. \end{aligned} \tag{3.11}$$

Hence, (3.6) and (3.11) yield

$$\left(\frac{u}{v}+\frac{v}{u}\right)\left(\left(\frac{u}{v}\right)^5+\left(\frac{v}{u}\right)^5\right)=\frac{4m(20m^4+80m^3+24m^2+400m+500)}{(-m^2+6m-5)^3}. \qquad (3.12)$$

Hence, multiplying (3.1) by $u/v+v/u$, we find that, by (3.6), (3.10), and (3.12), the new right side of (3.1) can be written in the form

$$\frac{16m(5m^4+20m^3+6m^2+100m+125)}{(-m^2+6m-5)^3} - \frac{128m(3m^3-2m^2+15m)}{(-m^2+6m-5)^2}+\frac{64m}{-m^2+6m-5}+4. \qquad (3.13)$$

In view of (3.1), combining (3.8) and (3.13), we find that it suffices to prove that

$$\begin{aligned} m(m-5)^5-125(m-1)^5 &= -4m(5m^4+20m^3+6m^2+100m+125) \\ &\quad -32m(3m^2-2m+15)(m^2-6m+5) \\ &\quad -16m(m^2-6m+5)^2+(m^2-6m+5)^3. \end{aligned}$$

This last equality is easily verified via *Mathematica*, and this completes the proof. □

Entry 4 *If* $n=5$ *in* (1.1), *then the functions* u *and* v *satisfy the modular equation of degree* 25,

$$(uv)^2+\left(\frac{5}{uv}\right)^2+5\left(uv+\frac{5}{uv}\right)+15=\left(\frac{v}{u}\right)^3. \qquad (4.1)$$

Proof: In Part III [1, p. 268, eq. (11.8)], we proved that

$$v^6 := \frac{f^6(-q^5)}{q^5 f^6(-q^{25})}=(uv)^5+5(uv)^4+15(uv)^3+25(uv)^2+25uv,$$

where we have replaced q by q^5 in the cited formulation. Dividing the equality above by $(uv)^3$ and rearranging the terms, we easily deduce (4.1). □

Entry 5 *If* $n=7$ *in* (1.1), *the functions* u *and* v *obey the modular equation of degree* 35,

$$\begin{aligned} (uv)^3+\left(\frac{5}{uv}\right)^3 &= -\left\{\left(\frac{u}{v}\right)^4-\left(\frac{v}{u}\right)^4\right\}-7\left\{\left(\frac{u}{v}\right)^3+\left(\frac{v}{u}\right)^3\right\} \\ &\quad -7\left\{\left(\frac{u}{v}\right)^2-\left(\frac{v}{u}\right)^2\right\}+14\left(\frac{u}{v}+\frac{v}{u}\right). \end{aligned} \qquad (5.1)$$

Proof: We will use the theory of modular forms and employ the theory developed by the author and L.-C. Zhang in Part IV [2, pp. 237–239].

Let $q=\exp(2\pi iz)$, where $\operatorname{Im} z>0$, and recall that $f(-q)=q^{-1/24}\eta(z)$, where η denotes the Dedekind eta-function. In the notation of Part IV [2, p. 237],

$$uv=R_{5,7}(z) \qquad \text{and} \qquad v/u=S_{5,7}(z).$$

By Lemmas 68.1 and 68.2 in Part IV [2, pp. 237, 238], we deduce that

$$R^3_{5,7}(z),\ S_{5,7}(z) \in \{\Gamma_0(35), 0, 1\},$$

where $\{\Gamma_0(n), 0, 1\}$ is the space of modular forms on $\Gamma_0(n)$ of weight 0 and multiplier system identically equal to 1.

From Part IV [2, p. 239], if r/s denotes a cusp with $(r, s) = 1$, then, for any pair of positive integers m, n,

$$\operatorname{ord}\left(\eta(mnz); \frac{r}{s}\right) = \frac{(mn, s)^2}{24mn}, \tag{5.2}$$

where (a, b) denotes the greatest common divisor of a and b. A complete set of inequivalent cusps for $\Gamma_0(35)$ is $\{0, \infty, \frac{1}{5}, \frac{1}{7}\}$. Using (5.2) repeatedly, we compose the following table summarizing the information that we need about the orders of certain functions at these cusps. We have abbreviated the left and right sides of (5.1) by $L(5.1)$ and $R(5.1)$, respectively.

cusp/order	u	v	uv	u/v	$L(5.1)$	$R(5.1)$
0	$\frac{1}{30}$	$\frac{1}{210}$	$\frac{4}{105}$	$\frac{1}{35}$	$-\frac{4}{35}$	$-\frac{4}{35}$
$\frac{1}{5}$	$-\frac{1}{6}$	$-\frac{1}{42}$	$-\frac{4}{21}$	$-\frac{1}{7}$	$-\frac{4}{7}$	$-\frac{4}{7}$
$\frac{1}{7}$	$\frac{1}{30}$	$\frac{7}{30}$	$\frac{4}{15}$	$-\frac{1}{5}$	$-\frac{4}{5}$	$-\frac{4}{5}$

If $F(z)$ denotes the difference of the left and right sides of (5.1), and if $\sum_\zeta$ denotes the sum over a complete set of inequivalent cusps for $\Gamma_0(35)$, then, by the valence formula (Part IV [2, p. 239]),

$$0 = \sum_\zeta \operatorname{ord}(F; \zeta) \geq \operatorname{ord}(F; \infty) - \tfrac{4}{35} - \tfrac{4}{7} - \tfrac{4}{5} = \operatorname{ord}(F; \infty) - \tfrac{52}{35}. \tag{5.3}$$

Thus, if we can show that $F(z) = O(q^2)$ as $q \to 0$ $(z \to i\infty)$, then we will have obtained a contradiction to (5.3), unless $F(z) \equiv 0$, which is what we want to prove. In fact, using *Mathematica*, we find that

$$L(1.4) = \frac{1}{q^4} - \frac{3}{q^3} + \frac{5}{q} + 3q - 16q^2 + \cdots = R(1.4).$$

This then completes the proof of (5.1). □

In the second set of eta-function identities, Ramanujan sets

$$u^2 = \frac{f(-q^{1/5})}{q^{1/5} f(-q^5)} \quad \text{and} \quad v^2 = \frac{f(-q^{n/5})}{q^{n/5} f(-q^{5n})}. \tag{6.1}$$

It is not clear why Ramanujan did not write u and v for u^2 and v^2, respectively. There are just two modular equations in the second set.

Entry 6 *For* $n = 2$ *in* (6.1), *the functions* u *and* v *satisfy the modular equation of degree* 25,

$$uv + \frac{5}{uv} = \left(\frac{u}{v}\right)^3 + \left(\frac{v}{u}\right)^3 - 2\left(\frac{u}{v} + \frac{v}{u}\right). \tag{6.2}$$

Proof: We rewrite Entry 58 of Chapter 25 in Part IV [2, pp. 212, 213] in Ramanujan's notation (6.1). Thus, since $P = u^2$ and $Q = v^2$, we find that

$$u^2v^2 + \frac{25}{u^2v^2} = \left(\frac{u}{v}\right)^6 + \left(\frac{v}{u}\right)^6 - 4\left\{\left(\frac{u}{v}\right)^4 + \left(\frac{v}{u}\right)^4\right\}. \tag{6.3}$$

In (6.2), observe that, for sufficiently small and positive q, each side is positive. It thus suffices to show that the squares of both sides of (6.2) are equal, i.e., after slight simplification, we want to prove that

$$\begin{aligned} u^2v^2 + \frac{25}{u^2v^2} &= \left(\frac{u}{v}\right)^6 + \left(\frac{v}{u}\right)^6 + 4\left\{\left(\frac{u}{v}\right)^2 + \left(\frac{v}{u}\right)^2\right\} \\ &\quad -4\left(\frac{u}{v} + \frac{v}{u}\right)\left\{\left(\frac{u}{v}\right)^3 + \left(\frac{v}{u}\right)^3\right\}. \end{aligned} \tag{6.4}$$

In comparing (6.3) with (6.4), we see that it remains to prove that

$$-\left(\frac{u}{v}\right)^4 - \left(\frac{v}{u}\right)^4 = \left(\frac{u}{v}\right)^2 + \left(\frac{v}{u}\right)^2 - \left(\frac{u}{v} + \frac{v}{u}\right)\left\{\left(\frac{u}{v}\right)^3 + \left(\frac{v}{u}\right)^3\right\}.$$

Since the last equality is trivial, the proof is complete. □

Entry 7 *If* $n = 3$ *in* (6.1), *the functions* u *and* v *satisfy the modular equation of degree* 75,

$$\begin{aligned} u^2v^2 &+ \frac{25}{u^2v^2} + 3\left(\frac{u}{v} + \frac{v}{u}\right)\left(uv + \frac{5}{uv}\right) \\ &= \left(\frac{u}{v}\right)^2 + \left(\frac{v}{u}\right)^2 - 6\left(\frac{u}{v} + \frac{v}{u}\right) - 9. \end{aligned} \tag{7.1}$$

Proof: From Schoeneberg's book [10, p. 102], if σ_∞ denotes the number of inequivalent cusps of $\Gamma_0(N)$, then

$$\sigma_\infty = \sum_{d|N} \varphi\left((d, N/d)\right),$$

where φ denotes Euler's φ-function, and (a, b) denotes the greatest common divisor of a and b. If $N = 75$, then $\sigma_\infty = 12$, and a complete set of inequivalent cusps is given by $\{0, \infty, \frac{1}{3}, \frac{1}{5}, \frac{1}{10}, \frac{1}{15}, \frac{1}{20}, \frac{1}{25}, \frac{1}{30}, \frac{1}{45}, \frac{1}{50}, \frac{1}{60}\}$.

Set $U(q) = u(q^5)$ and $V(q) = v(q^5)$. In the notation of Part IV [2, pp. 237, 238],

$$U^2V^2 = R_{25,3}(z) \quad \text{and} \quad V^2/U^2 = S_{25,3}(z),$$

where $q = \exp(2\pi i z)$. By Lemmas 68.1 and 68.2 in Part IV [2, pp. 237, 238],

$$R_{25,3}(z), S_{25,3}(z) \in \{\Gamma_0(75, 0, 1)\}.$$

Using (5.2), we compose the following table for orders of cusps.

cusp/order	u^2	v^2	uv	u/v	$L(7.1)$	$R(7.1)$
0	$\frac{1}{25}$	$\frac{1}{75}$	$\frac{2}{75}$	$\frac{1}{75}$	$-\frac{4}{75}$	$-\frac{4}{75}$
$\frac{1}{3}$	$\frac{1}{25}$	$\frac{3}{25}$	$\frac{2}{25}$	$-\frac{1}{25}$	$-\frac{4}{25}$	$-\frac{4}{25}$
$\frac{1}{5}$	0	0	0	0	0	0
$\frac{1}{10}$	0	0	0	0	0	0
$\frac{1}{15}$	0	0	0	0	0	0
$\frac{1}{20}$	0	0	0	0	0	0
$\frac{1}{25}$	-1	$-\frac{1}{3}$	$-\frac{2}{3}$	$-\frac{1}{3}$	$-\frac{4}{3}$	$-\frac{4}{3}$
$\frac{1}{30}$	0	0	0	0	0	0
$\frac{1}{45}$	0	0	0	0	0	0
$\frac{1}{50}$	-1	$-\frac{1}{3}$	$-\frac{2}{3}$	$-\frac{1}{3}$	$-\frac{4}{3}$	$-\frac{4}{3}$
$\frac{1}{60}$	0	0	0	0	0	0

If $F(z)$ denotes the difference of the left and right sides of (7.1), and if $\sum_\zeta$ denotes the sum over a complete set of inequivalent cusps, then, by the valence formula and the tables above,

$$0 = \sum_\zeta \operatorname{ord}(F;\zeta) \geq \operatorname{ord}(F;\infty) - \tfrac{4}{75} - \tfrac{4}{25} - \tfrac{4}{3} - \tfrac{4}{3} = -\tfrac{216}{75}. \tag{7.2}$$

Thus, if we can show that $F(z) = O(q^3)$ as q tends to 0, or z tends to $i\infty$, then we will have shown a contradiction to (7.2) unless $F(z) \equiv 0$, which is what we want to prove. In fact, using *Mathematica*, we find that

$$L(7.1) = \frac{1}{q^4} + \frac{2}{q^3} - \frac{1}{q^2} + \frac{2}{q} - 5 + 14q + 14q^2 + 44q^3 + \cdots = R(7.1).$$

This then completes the proof. □

The third set of eta-function identities comprises five modular equations. For these, Ramanujan sets

$$u = \frac{f(-q)}{q^{(n-1)/24} f(-q^n)} \quad \text{and} \quad v = \frac{f(-q^5)}{q^{5(n-1)/24} f(-q^{5n})}. \tag{8.1}$$

Entry 8 *For $n = 2$ in (8.1), the functions u and v satisfy the modular equation of degree 5,*

$$(uv)^2 + \left(\frac{2}{uv}\right)^2 = \left(\frac{v}{u}\right)^3 - \left(\frac{u}{v}\right)^3. \tag{8.2}$$

Proof: We will prove that (8.2) is equivalent to Entry 13(xiv) in Chapter 19 [1, p. 282]. To that end, first set

$$U = \frac{f(q)}{q^{1/24} f(-q^2)} \quad \text{and} \quad V = \frac{f(q^5)}{q^{5/24} f(-q^{10})}.$$

If we replace q by $-q$ in (8.2), we then find that (8.2) is equivalent to the identity

$$(UV)^2 - \left(\frac{2}{UV}\right)^2 = \left(\frac{V}{U}\right)^3 + \left(\frac{U}{V}\right)^3. \tag{8.3}$$

We now apply Entries 12(i) and 12(iii) in Chapter 17 [1, p. 124] to deduce that

$$UV = 2^{1/3}\{\alpha\beta(1-\alpha)(1-\beta)\}^{-1/24} \quad \text{and} \quad \frac{U}{V} = \left(\frac{\beta(1-\beta)}{\alpha(1-\alpha)}\right)^{1/24}.$$

Thus, (8.3) is equivalent to the modular equation of degree 5,

$$\frac{2^{2/3}}{\{\alpha\beta(1-\alpha)(1-\beta)\}^{1/12}} - 2^{4/3}\{\alpha\beta(1-\alpha)(1-\beta)\}^{1/12} = \left(\frac{\alpha(1-\alpha)}{\beta(1-\beta)}\right)^{1/8} + \left(\frac{\beta(1-\beta)}{\alpha(1-\alpha)}\right)^{1/8}. \tag{8.4}$$

But with

$$P := \{16\alpha\beta(1-\alpha)(1-\beta)\}^{1/12} \quad \text{and} \quad Q := \left(\frac{\beta(1-\beta)}{\alpha(1-\alpha)}\right)^{1/8},$$

(8.4) may be rewritten in the form

$$\frac{2}{P} - 2P = \frac{1}{Q} + Q,$$

which is Entry 13(xiv) in Chapter 19 [1, p. 282]. □

Entry 9 *With* $n = 3$ *in* (8.1), *the functions* u *and* v *satisfy the modular equation of degree* 15,

$$(uv)^2 + \left(\frac{3}{uv}\right)^2 + 5 = \left(\frac{v}{u}\right)^3 - \left(\frac{u}{v}\right)^3.$$

Entry 9 is identical to Entry 62 in Chapter 25 [2, p. 221].

Entry 10 *With* $n = 4$ *in* (8.1), *the functions* u *and* v *satisfy the modular equation of degree* 5,

$$(uv)^2 + \left(\frac{4}{uv}\right)^2 = \left(\frac{v}{u}\right)^3 + \left(\frac{u}{v}\right)^3 - 5\left(\frac{v}{u} + \frac{u}{v}\right). \tag{10.1}$$

Proof: By Entries 12(ii) and 12(iv) in Chapter 17 [1, p. 124],

$$u := \frac{f(-q)}{q^{1/8}f(-q^4)} = \sqrt{2}\left(\frac{1-\alpha}{\alpha}\right)^{1/8}$$

and

$$v := \frac{f(-q^5)}{q^{5/8}f(-q^{20})} = \sqrt{2}\left(\frac{1-\beta}{\beta}\right)^{1/8},$$

where β has degree 5 over α. It follows that

$$uv = 2\left(\frac{(1-\alpha)(1-\beta)}{\alpha\beta}\right)^{1/8} \quad \text{and} \quad \frac{u}{v} = \left(\frac{\beta(1-\alpha)}{\alpha(1-\beta)}\right)^{1/8}. \tag{10.2}$$

Thus, using (10.2), we see that, to prove (10.1), it suffices to prove the fifth degree modular equation

$$4\left(\frac{(1-\alpha)(1-\beta)}{\alpha\beta}\right)^{1/4} + 4\left(\frac{\alpha\beta}{(1-\alpha)(1-\beta)}\right)^{1/4} = \left(\frac{\alpha(1-\beta)}{\beta(1-\alpha)}\right)^{3/8}$$
$$+ \left(\frac{\beta(1-\alpha)}{\alpha(1-\beta)}\right)^{3/8} - 5\left(\frac{\alpha(1-\beta)}{\beta(1-\alpha)}\right)^{1/8} - 5\left(\frac{\beta(1-\alpha)}{\alpha(1-\beta)}\right)^{1/8}. \tag{10.3}$$

Recall that ρ is defined by (3.3). From Part III [1, pp. 285, 286, eqs. (13.10), (13.11)], we find that

$$\left(\frac{(1-\alpha)(1-\beta)}{\alpha\beta}\right)^{1/4} = \left(\frac{(\rho-3m+5)(\rho-m^2+3m)}{(\rho+3m-5)(\rho+m^2-3m)}\right)^{1/2},$$

where m is the multiplier of degree 5. Also, from Part III [1, p. 286, eq. (13.12)],

$$\left(\frac{\alpha(1-\beta)}{\beta(1-\alpha)}\right)^{1/8} = \left(\frac{2m+\rho}{2m-\rho}\right)^{1/2}.$$

Thus, (10.3) may be recast in the form

$$4\left(\frac{(\rho-3m+5)(\rho-m^2+3m)}{(\rho+3m-5)(\rho+m^2-3m)}\right)^{1/2} + 4\left(\frac{(\rho+3m-5)(\rho+m^2-3m)}{(\rho-3m+5)(\rho-m^2+3m)}\right)^{1/2}$$
$$= \left(\frac{2m+\rho}{2m-\rho}\right)^{3/2} + \left(\frac{2m-\rho}{2m+\rho}\right)^{3/2} - 5\left(\frac{2m+\rho}{2m-\rho}\right)^{1/2} - 5\left(\frac{2m-\rho}{2m+\rho}\right)^{1/2},$$

or

$$\frac{4\{(\rho-3m+5)(\rho-m^2+3m)+(\rho+3m-5)(\rho+m^2-3m)\}}{\{\rho^2-(3m-5)^2\}^{1/2}\{\rho^2-(m^2-3m)^2\}^{1/2}}$$
$$= \frac{(2m+\rho)^3+(2m-\rho)^3}{(4m^2-\rho^2)^{3/2}} - \frac{20m}{(4m^2-\rho^2)^{1/2}}. \tag{10.4}$$

Expanding all the numerators above, employing (3.3), putting the right side under one denominator, and omitting a goodly amount of elementary algebra, we find that (10.4) reduces to the equation

$$\frac{1}{(m^3-11m^2+35m-25)^{1/2}(-m^3+7m^2-11m+5)^{1/2}} = \frac{1}{(-m^2+6m-5)^{3/2}}. \tag{10.5}$$

However, (10.5) is easily shown by factoring all the polynomials in it, and so this completes the proof. □

Entry 11 *With $n = 5$, the functions u and v in (8.1) satisfy the modular equation of degree 25,*

$$(uv)^2 + \left(\frac{5}{uv}\right)^2 + 5\left(uv + \frac{5}{uv}\right) + 15 = \left(\frac{v}{u}\right)^3.$$

Entry 11 is identical to Entry 4 above.

Entry 12 *With $n = 7$, the functions u and v in (8.1) satisfy the modular equation of degree 35,*

$$(uv)^2 + \left(\frac{7}{uv}\right)^2 - 5 = \left(\frac{v}{u}\right)^3 - \left(\frac{u}{v}\right)^3 - 5\left\{\left(\frac{v}{u}\right)^2 + \left(\frac{u}{v}\right)^2\right\}.$$

Entry 12 is the same as Entry 71 in Chapter 25 [2, p. 236].

The fourth and last set of eta-function identities contains two modular equations featuring

$$u = \frac{f(-q^n)}{q^{(5-n)/24} f(-q^5)} \quad \text{and} \quad v = \frac{f(-q)}{q^{(5n-1)/24} f(-q^{5n})}. \tag{13.1}$$

Entry 13 *With $n = 2$ in (13.1), the two functions u and v satisfy the modular equation of degree 5,*

$$uv - \frac{5}{uv} = \left(\frac{v}{u}\right)^2 - \left(\frac{2u}{v}\right)^2.$$

Entry 13 is identical to Entry 54 of Chapter 25 [2, p. 207].

Entry 14 *With $n = 3$ in (13.1), the functions u and v satisfy the modular equation of degree 15,*

$$(uv)^3 - \left(\frac{5}{uv}\right)^3 = \left(\frac{v}{u}\right)^4 - \left(\frac{3u}{v}\right)^4 + \left(\frac{v}{u}\right)^2 - \left(\frac{3u}{v}\right)^2.$$

Entry 14 is identical to Entry 64 in Chapter 25 [2, p. 226].

For another approach, using the Atkin-Lehner involution, to deriving modular equations of the type considered in this section, see a paper by H.H. Chan and M.L. Lang [4].

2 Summary of Modular Equations of Six Kinds

We reproduce Ramanujan's summary of several of his modular equations found in a fragment on pages 350–352 of [9]. The modular equations are grouped into six types. It is interesting that, in contrast to his work in the notebooks [7] and lost notebook [9], Ramanujan used the more standard notations of k and ℓ to denote the moduli. Since most of the modular equations in this section have been given elsewhere by Ramanujan, so that readers may more easily compare the results stated here with Ramanujan's other work, we have put all the modular equations in Ramanujan's original notation.

There are three modular equations in the first set.

Entry 15 *If β has degree 2 over α, then*

$$(1-\sqrt{1-\alpha})(1-\sqrt{\beta})=2\sqrt{\beta(1-\alpha)}.$$

Entry 16 *If β has degree 4 over α, then*

$$(1-\sqrt[4]{1-\alpha})(1-\sqrt[4]{\beta})=2\sqrt[4]{\beta(1-\alpha)}.$$

Entry 17 *If β has degree 8 over α, then*

$$(1-\sqrt[4]{1-\alpha})(1-\sqrt[4]{\beta})=2\sqrt[8]{2\beta(1-\alpha)}.$$

After some elementary algebraic manipulation, it is easily seen that Entry 15 is equivalent to part of equation (24.12) in Chapter 18 of Part III [1, p. 213] and that Entry 16 is equivalent to (24.22) in Chapter 18 [1, p. 215]. Entry 17 is the equation just before Entry 24(vi) in Part III [1, p. 217]. Unfortunately, we erroneously claimed [1, pp. 216, 217] that two of Ramanujan's modular equations with degrees 8 and 16 are incorrect. It was the author, not Ramanujan, who was incorrect, and our work was corrected in Part V [3]. Modular equations of degree 2^n can be obtained from classical theta-function identities by iterating modular equations of degree 2^{n-1}. However, the complexity of these modular equations increases rapidly with n.

There are three sets of modular equations in the second and third groups as well.

Entry 18 *If m denotes the multiplier of degree 2 and β has degree 2 over α, then*

$$\tfrac{1}{2}m^2=\frac{1+\sqrt{\beta}}{1+\sqrt{1-\alpha}}=\frac{1+\beta}{1+(1-\alpha)}.$$

Entry 19 *If m denotes the multiplier of degree 4 and β has degree 4 over α, then*

$$\tfrac{1}{2}m=\frac{1+\sqrt[4]{\beta}}{1+\sqrt[4]{1-\alpha}}=\frac{1+\sqrt{\beta}}{1+\sqrt{1-\alpha}}.$$

Entry 20 *If m denotes the multiplier of degree 16 and β has degree 16 over α, then*

$$\tfrac{1}{2}\sqrt{m}=\frac{1+\sqrt[4]{\beta}}{1+\sqrt[4]{1-\alpha}}.$$

The two equalities of Entry 18 are given in (24.17), as part of Entry 24(ii) in Chapter 18 [1, p. 214]. The two equalities of Entry 19 are given in (24.20), as part of Entry 24(iii) in Chapter 18 [1, p. 215]. Lastly, Entry 20 is given in the middle of page 216 of [1] and is part of Entry 24(iv) of Chapter 18.

Entry 21 *If m is the multiplier of degree 2 and β has degree 2 over α, then*

(a) $m\sqrt{1-\alpha}+\sqrt{\beta}=1,$

(b) $\frac{2}{m}\sqrt{\beta}+\sqrt{1-\alpha}=1,$

(c) $m^2\sqrt{1-\alpha}+\beta=1,$

and

(d) $\frac{4}{m^2}\sqrt{\beta}+(1-\alpha)=1.$

Parts (a) and (c) are parts of Entry 24(ii) in Chapter 18 [1, p. 214, eqs. (24.15), (24.16)]. The equation in part (b) is the reciprocal of that of (a), and the equation in part (d) is the reciprocal of that of (c). (For the definition of the *reciprocal* of a modular equation, see Part III [1, p. 216, Entry 24(v)].)

Entry 22 *If m is the multiplier of degree 4 and β has degree 4 over α, then*

(a) $\sqrt{m}\sqrt[4]{1-\alpha}+\sqrt[4]{\beta}=1,$

(b) $\frac{2}{\sqrt{m}}\sqrt[4]{\beta}+\sqrt[4]{1-\alpha}=1,$

(c) $m\sqrt[4]{1-\alpha}+\sqrt{\beta}=1,$

and

(d) $\frac{4}{m}\sqrt[4]{\beta}+\sqrt{1-\alpha}=1.$

Parts (a) and (c) are parts of Entry 24(iii) in Chapter 18 [1, pp. 214, 215, eqs. (24.18), (24.19)]. The modular equations in parts (b) and (d) are the reciprocals of those in parts (a) and (c), respectively.

Entry 23 *If m is the multiplier of degree 8 and β has degree 8 over α, then*

(a) $\sqrt{m}\sqrt[8]{1-\alpha}+\sqrt[4]{\beta}=1,$

and

(b) $2\sqrt{\frac{2}{m}}\sqrt[8]{\beta}+\sqrt[4]{1-\alpha}=1.$

Part (a) is the same as equation (24.24) on page 216 of Part III [1], while the equation of part (b) is the reciprocal of part (a).

Ramanujan records four modular equations in his fourth set. The first, due to Legendre, is historically the first modular equation of a degree which is not a power of two. As we emphasized in Part III [1, Chaps. 19, 20], H. Schröter derived several modular equations of this sort. Many modular equations of this kind also were derived by R. Russell, but his methods are not completely rigorous. Russell's method has been put on a firm foundation by H.H. Chan and W.-C. Liaw [5].

Entry 24 *If β has degree 3 over α, then*

$$\{\alpha\beta\}^{1/4}+\{(1-\alpha)(1-\beta)\}^{1/4}=1.$$

Entry 24 is also Entry 5(ii) of Chapter 19 [1, p. 230].

Entry 25 *If β has degree 7 over α, then*

$$\{\alpha\beta\}^{1/8}+\{(1-\alpha)(1-\beta)\}^{1/8}=1.$$

Entry 25 is identical to part of Entry 19(i) of Chapter 19 [1, p. 314].

Entry 26 *If β has degree* 15 *over α, then*

$$(\alpha\beta)^{1/16}\left(\{(1+\sqrt{\alpha})(1+\sqrt{\beta})\}^{1/4}+\{(1-\sqrt{\alpha})(1-\sqrt{\beta})\}^{1/4}\right)$$
$$+\{(1-\alpha)(1-\beta)\}^{1/16}\left(\{(1+\sqrt{1-\alpha})(1+\sqrt{1-\beta})\}^{1/4}\right.$$
$$\left.+\{(1-\sqrt{1-\alpha})(1-\sqrt{1-\beta})\}^{1/4}\right)=\sqrt{2}.$$

Entry 26 is identical to Entry 20(vi) in Chapter 20 [1, p. 384].

Entry 27 *If β has degree* 31 *over α, then*

$$(\alpha\beta)^{1/32}\left(\{(1+\sqrt{\alpha})(1+\sqrt{\beta})\}^{1/8}\sqrt{1+\{\alpha\beta\}^{1/4}+\{(1-\sqrt{\alpha})(1-\sqrt{\beta})\}^{1/4}}\right.$$
$$\left.+\{(1-\sqrt{\alpha})(1-\sqrt{\beta})\}^{1/8}\sqrt{1+\{\alpha\beta\}^{1/4}+\{(1+\sqrt{\alpha})(1+\sqrt{\beta})\}^{1/4}}\right)$$
$$+\{(1-\alpha)(1-\beta)\}^{1/32}\left(\{(1+\sqrt{1-\alpha})(1+\sqrt{1-\beta})\}^{1/8}\right.$$
$$\times\sqrt{1+\{(1-\alpha)(1-\beta)\}^{1/4}+\{(1-\sqrt{1-\alpha})(1-\sqrt{1-\beta})\}^{1/4}}$$
$$+\{(1-\sqrt{1-\alpha})(1-\sqrt{1-\beta})\}^{1/8}$$
$$\left.\times\sqrt{1+\{(1-\alpha)(1-\beta)\}^{1/4}+\{(1+\sqrt{1-\alpha})(1+\sqrt{1-\beta})\}^{1/4}}\right)=2^{3/4}.$$

Entry 27 is the same as Entry 22(i) in Chapter 20 [1, p. 439].

The fifth set in this fragment contains seven results. These results are similar to modular equations of Russell-type. However, Ramanujan focuses on the algebraic expression,

$$\sqrt{\frac{1+\sqrt{\alpha\beta}+\sqrt{(1-\alpha)(1-\beta)}}{2}},$$

which has not been prominent in the work of any other mathematician on modular equations. In the next section, we will see how useful this expression becomes in simplifying modular equations.

Entry 28 *If β has degree* 7 *over α, then*

$$\sqrt{\frac{1+\sqrt{\alpha\beta}+\sqrt{(1-\alpha)(1-\beta)}}{2}}=\{\alpha\beta\}^{1/8}+\{(1-\alpha)(1-\beta)\}^{1/8}$$
$$-\{\alpha\beta(1-\alpha)(1-\beta)\}^{1/8}.$$

By combining both parts of Entry 19(i) in Chapter 19 [2, p. 314], we easily deduce Entry 28.

Ramanujan claimed that the modular equation of Entry 28 also holds for $n=\frac{5}{3}$. At first, this is somewhat puzzling, since modular equations have been defined for only *integral n*.

However, Ramanujan evidently had in mind modular equations of degree 15, where, in general, there are four moduli of degrees 1, 3, 5, and 15. Thus, Ramanujan asserted that the moduli of degrees 3 and 5 satisfy the modular equation above. In the proof below, we depend heavily on the parametrizations used in Part III [1, pp. 385–387] to establish many modular equations of degree 15.

Entry 29 *If α and β have degrees* 3 *and* 5, *respectively, then*

$$\sqrt{\frac{1+\sqrt{\alpha\beta}+\sqrt{(1-\alpha)(1-\beta)}}{2}} = \{\alpha\beta\}^{1/8}+\{(1-\alpha)(1-\beta)\}^{1/8} - \{\alpha\beta(1-\alpha)(1-\beta)\}^{1/8}. \tag{29.1}$$

Proof: Using the notation in Part III [1, p. 385], but with β and γ there replaced by α and β here, we set

$$B=\{\alpha\beta\}^{1/8} \quad \text{and} \quad B' = \{(1-\alpha)(1-\beta)\}^{1/8}.$$

Thus, (29.1) takes the form

$$\sqrt{\frac{1+B^4+B'^4}{2}} = B+B'-BB'. \tag{29.2}$$

Also set [1, p. 386]

$$B=\tfrac{1}{2}(M-\rho) \quad \text{and} \quad B'=\tfrac{1}{2}(M+\rho), \tag{29.3}$$

where

$$M=\sqrt{\frac{z_1 z_{15}}{z_3 z_5}}$$

and where [1, eq. (11.4), p. 386],

$$\rho^2=\frac{1+M-M^2}{M}. \tag{29.4}$$

Also [1, middle of p. 387],

$$\sqrt{\frac{1+B^4+B'^4}{2}} = \frac{1+M+3M^2-M^3}{4M}. \tag{29.5}$$

On the other hand, by (29.3) and (29.4),

$$\begin{aligned} B+B'-BB' &= M-\tfrac{1}{4}(M^2-\rho^2) \\ &= M-\tfrac{1}{4}M^2+\frac{1+M-M^2}{4M} \\ &= \frac{1+M+3M^2-M^3}{4M}. \end{aligned} \tag{29.6}$$

Comparing (29.5) and (29.6), we complete the proof of (29.2) and so also of Entry 29. □

Entry 30 *If β has degree 15 over α, then*

$$\sqrt{\frac{1+\sqrt{\alpha\beta}+\sqrt{(1-\alpha)(1-\beta)}}{2}} = \{\alpha\beta\}^{1/8}+\{(1-\alpha)(1-\beta)\}^{1/8} + \{\alpha\beta(1-\alpha)(1-\beta)\}^{1/8}. \tag{30.1}$$

Proof: The proof is similar to that above, and again we rely heavily on the notation and calculations from Part III. Set, as in Part III [1, p. 385], but with δ there replaced by β here,

$$A=\{\alpha\beta\}^{1/8} \quad \text{and} \quad A'=\{(1-\alpha)(1-\beta)\}^{1/8}.$$

Thus, (30.1) takes the form

$$\sqrt{\frac{1+A^4+A'^4}{2}}=A+A'+AA'. \tag{30.2}$$

With (Part III [1, p. 386])

$$A=\tfrac{1}{2}(M^{-1}-\rho) \quad \text{and} \quad A'=\tfrac{1}{2}(M^{-1}+\rho), \tag{30.3}$$

we find that (Part III [1, near bottom of p. 386])

$$\sqrt{\frac{1+A^4+A'^4}{2}}=\frac{1+3M-M^2+M^3}{4M^2}. \tag{30.4}$$

On the other hand, by using (30.3), (29.4), and a calculation similar to that in (29.6), we find that

$$A+A'+AA'=\frac{1+3M-M^2+M^3}{4M^2}. \tag{30.5}$$

Comparing (30.4) and (30.5), we see that we have established (30.2), and the proof is complete. □

Entry 31 *If β has degree 23 over α, then*

$$\sqrt{\frac{1+\sqrt{\alpha\beta}+\sqrt{(1-\alpha)(1-\beta)}}{2}} = \tfrac{1}{2}(1+\{\alpha\beta\}^{1/4}+\{(1-\alpha)(1-\beta)\}^{1/4}) + 2^{1/3}\{\alpha\beta(1-\alpha)(1-\beta)\}^{1/12}. \tag{31.1}$$

Entry 31 is identical to Entry 15(ii) in Chapter 20 [1, p. 411]. In fact, Ramanujan's formulation in the lost notebook is erroneous, since the last term on the right side of (31.1) was replaced by

$$2^{2/3}\{\alpha\beta(1-\alpha)(1-\beta)\}^{1/6}.$$

Entry 32 *If β has degree* 31 *over α, then*

$$\sqrt{\frac{1+\sqrt{\alpha\beta}+\sqrt{(1-\alpha)(1-\beta)}}{2}} = 1+\{\alpha\beta\}^{1/4}+\{(1-\alpha)(1-\beta)\}^{1/4} - \{\alpha\beta\}^{1/8}-\{(1-\alpha)(1-\beta)\}^{1/8} - \{\alpha\beta(1-\alpha)(1-\beta)\}^{1/8}.$$

Entry 32 is the same as Entry 22(iii) in Chapter 20 [1, p. 439].

Entry 33 *If β has degree* 47 *over α, then*

$$\sqrt{\frac{1+\sqrt{\alpha\beta}+\sqrt{(1-\alpha)(1-\beta)}}{2}} = \tfrac{1}{2}\left(1+\{\alpha\beta\}^{1/4}+\{(1-\alpha)(1-\beta)\}^{1/4}\right) + \left(\tfrac{1}{256}\alpha\beta(1-\alpha)(1-\beta)\right)^{1/24}\left(1+\{\alpha\beta\}^{1/8}+\{(1-\alpha)(1-\beta)\}^{1/8}\right).$$

Entry 33 is the same as Entry 23(i) in Chapter 20 [1, p. 444].

Entry 34 *If β has degree* 71 *over α, then*

$$\sqrt{\frac{1+\sqrt{\alpha\beta}+\sqrt{(1-\alpha)(1-\beta)}}{2}} = 1+\{\alpha\beta\}^{1/4}+\{(1-\alpha)(1-\beta)\}^{1/4} - \{\alpha\beta\}^{1/8}-\{(1-\alpha)(1-\beta)\}^{1/8}+\{\alpha\beta(1-\alpha)(1-\beta)\}^{1/8} + 2^{2/3}\{\alpha\beta(1-\alpha)(1-\beta)\}^{1/24}(\{\alpha\beta\}^{1/8}+\{(1-\alpha)(1-\beta)\}^{1/8}-1).$$

Entry 34 is identical to Entry 23(ii) of Chapter 20 [1, p. 444].

The sixth and last group of modular equations in this fragment contains seven pairs of formulas for moduli. Formulas of this sort seem to have originated with Ramanujan, and Ramanujan's methods for deriving these equations are unknown. From the definition of a modular equation, a formula for a modulus yields a transformation between two hypergeometric functions. It appears likely that such formulas are, in fact, special cases of more general transformation formulas for hypergeometric functions involving one or more parameters. It would be worthwhile to investigate such possibilities.

Entry 35 *If β and the multiplier m have degree* 3, *then*

$$m^2 = \sqrt{\frac{\beta}{\alpha}}+\sqrt{\frac{1-\beta}{1-\alpha}}-\sqrt{\frac{\beta(1-\beta)}{\alpha(1-\alpha)}},$$

$$\frac{9}{m^2} = \sqrt{\frac{\alpha}{\beta}}+\sqrt{\frac{1-\alpha}{1-\beta}}-\sqrt{\frac{\alpha(1-\alpha)}{\beta(1-\beta)}}.$$

Entry 36 *If β and the multiplier m have degree 5, then*

$$m = \left(\frac{\beta}{\alpha}\right)^{1/4} + \left(\frac{1-\beta}{1-\alpha}\right)^{1/4} - \left(\frac{\beta(1-\beta)}{\alpha(1-\alpha)}\right)^{1/4},$$
$$\frac{5}{m} = \left(\frac{\alpha}{\beta}\right)^{1/4} + \left(\frac{1-\alpha}{1-\beta}\right)^{1/4} - \left(\frac{\alpha(1-\alpha)}{\beta(1-\beta)}\right)^{1/4}.$$

Entry 37 *If β and the multiplier m have degree 7, then*

$$m^2 = \left(\frac{\beta}{\alpha}\right)^{1/2} + \left(\frac{1-\beta}{1-\alpha}\right)^{1/2} - \left(\frac{\beta(1-\beta)}{\alpha(1-\alpha)}\right)^{1/2} - 8\left(\frac{\beta(1-\beta)}{\alpha(1-\alpha)}\right)^{1/3},$$
$$\frac{49}{m^2} = \left(\frac{\alpha}{\beta}\right)^{1/2} + \left(\frac{1-\alpha}{1-\beta}\right)^{1/2} - \left(\frac{\alpha(1-\alpha)}{\beta(1-\beta)}\right)^{1/2} - 8\left(\frac{\alpha(1-\alpha)}{\beta(1-\beta)}\right)^{1/3}.$$

Entries 35, 36, and 37 are the same as Entries 5(vii), 13(xii), and 19(v), respectively, in Chapter 19 [1, pp. 230, 281–282, 314].

Entry 38 *If β and the multiplier m have degree 9, then*

$$\sqrt{m} = \left(\frac{\beta}{\alpha}\right)^{1/8} + \left(\frac{1-\beta}{1-\alpha}\right)^{1/8} - \left(\frac{\beta(1-\beta)}{\alpha(1-\alpha)}\right)^{1/8},$$
$$\frac{3}{\sqrt{m}} = \left(\frac{\alpha}{\beta}\right)^{1/8} + \left(\frac{1-\alpha}{1-\beta}\right)^{1/8} - \left(\frac{\alpha(1-\alpha)}{\beta(1-\beta)}\right)^{1/8}.$$

Entry 39 *If β and the multiplier m have degree 13, then*

$$m = \left(\frac{\beta}{\alpha}\right)^{1/4} + \left(\frac{1-\beta}{1-\alpha}\right)^{1/4} - \left(\frac{\beta(1-\beta)}{\alpha(1-\alpha)}\right)^{1/4} - 4\left(\frac{\beta(1-\beta)}{\alpha(1-\alpha)}\right)^{1/6},$$
$$\frac{13}{m} = \left(\frac{\alpha}{\beta}\right)^{1/4} + \left(\frac{1-\alpha}{1-\beta}\right)^{1/4} - \left(\frac{\alpha(1-\alpha)}{\beta(1-\beta)}\right)^{1/4} - 4\left(\frac{\alpha(1-\alpha)}{\beta(1-\beta)}\right)^{1/6}.$$

Entry 40 *If β and the multiplier m have degree 17, then*

$$\begin{aligned} m = {} & \left(\frac{\beta}{\alpha}\right)^{1/4} + \left(\frac{1-\beta}{1-\alpha}\right)^{1/4} + \left(\frac{\beta(1-\beta)}{\alpha(1-\alpha)}\right)^{1/4} \\ & -2\left(\frac{\beta(1-\beta)}{\alpha(1-\alpha)}\right)^{1/8}\left\{1 + \left(\frac{\beta}{\alpha}\right)^{1/8} + \left(\frac{1-\beta}{1-\alpha}\right)^{1/8}\right\}, \end{aligned}$$
$$\begin{aligned} \frac{17}{m} = {} & \left(\frac{\alpha}{\beta}\right)^{1/4} + \left(\frac{1-\alpha}{1-\beta}\right)^{1/4} + \left(\frac{\alpha(1-\alpha)}{\beta(1-\beta)}\right)^{1/4} \\ & -2\left(\frac{\alpha(1-\alpha)}{\beta(1-\beta)}\right)^{1/8}\left\{1 + \left(\frac{\alpha}{\beta}\right)^{1/8} + \left(\frac{1-\alpha}{1-\beta}\right)^{1/8}\right\}. \end{aligned}$$

Entries 38, 39, and 40 are, respectively, Entries 3(x), (xi), Entries 8(iii), (iv), and Entries 12(iii), (iv) in Chapter 20 [1, pp. 352, 376, 397–398].

Entry 41 *If β and the multiplier m have degree* 25, *then*

$$\sqrt{m} = \left(\frac{\beta}{\alpha}\right)^{1/8} + \left(\frac{1-\beta}{1-\alpha}\right)^{1/8} - \left(\frac{\beta(1-\beta)}{\alpha(1-\alpha)}\right)^{1/8} - 2\left(\frac{\beta(1-\beta)}{\alpha(1-\alpha)}\right)^{1/12},$$

$$\frac{5}{\sqrt{m}} = \left(\frac{\alpha}{\beta}\right)^{1/8} + \left(\frac{1-\alpha}{1-\beta}\right)^{1/8} - \left(\frac{\alpha(1-\alpha)}{\beta(1-\beta)}\right)^{1/8} - 2\left(\frac{\alpha(1-\alpha)}{\beta(1-\beta)}\right)^{1/12}.$$

Entry 41 is identical to Entries 15(i), (ii) in Chapter 19 [1, p. 291].

3 A Fragment on Page 349

By introducing a new parameter, Q, Ramanujan found simpler forms for some old modular equations and found some new ones as well. Each degree n satisfies the congruence $n \equiv 7 \pmod{16}$. Set

$$\begin{cases} P = 1 - \{\alpha\beta\}^{1/8} - \{(1-\alpha)(1-\beta)\}^{1/8}, \\ Q = \sqrt{\dfrac{1+\sqrt{\alpha\beta}+\sqrt{(1-\alpha)(1-\beta)}}{2}} - \{\alpha\beta\}^{1/8} - \{(1-\alpha)(1-\beta)\}^{1/8} \\ \qquad\qquad + \{\alpha\beta(1-\alpha)(1-\beta)\}^{1/8}, \\ R = 4\{\alpha\beta(1-\alpha)(1-\beta)\}^{1/8}. \end{cases} \tag{42.1}$$

Entry 42 *If β has degree* 7 *and P, Q, and R are defined by* (42.1), *then*

$$P^2 = Q = 0.$$

Both equations above are in Entry 19(i) of Chapter 19 [1, p. 314].

Entry 43 *If β has degree* 23 *and P, Q, and R are defined by* (42.1), *then*

$$P^2 = Q = R^{2/3}.$$

The equality $P = R^{1/3}$ is Entry 15(i) in Chapter 20 [1, p. 411], while the equality $P^2 = Q$, after some elementary manipulation, can be shown to be equivalent to Entry 15(ii) in Chapter 20 [1, p. 411].

Entry 44 *If P, Q, and R are defined by* (42.1) *and $n = 39$, then*

$$Q(P^2 - Q) = PR.$$

Entry 45 *If P, Q, and R are defined by* (42.1) *and $n = 55$, then*

$$Q(P^2 - Q)^2 = R(P^3 - R).$$

Entry 46 *If β has degree* 71 *and P, Q, and R are defined by* (42.1), *then*

$$P^2 - Q = PR^{1/3}.$$

Entry 46 is identical to Entry 23(ii) in Chapter 20 [1, p. 444].

Entry 47 *If P, Q and R are defined by* (42.1) *and $n = 119$, then*

$$(P^2 - Q)^2 = QR^{1/3}(P - R^{1/3}).$$

At this moment, we only have proofs of Entries 44, 45, and 47 using the theory of modular forms, which we do not give here. Hopefully, proofs in the spirit of Ramanujan can be given at a later date.

Ramanujan also listed the numbers 103 and 167, but he did not give modular equations for these degrees.

References

1. B.C. Berndt, *Ramanujan's Notebooks, Part III*, Springer-Verlag, New York, 1991.
2. B.C. Berndt, *Ramanujan's Notebooks, Part IV*, Springer-Verlag, New York, 1994.
3. B.C. Berndt, *Ramanujan's Notebooks, Part V*, Springer-Verlag, New York, 1998.
4. H.H. Chan and M.L. Lang, Ramanujan's modular equations and Atkin-Lehner involutions, *Israel J. Math.* **103** (1998), 1–16.
5. H.H. Chan and W.-C. Liaw, On Russell-type modular equations, *Canad. d. Math.* (*to appear*).
6. K.G. Ramanathan, Ramanujan's modular equations, *Acta Arith.* **53** (1990), 403–420.
7. S. Ramanujan, *Notebooks* (*2 volumes*), Tata Institute of Fundamental Research, Bombay, 1957.
8. S. Ramanujan, *Collected Papers*, Chelsea, New York, 1962.
9. S. Ramanujan, *The Lost Notebook and Other Unpublished Papers*, Narosa, New Delhi, 1988.
10. B. Schoeneberg, *Elliptic Modular Functions*, Springer-Verlag, Berlin, 1974.

Department of Mathematics
University of Illinois at Urbana-Champaign
Urbana IL 61801
USA

The *abc*-conjecture

Jerzy Browkin

In the present paper we give a survey of the *abc*-conjecture and of its modifications and generalizations. We discuss several consequences of the conjecture. At the end of the paper there are given numerical examples giving some evidence for the conjecture.

1 Introduction

1.1 The abc-conjecture

In the present paper we discuss the *abc*-conjecture on an elementary level, we do not say much about connections of the conjecture with more advanced theories.

We shall give the precise form of the *abc*-conjecture below (see 2.1), here we state it in a preliminary vague form, to explain the main idea lying behind it.

For any natural number n, let $r(n) = \prod_{p|n} p$ be the product of all distinct prime divisors of n. Thus $r(n)$ is the maximal squarefree divisor of n, we call it the *radical* of n.

We say that a positive integer n is very composite, if $r(n)$ is "small" with respect to n, or equivalently, if $\frac{\log n}{\log r(n)}$ is "large".

E.g. if

$$n_1 = 2^{41} \cdot 3^{15} \cdot 7 \cdot 11^3 \cdot 97^{10} = 2167914155712082356500105676431538626179891 2,$$

then $r(n_1) = 2 \cdot 3 \cdot 7 \cdot 11 \cdot 97 = 44814$ and $\frac{\log n_1}{\log r(n_1)} = 9.316746$.

Similarly, if

$$n_2 = 2^3 \cdot 3 \cdot 7^2 \cdot 11 \cdot 97 = 1254792,$$

then $r(n_2) = 2 \cdot 3 \cdot 7 \cdot 11 \cdot 97 = r(n_1)$ and $\frac{\log n_2}{\log r(n_2)} = 1.311122$.

Therefore n_1 is very composite, and n_2 is not.

The *abc*-conjecture (a preliminary form). *If $a + b = c$, where a, b are relatively prime natural numbers, then the number abc cannot be very composite. In particular, the sum of two relatively prime very composite numbers cannot be very composite.*

Thus the *abc*-conjecture expresses some relation between the addition and the multiplication in the ring of integers.

1991 *Mathematics Subject Classification.* 11D04, 11D41, 11C08.

Key words and phrases. *abc*-conjecture, Diophantine equations, algebraic functions.

1.2 Consequences of the abc-conjecture

It turns out that many deep theorems concerning integers, and some known conjectures in Number Theory follow easily from the *abc*-conjecture. We shall discuss this in 2.3. Let us mention here the following areas where there are known several interesting consequences of the *abc*-conjecture:

a) Diophantine equations and inequalities,
b) Elliptic curves,
c) Polynomials.

Therefore it seems to be difficult to prove the *abc*-conjecture, one may even doubt if it is true. We can rather suggest to replace the *abc*-conjecture by some weaker one, which also implies many statements in Number Theory, but which could be easier to prove. We shall discuss weak forms of the *abc*-conjecture in 2.4 and 2.5.

1.3 Modifications and Generalizations of the abc-conjecture

Since it seems to be hopeless to attack the *abc*-conjecture directly, one can try to understand it better by considering some of its modifications and generalizations.

E.g. we can replace the ring $\mathbb{Z}$ of rational integers by the ring of algebraic integers of an algebraic number field of a finite degree over the field $\mathbb{Q}$ of rational numbers, or by the ring of polynomials $k[x]$ over a field k of characteristic zero, or, more generally, by a ring of algebraic functions of one variable.

Another generalization is the following. In the *abc*-conjecture we consider sums of two relatively prime integers; we can consider sums of n (where $n \geq 3$ is fixed) relatively prime integers, or of n pairwise relatively prime integers.

We shall discuss these generalizations in part 3.

1.4 Limit Points

One of the forms of the *abc*-conjecture states that the maximal limit point of the set $\mathcal{L}$ of real numbers (defined below, see 2.1) is equal to 1. One may ask, more generally, about all limit points of $\mathcal{L}$.

It has been proved recently (see [BFGC], [GN]) that every point in the interval $[\frac{1}{3}, \frac{35}{36}]$ is the limit point of the set $\mathcal{L}$, but still we cannot prove that 1 is a limit point of $\mathcal{L}$ (not necessarily the maximal one!).

We shall discuss this topic in 4.1.

Of course, one can consider simultaneously some different points of view mentioned above (e.g. to investigate limit points in the case of the ring of polynomials etc.).

1.5 Computations

We cannot prove that 1 is the maximal limit point of the set $\mathcal{L}$, but we can give numerically many elements of $\mathcal{L}$ (but not infinitely many!). Therefore we can look for large numbers belonging to $\mathcal{L}$, to get some evidence for the *abc*-conjecture.

At the end of the paper we give the table of all known numbers > 1.4 belonging to $\mathcal{L}$, and the corresponding values of a, b, c.

1.6 References

There are several expository papers devoted to the *abc*-conjecture (see e.g. [La], [Ni]) containing long lists of references. In the references at the end of the paper we include a large number of important papers on the subject, but evidently our list is not complete.

Let us mention that there exist (see e.g. [Vo]) more general conjectures, from which the *abc*-conjecture follows. One can interpret this in two ways: either it suggests that the *abc*-conjecture is true, or that these more general conjectures are false.

1.7 Acknowledgements

I should like to express my gratitude for invitation and hospitality to the Abdus Salam International Centre for Theoretical Physics in Trieste. The first draft of this paper has been written during my stay there.

Paulo Ribenboim suggested me to look for weaker forms of the *abc*-conjecture. The table of the extremal examples at the end of the paper is a result of the collaboration with Juliusz Brzeziński and Abderrahmane Nitaj. I also thank A. Granville and A. Nitaj for informing me on their recent results. Discussions with many participants of the Zakopane Conference (1997) have also influenced on the contents of the present paper.

2 The *abc*-conjecture and some of its Consequences

2.1 Statement of the Conjecture

For relatively prime natural numbers a, b denote

$$L = L(a,b) = \frac{\log c}{\log r(abc)},$$

where $c = a + b$, and $r(n)$ is the radical of n defined above.

Let

$$\mathcal{L} = \{L(a,b) \ : \ a, b \in \mathbb{N}, \ \gcd(a,b) = 1\}.$$

The *abc*-conjecture *The maximal limit point of the set $\mathcal{L}$ is equal to* 1:

$$\limsup_{\gcd(a,b)=1} L(a,b) = 1.$$

Equivalently

$$\bigwedge_{\varepsilon>0} \bigvee_{N} \bigwedge_{\substack{a,b>N \\ \gcd(a,b)=1}} c < r(abc)^{1+\varepsilon},$$

or

$$\bigwedge_{\varepsilon>0} \bigvee_{C_\varepsilon} \bigwedge_{\substack{a,b\in\mathbb{N} \\ \gcd(a,b)=1}} c < C_\varepsilon r(abc)^{1+\varepsilon}.$$

2.2 The Present Status of the abc-conjecture (August 1999)

1. The largest known number belonging to $\mathcal{L}$ has been found by E. Reyssat [We]. It corresponds to the equality $2 + 3^{10} \cdot 109 = 23^5$. Namely

$$L(2, 3^{10} \cdot 109) = \frac{\log(23^5)}{\log(2 \cdot 3 \cdot 23 \cdot 109)} = 1.629912$$

Here $r(abc) = 2 \cdot 3 \cdot 23 \cdot 109 = 15042$ is "small" with respect to $23^5 = 6436343$.

2. We have

$$\limsup_{\gcd(a,b)=1} L(a, b) \geq 1.$$

In fact, for every $n \geq 1$, we have $9^n = 1 + 8t_n$, for some $t_n \geq 1$. Consequently, if $a = 1,\ b = 8t_n,\ c = 9^n$, then $r(abc) \leq 6t_n < 9^n = c$, and hence

$$L(a, b) = \frac{\log c}{\log r(abc)} > 1.$$

3. The table of all known pairs (a, b) such that $L(a, b) > 1.4$ is given at the end of the paper. Some of them have been found recently (compare the earlier table in [BB]).
4. J. Kanapka [Ka] determined all pairs (a, b) such that $\gcd(a, b) = 1,\ L(a, b) > 1.2$, and $a + b = c < 2^{36}$.
5. C.L. Stewart and Yu Kunrui [SY1], [SY2] proved that, for some effective constant k and every pair (a, b) of relatively prime natural numbers with $c = a + b$ we have

$$\log c < r(abc)^{\frac{1}{3} + \frac{k}{\log\log r(abc)}}.$$

The inequality in the *abc*-conjecture is much stronger.

6. On the other hand C.L. Stewart and R. Tijdeman [ST] proved that, for every $\delta > 0$, we have

$$\log c > \log r(abc) + (4 - \delta)\frac{\sqrt{\log r(abc)}}{\log\log r(abc)},$$

for infinitely many pairs (a, b) of relatively prime natural numbers and $c = a + b$.

2.3 Some Consequences of the abc-conjecture

We state here some theorems and conjectures which follow from the *abc*-conjecture. For most proofs we give references in the literature. We shall include some proofs later, see 2.4 and 2.5, when we shall discuss the weak forms of the *abc*-conjecture.

2.3.1 Fermat's Last Theorem

From the *abc*-conjecture it follows that there is only a finite number of positive integers n, x, y, z satisfying $n > 3,\ \gcd(x, y) = 1$ and

$$x^n + y^n = z^n.$$

Proof: Let $a = x^n,\ b = y^n,\ c = z^n$. Then

$$L(a,b) = \frac{n \log z}{\log r(xyz)} > \frac{n \log z}{\log z^3} = \frac{n}{3} \geq \frac{4}{3}.$$

The finiteness follows from $\limsup L(a,b) = 1$. □

2.3.2 The Equation $n! + 1 = m^2$

It is easy to verify that $4! + 1 = 5^2$ and $5! + 1 = 11^2$.

Theorem 2.1 (M. Overholt). *From the abc-conjecture it follows that the equation*

$$n! + 1 = m^2$$

has only a finite number of solutions.

Proof: See [Ov] and 2.5.2. □

2.3.3 Hall's Conjecture

M. Hall, Jr. [Ha] conjectured that

$$|x^2 - y^3| > Cy^{1/2},$$

for positive integers x, y with $x^2 \neq y^3$, where C is a positive constant.

On the other hand the result of L.V. Danilov [Da] improved by A. Schinzel [Si] says that for infinitely many positive integers x, y satisfying $x^2 \neq y^3$ we have

$$|x^2 - y^3| < \frac{54}{25\sqrt{5}} y^{1/2}.$$

From the *abc*-conjecture it follows the weak Hall conjecture: For every $\varepsilon > 0$ there exists $C(\varepsilon) > 0$ such that

$$|x^2 - y^3| > C(\varepsilon) y^{1/2-\varepsilon},$$

for every positive integers x, y satisfying $x^2 \neq y^3$.

Proof: See [Ni] and 2.5.3. □

2.3.4 The Congruence $a^{p-1} \equiv 1 \pmod{p^2}$

There are known only two prime numbers p satisfying

$$2^{p-1} \equiv 1 \pmod{p^2},$$

namely $p = 1093$ and $p = 3511$.

Nevertheless we cannot prove that the number of such primes is finite. A weaker result in this direction follows from the *abc*-conjecture:

Theorem 2.2 (J.H. Silverman). *The abc-conjecture implies that for every $a \geq 2$ there are infinitely many primes p satisfying $a^{p-1} \not\equiv 1 \pmod{p^2}$.*

Moreover the following quantitative result holds: From the abc-conjecture it follows that

$$\#\{p \leq X \ : \ a^{p-1} \not\equiv 1 \pmod{p^2}\} > C(a) \log X,$$

for some positive constant $C(a)$.

Proof: See [Sil]. □

2.3.5 Mordell's Conjecture

Mordell's conjecture proved by G. Faltings states that on every algebraic curve of genus $g \geq 2$ defined over an algebraic number field K there are only finitely many K-rational points.

N. Elkies [El] proved that Mordell's conjecture follows from the *abc*-conjecture for algebraic number fields (see 3.1 below).

2.3.6 Class Numbers of Imaginary Quadratic Fields

Let $-d$ be a negative fundamental discriminant. A. Granville and H. Stark [GS] proved that the uniform *abc*-conjecture (see 3.2. below) implies the following estimation from below for the class number of the field $\mathbb{Q}(\sqrt{-d})$.

$$h(-d) \geq \left(\frac{\pi}{3} + o(1)\right) \frac{\sqrt{d}}{\log d} \sum \frac{1}{\alpha},$$

where the sum runs over all reduced quadratic forms $\alpha x^2 + \beta xy + \gamma y^2$ of discriminant $-d$.

Much weaker result has been proved unconditionally (see [GZ]):

$$h(-d) > C(\varepsilon)\,(\log d)^{1-\varepsilon},$$

for every $\varepsilon > 0$ and an effectively computable constant $C(\varepsilon) > 0$.

2.3.7 Squarefree Values of Polynomials

It is not known any polynomial $f \in \mathbb{Z}[x]$ of degree > 4 such that $f(m)$ is squarefree for infinitely many $m \in \mathbb{Z}$.

On the other hand from the *abc*-conjecture it follows that the polynomials $(x^n - 1)/(x - 1)$ and the n-th cyclotomic polynomial $\Phi_n(x)$, for $n \geq 1$, take infinitely many times squarefree values, see [BFGS].

This result has been strengthened considerably by A. Granville ([Gr]): Assume that the *abc*-conjecture is true. Suppose that $f(x) \in \mathbb{Z}[x]$ does not have multiple roots. Let

$$B = \gcd\{f(n) \ : \ n \in \mathbb{Z}\},$$

and let B' be the minimal divisor of B such that B/B' is squarefree.

Then $f(m)/B'$ is squarefree for infinitely many $m \in \mathbb{Z}$.

2.3.8 Elliptic Curves

Let E be an elliptic curve defined over $\mathbb{Q}$. Denote by N its conductor, and by Δ its minimal discriminant.

Then the Szpiro conjecture (see [Oe], [Sz]) says that

$$|\Delta| < C(\varepsilon)\, N^{6+\varepsilon}.$$

Moreover there is the Goldfeld-Szpiro conjecture (see [GS]) giving an estimate of the order of Tate-Shafarevich group Ш of the curve E

$$\#Ш < C(\varepsilon)\, N^{1/2+\varepsilon}.$$

It is known (see [GS]) that both these conjectures are equivalent provided the Birch-Swinnerton-Dyer conjecture is true.

Now let $a, b \in \mathbb{N}$ be relatively prime integers, and let $a + b = c$. We consider the elliptic curve

$$y^2 = x(x-a)(x+b).$$

One can prove (assuming some mild conditions on a, b) that its conductor N equals $r(abc)$, and its minimal discriminant Δ equals $(abc)^2$ (up to absolutely bounded powers of 2).

The Szpiro conjecture (for these curves only!) follows easily from the *abc*-conjecture. Namely we have

$$\Delta = \mu(abc)^2 < \mu c^6, \quad N = \nu r(abc),$$

where μ, ν are powers of 2 bounded by absolute constants. Hence from the *abc*-conjecture $c < C(\varepsilon)\, r(abc)^{1+\varepsilon}$ we get

$$\Delta < \mu c^6 < \mu C(\varepsilon)^6 r(abc)^{6(1+\varepsilon)} = \mu C(\varepsilon)^6\, N^{6+6\varepsilon}/\nu^{6(1+\varepsilon)} \le C'(\varepsilon')N^{6+\varepsilon'},$$

where $\varepsilon' = 6\varepsilon$.

Therefore one may expect that extremal examples for the *abc*-conjecture give elliptic curves with Δ and #Ш large with respect to N. In fact B.M.M. de Weger [We1] has computed many examples which support this expectation.

E.g. the best known example for the *abc*-conjecture found by E. Reyssat ($a = 2$, $b = 3^{10} \cdot 109$) gives the elliptic curve with $\#Ш = 361 = 19^2$ and $\#Ш/N^{1/2} = 0.7359$, but the 87. example in the table at the end of the present paper corresponding to $a = 7^3$, $b = 5^{13} \cdot 181$ defines the curve isogeneus to a curve with $\#Ш = 50176 = 224^2$ and $\#Ш/N^{1/2} = 6.983$. It was the largest known value of $\#Ш/N^{1/2}$. On the other hand de Weger conjectured that for every $\varepsilon > 0$ there is a $C(\varepsilon) > 0$ such that there exist infinitely many elliptic curves defined over $\mathbb{Q}$ satisfying

$$\#Ш > C(\varepsilon)\, N^{1/2-\varepsilon}.$$

Recently A. Nitaj [Ni1] has found the elliptic curve

$$\begin{aligned} Y^2 + XY + Y &= X^3 + X^2 - 35329784286944646114547686 01X \\ &\quad - 8082758297957430129953722293855558299 2327, \end{aligned}$$

with $\#Ш = 1832^2$ and $\#Ш/N^{1/2} = 42.265$. It corresponds in some way to the example 18. in the table at the end of the present paper.

2.4 Weak Forms of the abc-conjecture

Now we shall discuss some weaker forms of the *abc*-conjecture which also imply many theorems and conjectures in number theory.

Namely, we replace the radical $r(n) = \prod_{p|n} p$ by some larger number, denote it by $r'(n)$ for a moment. Then we consider the corresponding function

$$L'(a,b) = \frac{\log c}{\log r'(abc)},$$

where a, b run over relatively prime natural numbers, and $c = a + b$.

The weak form of the *abc*-conjecture says that

$$\limsup L'(a,b) \leq 1.$$

We look for such functions $r'(n)$ that the weak *abc*-conjecture implies the same statements as the *abc*-conjecture itself. Of course, proofs of such implications are not so easy in general as the proofs in the case of the *abc*-conjecture.

To make our discussion more precise, we introduce the following notation. Let $f : \mathbb{N} \to \mathbb{R}$ be a function satisfying

$$1 \leq f(k) \leq k, \quad \text{for every} \quad k \in \mathbb{N},$$

and a condition on the growth at infinity:

$$\lim_{k\to\infty} f(k)/k = \alpha, \quad \text{where} \quad \alpha \geq 0.$$

Then we define the multiplicative function $r_f : \mathbb{N} \to \mathbb{R}$ by the condition

$$r_f(p^k) = p^{f(k)}, \quad \text{for every prime number } p, \text{ and } k \geq 1.$$

For relatively prime natural numbers a, b let $c = a + b$ and

$$L_f(a,b) = \frac{\log c}{\log r_f(abc)}.$$

The weak *abc*-conjecture (with a given function f as above) says that

$$\limsup L_f(a,b) \leq 1, \tag{2.1}$$

where a, b run over all relatively prime natural numbers.

2.4.1 Basic Properties of L_f

Lemma 1 *If* $\lim_{k\to\infty} f(k)/k = 0$, *then* $\limsup L_f(a,b) \geq 1$, *i.e. in the weak abc-conjecture* (2.1) *we can replace the inequality by the equality.*

Proof: For every natural number n we have

$$3^{2^n} = 1 + 2^{n+2}t, \quad \text{where} \quad (t, 6) = 1.$$

Put $a = 1,\ b = 2^{n+2}t,\ c = 3^{2^n}$. Since $r_f(t) \le t \le 3^{2^n}/2^{n+2}$, we get

$$r_f(abc) = r_f(2^{n+2})r_f(t)r_f(3^{2^n}) \le 2^{n+2}t\, r_f(3^{2^n}) \le 3^{2^n} \cdot 3^{f(2^n)}.$$

Hence

$$L_f(a,b) = \frac{\log c}{\log r_f(abc)} \ge \frac{2^n \log 3}{2^n \log 3 + f(2^n)\log 3} = \frac{1}{1 + \frac{f(2^n)}{2^n}} \longrightarrow 1$$

as $n \to \infty$. □

Lemma 2 *If, for some $0 < \beta \le 1$, we have $f(k)/k \ge \beta$, for every $k \in \mathbb{N}$, then*

$$\limsup L_f(a,b) \le \frac{1}{2\beta}.$$

In particular, if $f(k)/k \ge \frac{1}{2}$ for $k \in \mathbb{N}$, then $\limsup L_f(a,b) \le 1$, i.e. the weak abc-conjecture is a theorem.

Proof: From $f(k) \ge k\beta$ it follows that $r_f(n) \ge n^\beta$ for every $n \in \mathbb{N}$. Therefore assuming that $a \le b$ we get

$$L_f(a,b) = \frac{\log(a+b)}{\log r_f(ab(a+b))} \le \frac{\log(a+b)}{\beta \log(ab(a+b))} \le \frac{\log 2b}{\beta \log b^2} \longrightarrow \frac{1}{2\beta}$$

as $b \to \infty$. Hence $\limsup L_f(a,b) \le \dfrac{1}{2\beta}$. □

2.5 Some Consequences of the Weak abc-conjecture

2.5.1 Fermat's Last Theorem for Large Exponents

Theorem 2.3 *The weak abc-conjecture with f satisfying $\lim_{k\to\infty} f(k)/k = \alpha < 1/3$ implies Fermat's Last Theorem for large exponents.*

Proof: For every β satisfying $\alpha < \beta < 1/3$, from the assumption we get

$$f(k)/k < \beta \quad \text{for} \quad k > k_0.$$

Hence for every $m = \prod_{p|m} p^{m_p}$ and every $n > k_0$ we have

$$r_f(m^n) = \prod_{p|m} p^{f(m_p n)} < \prod_{p|m} p^{m_p n\beta} = m^{n\beta}. \tag{2.2}$$

Suppose that relatively prime natural numbers x, y, z satisfy $x^n + y^n = z^n$, for some $n > k_0$. Put $a = x^n,\ b = y^n,\ c = z^n$.

From (2.2) we get

$$r_f(abc) = r_f(x^n)r_f(y^n)r_f(z^n) \le (xyz)^{n\beta} < z^{3n\beta}.$$

Consequently

$$L_f(a,b) = \frac{\log c}{\log r_f(abc)} \ge \frac{\log z^n}{\log z^{3n\beta}} = \frac{1}{3\beta},$$

and $\limsup L_f(a,b) \ge \dfrac{1}{3\beta} > 1$. Contradiction. □

2.5.2 The Equation $n!+1=m^2$ (see [Ov])

Theorem 2.4 *Assume that $f(2k) \le 2kq$ for some $q < 1$ and every $k \in \mathbb{N}$, and that $\lim_{k\to\infty} f(k)/k = 0$. Then from the weak abc-conjecture with the function f it follows that the equation $n!+1=m^2$ has only a finite number of solutions in natural numbers m, n.*

Proof: For every natural number $m = \prod_{p|m} p^{m_p}$ from the assumption we get

$$r_f(m^2) = \prod_{p|m} p^{f(2m_p)} \le \prod_{p|m} p^{2m_p q} = m^{2q}. \tag{2.3}$$

To prove the theorem it will be sufficient to prove that

$$\lim_{n\to\infty} \frac{\log(r_f(n!))}{\log n!} = 0. \tag{2.4}$$

In fact, if $n!+1=m^2$, then

$$\begin{aligned} L_f(n!,1) &= \frac{\log(m^2)}{\log r_f(m^2) + \log r_f(n!)} \ge \frac{\log(m^2)}{\log(m^{2q}) + \log r_f(n!)} \\ &= \frac{1}{q + \frac{\log r_f(n!)}{\log(n!+1)}} \longrightarrow \frac{1}{q} \end{aligned}$$

as $n \to \infty$ in view of (2.3) and (2.4). Consequently $\limsup L_f(n!,1) \ge \frac{1}{q} > 1$, where n runs over all solutions of the equation $n!+1=m^2$, provided the number of solutions is infinite. Contradiction.

To prove (2.4) we shall use the Stirling formula

$$\log n! = n\log n + O(n),$$

and we shall estimate $\log r_f(n!)$ from above.

It is well known that

$$n! = \prod_{p\le n} p^{\alpha_p(n)},$$

where

$$\alpha_p(n) = \frac{n - s_p(n)}{p - 1} \tag{2.5}$$

and $s_p(n)$ is the sum of digits of n written in base p.

We shall write $\log r_f(n!)$ as the sum of two summands, and we shall estimate them separately.

$$\log r_f(n!) = \sum_{p \le n} f(\alpha_p(n)) \log p = \sum_{p \le n/\log n} f(\alpha_p(n)) \log p$$
$$+ \sum_{\frac{n}{\log n} < p \le n} f(\alpha_p(n)) \log p.$$

For $p \le n/\log n$, in view (2.5) we have

$$\alpha_p(n) = \frac{n - s_p(n)}{p - 1} \ge \frac{n - \log_2 n}{n/\log n} = \log n - \frac{(\log n)^2}{n \log 2} \longrightarrow \infty$$

as $n \to \infty$.

Then

$$g(n) := \max_{p \le n/\log n} \frac{f(\alpha_p(n))}{\alpha_p(n)} \longrightarrow 0$$

as $n \to \infty$ in view of the assumption of the theorem.

To get the estimation of the first summand we use the inequality

$$\alpha_p(n) = \frac{n - s_p(n)}{p - 1} < \frac{2n}{p} \tag{2.6}$$

and the well known formula (see [Pr])

$$\sum_{p \le x} \frac{\log p}{p} = \log x + O(1). \tag{2.7}$$

We have

$$\sum_{p \le n/\log n} f(\alpha_p(n)) \log p \le \sum_{p \le n/\log n} g(n)\alpha_p(n) \log p < 2ng(n) \sum_{p \le n/\log n} \frac{\log p}{p}$$
$$= 2ng(n)\left(\log(n/\log n) + O(1)\right) \le 2ng(n) \log n + O\left(ng(n) \log\log n\right)$$
$$= o(n \log n). \tag{2.8}$$

To estimate the second summand we use (2.6) and (2.7) once more. We have

$$\sum_{\frac{n}{\log n} < p \le n} f(\alpha_p(n)) \log p \le \sum_{\frac{n}{\log n} < p \le n} \frac{2n}{p} \log p$$
$$= 2n(\log n - \log(n/\log n) + O(1)) = 2n \log\log n + O(n) = o(n \log n). \tag{2.9}$$

From (2.8) and (2.9) we get (2.4). □

2.5.3 Hall's Conjecture

We shall prove that the weak Hall's conjecture stated in 2.3.3. follows also from an appropriate weak *abc*-conjecture.

Theorem 2.5 *Assume that f is not decreasing and satisfies*

$$f(2k) \leq k \quad \text{and} \quad f(3k) \leq k, \quad \text{for} \quad k \in \mathbb{N}.$$

Then the weak abc-conjecture with this function f implies the weak Hall's conjecture.

Proof: For natural numbers x, y satisfying $x^2 \neq y^3$ let $d = \gcd(x^2, y^3)$, and put

$$a = \frac{1}{d}|x^2 - y^3|, \quad b = \frac{1}{d}\min(x^2, y^3), \quad c = \frac{1}{d}\max(x^2, y^3).$$

Then $a + b = c$, and $\gcd(a, b) = 1$.

Since f is not decreasing, from the divisibility $m|n$ it follows that $r_f(m) \leq r_f(n)$. Moreover the assumption of the theorem implies that

$$r_f(x^2) \leq x \quad \text{and} \quad r_f(y^3) \leq y.$$

Consequently

$$r_f\left(\frac{1}{d}x^2\right) \leq r_f(x^2) \leq x$$

and

$$r_f\left(\frac{1}{d}y^3\right) \leq r_f(y^3) \leq y.$$

Hence $r_f(b)r_f(c) \leq xy$.

From the weak *abc*-conjecture we get, for some $C_1(\varepsilon) > 0$,

$$c \leq C_1(\varepsilon)r_f(abc)^{1+\varepsilon} \leq C_1(\varepsilon)\left(xy\frac{1}{d}|x^2 - y^3|\right)^{1+\varepsilon},$$

and consequently

$$x^2 \leq C_1(\varepsilon)(xy|x^2 - y^3|)^{1+\varepsilon},$$
$$y^3 \leq C_1(\varepsilon)(xy|x^2 - y^3|)^{1+\varepsilon}.$$

Multiplying these inequalities we obtain

$$x^2y^3 \leq C_1(\varepsilon)^2(xy|x^2 - y^3|)^{2+2\varepsilon},$$

and hence

$$|x^2 - y^3|^{2+2\varepsilon} \geq C_2(\varepsilon)x^{-2\varepsilon}y^{1-2\varepsilon}.$$

For $x \leq y^2$ it follows that

$$|x^2 - y^3|^{2+2\varepsilon} \geq C_2(\varepsilon)y^{1-6\varepsilon}.$$

Consequently

$$|x^2 - y^3| \geq C_3(\varepsilon) y^{1/2-\varepsilon'},$$

with $\varepsilon' = \frac{7}{2(1+\varepsilon)} \cdot \varepsilon$.

If $x > y^2$ we have obviously

$$|x^2 - y^3| > y^4 - y^3 > y > y^{1/2-\varepsilon}$$

for $y > 1$. □

3 Modifications and Generalizations of the *abc*-conjecture

3.1 The abc-conjecture for Algebraic Number Fields (see [El])

For an algebraic number field K of a finite degree over $\mathbb{Q}$ let $\|\cdot\|_v$ run over all normalized valuations of K, in particular let $\|\cdot\|_\mathfrak{p}$ be the non-archimedean valuation corresponding to a prime ideal $\mathfrak{p}$ of the ring of algebraic integers of K.

For $a, b, c \in K^*$ satisfying $a + b = c$ let

$$h_K(a, b, c) = \prod_v \max(\|a\|_v, \|b\|_v, \|c\|_v),$$

and let

$$r_K(a, b, c) = \prod_{\mathfrak{p} \in I_K(a,b,c)} N_{K/\mathbb{Q}}(\mathfrak{p}),$$

where $I_K(a, b, c)$ is the set of all prime ideals $\mathfrak{p}$ of K such that not all numbers $\|a\|_\mathfrak{p}$, $\|b\|_\mathfrak{p}$, $\|c\|_\mathfrak{p}$ are equal, and $N_{K/\mathbb{Q}}$ is the absolute norm.

It is easy to see that for $K = \mathbb{Q}$ and $a, b, c \in \mathbb{N}$ with $\gcd(a, b) = 1$ and $a + b = c$ we have

$$\begin{aligned} h_K(a, b, c) &= c \\ r_K(a, b, c) &= r(abc). \end{aligned}$$

This suggests the following definitions. For an algebraic number field K and $a, b, c \in K^*$ with $a + b = c$, where not all a, b, c are units in K, let

$$L_K(a, b) = \frac{\log h_K(a, b, c)}{\log r_K(a, b, c)},$$

and let

$$\mathcal{L}_K = \left\{ L_K(a, b) \ : \ a, b, c \in K^* \right\}.$$

The *abc*-conjecture for an algebraic number field K. *The maximal limit point of the set $\mathcal{L}_K$ is equal to 1, or equivalently, for every $\varepsilon > 0$ there exists a constant $C_{K,\varepsilon}$ depending on the field K and ε only, such that*

$$h_K(a, b, c) < C_{K,\varepsilon} r_K(a, b, c)^{1+\varepsilon},$$

for every $a, b, c \in K^*$ satisfying $a + b = c$.

Now we simplify the definition of the set $\mathcal{L}_K$. From the product formula:

$$\prod_v \|d\|_v = 1$$

for every $d \in K^*$, and from the definition of the set $I_K(a, b, c)$ of prime ideals of K it follows that the functions h_K and r_K defined above are scaling invariant:

$$h_K(da, db, dc) = h_K(a, b, c), \quad r_K(da, db, dc) = r_K(a, b, c),$$

for every $d \in K^*$. Therefore we can assume e.g. that $b = -1$. Then for $a \in K^*$, $a \neq 1$ we get

$$h_K(a, b, c) = h_K(a, -1, a-1) = \prod_{v\text{-arch.}} \max(\|a\|_v, 1, \|1-a\|_v) \cdot \prod_{\substack{\mathfrak{p} \\ \|a\|_\mathfrak{p} > 1}} \|a\|_\mathfrak{p}.$$

Similarly

$$r_K(a, b, c) = r_K(a, -1, a-1) = \prod_{\mathfrak{p} \in I_K(a)} N_{K/\mathbb{Q}}(\mathfrak{p}),$$

where $I_K(a)$ is the set of prime ideals $\mathfrak{p}$ of K such that $\|a\|_\mathfrak{p} \neq 1$ or $\|1-a\|_\mathfrak{p} < 1$. It follows that $r_K(a, -1, a-1) = 1$ iff $a(1-a)$ is a unit.

To simplify the notation we denote by K° the set of all $a \in K^*$ such that $a \neq 1$ and $a(1-a)$ is not a unit.

Therefore it is natural to consider the following functions of one variable

$$h_K(a) = h_K(a, -1, a-1), \quad r_K(a) = r_K(a, -1, a-1)$$

and

$$L_K(a) = \frac{\log h_K(a)}{\log r_K(a)},$$

where $a \in K^\circ$.

Then we get

$$\mathcal{L}_K = \{L_K(a) \, : \, a \in K^\circ\}.$$

Let us note that

$$h_K(a) = h_K(1-a) = h_K(1/a) \text{ and } r_K(a) = r_K(1-a) = r_K(1/a).$$

3.2 The Uniform abc-conjecture

Now we state the *abc*-conjecture for the field of all algebraic numbers.

If $K \subset M$ are algebraic number fields (of finite degrees over $\mathbb{Q}$, as usual) and $a \in K^\circ$, then in general $L_K(a)$ defined above is not equal to $L_M(a)$.

Moreover, it is easy to see that the set $\cup_K \mathcal{L}_K$, where K runs over all algebraic number fields, is not bounded above.

E.g. for $K = \mathbb{Q}(\sqrt[n]{2})$ and $a = -1$ we have $L_K(-1) = n$. In fact,

$$h_K(-1) = \prod_v \max(1, \|2\|_v) = \prod_{v-\text{arch.}} \|2\|_v = 2^n.$$

and $r_K(-1) = 2$, since $\|-1\|_\mathfrak{p} = \|1\|_\mathfrak{p} = \|2\|_\mathfrak{p}$ does not hold iff $\mathfrak{p}$ is the unique prime ideal of K dividing 2, then $N_{K/\mathbb{Q}}(\mathfrak{p}) = 2$.

Therefore it is desired to modify the function L_K in such a way that the new function, say $\widetilde{L}_K$, will be independent on K, i.e.

$$\widetilde{L}_K(a) = \widetilde{L}_M(a), \quad \text{for } a \in K^\circ.$$

We should take into account the ramification of prime ideals in $\mathbb{Q}(a)$. We modify the definition of $I_K(a)$ and $r_K(a)$ as follows.

For $a \in K$, let $J_K(a)$ be the set of all prime ideals of K which divide prime ideals of $\mathbb{Q}(a)$ ramified in $\mathbb{Q}(a)$.

Let

$$\widetilde{I}_K(a) = I_K(a) \cup J_K(a).$$

From this definition it follows that for $K \subset M$ and $a \in K$, and for prime ideals $\mathfrak{p}$ and $\mathfrak{P}$ of K and M respectively such that $\mathfrak{P} \mid \mathfrak{p}$ we have

$$\mathfrak{p} \in J_K(a) \text{ iff } \mathfrak{P} \in J_M(a).$$

Similarly

$$\mathfrak{p} \in I_K(a) \text{ iff } \mathfrak{P} \in I_M(a),$$

and hence

$$\mathfrak{p} \in \widetilde{I}_K(a) \text{ iff } \mathfrak{P} \in \widetilde{I}_M(a).$$

Let us observe that $\widetilde{I}_K(a)$ is not empty for every $a \in K^*$, $a \neq 1$.

Define

$$\widetilde{r}_K(a) = \prod_{\mathfrak{p} \in \widetilde{I}_K(a)} N_{K/\mathbb{Q}}(\mathfrak{p}^{e_{K/\mathbb{Q}}(\mathfrak{p})}) \quad \text{for} \quad a \in K^*,\ a \neq 1,$$

where $e_{K/\mathbb{Q}}(\mathfrak{p})$ is the ramification index of the ideal $\mathfrak{p}$. Then $\widetilde{r}_K(a) > 1$.

Denote

$$\widetilde{L}_K(a) = \frac{\log h_K(a)}{\log \widetilde{r}_K(a)}, \quad \text{for} \quad a \in K^*,\ a \neq 1.$$

Theorem 3.1 *If $K \subset M$ are algebraic number fields and $a \in K^*$, $a \neq 1$, then*

$$\widetilde{L}_K(a) = \widetilde{L}_M(a).$$

Proof: From the definition of normalized valuations it follows that

$$h_M(a) = h_K(a)^{(M:K)}.$$

For prime ideals $\mathfrak{p}$ and $\mathfrak{P}$ of K and M respectively, such that $\mathfrak{P} \mid \mathfrak{p}$ we have

$$\mathfrak{p} \in \widetilde{I}_K(a) \quad \text{iff} \quad \mathfrak{P} \in \widetilde{I}_M(a)$$

as observed above.

Moreover, for fixed $\mathfrak{p}$,

$$\prod_{\mathfrak{P}\mid\mathfrak{p}} N_{M/K}(\mathfrak{P})^{e_{M/K}(\mathfrak{P})} = \prod_{\mathfrak{P}\mid\mathfrak{p}} \mathfrak{p}^{f_{M/K}(\mathfrak{P})e_{M/K}(\mathfrak{P})} = \mathfrak{p}^{(M:K)},$$

where $f_{M/K}(\mathfrak{P})$ is the residue field degree, since

$$\sum_{\mathfrak{P}\mid\mathfrak{p}} f_{M/K}(\mathfrak{P})e_{M/K}(\mathfrak{P}) = (M : K).$$

Consequently

$$\begin{aligned}
\widetilde{r}_M(a) &= N_{M/\mathbb{Q}}\left(\prod_{\mathfrak{P}\in\widetilde{I}_M(a)} \mathfrak{P}^{e_{M/\mathbb{Q}}(\mathfrak{P})}\right) \\
&= N_{K/\mathbb{Q}}\left(N_{M/K}\left(\prod_{\mathfrak{p}\in\widetilde{I}_K(a)} \prod_{\mathfrak{P}\mid\mathfrak{p}} \mathfrak{P}^{e_{M/\mathbb{Q}}(\mathfrak{P})}\right)\right) \\
&= N_{K/\mathbb{Q}}\left(\prod_{\mathfrak{p}\in\widetilde{I}_K(a)} \mathfrak{p}^{e_{K/\mathbb{Q}}(\mathfrak{p})}\right)^{(M:K)} \\
&= \widetilde{r}_K(a)^{(M:K)},
\end{aligned}$$

since $e_{M/\mathbb{Q}}(\mathfrak{P}) = e_{M/K}(\mathfrak{P}) \cdot e_{K/\mathbb{Q}}(\mathfrak{p})$.

The result follows from the definition of $\widetilde{L}_K$. □

We denote the function $\widetilde{L}_K$ simply by L since it does not depend on K. It is defined for all algebraic numbers $a \neq 0, 1$.

The uniform *abc*-conjecture.

$$\limsup L(a) = 1,$$

where a runs over all algebraic numbers $\neq 0, 1$.

Example: The maximal known values of L have been discovered by N. Broberg [Bro]. They correspond to the equalities

1. $(1-\sqrt{2})^2 + (\sqrt{2})^5 = (1+\sqrt{2})^2$ $\quad L = 2.543107$
2. $(\sqrt{2}-1)^5(3-\sqrt{2}) + (\sqrt{2}+1)^5(3+\sqrt{2}) = (\sqrt{2})^{17}$ $\quad L = 2.232501$
3. $(1-\sqrt{2})^4 + (\sqrt{2})^7 \cdot 3 = (1+\sqrt{2})^4$ $\quad L = 1.967616$
4. $1 + (5+2\sqrt{6})^3 = (\sqrt{6}-2)^2(3+\sqrt{6})^5$ $\quad L = 1.919721$
5. $(\sqrt{2}+1)^7 + (\sqrt{2}-1)^7 = (\sqrt{2})^3 \cdot 13^2$ $\quad L = 1.893627$
6. $(5-2\sqrt{6})^8 + (5+2\sqrt{6})(\sqrt{6}-2)^9(3-\sqrt{6})(1+\sqrt{6})(\sqrt{6}-1) \cdot 7^2 = 1$ $\quad L = 1.714895$
7. $(\frac{1+\sqrt{-7}}{2})^{13} + \sqrt{-7} = (\frac{1-\sqrt{-7}}{2})^{13}$ $\quad L = 1.707222$
8. $(5+2\sqrt{6})^2(\sqrt{6}-2)^5(3-\sqrt{6})^5 \cdot 11 + (5-2\sqrt{6})^6 = 1$ $\quad L = 1.641494$

3.3 The abc-conjecture for Fields of Algebraic Functions of One Variable

3.3.1 The Field of Rational Functions

The above considerations concerning fields of algebraic numbers can be repeated with appropriate changes in the case of any global field of characteristic zero.

We begin with the field $k(x)$ of rational functions over a field k of characteristic zero. All normalized valuations $\|\cdot\|_v$ trivial on k have the form $\|\cdot\|_p$, where p runs over all irreducible polynomials in $k[x]$, and $\|\cdot\|_{1/x}$. For every $a \in k(x)^*$ we have $\|a\|_p = \exp(-v_p(a)\cdot\deg p)$, where $v_p(a)$ is the p-adic valuation of a, and $\|a\|_{1/x} = \exp(-v_{1/x}(a)) = \exp(\deg a)$.

Then, for $a, b, c \in k(x)^*$, satisfying $a + b = c$ we define

$$h(a,b,c) = \prod_v \max(\|a\|_v, \|b\|_v, \|c\|_v).$$

Moreover let

$$r(a,b,c) = \prod_{p\in I(a,b,c)} \|p\|_p^{-1} = \exp\left(\deg\left(\prod_{p\in I(a,b,c)} p\right)\right),$$

where $I(a,b,c)$ is the set of all irreducible polynomials $p \in k[x]$ or $p = 1/x$ such that not all numbers $\|a\|_p, \|b\|_p, \|c\|_p$ are equal.

Since $h(a,b,c)$ and $r(a,b,c)$ are scaling invariant, we may assume that a, b, c are relatively prime polynomials.

Then

$$h(a,b,c) = \exp(\max(\deg a, \deg b, \deg c))$$

and

$$r(a,b,c) = \exp(\deg(\mathrm{rad}(abc)) + \delta), \quad \delta = 0 \text{ or } 1,$$

where $\mathrm{rad}(f)$ is the maximal squarefree factor of the polynomial $f \in k[x]$. In other words, $\mathrm{rad}(f)$ is the number of distinct zeros of f in the algebraic closure of the field k.

Define

$$L(a,b,c) = \frac{\log h(a,b,c)}{\log r(a,b,c)}$$

provided not all a, b, c belong to k.

Then for relatively prime polynomials a, b, c we have

$$L(a,b,c) = \frac{\max(\deg a, \deg b, \deg c)}{\deg \mathrm{rad}(abc) + \delta}.$$

In this situation $L(a,b,c) < 1$ always holds. Namely

Theorem 3.2 ([Sto], [Mas]). *For relatively prime polynomials a, b over a field k of characteristic zero, not both belonging to k, and $c = a + b$, we have*

$$\max(\deg a, \deg b, \deg c) \le \deg \mathrm{rad}(abc) - 1.$$

Proof: See e.g. [La]. □

3.3.2 The Field of Algebraic Functions of One Variable

Let K be a field of algebraic functions of one variable over the field of constants k of characteristic zero. We assume for simplicity that the field k is algebraically closed.

For $a \in K \setminus k$ let $\deg a$ be the degree of the divisor of zeros of a, then $\deg a = (K : k(a))$. For $a \in k^*$ we put $\deg a = 0$.

For a prime divisor $\mathfrak{p}$ of K we define the normalized valuation $\|a\|_{\mathfrak{p}} = \exp(-v_{\mathfrak{p}}(a))$, for $a \in K^*$, where $v_{\mathfrak{p}}(a)$ is the multiplicity of zero of a at $\mathfrak{p}$.

Then, for $a \in K^*$, $a \neq 1$, we put

$$h_K(a) = \prod_{\mathfrak{p}} \max(\|a\|_{\mathfrak{p}}, \|1\|_{\mathfrak{p}}, \|1-a\|_{\mathfrak{p}}) = \exp(\deg a).$$

Let $I_K(a)$ be the set of prime divisors $\mathfrak{p}$ of K such that not all numbers $v_{\mathfrak{p}}(a)$, 1, $v_{\mathfrak{p}}(1-a)$ are equal. Then $I_K(a)$ is the set of zeros and of poles of a, and of zeros of $1-a$.

We define

$$r_K(a) = \exp(\#I_K(a)) \quad \text{for} \quad a \in K^*, a \neq 1,$$

and

$$L_K(a) = \frac{\log h_K(a)}{\log r_K(a)} = \frac{\deg a}{\#I_K(a)}, \quad \text{for} \quad a \in K \setminus k.$$

In this situation L_K is bounded from above by a constant depending only on the genus g_K of K. Namely

Theorem 3.3 *If K is a field of algebraic functions of one variable over an algebraically closed field of constants k of characteristic zero, and $a \in K \setminus k$, then*

$$\deg a \leq \#I_K(a) + 2g_K - 2.$$

Consequently

$$\begin{aligned} L_K(a) &< 1, && \text{if } g_K = 0, \\ &\leq 1, && \text{if } g_K = 1, \\ &\leq 1 + \tfrac{2}{3}(g_K - 1), && \text{if } g_K > 1. \end{aligned}$$

Proof: (cf. [El]). For $a \in K \setminus k$ and a prime divisor $\mathfrak{p}$ of K denote by $a(\mathfrak{p})$ the residue of a modulo $\mathfrak{p}$ if $\mathfrak{p}$ is not a pole of a, and put $a(\mathfrak{p}) = \infty$ otherwise.

For every $\alpha \in k \cup \{\infty\}$ we have

$$\#\{\mathfrak{p} : a(\mathfrak{p}) = \alpha\} = \deg a - \sum_{\substack{\mathfrak{p} \\ a(\mathfrak{p})=\alpha}} (e_{K/k(a)}(\mathfrak{p}) - 1),$$

where $e_{K/k(a)}(\mathfrak{p})$ is the ramification index of $\mathfrak{p}$.

Therefore

$$\#I_K(a) = \#\{\mathfrak{p} : a(\mathfrak{p}) \in \{0, 1, \infty\}\} = 3\deg a - \sum_{\substack{\mathfrak{p} \\ a(\mathfrak{p}) \in \{0,1,\infty\}}} (e_{K/k(a)}(\mathfrak{p}) - 1) \tag{3.1}$$

On the other hand we have the Riemann-Hurwitz formula

$$2(g_K - 1) = 2(K : k(a))(g_{k(a)} - 1) + \sum_{\mathfrak{p}} (e_{K/k(a)}(\mathfrak{p}) - 1),$$

where $\mathfrak{p}$ runs over all prime divisors of K.

Since $g_{k(a)} = 0$ and $(K : k(a)) = \deg a$ we get

$$2 \deg a = 2(1 - g_K) + \sum_{\mathfrak{p}} (e_{K/k(a)}(\mathfrak{p}) - 1).$$

Then from (3.1) it follows that

$$\begin{aligned} \# I_K(a) &= \deg a + 2 \deg a - \sum_{\substack{\mathfrak{p} \\ a(\mathfrak{p}) \in \{0,1,\infty\}}} (e_{K/k(a)}(\mathfrak{p}) - 1) \\ &= \deg a + 2(1 - g_K) + \sum_{\substack{\mathfrak{p} \\ a(\mathfrak{p}) \notin \{0,1,\infty\}}} (e_{K/k(a)}(\mathfrak{p}) - 1) \\ &\geq \deg a + 2(1 - g_K). \end{aligned}$$

Thus

$$\begin{aligned} \deg a &\leq \# I_K(a) - 2, && \text{if } g_K = 0, \\ &\leq \# I_K(a), && \text{if } g_K = 1, \\ &\leq \# I_K(a) + 2(g_K - 1) && \text{if } g_K \geq 2. \end{aligned}$$

Since each of a, $1/a$, and $1 - a$ has at least one zero, we have $\# I_K(a) \geq 3$. Therefore

$$L_K(a) = \frac{\deg a}{\# I_K(a)} \leq 1 + \frac{2}{3}(g_K - 1). \qquad \square$$

We shall not discuss here the uniform version of the *abc*-conjecture for the field of all algebraic functions of one variable. We remark only that if $K \subset M$ are fields of algebraic functions of one variable, and $a \in K \setminus k$ as above, then $\widetilde{L}_K(a) = \widetilde{L}_M(a)$, for some function $\widetilde{L}$ defined analogously as in the number field case.

3.4 The n-conjecture

In the *abc*-conjecture we consider all sums $a + b + (-c)$ equal to zero with non-zero and relatively prime summands. We can consider analogous sums with n summands, where $n \geq 3$.

3.4.1 The n-conjecture for the Ring $\mathbb{Z}$ of Integers

Let $a_1, \ldots, a_n \in \mathbb{Z}$, $n \geq 3$ satisfy

(i) $\gcd(a_1, \ldots, a_n) = 1$,
(ii) $a_1 + a_2 + \cdots + a_n = 0$,
(iii) No proper subsum of (ii) is equal to zero.

Define

$$L(a_1, \ldots, a_n) = \frac{\log \max_{1 \le j \le n} |a_j|}{\log r(a_1 \cdots a_n)},$$

where $r(m)$ is the maximal squarefree divisor of m.

The n-conjecture for $\mathbb{Z}$.

$$\limsup L(a_1, \ldots, a_n) = 2n - 5,$$

where $(a_1, \ldots, a_n)$ runs over all n-tuples of integers satisfying (i)–(iii).

Theorem 3.4 ([BB]). *Under the above assumptions (i)–(iii)*

$$\limsup L(a_1, \ldots, a_n) \ge 2n - 5, \quad \text{for } n \ge 3.$$

Proof: The polynomial

$$f_k(z) = \sum_{j=0}^{k} s_{k,j} z^j,$$

where

$$s_{k,j} = \frac{2k+1}{k+j+1} \binom{k+j+1}{2j+1} \in \mathbb{Z},$$

satisfies

$$x^k f_k\left(\frac{(x-1)^2}{x}\right) = \frac{x^{2k+1} - 1}{x - 1}. \tag{3.2}$$

Substituting $k = n - 3,\ x = 2^t$ in (3.2) we get

$$2^{t(2n-5)} - 1 - \sum_{j=0}^{n-3} s_{n-3,j} (2^t - 1)^{2j+1} 2^{t(n-3-j)} = 0.$$

It is the sum with n summands satisfying (i)–(iii). The summand with the maximal absolute value is $2^{t(2n-5)}$, and the product of distinct prime divisors of all summands does not exceed

$$2(2^t - 1) \cdot s, \quad \text{where } s = \prod_{j=0}^{n-3} s_{n-3,j}.$$

Consequently

$$L \ge \frac{\log(2^{t(2n-5)})}{\log(2(2^t - 1)s)} \ge \frac{t(2n-5)\log 2}{t \log 2 + \log 2s} \longrightarrow 2n - 5,$$

as $t \to \infty$. □

Let us remark that in the case of the n-conjecture for $\mathbb{Z}$ and $n > 3$ no estimations from below and from above of $\max(|a_1|, \ldots, |a_n|)$ by means of a function of $r(a_1 \cdots a_n)$ are known. I.e. the results of C.L. Stewart and Yu Kunrui, and of C.L. Stewart and R. Tijdeman mentioned in 2.2. above have not been generalized to any $n > 3$.

3.4.2 The n-conjecture for the Ring of Polynomials

Let $a_1, \ldots, a_n \in k[x]$, $n \geq 3$, where k is a field of characteristic zero, satisfy the conditions (i)–(iii) given above.

Define

$$L(a_1, \ldots, a_n) = \frac{\max_{1\leq j\leq n} \deg a_j}{\deg r(a_1 \cdots a_n)},$$

where $r(m)$ is the maximal squarefree factor of the polynomial m. We assume here that not all $a_1, \ldots, a_n$ belong to k.

The n-conjecture for $k[x]$.

$$\limsup L(a_1, \ldots, a_n) = 2n - 5,$$

where $(a_1, \ldots, a_n)$ runs over all n-tuples of polynomials in $k[x]$, not all constant satisfying (i)–(iii).

Theorem 3.5 ([BB], [BM]). *Under the above assumptions (i)–(iii)*

$$2n - 5 \leq \limsup L(a_1, \ldots, a_n) \leq \binom{n-1}{2}, \quad \text{for } n \geq 3.$$

Moreover, for $n = 3$ and 4 the above inequalities can be replaced by the equalities.

Proof: Substituting $k = n - 3$, $x = y^t$ in (3.2) we get analogously as before the sum of n summands satisfying the conditions (i)–(iii).

The summand with the maximal degree is $y^{t(2n-5)}$, and the product of distinct irreducible factors of all summands equals $y(y^t - 1)$. Consequently

$$L(a_1, \ldots, a_n) = \frac{t(2n-5)}{t+1} \longrightarrow 2n - 5,$$

as $t \to \infty$.

This proves the first inequality. The second one has been proved in [BM], see also [V] and [Za].

The last part of the theorem follows from the observation that $2n - 5 = \binom{n-1}{2}$, for $n = 3$ and $n = 4$. □

3.4.3 The Strong n-conjecture

Now we replace the condition (i) above by the stronger one:

(i′) $\gcd(a_i, a_j) = 1$ for all $1 \leq i < j \leq n$,

and we remove the condition (iii). Then we can state the corresponding conjecture as follows.

The strong n-conjecture: For every $n \geq 3$ there is a real number A_n such that

$$\limsup L(a_1, \ldots, a_n) = A_n,$$

where $(a_1, \ldots, a_n)$ runs over all n-tuples satisfying (i′) and (ii).

For $n = 3$ and $A_3 = 1$ the strong n-conjecture is simply the *abc*-conjecture.

In the case of the ring of integers we have

Theorem 3.6 (S. Konyagin [Ko]).

$$\limsup L(a_1, \dots, a_n) \geq \begin{cases} 3/2, & \text{if } n \geq 3 \text{ is odd}, \\ 1, & \text{if } n \geq 4 \text{ is even}, \end{cases}$$

where $(a_1, \dots, a_n)$ runs over all n-tuples of integers satisfying (i′) and (ii).

Proof: For $n = 5$ we use the identity

$$(6^m + 1)^3 - (6^m - 1)^3 = 6^{2m+1} + 1 + 1, \quad m \geq 1,$$

i.e. we put $a_1 = (6^m + 1)^3$, $a_2 = -(6^m - 1)^3$, $a_3 = -6^{2m+1}$, $a_4 = a_5 = -1$. Evidently the conditions (i′) and (ii) are satisfied.

Moreover

$$L(a_1, \dots, a_5) = \frac{3\log(6^m + 1)}{\log \operatorname{rad}(6(6^{2m} - 1))} \geq \frac{3\log(6^m + 1)}{\log(6^{2m+1})} \longrightarrow \frac{3}{2}$$

as $m \to \infty$.

For any other n odd, $n = 2k + 1$, it is sufficient to take $a_1, \dots, a_5$ as above and to put $a_6 = a_8 = \dots = a_{2k} = 1$, and $a_7 = a_9 = \dots = a_{2k+1} = -1$.

If $n \geq 4$ is even, then it is sufficient to choose a prime number $p > n - 2$ and to define $a_1 = p^m$, $a_2 = -p^m + (n - 2)$, $a_3 = \dots = a_n = -1$, where $m \geq 1$. Then evidently the conditions (i′) and (ii) are satisfied, and

$$L(a_1, \dots, a_n) = \frac{m \log p}{\log \operatorname{rad}(p\,(p^m - n + 2))} \geq \frac{m \log p}{(m + 1)\log p} \longrightarrow 1$$

as $m \to \infty$. □

In the case of the ring of polynomials we have

Theorem 3.7 (see [SS]). *If $n \geq 3$ and $a_1, \dots, a_n$ are polynomials in $k[x_1, \dots, x_r]$ not all constant satisfying (i′) and (ii), then*

$$\max_{1 \leq j \leq n} \deg a_j \leq (n - 2)(\deg r(a_1 \cdots a_n) - 1),$$

where $r(m)$ is the maximal squarefree divisor of a polynomial m.

Consequently

$$\limsup L(a_1, \dots, a_n) \leq n - 2.$$

We have the following

Example: (D. Davies). One can easily verify the polynomial identity

$$(x^m + 2)^3 - x^{3m} - 6(x^m + 1)^2 - 2 = 0,$$

for $m \in \mathbb{N}$. The summands on the l.h.s. are relatively prime by pairs.

Therefore if we take, for a fixed $n \geq 3$,

$$a_1 = (x^m + 2)^3,\ a_2 = -x^{3m},\ a_3 = -6(x^m + 1)^2,\ a_4 = 2 - n,\ a_5 = \cdots = a_n = 1,$$

then conditions (i$'$) and (ii) are satisfied, and

$$L(a_1, \ldots, a_n) = \frac{3m}{2m+1} \longrightarrow \frac{3}{2}$$

as $m \to \infty$.

From the example it follows that in Theorem 3.7 the constant $n - 2$ on the r.h.s. cannot be replaced by a constant less than $\frac{3}{2}$.

4 The Set of all Limit Points

4.1 Limit Points in the Case of the abc-conjecture for the Ring $\mathbb{Z}$ of Integers

For $a, b, c \in \mathbb{N}$, where $\gcd(a, b) = 1$, and $c = a + b$, we defined the set $\mathcal{L} = \{L(a, b)\}$ where

$$L(a, b) = \frac{\log c}{\log r(abc)},$$

and $r(n)$ is the maximal squarefree divisor of a natural number n.

The *abc*-conjecture (for the ring $\mathbb{Z}$ of rational integers) claims that

$$\limsup \mathcal{L} = 1.$$

It is known only that $1 \leq \limsup \mathcal{L} \leq \infty$.

We shall consider the related more general question: What can be proved on the set $\mathcal{L}'$ of all limit points of the set $\mathcal{L}$?

In the above notation we have $\max(a, b) < c$ and $r(abc) \leq abc < c^3$. Hence

$$L(a, b) > \frac{\log c}{\log c^3} = \frac{1}{3}.$$

Therefore $\mathcal{L}' \subset [\frac{1}{3}, \infty]$.

We shall prove the following weak theorem concerning the set $\mathcal{L}'$ to present the main ideas of proofs of similar stronger theorems.

Theorem 4.1 *In the case of the abc-conjecture for $\mathbb{Z}$ we have $[\frac{1}{3}, \frac{1}{2}] \subset \mathcal{L}'$.*

Proof: Let $\alpha > 1$ be a fixed real number and let $a \in [X, 2X]$ be a fixed prime number, where X is large.

Let b run over squarefree numbers in the interval $[X^\alpha, 2X^\alpha]$ prime to a. Then the number of such b's is at least

$$\left(\frac{6}{\pi^2} - \frac{1}{X}\right) X^\alpha + O(\sqrt{X^\alpha}). \tag{4.1}$$

Then $c = a + b$ runs over some numbers in the interval $[X + X^\alpha, 2(X + X^\alpha)]$.

Since the number of not squarefree numbers in this interval is

$$\left(1 - \frac{6}{\pi^2}\right)(X + X^\alpha) + O(\sqrt{X^\alpha})$$

and it is less than (4.1) for large values of X, we conclude that for some b in question the number c is also squarefree.

Then for these a, b, c we have $r(abc) = abc$.

Since $X^\alpha \le c \le 4X^\alpha$ and $X^{1+2\alpha} \le abc \le 16X^{1+2\alpha}$, we get $c = \lambda X^\alpha,\ abc = \mu X^{1+2\alpha}$, where λ, μ are bounded by absolute constants. Consequently

$$L(a,b) = \frac{\alpha \log X + \log \lambda}{(1+2\alpha)\log X + \log \mu} \longrightarrow \frac{\alpha}{1+2\alpha} = \frac{1}{2 + 1/\alpha}$$

as X tends to infinity.

The set $\{\frac{1}{2+1/\alpha} : \alpha > 1\}$ equals to $(\frac{1}{3}, \frac{1}{2})$ and the set $\mathcal{L}'$ is closed. Therefore $[\frac{1}{3}, \frac{1}{2}] \subset \mathcal{L}'$. □

The best known result in this direction is

Theorem 4.2 ([BFGS], [GN]). *For the set $\mathcal{L}$ for the ring $\mathbb{Z}$ of integers we have*

$$\left[\frac{1}{3}, \frac{36}{37}\right] \subset \mathcal{L}'.$$

The proof of this theorem consists of several steps. In every step it is proved that some subinterval of $[\frac{1}{3}, \frac{36}{37}]$ is contained in $\mathcal{L}'$.

E.g. to prove that $[\frac{6}{7}, \frac{12}{13}] \subset \mathcal{L}'$we consider the polynomial identity

$$x^9(x^3 + 2y^3) + (x^3 + y^3)^3(y^3 - x^3) = y^9(2x^3 + y^3).$$

It can be proved that for fixed $\alpha > 1$ and X large we can choose $x \in [X, 2X]$ and $y \in [X^\alpha, 2X^\alpha]$ such that

$$xy(2x^3 + y^3)(x^3 + y^3)(y^3 - x^3)(x^3 + 2y^3)/2 \tag{4.2}$$

is squarefree; it is the most difficult part of the proof.

Then taking

$$a = x^9(x^3 + 2y^3),\ b = (x^3 + y^3)^3(y^3 - x^3),\ c = y^9(2x^3 + y^3)$$

we get $\gcd(a,b) \le 2$ and $c = \lambda X^{12\alpha},\ r(abc) = \nu abc = \mu X^{13\alpha+1}$, where λ, μ, ν are bounded by absolute constants. Therefore (dividing by $\gcd(a,b)$ if necessary) we obtain from (4.2)

$$L(a,b) = \frac{\log c}{\log r(abc)} = \frac{12\alpha \log X + \log \lambda}{(13\alpha + 1)\log X + \log \mu} \longrightarrow \frac{12\alpha}{13\alpha + 1} = \frac{12}{13 + 1/\alpha}$$

as X tends to infinity.

Consequently

$$\left\{\frac{12}{13+1/\alpha} : \alpha > 1\right\} = \left(\frac{12}{14}, \frac{12}{13}\right) \subset \mathcal{L}'.$$

and hence $[\frac{6}{7}, \frac{12}{13}] \subset \mathcal{L}'$.

The last step of the proof invokes in a similar way a polynomial identity of degree 48.

On the other hand a result of M. Filaseta and S. Konyagin ([FK]) gives some information on larger limit points:

Theorem 4.3 *For every $\varepsilon \in [0, 1)$ there is a limit point of the set $\mathcal{L}$ belonging to the interval $[\frac{3}{3+\varepsilon}, \frac{3}{2+\varepsilon}]$. In particular there is such a limit point in the interval $[1, \frac{3}{2}]$.*

It is not known if any point of the interval $(\frac{36}{37}, \frac{36}{25}]$ belongs to $\mathcal{L}'$. Nevertheless the following conditional result holds.

Theorem 4.4 ([BFGS]). *The abc-conjecture implies that $\mathcal{L}' = [\frac{1}{3}, 1]$.*

4.2 Limit Points for Modified and Generalized abc-conjectures

1. In the case of the n-conjecture for the ring $\mathbb{Z}$ of integers there is a limit point of the set $\mathcal{L}$ belonging to the interval $[2n-5, \infty]$, $n = 3, 4, \dots$, see Theorem 3.4 above. Moreover from the results given above and from the identity (3.2) it follows immediately that

$$\left[\frac{1}{n}, (2n-5)\frac{36}{37}\right] \subset \mathcal{L}', \quad n \geq 3$$

in the case of the n-conjecture.

Similarly the result of [FK] implies that there is a limit point of the set $\mathcal{L}$ belonging to the interval $[2n-5, 3n-\frac{15}{2}]$ in the case of the n-conjecture.

2. In the case of the n-conjecture for the ring of polynomials $k[x]$ over a field k of characteristic zero D. Davies [D] proved that

$$\left[\frac{1}{n}, 2n-5\right] \subset \mathcal{L}'.$$

Hence $\mathcal{L}' = [\frac{1}{3}, 1]$ for $n = 3$ and $\mathcal{L}' = [\frac{1}{4}, 3]$ for $n = 4$ since for $n = 3$ and 4 the n-conjecture is a theorem.

3. In the case of the strong n-conjecture for the ring $\mathbb{Z}$ of integers we know only that there is a limit point of the set $\mathcal{L}$ belonging to the interval $[\frac{3}{2}, \infty]$ provided n is odd and to $[1, \infty]$ if n is even (see Theorem 3.6).

4. In the case of the strong n-conjecture for the ring $k[x_1, \dots, x_r]$ of polynomials of many variables over a field of characteristic zero H.N. Shapiro and G.H. Sparer [SS] proved that

$$\mathcal{L}' \subset \left[\frac{1}{n}, n-2\right].$$

I do not know if this interval can be replaced by a smaller one.

5 Numerical Results

At the end of the paper there is the table of all known triples (a, b, c) of relatively prime natural numbers such that $L(a, b) > 1.4$.

Now we make a remark on the examples in the table. Let N be the number of triples (a, b, c) in the table satisfying $c < 2^n$. We give below the values of N for $15 \leq n \leq 60$, $3|n$, and for $n = 80$. For $n \leq 36$ all such examples are known ([Ka]), thus the corresponding values of N are correct.

$\log_2 c <$	15	18	21	24	27	30	33	36	39	42	45	48	51	54	57	60	80
N	3	8	12	18	26	34	51	62	71	78	90	100	107	115	125	129	140

Let us observe that N behaves like a linear function of $\log_2 c$ (approximately $N = 3 \log_2 c - 49$, for $2^{15} < c < 2^{60}$). If it would be true in general, we would have a contradiction with the *abc*-conjecture.

There are known few triples (a, b, c) with $c > 2^{60}$ and $L > 1.4$. This may be caused by difficulties of factoring of large numbers.

On the other hand, no contradiction with the known examples occurs, if we weaken the *abc*-conjecture as follows:

$$\limsup L(a, b) \leq 2,$$

where a, b run over all relatively prime natural numbers. From this weak form of the *abc*-conjecture one can also deduce many other theorems and conjectures like discussed above.

References

[Bro] N. Broberg, Some examples related to the *abc*-conjecture for algebraic number fields, *Math. Comp.* (to appear).

[BB] J. Browkin and Brzeziński, J., Some remarks on the *abc*-conjecture, *Math. Comp.* **62** (1994), no. 206, 931–939.

[BFGS] J. Browkin, Filaseta, M., Greaves, G. and Schinzel, A., *Squarefree values of polynomials and the abc-conjecture*, in: Sieve methods, exponential sums, and their applications in number theory (G.R.H. Greaves, G. Harman, M.N. Huxley, eds.), Cambridge University Press, Cambridge 1996, pp. 65–85.

[BM] W.D. Brownawell and Masser, D.W., Vanishing sums in function fields, *Math. Proc. Cambridge Philos. Soc.* **100** (1986), 427–434.

[Da] L.V. Danilov, Letter to the editors, *Mat. Zam.* **36** (1984), 457–459. (Russian)

[D] D. Davies, A note on the limit points associated with the generalized *abc*-conjecture for $\mathbb{Z}[t]$, *Coll. Math.* **71** (1996), 329–333.

[El] N.D. Elkies, ABC implies Mordell, *Internat. Math. Res. Notices* **7** (1991), 99–109; in: *Duke Math. Journ.* **64** (1991).

[FK] M. Filaseta and Konyagin, S., On a limit point associated with the *abc*-conjecture. *Coll. Math.* **76** (1998), 265–268.

[GS] D. Goldfeld and Szpiro, L., Bounds for the order of the Tate-Shafarevich group, *Compositio Math.* **97** (1995), 71–87.

[Gr] A. Granville, *ABC allows us to count squarefrees*, to appear.
[GS] A. Granville and Stark, H.M., *ABC implies no 'Siegel zero'*, preprint.
[GN] G. Greaves and Nitaj, A., Some polynomial identities related to the *abc*-conjecture, in: *Number Theory in Progress* (K. Györy, H. Iwaniec, J. Urbanowicz, eds.), W. de Gruyter, Berlin, 1999, pp. 229–236.
[GZ] B.H. Gross and Zagier, D.B., Heegner points and derivatives of L-series, *Invent. Math.* **84** (1986), 225–320.
[Ha] M. Hall, Jr., The diophantine equation $x^3 - y^2 = k$, *Computers in Number Theory* (A.O.L. Atkin, B.J. Birch, eds.), Academic Press, London 1971, pp. 173–198.
[Ka] J. Kanapka, private communication.
[Ko] S. Konyagin, private communication.
[La] S. Lang, Old and new conjectured diophantine inequalities, *Bull. Amer. Math. Soc.* **23** (1990), 37–75.
[Mas] R.C. Mason, Equations over function fields, *Lecture Notes in Math.* **1068** (1984), 149–157.
[Mass] D.W. Masser, *Note on a conjecture of Szpiro*, Astérisque **183** (1990), 19–23.
[Ni] A. Nitaj, La conjecture abc, *L'Enseign. Math.* **42** (1996), 3–24.
[Ni1] A. Nitaj, *Invariants des courbes de Frey-Hellegouarch et grands groupes de Tate-Shafarevich*, preprint, 1998.
[Oe] J. Oesterlé, *Nouvelles approaches du "théorème" de Fermat.*, Sém. Bourbaki 1987–1988, Astérisque **161–162** (1988), no. 694, 165–186.
[Ov] M. Overholt, The diophantine equation $n! + 1 = m^2$, *Bull. London Math. Soc.* **25** (1993), 104.
[Pr] K. Prachar, *Primzahlverteilung*, Springer, Berlin, 1957.
[SS] H.N. Shapiro and Sparer, G.H., Extension of a theorem of Mason, *Comm. Pure and Appl. Math.* **47** (1994), 711–718.
[Si] W. Sierpiński, *Elementary Theory of Numbers*, Elsevier, 1988, p. 105.
[Sil] J.H. Silverman, Wieferich's criterion and the *abc*-conjecture, *J. Number Theory* **30** (1988), 226–237.
[ST] C.L. Stewart and Tijdeman, R., On the Oesterlé-Masser conjecture, *Monatsh. Math.* **102** (1986), 251–257.
[SY1] C.L. Stewart and Kunrui Yu, On the *abc*-conjecture, *Math. Ann.* **291** (1991), 225–230.
[SY2] C.L. Stewart and Kunrui Yu, *Lecture given by C.L. Stewart during the Conference in Zakopane*, (1997).
[Sto] W.W. Stothers, Polynomial identities and Hauptmoduln, *Quart. J. Math. Oxford* (2), **32** (1981), 349–370.
[Sz] L. Szpiro, *Discriminant et conducteur*, Astérisque **183** (1990), 7–17.
[Vo] P. Vojta, Diophantine approximation and value distribution theory, *Lecture Notes in Math.* vol. **1239**, Springer-Verlag, 1987.
[V] J.F. Voloch, Diagonal equations over function fields, *Bol. Soc. Brasil. Mat.* **16** (1985), 29–39.
[We] B.M.M. de Weger, Algorithms for diophantine equations, CWI Tract, *Centr. Math. Comput. Sci.*, Amsterdam, 1989.
[We1] B.M.M. de Weger, $A + B = C$ and big Ш's, *Quart. J. Math. Oxford* (2), **49** (1998), no. 193, 105–128.
[Za] U. Zannier, Some remarks on the S-unit equation in function fields, *Acta Arith.* **64** (1993), 87–98.

Table
Of the extremal *abc* examples - version of June 1999

No.	$L(a,b,c)$	$a+b$	$=$	c	*Author(s)*
1.	1.629912	$2+3^{10}\cdot 109$	$=$	23^5	*E.R.*
2.	1.625991	$11^2+3^2\cdot 5^6\cdot 7^3$	$=$	$2^{21}\cdot 23$	*B.W.*
3.	1.623490	$19\cdot 1307+7\cdot 29^2\cdot 31^8$	$=$	$2^8\cdot 3^{22}\cdot 5^4$	*J.B. – J.B.*
4.	1.580756	$283+5^{11}\cdot 13^2$	$=$	$2^8\cdot 3^8\cdot 17^3$	*J.B. – J.B., A.N.*
5.	1.567887	$1+2\cdot 3^7$	$=$	$5^4\cdot 7$	*B.W.*
6.	1.547075	7^3+3^{10}	$=$	$2^{11}\cdot 29$	*B.W.*
7.	1.544434	$7^2\cdot 41^2\cdot 311^3+11^{16}\cdot 13^2\cdot 79$	$=$	$2\cdot 3^3\cdot 5^{23}\cdot 953$	*A.N.*
8.	1.536714	$5^3+2^9\cdot 3^{17}\cdot 13^2$	$=$	$11^5\cdot 17\cdot 31^3\cdot 137$	*H.R. – P.M.*
9.	1.526999	$13\cdot 19^6+2^{30}\cdot 5$	$=$	$3^{13}\cdot 11^2\cdot 31$	*A.N.*
10.	1.522160	$3^{18}\cdot 23\cdot 2269+17^3\cdot 29\cdot 31^8$	$=$	$2^{10}\cdot 5^2\cdot 7^{15}$	*A.N.*
11.	1.502839	$239+5^8\cdot 17^3$	$=$	$2^{10}\cdot 37^4$	*J.B. – J.B., A.N.*
12.	1.497621	$5^2\cdot 7937+7^{13}$	$=$	$2^{18}\cdot 3^7\cdot 13^2$	*B.W.*
13.	1.492432	$2^2\cdot 11+3^2\cdot 13^{10}\cdot 17\cdot 151\cdot 4423$	$=$	$5^9\cdot 139^6$	*A.N.*
14.	1.491590	$73+2^{13}\cdot 7^7\cdot 941^2$	$=$	$3^{16}\cdot 103^3\cdot 127$	*A.N.*
15.	1.489246	$2^{24}+11^7\cdot 19\cdot 29^2$	$=$	$3^{11}\cdot 5^3\cdot 7^3\cdot 41$	*A.N.*
16.	1.488865	$11^2+3^9\cdot 13$	$=$	$2^{11}\cdot 5^3$	*B.W.*
17.	1.482910	$37+2^{15}$	$=$	$3^8\cdot 5$	*B.W.*
18.	1.481322	$5^{14}\cdot 19+2^5\cdot 3\cdot 7^{13}$	$=$	$11^7\cdot 37^2\cdot 353$	*A.N.*
19.	1.474450	$1+3^{16}\cdot 7$	$=$	$2^3\cdot 11\cdot 23\cdot 53^3$	*J.B. – J.B., A.N.*
20.	1.474137	$7^2+2^{10}\cdot 11\cdot 53^2$	$=$	$3^4\cdot 5^8$	*J.B. – J.B., A.N.*
21.	1.471298	$3^4\cdot 199+11^8$	$=$	$2^3\cdot 5^7\cdot 7^3$	*J.B. – J.B., A.N.*
22.	1.465676	$17^4\cdot 67+2^{19}\cdot 137^4$	$=$	$3^{15}\cdot 5^3\cdot 13\cdot 89^2$	*H.R. – P.M.*
23.	1.465520	$7^{12}+2^{14}\cdot 67^3\cdot 461$	$=$	$3^{13}\cdot 11\cdot 19^4$	*A.N.*
24.	1.461924	$2^7\cdot 5^2+7^6\cdot 41$	$=$	13^6	*B.W.*
25.	1.459425	$5^{11}\cdot 31\cdot 191+2^8\cdot 7^{13}\cdot 89\cdot 859^2$	$=$	$3^{30}\cdot 13^4\cdot 277$	*K.V.*
26.	1.457794	$5^{12}\cdot 17^2\cdot 31^2\cdot 1699+23^{14}\cdot 29$	$=$	$2^{19}\cdot 3^2\cdot 11\cdot 13^{10}\cdot 47$	*A.N.*
27.	1.457790	$3^6\cdot 5^{12}+2^{16}\cdot 13\cdot 59^4$	$=$	$7^{11}\cdot 47\cdot 113$	*A.N.*
28.	1.457066	$3^2\cdot 5^2+2^4\cdot 17^3\cdot 31^4$	$=$	$7^{10}\cdot 257$	*J.B. – J.B., A.N.*
29.	1.456203	$2^{25}\cdot 19+3\cdot 5^{15}\cdot 1033$	$=$	$11^7\cdot 13^3\cdot 47^2$	*A.N.*
30.	1.455673	$1+2^5\cdot 3\cdot 5^2$	$=$	7^4	*B.W.*
31.	1.455126	$3^2\cdot 11^6+2^{35}$	$=$	$19^5\cdot 13883$	*J.B. – J.B.*
32.	1.453343	$13^6+2\cdot 3^4\cdot 7^4\cdot 11^9\cdot 23$	$=$	$5^7\cdot 103^4\cdot 2399$	*A.N.*
33.	1.452613	$2^{19}\cdot 13\cdot 103+7^{11}$	$=$	$3^{11}\cdot 5^3\cdot 11^2$	*B.W.*
34.	1.451344	$3^5\cdot 7+5^6\cdot 67$	$=$	2^{20}	*J.B. – J.B., A.N.*
35.	1.450858	$3^5\cdot 7^3+2^{13}\cdot 23^3\cdot 59$	$=$	$5^3\cdot 19^6$	*J.B. – J.B.*
36.	1.450026	$1+3^3\cdot 5^3\cdot 7^7\cdot 23$	$=$	$2^{13}\cdot 11^4\cdot 13\cdot 41$	*A.N.*
37.	1.449651	$1+3\cdot 5^5\cdot 47^2$	$=$	$2^{18}\cdot 79$	*G.F.*
38.	1.447977	$11^2\cdot 43+5^9\cdot 7^2\cdot 13^4\cdot 97$	$=$	$2^3\cdot 3\cdot 73^7$	*A.N.*
39.	1.447743	$89+7\cdot 11^8$	$=$	$2^{20}\cdot 3^3\cdot 53$	*J.B. – J.B., A.N.*
40.	1.446246	$3^2\cdot 5^7\cdot 79+2^{29}\cdot 13$	$=$	$11^7\cdot 19^2$	*A.N.*
41.	1.445064	$2\cdot 13^2+5^8$	$=$	$3\cdot 19^4$	*J.B. – J.B., A.N.*
42.	1.444199	$2^{19}\cdot 263+83\cdot 167^5$	$=$	$5^4\cdot 29^7$	*H.R. – P.M.*
43.	1.443502	$2^2\cdot 11^4\cdot 17+5^{17}\cdot 13577$	$=$	$3^4\cdot 23^9\cdot 71$	*A.N.*
44.	1.443307	$1+2^{12}\cdot 5^3$	$=$	$3^5\cdot 7^2\cdot 43$	*B.W.*
45.	1.443284	$3^2\cdot 19^3+5^{11}$	$=$	$2^{17}\cdot 373$	*J.B. – J.B., A.N.*

(Table Contd.)

No.	$L(a,b,c)$	$a+b$	$=$	c	*Author(s)*
46.	1.442014	$2^{5} \cdot 11^{2} \cdot 19^{9} + 5^{15} \cdot 37^{2} \cdot 47$	$=$	$3^{7} \cdot 7^{11} \cdot 743$	*A.N.*
47.	1.441814	$3^{16} \cdot 23^{2} + 2^{13} \cdot 29^{2} \cdot 37^{3}$	$=$	$5^{9} \cdot 11^{4} \cdot 13$	*A.N.*
48.	1.441519	$7^{3} \cdot 29^{5} \cdot 151^{2} + 2^{4} \cdot 5^{16} \cdot 97 \cdot 919$	$=$	$3^{27} \cdot 13^{4}$	*A.N.*
49.	1.441441	$31^{3} + 2 \cdot 17 \cdot 41^{5}$	$=$	$3 \cdot 5^{7} \cdot 7^{5}$	*J.B. – J.B., A.N.*
50.	1.440969	$3^{4} \cdot 23^{2} + 31^{5}$	$=$	$2^{15} \cdot 5^{3} \cdot 7$	*J.B. – J.B., A.N.*
51.	1.440264	$2^{35} \cdot 7^{2} \cdot 17^{2} \cdot 19 + 3^{27} \cdot 107^{2}$	$=$	$5^{15} \cdot 37^{2} \cdot 2311$	*A.N.*
52.	1.439063	$1 + 2^{4} \cdot 3^{7} \cdot 547$	$=$	$5^{8} \cdot 7^{2}$	*B.W.*
53.	1.438357	$1 + 19 \cdot 509^{3}$	$=$	$2^{19} \cdot 3^{4} \cdot 59$	*J.B. – J.B.*
54.	1.436180	$2 \cdot 13^{5} + 7^{6} \cdot 173^{2}$	$=$	$3^{13} \cdot 47^{2}$	*A.N.*
55.	1.435006	$2^{10} \cdot 7 + 5^{7}$	$=$	$3^{8} \cdot 13$	*B.W.*
56.	1.433956	$11^{9} \cdot 43 + 2^{4} \cdot 23^{6} \cdot 47 \cdot 277^{2}$	$=$	$5^{14} \cdot 7^{2} \cdot 13^{4}$	*A.N.*
57.	1.433464	$2^{5} \cdot 3^{18} + 5^{6} \cdot 7^{10} \cdot 23^{2}$	$=$	$11^{9} \cdot 691 \cdot 1433$	*A.N.*
58.	1.433452	$5^{3} \cdot 8111 + 19^{12} \cdot 29$	$=$	$2^{19} \cdot 3^{3} \cdot 17^{4} \cdot 233^{2}$	*A.N.*
59.	1.433043	$31^{2} + 3^{5} \cdot 5^{9}$	$=$	$2^{5} \cdot 23^{4} \cdot 53$	*J.B. – J.B., A.N.*
60.	1.432904	$2^{21} + 7^{6} \cdot 17 \cdot 8209^{2}$	$=$	$5^{12} \cdot 743^{2}$	*A.N.*
61.	1.431623	$17^{4} \cdot 79^{3} \cdot 211 + 2^{29} \cdot 23 \cdot 29^{2}$	$=$	5^{19}	*A.N.*
62.	1.431260	$2^{27} \cdot 7^{5} + 3^{26} \cdot 11 \cdot 19 \cdot 139$	$=$	$5^{2} \cdot 13^{6} \cdot 43^{4} \cdot 179$	*A.N.*
63.	1.431183	$2^{11} + 3^{9} \cdot 7^{3} \cdot 11^{3} \cdot 19$	$=$	$29 \cdot 277^{4}$	*K.V.*
64.	1.431092	$2^{9} \cdot 19^{2} + 59^{6} \cdot 73$	$=$	$3^{3} \cdot 5^{7} \cdot 7^{2} \cdot 31^{3}$	*A.N.*
65.	1.430418	$193 + 2 \cdot 5^{6} \cdot 19^{2} \cdot 1193^{2}$	$=$	$3^{9} \cdot 13^{8}$	*J.B. – J.B., A.N.*
66.	1.430176	$3^{6} \cdot 7^{2} \cdot 13 \cdot 127^{2} + 2^{38} \cdot 61 \cdot 137$	$=$	$5^{11} \cdot 19^{6}$	*J.B. – J.B.*
67.	1.429873	$2^{9} \cdot 37 \cdot 97^{5} + 5^{5} \cdot 7 \cdot 89^{7}$	$=$	$3^{20} \cdot 17^{4} \cdot 3323$	*A.N.*
68.	1.429552	$3^{9} \cdot 29 + 7^{6} \cdot 43^{2}$	$=$	$2^{24} \cdot 13$	*A.N.*
69.	1.429007	$3^{21} + 7^{2} \cdot 11^{6} \cdot 199$	$=$	$2 \cdot 13^{8} \cdot 17$	*A.N.*
70.	1.428908	$73^{2} + 2^{11} \cdot 11^{4} \cdot 13^{3}$	$=$	$3^{11} \cdot 5^{5} \cdot 7 \cdot 17$	*J.B. – J.B.*
71.	1.428402	$5^{14} \cdot 11 + 3^{6} \cdot 7^{5} \cdot 13^{2} \cdot 251$	$=$	$2^{21} \cdot 23^{4}$	*A.N.*
72.	1.428323	$11 + 7^{3} \cdot 167^{2}$	$=$	$2 \cdot 3^{14}$	*J.B. – J.B., A.N.*
73.	1.427566	$73 + 11^{5} \cdot 157^{2}$	$=$	$2^{2} \cdot 3^{10} \cdot 7^{5}$	*J.B. – J.B., A.N.*
74.	1.427488	$61^{4} + 2^{20} \cdot 41^{3} \cdot 83^{2}$	$=$	$3^{22} \cdot 5 \cdot 19 \cdot 167$	*A.N.*
75.	1.427115	$3^{10} + 7^{8} \cdot 23$	$=$	$2^{9} \cdot 509^{2}$	*A.N.*
76.	1.426753	$31 + 2^{5} \cdot 5^{10} \cdot 19^{2}$	$=$	$3 \cdot 7^{5} \cdot 11^{3} \cdot 41^{2}$	*J.B. – J.B., A.N.*
77.	1.426565	$3 + 5^{3}$	$=$	2^{7}	*B.W.*
78.	1.423381	$5^{2} \cdot 11 + 13^{3} \cdot 1483^{2}$	$=$	$2^{29} \cdot 3^{2}$	*J.B. – J.B., A.N.*
79.	1.422083	$17 \cdot 19^{4} + 3^{3} \cdot 5^{10} \cdot 7^{2} \cdot 29^{3}$	$=$	$2^{13} \cdot 13^{7} \cdot 613$	*K.V.*
80.	1.421828	$2^{4} \cdot 59 + 5^{12} \cdot 19$	$=$	$3^{3} \cdot 11^{2} \cdot 17^{5}$	*J.B. – J.B., A.N.*
81.	1.421575	$5^{7} + 11^{5} \cdot 13^{2}$	$=$	$2^{15} \cdot 7^{2} \cdot 17$	*J.B. – J.B., A.N.*
82.	1.421008	$2^{9} \cdot 37^{3} \cdot 89 + 3^{9} \cdot 5^{9} \cdot 31$	$=$	103^{6}	*J.B. – J.B., A.N.*
83.	1.420437	$7^{8} \cdot 19 + 2^{15} \cdot 5^{2} \cdot 37^{2}$	$=$	$3 \cdot 17^{7}$	*A.N.*
84.	1.420320	$3^{13} + 2^{21} \cdot 5^{4} \cdot 199^{2}$	$=$	$7^{8} \cdot 83^{2} \cdot 1307$	*A.N.*
85.	1.420036	$23^{3} + 3^{9} \cdot 5^{7} \cdot 31$	$=$	$2^{7} \cdot 7^{3} \cdot 13 \cdot 17^{4}$	*A.N.*
86.	1.419292	$19^{4} \cdot 37^{2} + 3^{4} \cdot 5^{14} \cdot 79$	$=$	$2^{8} \cdot 31^{5} \cdot 73^{2}$	*A.N.*
87.	1.418919	$7^{2} + 2^{17} \cdot 181^{2}$	$=$	$3^{8} \cdot 809^{2}$	*J.B. – J.B., A.N.*
88.	1.418233	$13 \cdot 3499 + 2^{39}$	$=$	$3^{4} \cdot 5^{11} \cdot 139$	*J.B. – J.B.*
89.	1.417633	$5^{6} \cdot 1609 + 2^{9} \cdot 3^{14} \cdot 13^{3}$	$=$	1523^{4}	*J.B. – J.B.*
90.	1.416793	$3^{9} \cdot 43^{3} + 5^{13} \cdot 5323$	$=$	$2^{7} \cdot 7^{3} \cdot 23^{6}$	*A.N.*
91.	1.416438	$41^{4} \cdot 33941 + 3^{12} \cdot 19^{7}$	$=$	$2^{23} \cdot 5^{9} \cdot 29$	*J.B. – J.B.*
92.	1.416078	$^{22} \cdot 37 \cdot 204749 + 2^{8} \cdot 5^{2}7$	$=$	$13^{2} \cdot 31^{6} \cdot 103^{4} \cdot 113$	*A.N.*
93.	1.416051	$3 \cdot 5^{4} \cdot 599 + 11 \cdot 23^{8}$	$=$	$2^{22} \cdot 59^{3}$	*J.B. – J.B., A.N.*

(Table Contd.)

No.	$L(a,b,c)$	$a+b$	=	c	Author(s)
94.	1.415633	$2^{46}\cdot 23+3^9\cdot 5^5\cdot 11^7\cdot 31^2\cdot 43$	=	$19^{11}\cdot 59\cdot 7207$	A.N.
95.	1.415561	$7^3+5^{13}\cdot 181$	=	$2^4\cdot 3\cdot 11\cdot 13^2\cdot 19^5$	A.N.
96.	1.415273	$3\cdot 23^4+5^{13}\cdot 31$	=	$2\cdot 7^4\cdot 199^3$	H.R. – P.M.
97.	1.415090	$2^6\cdot 5^2\cdot 7^{13}\cdot 13^2\cdot 463+3^4\cdot 43^{12}$	=	$11^{12}\cdot 389^2\cdot 6841$	A.N.
98.	1.414503	$3^{11}\cdot 5^4+7\cdot 11^6\cdot 43$	=	$2^{17}\cdot 17^3$	X.G.
99.	1.414352	$3^7\cdot 5^{14}\cdot 7^2+2^{51}\cdot 11^2$	=	$29^5\cdot 73\cdot 419^2\cdot 1039$	A.N.
100.	1.413698	$2^6\cdot 5\cdot 137+3^{14}$	=	13^6	J.B. – J.B., A.N.
101.	1.413279	$5^2+3^7\cdot 13^3$	=	$2^8\cdot 137^2$	J.B. – J.B., A.N.
102.	1.413166	$3^6\cdot 157^3\cdot 283+23^{10}$	=	$2^{30}\cdot 5^2\cdot 11^2\cdot 13$	J.B. – J.B., A.N.
103.	1.412893	$13\cdot 733+3^9\cdot 5^5\cdot 89^5$	=	$2^{19}\cdot 7^2\cdot 31^5\cdot 467$	K.V.
104.	1.412681	$5+3^{11}$	=	$2^{10}\cdot 173$	B.W.
105.	1.411682	$79^3+3^6\cdot 7\cdot 11\cdot 13^5$	=	$2^{18}\cdot 43^3$	A.N.
106.	1.411615	$3\cdot 13^2\cdot 1049+2^{39}\cdot 29^2\cdot 107$	=	$19^3\cdot 139^6$	J.B. – J.B., A.N.
107.	1.410830	$13\cdot 29^4+3\cdot 7^{10}\cdot 19^4$	=	$2^5\cdot 5\cdot 43^2\cdot 139^4$	A.N.
108.	1.410683	$67^2\cdot 2399+3^{13}\cdot 107^3$	=	$2^6\cdot 5^{15}$	J.B. – J.B.
109.	1.410044	$2^{13}\cdot 3^{13}\cdot 11^3+13\cdot 29\cdot 43^6\cdot 673$	=	$5^{20}\cdot 17$	A.N.
110.	1.409742	$5^{12}+2^2\cdot 3^{21}\cdot 43^2\cdot 52859$	=	$7^{10}\cdot 13^3\cdot 17^2\cdot 151^2$	A.N.
111.	1.408973	7^2+83^5	=	$2^2\cdot 3^{12}\cdot 17\cdot 109$	J.B. – J.B., A.N.
112.	1.408866	$2^{12}+3^{15}\cdot 19^2\cdot 73^2\cdot 3343$	=	$5\cdot 41^3\cdot 193^5$	A.N.
113.	1.407787	$2^2\cdot 13+7^3\cdot 41^5\cdot 181$	=	$3^{14}\cdot 5\cdot 67^3$	A.N.
114.	1.407404	$3^2\cdot 233+23^7\cdot 293^2$	=	$2^{15}\cdot 5^2\cdot 13^5\cdot 31^2$	A.N.
115.	1.407208	$241+2^{12}\cdot 3^4\cdot 5^6\cdot 1181$	=	$11^8\cdot 13^4$	J.B. – J.B.
116.	1.407051	$3^9\cdot 163+2^3\cdot 11^6\cdot 17$	=	5^{12}	J.B. – J.B., A.N.
117.	1.406524	$7^9+3^2\cdot 5^7\cdot 13^3$	=	$2^{16}\cdot 19^2\cdot 67$	J.K.
118.	1.406420	$2^{19}\cdot 367^3+5^{17}\cdot 197\cdot 281$	=	$13^2\cdot 251^6$	A.N.
119.	1.406097	$2^{16}\cdot 41\cdot 71+3^{15}\cdot 7^2$	=	19^7	A.N.
120.	1.406080	$13^5\cdot 19^3+2\cdot 11^{12}\cdot 1123\cdot 76081$	=	$3^{38}\cdot 397$	A.N.
121.	1.406079	$5\cdot 7^2+13^2\cdot 43^3$	=	$2^{11}\cdot 3^8$	J.B. – J.B., A.N.
122.	1.405785	$13^3+2^9\cdot 37^2$	=	$3^2\cdot 5^7$	J.B. – J.B., A.N.
123.	1.405443	$2^{24}\cdot 3^5+5\cdot 19^5\cdot 59^2$	=	$7^{10}\cdot 167$	A.N.
124.	1.404484	$631+2^{26}\cdot 5\cdot 29^2$	=	$3^3\cdot 7^{10}\cdot 37$	J.B. – J.B., A.N.
125.	1.404264	$1+3^9\cdot 7^2\cdot 197$	=	$2^7\cdot 5^7\cdot 19$	J.B. – J.B., A.N.
126.	1.403980	$5^{12}\cdot 227+2^8\cdot 3\cdot 7^3\cdot 23^7\cdot 41$	=	$11\cdot 19^5\cdot 67^5$	A.N.
127.	1.403958	$3^9\cdot 103+2^8\cdot 11^2\cdot 13^5\cdot 41^2\cdot 47$	=	$5^{14}\cdot 53^3$	A.N.
128.	1.403482	$3^3\cdot 13+2^5\cdot 11\cdot 19^2\cdot 73^3$	=	$5^2\cdot 7^{11}$	A.N.
129.	1.402864	$5\cdot 67^3\cdot 127^2\cdot 19219+13^{18}\cdot 37\cdot 277$	=	$2\cdot 3^{15}\cdot 7^2\cdot 31^{10}$	A.N.
130.	1.402737	$3^4\cdot 19\cdot 61\cdot 173^2+2^{44}\cdot 7^{10}$	=	$5^2\cdot 149^4\cdot 503\cdot 929^3$	K.V.
131.	1.402183	$3^{12}\cdot 5^6+7^9\cdot 31^2$	=	$2^9\cdot 11^5\cdot 571$	A.N.
132.	1.401993	$3\cdot 5^{14}\cdot 199+7^2\cdot 11^5\cdot 17^4\cdot 41$	=	$2^{30}\cdot 13^4$	A.N.
133.	1.401979	$2^{33}\cdot 5+3^9\cdot 7^6\cdot 31^2\cdot 97$	=	$11^2\cdot 19^3\cdot 127^4$	A.N.
134.	1.401419	$3^{10}\cdot 5^4\cdot 401+13^6\cdot 47^3$	=	$2^{29}\cdot 31^2$	J.K.
135.	1.401261	$3^6\cdot 11^2\cdot 47\cdot 359^2+17^{13}$	=	$2^{21}\cdot 5^4\cdot 2749^2$	K.V.
136.	1.401156	$3^2\cdot 31^2\cdot 73^4\cdot 349+2^{29}\cdot 7$	=	$5^{15}\cdot 53^2$	A.N.
137.	1.400812	$23^4\cdot 71^2+7^{14}\cdot 1231$	=	$2^4\cdot 5^2\cdot 11^2\cdot 29^7$	A.N.
138.	1.400588	$13^4+17^6\cdot 463$	=	$2^{21}\cdot 73^2$	J.K.
139.	1.400317	$2^{14}\cdot 3^{13}\cdot 5+7\cdot 29^6\cdot 71^2$	=	$11^9\cdot 13^2\cdot 53$	A.N.
140.	1.400262	$5^{18}\cdot 6359+3^2\cdot 47^6\cdot 73^3$	=	$2^7\cdot 19^{10}\cdot 79$	A.N.

The authors:

J.B.– J.B. : Jerzy Browkin and Juliusz Brzeziński
G.F. : Gerhard Frey,
J.K. : Joseph Kanapka,
A.N. : Abderrahmane Nitaj,
E.R. : Eric Reyssat,
H.R.– P.M. : Herman te Riele and Peter Montgomery,
K.V. : Kees Visser,
B.W. : Benjamin M.M. de Weger,
X.G. : Xiao Gang.

Institute of Mathematics
University of Warsaw
ul. Banacha 2
PL-02-097 Warsaw
Poland
E-Mail: bro@mimuw.edu.pl

On Values of Linear and Quadratic Forms at Integral Points

S.G. Dani

The aim of this article is to give an exposition of certain applications of the study of the homogeneous space $SL(n, \mathbb{R})/SL(n, \mathbb{Z})$ and the flows on it induced by subgroups of $SL(n, \mathbb{R})$, to problems on values of linear and quadratic forms at integral points. Also, some complements to Margulis's theorem on Oppenheim's conjecture are proved.

Let $n \geq 2$ and let $\mathbb{R}^n$ be the n-dimensional euclidean space, viewed as the space of n-rowed column vectors with real entries. We denote by $\mathbb{Z}^n$ the usual lattice in $\mathbb{R}^n$ consisting of column vectors with integral entries. We recall that the group $GL(n, \mathbb{R})$ of $n \times n$ nonsingular real matrices is a locally compact topological group and the integral matrices with determinant ± 1 form a discrete subgroup, which is denoted by $GL(n, \mathbb{Z})$. The following fact is basic to many applications. The space of lattices in $\mathbb{R}^n$ can be identified canonically with the homogeneous space $GL(n, \mathbb{R})/GL(n, \mathbb{Z})$, by identifying the lattice $g\mathbb{Z}^n$, $g \in GL(n, \mathbb{R})$, with the coset $gGL(n, \mathbb{Z})$ (we note that any lattice in $\mathbb{R}^n$ is of the form $g\mathbb{Z}^n$ for some $g \in GL(n, \mathbb{R})$). The correspondence has many interesting properties. Firstly, the (intrinsically defined) topology on the space of lattices[3,14] coincides, under the correspondence as above, with the quotient topology on $GL(n, \mathbb{R})/GL(n, \mathbb{Z})$. If we consider the subspace consisting of lattices with a given discriminant (volume of a fundamental parallelopiped for the lattice), then it corresponds to an orbit of $SL(n, \mathbb{R})$, the subgroup of $GL(n, \mathbb{R})$ consisting of matrices of determinant 1; the orbit can be realised as $SL(n, \mathbb{R})/SL(n, \mathbb{Z})$, $SL(n, \mathbb{Z})$ being the subgroup of $SL(n, \mathbb{R})$ consisting of matrices with integer entries.

1 The Homogeneous Space $SL(n, \mathbb{R})/SL(n, \mathbb{Z})$

It will be convenient from this point on to consider $SL(n, \mathbb{R})/SL(n, \mathbb{Z})$, which corresponds to the space of lattices with a given discriminant, rather than the whole space $GL(n, \mathbb{R})/GL(n, \mathbb{Z})$, which is in fact a continuous union of copies of the former, parametrised by the discriminant. We shall denote by $\mathcal{L}^1$ the space of lattices in $\mathbb{R}^n$ with discriminant 1 (for convenience we shall exclude n from the notation, as there is no possibility of confusion on account of it).

The space $SL(n, \mathbb{R})/SL(n, \mathbb{Z})$ is noncompact (this follows for instance from the following theorem). However, it carries a finite measure invariant under the action of $SL(n, \mathbb{R})$ by translations on the left (cf. [21]). This means that though noncompact it is still a 'small' space; it has a compact part together with a narrow cusp (end); it may be observed that 'tending to infinity' in such a space corresponds to going further and further into the cusp. We next recall the Mahler criterion which relates the geometric asymptotics in the space $SL(n,\mathbb{R})/SL(n,\mathbb{Z})$ with Diophantine asymptotics of lattices[21].

Theorem 1.1 *A sequence $\{\Lambda_i\}$ of lattices in $\mathcal{L}^1$ tends to infinity if and only if there exists a sequence $\{x_i\}$ of nonzero vectors in $\mathbb{R}^n$ such that $x_i \in \Lambda_i$ for all i and $x_i \to 0$ as $i \to \infty$.*

The correspondence is also amenable to analytical techniques. Let f be a measurable function on $\mathbb{R}^n$ vanishing outside a compact set (f need not be continuous). Following Siegel we associate to f a function $\tilde{f}$ on $\mathcal{L}^1$ by setting

$$\tilde{f}(\Lambda) = \sum_{x\in\Lambda-(0)} f(x) \quad \text{for all } \Lambda \in \mathcal{L}^1;$$

observe that since f vanishes outside a compact set and Λ is discrete, the sum on the right hand side is in fact a finite sum. It turns out that via the transform the Lebesgue integral corresponds to the integral on $\mathcal{L}^1 = SL(n,\mathbb{R})/SL(n,\mathbb{Z})$ with respect to the $SL(n,\mathbb{R})$-invariant probability measure (that is, the invariant measure normalised to have total measure 1). Specifically we have the following:

Theorem 1.2 (C.L. Siegel[24]). *Let f be an integrable function on $\mathbb{R}^n$ vanishing outside a compact subset of $\mathbb{R}^n$. Then $\tilde{f}$ is integrable and*

$$\int \tilde{f}\,dm = \int f\,dl,$$

where l is the Lebesgue measure on $\mathbb{R}^n$ and m is the $SL(n,\mathbb{R})$-invariant probability measure on $\mathcal{L}^1$.

Siegel deduced from this that for any star-shaped body S in $\mathbb{R}^n$, centered at the origin, with volume less that $\zeta(n)$ there exists a lattice Λ in $\mathbb{R}^n$ such that $S \cap \Lambda = (0)$, a statement which was formulated earlier by Minkowski and first proved by Hlawka. The theorem plays an important role in the recent results on the asymptotics of the solutions of quadratic inequalities, described in §8.

2 Integral Points in Convex Sets

It may be recalled that by Minkowski's theorem any convex symmetric body in $\mathbb{R}^n$ with volume exceeding 2^n contains a nonzero point of any lattice $\Lambda \in \mathcal{L}^1$ (see [3] or [14]). The result applies in particular to regions defined by inequalities of the form $|L_i(v)| < a_i$, $i = 1, \ldots, k$, for some $k \geq 1$, where L_i's are linear forms and a_i's are positive real numbers. Similarly, it also applies to sets of the form $Q(v) < a$ where Q is a positive definite quadratic form; (one can also consider mixed systems, so long as the volumes of the regions can be computed). When $k = n$ and $L_1, \ldots, L_n$ are linearly independent (in the vector space of linear forms) the regions as above are parallelopipeds and Minkowski's theorem is optimal in this case as, for example, the open cube of side 2 centered at the origin (which has volume 2^n) does not contain any nonzero integral point.

On the other hand for balls (and ellipsoids) centered at 0, the classical theorem of Hermite gives a better bound for existence of nonzero lattice points, than the Minkowski theorem for general convex symmetric bodies, namely, a ball contains a nonzero point of every lattice in $\mathcal{L}^1$ if its radius exceeds $(4/3)^{\frac{n-1}{4}}$. This can be deduced from considerations of a fundamental domain for the homogeneous space $SL(n, \mathbb{R})/SL(n, \mathbb{Z})$, which we shall discuss next.

Let D denote the subgroup of $SL(n, \mathbb{R})$ consisting of all diagonal matrices with positive entries. We shall write the diagonal matrices in the form diag $(d_1, \ldots, d_n)$, where $d_1, \ldots, d_n$ are the diagonal entries (in the natural order of the entries). For $\sigma > 0$ let D_σ denote the subset of D consisting of those $d = \mathrm{diag}\,(d_1, \ldots, d_n)$ for which $(d_i/d_{i+1}) \leq \sigma$, for all $i = 1, \ldots, n-1$. We note that for $d = \mathrm{diag}\,(d_1, \ldots, d_n) \in D_\sigma$, since $d_1 d_2, \ldots, d_n = 1$ we have $d_1^n = (d_1/d_2)^{n-1}(d_2/d_3)^{n-2} \cdots (d_{n-1}/d_n) \leq \sigma^{n(n-1)/2}$ and hence $d_1 \leq \sigma^{(n-1)/2}$. Let N be the subgroup of $SL(n, \mathbb{R})$ consisting of all upper triangular matrices with 1's on the diagonal and for any $\tau > 0$ let N_τ denote the subset of N consisting of all those matrices for which every off-diagonal entry is of absolute value at most τ. Also let K denote the subgroup of $SL(n, \mathbb{R})$ consisting of orthogonal matrices of determinant 1. A set of the form $KD_\sigma N_\tau = \{kdu \mid k \in K, d \in D_\sigma, u \in N_\tau\}$ is called a *Siegel set.* It can be verified that any Siegel set is of finite Haar measure in $SL(n, \mathbb{R})$. It is a crucial fact that for suitable values of σ and τ the corresponding Siegel set is a fundamental domain for $SL(n, \mathbb{Z})$ in $SL(n, \mathbb{R})$, namely we have the following.

Theorem 2.1 (cf. [21], Ch. X). *Let $F = KD_{2/\sqrt{3}}N_{1/2}$. Then*

$$SL(n, \mathbb{R}) = F(SL(n, \mathbb{Z})).$$

Concerning lattice points this implies that if e_1 is the column vector in which the first row entry is 1 and the others are 0 then the set $Fe_1 = \{ge_1 \mid g \in F\}$ contains a nonzero point of any lattice $\Lambda \in \mathcal{L}^1$; (though the argument could be applied to other integral points in the place of e_1, it does not lead to any useful information when the point is not a multiple of e_1). Now, Fe_1 can be seen to be the complement of $\{0\}$ in the closed ball of radius $(4/3)^{\frac{n-1}{4}}$, namely the maximum possible value for the first entry of any element of $D_{2/\sqrt{3}}$. This establishes the result of Hermite mentioned earlier, that a ball centered at 0 contains a nonzero point of any lattice from $\mathcal{L}^1$ if its radius exceeds $(4/3)^{\frac{n-1}{4}}$; for a closed ball conclusion holds for radius $(4/3)^{\frac{n-1}{4}}$ as well. Since any positive definite quadratic form Q on $\mathbb{R}^n$ is equivalent to the quadratic form Q_0 given by the square of the usual norm (namely there exists a $g \in GL(n, \mathbb{R})$ such that $Q(v) = Q_0(gv)$ for all $v \in \mathbb{R}^n$) this implies the following.

Corollary 2.2 *Let Q be a positive definite quadratic form on $\mathbb{R}^n$. Then there exists $x \in \mathbb{Z}^n$ such that $Q(x) \leq (4/3)^{(n-1)/2} d(Q)^{1/n}$, where $d(Q)$ denotes the discriminant of Q.*

3 A Theorem of Howe and Moore

In this section, we recall a theorem of Howe and Moore, specialised to the case of the homogeneous space $SL(n, \mathbb{R})/SL(n, \mathbb{Z})$, and discuss its implications to lattices in $\mathbb{R}^n$.

As before we shall denote by m the normalised $SL(n, \mathbb{R})$-invariant measure on $\mathcal{L}^1 = SL(n, \mathbb{R})/SL(n, \mathbb{Z})$.

Theorem 3.1 (Howe and Moore[15]). *Let $\{g_i\}$ be a divergent sequence in $SL(n, \mathbb{R})$ (namely with no limit point in $SL(n, \mathbb{R})$). Then for any square-integrable functions f and ϕ on $\mathcal{L}^1$ we have*

$$\int_{\mathcal{L}^1} f(g_i x)\phi(x)\, dm \longrightarrow \int_{\mathcal{L}^1} f\, dm \int_{\mathcal{L}^1} \phi\, dm, \quad \text{as } i \longrightarrow \infty.$$

This implies the following corollary; it may be mentioned here that the corollary generalises a theorem proved in a number-theoretic context by W.M. Schmidt[23]; the result of Schmidt is the particular case of the following corollary when the sequence $\{g_i\}$ is chosen to be $\{\mathrm{diag}\,(i, i, \ldots, i, i^{-n+1})\}$.

Corollary 3.2 *Let $\{g_i\}$ be a divergent sequence in $SL(n, \mathbb{R})$. Then for almost every lattice $\Lambda \in \mathcal{L}^1$ the set of lattices $\{g_i \Lambda\}$ is dense in $\mathcal{L}^1$.*

For convenience we shall include here a proof of the corollary. Before going over to the proof let me however note the following. In the corollary '*almost every*' is meant in the sense that the set of lattices $\Lambda \in \mathcal{L}^1$ for which the statement does not hold (for a given sequence $\{g_i\}$ as in the hypothesis) is a set of m-measure 0, where m is the $SL(n, \mathbb{R})$-invariant measure on $\mathcal{L}^1$. In the set of all lattices (not only from $\mathcal{L}^1$) one can also interpret 'almost every' to mean a lattice generated by n linearly independent vectors $v_1, \ldots, v_n$ which may be chosen from a set of full Lebesgue measure. It can be seen that the corollary means also that for almost every lattice Λ in this sense $\{g_i \Lambda\}$ is dense in the set of lattices whose discriminant coincides with that of Λ (we shall however not go into the details of this).

Proof of the Corollary: Firstly we observe that as a consequence of the theorem, for any measurable subset E of $\mathcal{L}^1$ such that $m(E) > 0$ we have $m(\cup_i g_i^{-1}(E)) = 1$; for if E is such a set and f and ϕ are the characteristic functions of E and $\cup_i g_i^{-1}(E)$ respectively then we have $f(g_i x)\phi(x) = f(g_i x)$ for all $x \in \mathcal{L}^1$, for all i. Hence, by the theorem $(\int f dm)(\int \phi dm) = \int f(g_i x) dm(x) = \int f dm$ and since $\int f dm = m(E) > 0$ this implies that $m(\cup_i g_i^{-1}(E)) = \int \phi dm = 1$, as claimed. Now let $\{\Omega_j\}_{j=1}^{\infty}$ be a countable basis for the topology of $\mathcal{L}^1$. Then for all j we have $m(\Omega_j) > 0$ and hence $m(\cup_i g_i^{-1}(\Omega_j)) = 1$. Put $Y = \cap_j \cup_i g_i^{-1}(\Omega_j)$. Then $m(Y) = 1$ and for any $\Lambda \in Y$, $\{g_i \Lambda\}$ is dense in $\mathcal{L}^1$.

The corollary implies in particular that for any closed noncompact subgroup H of $SL(n, \mathbb{R})$, for almost all $\Lambda \in \mathcal{L}^1$ the H-orbit of Λ is dense in $\mathcal{L}^1$. The theorem of Howe and Moore implies further that for such an H a measurable function f on $\mathcal{L}^1$ is H-invariant (i.e. $f(h\Lambda) = f(\Lambda)$ for all $h \in H$) if and only if it is constant almost everywhere. In particular all H-invariant measurable sets are of measure 0 or 1, namely the action is 'ergodic'.

4 Duality and Values of Linear Forms

Let H be a closed noncompact subgroup of $SL(n, \mathbb{R})$ and consider the action of $SL(n, \mathbb{R})$ on the quotient space $SL(n, \mathbb{R})/H$. By restriction we get an action of $SL(n, \mathbb{Z})$ on

$SL(n,\mathbb{R})/H$. We note that for a $g \in SL(n,\mathbb{R})$ the H-orbit of $gSL(n,\mathbb{Z})$ is dense in $SL(n,\mathbb{R})/SL(n,\mathbb{Z})$ if and only if the $SL(n,\mathbb{Z})$-orbit of $g^{-1}H$ is dense in $SL(n,\mathbb{R})/H$, as either of them holds if and only if $HgSL(n,\mathbb{Z})$ is a dense subset of $SL(n,\mathbb{R})$. Since almost all H-orbits are dense (as seen in the previous section) it follows that almost all orbits of $SL(n,\mathbb{Z})$ are dense in $SL(n,\mathbb{R})/H$.

The correspondence as above, known as duality, can be used together with results about orbits of subgroups H of $SL(n,\mathbb{R})$ on $SL(n,\mathbb{R})/SL(n,\mathbb{Z})$ to deduce results about orbits of $SL(n,\mathbb{Z})$ under various linear actions which are of significance in terms of Diophantine approximation. For many subgroups, especially those generated by unipotent elements, the closure of orbits is well-understood, thanks to the work of M. Ratner on Raghunathan's conjecture (see [22] and [7] for details). For the so called horospherical subgroups the Raghunathan conjecture was proved earlier. We next recall some results on values of linear forms which follow from the orbit behaviour of the horospherical subgroups.

For $1 \le p \le (n-1)$ let E_p denote the p-fold Cartesian product $\mathbb{R}^n \times \cdots \times \mathbb{R}^n$. An element $(v_1,\ldots,v_p) \in E_p$ is called a *Euclidean p-frame* if $v_1,\ldots,v_p$ are linearly independent. For $f = (v_1,\ldots,v_p) \in E_p$ we shall denote by $\langle f\rangle$ the subspace of $\mathbb{R}^n$ spanned by $\{v_1,\ldots,v_p\}$. The duality argument as above implies in particular the following result, first proved in [11].

Theorem 4.1 (cf. [11]). *Consider the action of $SL(n,\mathbb{Z})$ on E_p given by the componentwise action on each copy of $\mathbb{R}^n$. Let $f = (v_1,\ldots,v_p)$ be a Euclidean p-frame. Then the $SL(n,\mathbb{Z})$ orbit of f is dense in E_p if and only if $\langle f\rangle$ contains no nonzero integral vector (namely, $\langle f\rangle \cap \mathbb{Z}^n = (0)$).*

More generally, if $x_1,\ldots,x_q \in \mathbb{Z}^n$, $q \ge 1$, are such that

$$\langle(v_1,\ldots,v_p,x_1,\ldots,x_q)\rangle \cap \mathbb{Z}^n \subseteq \langle(x_1,\ldots,x_q)\rangle,$$

(when x_i's are 0 we get the special case as above) and Γ is the subgroup $\{\gamma \in SL(n,\mathbb{Z}) \mid \gamma(x_j) = x_j \text{ for all } j = 1,\ldots,q\}$, then the Γ-orbit of f is dense in E_p.

The second assertion in the theorem follows from the proof of Proposition 4.3 in [11]; in the statement of the proposition there, the conclusion as above is claimed under a weaker hypothesis, which however is incorrect.

Theorem 4.1 can be interpreted as a result on values of linear forms at integral forms, by viewing each vector in $\mathbb{R}^n$ as the coefficients of a linear form. Let $\mathcal{F}_n$ denote the space of linear forms on $\mathbb{R}^n$, namely the dual space. We consider the action of $SL(n,\mathbb{Z})$ on $\mathcal{F}_n$ given by $(\gamma, L) \mapsto L\circ\gamma^{-1}$ for all $\gamma \in SL(n,\mathbb{Z})$ and $L \in \mathcal{F}_n$ (the so called contragradient action). Via the duality of $\mathcal{F}_n$ and $\mathbb{R}^n$ as vector spaces the theorem implies the following:

Corollary 4.2 *Let $L_1,\ldots,L_p$ be linear forms on $\mathbb{R}^n$, where $1 \le p \le n-1$. Let $M_1,\ldots,M_q$, $q \ge 0$, be rational linear forms on $\mathbb{R}^n$ such that a linear combination of the form $\Sigma_{i=1}^p \lambda_i L_i + \Sigma_{j=1}^q \mu_j M_j$, where λ_i's and μ_j's are real numbers, is not a rational form unless $\lambda_i = 0$ for all i. Then*

$$\{(L_1\circ\gamma,\ldots,L_p\circ\gamma) \mid \gamma \in SL(n,\mathbb{Z}),\ M_j\circ\gamma = M_j \text{ for all } j = 1,\ldots,q\}$$

is dense in $\mathcal{F}_n \times \cdots \times \mathcal{F}_n$ (p copies). In particular for any $t_1,\ldots,t_q \in \mathbb{R}$ for which there exists $x_0 \in \mathbb{Z}^n$ such that $M_j(x_0) = t_j$ for all $j = 1,\ldots,q$, any $a_1,\ldots,a_p \in \mathbb{R}$ and $\epsilon > 0$

there exists $x \in \mathcal{P}(\mathbb{Z}^n)$ *such that*

$$|L_i(x) - a_i| < \epsilon \text{ and } M_j(x) = t_j, \text{ for all } i = 1, \dots, p \text{ and } j = 1, \dots, q.$$

Similarly, one can also obtain results involving approximation of vectors and forms simultaneously. In this respect we note the following (see [4]):

Theorem 4.3 *Let* $k, l \in \{1, \dots, n\}$ *be such that* $k + l \leq n - 1$ *and let* X *denote the space of* $(k+l)$*-tuples of the form* $(v_1, \dots, v_k, L_1, \dots, L_l)$ *where* $v_i \in \mathbb{R}^n$, $L_j \in \mathcal{F}_n$ *and* $L_j(v_i) = 0$ *for all* $i = 1, \dots, k$ *and* $j = 1, \dots, l$*; (we realise* X *as a subset of* $\mathbb{R}^n \times \cdots \times \mathbb{R}^n \times \mathcal{F}_n \times \cdots \times \mathcal{F}_n$, k *and* l *copies respectively). Consider the* $SL(n, \mathbb{Z})$*-action on* X*, defined componentwise. Then for* $(v_1, \dots, v_k, L_1, \dots, L_l) \in X$ *the orbit is dense in* X *if and only if* $v_1, \dots, v_k$ *and* $L_1, \dots, L_l$ *are linearly independent in the respective spaces and the subspaces spanned by them do not contain any nonzero rational elements (vector or linear form respectively).*

5 The Oppenheim Conjecture

We have seen in Section 1 that study of the space $SL(n, \mathbb{R})/SL(n, \mathbb{Z})$ enables to understand integral solutions of Diophantine inequalities of the form $Q(x) < a$, where Q is any positive definite quadratic form. Study of certain flows on the space is applied to similar questions in respect of indefinite quadratic forms. In fact in this case the only known complete results are arrived at from the study of flows on the homogeneous space as above.

Let Q be an indefinite quadratic form on $\mathbb{R}^n$. Then the sets of the form $\{v \in \mathbb{R}^n \mid 0 < |Q(v)| < a\}$ have infinite volume for every $a > 0$. In the light of this one may ask whether they all contain nonzero integral points. For this to hold (for all $a > 0$) one would have to assume however that Q is not a multiple of a form with rational coefficients, since otherwise the set of values of Q at integral points would be discrete. Even with this additional condition the conclusion does not hold for certain forms in two variables. It is well-known that there exist numbers $\lambda \in \mathbb{R}$ such that $q^2|\lambda - \frac{p}{q}|$ is bounded below by a positive number, as p, q vary over integers, $q \neq 0$ (the so called badly approximable numbers; see [19], where they are actually called 'numbers of constant type') and this means that for the quadratic form $Q(x_1, x_2) = x_2(\lambda x_2 - x_1)$ there exists $a > 0$ such that there is no integral point x with $0 < |Q(x)| < a$. There was a conjecture, originating from an observation by Oppenheim that for $n \geq 3$ a nondegenerate indefinite quadratic form Q on $\mathbb{R}^n$, the set of values at integral points is dense in $\mathbb{R}$, whenever Q is not a scalar multiple of a form with integer coefficients. The conjecture was proved by Margulis in the mid eighties; see [20] for a detailed exposition, including on historical aspects. Later, the following relatively stronger assertion was proved in a joint paper of Margulis with the present author and subsequently an elementary proof of the same result was given in [9]; see also [5] for another elementary proof and [6] for an exposition of the proof of a weaker result.

We recall that an element $x \in \mathbb{Z}^n$ is said to be *primitive* if $\frac{1}{k}x$ is not contained in $\mathbb{Z}^n$ for any integer $k \geq 2$. We denote the set of all primitive integral elements by $\mathcal{P}(\mathbb{Z}^n)$.

Theorem 5.1 *Let Q be a nondegenerate indefinite quadratic form on $\mathbb{R}^n$, $n \geq 3$, which is not a multiple of a form with integer coefficients. Then $Q(\mathcal{P}(\mathbb{Z}^n))$ is dense in $\mathbb{R}$, that is, for any $t \in \mathbb{R}$ and $\epsilon > 0$ there exists a primitive integral vector x such that $|Q(x) - t| < \epsilon$.*

The argument of Margulis and that in the subsequent papers referred to above depends on the study of orbits of the special orthogonal groups of indefinite quadratic forms, on $SL(n, \mathbb{R})/SL(n, \mathbb{Z})$. Let Q be a nondegenerate indefinite quadratic form on $\mathbb{R}^n$ and let $H = SO(Q)$ be the associated special orthogonal group, namely $\{g \in SL(n, \mathbb{R}) \mid Q(gv) = Q(v)$ for all $v \in \mathbb{R}^n\}$. Then H is a closed noncompact subgroup of $SL(n, \mathbb{R})$. Hence, Corollary 3.2 implies that almost all orbits of H on $\mathcal{L}^1 = SL(n, \mathbb{R})/SL(n, \mathbb{Z})$ are dense. In fact we have the following:-

Theorem 5.2 *Let the notation be as above; in particular, we suppose that $n \geq 3$. Then every orbit of H on $\mathcal{L}^1$ is either closed or dense.*

Theorem 5.2 was proved by Margulis and the present author for $n = 3$ and applied to deduce Theorem 5.1 (for all $n \geq 3$). For a general $n \geq 3$, it is a consequence of a celebrated theorem of M. Ratner, proving a conjecture of Raghunathan (see [22] and [7] for some details and references). Theorem 5.2 also implies the following generalisation of Theorem 5.1, noted by A. Borel and G. Prasad[1].

Theorem 5.3 (Borel and Prasad[1]). *Let Q be a nondegenerate indefinite quadratic form on $\mathbb{R}^n$, $n \geq 3$, which is not a multiple of a form with integer coefficients and let B be the corresponding bilinear form (defined by $B(v, w) = \frac{1}{4}(Q(v + w) - Q(v - w))$ for all $v, w \in \mathbb{R}^n$). Then for any $v_1, \ldots, v_{n-1} \in \mathbb{R}^n$ and $\epsilon > 0$ there exist $x_1, \ldots, x_{n-1} \in \mathcal{P}(\mathbb{Z}^n)$ such that*

$$|B(v_i, v_j) - B(x_i, x_j)| < \epsilon, \quad \textit{for all } i, j = 1, \ldots, n-1.$$

In respect of deduction of Theorem 5.3 from Theorem 5.2 it may be worthwhile to note the following. Let Q be a nondegenerate indefinite quadratic form on $\mathbb{R}^n$. Let Q_0 be a rational quadratic form and let $g \in SL(n, \mathbb{R})$ be such that for a suitable $\lambda \in \mathbb{R}$ (depending on the discriminants of Q and Q_0), $Q(v) = \lambda Q_0(gv)$ for all $v \in \mathbb{R}^n$; such Q_0 and g exist and in fact Q_0 can be chosen to be one of the standard quadratic forms. Then H-orbit of $gSL(n, \mathbb{Z})$ is closed $\mathcal{L}^1$ if and only if and Q is a multiple of a form with integer coefficients (see [9]). In the alternative case, namely when the orbit is dense, the conclusion as in Theorem 5.3 follows from a continuity argument.

It may be mentioned here that Borel and Prasad[1] consider also similar problems in the context of nonarchimedian valuations (in the place of the usual absolute value); we shall however not go into the details.

6 Complements to the Oppenheim Conjecture

We shall now discuss analogous questions about values of quadratic forms at integral points, in some cases complementary to the results on Oppenheim conjecture.

Let us first consider the nondegeneracy condition in the Oppenheim conjecture, as in Theorem 5.1. Though it has been traditional to assume nondegeneracy in the context of the problem, unlike in some other problems the general case here does not readily reduce to proving the result in the case of nondegenerate forms (the reader may convince himself of this by perusing the cases arising in the proof of Theorem 6.1 below).

Let Q be a (possibly degenerate) quadratic form. We recall that the *radical* of Q is the subspace $W = \{w \in \mathbb{R}^n \mid Q(v+tw) = Q(v)$ for all $v \in \mathbb{R}^n$ and $t \in \mathbb{R}\}$. Suppose that W is a rational subspace of $\mathbb{R}^n$ (namely, it is defined by linear equations with rational coefficients). Then the image $(\mathbb{Z}^n + W)/W$ of $\mathbb{Z}^n$ in R^n/W is a lattice and $\mathbb{R}^n/W$ can be realised as $\mathbb{R}^m$, where m is the codimension of W in $\mathbb{R}^n$, with $(\mathbb{Z}^n + W)/W$ corresponding to $\mathbb{Z}^m$. Hence Q factors to a quadratic form, say Q', on $\mathbb{R}^m$ and $Q(\mathbb{Z}^n) = Q'(\mathbb{Z}^m)$. Further, it can be seen that there exists a positive integer r such that $Q'(\mathcal{P}(\mathbb{Z}^m)) \subseteq Q(\mathcal{P}(\mathbb{Z}^n)) \subseteq \frac{1}{r}Q'(\mathcal{P}(\mathbb{Z}^m))$. This shows in particular that if the radical of a quadratic form Q is a *rational* subspace of codimension 2 then the study of its values reduces to the 2-variable case; (see below for a discussion on the 2-variables case). We now note the following:-

Theorem 6.1 *Let Q be a quadratic form on $\mathbb{R}^n$, $n \geq 3$, which is indefinite (that is, there exist $v, w \in \mathbb{R}^n$ such that $Q(v) < 0 < Q(w)$) and not a multiple of a form with rational coefficients. Then at least one of the following conditions holds:-*

i) *the radical of Q is a rational subspace of codimension* 2; *or*
ii) *$Q(\mathcal{P}(\mathbb{Z}^n))$ is dense in $\mathbb{R}$.*

Proof: If the rank of Q is at least 3 then the argument as in the proof of Theorem 1 in [8] (reducing the number of variables to 3, in which case Q would be nondegenerate) shows, together with Theorem 5.1, that condition (ii) holds. Note that since the form is indefinite, the rank is at least 2. Now suppose that Q is of rank 2. Then there exist linear forms L_1 and L_2 on $\mathbb{R}^n$ such that $Q(v) = L_1(v)L_2(v)$ for all $v \in \mathbb{R}^n$. Since Q is indefinite L_1 and L_2 are linearly independent (as linear forms). Suppose first that no linear combination of the form $\lambda_1 L_1 + \lambda_2 L_2$, with $\lambda_1, \lambda_2 \in \mathbb{R}$, is a rational linear form. Let $\mathcal{F}_n$ be the space of linear forms on $\mathbb{R}^n$, and consider the contragradient action of $SL(n, \mathbb{Z})$ on $\mathcal{F}_n$ as in §4. Hence, by Corollary 4.2 for L_1, L_2 as above the set of pairs $\{(L_1(x), L_2(x)) \mid x \in \mathcal{P}(\mathbb{Z}^n)\}$ is dense in $\mathbb{R}^2$. Since $Q = L_1 L_2$ this implies that $Q(\mathcal{P}(\mathbb{Z}^n))$ is dense in $\mathbb{R}$.

Next suppose that there exist $\lambda_1, \lambda_2 \in \mathbb{R}$ such that $\lambda_1 L_1 + \lambda_2 L_2 = L_0$, say, is a rational linear form. If the span of L_1 and L_2 contains a rational form which is not a multiple of L_0 then the radical of Q, which is given by $\{v \in \mathbb{R}^n \mid L_1(v) = L_2(v) = 0\}$, is a rational subspace of codimension 2, so condition (i) is satisfied. We may, therefore, suppose that any rational form in the span of L_1 and L_2 is a multiple of L_0. Let Γ_0 be the subgroup $\{\gamma \in SL(n, \mathbb{Z}) \mid L_0 \circ \gamma = L_0\}$. Since Q is not a multiple of a form with integral coefficients, at least one of L_1 and L_2 is not a multiple of a linear form with integral coefficients; for definiteness assume that this holds for L_1. Then in view of our assumption as above, any rational linear form in the span of L_1 and L_0 is a multiple of L_0. Now let $t \in L_0(\mathcal{P}(\mathbb{Z}^n))$, $t \neq 0$. Then by Corollary 4.2 the set of pairs $\{(L_1(x), t) \mid x \in \mathcal{P}(\mathbb{Z}^n), L_0(x) = t\}$ is dense in $\{(s, t) \mid s \in \mathbb{R}\}$. We note that $\lambda_2 \neq 0$, since L_0 is a rational form while L_1 is not a multiple of any rational form. Now, for any $x \in \mathbb{R}^n$, $Q(x) = L_1(x)L_2(x) = L_1(x)(L_0(x) - \lambda_1 L_1(x))/\lambda_2$. Hence, the closure of $Q(\mathcal{P}(\mathbb{Z}^n))$ contains the set $\{s(t - \lambda_1 s)/\lambda_2 \mid s \in \mathbb{R}\}$, which is the whole of $\mathbb{R}$ if $\lambda_1 = 0$ and contains $[-t^2/4|\lambda_1\lambda_2|, t^2/4|\lambda_1\lambda_2|]$ if $\lambda_1 \neq 0$ (it extends to infinity on one side, depending on the sign of λ_1/λ_2). Since $L_0(\mathcal{P}(\mathbb{Z}^n))$ is unbounded this means that $Q(\mathcal{P}(\mathbb{Z}^n))$ is dense in $\mathbb{R}$, thus completing the proof. □

We now consider the case with $n = 2$. It was noted earlier that there exist binary quadratic forms which are not multiples of forms with integer coefficients but nevertheless their values at integer points do not intersect some interval of the form $(0, \delta)$ for some $\delta > 0$. Now, let Q be any indefinite binary quadratic form. Then $Q(v, w)$ can be expressed as $(av + bw)(cv + dw)$ for all $v, w \in \mathbb{R}$ (with respect to the standard basis), where $a, b, c, d \in \mathbb{R}$ and $ad - bc \neq 0$. An elementary argument then shows that for every $\epsilon > 0$ there exist $x, y \in \mathbb{Z}$, not both 0, such that $|Q(x, y)| < \epsilon$, if and only if either a/b or c/d is not badly approximable. The question of taking values arbitrarily close to nonzero real numbers is however more intricate (see [2]). It is closely related to the dynamics of the flow induced by the one parameter subgroup of diagonal matrices on the space $SL(2, \mathbb{R})/SL(2, \mathbb{Z})$. We shall, however, not go into the details of this here.

7 Simultaneous Solution of Quadratic and Linear Inequalities

In the earlier sections, we considered inequalities involving linear and quadratic forms separately. In this section we shall briefly consider mixed systems involving a quadratic and a linear form, on $\mathbb{R}^3$. The following result can be deduced from Corollary 2 in [10], using rotational symmetry, with respect to the maximal compact subgroup of the special orthogonal group.

Theorem 7.1 *Let Q be a nondegenerate indefinite quadratic form on $\mathbb{R}^3$ and L be a linear form on $\mathbb{R}^3$. Suppose that the plane $\{v \in \mathbb{R}^3 \mid L(v) = 0\}$ is tangential to the (double) cone $\{v \in \mathbb{R}^3 \mid Q(v) = 0\}$ and that no linear combination (with real coefficients) of Q and L^2 is a rational quadratic form. Then for any $a, b \in \mathbb{R}$ and $\epsilon > 0$ there exists $x \in \mathcal{P}(\mathbb{Z}^3)$ such that*

$$|Q(x) - a| < \epsilon \text{ and } |L(x) - b| < \epsilon.$$

The result is obtained by proving that for a unipotent one-parameter subgroup $\{u_t\}$ such that $u_t - I$ is of rank 2 for $t \neq 0$, (I being the identity matrix) the closure of any orbit of the flow induced by it on $SL(3, \mathbb{R})/SL(3, \mathbb{Z})$ is an orbit of a closed subgroup of $SL(3, \mathbb{R})$; this is a particular case of Raghunathan's conjecture, which was subsequently proved by M. Ratner (see [22] and [7] for details).

Remark 7.2 In the notation of Theorem 7.1 it can be seen that when some linear combination of Q and L^2 is a rational quadratic form then the (system of) inequalities as in the conclusion cannot admit nonzero integral solutions for all $a, b \in \mathbb{R}$ and $\epsilon > 0$; (there could however exist solutions for $a = b = 0$, for all $\epsilon > 0$).

It turns out that the condition in the hypothesis of Theorem 7.1 that the plane $\{v \in \mathbb{R}^3 \mid L(v) = 0\}$ be tangential to the cone $\{v \in \mathbb{R}^3 \mid Q(v) = 0\}$ cannot be weakened to the intersection of the two being nonzero; namely, intersection in a pair of straight lines may not suffice. Specifically we note the following result; the main idea of the proof is due to Margulis (oral communication) and the author would like to thank him for pointing it out.

Theorem 7.3 *Let Q_0 and L_0 be the forms (quadratic and linear respectively) on $\mathbb{R}^3$ defined by $Q_0(x_1e_1 + x_2e_2 + x_3e_3) = 2x_1x_3 - x_2^2$ and $L_0(x_1e_1 + x_2e_2 + x_3e_3) = x_2$ for all $x_1, x_2, x_3 \in \mathbb{R}$, e_1, e_2, e_3 being the standard basis of $\mathbb{R}^3$. Then there exist $g \in SL(3, \mathbb{R})$ and $\epsilon > 0$ such that the following conditions are satisfied for the forms Q and L defined by $Q(v) = Q_0(gv)$ and $L(v) = L_0(gv)$ for all $v \in \mathbb{R}^3$:*

i) *no linear combination of Q and L^2 (with real coefficients) is a rational quadratic form; and*
ii) *there does not exist any nonzero integral point $x \in \mathbb{Z}^3$ such that $|Q(x)| < \epsilon$ and $|L(x)| < \epsilon$.*

Further, the set of $g \in SL(3, \mathbb{R})$ for which the conditions are satisfied (for some $\epsilon > 0$) is of Hausdorff dimension 8, namely, same as the manifold dimension of $SL(3, \mathbb{R})$; in particular there are uncountably many pairs of forms (Q, L) arising as above for which conditions (i) and (ii) are satisfied.

Proof: For $t \in \mathbb{R}$ let d_t denote the diagonal matrix diag $(e^{-t}, 1, e^t)$ and let D be the one-parameter subgroup $\{d_t\}$. Let $\mathcal{B}$ denote the set of lattices Λ in $\mathcal{L}^1$ such that the D-orbit of Λ is bounded (has compact closure) in $\mathcal{L}^1$. By a theorem of Kleinbock and Margulis[17] the set $\{g \in SL(3, \mathbb{R}) \mid g\mathbb{Z}^3 \in \mathcal{B}\}$ intersects every nonempty subset of $SL(3, \mathbb{R})$ in a set of Hausdorff dimension 8; in particular it is not contained in any countable union of proper algebraic subvarieties of $SL(3, \mathbb{R})$; (we recall that an 'algebraic subvariety' is a set of the form $\{g \in SL(3, \mathbb{R}) \mid P(g) = 0\}$, where P is a polynomial function on $SL(3, \mathbb{R})$ in the coordinate variables).

We first show that if $\Lambda \in \mathcal{B}$ then there exists $\epsilon > 0$ such that Λ contains no nonzero point p with $|Q_0(p)| < \epsilon$ and $|L_0(p)| < \epsilon$; this implies that if $g \in SL(3, \mathbb{R})$ is such that $\Lambda = g\mathbb{Z}^3$ then condition (ii) as in the conclusion of the theorem is satisfied for the forms Q and L defined by $Q(v) = Q_0(gv)$ and $L(v) = L_0(gv)$ for all $v \in \mathbb{R}^3$.

Let $\Lambda \in \mathcal{B}$ and suppose, if possible, that for all $\epsilon > 0$ there exists $p \in \Lambda$, $p \neq 0$, such that $|L_0(p)| < \epsilon$ and $|Q_0(p)| < \epsilon^2$ (the square in the latter inequality is chosen for computational convenience). Consider any $\epsilon > 0$ small enough so that the open ball of radius 3ϵ centered at 0 contains no nonzero point of Λ and let p be a point satisfying the above conditions. If $p = x_1e_1 + x_2e_2 + x_3e_3$, where e_1, e_2, e_3 is the standard basis, then we have $|x_2| < \epsilon$ and $|2x_1x_3 - x_2^2| < \epsilon^2$. The conditions imply also that $|x_1x_3| < \epsilon^2$. Let $\alpha = \max\{|x_1|, |x_3|\}$. In view of the condition on ϵ as above it follows that α is positive and coincides with only one of $|x_1|$ or $|x_3|$. Now let $t = \sigma \log(\alpha/\epsilon)$, where $\sigma = 1$ if $\alpha = |x_1|$ and $\sigma = -1$ if $\alpha = |x_3|$. Then $d_t p = e^{-t}x_1e_1 + x_2e_2 + e^t x_3e_3$. If $\alpha = |x_1|$ then we have $|e^{-t}x_1| = \epsilon$ and $|e^t x_3| = |x_1x_3|/\epsilon \leq \epsilon$ and similarly, if $\alpha = |x_3|$ we get that $|e^t x_3| = \epsilon$ and $|e^{-t}x_1| \leq \epsilon$. Varying ϵ along a sequence converging to 0 we see from this that there exist a sequence of lattices $\{\Lambda_j\}$ of the form $d_t\Lambda$ and points $p_j \in \Lambda_j$ such that $p_j \to 0$ as $j \to \infty$. By Theorem 1.1 this means that $\{d_t\Lambda\}$ is not contained in a compact subset of $\mathcal{L}^1$, which contradicts Λ being from $\mathcal{B}$. The contradiction shows that the assertion formulated above must be true and hence condition (ii) holds for any $g \in SL(3, \mathbb{R})$ such that $g\mathbb{Z}^3 \in \mathcal{B}$.

To find a g such that condition (i) is also satisfied we proceed as follows: We note that for any rational quadratic form R the set of $g \in SL(3, \mathbb{R})$ for which the form defined by $v \mapsto R(g^{-1}v)$ (for all $v \in \mathbb{R}^3$) is a linear combination of the forms Q_0 and L_0^2 is an algebraic subvariety of $SL(3, \mathbb{R})$. Since $\{g \in SL(3, \mathbb{R}) \mid g\mathbb{Z}^3 \in \mathcal{B}\}$ is not contained in a countable union of algebraic varieties, we get that there exists a $g \in SL(3, \mathbb{R})$ such that $g\mathbb{Z}^3 \in \mathcal{B}$ and no linear combination of Q_0 and L_0^2 is of the form $v \mapsto R(g^{-1}v)$ for any rational quadratic form R. The latter condition means that for Q and L defined as in the hypothesis of the theorem, for this g, condition (i) is satisfied. This proves the existence of a g as claimed. The more general assertion in the theorem is clear from the proof as above. □

Remark 7.4 In the context of Theorem 6.1 one may look for analogues of Theorems 7.1 and 7.3 for quadratic forms of rank 2. Let $Q = L_1L_2$, where L_1 and L_2 are two linearly independent linear forms on $\mathbb{R}^3$ and let $W = \{v \in \mathbb{R}^3 \mid L_1(v) = L_2(v) = 0\}$, the line of intersection of the planes defined by L_1 and L_2. Let L be a linear form such that $L(w) = 0$ for all $w \in W$ (this corresponds to the tangential intersection condition in Theorem 7.1). Then L is a linear combination of L_1 and L_2 and hence $|Q(v)|$ and $|L(v)|$ can be small simultaneously only if $|L_1(v)|$ and $|L_2(v)|$ are small. This shows that for small a and b the inequalities $|Q(x) - a| < \epsilon$ and $|L(x) - b| < \epsilon$ admit integral solutions for all $\epsilon > 0$ only if no linear combination of L_1 and L_2 is a rational form. An argument as in the proof of Theorem 6.1 shows that the latter condition is also sufficient to ensure that the inequalities admit primitive integral solutions, for all $a, b \in \mathbb{R}$ and $\epsilon > 0$. (It may be noted that the condition is not equivalent to the condition in Theorem 7.1 that no linear combination of Q and L^2 is rational).

On the other hand, corresponding to Theorem 7.3 one can show by an argument analogous to the one there, using the theorem of Kleinbock and Margulis, that there exist linearly independent linear forms L_1, L_2 and L such that for some $\epsilon > 0$ there is no $x \in \mathbb{Z}^3 - \{0\}$ such that $|L_1(x)L_2(x)| < \epsilon$ and $|L(x)| < \epsilon$, and no linear combination of L_1 and L_2 is rational (one can also additionally arrange so that no linear combination of L_1L_2 and L^2 is rational).

8 Quantitative Results

Apart from the problems of existence of solutions of Diophantine inequalities, results on the modular homogeneous space and the flows on it can be applied to get quantitative results on the number of solutions of such inequalities, involving quadratic forms; (for linear forms the corresponding statements follow from results on uniform distribution of sequences of the form $\{i(v_1, \ldots, v_n)\}$ mod 1, for fixed $v_1, \ldots, v_n \in \mathbb{R}$, - see [2] - and will not be considered here). The classification of invariant measures of flows induced by unipotent one-parameter subgroups, due to M. Ratner, plays an important role in this respect. It is applied to obtain results on the distribution of the flows which, in turn, together with Theorem 1.2 are applied to deduce the asymptotics of the number of integral solutions of the inequalities, in balls centered at 0, as the radius tends to infinity. We mention here the following result in this regard; the purpose being only to give an idea of the results, we shall not strive for generality or completeness (see [12] or [20] for details).

Theorem 8.1 (Eskin, Margulis and Mozes[12]). *Let Q be a nondegenerate indefinite quadratic form on $\mathbb{R}^n$, $n \geq 5$, which is not a multiple of a rational form. Let $B(r)$ denote the* *ball of radius r in $\mathbb{R}^n$ centered at* 0. *Then there exists a constant λ such that for any open interval (a,b) in $\mathbb{R}$,*

$$\frac{1}{r^{n-2}}\#\{x \in B(r) \cap \mathbb{Z}^n \mid a < Q(x) < b\} \longrightarrow \lambda(b-a) \ \text{as}\ r \longrightarrow \infty.$$

The constant λ in the theorem also turns out to be such that for any open interval (a,b) the volumes of $\{v \in B(r) \mid a < Q(v) < b\}$ are asymptotic to $\lambda(b-a)r^{n-2}$, so the result signifies that the number of solutions of the inequalities in question is asymptotic to the volumes of the regions they describe in $\mathbb{R}^n$.

It turns out that similar asymptotics do not hold in general for $n = 3$ or 4. In these cases there are lower estimates comparable to the one above and upper estimates which are higher by a factor of $\log r$ (see [20] for details).

9 Comments in Conclusion

The study of flows on $SL(n,\mathbb{R})/SL(n,\mathbb{Z})$, and other more general homogeneous spaces of Lie groups, is also applied to various other Diophantine problems. The reader is referred to [7] for an exposition of the general theory of flows on homogeneous spaces as well as various applications. We mention here a only couple of recent applications of the theory to certain Diophantine problems. In [13], the authors study the asymptotics of the number of integral points on certain subvarieties, within distance r of the origin, as $r \to \infty$. In particular they obtain the following result, which may be of independent interest.

Theorem 9.1 (Eskin, Mozes and Shah[13]). *Let P be a monic polynomial with integral coefficients which is irreducible over the rationals and has degree at least two. For $r > 0$ let N_r be the number of integral $n \times n$ matrices $X = (x_{ij})$ with $\Sigma_{i,j}\, x_{ij}^2 \le r^2$, whose characteristic polynomial is P. Then there exists a constant $c > 0$ such that $N_r \sim c r^{n(n-1)/2}$ as $r \to \infty$.*

Interesting results have also been proved recently on 'approximability properties' for linear forms and vectors in [18] and [16]. In the former, a conjecture of Sprindzuk is verified and in the latter the theory of badly approximable and singular systems of linear forms is generalised.

Acknowledgement

The author would like to thank Nimish Shah for his valuable comments on an earlier version of the article.

References

1. A. Borel and Prasad, G., Values of isotropic quadratic forms at S-integral points, *Compositio Math.* **83** (1992), 347–372.
2. J.W.S. Cassels, *An Introduction to Diophantine Approximation*, Cambridge University Press, 1957.

3. J.W.S. Cassels, *An Introduction to Geometry of Numbers*, Springer-Verlag, 1959.
4. S.G. Dani, Approaching new points by application of linear transformations, *Curr. Sci.* **56** (1987), 507–513.
5. S.G. Dani, A proof of Margulis' theorem on values of quadratic forms, independent of the axiom of choice, *L'enseignement Math.* **40** (1994), 49–58.
6. S.G. Dani, Small values of quadratic forms at integral points, *Res. Bull. Panjab Univ.* **44** (1994), 271–294.
7. S.G. Dani, Flows on homogeneous spaces; a review, In: *Ergodic Theory of $\mathbb{Z}^n$-actions (Proc. Warwick Symp.*, 1993–94), pp. 63–112, *London math. Soc. LNS*, **228**, Cambridge University Press, 1996.
8. S.G. Dani and Margulis, G.A., Values of quadratic forms at primitive integral points, *Invent. Math.* **98** (1989), 405–424.
9. S.G. Dani and Margulis, G.A., Values of quadratic forms at integral points: an elementary approach, *L'Enseignement Math.* **36** (1990), 143–174.
10. S.G. Dani and Margulis, G.A., Orbit closures of generic unipotent flows on homogeneous spaces of $SL(3, \mathbb{R})$, *Math. Ann.* **286** (1990), 101–128.
11. S.G. Dani and Raghavan, S., Orbits of euclidean frames under discrete linear groups, *Israel J. Math.* **36** (1980), 300–320.
12. A. Eskin, Margulis, G.A. and Mozes, S., Upper bounds and asymptotics in a quantitative version of Oppenheim conjecture, *Ann. Math.* **147** (1998), 93–141.
13. A. Eskin, Mozes, S. and Shah, N.A., Unipotent flows and counting lattice points on homogeneous varieties, *Ann. Math.* **143** (1996), 253–299.
14. P.M. Gruber and Lekkerkerkar, C.G., *Geometry of Numbers*, North-Holland Publishing Co., Amsterdam - New York, 1987.
15. R. Howe and Tan, E.C., *Non-Abelian Harmonic Analysis*, Springer-Verlag, 1992.
16. D. Kleinbock, Flows on homogeneous spaces and Diophantine properties of matrices, *Duke Math. J.* **95** (1998), 107–124.
17. D. Kleinbock and Margulis, G.A., Bounded orbits of nonquasiunipotent flows on homogeneous spaces, *Amer. Math. Soc. Trans.* **171** (1996), 141–172.
18. D. Kleinbock and Margulis, G.A., Flows on homogeneous spaces and Diophantine approximation on manifolds, *Ann. Math.* **148** (1998), 339–360.
19. S. Lang, *Introduction to Diophantine Approximation*, Springer-Verlag, 1995.
20. G.A. Margulis, Oppenheim conjecture, In: *Fields Medalists Lectures*, pp. 272–327, World Scientific, 1997.
21. M.S. Raghunathan, *Discrete Subgroups of Lie Groups*, Springer-Verlag, 1972.
22. M. Ratner, Interactions between ergodic theory, Lie groups and number theory, In: *Proc. Int. Cong. Mathematicians*, Zurich-1994, pp. 157–182, Birkhauser, Basel, 1995.
23. W.M. Schmidt, Diophantine approximation and certain sequences of lattices, *Acta Arith.* **18** (1971), 165–178.
24. C.L. Siegel, A mean value property in geometry of numbers, *Ann. Math.* **46** (1945), 340–347.

School of Mathematics
Tata Institute of Fundamental Research
Homi Bhabha Road, Colaba
Mumbai 400 005
E-Mail: dani@math.tifr.res.in

Variants of the Second Borel-Cantelli Lemma and their Applications in Metric Number Theory

Glyn Harman

§1 Introduction

In Metric Number Theory we are concerned with the arithmetical properties of almost all numbers (in the first case with respect to Lebesgue measure on the real line, but the ideas generalise to $\mathbb{R}^k$ and other situations). Investigations therefore involve both analysis and number theory. It is the purpose of this paper to review the important contribution made by variants of the second Borel-Cantelli Lemma (also known as the *divergence part* of the Borel-Cantelli Lemma). While doing this we shall prove sharper variants than have previously appeared and give their applications. To illustrate the types of result which fall within the ambit of our approach we state the following:-

Theorem 1 *Let $\psi(n)$ be a monotonic decreasing function with $\psi(n) \in (0, \frac{1}{2})$, and write $P(\alpha, N)$ for the number of solutions to*

$$|\alpha p - q| < \psi(p), \quad p, q \text{ prime}, \quad p \le N, \tag{1.1}$$

and put

$$\Psi(N) = 2 \sum_{p \le N} \frac{\psi(p)}{\log p}.$$

Then, if $\Psi(\infty)$ diverges, we have, for almost all $\alpha > 0$,

$$\limsup_{N\to\infty} \frac{P(\alpha, N)}{\Psi(N)} \ge 1, \quad \text{and} \quad \liminf_{N\to\infty} \frac{P(\alpha, N)}{\Psi(N)} \le 1. \tag{1.2}$$

Remarks: In [8] (see also Chapter 6 in [11]), the author showed that if $\Psi(\infty)$ diverges then $P(\alpha, N) \to \infty$ as $N \to \infty$ for almost all positive α. The quantitative results (1.2) are new. One would expect that, for almost all $\alpha > 0$,

$$\lim_{N\to\infty} \frac{P(\alpha, N)}{\Psi(N)}$$

exists and equals 1, but this result appears to be out of reach at present unless one puts certain growth conditions on $\Psi(N)$.

In §2, we shall prove the Borel-Cantelli lemmas and two important variants. In §3 we consider the zero-one laws which often hold for problems in metric number theory. We then apply the results to Khintchine and Duffin-Schaeffer type problems in §4, and to two restricted variable problems (like Theorem 1 above) in §5. We then consider a different variant of the second Borel-Cantelli lemma in §6 which provides an asymptotic formula for the number of solutions to Diophantine inequalities. We end in §7 with open problems and other generalizations of the ideas.

§2 The Borel-Cantelli Lemmas

Let $(\Omega, \mathcal{M}, \mu)$ be a probability space, and suppose that $\mathcal{A}_1, \mathcal{A}_2, \ldots$ is an infinite sequence of subsets (events) in $\mathcal{M}$. Write

$$F_N(\alpha, \mathcal{A}_n) = |\{n \leq N : \alpha \in \mathcal{A}_n\}|.$$

Lemma 1 *(First Borel-Cantelli Lemma). If*

$$\sum_{n=1}^{\infty} \mu(\mathcal{A}_n) \tag{2.1}$$

converges, then $F_N(\alpha, \mathcal{A}_n)$ is finite for almost all $\alpha \in \Omega$ as $N \to \infty$.

Proof: Clearly the set on which $F_N(\alpha, \mathcal{A}_n) \to \infty$ has measure not exceeding

$$\sum_{n=M}^{\infty} \mu(\mathcal{A}_n) \tag{2.2}$$

for every M. From the convergence of (2.1) we deduce that (2.2) can be made as small as possible by taking M suitably large. This completes the proof. □

Lemma 2 *(Second Borel-Cantelli Lemma). Suppose that* (2.1) *diverges, and that*

$$\mu(\mathcal{A}_j \cap \mathcal{A}_k) = \mu(\mathcal{A}_j)\mu(\mathcal{A}_k) \quad \textit{for } j \neq k. \tag{2.3}$$

Then

$$F_N(\alpha, A_N) \longrightarrow \infty \quad \textit{as} \quad N \longrightarrow \infty \quad \textit{for almost all} \quad \alpha \in \Omega. \tag{2.4}$$

Remarks: The condition (2.3) is often called *independence*. It is trivial to observe that the divergence of (2.1) is not sufficient to obtain (2.4). For example, if $\Omega = [0, 1)$, $\mathcal{A} = (0, 1/n)$, and $\mu(\cdot)$ denotes Lebesgue measure, then (2.1) diverges, yet (2.4) is not true for any α.

Proof: Let $\chi_j(\alpha)$ be the characteristic function of $\mathcal{A}_j$, and write

$$f_N(\alpha) = \frac{\sum_{j=1}^{N}(\chi_j(\alpha) - \mu(\mathcal{A}_j))}{\sum_{j=1}^{N} \mu(\mathcal{A}_j)}.$$

Since

$$\int_\Omega \chi_j(\alpha)d\mu = \mu(\mathcal{A}_j),$$

and, for $j \neq k$,

$$\int_\Omega \chi_j(\alpha)\chi_k(\alpha)d\mu = \mu(\mathcal{A}_j \cap \mathcal{A}_k) = \mu(\mathcal{A}_j)\mu(\mathcal{A}_k),$$

we have

$$\int_\Omega f_N(\alpha)^2 d\mu = \frac{\sum_{j=1}^N \mu(\mathcal{A}_j)(1-\mu(\mathcal{A}_j))}{\left(\sum_{j=1}^N \mu(\mathcal{A}_j)\right)^2} \leq \left(\sum_{j=1}^N \mu(\mathcal{A}_j)\right)^{-1}.$$

Since (2.1) diverges this gives

$$\lim_{N\to\infty} \int_\Omega f_N(\alpha)^2 a\mu = 0.$$

Hence, by Fatou's lemma,

$$\liminf_{N\to\infty} f_N(\alpha) = 0 \quad \text{for almost all } \alpha$$

Thus

$$\limsup_{N\to\infty} \frac{\sum_{j=1}^N \chi_j(\alpha)}{\sum_{j=1}^N \mu(\mathcal{A}_j)} = 1 \quad \text{for almost all } \alpha,$$

from which (2.4) follows. □

The second Borel-Cantelli lemma is of limited use in metric number theory since (2.3) only holds for certain types of problems. The following variant, which is a simple consequence of Cauchy's inequality, has been much more useful, however. The basic idea goes back to a lemma of Paley and Zygmund[18].

Lemma 3 *Suppose that* (2.1) *diverges. Then the set* $\mathcal{D}$ *on which* (2.4) *holds satisfies* $\mu(d) \geq c$, *where*

$$c = \limsup_{N\to\infty} \left(\sum_{j=1}^N \mu(\mathcal{A}_j)\right)^2 \left(\sum_{j,k=1}^N \mu(\mathcal{A}_j \cap \mathcal{A}_k)\right)^{-1}. \tag{2.5}$$

Proof: Write

$$\mathcal{A}_m^n = \bigcup_{j=m}^n \mathcal{A}_j, \quad M(m,n) = \sum_{j=m}^n \mu(\mathcal{A}_j), \quad V(m,n) = \sum_{j,k=m}^n \mu(\mathcal{A}_j \cap \mathcal{A}_k).$$

We take $\chi_j(\alpha)$ as in Lemma 2, and let $\chi_{m,n}(\alpha)$ be the characteristic function of $\mathcal{A}_m^n$. To prove the lemma it suffices to show that

$$\lim_{m\to\infty}(\lim_{n\to\infty}\mu(\mathcal{A}_m^n)) \geq \limsup_{N\to\infty}\frac{M(0,N)^2}{V(0,N)}. \tag{2.6}$$

Since (2.1) diverges, we have

$$M(m,n) = M(0,n) + O(m), \quad V(m,n) = V(0,n) + O(m).$$

Hence, since $\mu(\mathcal{A}_m^n)$ is a non-decreasing function of n, we need only show that

$$\mu(\mathcal{A}_m^n) \geq \frac{M(m,n)^2}{V(m,n)}. \tag{2.7}$$

This is a simple consequence of the Cauchy-Schwarz inequality as we now show. We have

$$M(m,n) = \int_\Omega \sum_{j=m}^n \chi_j(\alpha)\,d\alpha = \int_\Omega \chi_{m,n}(\alpha)\sum_{j=m}^n \chi_j(\alpha)\,d\alpha.$$

Thus, by the Cauchy-Schwarz inequality,

$$\begin{aligned} M(m,n)^2 &\leq \left(\int_\Omega \chi_{m,n}(\alpha)\,d\alpha\right)\left(\int_\Omega \sum_{j,k=m}^n \chi_j(\alpha)\chi_k(\alpha)\,d\alpha\right) \\ &= \mu(\mathcal{A}_m^n)V(m,n). \end{aligned}$$

This gives (2.7) and completes the proof. □

The above lemma usually gives the stronger conclusion that $\mu(\mathcal{D}) = 1$, since the sets dealt with frequently satisfy a zero-one law, that is $\mu(\mathcal{D})$ can only be 0 or 1. Hence, we only require $c > 0$ in order to establish that $\mu(\mathcal{D}) = 1$. Even when such a law is not known *a priori* it may still be possible to reach the same conclusion as we show in §4.

Sullivan[24] appears to have been the first to state that the hypotheses of Lemma 3 lead to a quantitative conclusion, although, with different notation, we note that the result is essentially due to Cassels (Lemma 1 in [3]), following Paley and Zygmund[18]. We now give a strengthened version of Sullivan's result.

Theorem 2 *Given the hypotheses of Lemma 3, write, for* $\epsilon > 0$,

$$\begin{aligned} \mathcal{E} &= \left\{\alpha : \limsup_{N\to\infty}\frac{F_N(\alpha,\mathcal{A}_n)}{V(N)} \geq 1-\epsilon\right\}, \\ \mathcal{J} &= \left\{\alpha : \liminf_{N\to\infty}\frac{F_N(\alpha,\mathcal{A}_n)}{V(N)} \leq 1+\epsilon\right\}, \end{aligned}$$

where

$$V(N) = \sum_{j=1}^N \mu(\mathcal{A}_j).$$

Then

$$\mu(\mathcal{E}) \geq c\epsilon^2, \quad \mu(\mathcal{J}) \geq \frac{\epsilon}{1+\epsilon}. \tag{2.8}$$

Remark: Note that the second statement of (2.8) is independent of c.

Proof: Let $\mathcal{F}_N$ be the set on which

$$F_N(\alpha, \mathcal{A}_n) \geq (1-\epsilon)V(N),$$

and write $\mathcal{G}_N$ for its complement in Ω. We then have

$$V(N) = \int_\Omega \sum_{n=1}^N \chi_n(\alpha) d\mu = \int_{\mathcal{G}_N} \sum_{n=1}^N \chi_n(\alpha) d\mu + \int_{\mathcal{F}_N} \sum_{n=1}^N \chi_n(\alpha) d\mu.$$

Thus

$$\begin{aligned} \int_{\mathcal{F}_N} \sum_{n=1}^N \chi_n(\alpha) d\mu &= V(N) - \int_{\mathcal{G}_N} \sum_{n=1}^N \chi_n(\alpha) d\mu \\ &\geq V(N) - \mu(\mathcal{G}_N)(1-\epsilon)V(N) \geq \epsilon V(N). \end{aligned}$$

So, by Cauchy's inequality,

$$\mu(\mathcal{F}_N) \sum_{1\leq j,k\leq N} \mu(\mathcal{A}_j \cap \mathcal{A}_k) = \mu(\mathcal{F}_N) \int_\Omega \sum_{1\leq j,k \leq N} \chi_j(\alpha)\chi_k(\alpha) d\mu \geq \epsilon^2 V^2(N).$$

Hence, by (2.5), there is a sequence $\eta_N \to 0$ such that

$$\mu(\mathcal{F}_N) \geq c\epsilon^2 - \eta_N,$$

for infinitely many N, say $N_1, N_2, \ldots$. Write $\mathcal{F}_j^* = \mathcal{F}_{N_j}$. Then

$$\sum_{1\leq j,k\leq N} \mu(\mathcal{F}_j^* \cap \mathcal{F}_k^*) \leq N \sum_{1\leq n\leq N} \mu(\mathcal{F}_n^*), \quad \sum_{1\leq n\leq N} \mu(\mathcal{F}_n^*) \geq c\epsilon^2 N + o(N),$$

and so

$$\limsup_{N\to\infty} \left(\sum_{n=1}^N \mu(\mathcal{F}_n^*)\right)^2 \left(\sum_{1\leq j,k\leq N} \mu(\mathcal{F}_j^* \cap \mathcal{F}_k^*)\right)^{-1} \geq c\epsilon^2.$$

It follows from Lemma 3 that

$$\mu(\{\alpha : \alpha \in \mathcal{F}_j^* \text{ for infinitely many } j\}) \geq c\epsilon^2,$$

which establishes the first statement in (2.8).

Now let $\mathcal{H}_N$ be the set on which

$$F_N(\alpha, \mathcal{A}_n) \leq (1+\epsilon)V(N),$$

and write $\mathcal{I}_N$ for its complement in Ω. Then

$$(1+\epsilon)V(N)\mu(\mathcal{I}_N) \le \int_{\mathcal{I}_N} F_N(\alpha, \mathcal{A}_n)d\mu \le \int_{\Omega} F_N(\alpha, \mathcal{A}_n)d\mu = V(N).$$

Hence $\mu(\mathcal{H}_N) = 1 - \mu(\mathcal{I}_N) \ge 1 - \frac{1}{1+\epsilon} = \frac{\epsilon}{1+\epsilon}$. It is now straightforward, as above, to deduce that the set of α belonging to infinitely many $\mathcal{H}_N$ also has measure at least $\frac{\epsilon}{1+\epsilon}$. □

One might wonder whether a stronger conclusion than (2.8) were obtainable with the given hypotheses, say with the *liminf* in $\mathcal{J}_N$ bounded below by a constant. That this is not the case is shown clearly by the following example. Take $\Omega = \{0, 1\}$, $\mu(\{0\}) = \mu(\{1\}) = \frac{1}{2}$,

$$\mathcal{A}_N = \begin{cases} \{1\} & \text{if } 2^{j^2} \le n < 2^{(j+1)^2} \text{ with } j \text{ even} \\ \{0\} & \text{if } 2^{j^2} \le n < 2^{(j+1)^2} \text{ with } j \text{ odd.} \end{cases}$$

Then

$$\limsup_{N\to\infty} \left(\sum_{j=1}^{N} \mu(\mathcal{A}_j)\right)^2 \left(\sum_{j,k=1}^{N} \mu(\mathcal{A}_j \cap \mathcal{A}_k)\right)^{-1} \ge \frac{1}{4}$$

for all N, $\mu(\mathcal{F}) = \mu(\mathcal{J}) = 1$, yet for *all* α

$$\liminf_{N\to\infty} \frac{F_N(\alpha, \mathcal{A}_N)}{V(N)} = 0. \tag{2.9}$$

There is no benefit imposing the condition $\mu(\mathcal{A}_n) \to 0$. We could take $\Omega = [0, 1)$, $\mu(\cdot)$ as Lebesgue measure, and

$$\mathcal{A}_n = \mathcal{I}_n \cap \bigcup_{(r,n)=1} \left(\frac{r}{n} - \frac{1}{n^2}, \frac{r}{n} + \frac{1}{n^2}\right),$$

where

$$\mathcal{I}_n = \begin{cases} [0, \frac{1}{2}) & \text{if } 2^{j^2} \le V(n) < 2^{(j+1)^2} \text{ with } j \text{ even} \\ [\frac{1}{2}, 1) & \text{if } 2^{j^2} \le V(n) < 2^{(j+1)^2} \text{ with } j \text{ odd} \end{cases}$$

Then, writing $\phi(n)$ for Euler's function, we have

$$\mu(\mathcal{A}_n) = \frac{\phi(n)}{n^2}, \quad \sum_{n=1}^{N} \frac{\phi(n)}{n^2} \sim \frac{6}{\pi^2} \log N,$$

(2.5) holds with $c > 0$, $\mu(\mathcal{E}) = \mu(\mathcal{J}) = 1$, but (2.9) also holds for almost all α. Of course, it is the intersection with $\mathcal{I}_n$ which makes the problem unnatural in Number Theory. If we removed that we would get an asymptotic formula for $F_N(\alpha, \mathcal{A}_n)$ for almost all α (see Theorem 4.4 in [11]).

§3 Zero-One Laws

We now consider how the conclusion of positive measure can be converted to a result of full measure. One approach is to define a suitable mapping T such that $\mathcal{D}T = \mathcal{D}$, where $\mathcal{D}$ is the set on which our required property holds. If T is *ergodic*, that is the only invariant sets have measure 0 or 1, then we have achieved our goal (see Section 2.2 of [11] for example). Another approach is to apply the results of the previous section to short intervals after a suitable scaling. In this way we obtain in place of (2.8):

$$\mu(\mathcal{E}) \geq c\epsilon^2, \quad \mu(\mathcal{J}) \geq \frac{\epsilon}{1+\epsilon}\mu(\Omega). \tag{3.1}$$

Note that the first statement remains unchanged, but one should think of c now as being of size $c'\mu(\Omega)$. If we are dealing with Lebesgue measure then the following consequence of the Lebesgue density theorem is crucial. Throughout this section $\mu(\cdot)$ denotes Lebesgue measure on $\mathbb{R}$.

Lemma 4 *Let $\mathcal{K}$ be a given subset of $\mathbb{R}$ such that*

$$\mu(\mathcal{K} \cap \mathcal{I}) \geq \delta\mu(\mathcal{I}) \tag{3.2}$$

for every finite subinterval $\mathcal{I}$ of $\mathbb{R}$, where δ is a positive constant. Then almost all real α belong to $\mathcal{K}$.

Proof: Write $\mathcal{C} = \mathbb{R}\backslash\mathcal{K}$ and suppose that $\mu(\mathcal{K}) > 0$. Then, by the Lebesgue density theorem, $\mathcal{K}$ has a point of metric density, say α, that is to say

$$\mu((\alpha - \epsilon, \alpha + \epsilon) \cap \mathcal{C}) \geq 2\epsilon(1 - \delta/2) \tag{3.3}$$

for all sufficiently small ϵ. Let $\mathcal{I} = (\alpha - \epsilon, \alpha - \epsilon)$. Then, by (3.2) and (3.3) we have

$$\mu(\mathcal{I}) = \mu(\mathcal{I} \cap \mathcal{K}) + \mu(\mathcal{I} \cap \mathcal{C}) \geq \delta\mu(\mathcal{I}) + \mu(\mathcal{I})\left(1 - \frac{\delta}{2}\right) = \left(1 + \frac{\delta}{2}\right)\mu(\mathcal{I}).$$

This contradiction establishes $\mu(\mathcal{C}) = 0$ as required. □

In the above we could have replaced $\mathbb{R}$ by any subinterval $\mathcal{J}$ of $\mathbb{R}$, and the conclusion would have been that almost all $\alpha \in \mathcal{J}$ belong to $\mathcal{K}$. Combining Lemma 4 with Theorem 2 having replaced (2.8) by (3.1) then gives the following result.

Theorem 3 *Let $\mathcal{J}$ be a subinterval of $\mathbb{R}$, and suppose that we have, for any finite subinterval $\mathcal{I} \subset \mathcal{J}$,*

$$\lim_{N\to\infty} \frac{1}{V(N)} \sum_{n=1}^{N} \mu(\mathcal{I} \cap A_n) = \lambda(\mathcal{I}) \tag{3.4}$$

with $V(N)$ as Theorem 2 and $V(\infty)$ divergent, and

$$\limsup_{N\to\infty} \left(\sum_{n=1}^{N} \mu(A_n \cap \mathcal{I})\right)^2 \left(\sum_{1\leq j,k\leq N} \mu(A_j \cap A_k \cap \mathcal{I})\right)^{-1} \geq c\lambda(\mathcal{I}) > 0. \tag{3.5}$$

Then, for almost all $\alpha \in \mathcal{J}$ we have

$$\limsup_{N\to\infty} \frac{F_N(\alpha, \mathcal{A}_n)}{V(N)} \geq 1, \text{ and } \liminf_{N\to\infty} \frac{F_N(\alpha, \mathcal{A}_n)}{V(N)} \leq 1. \tag{3.6}$$

In particular, given $\epsilon > 0$, for almost all $\alpha \in \mathcal{J}$ there are infinitely many N with

$$1 - \epsilon < \frac{F_N(\alpha, \mathcal{A}_n)}{V(N)} < 1 + \epsilon. \tag{3.7}$$

The example given at the end of §2 can be modified to show that the hypotheses of Theorem 3 do not imply

$$\liminf_{N\to\infty} \frac{F_N(\alpha, \mathcal{A}_n)}{V(N)} > 0. \tag{3.8}$$

In this case we take

$$\mathcal{I}_n = \begin{cases} B_j & \text{if } 2^{(2j)^2} \leq V(n) < 2^{(2j+1)^2} \\ C_j & \text{if } 2^{(2j+1)^2} \leq V(n) < 2^{(2j+2)^2}, \end{cases}$$

where

$$B_j = \bigcup_{k \text{ odd}} \left[\frac{k}{2^j}, \frac{k+1}{2^j}\right), \quad C_j = \bigcup_{k \text{ even}} \left[\frac{k}{2^j}, \frac{k+1}{2^j}\right).$$

Again, the intersection with $\mathcal{I}_n$ is unnatural in number-theoretic problems, but this example shows that some other information is needed to obtain (3.8).

§4 Application to Khintchine and Duffin-Schaeffer Type Problems

In 1924, Khintchine[16] proved the following result:-

Let $\psi(n)$ be a function such that $n\psi(n)$ is decreasing and

$$\sum_{n=1}^{\infty} \psi(n) = \infty. \tag{4.1}$$

Then, for almost all α, there exist infinitely many rationals a/q which satisfy

$$\left|\alpha - \frac{a}{q}\right| < \frac{\psi(q)}{q}. \tag{4.2}$$

In this situation we can take $\Omega = [0, 1)$, and our first thought would be to put

$$\mathcal{A}_n = \bigcup_{a=1}^{n-1} \left(\frac{a}{n} - \frac{\psi(n)}{n}, \frac{a}{n} + \frac{\psi(n)}{n}\right).$$

In fact we run into serious trouble with this choice for $\mathcal{A}_n$, and we need to impose the extra condition $(a, n) = 1$ (or else our overlap estimates become too large). This gives

$$\sum_{n=1}^{N} \mu(\mathcal{A}_n) = 2\sum_{n=1}^{N} \frac{\psi(n)\phi(n)}{n}. \tag{4.3}$$

Now, if $n\psi(n)$ is decreasing, then the divergence of (4.1) implies the divergence of (4.3) by partial summation. The crucial feature is the consideration of the overlaps $\mathcal{A}_m \cap \mathcal{A}_n$. To be precise, we must investigate

$$\sum_{1\le j,k\le N} \mu(\mathcal{A}_j \cap \mathcal{A}_k) = 2 \sum_{1\le j<k\le N} \mu(\mathcal{A}_j \cap \mathcal{A}_k) + \sum_{n=1}^{N} \mu(\mathcal{A}_n). \tag{4.4}$$

The second term on the right of (4.4) is of smaller order than the other terms in view of the divergence of (4.1). Also

$$\sum_{1\le j<k\le N} \mu(\mathcal{A}_j \cap \mathcal{A}_k) \le 2 \sum_{1\le j<k\le N} \frac{\psi(k)}{k} \sum\nolimits^{*} 1.$$

Where * indicates the summation conditions:-

$$|aj - bk| < k\psi(j), \quad (j,b) = (a,k) = 1, \quad 1 \le a < k, 1 \le b < j.$$

Since $aj - bk = 0$ has no solution with the additional constraints, and $aj - bk = h$ has no solution if $\frac{h}{(j,k)}$ has any factor in common with j or k, and (j,k) solutions in a otherwise, we obtain

$$\begin{aligned}\sum_{1\le j<k\le N} \mu(\mathcal{A}_j \cap \mathcal{A}_k) &\le 4 \sum_{1\le j<k\le N} \left(\frac{\psi(k)}{k}\right) k\psi(j)\\ &\le 4\left(\sum_{n=1}^{N}\psi(n)\right)^2 \le 16\left(\sum_{n=1}^{N}\frac{\psi(n)\phi(n)}{n}\right)^2\end{aligned}$$

for all large N. Since it is not difficult to prove a zero-one law for this situation, this proves Khintchine's result in view of Lemma 3.

The above proof is essentially due to Duffin and Schaeffer[5] who observed that the hypothesis '$n\psi(n)$ decreasing' could be replaced with

$$\limsup_{N\to\infty} \left(\sum_{n=1}^{N} \frac{\psi(n)\phi(n)}{n}\right)\left(\sum_{n=1}^{N}\psi(n)\right)^{-1} > 0,$$

and the conclusion (4.2) holds with $(a,q) = 1$. It is, in fact, the repeat solutions $b/r = a/q$ that lead to problems in estimating the overlaps, and these are excluded by imposing $(a,n) = 1$ in $\mathcal{A}_n$. Duffin and Schaeffer made the following conjecture which is an important unsolved problem in this field.

Conjecture: Let $\psi(n) \in [0,1)$ be any function such that

$$\sum_{n=1}^{\infty} \frac{\psi(n)\phi(n)}{n} = \infty.$$

Then there are infinitely many solutions to

$$|a\alpha - q| < \psi(q), \quad (a,q) = 1$$

for almost all real α.

See Chapters 2 and 3 of [11] for the known cases of this conjecture.

Now let us suppose that we want to establish, for almost all real α, a result of the form

$$\limsup_{N\to\infty} \frac{S(\alpha, N)}{\Psi(N)} \geq c \tag{4.5}$$

where

$$\Psi(N) = 2\sum_{n=1}^{N} \psi(n),$$

and $S(\alpha, N)$ denotes the number of solutions to (4.2) with $q \leq N$. If we assume that $\psi(n)$ is monotonic decreasing, or that

$$c_1 < \frac{\psi(m)}{\psi(n)} < c_2 \text{ for all } m, n \text{ with } n < m < 2n, \tag{4.6}$$

for positive constants c_1, c_2, then it is possible to use Theorem 3 as we now show.

First we replace $(a, n) = 1$ in the definition of $\mathcal{A}_n$ by $(a, n) \leq D$ (this idea is due to W.M. Schmidt [22]). This gives, if $\mathcal{J} = (\beta, \beta + \delta)$,

$$\sum_{n=1}^{N} \mu(\mathcal{A}_n \cap \mathcal{J}) = \delta \sum_{n=1}^{N} \psi(n) \frac{\phi(D, n)}{n} (1 + o(1)) \text{ as } N \longrightarrow \infty,$$

where

$$\phi(D, n) = \sum_{\substack{d \leq N \\ (d,n) \leq D}} 1.$$

Also,

$$\sum_{1 \leq m < n \leq n} \mu(\mathcal{A}_m \cap \mathcal{A}_n \cap \mathcal{J}) \leq K_1 \sum_{M,R} \min\left(\frac{\psi(R)}{R}, \frac{\psi(M)}{M}\right) \sum_{a,b,m,n} 1, \tag{4.7}$$

where M and R satisfy

$$M = (1 + \delta)^s, \quad R = (1 + \delta)^r, \quad M \leq R \leq N,$$

and a, b, m, n are restricted by

$$|am - bn| < K_2 \max(\psi(R)M, \psi(M)R) = H, \text{ say}, \tag{4.8}$$

and

$$(m, b) \leq D, \ (n, a) \leq D, \ m : M, \ n : R, \ B :: \beta M, \ a :: \beta R,$$

where $x : X$ indicates $X \leq x < (1 + \delta)X$, while $x :: X$ stands for $X \leq x < (1 + \delta)^2 X$. Now we can replace (4.8) by

$$am' - b'n = h, \quad 1 \leq |h| \leq H', \tag{4.9}$$

where

$$m' = \frac{m}{(m,b)}, \quad b' = \frac{b}{(m,b)} \quad H' = \frac{H}{(m,b)},$$

since the terms with $|am - bn| = 0$ contribute to (4.7)

$$\leq D^2 \sum_{n=1}^{N} \psi(n).$$

For each m, b and h there is exactly one solution to (4.9) in n per set of residues (mod m'). Hence the number of solutions all told for each pair m, b is no more than

$$\frac{H}{(m,b)} \left(1 + \frac{\delta R(b,m)}{m}\right).$$

Summing over m and b thus gives a bound

$$\leq 3\delta^2 M^2 h \left(1 + \frac{\delta R}{M}\right).$$

Substituting this into (4.7) gives a bound

$$\begin{aligned}
&\ll \sum_{M,R} \min\left(\frac{\psi(R)}{R}, \frac{\psi(M)}{M}\right) (\delta^3 RM + \delta^2 M^2) \max(\psi(R)M, \psi(M)R) \\
&\ll \sum_{M,R} \delta^3 RM \psi(R)\psi(M) + \sum_{R} \delta\psi(R) \sum_{m \leq R} m\psi(m) \\
&\ll \delta \left(\sum_{n=1}^{N} \psi(n)\right)^2 + \sum_{n=1}^{N} \psi(n)
\end{aligned}$$

provided that $\psi(n) \ll n^{-1}$. Indeed, we could also have obtained a satisfactory bound if

$$g(n)(\log n)^{A-\frac{1}{2}} \ll n\psi(n) \ll (\log n)^A$$

for some $A > 0$ and some function $g(n) \to \infty$ as $n \to \infty$. Under these conditions we obtain, for almost all α, by Theorem 3 that

$$\limsup_{N \to \infty} \frac{S(\alpha, D, N)}{\Psi(D, N)} \geq 1,$$

with

$$\Psi(D, N) = 2 \sum_{n=1}^{N} \frac{\phi(D,n)}{n} \psi(n),$$

and where $S(\alpha, D, N)$ is the number of solutions to

$$|\alpha n - m| < \psi(n), \; n \leq n, \; (m,n) \leq D.$$

Since

$$\lim_{D\to\infty}\lim_{N\to\infty}\frac{\Psi(D,N)}{\Psi(N)}=1$$

for many sequences $\psi(n)$, in particular if $\psi(n)$ is monotonic decreasing, this gives, in many cases, the expected lower bound for almost all α:

$$\limsup_{N\to\infty}\frac{S(\alpha,N)}{\Psi(N)}\geq 1.$$

In fact, it is possible to take D increasing with n in the above, say $D=\Psi(n)^{\frac{1}{4}}$.

We now consider the weak point in the above proof. We required a bound on $\psi(n)$ to deal with a sum which arose because n might not run over a complete set of residues (mod m) (the problem arises from the need to keep our results uniform in δ). We did not make use of all the variables in (4.8)/(4.9). We can reduce the number of variables to 3 by recasting (4.8) as

$$\left\{\frac{-h\bar{n}}{m}\right\}\in[\beta,(1+\delta^2)\beta)\quad(\text{mod } 1). \tag{4.10}$$

Here we have assumed $(n,m)=1$, written $\bar{n}$ for the inverse of n (mod m), denoted fractional part by $\{\cdot\}$, and the conditions on the variables are

$$1\leq|h|\leq H,\ m:M,\ n:R.$$

The requirement $(m,n)=1$ is overcome by considering all $d|m$ and restricting attention to all n with $(n,m)=d$. After summing over d this deals with all n. To bound the number of solutions to (4.10) first fix m, then by standard techniques (see Lemma 5.1) in [11], writing $e(x)=\exp(2\pi ix)$, the number of solutions is

$$\leq 3\delta^2RH+6\delta\sum_{l\delta<1}\left|\sum_{h=1}^{H}\sum_{n:R}e\left(\frac{lh\bar{n}}{m}\right)\right|\leq 3\delta^2RH+H\left(M^{\frac{1}{2}+\epsilon}+\frac{R}{M}\right)$$

using the well-known bound for an incomplete Kloostermann sum (see Lemma 3.6 in [14] for example). Summing over m then gives a bound

$$3\delta^3RHM+M^{\frac{3}{2}+\epsilon}H+RH.$$

Since

$$\sum_{M,R}M^{\frac{3}{2}+\epsilon}\psi(R)\psi(M)=o(\Psi^2(N))\ \text{ and }\ \sum_{M,R}R\psi(R)\psi(M)=o(\Psi^2(N)),$$

this establishes a suitable bound for the overlap estimates without any restriction on the size of $\psi(n)$ except (4.6), or monotonicity. In this way we obtain more precise results than Sullivan (Theorem 4 in [24]) with weaker hypotheses. For example we can easily modify the above to obtain the following quantitative form of the Duffin and Schaeffer result.

Theorem 4 *Suppose that $\psi(n) \in (0, \frac{1}{2})$, and is either monotonic or satisfies* (4.6). *Also suppose that $\Psi'(\infty) = \infty$ where*

$$\Psi'(N) = 2\sum_{n=1}^{N} \frac{\phi(n)}{n}\psi(n).$$

Then, writing $\Sigma(\alpha, N)$ for the number of solutions to

$$|\alpha n - m| < \psi(n), \quad (m, n) = 1, \ n \leq N,$$

we obtain

$$\limsup_{N\to\infty} \frac{\Sigma(\alpha, N)}{\Psi'(N)} \geq 1$$

for almost all α.

§5 Application to Two Restricted Variable Problems

In [6–9] (see also Chapter 6 of [11]) the author has considered the problem of showing whether infinitely many solutions exist to

$$|\alpha n - m| < \psi(n), \quad n \in \mathcal{A}, m \in \mathcal{B},$$

with $\mathcal{A}$ and $\mathcal{B}$ given sets of integers. One expects infinitely many solutions for almost all α if the sum

$$\sum_{n\in\mathcal{A}} \rho(n)\psi(n)$$

diverges, where $\rho(n)$ is a 'smooth' function representing the probability that $n \in \mathcal{B}$ (for example, if $\mathcal{B}$ is the set of primes then $\rho(n) = (\log n)^{-1}$). That convergence of the sum gives only finitely many solutions for almost all α is easy to establish from the first Borel-Cantelli lemma. Many of the divergence cases were settled by combining Lemmas 3 and 4 above. However, the arithmetical information obtained in some of the cases discussed is strong enough to apply Theorem 3 instead. One example is given above as Theorem 1. The most delicate part of the proof, as expected, is to obtain the correct order size for the overlap estimates $\mu(\mathcal{A}_p \cap \mathcal{A}_q \cap \mathcal{J})$ with p, q primes and

$$\mathcal{A}_p = \bigcup_{r\neq p} \left(\frac{r - \psi(p)}{p}, \frac{r + \psi(p)}{p}\right),$$

with r also denoting a prime. Here Kloostermann sums together with a three-dimensional sieve were needed to bound the number of solutions to the resulting inequalities:

$$|pr - qs| < A, \quad p, q, r, s \text{ primes.}$$

In the same way, several of the results in [9] also can now be restated to give a conclusion like (1.2).

As another application of the general technique we mention the following result which can be obtained by combining the present Theorem 3 with the overlap estimates in [10].

Theorem 5 *Let a_n be an increasing sequence of reals with $a_{n+1} \geq a_n + 1, a_1 \geq 2$. Write $P(\alpha, n)$ for the number of primes of the form $[a_n\alpha]$, $n \leq N$, and put*

$$V(N) = \sum_{n=1}^{N} \frac{1}{\log a_n}.$$

Then, if $V(\infty)$ diverges, for almost all α we have

$$\limsup_{N\to\infty} \frac{P(\alpha, N)}{V(N)} \geq 1, \qquad \liminf_{N\to\infty} \frac{P(\alpha, N)}{V(N)} \leq 1.$$

With some tedious, but straightforward, modifications the method of [12] can be combined with Theorem 3 to produce the following:-

Theorem 6 *Let $k \geq 2$. Write $\pi_k^*(\alpha, N)$ for the number of k-tuplets of primes $(p_1, \ldots, p_k)$ such that*

$$p_j = [p_{j-1}^{\alpha}], \quad k \geq j \geq 2.$$

Then, for almost all $\alpha > 2$, we have

$$\limsup_{N\to\infty} \frac{\pi_k^*(\alpha, N) \log^k N}{N} \geq \alpha^{-k(k-1)/2}, \quad \liminf_{N\to\infty} \frac{\pi_k^*(\alpha, N) \log^k N}{N} \leq \alpha^{-k(k-1)/2}.$$

§6 Asymptotic Formula Variants

In the applications given above it would be expected that instead of inequalities for *limsup* and *liminf* we should be able to get a result

$$\lim_{N\to\infty} \frac{H(\alpha, N)}{V(N)} = 1,$$

or even an explicit error term:

$$H(\alpha, N) = V(N) + O(V'(N)) \text{ as } N \longrightarrow \infty.$$

Here $H(\alpha, N)$ counts the number of solutions to an inequality, $V(N)$ is the expected number of solutions, and $V'(N) = o(V(N))$. Weyl[26] presented a technique which has been refined by subsequent authors[20–22] to give the following.

Theorem 7 *Let Ω be a measure space with measure μ such that $0 < \mu(\Omega) < \infty$. Let $f_k(\alpha)$ $(k = 1, 2, \ldots)$ be a sequence of non-negative μ-measurable functions, and let f_k, ϕ_k be sequences of real numbers such that*

$$0 \leq f_k \leq \phi_k \quad (k = 1, 2, \ldots).$$

Write

$$\Phi(N) = \sum_{k=1}^{N} \phi_k,$$

and suppose that $\Phi(N) \to \infty$ *as* $N \to \infty$. *Suppose that for arbitrary integers* m, n $(1 \leq m < n)$ *we have*

$$\int_{\Omega} \left(\sum_{m \leq k < n} (f_k(\alpha) - f_k) \right)^2 d\mu \leq K \sum_{m \leq k < n} \phi_k \tag{6.1}$$

for an absolute constant K. *Then, given* $\epsilon > 0$, *for almost all* $\alpha \in \Omega$ *we have, as* $N \to \infty$,

$$\sum_{k=1}^{N} f_k(\alpha) = \sum_{k=1}^{N} f_k + O\left(\Phi^{\frac{1}{2}}(N)(\log(\Phi(N)+2))^{\frac{3}{2}+\epsilon} + \max_{1 \leq k \leq N} f_k \right). \tag{6.2}$$

Proof: For a detailed proof see Lemma 1.5 in [11]. We sketch the basic ideas here. First write

$$n_j = \max\{n : \Phi(n) < j\}.$$

It then suffices to prove (6.2) for $N = n_j$. Next express j in binary as

$$j = \sum_{v} b(j, v) 2^v,$$

and write

$$B(j) = \{(i, s) : i = \sum\nolimits_{v=s+1}^{r} b(j, v) 2^{v-s}, b(j, s) = 1, 0 \leq s \leq r\}$$

where $r = r(j) = [\log_2 j]$, and put

$$F(i, s, \alpha) = \sum_{u_0 < k \leq u_1} (f_k(\alpha) - f_k),$$

with

$$u_t = u_t(i, s) = \max\{n > 0 : \Phi(n) < (i+t)2^s\},$$

with the convention $\max \emptyset = 0$. Thus

$$\sum_{k=1}^{n_j} f_k(\alpha) = \sum_{k=1}^{n_j} f_k + \sum_{(i,s) \in B(j)} F(i, s, \alpha).$$

It thus suffices to show that, for almost all α,

$$\sum_{(i,s) \in B(j)} |F(1, s, \alpha)| = O(j^{\frac{1}{2}}(\log(j+2))^{\frac{3}{2}+\epsilon}).$$

We have

$$\sum_{(i,s) \in B(j)} |F(1, s, \alpha)| \leq |B(j)|^{\frac{1}{2}} G^{\frac{1}{2}}(r, \alpha) \leq r^{\frac{1}{2}} G^{\frac{1}{2}}(r, \alpha)$$

with

$$G(r, \alpha) = \sum_{\substack{0 \leq s \leq r \\ i < 2^{r-s}}} F^2(i, s, \alpha).$$

The required bound for $G(r,\alpha)$ for almost all α, namely $G(r,\alpha) < R^{2+\epsilon}2^r$ for $r > r(\alpha)$ may then be deduced from (6.1) with the first Borel-Cantelli lemma since

$$\sum_{r=1}^{\infty} \mu(\{\alpha \in \Omega : G(r,\alpha) > r^{2+\epsilon}2^r\})$$

converges. This completes the proof. □

There are examples which show that the error term in (6.2) can sometimes exceed $\Phi^{\frac{1}{2}}(N)(\log(\Phi(N)+2))^{\frac{1}{2}}$ infinitely often (see Chapters One and Five in [11]). To apply Theorem 7 to the type of problems considered so far, let f_k be the characteristic function of $\mathcal{A}_k$ and put $f_k = \mu(\mathcal{A}_k)$. The left hand side of (6.1) is thus

$$\int_\Omega \sum_{m\le j,k<n} f_k(\alpha)f_j(\alpha)d\mu - 2\int_\Omega \sum_{m\le k<n} f_k(\alpha)d\mu \sum_{m\le j<n} f_j + \mu(\Omega)\left(\sum_{m\le k<n} f_k\right)^2$$

$$= \sum_{m\le j,k<n} \mu(\mathcal{A}_j \cap \mathcal{A}_k) - \left(\sum_{m\le k<n} \mu(\mathcal{A}_k)\right)^2 \tag{6.3}$$

assuming that $\mu(\Omega) = 1$. We thus see that to obtain an asymptotic formula we require much more than

$$\limsup_{N\to\infty} \frac{\left(\sum_{k=1}^N \mu(\mathcal{A}_k)\right)^2}{\sum_{1\le j,k\le N} \mu(\mathcal{A}_j \cap \mathcal{A}_k)} > 0;$$

we actually require

$$\sum_{m\le j,k<n} \mu(\mathcal{A}_j \cap \mathcal{A}_k) = \left(\sum_{m\le k<n} \mu(\mathcal{A}_k)\right)^2 + O\left(\sum_{m\le k<n} \phi_k\right),$$

where

$$\Phi^{\frac{1}{2}}(N)(\log(\Phi(N)+2))^{\frac{3}{2}+\epsilon} = o\left(\sum_{k=1}^N \mu(\mathcal{A}_k)\right).$$

Theorem 7 has had many applications in metric number theory. Often it is easier to work directly with the left hand side of (6.1) rather than the expression in (6.3). We give three examples by way of illustration, pointing out the distinctive features of their proofs..

Theorem 8 *(W.M. Schmidt)*[22]. *Let $\psi(n)$ be a monotonic decreasing function with $\psi(n) \in (0, \frac{1}{2})$. Write*

$$\Psi(N) = \sum_{n=1}^N \psi(n),$$

and suppose that $\Psi(\infty)$ diverges. Then the number of solutions to

$$|\alpha n - m| < \psi(n), \quad n \leq N \tag{6.4}$$

is, for almost all α,

$$2\Psi(N) + O(\Psi^{\frac{1}{2}}(\log(\Psi(N)+2))^{2+\epsilon}). \tag{6.5}$$

Proof: For full details see [22], or Theorem 4.1 in [11]. One very important ingredient in the proof is to impose the condition $(m, n) < D$ in (6.4) (as we did in §4). Now D is taken to vary with n, to be precise $D = \Psi^2(n) + 1$. It can then be shown that the number of solutions excluded by this condition is relatively small for almost all α. Also (compare §4) the difference between $\Psi(N)$ and

$$\sum_{n \leq N} \psi(n)\phi(D(n), n)/n$$

is not larger than the error term in (6.5). It then remains to show that the sets $\mathcal{A}_j$, $\mathcal{A}_k$ overlap in the way expected on average. In §4 the worst case estimates were taken, but these can be improved for monotonic $\psi(n)$ to give the required result. The crucial feature is that the length of the overlap is $(j > k)$

$$\min\left(\frac{2\psi(j)}{j}, \frac{\psi(j)}{j} + \frac{\psi(k)}{k} - \frac{|h|}{jk}\right)$$

when

$$|ju - kv| = h, \quad |h| \leq j\psi(k) + k\psi(j), \quad (j, v) \leq D(j), (k, u) \leq D(k). \tag{6.6}$$

Since one only needs an upper bound on the overlap (as (6.3) cannot be negative), except for $h = 0$ one removes the conditions $(j, v) \leq D(j)$, $(k, u) \leq D(k)$, and thereby we obtain the required estimate. The fact that D increases with n only worsens the power of log in the error term from $\frac{3}{2}$ to 2. □

Theorem 9 *(W. Philipp)*[20]. *Let T be the transformation associated with the continued fraction expansion of a real number, namely*

$$T\alpha = \begin{cases} 0 & \text{if } \alpha = 0 \\ \{\alpha^{-1}\} & \text{otherwise.} \end{cases}$$

Let $\mathcal{I}_n$ be an arbitrary sequence of intervals contained in $[0, 1)$. *Write $A(N, \alpha)$ for the number of solutions to $T^n\alpha \in \mathcal{I}_n$, $1 \leq n \leq N$, and put*

$$\Phi(N) = \sum_{n \leq N} \mu(\mathcal{I}_n)$$

where

$$\mu(\mathcal{I}_n) = \frac{1}{\log 2} \int_{\mathcal{I}_n} \frac{dx}{1+x}.$$

Then, given $\epsilon > 0$, for almost all α, we have

$$A(N, \alpha) = \Psi(N) + O(\Psi^{\frac{1}{2}}(N)(\log(\Psi(N)+2))^{\frac{3}{2}+\epsilon}).$$

Remark: Philipp actually proved this result for other transformations of number-theoretic interest with differing choices for μ appropriate to each transformation. The above theorem has an interesting corollary in providing information on blocks of numbers among consecutive partial quotients in the continued fraction expansion for almost all α (see [13]).

Proof: See [20] for details. As usual we need to estimate certain overlaps. If we put $\mathcal{E}_n = T^{-n}\mathcal{I}_n$, it is shown that, for $n > m$,

$$\mu(\mathcal{E}_n \cap \mathcal{E}_m) \le \mu(\mathcal{E}_n)\mu(\mathcal{E}_m) + O(\mu(\mathcal{E}_n)q^{\sqrt{(n-m)}})$$

for some $q < 1$. Since

$$\sum_{m \le n} q^{\sqrt{(n-m)}} = O(1)$$

we obtain Theorem 9 from Theorem 7. □

Theorem 10 *(G. Harman)*[10,11] *Let $a_n \ge 2$ be a lacunary sequence, and write*

$$S_N(\alpha) = |\{[\alpha a_n] = p : n \le N\}|, \quad \Psi(N) = \sum_{n=1}^{N} \frac{1}{\log a_n}.$$

Suppose that $\Psi(N) \to \infty$ as $N \to \infty$. Then, for almost all $\alpha > 0$ and for any $\epsilon > 0$ we have

$$S_N(\alpha) = \Psi(N) + O(\Psi(N)^{\frac{1}{2}}(\log(\Psi(N) + 2))^{\frac{3}{2}+\epsilon}).$$

Remark: If we take $a_n = 10^n$ this has an interesting corollary for primes appearing in the decimal expansions of almost all positive α.

Proof: Consider the interval $[A, A+1]$, $A > 0$, with

$$\mathcal{A}_n = [A, A+1] \cap \bigcup_p \left[\frac{p}{a_n}, \frac{p+1}{a_n}\right).$$

It is relatively straightforward to obtain

$$\begin{aligned}
&\sum_{\substack{M \le m < n \le N \\ a_m^3 \le a_n}} \mu(\mathcal{A}_m \cap \mathcal{A}_n) \\
&= \sum_{\substack{M \le m < n \le N \\ a_m^3 \le a_n}} \frac{1}{(\log a_n)(\log a_m)} + O\left(\sum_{M \le n \le N} \frac{1}{\log a_n}\right),
\end{aligned}$$

using known results on primes in short intervals together with the convergence of

$$\sum_n \frac{1}{\log^2 a_n}.$$

For the overlap sum with $a_m^3 > a_n$ it suffices to obtain a rough upper bound for

$$\sum_{\substack{M \leq m < n \leq N \\ a_m^3 > a_n}} \mu(\mathcal{A}_m \cap \mathcal{A}_n)$$

using a sieve method. □

§7 Further Applications

Most of the results given above generalize to $\mathbb{R}^k$ for $k > 1$. For example, the present author's research student Huw Jones[15] has recently obtained the following generalization of Theorem 1.

Theorem 11 *Let $\psi_j(n)$, $j = 1, \ldots, k$ be monotonic decreasing functions with $\psi_j(n) \in (0, \frac{1}{2})$, and write $P(\boldsymbol{\alpha}, N)$ ($\boldsymbol{\alpha} = (\alpha_1, \ldots, \alpha_k)$) for the number of solutions to*

$$|p\alpha_j \pm q_j| < \psi_j(p), \quad p, q_j \text{ primes}, \ p \leq N. \tag{7.1}$$

Also, put

$$\Psi(N) = 2^k \sum_{p \leq N} \frac{\psi_1(p) \cdots \psi_k(p)}{\log^k p}.$$

Then, if $\Psi(\infty)$ diverges, we have, for almost all $\boldsymbol{\alpha} \in \mathbb{R}^k$,

$$\limsup_{N \to \infty} \frac{P(\boldsymbol{\alpha}, N)}{\Psi(N)} \geq 1, \quad \liminf_{N \to \infty} \frac{P(\boldsymbol{\alpha}, N)}{\Psi(N)} \leq 1. \tag{7.2}$$

Several authors have generalized Khintchine's theorem to other situations. In many of these generalizations (for example [1], [19], [23], [24], [25]) the reals with Lebesgue measure are replaced with some group on which an appropriate measure is defined, while the rôle played by the integers is now taken by a discrete subgroup (or the rôle of the rationals is taken by the orbit of some element under a group action). The result we stated as Lemma 3 features in these works, but the approach to the overlap estimates is more geometrical.

Work on submanifolds of $\mathbb{R}^k$ has mainly concentrated on proving the convergence/finitely many solutions result which for the whole of $\mathbb{R}^k$ is very easy, but is far from non-trivial for submanifolds (see Chapter 9 of [11]). We would seem some way from proving a complete Khintchine type result in this situation, although Khintchine himself proved one result [17], and some authors have shown that one can obtain an asymptotic formula result in the divergence case for certain manifolds[2,4].

An outstanding problem in the area we have discussed is to obtain asymptotic formulae for those problems where only lim sup and lim inf results are known. Indeed, it would be an advance just to obtain a non-trivial upper bound for lim sup and a non-trivial lower bound for lim inf for almost all $\alpha \in \Omega$. In view of the examples we gave above it would seem that more arithmetical information must be fed into the method.

Perhaps the major problem, however, is the stubborn Duffin and Schaeffer conjecture. If we attack it by Lemma 3 we are lead to certain overlap estimates, but we seem unable to obtain satisfactory bounds in the most general case (see the discussion in Chapter 2 of [11]).

References

[1] L. Auslander, Brezin, J. and Sacksteder, R., *J. different. Geomet.* **6** (1972), 479–496.
[2] V.I. Bernik, *Vesci Akad. Navuk BSSR Ser. Fiz.-Mat. Navuk* **1** (1973), 10–17, 134.
[3] J.W.S. Cassels, *Proc. Camb. phil. Soc.* **46** (1950), 209–218.
[4] M.M. Dodson, Rynne, B.P. and Vickers, J.A.G., *J. N. Theory* **58** (1996), 298–316.
[5] R.J. Duffin and Schaeffer, A.C., *Duke Math. J.* **8** (1941), 243–255.
[6] G. Harman, *Math. Proc. Camb. phil. Soc.* **103** (1988), 197–206.
[7] G. Harman, *Mathematika* **35** (1988), 59–68.
[8] G. Harman, *J. N. Theor.* **29** (1988), 364–375.
[9] G. Harman, *Acta Arith.* **53** (1989), 207–216.
[10] G. Harman, *Proc London math. Soc.* **75** (3) (1997), 481–496.
[11] G. Harman, *Metric Number Theory*, Oxford University Press, 1998.
[12] G. Harman, *Mathematika* **45** (1998), 351–357.
[13] G. Harman and Wong, K.C., *Preprint* Cardiff (1997).
[14] C. Hooley, *Applications of Sieve Methods to the Theory of Numbers*, Cambridge University Press, 1976.
[15] H.R. Jones, *Preprint* Cardiff (1998).
[16] A. Khintchine, *Math. Ann.* **92** (1924), 115–125.
[17] A. Khintchine, *Math. Zeit.* **22** (1925), 274–284.
[18] R.E.A.C. Paley and Zygmund, A., *Proc. Camb. phil. Soc.* **26** (1930), 337–357, 458–474 and **28** (1932), 190–205.
[19] S.J. Patterson, *Number Theory and Dynamical Systems* (*Eds*. M.M. Dodson and J.A.G. Vickers) Cambridge University Press, 1987, pp. 37–48.
[20] W. Philipp, *Pacific J. Math.* **20** (1967), 109–127.
[21] H. Rademacher, *Math. Ann.* **87** (1922), 112–138.
[22] W.M. Schmidt, *Canad. J. Math.* **12** (1960), 619–631.
[23] B. Stratmann and Velani, S.L., *Proc. London math. Soc.* (3) **71** (1995), 197–220.
[24] D. Sullivan, *Acta Math.* **144** (1983), 215–237.
[25] S.L. Velani, *Math. Proc. Camb. phil. Soc.* **120** (1996), 647–662.
[26] H. Weyl, *Math. Ann.* **77** (1916), 313–352.

Department of Mathematics
Royal Holloway
University of London
Egham
Surrey TW20 0EX
UK
E-Mail: G.Harman@rhbnc.ac.uk

Pythagorean Triples

Edmund Hlawka

Historical Introduction

The investigation of Pythagorian triples has a very long history. For the first hundred years I refer to the famous book [DIC01]. Triangles of this type were given by Greek and Indian mathematicians. Arithmetically these are the solutions of the diophantine equation

$$x^2 + y^2 = z^2$$

in rational numbers. The general solution is given by the formulas

$$\begin{aligned} x &= l(m^2 - n^2) \\ y &= l \cdot 2mn \\ z &= l(m^2 + n^2) \qquad l\ (l \neq 0),\ m, n \text{ arbitrary.} \end{aligned}$$

These formulas are already contained in the works of Euclid and Brahmegupta. We also mention Bháscara, Pisano, Vieta,[1] Euler and Kronecker.

In early days the case $|x - y| = 1$ has been studied (example: $x = 3, y = 4, z = 5$). This leads to the Pell-Fermat equation

$$x^2 - 2z^2 = \pm 1.$$

It has been treated in the antiquity but is still of interest (In [RUN01] further references can be found. The paper [PAR01], unfortunately, was not accessible in the original version).

In this paper we will refer to further historic articles.

The intimate connection with the right-angled triangles and also with the unit circle given by

$$\xi = \cos\omega = \frac{1-t^2}{1+t^2}\,, \quad \eta = \sin\omega = \frac{2t}{1+t^2}\,, \quad t = \frac{m}{n}$$

is well-known.

The first important progress, namely that $\frac{\omega}{\pi}$ is irrational, was made by Scherrer and Hadwiger. I refer to my article [HLA01]. This article contains new results which were presented in a lecture on Dec 9, 1977.

[1] F. Vieta, "Genesis triangolarum".

§1

Consider the Gaussian plain of numbers $\alpha = A+Bi$, where $i = \sqrt{-1}$ and A, B are integers. Here the norm $N(\alpha) = A^2 + B^2$ *satisfies* $N(\alpha) \geq 1$. The pair (A, B) is considered to be the point with the integer coordinates A and B. The set of these points forms a lattice with the unit sqare forming its fundamental area. This lattice is invariant under transformations T of the type

$$\begin{aligned} x' &= Ex + Fy \\ y' &= Gx + Hy, \end{aligned}$$

where E, F, G, H are integers with determinant 1. The corresponding inverse transformation is given by

$$\begin{aligned} x &= Hx' - Fy' \\ y &= -Gx' + Ey'. \end{aligned}$$

Together with α we consider its conjugate $\bar{\alpha} = A - iB$ and define

$$p(\alpha, \varepsilon) = \frac{\alpha}{\bar{\alpha}}\varepsilon = \frac{A+iB}{A-iB}\varepsilon, \tag{1}$$

where ε is one of the four different powers of i: $1, i, i^2 = -1, i^3 = -i$.

The set of all $p(\alpha, \varepsilon)$ forms a group with respect to multiplication with unit $E = p(1, 1)$ and inverse $p(\alpha, \varepsilon)^{-1} = p(\bar{\alpha}, \bar{\varepsilon})$. We call this group the pythagorian group P. All its members have norm 1. We call all $p(\alpha, \varepsilon)$ with $p(\alpha, 1)$ associated. If we choose one of them we write $p(\alpha)$. Note that among these numbers

$$\alpha = A + iB, \quad \alpha i = -B + Ai, \quad \alpha i^2 = -(A + Bi), \quad \alpha i^3 = B - Ai \tag{2}$$

resp. among the numbers

$$\bar{\alpha} = A - iB, \quad \bar{\alpha} i = B + Ai, \quad \bar{\alpha} i^2 = -A + iB, \quad \bar{\alpha} i^3 = -B - Ai$$

exactly one has positive real and imaginary part which we call the principal number of these four numbers. Here we may assume $A \geq B$, otherwise exchange A and B.

Multiplication of numerator and denominator in the fraction (1) with $\bar{\alpha}$, yields

$$p(\alpha) = X(\alpha) + iY(\alpha), \tag{3}$$

with

$$X(\alpha) = \frac{A^2 - B^2}{A^2 + B^2}, \quad Y(\alpha) = \frac{2AB}{A^2 + B^2}. \tag{4}$$

Putting

$$x = X(\alpha), \quad y = Y(\alpha), \quad z = p(\alpha) = x + iy, \tag{5}$$

we get

$$z\bar{z} = x^2 + y^2 = 1, \tag{6}$$

i.e. (x, y) is on the unit circle and the triple

$$(A^2 - B^2, 2AB, A^2 + B^2) = (a, b, c) \tag{7}$$

fulfills the diophantine equation

$$a^2 + b^2 = c^2. \tag{8}$$

We have

$$z = \frac{z+1}{\bar{z}+1}, \tag{9}$$

if $\bar{z} + 1 \neq 0$, hence $x + 1 \neq 0$.

Putting

$$t = \frac{y}{x+1} = \frac{B}{A}, \tag{10}$$

we get for $A \neq 0$

$$z = x + iy = \frac{1+it}{1-it} = \frac{(1+it)^2}{1+t^2}, \tag{11}$$

hence

$$x = \frac{1-t^2}{1+t^2}, \quad y = \frac{2t}{1+t^2}, \tag{12}$$

the well known parameter representation of the unit circle, where the point $(-1, 0)$ is deleted.

The parameter representation in (4) is given in the homogeneous parameters (A, B) containing the point $(-1, 0)$ corresponding to $A = 0$, $B = 1$.

In (7) we suppose all multiples $(\lambda a, \lambda b, \lambda c)$, $\lambda \neq 0$ integer, to be contained.

In the author's paper [HLA01] we have gone the converse direction (here we refer to it briefly as "Pythagoräische Tripel I" or "PT I", while for the paper in front of you we write "PT II"). PT II can be read independently. Some results from PT I will be quoted and proofs will be sketched.

We want to describe the group P explicitly: If

$$\alpha_1 = A_1 + B_1 i, \quad \alpha_2 = A_2 + B_2 i,$$

then

$$\alpha_3 = \alpha_1\alpha_2 = A_3 + iB_3, \tag{13}$$

$$A_3 = A_1A_2 - B_1B_2, \quad B_3 = A_2B_1 + A_1B_2 \tag{14}$$

$$P(\alpha_3) = P(\alpha_1)P(\alpha_2).$$

Hence the triple (7) is given by

$$(a_3, b_3, c_3)$$

with

$$a_3 = A_3^2 - B_3^2, \quad b_3 = 2A_3B_3, \quad c_3 = A_3^2 + B_3^2 = (A_1^2 + B_1^2)(A_2^2 + B_2^2).$$

If $\alpha = A + Bi$, then (L integer)

$$\alpha^L = A_L + B_L i.$$

Let first L denote natural number, then

$$A_L + iB_L = (A + iB)^L = \sum_{k=0}^{L} \binom{L}{k} A^{L-k}(iB)^k. \tag{15}$$

We distinguish the case $k = 2r$ even (to get A_L) and the case $k = 2r + 1$ odd (to get B_L):

$$\begin{aligned} A_L &= \sum_r \binom{L}{2r}(-1)^r A^{L-2r} B^{2r} \\ B_L &= \sum_r \binom{L}{2r+1}(-1)^r A^{L-2r-1} B^{2r+1} \\ C_L &= (A^2 + B^2)^L, \end{aligned} \tag{16}$$

Here r takes all nonnegative integers with $r \leq \frac{1}{2}L$.

For negative $L = -m$, let $A_L = A_m$, $B_L = -B_m$, hence

$$\alpha^L = A_L + iB_L = A_m - iB_m = \bar{\alpha}^m,$$

In P we introduce a further operation,[2] which we denote by $\oplus$,

$$p(\alpha_0) = p(\alpha_1) \oplus p(\alpha_2).$$

For $\alpha_j = A_j + iB_j$, $j = 1, 2$, define

$$A_0 = A_1A_2, \quad B_0 = B_1B_2.$$

For all $j = 0, 1, 2$ we put

$$N_j = A_j^2 + B_j^2, \quad D_j = A_j^2 - B_j^2, \quad L_j = 2A_jB_j.$$

We get

$$p(\alpha_j) = X_j + iY_j = \frac{D_j + iL_j}{N_j}.$$

and $N_0 = |\alpha_0|^2 = A_0^2 + B_0^2$. An easy computation shows

$$N_0 = \frac{1}{2}(N_1N_2 + D_1D_2). \tag{$\circ$}$$

---

[2] which seems not to occur in the literature explicitly.

Furthermore we have

$$\alpha_0^2 = A_0^2 - B_0^2 + 2iA_0B_0 = D_0 + iL_0.$$

and

$$D_0 = \frac{1}{2}(D_1N_2 + D_2N_1), \quad L_0 = \frac{1}{2}L_1L_2, \tag{$\circ\circ$}$$

hence

$$p(\alpha_0) = \frac{\alpha_0}{\bar{\alpha}_0} = \frac{\alpha_0^2}{|\alpha_0|^2} = X_0 + iY_0.$$

Since

$$X_0 = \frac{D_0}{N_0}, \quad Y_0 = \frac{L_0}{N_0},$$

($\circ$) and ($\circ\circ$) yield

$$X_0 = \frac{x_1 + x_2}{1 + x_1x_2}, \quad Y_0 = \frac{y_1y_2}{1 + x_1x_2},$$

and therefore

$$\mathrm{Re}(p(\alpha_1) \oplus p(\alpha_2)) = \frac{\mathrm{Re}(p(\alpha_1) + p(\alpha_2))}{1 + \mathrm{Re}(p(\alpha_1))\,\mathrm{Re}(p(\alpha_2))}, \tag{17}$$

$$\mathrm{Im}(p(\alpha_1) \oplus p(\alpha_2)) = \frac{\mathrm{Im}(p(\alpha_1))\,\mathrm{Im}(p(\alpha_2))}{1 + \mathrm{Re}(p(\alpha_1))\,\mathrm{Re}(p(\alpha_2))}. \tag{18}$$

We want to bring out a consequence:

$$X_i^2 + Y_i^2 = 1,$$

holds for all i, since

$$|Y_i| = \sqrt{1 - X_i^2}\,.$$

Hence (18′) gets

$$\sqrt{1 - X_0^2} = \frac{\sqrt{1 - X_1^2}\sqrt{1 - X_2^2}}{1 + X_1X_2}, \tag{18′}$$

yielding

$$X_0^2 = 1 - (1 - X_0^2) = 1 - \frac{(1 - X_1^2)(1 - X_2^2)}{(1 + X_1X_2)^2}. \tag{18″}$$

We want to write (18′) in a different form: Define

$$\gamma(X_i) = \frac{1}{\sqrt{1 - x_i^2}}, \tag{19}$$

then (18″) can be written as

$$\gamma(X_0) = \gamma(X_1)\gamma(X_2)(1 + X_1X_2). \tag{20}$$

Now we involve geometric aspects. There is a unique angle φ_0 with $0 \leq \varphi_0 < 2\pi$, such that for all φ of the type $\varphi_0 + 2\pi k$ (k takes all integer values, we write $\varphi \equiv \varphi_0 \pmod{2\pi}$)

$$p(\alpha) = e^{i\varphi} = \cos\varphi + i\sin\varphi,$$

hence

$$\cos\varphi = \frac{A^2 - B^2}{A^2 + B^2}, \quad \sin\varphi = \frac{2AB}{A^2 + B^2}.$$

More explicitly we write

$$\varphi = \operatorname{arc} p(\alpha).$$

Note

$$\operatorname{arc}(p(\alpha_1\alpha_2)) = \operatorname{arc} p(\alpha_1) + \operatorname{arc} p(\alpha_2) \pmod{2\pi} \tag{21}$$

$$\operatorname{arc}(p(\alpha)) = L \operatorname{arc} p(\alpha) \pmod{2\pi}. \tag{22}$$

Again we have

$$X = \frac{A^2 - B^2}{A^2 + B^2}, \qquad X = \frac{2AB}{A^2 + B^2}.$$

Considering the determinants

$$T_1 = \begin{vmatrix} X & 1 \\ Y & i \end{vmatrix} \qquad T_2 = \begin{vmatrix} X & 1 \\ Y & -i \end{vmatrix}$$

$$T_3 = \begin{vmatrix} 1 & 1 \\ 0 & i \end{vmatrix} \qquad T_4 = \begin{vmatrix} 1 & 1 \\ 0 & -i \end{vmatrix}$$

and their cross ratio we get

$$CR(T_1, T_3; T_2, T_4) = \frac{T_1}{T_3} : \frac{T_2}{T_4} = \left(\frac{A + iB}{A - iB}\right)^2,$$

hence (due to Laguerre)

$$\varphi_0 = \frac{1}{2i} \log CR(T_1, T_3; T_2, T_4).$$

This interpretation is often useful for applications of pythagorean triples.

We put

$$\varphi_0 = \pi\chi.$$

The Swiss Mathematicians Scherrer and Hadwiger have proved:

If $\alpha = A + iB$, $AB \neq 0$ and $A^2 \neq B^2$, then χ is irrational.[3]

---

[3] [SCH01], [HAD01].

Proof: [4] Assume, by contradiction, that $\chi = \frac{m}{L}$ is rational. Then

$$\frac{\alpha}{\bar{\alpha}} = e^{\frac{i\pi m}{L}}$$

and

$$\left(\frac{\alpha}{\bar{\alpha}}\right)^L = e^{i\pi m} = \pm 1.$$

Since

$$\left(\frac{\alpha}{\bar{\alpha}}\right)^L = \frac{A_L + iB_L}{A_L - iB_L},$$

we get

$$A_L + iB_L = \pm(A_L - iB_L). \tag{$*$}$$

We may assume that A_L and B_L have no common prime divisors. Let us make the assumption that

$$L \equiv 1 \pmod 2, \quad A \equiv 1 \pmod 2, \quad B \equiv 0 \pmod 2. \tag{$**$}$$

Consider first the case of negative sign in ($*$), thus

$$A_L = 0.$$

Considering (16) modulo 2 leads to a contradiction, since with the exception of A (which is odd by assumption ($**$)) all terms have the even divisor B.

Suppose now that L is even:
Let 2^ρ the maximal power of 2 in L, i.e.

$$L = 2^\rho L_1,$$

L_1 odd. We put $\beta = \alpha^{2^\rho}$, then

$$\left(\frac{\alpha}{\bar{\alpha}}\right)^L = \left(\frac{\beta}{\bar{\beta}}\right)^{L_1}.$$

Let

$$e^{i\pi\chi_1} = \frac{\beta}{\bar{\beta}}.$$

Maintaining all other assumptions from ($**$), we see from the preceeding argument that χ_1 and, since $\chi_1 = 2^\rho\chi$, χ are irrational. Now cancel condition ($**$)!

Assume that A and B have no common prime divisors. It suffices to treat the case

$$A \equiv B \equiv 1 \pmod 2. \tag{$***$}$$

Take α^2 instead of α. The corresponding triple is $(a_2, b_2, c_2)a_2 = A^2 - B^2$, $b_2 = 2AB$, $c_2 = (A^2 + B^2)^2$. ($***$) implies

$$A_2 \equiv 0 \pmod 4, \quad B_2 \equiv 0 \pmod 2,$$

[4]The proof presented here follows the book [MES01].

thus

$$A_3 = \frac{1}{2}A_2 \equiv 0 \pmod 2, \quad B_3 = \frac{1}{2}B_2 \equiv AB \equiv 1 \pmod 2.$$

Thus the triple (B_3, A_3, C_3) fulfills our conditions and the proof is complete. □

Consider the sequence $\omega = (2k\chi)$. Since 2χ is irrational, a result of the theory of uniform distribution (confer the textbooks [HLA02] and [KUI01]) tells us that the sequence is uniformly distributed modulo 1.

Now we compute the discrepancy D_N of the sequence. By Erdös-Turan's inequality we have

$$D_N \leq C\left(\frac{1}{M} + \sum_{|h|=1}^{M} \frac{1}{|h|}|W_N(h)|\right).$$

Here $W_N(h)$ is the Weyl sum

$$W_N(h) = \frac{1}{N}\sum_{k=1}^{N} e^{2\pi i h k \chi}.$$

We recall the easy computation. Note

$$|NW_N(h)| \leq \frac{1}{|1 - e^{2\pi i h \chi}|} = \frac{1}{|\sin^2 \pi h \chi|}$$

and

$$\sin \pi h \chi = \frac{B_h}{C^h},$$

with $C = A^2 + B^2$ ist.

B_h is an integer and, since χ is irrational, we have $B_h \neq 0$. Thus we get

$$|\sin \pi h \chi| \geq \frac{1}{C^h}$$

and we have shown that

$$|W_N(h)| \leq C^h.$$

The choice $M = [\frac{\log N}{\log C}] + 1$ leads to

$$D_N \leq \frac{20 \log C}{\log N}. \tag{23}$$

We will use this formula several times.

If f is integrable in the Riemann sense and with a period 1, then

$$\frac{1}{N}\sum_{k=1}^{N} f(k\chi) = \int_0^1 f(x)dx + \vartheta\, \sigma(D_N^{\frac{1}{3}}, f),$$

where $|\vartheta| \leq 1$ and $\sigma(\varepsilon, f)$ denotes the integrability module ε of f. If f has bounded variation $V(f)$, then we can give the more concrete estimation

$$\lambda_N(f) = \frac{1}{N}\sum_{k=1}^{N} f(kx) = \int_0^1 f(x)dx + \vartheta V(f)D_N.$$

If f has the form $G(\cos 2\pi x, \sin 2\pi x)$ (G integrable in E^2), then, provided that $G(\cos x, \sin x)$ has bounded variation $V(G)$, we get

$$\lambda_N(G) = \int_0^1 G(\cos 2\pi x, \sin 2\pi x)dx + \vartheta V(f)D_N,$$

where

$$\lambda_N(G) = \frac{1}{N}\sum G(a_{2L}, b_{2L})$$

with

$$a_{2L} = \frac{A_{2L}^2 - B_{2L}^L}{A_{2L}^2 + B_{2L}^L}, \quad b_{2L} = \frac{2A_{2L}B_{2L}}{A_{2L}^2 + B_{2L}^L}$$

and

$$A_{2L}^2 + B_{2L}^2 = (A^2 + B^2)^{2L}.$$

For differentiable G we have

$$V(G) \leq \operatorname*{Max}_{E^2} \sqrt{\left(\frac{\partial G}{\partial x}\right)^2 + \left(\frac{\partial G}{\partial y}\right)^2}.$$

§2

It is useful to transfer the uniformly distributed sequences considered above to the higher dimensional case. Consider a prime of the type $p = 4k + 1 = A^2 + B^2$ where the representation as a sum of two integer squares is unique up to the sign and order of A and B. Using complex numbers we have the representation

$$p = (A + iB)(A - iB) = \pi(p) \cdot \bar{\pi}(p),$$

where $\pi(p), \bar{\pi}(p)$ are different primes in the number field $Z(i)$. We have

$$\frac{\pi(p)}{\bar{\pi}(p)} = e^{i\pi\chi(p)}, \tag{1}$$

where χ is irrational and the sequence $(k\chi)$ is uniformly distributed mod 1.

If $p_1, p_2, \ldots, p_s$ are pairwise different primes of the type $4k + 1$ with corresponding $(\pi_1, \bar{\pi}_1), \ldots, (\pi_s, \bar{\pi}_s)$ and angles $\chi_1, \ldots, \chi_s$, then these angles are linearly independent in the sense of uniform distribution. The reason is that the prime decomposition in $Z(i)$ is unique. A consequence is that the sequence $(k\chi_1, \ldots, k\chi_s)$ is uniformly distributed mod 1 in E^s. In the author's paper ([HLA03]) the proof of the estimation

$$D_N^s(p_1, \ldots, p_s) \leq 4^s C_s (\log P_s) \frac{(\log \log N)^s}{\log N} \tag{2}$$

for $P_s = p_1 \cdot \ldots \cdot p_s$ can be found.

In

$$p_j = (A_j + iB_j)(A_j - iB_j) = A_j^2 + B_j^2 \tag{3}$$

the numbers A_j and B_j are given by (this representation is due to Jacobsthal)

$$A_j = -\frac{1}{2} \sum_{k=1}^{p_j} \left(\frac{k}{p_j}\right) \left(\frac{k^2 + r_j}{p_j}\right) \tag{4}$$

$$B_j = -\frac{1}{2} \sum_{k=1}^{p_j} \left(\frac{k}{p_j}\right) \left(\frac{k^2 + u_j}{p_j}\right), \tag{5}$$

where the r_j are the quadratic residues modulo p_j, and u_j the quadratic non-residues modulo p_j. For r_j one may independently take the residue $= -1$. For the non-residues one has to look among the numbers $1, \ldots, 5\sqrt{p_j}$ (confer [VIN01]).

There are many results on the A_j and B_j, even recent investigations. For instance, the case that also A_j is a prime is treated in [FOU01].

As an example we give

$$p_1 = 5,\ \pi_1 = 1 + 2\sqrt{-1},\ p_2 = 13,\ \pi_2 = 3 + 2\sqrt{-1},\ p_3 = 17,\ \pi_3 = 4\sqrt{-1}.$$

§3 Applications

Applications to formulas of relativity theory. Consider the famous energy formula

$$E = \frac{m_0}{\sqrt{1 - v^2}}.$$

Expanding of $(1 - v^2)^{-k}$ in terms of powers of v^2 and taking only two terms one gets

$$(1 - v^2)^{-\frac{1}{2}} \sim 1 + \frac{1}{2} v^2,$$

and

$$E \sim m_0 \left(1 + \frac{1}{2} v^2\right). \tag{6}$$

Taking for v the value

$$v_L = \frac{A_L^2 - B_L^2}{A_L^2 + B_L^2}, \tag{7}$$

The **theorem of additivity** §1 (17) is

$$\left(v_L \oplus v_m = \frac{v_L + v_m}{1 + v_L v_m} \right)$$

we get

$$E_L = \frac{1}{2} m_0 \left(\frac{A_L}{B_L} + \frac{B_L}{A_L} \right) \tag{8}$$

and for the impulse $p = \frac{m_0 v}{\sqrt{1-v^2}}$,

$$p_L = \frac{m_0 v_L}{\sqrt{1 - v^2}} = \frac{1}{2} \left(\frac{A_L}{B_L} - \frac{B_L}{A_L} \right) m_0 . \tag{9}$$

We have

$$E_L^2 - p_L^2 = m_0^2 . \tag{10}$$

It is interesting to insert (2) in (1) to get

$$\hat{E}_L \sim m_0 \left(1 + \frac{1}{2} v_L^2 \right) = m_0 \left(1 + \frac{1}{2} \left(\frac{A_L^2 - B_L^2}{A_L^2 + B_L^2} \right)^2 \right) .$$

The difference

$$E_v - \hat{E}_v$$

shows the error.

Now we consider the change of the mass

$$\Delta m_0 = 2 m_0 \left(\frac{1}{\sqrt{1 - v^2}} - 1 \right)$$

of two particles of the same mass coming together. For $v = v_L$ we get

$$\Delta_L m_0 = m_0 \frac{(|A| - |B|)^2}{|AB|} . \tag{11}$$

Computing for instance

$$\frac{1}{N} \sum_{L=1}^{N} {}^{*} 1$$

under the assumption that v_L lies in the interval $J : 0 < \alpha < \beta < 1$, we get from the theory of uniform distribution

$$\int_\alpha^\beta \frac{dv}{\sqrt{1 - v^2}} + \text{error} = \arcsin \beta - \arcsin \alpha + \text{error}$$

and

$$\frac{1}{N}\sum_{L=1}^{N}{}^{*}(1-v_L^2)^{-1} = \operatorname{arctg}\gamma - \operatorname{arctg}\delta + \text{error},$$

where $\gamma = \arccos\beta,\ \delta = \arccos\alpha$.

The error terms can be estimated using the discrepancy.

Putting $v = v_L$ the Lorentz transformation

$$x' = \frac{x - vt}{\sqrt{1-v^2}}, \qquad t' = \frac{vx + t}{\sqrt{1-v^2}} \tag{12}$$

gets to

$$\begin{aligned} x'_L &= \frac{1}{2}\left(\frac{A_L}{B_L} + \frac{B_L}{A_L}\right)x + \frac{1}{2}\left(\frac{A_L}{B_L} - \frac{B_L}{A_L}\right)t \\ t'_L &= \frac{1}{2}\left(\frac{A_L}{B_L} - \frac{B_L}{A_L}\right)x + \frac{1}{2}\left(\frac{A_L}{B_L} + \frac{B_L}{A_L}\right)t. \end{aligned} \tag{13}$$

Important points

$$x_L = c\frac{1}{2}\left(\frac{A_L}{B_L} + \frac{B_L}{A_L}\right), \quad t_L = c\frac{1}{2}\left(\frac{A_L}{B_L} - \frac{B_L}{A_L}\right)$$

lie on the curve $x^2 - t^2 = c^2$.

In the three dimensional case the Lorentz transformation is given by

$$\begin{aligned} \hat{x}_L &= \mathbf{x} - (xo)\left(1 - \frac{1}{\sqrt{1-v_L^2}}\right) + \frac{v_L t}{\sqrt{1-v_L^2}} \\ \hat{t}_L &= \frac{t - v_L(xo)}{\sqrt{1-v_L^2}}, \end{aligned}$$

where

$$\begin{aligned} \mathbf{x} &= (x_1, x_2, x_3), \quad \hat{x} = (\hat{x}_1, \hat{x}_2, \hat{x}_3) \\ o &= (o_1, o_2, o_3) \quad \text{unit vector} \\ (xo) &= x_1 o_1 + x_2 o_2 + x_3 o_3. \end{aligned}$$

The case $\tilde{v}_L = -\frac{1}{v_L}$ ($|\tilde{v}_L| > 1$) (Dirac electron) is also important. We do not go further into this direction.

The sequence of the pairs $(v_L, \tilde{v}_L)$ für $l = 1, 2, \ldots$ forms a countable model of a 'Dirac sea' (maybe this is an interpretation of a remark of Hilbert 1934 reported by the physicist Sauter).

A little application to the general relativity theory: Using the equivalence principle we get the Schwarzschild radii r_k:

$$r_k = \frac{2GM}{v_k^2} = 2GM\left(\frac{A_k^2 - B_k^2}{A_k^2 + B_k^2}\right)^2 .$$

Application to the triangle: If a right-angled triangle has the legs x, y and the hypotenuse z then, by Pytharoras' theorem, $x^2 + y^2 = z^2$ and the triple $(A^2 - B^2, 2AB, A^2 + B^2)$ satisfies the equation. Frequently z and x are given and y has to be computed.

Starting with the relation $z^2 - y^2 = x^2$ we get the solution

$$z = \frac{1}{2}\left(\frac{A}{B} + \frac{B}{A}\right)x, \quad y = \frac{1}{2}\left(\frac{A}{B} - \frac{B}{A}\right)x.$$

We get triples of a type treated above.

The following idea is due to R. Lauffer:[5]

Consider an arbitrary triangle with vertices E, F, G arranged as in the picture.

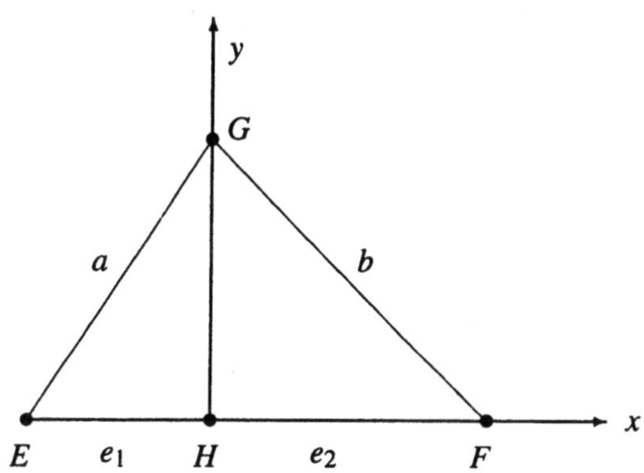

Let the coordinates be given by $E = (-e_1, 0)$, $F = (e_2, 0)$, $G = (0, h)$, $H = (0, 0)$ and the lengths of the edges be a for EG, b for FG. The length of EH ist e_1, that of HF is e_2.

Let p_1, p_2 be primes with the corresponding Gaussian primes $\pi_1 = A_1 + iB_1, \pi_2 = A_2 + iB_2$. For

$$a = \frac{1}{2}\left(\frac{A_1}{B_1} + \frac{B_1}{A_1}\right)h, \quad b = \frac{1}{2}\left(\frac{A_2}{B_2} + \frac{B_2}{A_2}\right)h,$$

we have

$$e_1 = \frac{1}{2}\left(\frac{A_1}{B_1} - \frac{B_1}{A_1}\right)h, \quad e_2 = \frac{1}{2}\left(\frac{A_2}{B_2} - \frac{B_2}{A_2}\right)h.$$

[5] Historical remark: Before Lauffer there were several mathematicians treating related problems: Brahmagupta, Euler, Gauß, Blichfeld and, mainly H. Schubert. We again mention Dickson. Unfortunately, Schubert's original papers were not accessible for me.

The total length EF is

$$\frac{1}{2}\left(\frac{A_1}{B_1}+\frac{A_2}{B_2}+\left(\frac{B_1}{A_1}-\frac{B_2}{A_2}\right)\right)h.$$

We compute the angles α in E, β in F and γ in G we get

$$\cos\alpha=\frac{A_1^2-B_1^2}{A_1^2+B_1^2},\qquad \sin\alpha=\frac{2A_1B_1}{A_1^2+B_1^2}, \tag{14}$$

$$\cos\beta=\frac{A_2^2-B_2^2}{A_2^2+B_2^2},\qquad \sin\beta=\frac{2A_2B_2}{A_2^2+B_2^2}, \tag{15}$$

(the formula for $\sin\alpha$ is due to H. Schubert).

To compute γ we use

$$\cos\gamma=\cos(\pi(-(\alpha+\beta)))=\cos(\alpha+\beta)=\cos\alpha\cos\beta-\sin\alpha\sin\beta$$

and

$$\sin\gamma=\sin(\alpha+\beta)=\sin\alpha\cos\beta+\cos\alpha\sin\beta,$$

which follows from above. Thus the trigonometric functions of the angles are rational numbers. If h is a rational number, then the same holds for the edges.

There are applications of the pythagorean triples to different branches of mathematics:

rotations
3-dimensional pythagorean triples
quadratures
interpolation on the unit circle
rational points on spheres
line geometry
Clifford surfaces
spherical and non-euclidean geometry
infinite series
partial differential equations
'natural geometry'.

These applications are part of the content of a manuscript which is to be published later on.

References

[DIC01] L.E. Dickson, *History of the Theory of Numbers,* Chelsea Publ. Comp. New York 1952, vol. II, Diophantic Analysis, chapter IV, Rational Right Triangles, pp. 165–172.

[FOU01] E. Fouvry and Iwaniec, H., Gaussian Primes, *Acta Arithm.* **79** (1997), 249–287.

[HAD01] H. Hadwiger, Über die rationalen Hauptwinkel der Goniometrie, *El. Math. I* (6) (1946), 98–100.

[HLA01] E. Hlawka, Approximation von Irrationalzahlen und pythagoräische Tripel, *Bonner Math. Schriften* **121** (1980), 1–32 and P.M. Gruber, W.M. Schmidt, Edmund Hlawka, Selecta, Springer, Berlin-Heidelberg 1990, pp. 431–455.

[HLA02] E. Hlawka, Theorie der Gleichverteilung, *Bibliograph. Institut Wissenschaftsverlag*, Mannheim-Wien-Zrich, 1979.

[HLA03] E. Hlawka, Über eine Klasse von gleichverteilten Folgen, *Acta Arithm.* **53** (1990), 389–402.

[KUI01] L. Kuipers u. H. Niederreiter, *Uniform Distribution of Sequences*, J. Wiley & Sons, New York, 1974.

[MES01] H. Meschkowski, *Ungelöste und unlösbare Probleme in der Geometrie*, Vieweg, Braunschweig, 1960.

[PAR01] S.A. Paramhans and Prasad, S., Recurrence formulae for the identities generating Pythagorean triples, *Bull. Number Theory Relat. Top.* **16** (1992), 113–116.

[RUN01] J. Rung, Pythagoreische Dreiecke und lineare Differentialgleichungen, *Praxis der Mathematik* **3/32** (1990), 102–106.

[SCH01] W. Scherrer, Die Einlagerung eines regulären Vielecks in ein Gitter, *El. Math. I* (6) (1946), 97–98.

[VIN01] I.M. Vinogradov, *Elemente der Zahlentheorie*, German edition, Oldenbourg, München, 1956.

TU Wien, Inst. 114
Wiedner Hauptstrasse 8
A-1040 Wien
Austria
E-Mail: elhlawka@osiris.tuwien.ac.at

Integer Points in Plane Regions and Exponential Sums

M.N. Huxley

How many integer points (m, n) lie inside a large circle, or in the annulus between two circles? There are approaches by real-variable approximation theory, or by Fourier analysis. The same ideas occur in both, and the latest results use both methods at different stages of the argument.

Analytic number theory is about counting the number of sets of integers satisfying certain conditions. There are two famous questions.

1. The number of prime numbers in a short interval.
2. The number of integer points inside a circle: (m, n) with

$$m^2 + n^2 \leq R^2.$$

In question 1, prime numbers are defined negatively by ruling out multiples of smaller prime numbers. One asks: how many solutions of

$$pq = N + h,$$

with p a prime number in some range, and h a small integer? This is a very hard problem. An easier version is to change from prime numbers to square-free numbers, defined negatively by not being multiples of the squares of prime numbers. Now the relation is

$$p^2 q = N + h, \quad p^2 q \simeq N.$$

A version with two terms,

$$p_1^2 q_1 - p_2^2 q_2 = h, \quad \frac{p_1}{p_2} \simeq \sqrt{\frac{q_2}{q_1}},$$

leads to an approximate equation involving rational numbers.

The two questions can be generalised to two problems.

Problem A Count the number of integer points inside a closed curve (as an asymptotic formula).

Problem B Count the number of integer points (m, n) close to an arc of a curve $y = f(x)$, so that

$$|n - f(m)| \leq \delta. \tag{1}$$

Upper or lower bounds are still interesting when there is no asymptotic formula.

Variants of these problems have integer points (m, n) replaced by rational points, either projectively (points $(\frac{m}{q}, \frac{n}{q})$ with $q \leq Q$ corresponding to an integer point (m, n, q) in projective space) or generally (points $(\frac{m}{n}, \frac{r}{q})$ with $n \leq N, q \leq Q$).

The tools of analytic number theory are:

1. *Discreteness*. An integer n with $|n| < 1$ is zero.
2. *Cauchy's inequality*. Mean squares are positive.
3. *Fourier analysis*, usually Poisson summation.

Problem A by Fourier analysis leads to exponential sums.

Problem C Estimate the sum

$$\sum_{M}^{M'} e(f(m)), \quad e(x) = \exp 2\pi i x, \quad M' \le 2M. \tag{2}$$

or the two-variable sum

$$\sum_{h=H}^{H'} \sum_{m=M}^{M'} e(hf(m)), \quad H' \le 2H, \quad M' \le 2M. \tag{3}$$

To compare the problems we suppose that $M \le m \le 2M$, with $f(m) = TF(\frac{m}{M})$ in Problems A and C (T has the dimensions of area), and $f(m) = NF(\frac{m}{M})$ in Problem B (N has the dimensions of length, so MN corresponds to T in Problem A). Here $F(x)$ is a bounded function defined on an interval containing [1,2]. Certain derivatives and determinants of derivatives will occur in the estimates. These derivatives are assumed to be bounded in modulus.

Standard case. Any derivative in the denominator is bounded away from zero.

Non-standard case. Some derivative which should be in the denominator becomes very small.

In a non-standard case some quantity is approximately constant. If its value has a good rational approximation a/q with q small, then strange things happen, and the argument may fail. If there is no good rational approximation, then the argument can be modified.

Here are some typical functions $F(x)$.

Problem A $F(x) = \sqrt{1-x^2}$ (Gauss's circle problem), $F(x) = 1/x$ (Dirichlet's divisor problem).

Problem B $F(x) = 1/x$ (prime numbers), $F(x) = 1/\sqrt{x}$ (square-free numbers).

Problem C As in Problems A and B, and also $F(x) = \log x$ (size of the Riemann zeta-function).

The first non-trivial results in each problem typically show a saving of 1/3 on some exponent in the trivial bound, leading to exponents with denominators 3 or 6 in the final result.

Circle Problem (Sierpinski 1906). The number of integer solutions of $x^2 + y^2 \le R^2$ is

$$\pi R^2 + O(R^{2/3}).$$

Divisor problem (integer points inside an hyperbola) (Voronoi 1903). The number of positive integer solutions of $mn \leq T$ is

$$T \log T + (2\gamma - 1)T + O(T^{1/3}(\log T)^c).$$

Size of the zeta function (Hardy and Littlewood 1921).

$$\zeta\left(\frac{1}{2} + it\right) = O(T^{1/6} \log T).$$

Integer points close to curves (Vinogradov and van der Corput independently about 1914–18).

$$R = 2\delta M + O((MN)^{1/3}(\log MN)^c).$$

Rational points close to a curve (Huxley 1994a). With (m, n) in (1) replaced by $(m/q, n/q)$ with highest common factor $(m, n, q) = 1$ and $q \leq Q$, $Mq \leq m \leq 2Mq$, and δ replaced by δ/Q, for $F''(x) \neq 0$

$$R = O\left(\delta M Q^2 \left(\frac{MN}{\delta}\right)^{\epsilon} + (MN)^{1/3} Q\right).$$

General case (Huxley 2000). With (m, n) in (1) replaced by $(m/n, r/q)$ with highest common factors $(m, n) = (r, q) = 1$, $n \leq M$, $q \leq Q$ and δ replaced by δ/Q^2, with $f(x) = TF(x)$, satisfying $F'(x)$, $F''(x)$ and $3F''(x)^2 - 2F'(x)F^{(3)}(x)$ all non-zero,

$$R = O(\delta^{1/4} M^2 + (M^2 Q^2 T(\delta M^2 + 1))^{1/3}(MQ^2T)^{\epsilon}).$$

In these results c is some fixed exponent, ϵ is an exponent which can be taken arbitrarily small. The constant factor suppressed in the $O(.)$ notation is constructed from the range of values of the derivatives of $F(x)$, and from ϵ where present.

Van der Corput (1920 and later papers) developed an iterative procedure for estimating exponential sums.

Step A First differences (or mean square over short intervals). This step keeps M the same, and replaces $F(x)$ by $F'(x)$ (to a first approximation).

Step B Poisson summation. This step is used if $M > \sqrt{T}$ and $F''(x)$ is non-zero. It keeps T the same, reduces M, and replaces $F(x)$ by the complementary function $G(y)$ satisfying $F'(G'(y)) = y$.

The iteration continues until M is small enough for the sum to be estimated trivially, or until the new function that replaces $F(x)$ becomes non-standard. The B step must be followed by an A step, but A steps can be iterated, so the iteration has a branched tree structure. Each A step introduces an extra variable, summed over a short range. A more complicated version of the iteration allows A and B steps with respect to these subsidiary variables, or in several variables at once. In practice the iteration is stopped after a few steps, either because error terms from a B step in one variable become large when summed over the other variables, or because large parts of the transformed sums are non-standard. Van der Corput's iteration can now be easily studied in Graham and Kolesnik (1991).

For exponential sums with $(\log M)/(\log T)$ very small, Vinogradov's mean value method using high moments gives a better estimate than repeated A steps. Vinogradov's mean value method is a single step that does not lead on to further iteration. In some ways it is analogous to taking r-th differences.

Huxley (1989) discovered an analogous iteration that gives upper bounds for Problem B, the number of integer points close to a curve.

Step A Differencing using the interpolation polynomial over short intervals. This step keeps M the same, reduces N, increases δ, and replaces $F(x)$ by $F^{(r)}(x)$ (to a first approximation), where r is the order of differencing.

Step B Interchange x and y. This step is used if $M > N$ and $F'(x)$ is non-zero. It interchanges M and N, reducing M and increasing δ. The function $F(x)$ is replaced by its inverse function.

The iteration ends with a trivial estimate, which is $O(M)$ if $\delta \leq 1/2$, $O(\delta M)$ if the previous step has made $\delta > 1/2$. The use of r-th differences makes this iteration more powerful than the van der Corput iteration for exponential sums. Recent work (Filaseta and Trifonov 1996, Huxley and Sargos 1995) uses essentially an A step followed by a B step. The B step is elaborated by focussing on short arcs of the curve. Sargos introduced the classification into major and minor arcs. A major arc is a region where there is a good rational approximation $y = g(x)$ (with small denominator) to the equation of the curve, such that any integer point (m, n) close to this arc of the curve must satisfy $n = g(m)$. Other regions of the curve are called minor arcs. Huxley and Sargos (1995) gave a simple upper bound in Problem B.

$$R = O\left(\Delta^{\frac{2}{r(r+1)}} M + \delta^{\frac{2}{r(r-1)}} M + \left(\frac{\delta}{\Delta}\right)^{\frac{1}{r}} + 1\right), \tag{4}$$

provided that $F^{(r)}(x)$ is non-zero and

$$\Delta = T/M^r < 1.$$

The first two terms on the right of (4) are the estimate from the minor arcs. The third term on the right of (4) is the possible contribution of a single major arc, and the term 1 covers trivial cases. An object of current research (Filaseta and Trifonov 1996, Huxley and Sargos 2000) is to reduce the first term in (4) under further conditions, such as the non-vanishing of $F^{(r-1)}(x)$ as well.

Lower bounds can be given in Problem B if δ is not too small (Huxley 1996b). The number of integer points found in this way is less than the expected number $2\delta M$, and they all lie on major arcs. When $(\log M)/(\log N)$ is large, the rational approximations $y = g(x)$ on the major arcs are constructed by an iterative process involving differencing. At present the denominator of approximation has to be a power of two for the iteration to proceed for more than two steps. This restriction permits an approximation argument in the 2-adic metric, and avoids looking in a short interval for a solution of a congruence to an arbitrary modulus q. The estimate that starts the iteration is

$$R \geq c \min\left(\delta M, \frac{\delta^4 M^3}{N}\right) \quad \text{when} \quad \delta \geq \frac{d}{\sqrt{M}}, \tag{5}$$

when $N \geq M$ and $F''(x)$ is non-zero and numerically less than 1. Here c and d are explicit positive constants constructed from the range of values of $F''(x)$. The iteration leads to similar but weaker lower bounds when the second derivative $F''(x)$ is replaced by $F^{(r)}(x)$ in the condition.

The van der Corput iteration for exponential sums shows that, for large δ, R is not only non-zero, but approximately equal to $2\delta M$. The lower bounds are still positive for δ a little smaller than this range.

For both problems B and C there is a deeper method in the middle range, where $(\log M)/(\log N)$ is near 1 for points close to a curve, or $(\log M)/(\log T)$ is near $1/2$ for exponential sums. For points close to a curve, Swinnerton Dyer's method uses a fourth moment short interval estimate on the minor arcs (Swinnerton-Dyer (1974) considered points on the curve, with $M = N$ and $\delta = 0$, so he had no major arcs). The bounds take the form

$$R = O((MN)^{3/10}(\log\ MN)^{1/10} + \text{terms in } \delta) \tag{6}$$

for $F''(x)$ and $F^{(3)}(x)$ non-zero, and $N \geq M$. For the latest version, see Huxley (1999). Similar arguments with higher moments come up against determinants which do not factor completely into linear factors.

For exponential sums the deeper method is that of Bombieri and Iwaniec (1986), which uses high moments of short exponential sums. Each short sum is labelled by a rational approximation a/q to $\frac{1}{2}f''(x)$, and is transformed by Poisson summation (incorporating the finite Fourier transform modulo q). On major arcs q is small, and the transformed sum can be estimated at once. The transformed minor arc sums are estimated in mean 2k-th power ($k = 4$ in Bombieri and Iwaniec (1986) and Huxley and Watt (1988), $k = 5$ in Watt (1989) and subsequent papers). After some simplification, the large sieve inequality is used to bound a bilinear form

$$\sum_h \sum_j \alpha_h \beta_j e(\mathbf{x}^{(h)} \cdot \mathbf{y}^{(j)}),$$

where h is the integer variable introduced by Poisson summation, and j indexes the rationals a_j/q_j that label the short sums on the minor arcs. The vectors $\mathbf{x}^{(h)}$ lie in a box **A** in four-dimensional space, and the vectors $\mathbf{y}^{(j)}$ lie in a box **B**. The upper bound for the bilinear form requires an estimate for the sum

$$\sum_h \sum_i |\alpha_h \alpha_i|$$

taken over a neighbourhood of the diagonal in $\mathbf{A} \times \mathbf{A}$. The size of the neighbourhood is determined by the size of the box **B**. It is sufficient to count the number of pairs of vectors

$\mathbf{x}^{(h)}, \mathbf{x}^{(i)}$ in the given neighbourhood. This is the First Spacing Problem; the Second Spacing Problem is the analogue for the vectors $\mathbf{y}^{(j)}$.

The First Spacing Problem involves counting the number of sets of integers $h_1, \ldots, h_{2k}$ with

$$h_1^i + \cdots + h_k^i \simeq h_{k+1}^i + \cdots + h_{2k}^i$$

simultaneously for $i = 2, 3/2, 1, 1/2$. Vinogradov's mean value method produces conditions like this, but with integer exponents and exact equality. Progress with the First Spacing Problem was made in Watt (1989) and in Huxley and Kolesnik (1991).

In the Second Spacing Problem, a pair of vectors $\mathbf{y}^{(j)}$, $\mathbf{y}^{(k)}$ in a neighbourhood of the diagonal in $\mathbf{B} \times \mathbf{B}$ corresponds to a relation between two minor arcs which we call resonance. The four coordinates give four coincidence conditions, whose strengths are measured by real numbers $\Delta_1, \Delta_2, \Delta_3, \Delta_4$, each less than unity. The first coincidence condition implies the existence of the 'magic matrix' M, which has small integer entries and determinant one, with

$$\begin{pmatrix} a_k \\ q_k \end{pmatrix} = M \begin{pmatrix} a_j \\ q_j \end{pmatrix}.$$

The factor Δ_1 has to be saved in the construction in order to get even a trivial estimate.

When the magic matrix is fixed, then the second coincidence condition holds on a block of consecutive minor arcs, the domain of the magic matrix. The factor Δ_2 saved leads to the non-trivial results of Bombieri and Iwaniec (1986), Huxley and Watt (1988), Watt (1989) and Huxley and Kolesnik (1991).

The next step is to assume the first and second coincidence conditions (the first, alas, with borrowed strength which must be repaid). Among the minor arcs in the domain of the magic matrix, those on which the third and fourth coincidence conditions also hold, they give rise to integer points close to a certain curve, the resonance curve. More precisely, a resonance curve corresponds to a block of U consecutive minor arcs, with $U = O(1/\Delta_4)$. The length of the resonance curve grows like $U^{3/2}$. The result of Huxley (1993) was obtained by choosing U so that the resonance curve had bounded length, and thus a bounded number of integer points close to it. This gives a saving of a factor $\Delta_4^{2/3}$ from the third and fourth coincidence conditions together.

Work in progress (Huxley 2001a, 2001b) uses a bound of the form (6). There is a possible saving by a factor $\Delta_4^{7/10} < \Delta_4^{2/3}$. However the typical resonance curve has a cusp, with the gradient on one branch tending to infinity, and on the other branch tending to zero. At the cusp and the ends of the resonance curve, the conditions for (6) do not hold, and a weaker argument must be found.

Further improvements would come if different resonance curves could be compared, even by showing that most resonance curves do not have an integer point very close to the cusp.

The usual test of exponential sum bounds is the size of the Riemann zeta function

$$\zeta\left(\tfrac{1}{2} + iT\right) = O(T^{\theta}(\log T)^c).$$

Hardy and Littlewood (1921) had $\theta = 1/6$. Rankin (1955) and Graham and Kolesnik (1991) calculated the limit of van der Corput's iteration in one variable as $\theta = 0.164\ldots$. Using the subsidiary variables in van der Corput's iteration gives exponents which are smaller, but which have $\theta > 0.1618\ldots$ (Graham and Kolesnik 1991). Bombieri and Iwaniec (1986) obtained $\theta = 0.1607\ldots$, and the latest form of their method (Huxley 1993a) has $\theta = 0.156140\ldots$. Huxley (2001a) uses (6) to reach $\theta = 0.156098\ldots$. The best possible estimates in the First and Second Spacing Problems in the Bombieri-Iwaniec method would reach $\theta = 3/20$.

The Bombieri-Iwaniec method works best when the ratio $(\log M)/(\log T)$ is close to $1/2$. Sargos (1995) has a modified form of the method which works well near $2/5$. Kolesnik has a different construction of resonance curves suited to the case when all magic matrices are upper triangular, which happens when the ratio is below $3/7$. Huxley and Kolesnik made extensive calculations of an iteration based on the Kolesnik resonance curve. Their results are summarised in chapter 19.3 of Huxley (1996a).

There seems to be no analogue of the Bombieri-Iwaniec method to exponential sums in two variables involving a function $F(x, y)$, because simultaneous Diophantine approximations to three second order partial derivatives with the same denominator cannot be as accurate as an approximation to one second derivative for a function of one variable. The form of the large sieve inequality allows a small saving where the second variable is used as a parameter that modifies the sum. A similar argument allows a coefficient $a(m)$ to be inserted in the sums (2), where $a(m)$ has some integer period $q < M$, and it can be used to extend the estimates from the zeta function to the Dirichlet L-functions (Huxley and Watt 1997).

The best treatment for exponential sums with $(\log M)/(\log T)$ not near $1/2$ is to apply some A steps of van der Corput's iteration (or a B step followed by some A steps) until the new parameters M and T are in the Bombieri-Iwaniec range. The extra variables from the A steps give parameters that modify the main sum, so there is a small extra saving.

The Bombieri-Iwaniec method does work if there is one function, but two integer variables occurring as $hf(m)$ as in (3) (Iwaniec and Mozzochi 1989) or as $f(m+h)-f(m-h)$ (Heath-Brown and Huxley 1990). The First Spacing Problem is different in these constructions, but the Second Spacing Problem is almost the same; the main difference is that $f(x)$ in (3) corresponds to $f'(x)$ in (2). This leads to an estimate for the generalised circle problem. Let C be a closed convex plane curve of area A for which the arc length s is three times continuously differentiable with respect to the tangent angle ψ (the circle is the case $s = \psi$, $A = \pi$). Let D be a plane domain bounded by a copy of C enlarged by a factor $M \geq 2$. Then the number of integer points (m, n) in D is

$$AM^2 + O(M^\theta (\log\ M)^c)$$

with $\theta = 46/73 = 0.6301\ldots$; the constant of proportionality depends on C but not on the orientation of the domain D (Huxley 1993b). In the Dirichlet divisor problem, the number of positive integer solutions of $mn \leq M^2$ is

$$2M^2 \log\ M + (2\gamma - 1)M^2 + O(M^\theta (\log\ M)^c)$$

with the same θ. The possible improvement in the Second Spacing Problem would give $\theta = 0.6298\ldots$. The analogous sums with $f(m+h) - f(m-h)$ in place of $hf(m)$ in (3) lead to short interval mean square bounds for exponential sums (Heath-Brown and Huxley 1990, Huxley 1994b).

For exponential sums in one variable, the form of the underlying function $F(x)$ does not matter very much. In two variables the non-standard regions of the double sum depend on the shape of $F(x, y)$, for instance on whether the second order terms are elliptic or hyperbolic. The functions $F(x)$ which arise from the famous problems satisfy algebraic differential equations. For problem B, points close to a curve, it is sometimes possible to use this. The relation

$$\frac{p_1}{p_2} \simeq \sqrt{\frac{q_2}{q_1}}$$

arises in the study of square-free numbers in a short interval. The curve $F(x) = 1/\sqrt{x}$ has a major arc around the point (1,1) with an approximation $y = g(x)$, an explicit Pade approximant rational function. This idea was introduced by Roth (1951). Filaseta and Trifonov combined this with a differencing argument to discuss the gaps between square-free numbers. Let $s_1, \ldots, s_N$ be the square-free numbers in the range $1, \ldots, M$, so that

$$N \simeq \frac{6M}{\pi^2}.$$

Then the maximum gap satisfies

$$s_{n+1} - s_n = O(M^{1/5}(\log M)^c)$$

(Filaseta and Trifonov 1992), and Erdös's asymptotic formula

$$\sum_{1}^{N-1} (s_{n+1} - s_n)^\gamma \simeq \beta(\gamma)M,$$

where $\beta(\gamma)$ is a constant determined by the exponent γ, holds for $0 \le \gamma < 59/16 = 3.6875$ (Huxley 1999). These arguments use the upper bounds in Problem B.

For the case $\delta = 0$ of Problem B, points actually on the curve, Bombieri and Pila (1989) have upper bounds using intersection theory in algebraic geometry, which are essentially best possible for algebraic curves. In general the constant of proportionality in the upper bound is shown to exist by a compactness argument, and it cannot be calculated explicitly or uniformly, which rules out some applications. For results of the quality of (6), the curve must be differentiable a large finite number of times. It seems to be possible to extend Bombieri and Pila's argument to non-zero δ, but the upper bound increases rapidly as δ moves away from zero.

The lower bounds in Huxley (1996b) imply but do not improve the famous result of Bambah and Chowla (1947) that the gap between numbers less than N which are sums of two squares is $O(N^{1/4})$.

References

1. R.P. Bambah, and Chowla, S., "On numbers which can be expressed as a sum of two squares", *Proc. Nat. Acad. Sci. India* **13** (1947), 101–103.
2. E. Bombieri, and Iwaniec, H., "On the order of $\zeta(1/2 + it)$", *Ann. Scuola Norm. Sup. Pisa Cl. Sci.* (4) **13** (1986), 449–472.

3. E. Bombieri, and Pila, J., "The number of integral points on arcs and ovals", *Duke Math. J.* **59** (1989), 337–357.
4. J.G. van der Corput, "Über Gitterpunkte in der Ebene", *Math. Annalen* **81** (1920), 1–20.
5. J.G. van der Corput, "Zahlentheoretische Abschätzungen", *Math. Annalen* **84** (1921), 53–79.
6. J.G. van der Corput, "Verschärfung der Abschätzung beim Teilerproblem", *Math. Annalen* **87** (1922), 39–65.
7. J.G. van der Corput, "Neue zahlentheoretische Abschätzungen", *Math. Annalen* **89** (1923), 215–254.
8. M. Filaseta, and Trifonov, O., "On the gaps between squarefree numbers II", *J. London Math. Soc.* (2) **45** (1992), 215–221.
9. M. Filaseta, and Trifonov, O., "The distribution of fractional parts with application to gap results in number theory", *Proc. London Math. Soc.* (3) **73** (1996), 241–78.
10. S.W. Graham, and Kolesnik, G., Van der Corput's Method for Exponential Sums, *London Math. Soc. Lecture Notes* **126** (1991) (University Press, Cambridge).
11. G.H. Hardy, and Littlewood, J.E., "The zeros of Riemann's zeta function on the critical line", *Math. Zeitschrift* **10** (1921), 283–317.
12. D.R. Heath-Brown, and Huxley, M.N., "Exponential sums with a difference", *Proc. London Math. Soc.* (3) **61** (1990), 227–250.
13. M.N. Huxley, "The integer points close to a curve", *Mathematika* **36** (1989), 198–215.
14. M.N. Huxley (1993a), "Exponential sums and the Riemann zeta function IV", *Proc. London Math. Soc.* (3) **66**, 1–40.
15. M.N. Huxley (1993b), "Exponential sums and lattice points II", *Proc. London Math. Soc.*, (3) **66** 279–301, Corrigenda *ibid.* (3) **68**, 264.
16. M.N. Huxley (1994a), "The rational points close to a curve", *Ann. Scuola Norm. Sup. Pisa Cl. Sci. Fis. Mat.* (4) **21** (1994), 357–375.
17. M.N. Huxley (1994b), "A note on exponential sums with a difference", *Bull. London Math. Soc.* **26**, 325–7.
18. M.N. Huxley (1996a), Area, Lattice Points and Exponential Sums, *London Math. Soc. Monographs* **13** (Oxford University Press).
19. M.N. Huxley (1996b), "The integer points close to a curve II", in *Analytic Number Theory, Proceedings of a Conference in Honor of Heini Halberstam* **2**, 487–516 (Birkhäuser, Boston).
20. M.N. Huxley, "The integer points close to a curve III", in *Number Theory in Progress* II (1999), 911–940 (de Gruyter, Berlin).
21. M.N. Huxley, "The rational points close to a curve II", submitted to *Acta Arithmetica* (2000).
22. M.N. Huxley, "Exponential sums and the Riemann zeta function V" (2001a), to appear.
23. M.N. Huxley, "Exponential sums and lattice points III" (2001b), to appear.
24. M.N. Huxley, and Kolesnik, G., "Exponential sums and the Riemann zeta function III", *Proc. London Math. Soc.* (3) **62**, 449–468, Corrigenda *ibid.* (3) **66** (1991), 302.
25. M.N. Huxley, and Sargos, P., "Points entiers au voisinage d'une courbe plane de classe C^n", *Acta Arithmetica* **59** (1995), 359–66.
26. M.N. Huxley, and Sargos, P., "Points entiers au voisinage d'une courbe plane de classe C^n II" (2000), to appear.
27. M.N. Huxley, and Watt, N., "Exponential sums and the Riemann zeta function", *Proc. London Math. Soc.* (3) **57** (1988), 1–24.
28. M.N. Huxley, and Watt, N., "Congruence families of exponential sums", in Analytic Number Theory (ed. Motohashi Y.), *London Math. Soc, Lecture Notes* **247** (1997), 127–138 (Cambridge University Press).
29. A. Ivić, *The Riemann Zeta Function* (Wiley, New York) (1985).

30. H. Iwaniec, and Mozzochi, C.J., "On the divisor and circle problems", *J. Number Theory* **29** (1988), 60–93.
31. R.A. Rankin, "Van der Corput's method and the theory of exponent pairs", *Quarterly J. Math. (Oxford)* **6** (2) (1955), 147–53.
32. K.F. Roth, "On the differences between squarefree numbers", *J. London Math. Soc.* (2) **26** (1951), 263–268.
33. P. Sargos, "Points entiers au voisinage d'une courbe, sommes trigonométriques courtes et paires d'exposants", *Proc. London Math. Soc.* (3) **70** (1995), 285–312.
34. W. Sierpiński, "Sur un problème du calcul des fonctions asymptotiques," *Prace Mat.-Fiz.* **17** (1906), 77–118.
35. H.P.F. Swinnerton-dyer, "The number of lattice points on a convex curve", *J. Number Theory* **6** (1974), 128–135.
36. E.C. Titchmarsh, *The theory of the Riemann zeta function* (University Press, Oxford) (1951).
37. I.M. Vinogradov (translated Popova A.) *An Introduction to the Theory of Numbers* (Interscience, Oxford) (1955).
38. G. Voronoi, "Sur un problème du calcul des fonctions asymptotiques", *J. reine angew. Math.* **126** (1903), 241–282.
39. N. Watt, "Exponential sums and the Riemann zeta function II", *J. London Math. Soc.* **39** (1989), 385–404.

School of Mathematics
Cardiff University
P.O. Box 926
Cardiff CF24 4YH
UK
E-Mail: huxley@cf.ac.uk

Artin's Conjecture for Polynomials Over Finite Fields

*Erik Jensen and M. Ram Murty**

1 Introduction

A classical conjecture of E. Artin[Ar] predicts that any integer $a \neq \pm 1$ or a perfect square is a primitive root (mod p) for infinitely many primes p. This conjecture is still open. In 1967, Hooley[H] proved the conjecture assuming the (as yet) unresolved generalized Riemann hypothesis for Dedekind zeta functions of certain number fields.

In 1983, R. Gupta and M.R. Murty[GM] made the first breakthrough by showing the following: given three prime numbers a, b, c, then at least one of the thirteen numbers

$$\{ac^2, a^3b^2, a^2b, b^3c, b^2c, a^2c^3, ab^3, a^3bc^c, bc^3, a^2b^3c, a^3c, ab^2c^3, abc\}$$

is a primitive root (mod p) for infinitely many primes p. This result has been further refined by R. Gupta, M.R. Murty, and V.K. Murty[GMM] to establish that at least one of the seven numbers

$$\{a, b, c, a^2b, ab^2, a^2c, ac^2\}$$

is a primitive root (mod p) for infinitely many primes p. Finally, Heath-Brown[HB] used the Chen-Iwaniec switching and a celebrated 1986 theorem of Bombieri, Friedlander, and Iwaniec[BFI] to derive the further refinement that at least one of $\{a, b, c\}$ is a primitive root for infinitely many primes p. The paper by Murty[M1] contains an overview of Artin's conjecture and its analogues for elliptic curves. (See also [M2]).

Although any undergraduate student can easily understand Artin's conjecture, to understand Hooley's results requires a strong background in both analytic and algebraic number theory, and thus is only possible at the senior or graduate level. The results of Gupta, M.R. Murty, V.K. Murty, and Heath-Brown require an even more formidable background in advanced sieve theory that is available only to the doctoral student or to the expert working in the field.

The purpose of this paper is twofold. Our first goal is to show that the sieve approach can be understood, at least conceptually, by the undergraduate student. Indeed, it was this kind of conceptual reasoning that led to the original breakthrough. Once our reasoning is in place, it is then just a matter of technical expertise to fine tune the argument to get a refined result.

*The first author, an undergraduate, would like to thank the second author for giving him the opportunity of being a part of this research project. Research of the second author was partially supported by NSERC.

Our second goal is to study the analogue of Artin's conjecture for polynomials mod p. More precisely, fix a prime p and let $a(x)$ be a non-constant polynomial which is not equal to the square of a polynomial mod p. Are there infinitely many irreducible polynomials $p(x)$ such that $a(x)$ generates the residue classes of $(\mathbf{F}_p[x]/(p(x)))^*$?

In 1937, Bilharz[B], a student of E. Artin, answered this in the affirmative, by assuming the truth of the so-called Riemann hypothesis for curves. Then, in 1948, A. Weil proved the Riemann hypothesis for curves as a consequence of his rigorous treatment of algebraic geometry (see [L] for a proof). Thus, as it stands, even this "function field analogue" is not accessible to the undergraduate student.

In this paper, we will show that for $a(x) = x^m + c$, Artin's conjecture for $\mathbf{F}_p[x]$ can be established very easily and almost from first principles. The fact that the junior author co-authored this paper is proof enough that it is accessible to the undergraduate.

2 Sieve Theory and Artin's Conjecture

Understanding this section requires a familiarity with quadratic residues. We direct the reader to chapter 5, section 1 of Ireland and Rosen[IR].

Suppose that we want to prove that 2 is a primitive root (mod p) for infinitely many primes p. First, observe that if $(p-1) = 4q$, with q an odd prime, then 2 is a primitive root (mod p). To see this, suppose that 2 has order r. Then $r|(p-1)$. Hence, $r|4q$. Therefore, $r \in \{1, 2, 4, q, 2q, 4q\}$. Let's examine the separate cases. Clearly the case $r = 1$ is not possible. If $r = 2$, we have that $p|(2^2 - 1)$, which implies that $p = 3$. However, $p = 4q+1$, which is clearly ≥ 13. So, we see that $r \neq 2$. Similarly, if $r = 4$ we have that $p|(2^4 - 1)$, which implies that $p = 3$ or $p = 5$. However, we have seen that $p \geq 13$, so it is clear that $r \neq 4$. If $r = q$ or $r = 2q$, then

$$2^{\frac{p-1}{2}} \equiv 1 \pmod{p}.$$

Hence, 2 is a quadratic residue (mod p). Therefore,

$$p \equiv \pm 1 \pmod{8}.$$

However, it is easily seen that

$$p = 4q + 1 = 4(2k + 1) + 1 = 8k + 5 \equiv 5 \pmod{8}.$$

Hence, $r = 4q = (p-1)$. It is at present unknown whether there are infinitely many primes p such that $(p-1) = 4q$, with q prime. Heuristic reasoning[IR, ch. 2, §4] suggests that the number of such primes $p \leq x$ is asymptotic to $\frac{cx}{\log^2 x}$ as $x \to \infty$, for some constant $c > 0$. So, our attempt at a proof stalls here. However, if we continue along this line of thought, and use the ideas of sieve theory, we can prove that one of 2, 3, 5 is a primitive root for infinitely many primes p. First, though, we will pause to prove a lemma which will be useful later on.

Lemma: *A natural number n cannot have more than* $\frac{\log n}{\log 2}$ *prime factors.*

Proof: Let $n = p_1^{\alpha_1} p_2^{\alpha_2} \cdots p_l^{\alpha_l}$, with $p_1, \ldots, p_l$ distinct primes. Then we see that

$$\log n = \sum_{i=1}^{l} \alpha_i \log p_i \geq l \log 2$$

and so clearly we have that

$$l \leq \frac{\log n}{\log 2}$$

and the lemma is proved. □

Ramanujan[R] observed that this estimate can be refined to

$$l = O\left(\frac{\log n}{\log \log n}\right)$$

in the following way: first, note that

$$\log n \geq \sum_{i=1}^{l} \log p_i \geq \sum_{i=1}^{l} \log q_i$$

where $2 = q_1 < q_2 < \cdots$ is the sequence of primes. Then, by the elementary Tchebychef theorem (see [IR, p. 25]), we obtain

$$\log n \geq cl \log l$$

for some constant $c > 0$. If $l \leq \frac{\log n}{\log \log n}$, then we are done, so assume that $l > \frac{\log n}{\log \log n}$. Then,

$$\log l > \log \log n - \log \log \log n$$

which gives

$$\log n \gg l \log \log n$$

as desired.

The seminal idea of Gupta and Murty was to use sieve theory to produce primes p such that $(p-1)$ has very few prime factors, thereby restricting the possibility for 2 to have order $< (p-1) \bmod p$. Indeed, sieve theory provides at least

$$\frac{cx}{\log^2 x}, \quad c > 0$$

primes $p \leq x$ such that $(p-1) = 2q$ or $(p-1) = 2q_1 q_2$, with q, q_1, q_2 prime and $(2/p) = (3/p) = (5/p) = -1$. Moreover, one can arrange

$$q_2 > x^{\frac{1}{2}-\delta} > q_1 > x^{0.26} \quad \text{for some } \delta > 0.$$

If $(p-1)=2q$, then as $(2/p)=-1$, we use the same reasoning as before and easily see that 2 has order $(p-1)$ mod p in this case. If $(p-1)=2q_1q_2$, then the order of 2 (mod p) is either $2q_1$, $2q_2$, or $2q_1q_2$. If the order of 2 (mod p) is $2q_1$, then $p|(2^{2q_1}-1)$ and hence

$$p \,\Bigg|\, \prod_{a<2x^{\frac{1}{2}-\delta}} (2^a-1)\,. \tag{1}$$

Using Lemma 1, we see that the number of primes $p \le x$ satisfying (1) cannot exceed

$$\sum_{a<2x^{\frac{1}{2}-\delta}} a \le 4x^{1-2\delta}.$$

Therefore, we have at least

$$\frac{cx}{\log^2 x} - 4x^{1-2\delta}$$

primes $p \le x$ such that $(p-1)=2q$ or $(p-1)=2q_1q_2$ where the order of 2 (mod p) is either $2q_2$ or $(p-1)$.

Among these primes, let us consider the order of 3 (mod p). If the order is $2q_1$, then, as above, we have that

$$p \,\Bigg|\, \prod_{a<2x^{\frac{1}{2}-\delta}} (3^a-1)\,.$$

and the number of such primes cannot exceed

$$\sum_{a<2x^{\frac{1}{2}-\delta}} \left(\frac{\log 3}{\log 2}\right) a \le \left(\frac{\log 3}{\log 2}\right) 4x^{1-2\delta}.$$

Similarly, the number of primes p for which the order of 5 (mod p) is $2q_1$ cannot exceed

$$\sum_{a<2x^{\frac{1}{2}-\delta}} \left(\frac{\log 5}{\log 2}\right) a \le \left(\frac{\log 5}{\log 2}\right) 4x^{1-2\delta}.$$

Therefore, we have at least

$$\frac{cx}{\log^2 x} - O(x^{1-2\delta})$$

primes $p \le x$ such that the order of 2, 3, and 5 (mod p) is one of $2q_2$ or $(p-1)$.

We now show that there are infinitely many primes p such that one of 2, 3, 5 is a primitive root mod p. If not, then 2, 3, and 5 generate a subgroup of order $2q_2$ (mod p). Observe that

$$2q_2 = \frac{p-1}{q_1} < \frac{x}{x^{0.26}} = x^{0.74} = z.$$

Notice that if the numbers in the set

$$\{2^a 3^b 5^c,\ 0 \le a, b, c \le x^\alpha\}, \quad \alpha = 0.247$$

are all distinct, they generate a subgroup of size $x^{3\alpha}$. This is clearly a contradiction, since $3\alpha = 0.741 \Rightarrow x^{3\alpha} > z$. Hence, for some a, b, c and a', b', c', we have that

$$2^a 3^b 5^c \equiv 2^{a'} 3^{b'} 5^{c'} \pmod p$$

which implies that

$$2^{a-a'} 3^{b-b'} 5^{c-c'} \equiv 1 \pmod p.$$

Thus, p divides the numerator of

$$\prod_{0 \le |a|, |b|, |c| \le 2x^\alpha} (2^a 3^b 5^c - 1). \tag{2}$$

Observe that if e_i, $1 \le i \le r$ are integers then the numerator of

$$\prod_{i=1}^{r} p_i^{e_i} - 1$$

is

$$\prod_{e_i > 0} p_i^{e_i} - \prod_{e_i < 0} p_i^{-e_i}.$$

Thus, for a given triple a, b, c in the product from (2), the numerator can have at most $O(x^\alpha)$ prime factors.

The number of such primes is $O(x^{4\alpha})$, which is $o(\frac{x}{\log^2 x})$, since $4\alpha < 1$. Hence, one of 2, 3, or 5 is a primitive root (mod p) for these primes.

3 Sifting Function for Primitive Roots

In this section, we will use the Ramanujan sum and Theorem 272 from Hardy and Wright [HW, p. 238]. The Ramanujan sum is defined as:

$$c_d(j) = \sum_{\substack{1 \le k \le d \\ (k,d)=1}} e^{\frac{2\pi i k j}{d}}.$$

Theorem 272 states that

$$c_d(j) = \frac{\mu(\frac{d}{a})\phi(d)}{\phi(\frac{d}{a})}, \quad \text{where } a = (d, j).$$

With these tools at our disposal, we will prove the following lemma:

Lemma: *Let G be a cyclic group of order n. Let*

$$f(g) = \frac{\phi(n)}{n}\left\{1 + \sum_{\substack{d|n\\ d>1}} \frac{\mu(d)}{\phi(d)} \sum_{\text{ord}\,\chi = d} \chi(g)\right\}$$

where the inner sum runs over characters χ of G which are of order d. Then

$$f(g) = \begin{cases} 1, & \text{if } g \text{ generates } G \\ 0, & \text{otherwise.} \end{cases}$$

Proof: Let ε be a generator of G. Let $g = \varepsilon^j$. Then g generates G if and only if $(j, n) = 1$. Notice that if $\Psi(\varepsilon) = e^{\frac{2\pi i}{n}}$, then $\Psi^{\frac{n}{d}}$ is a character of order d. In fact, all of the characters of order d are given by $\Psi^{\frac{n}{d}k}$, where $(k, d) = 1$ and $1 \leq k \leq d$. Then

$$\begin{aligned} f(g) &= \frac{\phi(n)}{n}\left\{1 + \sum_{\substack{d|n\\ d>1}} \frac{\mu(d)}{\phi(d)} \sum_{\text{ord}\,\chi = d} \chi(g)\right\} \\ &= \frac{\phi(n)}{n}\left\{1 + \sum_{\substack{d|n\\ d>1}} \frac{\mu(d)}{\phi(d)} \sum_{\substack{1\leq k\leq d\\ (k,d)=1}} \Psi^{\frac{n}{d}k}(\varepsilon^j)\right\} \\ &= \frac{\phi(n)}{n}\left\{1 + \sum_{\substack{d|n\\ d>1}} \frac{\mu(d)}{\phi(d)} \sum_{\substack{1\leq k\leq d\\ (k,d)=1}} \Psi(\varepsilon)^{\frac{nkj}{d}}\right\} \\ &= \frac{\phi(n)}{n}\left\{1 + \sum_{\substack{d|n\\ d>1}} \frac{\mu(d)}{\phi(d)} \sum_{\substack{1\leq k\leq d\\ (k,d)=1}} e^{\frac{2\pi ikj}{d}}\right\} \\ &= \frac{\phi(n)}{n}\left\{1 + \sum_{\substack{d|n\\ d>1}} \frac{\mu(d)}{\phi(d)} c_d(j)\right\} \end{aligned}$$

Notice that

$$\frac{\mu(1)}{\phi(1)} c_1(j) = 1.$$

We can use this fact to rewrite $f(g)$ more elegantly as

$$\frac{\phi(n)}{n}\left\{\sum_{d|n} \frac{\mu(d)}{\phi(d)} c_d(j)\right\}.$$

Now, we apply Theorem 272 to see that

$$\begin{aligned} f(g) &= \frac{\phi(n)}{n}\left\{\sum_{d|n}\frac{\mu(d)}{\phi(d)}\frac{\mu(\frac{d}{(d,j)})\phi(d)}{\phi(\frac{d}{(d,j)})}\right\} \\ &= \frac{\phi(n)}{n}\left\{\sum_{d|n}\frac{\mu(d)\mu(\frac{d}{(d,j)})}{\phi(\frac{d}{(d,j)})}\right\} \end{aligned}$$

Notice that

$$g = \varepsilon^j \text{ generates } G \Longrightarrow (j,n) = 1 \Longrightarrow (j,d) = 1\ \forall d|n.$$

So, let g generate G. We now have that

$$f(g) = \frac{\phi(n)}{n}\sum_{d|n}\frac{\mu^2(d)}{\phi(d)}.$$

Since both μ and ϕ are multiplicative functions, then $\frac{\mu^2}{\phi}$ is also a multiplicative function. So, when g generates G, we have

$$\begin{aligned} f(g) &= \frac{\phi(n)}{n}\left(\prod_{p|n}\left(1+\frac{1}{p-1}\right)\right) \\ &= \frac{\phi(n)}{n}\left(\prod_{p|n}\frac{p}{p-1}\right) \\ &= \left(\prod_{p|n}\left(1-\frac{1}{p}\right)\right)\left(\prod_{p|n}\frac{p}{p-1}\right) \\ &= \left(\prod_{p|n}\frac{p-1}{p}\right)\left(\prod_{p|n}\frac{p}{p-1}\right) \\ &= 1 \end{aligned}$$

So, when g generates G, $f(g) = 1$.

Now, let's consider the case when g does not generate G. So, let $(j,n) = \delta$, $\delta > 1$. Then $n = \delta t$. Observe also that $(j,t) = 1$ and $(\delta, t) = 1$, since $\delta | j$. We look at

$$\frac{\phi(n)}{n}\left\{\sum_{d|n}\frac{\mu(d)\mu(\frac{d}{(d,j)})}{\phi(\frac{d}{(d,j)})}\right\}$$

Since $(\delta, t) = 1$ and $d|n$, we have that $d = (d,\delta)(d,t)$. Let $(d,\delta) = d_1$ and $(d,t) = d_2$ so that $d = d_1 d_2$ Now, we can rewrite the above expression as

$$\frac{\phi(n)}{n}\left\{\sum_{\substack{d_1|\delta \\ d_2|t}} \frac{\mu(d_1d_2)\mu(\frac{d_1d_2}{(d_1d_2,j)})}{\phi(\frac{d_1d_2}{(d_1d_2,j)})}\right\}$$

Observe that since $(j,t) = 1$ and $(d,t) = d_2$, then $(d_2, j) = 1$, which implies that $(d_1d_2, j) = (d_1, j)$. So, we can again rewrite the expression for $f(g)$ as

$$\frac{\phi(n)}{n}\left\{\sum_{\substack{d_1|\delta \\ d_2|t}} \frac{\mu(d_1d_2)\mu(\frac{d_1d_2}{(d_1,j)})}{\phi(\frac{d_1d_2}{(d_1,j)})}\right\}$$

Using the facts that μ and ϕ are multiplicative, and that $(d_1, d_2) = 1$, we obtain

$$\begin{aligned}
f(g) &= \frac{\phi(n)}{n}\left\{\sum_{\substack{d_1|\delta \\ d_2|t}} \frac{\mu(d_1)\mu(d_2)\mu(\frac{d_1}{(d_1,j)})\mu(d_2)}{\phi(\frac{d_1}{(d_1,j)})\phi(d_2)}\right\} \\
&= \frac{\phi(n)}{n}\left(\sum_{d_1|\delta} \frac{\mu(d_1)\mu(\frac{d_1}{(d_1,j)})}{\phi(\frac{d_1}{(d_1,j)})}\right)\left(\sum_{d_2|t}\frac{\mu^2(d_2)}{\phi(d_2)}\right) \\
&= \frac{\phi(n)}{n}\left(\sum_{d_1|\delta} \frac{\mu(d_1)\mu(\frac{d_1}{(d_1,j)})}{\phi(\frac{d_1}{(d_1,j)})}\right)\left(\frac{t}{\phi(t)}\right)
\end{aligned}$$

But since $(j,n) = \delta$, and $(d,\delta) = d_1$, then $(d_1, j) = d_1$. So we have

$$\begin{aligned}
f(g) &= \frac{\phi(n)}{n}\left(\sum_{d_1|\delta}\mu(d_1)\right)\frac{t}{\phi(t)} \\
&= 0, \quad \text{since } \delta > 1. \text{ (See [IR, p. 19])}
\end{aligned}$$

So, we have shown that

$$f(g) = \begin{cases} 1, & \text{if } g \text{ generates } G \\ 0, & \text{otherwise.} \end{cases}$$

□

4 Reformulation and Solution of the Problem

Let $\mathbf{F}_q$ be a finite field with $q = p^n$ elements. Consider the polynomial ring $\mathbf{F}_p[x]$. Let $a(x)$ be a polynomial in $\mathbf{F}_p[x]$. We would like to know the number of irreducible polynomials

$p(x) \in \mathbf{F}_p[x]$ such that $a(x)$ generates $(\mathbf{F}_p[x]/(p(x)))^*$. Recall that if $\deg p(x) = n$ then

$$\mathbf{F}_p[x]/(p(x)) \simeq \mathbf{F}_{p^n}$$

Moreover, $\mathbf{F}_{p^n}^*$ is cyclic of order $p^n - 1$. Also recall that the isomorphism $\mathbf{F}_p[x]/(p(x)) \simeq \mathbf{F}_{p^n}$ is given explicitly as follows: for $g(x) \in \mathbf{F}_p[x]$, we write

$$g(x) = p(x)q(x) + r(x), \quad \text{with } r(x) = 0 \text{ or } 0 \le \deg r < \deg p = n.$$

Let $\theta \in \mathbf{F}_{p^n}$ be a root of $p(x)$. Then

$$g(\theta) = r(\theta) = a_0 + a_1\theta + \cdots + a_{n-1}\theta^{n-1}, \ a_i \in \mathbf{F}_p$$

describes all the elements of $\mathbf{F}_{p^n}$. Thus, $a(x)$ generating $(\mathbf{F}_p[x]/(p(x)))^*$ is equivalent to $a(\theta)$ generating $\mathbf{F}_{p^n}^*$.

Hence, to count the number of irreducible $p(x)$ of degree n for which $a(x)$ is a generator is tantamount to counting the number of θ of degree n for which $a(\theta)$ generates $\mathbf{F}_{p^n}^*$. Indeed, since each $p(x)$ has n roots, we find

$$\begin{aligned}&\#\{p(x) \in \mathbf{F}_p[x] : p(x) \text{ irreducible}, \ \deg p = n, \ a(x) \text{ generates } (\mathbf{F}_p[x]/(p(x)))^*\}\\ &= \frac{1}{n}\#\{\theta \in \mathbf{F}_{p^n} : \deg\theta = n, \ a(\theta) \text{ generates } \mathbf{F}_{p^n}^*\}\end{aligned}$$

We now apply the lemma of section 3 to see that the number in question is:

$$\begin{aligned}\frac{1}{n}\sum_{\substack{\theta\in\mathbf{F}_{p^n}\\ \deg\theta=n}} f(a(\theta)) &= \frac{1}{n}\sum_{\substack{\theta\in\mathbf{F}_{p^n}\\ \deg\theta=n}} \left(\frac{\phi(p^n-1)}{p^n-1}\left(1+\sum_{\substack{d\mid p^n-1\\ d>1}}\frac{\mu(d)}{\phi(d)}\sum_{\operatorname{ord}\chi=d}\chi(a(\theta))\right)\right)\\
&= \frac{1}{n}\sum_{\substack{\theta\in\mathbf{F}_{p^n}\\ \deg\theta=n}} \left(\frac{\phi(p^n-1)}{p^n-1} + \frac{\phi(p^n-1)}{p^n-1}\sum_{\substack{d\mid p^n-1\\ d>1}}\frac{\mu(d)}{\phi(d)}\sum_{\operatorname{ord}\chi=d}\chi(a(\theta))\right)\\
&= \frac{1}{n}\sum_{\substack{\theta\in\mathbf{F}_{p^n}\\ \deg\theta=n}} \frac{\phi(p^n-1)}{p^n-1} + \frac{1}{n}\sum_{\substack{\theta\in\mathbf{F}_{p^n}\\ \deg\theta=n}} \frac{\phi(p^n-1)}{p^n-1}\sum_{\substack{d\mid p^n-1\\ d>1}}\frac{\mu(d)}{\phi(d)}\\
&\quad \sum_{\operatorname{ord}\chi=d}\chi(a(\theta))\end{aligned}$$

So, the contribution from the main term (the first term in the expression above) is

$$\frac{p^n - p^{n-1}}{n} \cdot \frac{\phi(p^n-1)}{p^n-1}.$$

and the error term is

$$\frac{\phi(p^n-1)}{n(p^n-1)}\sum_{\substack{d\mid p^n-1\\ d>1}}\frac{\mu(d)}{\phi(d)}\sum_{\operatorname{ord}\chi=d}\sum_{\substack{\theta\in\mathbf{F}_{p^n}\\ \deg\theta=n}}\chi(a(\theta)).$$

We will show in the following sections that when $d > 1$, and $a(x)$ is of the form $x^m + c$, the sum

$$\left| \sum_{\substack{\theta \in \mathbf{F}_{p^n} \\ \deg \theta = n}} \chi(a(\theta)) \right| \leq m p^{\frac{n}{2}} \tag{3}$$

Hence, the contribution from the error term is $O(p^{\frac{n}{2}} \delta(p^n - 1))$, where $\delta(u)$ is the number of divisors of u.

The estimate in (3) is a consequence of A. Weil's celebrated work proving the analogue of the Riemann hypothesis for zeta functions over finite fields. However, one can also obtain this estimate in a more elementary manner using Gauss sums. In section 5, we recount some of the theory of Gauss sums over finite fields. Then, in section 6, we use that theory to obtain the estimate in (3).

Once we justify this estimate, it is clear that we have proven the function field analogue of Artin's conjecture in the special case where $a(x) = x^m + c$. To see this, notice that in our expression for the number of irreducible $p(x)$ of degree n such that $a(x)$ generates $(\mathbf{F}_p[x]/(p(x)))^*$, the contribution from the main term far outweighs the contribution from the error term.

5 Gauss Sums Over Finite Fields

In this section, we review some basic properties of Gauss sums over finite fields. All of this can be found in Davenport[D].

Let p be prime and consider the finite field $\mathbf{F}_{p^n}$. Let

$$e(\alpha) = \exp\left(\frac{2\pi i}{p} \mathrm{Tr}_{\mathbf{F}_{p^n}/\mathbf{F}_p}(\alpha)\right), \quad \text{for } \alpha \in \mathbf{F}_{p^n}.$$

Let χ be a character of $\mathbf{F}_{p^n}^*$. The Gauss sum is defined as

$$\tau(\chi) = \sum_{\alpha \in \mathbf{F}_{p^n}^*} \chi(\alpha) e(\alpha)$$

Observe that for any $\eta \neq 0$,

$$\chi(\eta)\tau(\overline{\chi}) = \sum_{\alpha \in \mathbf{F}_{p^n}^*} \overline{\chi}(\alpha)\chi(\eta)e(\alpha)$$

If we make the change of variables $\alpha = \eta\beta$, we have that

$$\chi(\eta)\tau(\overline{\chi}) = \sum_{\beta \in \mathbf{F}_{p^n}^*} \chi(\beta)e(\eta\beta)$$

since $\overline{\chi}(\eta)\chi(\eta) = |\chi(\eta)|^2 = 1$ and as β ranges over elements of $\mathbf{F}_{p^n}^*$, so does $\eta\beta$.

Theorem: $|\tau(\chi)| = p^{n/2}$, *for all non-trivial characters* χ.

Proof: If $\chi \neq \chi_0$ (the principal character), then

$$|\tau(\chi)|^2 = \sum_{\alpha,\beta\in\mathbf{F}^*_{p^n}} \chi(\alpha)\overline{\chi}(\beta)e(\alpha-\beta)$$

We set $\alpha = \beta\gamma$. Then we have that

$$|\tau(\chi)|^2 = \sum_{\gamma,\beta\in\mathbf{F}^*_{p^n}} \chi(\gamma)e(\beta(\gamma-1)).$$

Notice that if $w_1, \ldots, w_n$ is an $\mathbf{F}_p$-basis of $\mathbf{F}_{p^n}$, then

$$\begin{aligned}\sum_{\beta\in\mathbf{F}_{p^n}} e(\beta) &= \sum_{a_1,\ldots,a_n\in\mathbf{F}_p} e(a_1w_1+\cdots+a_nw_n)\\ &= \prod_{j=1}^{n}\left(\sum_{a_j\in\mathbf{F}_p} e(a_jw_j)\right)\end{aligned}$$

Since

$$\begin{aligned}\sum_{a_j\in\mathbf{F}_p} e(a_jw_j) &= \sum_{a_j=0}^{p-1}\exp\left(\frac{2\pi i a_j \mathrm{Tr}_{\mathbf{F}_{p^n}/\mathbf{F}_p}(w_j)}{p}\right)\\ &= \begin{cases} p, & \text{if } \mathrm{Tr}_{\mathbf{F}_{p^n}/\mathbf{F}_p}(w_j)=0\\ 0, & \text{otherwise.}\end{cases}\end{aligned}$$

we have that

$$\sum_{\beta\in\mathbf{F}_{p^n}} e(\beta) = \begin{cases} p^n, & \text{if } \mathrm{Tr}_{\mathbf{F}_{p^n}/\mathbf{F}_p}(w_1)=\cdots=\mathrm{Tr}_{\mathbf{F}_{p^n}/\mathbf{F}_p}(w_n)=0\\ 0, & \text{otherwise.}\end{cases}$$

In the first case, we deduce that

$$\mathrm{Tr}_{\mathbf{F}_{p^n}/\mathbf{F}_p}(\eta)=0 \quad \forall \eta\in\mathbf{F}_{p^n}.$$

In particular, if $1, \theta, \theta^2, \ldots, \theta^{n-1}$ is a basis, we find that

$$\begin{aligned}0 = a_{ij} &= \mathrm{Tr}_{\mathbf{F}_{p^n}/\mathbf{F}_p}(\theta^{i+j})\\ &= \sum_{k=1}^{n}\theta^{(k)i}\theta^{(k)j}\end{aligned}$$

and $\theta^{(1)}, \ldots, \theta^{(n)}$ are the conjugates of θ. The above equation can be rewritten as $\Omega\Omega^T = 0$, where $\Omega = (\theta^{(j)i})$ is a Vandermonde matrix. However, this is a contradiction since $\theta^{(1)}, \ldots, \theta^{(n)}$ are all distinct, which implies that $\det\Omega \neq 0$.

So, we have that

$$\sum_{\beta \in \mathbf{F}_{p^n}} e(\beta) = 0.$$

Now, if $\gamma \neq 1$, then

$$\sum_{\beta \in \mathbf{F}^*_{p^n}} e(\beta(\gamma - 1)) = -1.$$

If $\gamma = 1$, then

$$\sum_{\beta \in \mathbf{F}^*_{p^n}} e(\beta(\gamma - 1)) = p^n - 1.$$

Thus we have that

$$\begin{aligned} |\tau(\chi)|^2 &= p^n - 1 + (-1) \sum_{\substack{\gamma \neq 1 \\ \gamma \in \mathbf{F}^*_{p^n}}} \chi(\gamma) \\ &= (p^n - 1) + (-1)(-1), \quad \text{since } \chi \text{ is non-trivial} \\ &= p^n \end{aligned}$$

So we have shown that $|\tau(\chi)| = p^{\frac{n}{2}}$, for all non-trivial characters χ. □

6 Estimation

We will estimate

$$\sum_{\theta \in \mathbf{F}_{p^n}} \chi(a(\theta))$$

when $a(x) = x^m + c$. For this purpose, we can rewrite the above as

$$\sum_{\theta \in \mathbf{F}_{p^n}} \chi(\theta^m + c).$$

If $(m, p^n - 1) = 1$, then $\theta \mapsto \theta^m$ is an isomorphism, and the sum becomes

$$\sum_{\theta \in \mathbf{F}_{p^n}} \chi(\theta + c) = 0.$$

If $(m, p^n - 1) = t > 1$, then write $m = ts$. Then, $(s, p^n - 1) = 1$, and by the same reasoning,

$$\sum_{\theta \in \mathbf{F}_{p^n}} \chi(\theta^m + c) = \sum_{\theta \in \mathbf{F}_{p^n}} \chi(\theta^t + c). \tag{4}$$

Since $t|(p^n - 1)$, let Ψ be a character of $\mathbf{F}_{p^n}^*$ of order t. Then

$$\frac{1}{t}\sum_{i=0}^{t-1}\Psi^i(\alpha) = \begin{cases} 1, & \text{if } \alpha = \theta^t \text{ for some } \theta \\ 0, & \text{otherwise.} \end{cases}$$

Putting this in expression (4) gives

$$\begin{aligned}\sum_{\theta\in\mathbf{F}_{p^n}}\chi(\theta^t + c) &= t\sum_{\alpha\in\mathbf{F}_{p^n}}\chi(\alpha+c)\,\frac{1}{t}\sum_{i=0}^{t-1}\Psi^i(\alpha)\\ &= \sum_{i=0}^{t-1}\sum_{\alpha\in\mathbf{F}_{p^n}}\chi(\alpha+c)\Psi^i(\alpha)\end{aligned}$$

Now, we use Gauss sums to replace $\chi(\alpha + c)$ as follows:

$$\chi(\alpha+c) = \frac{1}{\tau(\overline{\chi})}\sum_{\beta\in\mathbf{F}_{p^n}^*}\overline{\chi}(\beta)e(\beta(\alpha+c)).$$

Thus interpreting $\chi(0)$ as zero, we have

$$\begin{aligned}\sum_{\alpha\in\mathbf{F}_{p^n}}\Psi^i(\alpha)\chi(\alpha+c) &= \frac{1}{\tau(\overline{\chi})}\sum_{\alpha\in\mathbf{F}_{p^n}}\Psi^i(\alpha)\sum_{\beta\in\mathbf{F}_{p^n}}\overline{\chi}(\beta)e(\beta(\alpha+c))\\ &= \frac{1}{\tau(\overline{\chi})}\sum_{\beta\in\mathbf{F}_{p^n}}\overline{\chi}(\beta)e(\beta c)\overbrace{\sum_{\alpha\in\mathbf{F}_{p^n}}\Psi^i(\alpha)e(\beta\alpha)}^{\text{also a Gauss sum}}\\ &= \frac{1}{\tau(\overline{\chi})}\sum_{\beta\in\mathbf{F}_{p^n}}\overline{\chi}(\beta)e(\beta c)\overline{\Psi}^i(\beta)\tau(\Psi^i)\\ &= \frac{\tau(\Psi^i)}{\tau(\overline{\chi})}\sum_{\beta\in\mathbf{F}_{p^n}}\overline{\chi}(\beta)\overline{\Psi}^i(\beta)e(\beta c)\end{aligned}$$

Observe that

$$\tau(\overline{\chi})\chi(n) = \sum_{\alpha\in\mathbf{F}_{p^n}}\overline{\chi}(\alpha)e(\alpha n).$$

So we have that

$$\sum_{\alpha\in\mathbf{F}_{p^n}}\Psi^i(\alpha)\chi(\alpha+c) = \frac{\tau(\Psi^i)}{\tau(\overline{\chi})}\tau(\overline{\chi\Psi^i})\chi(c).$$

Now, since $\chi, \overline{\chi}, \Psi^i, \overline{\Psi}^i$ are non-trivial characters, we apply the theorem of section 5 to see that

$$\left|\sum_{\alpha\in\mathbf{F}_{p^n}} \Psi^i(\alpha)\chi(\alpha+c)\right| = p^{n/2}.$$

Now, we conclude by showing that when $(m, p^n - 1) = t > 1$

$$\begin{aligned}\left|\sum_{\theta\in\mathbf{F}_{p^n}} \chi(\theta^m+c)\right| &= \left|\sum_{\theta\in\mathbf{F}_{p^n}} \chi(\theta^t+c)\right| \\ &= \left|\sum_{i=0}^{t-1}\sum_{\alpha\in\mathbf{F}_{p^n}} \Psi^i(\alpha)\chi(\alpha+c)\right| \\ &\leq \sum_{i=0}^{t-1}\left|\sum_{\alpha\in\mathbf{F}_{p^n}} \Psi^i(\alpha)\chi(\alpha+c)\right| \\ &\leq \sum_{i=0}^{t-1} p^{\frac{n}{2}} \\ &\leq mp^{\frac{n}{2}}.\end{aligned}$$

Thus, for all m and c, we have that

$$\left|\sum_{\theta\in\mathbf{F}_{p^n}} \chi(\theta^m+c)\right| \leq mp^{\frac{n}{2}}.$$

References

[Ar] E. Artin, *Collected Papers.* Addison-Wesley, 1965.

[B] H. Bilharz, Primdivisoren mit vorgeberger Primitivwurzel. *Math. Annalen* **114** (1937), 476–492.

[BFI] E. Bombieri, Friedlander, J. and Iwaniec, H., Primes in arithmetic progressions to large moduli. *Acta Math.* **370** (1986), 203–251.

[D] H. Davenport, On Primitive Roots in Finite Fields. *Quart. J. Math.* (2) **8** (1937), 308–312. (See also *Collected Papers*, vol. 4, p. 1557–1561.)

[GM] R. Gupta and Ram Murty, M., A remark on Artin's conjecture. *Invent. Math.* **101** (1990), 225–235.

[GMM] R. Gupta, Ram Murty, M. and Kumar Murty, V., The Euclidean algorithm for S-integers. In *Number Theory*, 189–201 (H. Kisilevsky and J. Labute, eds.). (Canadian Mathematical Society Conference Proceedings **7**) (1987).

[H] C. Hooley, On Artin's conjecture. *J. Reine. Angew. Math.* **225** (1967), 209–220.

[HB] D.R. Heath-Brown, Artin's conjecture for primitive roots. *Quart. J. Math. Oxford* (2) **37** (1986), 27–38.

[HW] G.H. Hardy and Wright, E.M., *An Introduction to the Theory of Numbers*, 5th ed. Oxford University Press, 1979.

[IR] K. Ireland and Rosen, M., *A Classical Introduction to Modern Number Theory*. Springer-Verlag, 1982.

[L] D. Lorenzini, *An Invitation to Arithmetic Geometry*, vol. 9. Graduate Studies in Mathematics, American Math Society, 1996, pp. 354–360.

[M1] M. Ram Murty, Artin's conjecture and elliptic analogues. In *Sieve Methods, Exponential Sums, and their Applications in Number Theory*, 325–344. (G.R.H. Greaves, G. Harman, and M.N. Huxley, eds.). Cambridge University Press, 1996.

[M2] M. Ram Murty, Artin's conjecture for primitive roots. *Math. Intelligencer* **10** (1988), 59–67.

Department of Mathematics
Queen's University
Kingston, Ontario
K7L 3N6, Canada
E-Mail: murty@mast.queensu.ca

Continuous Homomorphisms as Arithmetical Functions, and Sets of Uniqueness

I. Kátai

This is a survey paper on the characterization of continuous group homomorphisms as arithmetical functions, and on sets of uniqueness with respect to completely additive functions.

1 Introduction

Let, as usual $\mathbb{N}, \mathbb{Z}, \mathbb{Q}, \mathbb{R}, \mathbb{C}$ be the set of positive integers, integers, rational, real, and complex numbers, respectively. Let $\mathbb{Q}_\times, \mathbb{R}_\times$ be the multiplicative group of positive rationals, reals, respectively. Let $\mathcal{P}$ be the set of prime numbers.

For an arbitrary, additively written Abelian group G let $\mathcal{A}_G$, resp. $\mathcal{A}_G^*$ denote the classes of additive, resp. completely additive functions. A function $f : \mathbb{N} \to G$ belongs to $\mathcal{A}_G$ if $f(nm) = f(m) + f(n)$ holds for each pair of coprime m, n, and it belongs to $\mathcal{A}_G^*$ if the above equation holds for all pairs $m, n \in \mathbb{N}$. If G is written multiplicatively, then we shall write $\mathcal{M}_G, \mathcal{M}_G^*$ instead of $\mathcal{A}_G, \mathcal{A}_G^*$, and the corresponding functions are called multiplicative, completely multiplicative.

If $G = \mathbb{R}$, then we shall write simply $\mathcal{A}, \mathcal{A}^*$ instead of $\mathcal{A}_\mathbb{R}, \mathcal{A}_\mathbb{R}^*$.

If $f \in \mathcal{A}_G^*$, then its domain $\mathbb{N}$ can be extended to $\mathbb{Q}_\times$ by

$$f\left(\frac{m}{n}\right) := f(m) - f(n),$$

and the functional equation

$$f(r_1 r_2) = f(r_1) + f(r_2)$$

remains valid for every $r_1, r_2 \in \mathbb{Q}_\times$.

Let us assume that G is a topological group and $f : \mathbb{Q}_\times \to G$ is continuous at 1. Then for each $\alpha \in \mathbb{R}_\times$ there exists the limit

$$\lim_{r \to \alpha} f(r) =: \Phi(\alpha),$$

Φ is continuous everywhere in $\mathbb{R}_\times$, furthermore $\Phi(\alpha\beta) = \Phi(\alpha) + \Phi(\beta)$ valid for all $\alpha, \beta \in \mathbb{R}_\times$, i.e. Φ is a continuous homomorphism of $\mathbb{R}_\times$ into G.

On the other hand, if $\Phi : \mathbb{R}_\times \to G$ is a homomorphism, then its restriction to the domain $\mathbb{N}$ is a completely additive function.

Let S be an R-module, containing at least two elements, defined over an integral domain R which has an identity. Consider the set of all doubly infinite sequences $(\ldots s_{-1}, s_0, s_1, \ldots)$ of elements of S. We introduce the shift operator E whose action takes a typical sequence $\{s_n\}$ to the new sequence $\{s_{n+1}\}$. If

$$P(x) = \sum_{j=0}^{r} c_j x^j$$

is a polynomial with coefficients in R, we extend this definition by defining

$$P(E)s_n = \sum_{j=0}^{r} c_j s_{n+j}.$$

In this way we define a ring of operators which is isomorphic to the ring of polynomials with coefficients in R. Let I be the identity operator, and $\Delta := E - I$.

We shall say that an additive function f is of finite support, if it vanishes on the set of prime powers except possibly on the powers of finitely many primes.

For $z \in \mathbb{R}$ let $\|z\| := \min_{k\in\mathbb{Z}} |z-k|$.

2 Characterization of log as an Additive Arithmetical Function

The function $f(n) = \log n$ belongs to $\mathcal{A}^*$. Normally log is considered as a mapping $\mathbb{R}_\times \to \mathbb{R}$ and in this context it is wellknown that continuity along with the functional equation $f(xy) = f(x) + f(y)$ characterizes the logarithm up to a constant factor. Restricting the domain from $\mathbb{R}_\times$ to $\mathbb{N}$ creates an interesting question: What further properties along with the (complete) additivity will ensure that an arithmetic function f is in fact $c \log n$.

The first result of this type was proved by P. Erdös [1] in 1946.

Theorem 1 *If $f \in \mathcal{A}$ and $\Delta f(n) \geq 0$ for all n, or $f(n) \to \infty$ $(n \to \infty)$, then $f(n)$ is a constant multiple of* $\log n$.

In [2] we proved

Theorem 2 *If $f \in \mathcal{A}$ and* $\liminf \Delta^k f(n) \geq 0$ *with some $k \in \mathbb{N}$, then f is a constant multiple of* $\log n$.

An important progress has been achieved by E. Wirsing [3] proving the following conjecture of Erdös.

Theorem 3 *If $f \in \mathcal{A}$ and $\Delta f(n) \geq -K$ with some constant K, then $f(n) = c \log n + u(n)$, where $u(n)$ is bounded and c is a suitable constant.*

Another one of Erdös's conjecture was proved in [4].

Theorem 4 *If $f \in \mathcal{A}$ and*

$$\frac{1}{x} \sum_{n \leq x} |\Delta f(n)| \longrightarrow 0, \tag{2.1}$$

then $f = c$ log.

Somewhat later the condition (2.1) was weakened by E. Wirsing. Namely, he proved in [5]

Theorem 5 *Let $f \in \mathcal{A}$. Assume that there exists a constant $\gamma > 1$ and a sequence $x_1 < x_2 < \dots$ such that*

$$x_i^{-1} \sum_{x_i < n \leq \gamma x_i} |\Delta f(n)| \longrightarrow 0 \quad (i \longrightarrow \infty).$$

Then $f = c$ log.

By making use of very original new ideas and some deep results on the distribution of primes in arithmetical progressions, E. Wirsing [6] proved

Theorem 6 *If $f \in \mathcal{A}^*$ and $\Delta f(n) = o(\log n)$, then $f(n) = c \log n$.*

One can show easily that the following generalization of the preceding theorems hold true.

Theorem 7

(1) *Let $f, g \in \mathcal{A}$. If*

(a) *$g(n+1) - f(n) \to 0$, then $f = g = c$ log;*

(b) *$g(n+1) - f(n)$ is bounded, then $f(n) = c \log n + u(n)$, $g(n) = c \log n + v(n)$, and u, v are bounded.*

(2) *Let $f, g \in \mathcal{A}^*$. If $g(n+1) - f(n) = o(\log n)$, then $f(n) = g(n) = c \log n$.*

For the method of the proof of Theorem 7 see [7], [8].

In [9] and [10] I asked for a characterization of those additive functions which satisfy

$$f(an+b) - f(An+B) \longrightarrow C \quad \text{as} \quad n \longrightarrow \infty \tag{2.2}$$

for some integers $a > 0$, $A > 0$, b, B, and real constant C. I considered it with $B = 0$ and small values of a and b in [9] and [10].

With general a and b but still with $B = 0$ satisfactory results has been achieved by Mauclaire [11].

Elliott solved this problem completely. Namely he demonstrated in [12] that if (2.2) holds, then there is a constant F such that

$$f(m) = F \log m$$

holds for all m coprime to $aA\Delta$, where $\Delta = aB - Ab$, assuming $\Delta \neq 0$. Moreover he could give the values of f for those prime powers p^α for which $p | aA\Delta$.

Another important assertion proved by Elliott is formulated as

Theorem 8 *Assume that $aA\Delta \neq 0$. There exist positive constants c, c_1 so that*

$$\left| \frac{f(m)}{\log m} - \frac{f(n)}{\log n} \right| \leq c_1 \left(\frac{L(m)}{\log m} + \frac{L(n)}{\log n} \right)$$

holds uniformly for all integers m and n which satisfy $2 \le m \le n \le e^m$ *and are prime to* $aA\Delta$. *Here*

$$L(x) = \max_{n \le x^c} |f(an+b) - f(An+B)|.$$

The constants c, c_1 *may depend on* a, b, A, B.

The best source for the proof of this theorem and other important results is the excellent book of Elliott [13]. Theorem 8 generalizes a result of Wirsing [6] which sounds as follows:

Let $\beta(x)$ be a positive non-decreasing function so that $\beta(x^2) \le 2^{6/5}\beta(x)$. Let $f \in \mathcal{A}$ such that $f(2) \ge 0$ and $f(n+1) - f(n) \le \beta(n)$, for every $n \in \mathbb{N}$. Then, there is a suitable constant γ so that

$$\left|\frac{f(m)}{\log m} - \frac{f(n)}{\log n}\right| \le \gamma\left(\frac{\beta(m)}{\log m} + \frac{\beta(n)}{\log n}\right)$$

uniformly for $2 \le m \le n \le e^m$.

We shall say that a sequence of real numbers $t_n (n \in \mathbb{N})$ is tight if

$$\limsup_{x\to\infty} \frac{1}{x}\#\{n \le x, |t_n| > K\} =: c(K) \longrightarrow 0 \quad \text{as} \quad K \longrightarrow \infty. \tag{2.3}$$

A. Hildebrand [14] proved that $f(n+1) - f(n)$, $f \in \mathcal{A}$ has a limit distribution if and only if there exists a constant c such that $h(n) := f(n) - c\log n$ satisfies

$$\sum_p \frac{\min(1, h^2(p))}{p} < \infty. \tag{2.4}$$

Though explicitly it was not formulated but from this argument the following assertion follows immediately

Theorem 9 *Let* $f \in \mathcal{A}$. *Then* (2.3) *holds for* $t_n = f(n+1) - f(n)$, *if and only if* (2.4) *is satisfied.*

Later Elliott [15] went on to prove the following more general

Theorem 10 *Let* $a > 0$, $A > 0$, b, B *be integers which satisfy* $aB \ne Ab$, *and* $\eta(x)$ *a real-valued function defined for* $x \ge 2$. *Let* $f_1, f_2 \in \mathcal{A}$, *and* $\eta(x)$ *be an arbitrary function. Let*

$$F_x(z) := \frac{1}{x}\#\{n \le x \,|\, f_1(an+b) - f_2(An+B) - \eta(x) \le z\}.$$

The following three propositions are equivalent.

(1) *There is an* $\eta(x)$ *so that the frequencies* $F_x(z)$ *converge weakly to a distribution function as* $x \to \infty$.
(2) *There is an* $\eta(x)$ *so that*

$$\lim_{z\to\infty} \limsup_{x\to\infty} (1 - F_x(z) + F_x(-z)) = 0.$$

(3) *There are real numbers c_1, c_2 such that for $h_j(n) := f_j(n) - c_j \log n$ the conditions*

$$\sum_{p \in \mathcal{P}} \frac{\min(1, h_j^2(p))}{p} < \infty$$

hold.

Let $\varphi : [0, \infty) \to [0, \infty)$ be a so-called subadditive function, i.e. monotonically increasing, $\varphi(x) \to \infty$ as $x \to \infty$, and the condition

$$\varphi(x + y) \leq c_1(\varphi(x) + \varphi(y)) \quad \text{for} \quad x, y \geq 1 \tag{2.5}$$

holds with a suitable constant $c_1 > 0$.

We are interested in giving necessary and sufficient conditions for an additive f to satisfy

$$\sum_{n \leq x} \varphi(|f(n+1) - f(n)|) \ll x \quad (x \longrightarrow \infty) \tag{2.6}$$

Applying the argument we used in our paper [16] written jointly with Indlekofer, one gets

Theorem 11 *Let φ be a subadditive function. The relation* (2.6) *holds for an additive function f if and only if there exists a constant c such that $h(n) := f(n) - c \log n$ satisfies* (2.4) *and*

$$\sum_{|h(q^m)| \geq 1} \frac{\varphi(|h(q^m)|)}{q^m} < \infty, \tag{2.7}$$

where q^m runs over the set of prime powers.

Proof: Necessity. Assume that (2.6) holds. Then $\Delta f(n)$ is a tight sequence, and so, by Theorem 8 we obtain the fulfilment of (2.4). Since $\Delta f(n) = \Delta h(n) + o(1)$, therefore $\sum_{n \leq x} \varphi(|\Delta h(n)|) \ll x$. Let $h(n)$ be written as the sum of the additive functions $h_1(n), h_2(n)$, where h_1 is a strongly additive function defined for primes q such that

$$h_1(q) = \begin{cases} h(q) & \text{if } |h(q)| < 1, \text{ or if } q = 2 \\ 0 & \text{otherwise,} \end{cases}$$

and $h_2(n)$ is defined by $h_2(n) := h(n) - h_1(n)$.

From (2.5) one gets easily that $\varphi(x) \ll x^c$ for $x \geq 1$ with a suitable constant c. Furthermore, from the generalized Turán-Kubilius inequality due to Elliott (see Lemma 1.4.[13]), together with (2.4) we obtain that $\sum_{n \leq x} \varphi(|\Delta h_1(n)|) \ll \sum_{n \leq x} |\Delta h_1(n)|^c \ll x$, consequently, from the assumptions (2.6), (2.5), and $|\Delta h_2(n)| \leq |\Delta h_1(n)| + |\Delta h(n)|$ we obtain that

$$\sum_{n \leq x} \varphi(|\Delta h_2(n)|) \leq c_3 x. \tag{2.8}$$

with a suitable constant c_3, for all $x \geq 2$. From (2.8) we obtain (2.7) readily. Let $\tilde{\mathcal{P}}$ be the set of those primes q for which $|h(q)| \geq 1$. As we know (see (2.4)) $\sum_{q \in \tilde{\mathcal{P}}} \frac{1}{q} < \infty$. Let us choose an arbitrary $Y \geq 1$. For all β, $(2 <)2^\beta \leq Y$ consider those integers $n = 2^\beta \gamma$, γ odd for which $\gamma(n+1)$ is square-free and coprime to $\tilde{\mathcal{P}}$. By making use of the Eratosthenian sieve we can see that the density of these integers is $\frac{e}{2^\beta}$ with a positive constant e which may depend only on $\tilde{\mathcal{P}}$. Since $h_2(2^\beta\gamma) = h_2(2^\beta)$, and the sequences defined for different β are disjoint ones, from (2.8) we get that

$$\sum_{2^\beta \le Y} \varphi(|h_2(2^\beta)|)\frac{e}{2^\beta} < c_3. \tag{2.9}$$

Let now $\mathcal{B} = \{q^m, q > 2, m \ge 2\} \cup \{q \in \tilde{\mathcal{P}}, q \ne 2\}$. Let $Q \le Y$, $\mathcal{S}_Q$ be the set of those integers $n = 2Q\nu$ for which ν is odd, and $\nu(2Q\nu+1)$ is square free and coprime to $\tilde{\mathcal{P}}$. By the Eratosthenian sieve we obtain that the asymptotic density of $\mathcal{S}_Q$ is $\ge e_1/Q$ with a positive constant e_1. Since $h_2(2Q\nu) = h_2(2) + h_2(Q)$, and the sets $\mathcal{S}_Q$ are disjoint, we obtain that

$$\sum_{Q<Y} \varphi(|h_2(2) + h_2(Q)|)\frac{e_1}{Q} < c_3.$$

Hence we get (2.7) immediately.

Sufficiency. Assume that (2.4), (2.7) hold true. Since $\varphi(|\Delta f(n)|) \ll \varphi(|c\Delta \log n|) + \varphi(|\Delta h_1(n)|) + \varphi(|\Delta h_2(n)|)$, therefore summing over n up to x, the first two sums on the right hand side are bounded by x, it remains to prove that

$$\sum_{n\le x} \varphi(|\Delta h_2(n)|) \ll x,$$

which will follow if we show that

$$\sum_{n\le x} \varphi(|h_2(n)|) \ll x. \tag{2.10}$$

Let $\mathcal{T}$ denote the set of those integers D for which $p\|D$ implies that $h_2(p) \ne 0$. The left hand side of (2.10) is bounded by

$$x \sum_{\substack{D\in\mathcal{T}\\ D\le x}} \frac{\varphi(|h_2(D)|)}{D}. \tag{2.11}$$

Iterating (2.5) we obtain that

$$\varphi(|h_2(D)|) \le \sum_{q^m\|D} c^{\omega(\frac{D}{q^m})}\varphi(|h_2(q^m)|),$$

where $\omega(n)$ is the number of distinct prime divisors of n. Thus we have

$$\sum_{\substack{D\in\mathcal{T}\\ D\le x}} \frac{\varphi(|h_2(D)|)}{D} \le 1 + \sum_{D}\sum_{q^m\|D} \frac{\varphi(|h_2(q^m)|)}{q^m}\frac{c^{\omega(\frac{D}{q^m})}}{\frac{D}{q^m}}$$

$$\le 1 + \left(\sum_{q^m\in\mathcal{T}} \frac{\varphi(|h_2(q^m)|)}{q^m}\right)\left\{\sum_{D_1\in\mathcal{T}} \frac{c^{\omega(D_1)}}{D_1}\right\}.$$

On the right hand side both sums are convergent, and the proof is complete. □

As a special case we have the following

Corollary 1 *Let $f \in \mathcal{A}$. The inequality*

$$\sum_{n \le x} |\Delta f(n)|^{\alpha} \ll x$$

holds with some constant $\alpha > 0$, if and only if there is a suitable constant c such that for $h(n) := f(n) - c \log n$

$$\sum_{|h(p)|<1} \frac{h^2(p)}{p} < \infty,$$

and

$$\sum_{|h(q^m)| \ge 1} \frac{|h(q^m)|^{\alpha}}{q^m} < \infty$$

hold.

This assertion for $\alpha = 2$ was proved earlier by Elliott [17].

3 Characterization of n^s as a Multiplicative Function

In a series of papers ([18] I–VI) I considered functions $f \in \mathcal{M}$ under the conditions that $\Delta f(n)$ tends to zero in some sense. I could determine all those functions $f, g \in \mathcal{M}^*$ for which the relation

$$\sum_{n=1}^{\infty} \frac{1}{n} |g(n+k) - f(n)| < \infty \tag{3.1}$$

with some fixed $k \in \mathbb{N}$ holds. Namely I proved the following assertions.

Theorem 12 *If $f, g \in \mathcal{M}$ and* (3.1) *holds with $k = 1$, then either*

$$\sum_{n=1}^{\infty} \frac{|f(n)|}{n} < \infty, \quad \sum_{n=1}^{\infty} \frac{|g(n)|}{n} < \infty, \tag{3.2}$$

or

$$f(n) = g(n) = n^{\sigma + i\tau}, \quad \sigma, \tau \in \mathbb{R}, \quad 0 \le \sigma < 1. \tag{3.3}$$

Theorem 13 *Let $f, g \in \mathcal{M}^*$ and $k \ge 2$ be fixed. Assume that* (3.1) *holds, furthermore that $f(n) = g(n) = 0$ if $(n, k) > 1$ and $f(n) \ne 0, g(n) \ne 0$ if $(n, k) = 1$. Then either* (3.2) *is satisfied or there exist $F, G \in \mathcal{M}^*$ and $s \in \mathbb{C}$ with $\Re s < 1$, such that $f(n) = n^s F(n)$, $g(n) = n^s G(n)$, and*

$$G(n+k) = F(n) \quad (n \in \mathbb{N}) \tag{3.4}$$

holds.

In [18, IV.] I determined all the solutions of (3.4) for completely multiplicative pairs of F, G and in [19] even for $F, G \in \mathcal{M}$ under the additional condition that $F(n) \neq 0$ if $(n, k) = 1$. The above assertions are not obvious even in the case $g = f$.

An immediate consequence of Theorem 12 is that $\sum_{n=1}^{\infty} \frac{1}{n}|\lambda(n+1) - \lambda(n)| = \infty$, where λ is the Liouville function. This shows that the size of the integers n for which $\lambda(n) \neq \lambda(n+1)$ is not too small.

Recently in a joint paper with B.M. Phong [20] we proved

Theorem 14 *Let $k \in \mathbb{N}$ be fixed. Assume that $F, G \in \mathcal{M}$ and* (3.4) *is satisfied. Then either*

$$S_F := \{n | F(n) \neq 0\} \quad \text{and} \quad S_G := \{n | G(n) \neq 0\}$$

are finite sets, or $F(n) \neq 0$ for every n coprime to k.

A special case was treated earlier in [21].

In [22] I formulated the following

Conjecture 1 If $f \in \mathcal{M}$ and

$$\frac{1}{x} \sum_{n \le x} |\Delta f(n)| \longrightarrow 0, \tag{3.5}$$

then either

$$\frac{1}{x} \sum_{n \le x} |f(n)| \longrightarrow 0 \quad \text{as} \quad x \longrightarrow \infty, \tag{3.6}$$

or $f(n) = n^s$, $\Re s < 1$.

Towards this conjecture, a few partial results are known.

First, assuming that (3.6) does not hold, from (3.5) one can deduce that $f \in \mathcal{M}^*$. This assertion was explicitly proved by Mauclaire and Murata [23] for functions f of modulus 1, but their method can be applied to the general case.

The second observation is that either $|f(n)| \ge 1$ for every n, or (3.6) holds. Indeed, let $|f(q)| = \varrho < 1$, $S(x) := \sum_{n \le x} |f(n)|$. Then

$$S(x) \le \sum_{m=1}^{[x/q]+1} \sum_{j=0}^{q-1} |f(mq+j)| \le \sum_{m=1}^{[x/q]+1} q|f(mq)| + \sum_m \sum_j |f(mq+j) - f(mq)|.$$

According to (3.5), the second sum on the right hand side is smaller than ε, ($\varepsilon > 0$ arbitrary), if x is large enough, the first sum is $q\varrho S([\frac{x}{q}]+1)$, consequently,

$$\frac{S(x)}{x} \le \varepsilon + \varrho \frac{S([\frac{x}{q}]+1)}{x/q},$$

whence $S(x)/x \to 0$ immediately follows.

Moreover arguing similarly, one can deduce that if (3.6) does not hold, then $|f(n)| = n^\sigma$ with a constant $\sigma, 0 \le \sigma < 1$. Let $t(n) := f(n)n^{-\sigma}$, and assume that $\sigma > 0$. Since $t(n+1) - t(n) = f(n+1)((n+1)^{-\sigma} - n^{-\sigma}) + (\Delta f(n))n^{-\sigma}$, therefore

$$\sum_{n\le x} \frac{|\Delta t(n)|}{n} \ll \sum_{n\le x} \frac{1}{n^2} + \sum_{n\le x} \frac{|\Delta f(n)|}{n^{\sigma+1}}.$$

The right hand side is clearly convergent, therefore Theorem 12 can be applied, whence we obtain that $t(n) = n^{i\tau}, \tau \in \mathbb{R}$, i.e. $f(n) = n^s, 0 < \Re s < 1$.

The case, when $f(n)$ is of modulus 1 seems to be very hard. Hildebrand [23] proved

Theorem 15 *There exists a positive constant c with the following property. If $g \in \mathcal{M}^*$, $|g(n)| = 1$ for $n \in \mathbb{N}$ and for every $p \in \mathcal{P}, |g(p) - 1| \le c$, then either $g(n) = 1$ identically, or*

$$\liminf \frac{1}{x} \sum_{n\le x} |\Delta g(n)| > 0. \tag{3.7}$$

By using the ideas of Hildebrand and some of mine, I obtained [18, VI.]

Theorem 16 *Let $g \in \mathcal{M}^*$, $|g(n)| = 1$ for $n \in \mathbb{N}$. There exist positive constants $\beta < 1$ and δ such that*

$$\limsup \sum_{x^\beta < p < x} \frac{|g(p) - 1|}{p} < \delta \tag{3.8}$$

and

$$\liminf \frac{1}{x} \sum_{x/2 \le n \le x} |\Delta g(n)| = 0 \tag{3.9}$$

imply that $g(n) = 1$.

Let $\varrho : [1, \infty) \to [1, \infty)$ be a slowly varying function, i.e. such that

$$\lim_{x\to\infty} \max_{y\in[x/2,x]} \left| \frac{\varrho(y)}{\varrho(x)} - 1 \right| = 0.$$

Let Ω denote the set of all arithmetical functions having complex values. $f \in \Omega$ is considered as an infinite dimensional vector, the n'th coordinate of which is $f(n)$. Let $\alpha \ge 1$ be a constant and $\Omega_{\alpha,\varrho}$ be the subspace of Ω which consists of those $x \in \Omega$ for which

$$\sup_{y\ge 1} \frac{1}{y\varrho(y)^\alpha} \sum_{n\le y} |x_n|^\alpha$$

is finite.

Let $\mathcal{L}_{\alpha,\varrho} = \mathcal{M} \cap \Omega_{\alpha,\varrho}$, $\mathcal{L}^*_{\alpha,\varrho} = \mathcal{M}^* \cap \Omega_{\alpha,\varrho}$.

In a joint paper with Indlekofer [24] we proved

Theorem 17 *If* $f \in \mathcal{M}$, $P \in C[z]$, $P \neq 0$, $k = \deg P$, *and*

$$P(E)f \in \Omega_{\alpha,\varrho}$$

then either $f \in \mathcal{L}_{\alpha,\varrho}$, *or* $f(n) = n^s u(n)$, *where* $0 \leq \Re s \leq k$ *and*

$$P(E)u = 0$$

The next assertion was proposed by myself as a conjecture and proved by Wirsing in 1984.

Theorem 18 *If* $f \in \mathcal{M}$, $\Delta f(n) \to 0$ *as* $n \to \infty$, *then* $f(n) \to 0$ *as* $n \to \infty$ *or* $f(n) = n^s, 0 \leq \Re s < 1$.

This theorem has been proved some years later independently by Tang and Shao. The joint paper of Wirsing, Tang and Shao [25] contains two different proofs.

Wirsing's theorem can be formulated in the following way: If $F \in \mathcal{A}$ and $\|\Delta F(n)\| \to 0$, then with some suitable constant $\lambda \in \mathbb{R}$ we have that $F(n) - \lambda \log n$ is integer for every $n \in \mathbb{N}$.

In other words, if T is the group of the reals mod 1, and $F \in \mathcal{A}_T$, $\Delta F(n) \to 0$, then F is a restriction of a continuous homomorphism from $\mathbb{R}_\times$ to T.

B.M. Phong proved the following generalization of Wirsing's theorem.

Theorem 19 *Let* A, B *be positive integers and let* D *be a real constant. If* $h \in \mathcal{A}_T^*$ *and*

$$h(An + B) - h(n) - D \longrightarrow 0 \quad as \quad n \longrightarrow \infty,$$

then h *is the restriction of a continuous homomorphism:* $\mathbb{R}_\times \to T$.

For $A = 1$ this assertion was generalised by Tang [29]:

Theorem 20 *Let* B *be a fixed positive integer,* f *a multiplicative function defined on the set of the integers* n *coprime to* B, *such that* $|f(n)| = 1$ *and* $f(n + B) - f(n) \to 0, n \to \infty$. *Then there must be a* $\tau \in \mathbb{R}$ *such that* $f(n) = n^{i\tau}\chi_B(n)$, *where* $\chi_B(n)$ *is a Dirichlet-character mod* B.

By using this assertion one can completely characterize all those multiplicative functions f of modulus 1, for which $P(E)f(n) \to 0$, $(n \to \infty)$ holds. (For this see [18, I])

In a joint paper with N.L. Bassily [28] we proved

Theorem 21 *If* $f, g \in \mathcal{M}$ *and* $g(2n+1) - Cf(n) \to 0$ *with some nonzero constant* C, *then either* $f(n) \to 0$ *as* $n \to \infty$, *or* $C = f(2)$, $f(n) = n^s, 0 \leq \Re s < 1$, *and* $g(n) = f(n)$ *for every odd* n.

The complete description of those $f, g \in \mathcal{M}$ for which $g(An + B) - Cf(an + b) \to 0$ $(n \to \infty)$ is not given yet.

4 On Additive Functions mod 1

T is considered here as the additive group $\mathbb{R}/\mathbb{Z}$. We say that $F \in \mathcal{A}_T$ is of finite support if $F(p^\alpha) = 0$ holds for every large prime p, and every $\alpha \in \mathbb{N}$. For $F_\nu \in \mathcal{A}_T (\nu = 0, 1, \ldots, k-1)$ let

$$L_n(F_0, \ldots, F_{k-1}) := F_0(n) + \cdots + F_{k-1}(n+k-1). \tag{4.1}$$

Conjecture 2 Let $\mathcal{L}_0^{(k)}$ be the space of those k-tuples $(F_0, \ldots, F_{k-1})$ of $F_\nu \in \mathcal{A}_T$ for which

$$L_n(F_0, \ldots, F_{k-1}) = 0 \quad (n \in \mathbb{N}) \tag{4.2}$$

holds. Then each F_j is of finite support, and $\mathcal{L}_0^{(k)}$ is a finite dimensional $\mathbb{Z}$ module. Let $G_j(n) = \tau_j \log n \pmod 1$, $\tau_0 + \cdots + \tau_{k-1} = 0$. Then $L_n(G_0, G_1, \ldots, G_{k-1}) \to 0$ as $n \to \infty$.

Conjecture 3 If $F_\nu \in \mathcal{A}_T (\nu = 0, \ldots, k-1)$, and

$$L_n(F_0, \ldots, F_{k-1}) \longrightarrow 0 \quad (n \longrightarrow \infty),$$

then there exist suitable real numbers $\tau_0, \ldots, \tau_{k-1}$ such that $\tau_0 + \cdots + \tau_{k-1} = 0$, and for $H_j(n) := F_j(n) - \tau_j \log n$ we have

$$L_n(H_0, \ldots, H_{k-1}) = 0 \quad (n = 1, 2, \ldots).$$

Remarks:

1. Conjecture 3 for $k = 1$ can be deduced easily from Wirsing's theorem.
2. Conjecture 2 was proved for $k = 3$ under the more strict condition that $F_\nu \in \mathcal{A}_T^*$ in [30]. We obtained that (4.2) implies that $F_\nu = 0$ $(\nu = 0, 1, 2)$ identically.
3. Conjecture 2 for $k = 3$ was proved completely by R. Styer [31].
4. M. Wijsmuller treated similar problems for additive functions defined on the set of Gaussian integers taking values from T. See [32], [33].

Let $P(n)$ be the largest and $p(n)$ the smallest prime divisor of n.

Conjecture 4 For every integer $k (\geq 1)$ there exists a constant c_k such that for every prime p greater than c_k,

$$\min_{\substack{1 \leq j \\ P(j) < p}} \max_{\substack{i \in [-k,k] \\ i \neq 0}} P(jp + l) < p \tag{4.3}$$

holds.

We are unable to prove it even for $k = 2$.

Proposition 1 *Let $\mathcal{L}_0^{*(l)}$ be the space of those l-tuples $(F_0, \ldots, F_{l-1})$ of $F_\nu \in \mathcal{A}_T^*$ for which $L_n(F_0, \ldots, F_{l-1}) = 0$ $(n \in \mathbb{N})$. Assume that Conjecture 4 is true for $k = l$. Then $\mathcal{L}_0^{*(l)}$ is a finite dimensional space.*

Proof: Let $(\hat{F}_0, \ldots, \hat{F}_{l-1})$ be such an element of $\mathcal{L}_0^{*(l)}$ for which $\hat{F}_j(q) = 0$ for every $q \le \max(c_l, l)$ and $j = 0, \ldots, l-1$. We shall prove that $\hat{F}_j(n) = 0$ for every $n \in \mathbb{N}$, $j = 0, \ldots, l-1$. Assume the contrary, and let M be the smallest integer for which $\hat{F}_t(M) \ne 0$ for some $t \in \{0, \ldots, l-1\}$. Then M should be a prime. Since

$$L_{jM-t}(\hat{F}_0, \ldots, \hat{F}_{l-1}) = \sum_{i=0}^{l} \hat{F}_i(jM - T + i) = 0,$$

from (4.3), by choosing that j for which (4.3) is attained (with $M = p$), we obtain that $\hat{F}_t(M) = 0$.

Hence it follows that the initial values $F_j(q)$, $j = 0, \ldots, l-1$; $q \le \max(c_l, l)$ completely determine the functions F_j, if they are correlated according to (4.2).

The proof is complete. □

Let K be the closure of the set $\{L_n(F_0, \ldots, F_{k-1}) | n \in \mathbb{N}\}$.

Conjecture 5 If $F_0, \ldots, F_{k-1} \in \mathcal{A}_T^*$ and K contains an element of infinite order, then $K = T$.

This conjecture is obvious if $k = 1$, and it seems to be hard for $k \ge 2$. Recently, in our joint papers with M.V. Subbarao [34], [35] we obtained some partial results. This will be explained in the remaining part of this section.

Let $E_k = \{u/k | u = 0, 1, \ldots, k-1\}$, i.e. the group of those elements $\alpha \in T$ for which $k\alpha = 0$. A special case of Conjecture 5 would be

Conjecture 6 Let $f \in \mathcal{A}_T^*$, and $\mathcal{H} = \{\alpha_1, \ldots, \alpha_k\}$ be the set of the limit points of the sequence $f(n+1) - f(n) (n \in \mathbb{N})$. Then $\mathcal{H} = E_k$, and there exists a real number τ such that $f(n) = \tau \log n + U(n) \pmod 1$, $U(\mathbb{N}) = E_k$, and for every $\omega \in E_k$ there exists a subsequence n_ν of integers such that $U(n_\nu + 1) - U(n_\nu) = \omega$.

We proved

Theorem 22

1) *Conjecture* 6 *is true for* $k = 1, 2, 3$.
2) *Let* $k = 4$, *and assume that the conditions of Conjecture* 6 *are satisfied. Then there is a* $\tau \in \mathbb{R}$ *such that* $f(n) = \tau \log n + U(n) \pmod 1$ *and either* (A) *or* (B) *hold:*
   (A) $\mathcal{H} = E_4, U(\mathbb{N}) \subseteq E_4$
   (B) $\mathcal{H}$ *consists of four distinct elements of* E_5, *i.e.* $\mathcal{H} = \{\kappa^{l_1}, \kappa^{l_2}, \kappa^{l_3}, \kappa^{l_4}\}$, *where* κ *is any nonzero element of* E_5, *moreover* $U(\mathbb{N}) \subseteq E_5$ *and* $U(n+1) - U(n) \in \mathcal{H}$ *for every large* n.

Remark: We think that case (B) cannot hold, which would follow if we could prove that $U(\mathbb{N}) = E_5$ implies that for every $\alpha \in E_5$, $U(n+1) - U(n) = \alpha$ occurs infinitely often.

5 Characterization of Continuous Homomorphisms as Elements of $\mathcal{A}_G$ for Compact Groups

We investigated this topic in a series of papers written jointly with Z. Daróczy [36–41].

Assume in this section that G is a metrically compact Abelian group supplied with some translation invariant metric ϱ. An infinite sequence $\{x_n\}_{n=1}^{\infty}$ in G is said to belong to $\mathcal{E}_D$, if for every convergent subsequence $x_{n_1}, x_{n_2}, \ldots$ the "shifted subsequence" $x_{n_1+1}, x_{n_2+1}, \ldots$ is convergent, too. Let $\mathcal{E}_\Delta$ be the set of those sequences $\{x_n\}_{n=1}^{\infty}$ for which $\Delta x_n = x_{n+1} - x_n \to 0$ $(n \to \infty)$ holds. Then $\mathcal{E}_\Delta \subseteq \mathcal{E}_D$. We say that $f \in \mathcal{A}_G^*$ belongs to $\mathcal{A}_G^*(\Delta)$ (resp. $\mathcal{A}_G^*(D)$) if the sequence $\{f(n)\}_{n=1}^{\infty}$ belongs to $\mathcal{E}_\Delta$ (resp. $\mathcal{E}_D$).

We proved the following assertions.

(1) $\mathcal{A}_G^*(\Delta) = \mathcal{A}_G^*(D)$.

(2) If $f \in \mathcal{A}_G^*(D)$, then there exists a continuous homomorphism $\Phi : \mathbb{R}_\times \to G$ such that $f(n) = \Phi(n) (n \in \mathbb{N})$.
The proof of (2) was based upon the theorem of Wirsing (Theorem 18).
The set of all limit points of $\{f(n)\}_{n=1}^{\infty}$ form a compact subgroup in G which is denoted by S_f.

(3) $f \in \mathcal{A}_G^*(D)$ if and only if there exists a continuous function $H : S_f \to S_f$ such that $f(n+1) - H(f(n)) \to 0$ as $n \to \infty$.

(4) In [41] we characterized those $f \in \mathcal{A}_G^*$ for which with some continuous function $F : S_f \to S_f$ the relation $f(2n-1) - F(f(n)) \to 0$ $(n \to \infty)$ holds. For $G = T$ we obtained that either $f(n) = 0$ for every odd n, or there exists a nonzero $\lambda \in \mathbb{R}$ such that $f(n) = \lambda \log n (\mathrm{mod}\ 1)$ for every $n \in \mathbb{N}$.

(5) In [44] we solved the following problem. Let G_1, G_2 be metrically compact Abelian groups with some translation invariant metrics. Let $f \in \mathcal{A}_{G_1}^*, g \in \mathcal{A}_{G_2}^*$, and assume that with some continuous function $F : S_f \to S_g$ the relation $g(n-1) - F(f(n)) \to 0$ $(n \to \infty)$ holds. E.g. for $G_1 = T$ we proved: Under the above conditions, either $g(n) = 0$ identically, or there exist $\tau \in \mathbb{R}, M \in \mathbb{N}, u \in \mathcal{A}_{E_M}$ such that $f(n) = \frac{\tau}{M} \log n + u(n) \bmod 1$. Let $\lambda(n) := Mf(n)(n \in \mathbb{N})$. Then the correspondence $\lambda(n) \leftrightarrow g(n)(n \in \mathbb{N})$ generates a topological isomorphism between S_λ and S_g. The converse assertion is also true.

(6) Further interesting results were obtained by Phong [42], [43].

(7) The main problem we are interested in is the following one:
Let $f_j \in \mathcal{A}_{G_j} (j = 0, \ldots, k-1)$, G_j be compact groups, $e_n := \{f_0(n), f_1(n+1), \ldots, f_{k-1}(n+k-1)\}$. Then $e_n \in S_{f_0} \times \cdots \times S_{f_{k-1}} (=: U)$. What can we say about the functions f_j if the set of limit points is not everywhere dense in U? We shall formulate our guesses only for special cases.

Conjecture 7 Let $f \in \mathcal{A}_T^*, S_f = T, e_n = (f(n), \ldots, f(n+k-1))$. Then, either $f(n) = \lambda \log n \pmod 1$ with some $\lambda \in \mathbb{R}$, or $\{e_n | n \in \mathbb{N}\}$ is dense in $T_k = T \times \cdots \times T$.

Conjecture 8 Let $f, g \in \mathcal{A}_T^*, S_f = S_g = T, e_n := (f(n), g(n+1))$. If e_n is not everywhere dense in $T_2 = T \times T$, then f and g are rationally dependent continuous homomorphisms, i.e. there exist $\lambda \in \mathbb{R}, s \in \mathbb{Q}$ such that $g(n) \equiv sf(n) (\mathrm{mod}\ 1)$, $f(n) = \lambda \log n \pmod 1$.

Mauclaire proved in [45] that if G is an arbitrary locally compact group and $f \in \mathcal{A}_G$ satisfies $\Delta f(n) \to 0$ $(n \to \infty)$ then f is the restriction of a continuous homomorphism $\varphi : \mathbb{R}_\times \to G$. Ruzsa and Tijdeman proved [46] that it cannot be generalized for all groups.

6 Sets of Uniqueness for Completely Additive Functions

Definition: We say that $E \subseteq \mathbb{N}$ is a set of uniqueness for the functions belonging to $\mathcal{A}^*$ if $f \in \mathcal{A}^*$, $f(E) = 0$ implies that $f(\mathbb{N}) = 0$.

I introduced this notion in [47], and in [48] it was proved that if to the sequence of "prime + one"s we adjoin a finite set of integers then we obtain a set of uniqueness. My guess that the set of shifted primes itself is a set of uniqueness, was proved by Elliott [49].

It was proved by Wolke [49], and Dress and Volkman [50], that in order for a set E to be such a set of uniqueness, it is necessary and sufficient that every positive integer n has a multiplicative representation:

$$n^k = \prod_{i=1}^{k} a_{j_i}^{\varepsilon_i} \quad a_{j_i} \in E, \quad \varepsilon_i = \pm 1.$$

The h, k may vary with n. They used vector spaces over the field of rational numbers.

In [52] Elliott proved my further conjecture, namely that if $f \in \mathcal{A}^*$, $M(x) = \max_{n \le x} |f(n)|$, $E(x) = \max_{p \le x} |f(p+1)|$, then

$$M(x) \le AE(x^B) \quad x \ge 2 \tag{6.1}$$

holds with suitable numerical constants A, B. For the wider class $f \in \mathcal{A}$ he got a weaker result, namely that

$$M(x) \le AE(x^B) + AM((\log x)^C)$$

for some $C > 0$.

Wirsing extended (6.1) for $f \in \mathcal{A}$ [53]. He proved that every $n \in \mathbb{N}$ has a representation

$$n^h = \prod_{i=1}^{k} (p_i + 1)^{\varepsilon_i}$$

where h and k are bounded, $\varepsilon_i = \pm 1$, and the primes p_i lie in an interval $n < p_i \le n^B$. In particular, Wirsing's result showed that for the multiplicative group K generated by the "prime plus one"s $\mathbb{Q}_\times / K$ has bounded order.

Another interesting consequence of Wirsing's result is that $f \in \mathcal{A}$, $f(p+1) \to 0$ $(p \in \mathcal{P})$ implies that $f(n) = 0$.

My motivation with the investigation of the set of shifted primes was the following. In 1968 I proved [54] that $f \in \mathcal{A}$ has a limit distribution on the set of shifted primes if the three series

$$\sum_{|f(p)| \le 1} \frac{f(p)}{p}, \quad \sum_{|f(p)| \le 1} \frac{f^2(p)}{p}, \quad \sum_{|f(p)| \ge 1} \frac{1}{p} \tag{6.2}$$

are convergent. But the question of the necessity of these conditions remained open.

The necessity of the convergence of the series was proved by additional assumptions: a) if $f(p) \geq 0$, by Elliott [55]; b) if $f(p) = \mathcal{O}(1)$, by Kátai [56]. Finally it was proved without any other conditions by Hildebrand [58] in 1988. From his result it follows that, if $f \in \mathcal{A}$ satisfies

$$\frac{1}{\pi(x)} \#\{p \leq x : |f(p+1)| \geq \varepsilon\} \longrightarrow 0 \quad (x \longrightarrow \infty)$$

for every $\varepsilon > 0$, then $f(n) = 0$ identically.

The notion of sets of uniqueness can be extended into group valued arithmetical functions.

Definition 2 Let G be an arbitrary Abelian group. We say that $E \subseteq \mathbb{N}$ is a set of uniqueness for the class of functions in $\mathcal{A}_G^*$ if $f \in \mathcal{A}_G^*$, $f(E) = 0$ implies that $f(\mathbb{N}) = 0$.

For $G = T$ the following assertion has been proved by Meyer [58], Indlekofer [59], Dress and Volkman [51], see also Elliott [60]:

In order that E would be a set of uniqueness for the class $\mathcal{A}_T^*$ it is necessary and sufficient that every positive integer n has a representation

$$n = \prod_{j=1}^{s} a_j^{d_j}$$

with some integers d_j, positive, negative or zero.

Probably, the set of "prime plus one"s is a set of uniqueness for $\mathcal{A}_T^*$ but it does not seem to be easy. Presently it is not disproved even that $\frac{1}{2}\Omega(p+1) \equiv 0 \pmod 1$ for every large p.

In my paper [15] implicitly it was proved that there is a constant L such that every integer n has a representation

$$n = A \prod_{i=1}^{k} (p_i + 1)^{\varepsilon_i}, \quad \varepsilon_i = \pm 1,$$

where A is such a rational number in the reduced form of which all prime factors are less than L. The constant L was implicit, since I used the Bombieri-Vinogradov theorem. Later Elliott [61] proved that $L = 10^{387}$ is appropriate.

This bound is extremely large for computation. If we could reduce it to 10^{12}, say, then with a massive computer calculation perhaps we could prove that $K = \mathbb{Q}_\times$.

Recently Elliott [62] proved that the factor group $\mathbb{Q}_\times / K$ is either trivial or is of order 2, or 3.

Schinzel and Sierpinski in 1958 stated the conjecture [62], that every positive rational has infinitely many representations of the form $(p+1)(q+1)^{-1}$ with $p, q \in \mathcal{P}$. From this $K = \mathbb{Q}_\times$ would immediately follow.

By using the method of Chen [63] one can prove that every natural number n has infinitely many representations of the form $(P_2 + 1)(Q_2 + 1)^{-1}$, where P_2, Q_2 run over the integers the number of prime factors of which is at most 2. Consequently the multiplicative group K_1 generated by set $P_2 + 1$, where P_2 runs over the integers having at most two prime factors equals to $\mathbb{Q}$.

References

1. P. Erdös, On the distribution function of additive functions, *Ann. Math.* **47** (1946), 1–20.
2. I. Kátai, A remark on additive arithmetical functions, *Annales Univ. Sci. Budapest, Sectio Math.* **10** (1967), 81–83.
3. E. Wirsing, A characterization of log n as an additive arithmetic function, *Symposia Mathematica, Instituto Nationale de Alta Mathematica* IV (1970).
4. I. Kátai, On a problem of P. Erdös, *J. Number Theory* **2** (1970), 1–6.
5. E. Wirsing, Characterization of the logarithm as an additive function, Proceedings of the 1969 Summer Institute of number theory, prepared by the *Amer. Math. Soc.* (1971), 375–381.
6. E. Wirsing, Additive and completely additive functions with restricted growth, *Recent Progress in Analytic Number Theory*, vol. 2, London, 1981, pp. 231–280.
7. I. Kátai, On additive functions, *Publ. Math. Debrecen* **25** (1978), 251–257.
8. I. Kátai, Characterization of log n, Studies in Pure Mathematics (to the memory of Paul Turán), Akadémiai Kiadó, Budapest, 1984, pp. 415–421.
9. I. Kátai, Some results and problems in the theory of additive functions, *Acta Sci. Math. Szeged* **30** (1969), 305–311.
10. I. Kátai, On number-theoretical functions, Colloquia Mathematica Sociates János Bolyai, vol. 2., North-Holland, Amsterdam, 1970, pp. 133–136.
11. J.L. Mauclaire, *Sur la regularité des fonctions additives*, Seminaire Delange-Pisot-Poitou, Theorie des Nombres, Paris **15** (1973/74), no. 23.
12. P.D.T.A. Elliott, On additive arithmetic function $f(n)$ for which $f(an+b) - f(cn+d)$ is bounded, *J. Number Theory* **16** (1983), 285–310.
13. P.D.T.A. Elliott, Arithmetic functions and integer products, Springer V., New York, 1985.
14. A. Hildebrand, An Erdös-Wintner theorem for differences of additive functions, *Trans. Amer. Math. Soc.* **310** (1988), 257–276.
15. P.D.T.A. Elliott, The value distribution of differences of additive arithmetic functions, *J. Number Theory* **32** (1989), 339–370.
16. I. Kátai and Indlekofer, K.-H., Estimation of generalized moments of additive functions over the set of shifted primes, *Acta Sci. Math.* **56** (1992), 229–236.
17. P.D.T.A. Elliott, Sums and differences of additive arithmetic functions in mean square, *J. reine und angewandte Mathematik* **309** (1979), 21–54.
18. I. Kátai, Multiplicative functions with regularity properties, I–VI, *Acta Math. Hung.* **42** (1983), 295–308; **43** (1984), 105–130, 259–272; **44** (1984), 125–132; **45** (1985), 379–380; **58** (1991), 343–350.
19. I. Kátai, Arithmetical functions satisfying some relations, *Acta Sci. Math.* **55**, 249–268.
20. I. Kátai and Phong, B.M., On some pairs of mulitplicative functions correlated by an equation, New Trends in Probability and Statistics, Analytic and Probabilistic Methods in Number Theory, TEV, Vilnius, Lithuania **4**, 191–203.
21. J. Fehér, Kátai, I. and Phong, B.M., On multiplicative functions satisfying a special relation, *Acta Sci. Math.* (accepted).
22. I. Kátai, Some problems in number theory, *Studia Scient. Math. Hung.* **16** (1981), 289–295.
23. A. Hildebrand, Multiplicative functions at consecutive integers II., *Math. proc. Cambridge Phil. Soc.* **103** (1988), 389–398.
24. K.-H. Indlekofer and Kátai, I., Multiplicative functions with small increments III., *Acta Math. Hung.* **58** (1991), 121–132.
25. E. Wirsing, Tang Yuansheng and Shao Pintsung, On a conjecture of Kátai for additive functions, *J. Number Theory* **56** (1996), 391–395.
26. I. Kátai, Additive functions with regularity properties, *Acta Sci. Math.* **44** (1982), 299–305.

27. B.M. Phong, A characterization of some arithmetical multiplicative functions, *Acta Math. Hung.* **63** (1994), 29–43.
28. N.L. Bassily and Kátai, I., On the pairs of multiplicative functions satisfying some relations, *Aequationes Math.* **55** (1998), 1–14.
29. Tang Yuansheng, A reverse problem on arithmetic functions, *J. Number Theory* **58** (1996), 130–138.
30. I. Kátai, On additive functions satisfying a congruence, *Acta Sci. Math.* **47** (1984), 85–92.
31. R. Styer, A problem of Kátai on sums of additive functions, *Acta Sci. Math.* **55** (1991), 269–286.
32. M. Wijsmuller, Additive functions on the Gaussian integers, *Publ. Math. Debrecen* **38** (1991), 255–262.
33. I. Kátai and Wijsmuller, M., Additive functions satisfying congruences, *Acta Sci. Math.* **56** (1992), 63–72.
34. I. Kátai and Subbarao, M.V., The characterization of $n^{i\tau}$ as a multiplicative function, *Acta Math. Hung.* (accepted).
35. I. Kátai and Subbarao, M.V., On the multiplicative function $n^{i\tau}$, *Studia Sci. Math.* (accepted).
36. Z. Daróczy and Kátai, I., On additive arithmetical functions with values in the circle group, *Publ. Math. Debrecen* **34** (1987), 307–312.
37. Z. Daróczy and Kátai, I., On additive arithmetical functions with values in topological groups, *Publ. Math. Debrecen* **33** (1986), 287–292.
38. Z. Daróczy and Kátai, I., On additive number-theoretical functions with values in a compact Abelian group, *Aequationes Math.* **28** (1985), 288–292.
39. Z. Daróczy and Kátai, I., On additive arithmetical functions with values in topological groups II, *Publ. Math. Debrecen* **34** (1987), 65–68.
40. Z. Daróczy and Kátai, I., On additive functions taking values from a compact group, *Acta Sci. Math.* **53** (1989), 59–65.
41. Z. Daróczy and Kátai, I., Characterization of additive functions with values in the circle group, *Publ. Math. Debrecen* **36** (1989), 1–7.
42. B.M. Phong, Note on multiplicative functions with regularity properties, *Publ. Math. Debrecen* **41** (1992), 117–125.
43. B.M. Phong, Characterization of additive functions with values in a compact Abelian group, *Publ. Math. Debrecen* **40** (1992), 273–278.
44. Z. Daróczy and Kátai, I., Characterization of pairs of additive functions with some regularity property, *Publ. Math. Debrecen* **37** (1990), 217–221.
45. J.L. Mauclaire, On the regularity of group valued additive arithmetical functions, *Publ. Math. Debrecen* **44** (1994), 285–290.
46. I.Z. Ruzsa and Tijdeman, R., On the difference of integer-valued additive functions, *Publ. Math. Debrecen* **39** (1991), 353–358.
47. I. Kátai, On sets characterizing number-theoretical functions, *Acta Arithm.* **13** (1968), 315–320.
48. I. Kátai, On sets characterizing number-theoretical functions II (The set of "prime plus one"s is a set of quasi uniqueness), *Acta Arithm.* **16** (1968), 1–4.
49. P.D.T.A. Elliott, A conjecture of Kátai, *Acta Arithm.* **26** (1974), 11–20.
50. D. Wolke, Bemerkungen über Eindeutigkeitsmengen additiver Functionen, *Elem. der Math.* **33** (1978), 14–16.
51. F. Dress and Volkman, B., Ensembles d'unicité pour les fonctions arithemtiques additives ou multiplicatives, *C. R. Acad. Sci.* Paris Ser. A **287** (1978), 43–46.
52. P.D.T.A. Elliott, On two conjectures of Kátai, *Acta Arithm.* **30** (1976), 35–39.
53. E. Wirsing, Additive functions with restricted growth on the numbers of the form $p + 1$, *Acta Arithm.* **37** (1980), 345–357.

54. I. Kátai, On distribution of arithmetical functions on the set of prime plus one, *Compositio Math.* **19** (1968), 278–289.
55. P.D.T.A. Elliott, On the limiting distribution of $f(p+1)$ for non-negative additive functions, *Acta Arithm.* **25** (1974), 259–264.
56. I. Kátai, Some remarks on additive arithmetical functions, *Litovsk. Mat. Sb.* **9** (1969), 515–518.
57. A. Hildebrand, Additive and multiplicative functions on shifted primes, *Proc. London Math. Soc.* **53** (1989), 209–232.
58. J. Meyer, *Ensembles d'unicité pour les fonctions additives. Étude analogue dans les cas des fonstions multiplicatives*, Journées de Théorie Analytique et Élémentaire des Nombres, Orsay, 2 et 3 Juin, 1980. Publications Mathématiques d'Orsay, 50–66.
59. K.-H. Indlekofer, On sets characterizing additive and multiplicative arithmetical functions, *Illinois J. of Math.* **25** (1981), 251–257.
60. P.D.T.A. Elliott, On representing integers as products of integers of a prescribed type, *J. Australian Math. Soc.* (Series A) **35** (1983), 143–161.
61. P.D.T.A. Elliott, On representing integers as products of the $p+1$, *Monatschrifte für Math.* **97** (1984), 85–97.
62. P.D.T.A. Elliott, *The multiplicative group of rationals generated by the shifted primes*, I (manuscript).
63. J. Chen, On the representation of a larger even integer as the sum of a prime and the product of at most two primes, *Sci. Sinica* **16** (1973), 157–176.

Eotvos University
Math Institute
Budapest, Muzeum Krt. 6-8
1088, Hungary
E-Mail: katai@compalg.elte.hu
imre.katai@elte.hu

Hamburger's Theorem on $\zeta(s)$ and the Abundance Principle for Dirichlet Series with Functional Equations

Marvin I. Knopp

I Introduction

Ask any mathematician - indeed any number theorist - to state Hamburger's theorem; chances are the response will be something like, "Riemann's function $\zeta(s)$ is uniquely determined by its functional equation." In fact, this is correct, as far as it goes, but (as is often the case) closer examination show that it does not go nearly far enough.

Hecke grasped the subtleties inherent in Hamburger's theorem (1921) at least by 1944. In his final published paper [8], appearing that year, Hecke describes two versions of the theorem. I quote from the introduction to [8] (translation mine):

"The analytic function $\varphi(s)$ of the complex variable s is determined up to a constant by the following conditions: Put $R(s) = \pi^{-s}\Gamma(s)\varphi(s)$".

1. With a suitable polynomial $P(s)$ suppose that $P(s)\varphi(s)$ is an entire function of finite genus.
2. Suppose $\varphi(s)$ satisfies the equation

$$R(s) = R\left(\frac{1}{2} - s\right).$$

3(a). Suppose that not only $\varphi(s)$, but also $\varphi(s/2)$, can be expanded in a Dirichlet series convergent somewhere:[1] $\varphi(s) = \sum_{n=1}^{\infty} b(n)n^{-2s}$. This condition can also be replaced by

3(b). Suppose that the only pole allowed for $\varphi(s)$ is $s = 1/2^2$; but we assume only the expressibility of $\varphi(s)$ itself as a Dirichlet series $\varphi(s) = \sum_{n=1}^{\infty} b(n)n^{-s}$, not that of $\varphi(s/2)$.

"Mr. Hamburger first proved that $\varphi(s)$ is uniquely determined by $1, 2, 3$ (a) and thus $=$ const. $\zeta(2s)$" [5]; "that also $1, 2, 3$ (b) suffice I have proved within the framework of a general investigation, by means of reduction to the theory of certain automorphic functions [6]."

While Hamburger discovered and gave the first proof of the well-known theorem bearing his name, for our purposes Siegel's elegant proof, published one year after Hamburger's, has greater relevance. The two formulations described by Hecke are in some ways quite distinct, but Siegel's (and, indeed, Hamburger's) proof of the Hamburger version and Hecke's proof of his own version are closely linked by their common use of the Mellin transform or, more accurately, its inverse. The idea in both cases is to show that the inverse Mellin transform of the function $R(s)$ described by Hecke is a constant times $\vartheta(z)-1$, where ϑ is the classical Jacobi function, given by

$$\vartheta(z)=\sum_{n=-\infty}^{\infty} e^{\pi i n^2 z}=1+2\sum_{n=1}^{\infty} e^{\pi i n^2 z}, \tag{1.1}$$

for z in $\mathcal{H}=\{z\in\mathbb{C}\big|\operatorname{Im} z>0\}$. From this fact it follows immediately that $\varphi(s)=$ const. $\zeta(2s)$.

As we shall observe in §II.2 (see especially (2.6)), the Hamburger condition 3(a), above, implies immediately that the inverse Mellin transform of $R(s)$ has the form

$$\sum_{n=1}^{\infty} b(n)e^{\pi i n^2 z}, \tag{1.2}$$

and thus has the general shape of $\vartheta(z)-1$, even before the application of conditions 1 and 2. Siegel then shows that the two latter conditions lead to a "modular relation" for the series (1.2), not that of a modular form, but the more general relation of a "modular integral." (See (2.12), below, and §IV.2 for the definition.) As it turns out, in the presence of (1.2) this more general relation suffices to imply that the function defined by (1.2) equals const.$\{\vartheta(z)-1\}$, and thus to conclude the proof.

The Hecke condition 3(b), on the other hand, gives nothing like (1.2) (only that $F(z)$, the inverse Mellin transform of $R(s)$, has the form $\sum_{n=1}^{\infty}\beta(n)e^{\pi i n z}$), but the severe restriction on the singularities of $\varphi(s)$ in 3(b), together with condition 2, implies instead that with a_0 suitably chosen, $F(z)+a_0$ is a modular form of weight $\frac{1}{2}$, possessing precisely the same transformation properties as does $\vartheta(z)$. Hecke then invokes a familiar uniqueness result on modular forms to conclude that $F(z)+a_0=a_0\vartheta(z)$, and thus that $\varphi(s)=2a_0\zeta(2s)$.

With these contrasting versions of Hamburger's theorem in mind, it appears natural to relax both the expressibility of $\varphi(s/2)$ as a Dirichlet series in 3(a) and the restriction on the poles of $\varphi(s)$ in 3(b), to conjecture that $\varphi(s)$ is uniquely determined by 1, 2 and

3. Suppose (only) that $\varphi(s)$ can be expanded in a Dirichlet series convergent somewhere.

While appealing, this conjectured "strong Hamburger's theorem" fails spectacularly. Indeed, [12, Theorem 1] presents the

Abundance Principle for Dirichlet Series with Functional Equation. *There exist infinitely many linearly independent Dirichlet series satisfying the conditions* 1, 2 *and* 3.

There are generalizations of this Principle. For detailed statements see §V.1, below, [12, §§I & V], and [13, Theorem 1].

The proofs of the Principle and its generalizations fall into two steps. The first is an application of the Riemann-Hecke correspondence, as extended by Bochner [1], to translate the question of existence of the desired Dirichlet series into a question of existence of modular integrals with equivalent properties (§IV.4). The second step is the construction, by means of Eichler's generalized Poincaré series [4, 11], of infinitely many linearly independent modular integrals of the appropriate kind. (See §V.2 for further details.)

II Siegel's proof

1. Preliminary observations. We begin by outlining Siegel's celebrated proof of Hamburger's theorem [17], above all because of its relevance to our point of view. Indeed, his proof foreshadows our approach to the Abundance Principle, featuring an application of the Riemann-Hecke-Bochner correspondence, fourteen years before Hecke developed it as a systematic link between modular forms, on the one hand, and Dirichlet series with functional equations, on the other [6, 7], and twenty-nine years before Bochner's generalization [1, 3] to modular integrals. (We stop short of claiming that Siegel's work antedates Riemann's invention of the correspondence in the latter's derivation of the functional equation of $\zeta(s)$ from the transformation formula of $\vartheta(z)$ under $z \to -\frac{1}{z}$. See (4.1), below.)

Furthermore, aside from Hurwitz's construction of the Eisenstein series E_2 of weight 2 on the full modular group [9], Siegel's proof contains the first (to my knowledge) published example of a modular integral with log-polynomial period function. (See (2.12) and §IV.3, below, for the definition.) It is certainly the first occurrence of a modular integral within the context of the Riemann-Hecke-Bochner correspondence.[3] (Of course, Siegel proceeds to show that the modular integral is a multiple of $\vartheta(z) - 1$ and thus, in fact, a modular form; however, this small irony does not diminish the point.) A model of mathematical insight and elegance, this proof is relevant to research today, notwithstanding the passage of three-quarters of a century.

The statement that Siegel-like Hamburger before him-proves differs from Hecke's description of it in two respects. It posits the existence of two Dirichlet series

$$f(s) = \sum_{n=1}^{\infty} a_n n^{-s}, \quad g(s) = \sum_{n=1}^{\infty} b_n n^{-s} \tag{2.1}$$

and a polynomial $P(s)$, such that

(i) $P(s)f(s)$ is an entire function of finite genus;

(ii) $f(s)$ converges absolutely for $\sigma = \operatorname{Re} s > 2 - \theta$ (some $\theta > 0$);

(iii) $g(s)$ converges absolutely for $\sigma > 1 + \alpha$ (some $\alpha > 0$);

(iv) $$\pi^{-s/2}\Gamma(s/2)f(s) = \pi^{-(\frac{1-s}{2})}\Gamma\left(\frac{1-s}{2}\right)g(1-s). \tag{2.2}$$

The conclusion: $f(s) = g(s) = \text{const.}\zeta(s)$.

The two ways in which Hecke rephrased the hypotheses (2.2) can be read in condition (2.2, iv). In Hecke the two Dirichlet series $f(s)$, $g(s)$ have been replaced by the single $\varphi(s)$. However, this apparent loss of generality is not significant since (2.2, iv) implies immediately that $R_1(s) = \pi^{-s/2}\Gamma(s/2)\{f(s) + g(s)\}$ and $R_2(s) = \pi^{(-s/2)}\Gamma(s/2)\{f(s) - g(s)\}$ satisfy, respectively, $R_1(s) = R_1(1-s)$ and $R_2(s) = -R_2(1-s)$, functional equations with the same Dirichlet series on both sides.

Hecke's second change in the functional equation amounts to a replacement of s by $2s$ in (2.2, iv), which then becomes

$$\pi^{-s}\Gamma(s)\hat{f}(s) = \pi^{-(\frac{1}{2}-s)}\Gamma\left(\frac{1}{2}-s\right)\hat{g}\left(\frac{1}{2}-s\right), \tag{2.3}$$

where $\hat{f}(s) = f(2s)$ and $\hat{g}(s) = g(2s)$. With $\hat{f} = \hat{g} = \varphi$ and $R(s) = \pi^{-s}\Gamma(s)\varphi(s)$, (2.3) reduces to the Hecke formulation $R(s) = R(\frac{1}{2} - s)$. That $\varphi(s) = \hat{f}(s) = f(2s)$ accounts for the condition 3(a) in Hecke's description of Hamburger's version, the condition which, in his own formulation, he replaces by 3(b), the restriction on the poles of $\varphi(s)$.

2. Outline of the Proof. Siegel's proof is considerably more direct than that of Hamburger, using only the familiar formulae

$$e^{-y} = \frac{1}{2\pi i}\int_{1-i\infty}^{1+i\infty} y^{-s}\Gamma(s)ds,\ y > 0 \tag{2.4}$$

(the inverse Mellin transform of $\Gamma(s)$ is e^{-y}) and

$$\int_0^\infty e^{-a^2x-b^2/x}\frac{dx}{\sqrt{x}} = \frac{\sqrt{\pi}}{a}e^{-2ab}, \tag{2.5}$$

for $a > 0, b \geq 0$ (evaluation of the Bessel integral). The first step is the simple observation that (2.2, iv) implies $S_1 = S_2$ for $y > 0$, where

$$\begin{aligned} S_1 &= \frac{1}{2\pi i}\int_{2-i\infty}^{2+i\infty} f(s)\Gamma(s/2)\pi^{-s/2}y^{-s/2}ds, \\ S_2 &= \frac{1}{2\pi i}\int_{2-i\infty}^{2+i\infty} g(1-s)\Gamma\left(\frac{1-s}{2}\right)\pi^{-(\frac{1-s}{2})}y^{-s/2}ds. \end{aligned}$$

From (2.4) and an easily-justified interchange of sum and integral it follows that

$$S_1 = 2\sum_{n=1}^{\infty} a_n e^{-\pi n^2 y}. \tag{2.6}$$

Thus S_1 is the exponential series

$$2\sum_{n=1}^{\infty} a_n e^{\pi i n^2 z},\ \operatorname{Im} z > 0,$$

evaluated on the positive imaginary axis $z = iy$, $y > 0$. It is this series that must be proved equal to $\alpha(\vartheta(z) - 1)$, with $\alpha \in \mathbb{C}$. (This equation is equivalent to: $a_n = \alpha$ for all $n \geq 1$. The same is true of the equation $f(s) = \alpha\zeta(s)$.)

The conditions (2.2, i, ii), the functional equation (2.2, iv), Stirling's formula and the Phrágmen-Lindelöf principle combine to give an estimate on the growth of $g(1-s)$ in the vertical strip $-\alpha - 1 \leq \sigma \leq 2$, an estimate sufficiently strong to make possible an application of the residue theorem to S_2 in the infinite strip $-\alpha - 1 \leq \sigma \leq 2$, $|t| \geq T_0 > 0$ (T_0 sufficiently large). This yields

$$S_2 = \frac{1}{2\pi i}\int_{-\alpha-1-i\infty}^{-\alpha-1+i\infty} g(1-s)\Gamma\left(\frac{1-s}{2}\right)\pi^{-(\frac{1-s}{2})}y^{-s/2}ds + \sum_{\nu=1}^{m} R_\nu(y), \tag{2.7}$$

where the $R_\nu(y)$ are the residues of the integrand at the poles in the region $-\alpha - 1 < \sigma < 2$.

The Dirichlet series representation (2.1) of $g(s)$ and (2.4) together show that the integral in (2.7) is equal to

$$\frac{2}{\sqrt{y}}\sum_{n=1}^{\infty} b_n e^{-\pi n^2/y},\ y>0.$$

On the other hand, calculation of the residues of the integrand leads to

$$\sum_{\nu=1}^{m} R_\nu(y)=\sum_{\nu=1}^{m} y^{-s_\nu/2}Q_\nu(\log y)=Q(y), \tag{2.8}$$

where the Q_ν are polynomials and the s_ν are the poles of the integrand in $-\alpha-1<\sigma<2$. (The conditions (2.2, ii) and (2.2, iv) imply that Re $s_\nu\le 2-\theta$, for $1\le\nu\le m$.) Then, $S_1=S_2$, (2.6) and (2.7) together yield the fundamental transformation property

$$2\sum_{n=1}^{\infty} a_n e^{-\pi n^2 y}=\frac{2}{\sqrt{y}}\sum_{n=1}^{\infty} b_n e^{-\pi n^2/y}+Q(y). \tag{2.9}$$

At this juncture Siegel introduces another integral transform, multiplying both sides of (2.9) by $e^{-\pi t y^2}$, with fixed $t>0$, and integrating on y from 0 to ∞. Absolute convergence justifies termwise integration on both sides of (2.9); the application of (2.5) on the right-hand side leads to

$$\sum_{n=1}^{\infty} a_n\left(\frac{1}{t+ni}+\frac{1}{t-ni}\right)-\pi t H(t)=2\pi\sum_{n=1}^{\infty} b_n e^{-2\pi n t}, \tag{2.10}$$

where $H(t)=\sum_{\nu=1}^{m} t^{s_\nu-2}H_\nu(\log t)$, with H_ν a polynomial. Since the identity (2.10) clearly extends to the half-plane Re $t>0$, Siegel can exploit the periodicity of the right-hand side of (2.10) and the singularities of the left-hand side, to conclude that $a_n=a_1$ for all $n\ge 1$, thus that $f(s)=a_1\zeta(s)$. This completes the proof.

Remark: If we put

$$F(z)=2\sum_{n=1}^{\infty} a_n e^{\pi i n^2 z},\ G(z)=2\sum_{n=1}^{\infty} b_n e^{\pi i n^2 z},$$

then (2.9) becomes

$$F(z)=(z/i)^{-1/2}G(-1/z)+Q(z/i), \tag{2.11}$$

on the positive imaginary axis: $z=iy$, $y>0$. By the principle of analytic continuation, (2.11) holds in all of $\mathcal{H}$. Thus, in $\mathcal{H}$, $K(z)=F(z)+G(z)$ satisfies

$$K(z)=(z/i)^{-1/2}K(-1/z)+Q(z/i)-(z/i)^{-1/2}Q(i/z). \tag{2.12}$$

With $Q(z/i)$ of the form (2.8), the same is true of $(z/i)^{-1/2}Q(i/z)$, so that $Q(z/i)-(z/i)^{-1/2}Q(i/z)$ is a "log-polynomial sum." Consequently, $K(z)$ is a "modular integral with log-polynomial period function." (See §IV, 3 below.)

III Hecke's Proof

The beginning of the Hecke proof in essence follows that of Siegel, modified only to take account of the modified hypotheses. In place of the functional equation (2.2, iv) linking the Dirichlet series f and g, Hecke assumes

$$\pi^{-s}\Gamma(s)\varphi(s) = \pi^{-(\frac{1}{2}-s)}\Gamma\left(\frac{1}{2}-s\right)\varphi\left(\frac{1}{2}-s\right), \tag{3.1}$$

with φ a Dirichlet series: $\varphi(s) = \sum_{n=1}^{\infty} c_n n^{-s}$. In place of Siegel's assumption (2.2, i), which permits f to have an arbitrary finite number of poles in the s-plane, Hecke imposes the condition that $(s-\frac{1}{2})\varphi(s)$ can be continued to an entire function of finite genus. That is to say, in Hecke's formulation the polynomial $P(s) = s - 1/2$; this is what Hecke actually intends in his condition 3(b). (See §I, above.)

Siegel's technique, employing the inverse Mellin transform, is equally effective under Hecke's modified assumptions. Using this procedure, Hecke obtains, in place of Siegel's transformation law (2.9),

$$\sum_{n=0}^{\infty} c_n e^{-\pi n y} = \frac{1}{\sqrt{y}} \sum_{n=0}^{\infty} c_n e^{-\pi n/y}, \tag{3.2}$$

where $c_n, n \geq 1$, are the coefficients in the expression of $\varphi(s)$ as a Dirichlet series and $c_0 = \operatorname{Res}_{s=1/2} \varphi(s)$. (Comparison shows obvious changes from (2.9), upon which we shall comment shortly.)

Putting $L(z) = \sum_{n=0}^{\infty} c_n e^{\pi i n z}$ permits us to rewrite (3.2) as

$$L(z) = \frac{1}{\sqrt{-iz}} L(-1/z),\ z = iy,\ y > 0. \tag{3.3}$$

Since $\varphi(s)$ is assumed to converge in some right half-plane, the coefficients c_n have at worst polynomial growth in n. This implies that $L(z)$ is holomorphic in $\mathcal{H}$, so (3.3) holds in all of $\mathcal{H}$. The definition of $L(z)$ shows further that $L(z+2) = L(z)$, and this, together with (3.3) and the growth condition on the coefficients c_n, implies that $L(z)$ is an entire modular form of weight 1/2 on Γ_ϑ, the subgroup of index 3 in $SL(2, Z)$ generated by $z \to z+2$ and $z \to -1/z$. (See §IV.2, below. Recall that the full modular group $SL(2, Z)$ is generated by $z \to z+1$ and $z \to -1/z$.)

Since $L(z)$ has precisely the same transformation properties with respect to the generators of Γ_ϑ as does the Jacobi ϑ-function (1.1) (see (4.1), below), Hecke can complete his proof simply by comparing $L(z)$ with $\vartheta(z)$. It turns out that $L(z)/\vartheta(z)$ is entire modular *function* (i.e. modular form of weight 0) on Γ_ϑ, and thus a constant. But $L(z) = \text{const.}\ \vartheta(z)$ leads directly to $\varphi(s) = \text{const.}\zeta(s)$. This concludes the proof.

Remarks: The transformation law (3.3) differs from (2.12) in two essential respects:

(i) The series defining $K(z) = F(z) + G(z)$ is supported on integral squares, while the exponents in the series defining $L(z)$ are linear in n.

(ii) The "period function" $Q(z/i) - (z/i)^{-1/2}Q(i/z)$ appearing on the right-hand side of (2.12) is not present in (3.3)

That $K(z)$ is not *a priori* a modular form on Γ_ϑ, but only a modular integral with period function, makes Hecke's proof unavailable under the conditions Siegel imposes (notwithstanding that, as a consequence of his proof of $f(s) = g(s) = a_1\zeta(s)$, $K(z) = 2a_1(\vartheta(z) - 1)$, so $K(z) + 2a_1$ is indeed a modular form).

Equally, Siegel's proof fails if applied to Hecke's case. For, multiplying both sides of (3.2) by $e^{-\pi t^2 y}$ and integrating on y from 0 to ∞ yields

$$\sum_{n=0}^{\infty} \frac{c_n t}{\pi(t^2+n)} = \sum_{n=0}^{\infty} c_n e^{-2\pi t\sqrt{n}} \tag{3.4}$$

in place of (2.10). Siegel's derivation of Hamburger's theorem from (2.10) breaks down if we start instead with (3.4).

IV Modular Integrals and the Riemann-Hecke-Bochner Correspondence

The term "modular integral" has arisen several times above, most prominently in the discussion of Siegel's proof of Hamburger's theorem (§II). In this section we define the notion, important here for its role in explaining the origin of the Abundance Principle for Dirichlet Series (§I).

The ideas we shall introduce apply to the entire class of finitely generated discrete groups Γ acting on $\mathcal{H}$, of finite or infinite hyperbolic area. However, for the most part we restrict the discussion to $\Gamma = \Gamma_\vartheta$, the subgroup of $SL(2, Z)$ generated by $z \to z+2$, $z \to -1/z$. (See §III, above.) The group Γ_ϑ is so called because of its connection with the Jacobi function defined by (1.1): $\vartheta(z)$ is an entire modular form of weight $\frac{1}{2}$ on Γ_ϑ. That is, for z in $\mathcal{H}$,

$$\vartheta(z+2) = \vartheta(z),\ \vartheta(-1/z) = e^{-\pi i/4} z^{1/2}\vartheta(z), \tag{4.1}$$

and $\vartheta(z)$ has bounded behaviour at the two parabolic points of a fundamental region $\mathcal{R}$ for Γ_ϑ. ($\mathcal{R}$ can be so chosen that the two parabolic points are $i\,\infty$ and -1. The expansion (1.1) expresses the behaviour at $i\infty$; there is a similar expansion at -1. See [10, Theorem 13, 46].)

1. Multiplier systems and period cocycles for Γ_ϑ. Let k be a real number and v a "multiplier system" for the weight k and the group Γ_ϑ. This means that v is a function on the group Γ_ϑ - thought of as a *matrix group* - such that

$$\text{(i)}\ \ |v(M)| = 1 \text{ for all } M \text{ in } \Gamma_\vartheta;$$

$$\text{(ii)}\ \ v(M_3)(c_3z+d_3)^k = v(M_1)v(M_2)(c_1M_2z+d_1)^k(c_2z+d_2)^k. \tag{4.2}$$

The identity (4.2, ii) is required to hold for all z in $\mathcal{H}$ and M_1, M_2 in Γ_ϑ, with $M_3 = M_1M_2$ and $M_i = \begin{pmatrix} * & * \\ c_i & d_i \end{pmatrix}$, $1 \le i \le 3$. It is not too hard to show from (4.2, ii) that v is a character on the matrix group Γ_ϑ when $k \in Z$, and a character on the linear fractional transformation group Γ_ϑ when $k \in 2Z$.

With k real, v a fixed multiplier system in weight k for Γ_ϑ and f a function defined on $\mathcal{H}$, we introduce the *stroke* (or *slash*) *operator*

$$(f|_k^v M)(z) = \bar{v}(M)(cz+d)^{-k} f(Mz), M = \begin{pmatrix} * & * \\ c & d \end{pmatrix} \in \Gamma_\vartheta. \tag{4.3}$$

With this notation, condition (4.2, ii) is equivalent to

$$f|_k^v M_1 M_2 = (f|_k^v M_1)|_k^v M_2, \tag{4.4}$$

for M_1, M_2 in Γ_ϑ and any f defined on $\mathcal{H}$.

Suppose f is a function holomorphic in $\mathcal{H}$; define the functions $q_M(z) = q(M; z)$, $M \in \Gamma_\vartheta$, as follows:

$$\bar{v}(M)(cz+d)^{-k} f(Mz) = f(z) + q_M(z), \tag{4.5, i}$$

with $M = \begin{pmatrix} * & * \\ c & d \end{pmatrix} \in \Gamma_\vartheta$. This can be rewritten as

$$f|M = f + q_M, M \in \Gamma_\vartheta. \tag{4.5, ii}$$

(Note the abbreviation $f|M$ for $f|_k^v M$.) The q_M are called the *period functions* of f relative to $\{\Gamma_\vartheta, k, v\}$. From (4.4) the *cocycle condition* for $\{q_M | M \in \Gamma_\vartheta\}$ follows directly:

$$q_{MN} = q_M|N + q_N, \text{ for } M, N, \in \Gamma_\vartheta. \tag{4.6}$$

A collection of functions $\{q_M | M \in \Gamma_\vartheta\}$ satisfying (4.6) is called a *cocyle for* (or *relative to)* $\{\Gamma_\vartheta, k, v\}$.

The condition (4.5) above does not restrict f in any way, since q_M is defined to be $f|M - f$. To construct a meaningful theory we impose restrictions upon the q_M, suited to the purpose at hand. In the present context it is essential to assume that the q_M lie in $\mathcal{P}$, the collection of all functions f holomorphic in $\mathcal{H}$, subject to the growth condition

$$|f(z)| \le K(|z|^\alpha + y^{-\beta}), y = \operatorname{Im} z > 0, \tag{4.7}$$

for some constants $K, \alpha, \beta > 0$. Note that $\mathcal{P}$ is an algebra over the complex field $\mathbb{C}$; moreover, it is preserved under differentiation, integration and the stroke operator (4.3), with $M \in SL(2, \mathbb{R})$.

2. Modular integrals. Assume that $\{q_M\}$ is a cocycle relative to $\{\Gamma_\vartheta, k, v\}$ such that $q_M \in \mathcal{P}$, for all $M \in \Gamma_\vartheta$. Suppose that f is holomorphic in $\mathcal{H}$ and satisfies (4.5). Standard arguments using (4.5) imply the existence of "Fourier expansions" for f at the parabolic cusps $i\infty$ and -1, of Γ_ϑ. (See, for example, [10, Chapter 2, 17–23].) Let $S = \begin{pmatrix} 1 & 1 \\ 0 & 1 \end{pmatrix}$ and $T = \begin{pmatrix} 0 & -1 \\ 1 & 0 \end{pmatrix}$. The expansion at $i\infty$ has the form

$$f(z) = \rho_0(z) + \sum_{n=-\infty}^{\infty} a_n e^{\pi i(n+\kappa)z}, \operatorname{Im} z > 0, \tag{4.8a}$$

with complex coefficients a_n. Here ρ_0 is a function in $\mathcal{P}$ determined by $q(S^2; z)$ and κ derives from the multiplier system $v : v(S^2) = e^{2\pi i\kappa}, 0 \leq \kappa < 1$. The expansion at -1 has the form

$$f(z) = \rho_1(z) + (z+1)^{-k} \sum_{n=-\infty}^{\infty} b_n \exp\{-2\pi i(n+\mu)/(z+1)\}, \tag{4.8b}$$

with complex b_n and $\rho_1 \in \mathcal{P}$ determined by $q(S^{-2}T; z)$; μ is defined by $v(S^{-2}T) = e^{2\pi i\mu}$, $0 \leq \mu < 1$.

Assume that the expansions (4.8) are both left-finite. Then we call f a *modular integral with respect to Γ_ϑ, of weight k and multiplier system v, with period functions* (or *period cocycle*) $\{q_M | M \in \Gamma_\vartheta\}$. If no a_n, b_n with $n < 0$ occur in the expansions (4.8) we call f an *entire* modular integral. If $q_M = 0$ for all M in Γ_ϑ, f is a *modular form* (*entire* modular form) rather than a modular integral (*entire* modular integral).

In §IV.4 we shall introduce the Riemann-Hecke-Bochner (R-H-B) correspondence, which plays an essential role in the derivation of the Abundance Principle (§I, above). To gain a measure of flexibility we state the correspondence not merely for Γ_ϑ, but for the entire class of "Hecke triangle groups". With $\lambda > 0$ let $S_\lambda = \begin{pmatrix} 1 & \lambda \\ 0 & 1 \end{pmatrix}$ (hence $S_1 = S = \begin{pmatrix} 1 & 1 \\ 0 & 1 \end{pmatrix}$ and $S_2 = S^2$). Then the *Hecke group* G_λ is defined by

$$G_\lambda = \langle S_\lambda, T \rangle. \tag{4.9}$$

(Note that $G_1 = \Gamma(1) = SL(2, Z)$ and $G_2 = \Gamma_\vartheta$.)

Hecke [7] has shown that G_λ is discrete if and only if (i) $\lambda \geq 2$ or (ii) $\lambda = 2\cos \pi/p$, with $p \in Z$, $p \geq 3$. When $\lambda \geq 2$, G_λ has the single relation $T^2 = I$; when $\lambda = 2\cos\pi/p$, G_λ has the two relations $T^2 = (S_\lambda T)^p = I$. When $\lambda \leq 2$, G_λ has a fundamental region of finite hyperbolic area. For $\lambda > 2$, by contrast, this area is infinite.

One can define "period cocyle for G_λ" and "(entire) modular integral" by analogy with the definitions given above for $\Gamma_\vartheta (\lambda = 2)$. For general $\lambda > 0$, the expansion (4.8a) is replaced by

$$f(z) = \rho_0(z) + \sum_{-\infty}^{\infty} a_n e^{2\pi i(n+\kappa)/\lambda}, \quad \operatorname{Im} z > 0. \tag{4.10}$$

If $\lambda \neq 2$, the point -1 is not a parabolic cusp, so there is no analogue of (4.8b). The definitions of "modular integral" and "entire modular integral" of course entail the same restrictions on the expansion (4.10) as in the case $\lambda = 2$.

3. Modular integrals with log-polynomial period. Our application to the Abundance Principle for Dirichlet series requires period cocycles $\{q_M | M \in G_\lambda\}$ satisfying conditions far more stringent than $q_M \in \mathcal{P}$. To describe these conditions we introduce *log-polynomial sums*, functions of the form

$$q(z) = \sum_{1 \leq j \leq J} z^{\alpha_j} \sum_{0 \leq t \leq M(j)} \beta(j,t)(\log z)^t, \tag{4.11}$$

where the exponents α_j and the coefficients $\beta(j, t)$ are arbitrary complex numbers. The t are nonnegative integers. Note that a log-polynomial sum is holomorphic in $\mathbb{C}\backslash\{iy | y \leq 0\}$; in particular, $q(z)$ defined by (4.11) is holomorphic in $\mathcal{H}$.

We say that the modular integral f on Γ_ϑ (respectively, G_λ) has *log-polynomial* period, provided $q(S^2; z) \equiv 0$ (respectively, $q(S_\lambda; z) \equiv 0$) and $q(T; z)$ is a log-polynomial sum, for the generators S^2 (respectively, S_λ) and T, in the modular transformation law for f ((4.5) or its analogue for G_λ). Note that $q(T; z^2)$ and $q(S^2; z)(q(S_\lambda; z))$ are in $\mathcal{P}$. Since $\Gamma_\vartheta = \langle S^2, T\rangle$ (respectively, $G_\lambda = \langle S_\lambda, T\rangle$), it follows that $q_M \in \mathcal{P}$ for all $M \in \Gamma_\vartheta$ (respectively, $M \in G_\lambda$), by the closure properties of $\mathcal{P}$ and the cocycle condition (4.6) (or its analogue for G_λ). The cocycle condition implies as well that $q_T | T + q_T = 0$, since $T^2 = I$.

The significance for us of modular integrals with log-polynomial period is this: *By the Riemann-Hecke-Bochner correspondence* (§IV.4, below), *the Abundance Principle for Dirichlet Series with Functional Equation is equivalent to the existence of infinitely many linearly independent entire modular integrals on* Γ_ϑ, *with log-polynomial period.*

4. The Riemann-Hecke-Bochner correspondence. Before stating the correspondence we make a few observations about the expansion (4.8a). If f is an entire modular integral on Γ_ϑ, then by our definition the expansion has the form

$$f(z) = \rho_0(z) + \sum_{n=0}^{\infty} a_n e^{\pi i(n+\kappa)z}.$$

If, in addition, $q(S^2, z) \equiv 0$ (as is the case when f has log-polynomial period), the expansion has the form $f(z) = \sum_{n=0}^{\infty} a_n e^{\pi i(n+\kappa)z}$.

We note further that the only relation in Γ_ϑ, $T^2 = I$, does not involve the generator S^2. Thus, in any weight k we can determine a multiplier system v on Γ_ϑ by choosing $v(S^2) = e^{2\pi i\kappa}$, with $0 \leq \kappa < 1$, but κ otherwise arbitrary, and putting $v(T) = C$, with C chosen to respect the relation $T^2 = I$. By (4.2, ii), this means that $v(T)$ has one of the four values $\pm e^{-\pi ik/2}$, $\pm ie^{-\pi ik/2}$; however, only the two values $\pm e^{-\pi ik/2}$ give rise to nontrivial modular forms or modular integrals on Γ_ϑ. Now, given M in Γ_ϑ we write M as a word in S^2 and T, and determine $v(M)$ from $v(S^2)$, $v(T)$ by applying the consistency condition (4.2, ii).

In the statement of the correspondence we choose $\kappa = 0$, so $v(S^2) = 1$. Then the expansion at $i\infty$ of an entire modular integral f on Γ_ϑ assumes the form

$$f(z) = \sum_{n=0}^{\infty} a_n e^{\pi inz}, \ \operatorname{Im} z > 0. \tag{4.12}$$

R-H-B Correspondence. Let k be real and C complex. Suppose $F(z)$ is holomorphic in the upper half-plane $\mathcal{H}$, defined there by an exponential series of the form (4.12), where the complex coefficients a_n satisfy the polynomial growth condition

$$a_n = \mathcal{O}(n^\gamma), \gamma > 0, n \longrightarrow \infty. \tag{4.13}$$

Let $\Phi(s)$ be the Mellin transform of $F(iy) - a_0$:

$$\Phi(s) = \int_0^\infty \{F(iy) - a_0\} y^s \frac{dy}{y} = \pi^{-s}\Gamma(s) \sum_{n=1}^{\infty} a_n n^{-s}.$$

Then, we have the following.

1. The assertions (A) and (B) are equivalent:
   (A) $F(z)$ is an entire modular integral on Γ_ϑ, with log-polynomial period, of weight k and multiplier system v such that $v(S^2) = 1$, $v(T) = C$.
   (B) Φ can be continued analytically into the entire s-plane, except for a finite number of poles. Furthermore, Φ is bounded in each lacunary vertical strip,

   $$\sigma_1 \leq \operatorname{Re} s \leq \sigma_2, \, |\operatorname{Im} s| \geq t_0 > 0, \quad t_0 \text{ sufficiently large},$$

   and satisfies the functional equation

   $$\Phi(k - s) = e^{\pi i k/2} C \Phi(s). \tag{4.14}$$

2. Consider the modular relation included in condition (A):

   $$z^{-k} F(-1/z) = C F(z) + q_T(z), \tag{4.15}$$

   with $q_T(z)$ of the form (4.11). The term $\beta z^\alpha (\log z)^t$ ($\beta \neq 0, t \in Z, t \geq 0$) occurs in $q_T(z)$ if and only if $\Phi(s)$ has poles of order $t + 1$ at $s = \alpha + k$ and $s = -\alpha$. The only possible further singularities of Φ are simple poles at $s = 0$ and $s = k$.

Remarks: (i) By (4.14), $C = \pm e^{-\pi i k/2}$. (See the discussion immediately preceding the statement of the R-H-B correspondence.)

(ii) The correspondence, as stated here, differs somewhat from Bochner's original formulation in [1], which deals with generalized Dirichlet series rather than the ordinary Dirichlet series we have here; Bochner allows as well two distinct exponential series in the modular relation. Furthermore, Bochner's period function $q_T(z)$ in (4.15) is a "residual function" (in his terminology) rather than a sum of the form (4.11). (See also [3].) However, the sums (4.11) are residual in Bochner's sense, and a residual function appearing as a period function in a modular relation (4.15) necessarily has the form (4.11).

V The Abundance Principle for Dirichlet Series with Functional Equation

1. Detailed statement of results. In the Introduction we have stated the Abundance Principle in general terms, without explanatory details. We provide those now. Recall the conditions 1, 2, 3 of §I: Let $\varphi(s)$ be an analytic function and put $R(s) = \pi^{-s}\Gamma(s)\varphi(s)$. Assume the following concerning $\varphi(s)$ and $R(s)$.

1. There exists a polynomial $P(s)$ such that $P(s)\varphi(s)$ is an entire function of finite genus.

2. $R(s) = R(\frac{1}{2} - s)$.
3. $\varphi(s)$ can be expanded in a Dirichlet series convergent in some right half-plane.

Theorem 1 *Let σ_0 be a real number $\geq \frac{1}{4}$. Let $\mathcal{A}(\sigma_0)$ be the space of rational functions $A(s)$ with poles restricted to the strip $\frac{1}{2} - \sigma_0 \leq \text{Re } s \leq \sigma_0$ and satisfying the symmetry condition $A(\frac{1}{2} - s) = A(s)$. Let $\mathcal{A}_H(\sigma_0)$ be the subspace of A in $\mathcal{A}(\sigma_0)$ such that $R(s) - A(s)$ is entire for some $R(s) = \pi^{-s}\Gamma(s)\varphi(s)$ satisfying 1, 2 and 3. Then for $A_1, \ldots, A_n \in \mathcal{A}(\sigma_0)$, with $n > [\frac{\sigma_0}{2} + \frac{3}{8}] + 2$, some nontrivial linear combination of $A_1, \ldots, A_n$ is in $\mathcal{A}_H(\sigma_0)$.*

Theorem 1 can be generalized to

Theorem 2 (Weight k Abundance Principle). *Let k be an arbitrary real number. For $\sigma_0 \geq k/2$, let $\mathcal{A}(\sigma_0; k)$ denote the space of rational functions $A(s)$ with poles restricted to the strip $k - \sigma_0 \leq \text{Re } s \leq \sigma_0$ and such that $A(k - s) = A(s)$. Replace condition 2 by the functional equation*

$$R(s) = R(k - s). \tag{2_k}$$

Let $\mathcal{A}_H(\sigma_0; k)$ be the subspace of A in $\mathcal{A}(\sigma_0; k)$ such that $R(s) - A(s)$ is entire for some $R(s) = \pi^{-s}\Gamma(s)\varphi(s)$ satisfying $1, 2_k$ and 3. Then for $A_1, \ldots, A_n \in \mathcal{A}(\sigma_0; k)$, with $n > N(\sigma_0, k)$, some nontrivial linear combination of $A_1, \ldots, A_n$ lies in $\mathcal{A}_H(\sigma_0; k)$. Here, $N(\sigma_0, k)$ is an explicit constant dependent only upon σ_0 and k.

For $k \geq 2$, Theorem 2 can be strengthened considerably, to

Theorem 3 (Mittag-Leffler Principle). *Let $k \geq 2$ and let $A(s)$ be any rational function satisfying $A(k-s) = A(s)$. Then there exists $\varphi(s)$ such that $R(s) = \pi^{-s}\Gamma(s)\varphi(s)$ satisfies $1, 2_k$ and 3, and such that $R(s) - A(s)$ is entire.*

Finally, we can extend all of these results to the case in which 2_k is replaced by the functional equation

$$R(s) = -R(k - s). \tag{4_k}$$

Theorem 4 **(a)** *For k an arbitrary real number and $\sigma_0 \geq k/2$, let $\mathcal{B}(\sigma_0; k)$ be the space of rational functions $A(s)$ with poles restricted to the strip $k - \sigma_0 \leq \text{Re } s \leq \sigma_0$ and satisfying $A(k - s) = -A(s)$. Let $\mathcal{B}_H(\sigma_0; k)$ denote the subspace of A in $\mathcal{B}(\sigma_0; k)$ such that $R(s) - A(s)$ is entire for some $R(s) = \pi^{-s}\Gamma(s)\varphi(s)$ satisfying $1, 3$ and 4_k. Then for $A_1, \ldots, A_n$ in $\mathcal{B}(\sigma_0; k)$ with $n > M(\sigma_0, k)$, some nontrivial linear combination of $A_1, \ldots, A_n$ lies in $\mathcal{B}_H(\sigma_0; k)$. Here $M(\sigma_0, k)$ is an explicit constant determined by σ_0 and k.*

(b) *Let $k \geq 2$ and suppose $A(s)$ is a rational function with $A(k-s) = -A(s)$. Then there exists $\varphi(s)$ such that $R(s) = \pi^{-s}\Gamma(s)\varphi(s)$ satisfies $1, 3$ and 4_k, and such that $R(s) - A(s)$ is entire.*

Theorem 1 is to be found in [12, 362, Theorem 1]. Theorem 3 appeared as Theorem 2 of the same article (362–3).

2. Brief discussion of the proofs. The first step is the observation that the abundance of Dirichlet series with a functional equation is equivalent, by the R-H-B correspondence, to the abundance of entire modular integrals on Γ_ϑ, with log-polynomial period. In this equivalence the functional equation 2_k is associated with modular integrals of weight k and multiplier system v_k^+ determined by

$$v_k^+(S^2) = 1, v_k^+(T) = e^{-\pi ik/2}, \tag{5.1}$$

while 4_k corresponds to modular integrals of weight k and multiplier system v_k^- determined by

$$v_k^-(S^2) = 1, v_k^-(T) = -e^{-\pi ik/2}. \tag{5.2}$$

The proofs, then, entail the construction of "many" modular integrals of fixed weight k on Γ_ϑ, with log-polynomial period for both multiplier systems v_k^+, v_k^-.

The construction of these modular integrals relies upon Eichler's "generalized Poincaré series" (GPS), first introduced in [4], and later developed and applied in [14,11,12,13]. To construct a GPS we must have, at the outset, a period cocycle $\{q_M\}$ on Γ_ϑ, in weight k and connected with the appropriate multiplier system, either v_k^+ or v_k^- in the present situation. We require further that $q(S^2; z) \equiv 0$ and that $q(T; z)$ is a log-polynomial sum. (Recall the alternative notation $q(M; z)$ for $q_M(z)$.) Starting with these two restrictions, we wish to generate $\{q_M\}$ by applying the cocycle condition (4.6).

As it turns out, the necessary condition

$$q_T|T + q_T = 0 \tag{5.3}$$

of §IV.3 is sufficient for this process to yield a well-defined period cocycle $\{q_M\}$. Initially, let q be any log-polynomial sum whatsoever and put $q_T = q|T - q$, which is then a log-polynomial sum satisfying (5.3). We now define a period cocycle in Γ_ϑ by writing $M \in \Gamma_\vartheta$ as a word in S^2 and T and then applying (4.6) several times to define q_M. While the relation $T^2 = I$ gives rise to the complication that M is not given uniquely as a word in S^2 and T, the restriction (5.3) on q_T guarantees the uniqueness of q_M defined in this manner.

Next, let m be a positive even integer. With $\{q_M\}$ in hand, define Eichler's *generalized poincaré series* $\Psi(\{q_M\}; m; z) = \Psi(z)$ by

$$\Psi(z) = \sum_V q_V(z)(cz + d)^{-m}; \tag{5.4}$$

the summation is on all $V = \begin{pmatrix} * & * \\ c & d \end{pmatrix} \in \Gamma_\vartheta$ with distinct lower rows. (The condition $q(S^2; z) \equiv 0$ ensures that the individual terms of the series depend only upon the lower row c, d of $V \in \Gamma_\vartheta$.) Eichler shows that, for sufficiently large m, the series (5.4) converges absolutely for z in $\mathcal{H}$, and uniformly on compact subsets of $\mathcal{H}$ [4]. Eichler's proof assumes that the q_V are polynomials, but a slight elaboration of his method establishes convergence for the more general case $q_V \in \mathcal{P}$, $V \in \Gamma_\vartheta$ [11, 615–619]. Since we began with q_T a log-polynomial sum, it follows that $q_V \in \mathcal{P}$ for all V in Γ_ϑ, so the proof applies here.

As a consequence of absolute convergence, $\Psi(z)$ has the transformation property

$$(\Psi|M)(z) = (\gamma z + \delta)^m \Psi(z) - (\gamma z + \delta)^m E_m(z) q_M(z), \tag{5.5}$$

for $M = \begin{pmatrix} * & * \\ \gamma & \delta \end{pmatrix} \in \Gamma_\vartheta$, where $E_m(z)$ is the familiar Eisenstein series of weight m,

$$E_m(z) = \sum_V (cz+d)^{-m} \tag{5.6}$$

with summation conditions as in (5.4). Since m is an even integer, $E_m(z)$ does not vanish identically, so we may form the quotient $F(z) = -\Psi(z)/E_m(z)$. By (5.5) and the well-known modular transformation properties of $E_m(z)$, $F(z)$ has the formal behavior of a modular integral relative to $(\Gamma_\vartheta, k, v^\pm)$, with the preassigned cocycle $\{q_M\}$ of period functions:

$$F|M = F + q_M, \quad M \in \Gamma_\vartheta. \tag{5.7}$$

On the other hand, F may not be an entire modular integral (or a modular integral at all) in the sense of §IV.2, since the zeros of $E_m(z)$ in $\mathcal{H}$ may well be poles of F. There is the further issue that F may not behave appropriately at the parabolic cusps, -1 and $i\,\infty$, of Γ_ϑ. The proofs of the Theorems are largely procedures for modifying F to obtain entire modular integrals for $\{\Gamma_\vartheta, k, v^\pm\}$.

In all of the proofs the key point is this: the log-polynomial sum q_T is completely arbitrary except for the restriction (5.3). The proof of Theorem 4 employs the multiplier system v_k^-, given in (5.2), while the proofs of Theorems 1–3 utilize v_k^+, characterized by (5.1). This is the only distinction. The proofs of those results labelled "Mittag-Leffler" rely upon a "Mittag-Leffler" theorem for modular forms of weights $k \geq 2$. This theorem fails in weights $k < 2$. For detailed proofs see [12].

VI Conclusion

1. Extension to other Hecke groups. All of the above results rely upon considerations regarding the particular Hecke triangle group $\Gamma_\vartheta = \langle S^2, T\rangle$. There is a modification of these results, in which the function $R(s) = \pi^{-s}\Gamma(s)\varphi(s)$ is replaced by

$$R_\lambda(s) = \left(\frac{2\pi}{\lambda}\right)^{-s} \Gamma(s)\varphi_\lambda(s), \lambda > 2, \tag{6.1}$$

with $\varphi_\gamma(s)$ a Dirichlet series convergent in some right half-plane. (In the notation of (6.1), $R(s) = R_2(s)$.) Bringing to bear the Hecke triangle groups G_λ defined in (4.9) with $\lambda > 2$, leads to

Theorem 5 (Mittag-Leffler) (a) *Let k be an arbitrary real number and $A(s)$ any rational function such that $A(k-s) = A(s)$. Then there exists $\varphi_\lambda(s)$ such that $R_\lambda(s)$, defined in* (6.1), *satisfies* 1, 2_k *and* 3, *and such that $R_\lambda(s) - A(s)$ is entire.*

(b) *The same, with $A(k-s) = -A(s)$ and 2_k replaced by 4_k.*

The proof, based in part upon a Mittag-Leffler theorem for modular forms on G_λ *of all real weights*, can be found in [13].

2. Zeros. Ever since Riemann's path-breaking work on $\zeta(s)$ [16], there has been a great deal of interest in the zeros of Dirichlet series, especially those with Euler product. While we can assert nothing about Euler products for the Dirichlet series constructed here, the doctoral dissertation of Lekkerkerker [15] contains results yielding information about the distribution of their zeros in the plane. These generalize familiar results concerning the zeros of Riemann's $\zeta(s)$.

Specifically, Theorem 4 in [15, chapter II, 24] immediately implies the following for the Dirichlet series $\varphi(s)$ occurring in our Theorems 1–5: *Let $\mathcal{R}$ be a rectangle of the form*

$$|\mathrm{Re}\, s - k/2| \le \alpha, |\mathrm{Im}\, s| \le \beta,$$

with α, β real, and such that $\varphi(s)$ is holomorphic in the complement of $\mathcal{R}$. Let $N_1(T)$ ($N_2(T)$) denote the number of zeros s_0 of $\varphi(s)$ in the complement of $\mathcal{R}$ such that $0 < \mathrm{Im}\, s_0 < T (-T < \mathrm{Im}\, s_0 < 0)$. Then,

$$N_i(T) = \frac{1}{\pi} T \log T + \delta_\varphi T + \mathcal{O}(\log T),$$

$T \to \infty$ for $i = 1$ and 2. Here δ_φ is a real number determined by $\varphi(s)$.

Chapter IV of [15] presents results concerning the zeros on the "critical line" (in our notation, $\mathrm{Re}\, s = k/2$) of Dirichlet series with functional equations, but there are technical difficulties in applying them to the Dirichlet series constructed here. This matter bears further investigation.

Notes

1. (p. 1) Hecke's words are "$\varphi(s/2)$ soll in eine *irgendwo* konvergente Dirichlet-Reihe entwickelbar sein" (emphasis added). However Hamburger's proof [5] requires that $\varphi(s/2)$ be absolutely convergent for $\mathrm{Re}\, s > 1$. Siegel [17] relaxes this condition, assuming convergence for $\mathrm{Re}\, s > 2 - \theta, \theta > 0$. To my knowledge, the proof of the stronger formulation of Hamburger's version, asserted by Hecke, did not appear until 1956 [2, Theorem 7.1].
2. (p. 1) Hecke assumes that the pole at $s = 1/2$ is simple.
3. (p. 3) In hindsight, Hamburger's original proof [5] does suggest these ideas, but they are not actually present.

References

1. S. Bochner, Some properties of modular relations, *Ann. Math.* **53** (1951), 332–360.
2. S. Bochner and Chandrasekharan, K., On Riemann's functional equation, *Ann. Math.* **63** (1956), 336–360.
3. K. Chandrasekharan and Narashimhan, R., Hecke's functional equation and arithmetical identities, *Ann. Math.* **74** (1961), 1–23.
4. M. Eichler, Grenzkreisgruppen und kettenbruchartige Algorithmen, *Acta Arith.* **11** (1965), 169–180.
5. H. Hamburger, Über die Riemannsche Funktionalgleichung der ζ-Funktion, *Math. Zeit.* **10** (1921), 240–254.
6. E. Hecke, Über die Bestimmung Dirichletscher Reihen durch ihre Funktionalgleichung, *Math. Ann.* **112** (1936), 664–669.

7. E. Hecke, Lectures on Dirichlet series, modular functions and quadratic forms, Edwards, Ann Arbor, 1938.
8. E. Hecke, Herleitung des Euler-Produkts der Zetafunktion und einiger L-Reihen aus ihr Funktionalgleichung, *Math. Ann.* **119** (1944), 266–287.
9. A. Hurwitz, Grundlagen einer independenten Theorie der elliptischen Modulfunctionen und Theorie der Multiplicatorgleichungen erster Stufe, *Math. Ann.* **18** (1881), 528–591.
10. M. Knopp, *Modular functions in analytic number theory*, Markham, Chicago, 1970; 2nd ed., Chelsea, New York, 1993.
11. M. Knopp, Some new results on the Eichler cohomology of automorphic forms, *Bull. Amer. Math. Soc.* **80** (1974), 607–632.
12. M. Knopp, On Dirichlet series satisfying Riemann's functional equation, *Invent. Math.* **117** (1994), 361–372.
13. M. Knopp and Sheingorn, M., On Dirichlet series and Hecke triangle groups of infinite volume, *Acta. Arith.* **76** (1996), 227–244.
14. J. Lehner, Automorphic forms with preassigned periods, *J. Res. Nat. Bur. Standards Sect.* B **73** (1969), 153–161.
15. C.G. Lekkerkerker, *On the zeros of a class of Dirichlet series*, Doctoral dissertation, Utrecht, 1955, 64 pp.
16. B. Riemann, *Über die Anzhl der Primzahlen unter einer gegebenen Grösse*, Monatsberichte der Berliner Akad. (Nov., 1859), 671–680.
17. C.L. Siegel, Bemerkungen zu einem Satz von Hamburger über die Funktionalgleichung der Riemannschen Zetafunktion, *Math. Ann.* **86** (1922), 276–279.

Department of Mathematics
Temple University
Philadelphia, PA 19122
USA

A Survey of Number Theory and Cryptography

Neal Koblitz

Our purpose is to give an overview of the applications of number theory to public-key cryptography. We conclude by describing some tantalizing unsolved problems of number theory that turn out to have a bearing on the security of certain cryptosystems.

Key-words. Public-Key Cryptography, Primality, Factorization, Discrete Logarithm, Elliptic Curve.

1 Introduction

Cryptography, broadly defined, is the science that studies a wide range of issues in the transmission and safeguarding of information. The increasing importance of cryptography in the "information age" and the concomitant flourishing of cryptographic research have had a profound impact on number theory, causing certain algebraic, analytic, and computational questions unexpectedly to acquire a practical urgency. Most of the applications of number theory have arisen since 1976 as a result of the development of *public-key* cryptography.

This survey is devoted to the questions that one encounters in the study of public-key cryptosystems. After discussing the idea of public-key cryptography and its importance, we next describe certain prototypical public-key constructions that use number theory. Then we survey the progress that has been made in the three number-theoretic problems that are at the heart of these constructions — primality testing, factorization of integers, and discrete logarithms in a finite field. We also treat elliptic curve cryptosystems and the corresponding discrete logarithm problem. We conclude by describing some difficult unsolved problems of number theory that have cryptographic significance.

2 Early History

A cryptosystem for message transmission means a map from units of ordinary text called *plaintext message units* (each consisting of a letter or block of letters) to units of coded text called *ciphertext message units*. The idea of using arithmetic operations to construct such a map goes back at least to the Romans. In modern terminology, they used the operation of addition modulo N, where N is the number of letters in the alphabet, which we suppose has been put in one-to-one correspondence with $\mathbf{Z}/N\mathbf{Z}$. For example, if $N = 26$ (i.e., messages are in the usual Latin alphabet, with no additional symbols for punctuation, numerals, capital letters, etc.), the Romans might encipher single letter message units according to the formula $C \equiv P + 3 \pmod{26}$. This means that we replace each plaintext letter by the letter three positions farther down the alphabet (with the convention that $X \mapsto A$, $Y \mapsto B$, $Z \mapsto C$). It is not hard to see that the Roman system — or in fact any cryptosystem based on a permutation of single letter message units — is easy to break.

In the 16th Century, the French cryptographer Vigenère invented a variant on the Roman system that is not quite so easy to break. He took a message unit to be a block of k letters — in modern terminology, a k-vector over $\mathbf{Z}/N\mathbf{Z}$. He then shifted each block by a "code word" of length k; in other words, his map from plaintext to ciphertext message units was translation of $(\mathbf{Z}/N\mathbf{Z})^k$ by a fixed vector.

For the most part, until about 20 years ago only rather elementary number theory was used in cryptography. The first mention in print of a new type of cryptography seems to have been a passage in a book about time-sharing systems that was published in 1968 (p. 91–92 of [103]). In it, the author describes a new *one-way cipher* used by R. M. Needham in order to make it possible for a computer to verify passwords without storing information that could be used by an intruder to impersonate a legitimate user.

> In Needham's system, when the user first sets his password, or whenever he changes it, it is immediately subjected to the enciphering process, and it is the enciphered form that is stored in the computer. Whenever the password is typed in response to a demand from the supervisor for the user's identity to be established, it is again enciphered and the result compared with the stored version. It would be of no immediate use to a would-be malefactor to obtain a copy of the list of enciphered passwords, since he would have to decipher them before he could use them. For this purpose, he would need access to a computer and even if full details of the enciphering algorithm were available, the deciphering process would take a long time.

In [89] the first detailed description of such a *one-way function* was published. The original passwords and their enciphered forms are regarded as integers modulo a large prime p, and the "one-way" map $\mathbf{F}_p \to \mathbf{F}_p$ is given by a polynomial $f(x)$ which is not hard to evaluate by computer but which takes an unreasonably long time to invert. Purdy used $p = 2^{64} - 59$, $f(x) = x^{2^{24}+17} + a_1x^{2^{24}+3} + a_2x^3 + a_3x^2 + a_4x + a_5$, where the coefficients a_i were arbitrary 19-digit integers.

3 The Idea of Public-Key Cryptography

Until 1976, all cryptographic message transmission was by what can be called *private key*. This means that someone who has enough information to encrypt messages automatically has enough information to decipher messages as well. As a result, any two users of the system who want to communicate secretly must have exchanged keys in a safe way, e.g., using a trusted courier.

The face of cryptography was radically altered when Diffie and Hellman invented an entirely new type of cryptography, called *public-key* [28]. At the heart of this concept is the idea of using a one-way function for encryption. Speaking informally, we can define a public-key encryption function (also called a "trapdoor" function) as a map from plaintext message units to ciphertext message units that can be easily computed by anyone having the so-called "public" key but whose inverse function (which deciphers the ciphertext message units) cannot be computed in a reasonable amount of time without some additional information (the "private" key). This means that everyone can send a message to a given user using the same enciphering key, which they simply look up in a public directory. There is no need for the sender to have made any prior secret arrangement with the recipient; indeed, the recipient need never have had any prior contact with the sender at all.

It was the invention of public-key cryptography that led to a dramatic expansion of the role of number theory in cryptography. The reason is that number theory seems to be the best source of one-way functions. Later we shall discuss the most important examples.

3.1 Tasks for Public-Key Cryptography

The most common purposes for which public-key cryptography has been applied are:

(1) *confidential message transmission*;
(2) *authentication* (verification that the message was sent by the person claimed and that it has not been tampered with), often using *hash functions* and *digital signatures*; *password* and *identification* systems (proving authorization to have access to data or a facility, or proving that you are who you claim to be); *non-repudiation* (guarding against people claiming not to have agreed to something that they really agreed to);
(3) *key exchange*, where two people using the open airwaves want to agree upon a secret key for use in some private-key cryptosystem;
(4) *coin flip* (also called *bit commitment*); for example, two chess players in different cities want to determine by telephone (or e-mail) who plays white;
(5) *secret sharing*, where some secret information (such as the password to launch a missile) must be available to k subordinates working together but not to $k-1$ of them;
(6) *zero knowledge proof*, where you want to convince someone that you have successfully solved a number-theoretic or combinatorial problem (for example, you have found the square root of an integer modulo a large unfactored integer, or you have 3-colored a map) without conveying any knowledge whatsoever of what the solution is.

3.2 Probabilistic Encryption

Most of the number theory based cryptosystems for message transmission are *deterministic*, in the sense that a given plaintext will always be encrypted into the same ciphertext by anyone. However, deterministic encryption has two disadvantages: (1) if an eavesdropper knows that the plaintext message belongs to a small set (for example, the message is either "yes" or "no"), then she can simply encrypt all possibilities in order to determine which is the supposedly secret message; and (2) it seems to be very difficult to *prove* anything about the security of a system if the encryption is deterministic. For these reasons, *probabilistic encryption* was introduced. We will not discuss this further or give examples in the present paper. For more information, see the fundamental papers on the subject [33, 34].

We will now discuss two examples of public-key cryptosystems — RSA [93] and Diffie–Hellman [28] — that have been at the center of research in public key cryptography. Both are connected with fundamental questions in number theory — integer factorization and discrete logarithms, respectively. Although the systems can be modified to perform most or all of the six tasks listed above, we will describe a protocol for only one of these tasks in each case (message transmission in the case of RSA, and key exchange in the case of discrete log).

4 The RSA Cryptosystem

Suppose that we have a large number of users of our system, each of whom might want to send a secret message to any one of the other users. We shall assume that the message units m have been identified with integers in the range $0 \le m < N$. For example, a message might be a block of k letters in the Latin alphabet, regarded as an integer to the base 26 with the letters of the alphabet as digits; in that case $N = 26^k$. In practice, in the RSA system N is a number of between 200 and 300 decimal digits.

Each user A (traditionally named Alice) selects two extremely large primes p and q whose product n is greater than N. Alice keeps the individual primes secret, but she publishes the value of n in a directory under her name. She also chooses at random an exponent e that must have no common factor with $p - 1$ or $q - 1$ (and presumably has the same order of magnitude as n), and publishes that value along with n in the directory. Thus, her *public key* is the pair (n, e).

Suppose that another user B (Bob) wants to send Alice a message m. He looks up her public key in the directory, computes the least nonnegative residue of m^e modulo n, and sends Alice this *ciphertext* value (which we denote c). Bob can perform the modular exponentiation $c = m^e \pmod n$ very rapidly (in $O(\log^3 n)$ bit operations), using the so-called "repeated squaring method" (see §4.6.3 of [45]).

To decipher the message, Alice uses her secret *deciphering key* d, which is any integer with the property that $de \equiv 1 \pmod{p-1}$ and $de \equiv 1 \pmod{q-1}$. She can find such a d easily by applying the extended Euclidean algorithm to the two numbers e and l.c.m. $(p-1, q-1)$. It is easy to check (using Fermat's Little Theorem) that if Alice computes the least nonnegative residue of c^d modulo n, the result will be the original message m.

What would prevent an unauthorized person C (Cynthia) from using the public key (n, e) to decipher the message? The problem for Cynthia is that without knowing the factors p and q of n there is apparently no way to find a deciphering exponent d that inverts the operation $m \mapsto m^e \pmod n$. Nor does there seem to be any way of inverting the encryption other than through a deciphering exponent. Here I use the words "apparently" and "seem" because these assertions have not been proved. Thus, one can only say that *apparently* breaking the RSA cryptosystem is as hard as factoring n.

5 Diffie–Hellman Key Exchange

Another basic type of public-key cryptographic system is based on the discrete logarithm problem. First we define this problem.

Let $\mathbf{F}_q^*$ denote the multiplicative group of the finite field of q elements. Let $g \in \mathbf{F}_q^*$ be a fixed element ("base"). The *discrete log problem* in $\mathbf{F}_q^*$ to the base g is the problem, given $x \in \mathbf{F}_q^*$, of determining an integer y such that $x = g^y$ (if such y exists; otherwise, one must receive an output to the effect that x is not in the group generated by g).

The Diffie–Hellman key exchange works as follows. Suppose that Alice and Bob want to agree upon a large integer to serve as a key for some private-key cryptosystem. This must be done using open communication channels, i.e., any eavesdropper (Cynthia) knows everything Alice sends to Bob and everything Bob sends to Alice. Alice and Bob first agree on the field $\mathbf{F}_q$ and a base element g. They also agree on a map from $\mathbf{F}_q^*$ to the positive integers; for example, in the case $q = p$ it could be the obvious map $\mathbf{F}_p^* \to \{1, 2, \ldots, p-1\}$. All of this is agreed upon publicly, so that Cynthia also has this information at her disposal.

Next, Alice and Bob choose random secret integers k_A and k_B, respectively. Alice sends Bob the group element g^{k_A} (but not her secret integer k_A), and Bob sends Alice the element g^{k_B} (but not the integer k_B). The common key is then $g^{k_A k_B} \in \mathbf{F}_q$ (or rather, the integer associated to $g^{k_A k_B}$ under the agreed upon correspondence). Alice determines this key by taking the element g^{k_B} received from Bob and raising it to the k_A-th power; Bob determines the key by taking the element g^{k_A} that he received from Alice and raising it to the k_B-th power. But Cynthia is in the unfortunate position of having to determine the element $g^{k_A k_B}$ knowing only g^{k_A} and g^{k_B} but not k_A or k_B.

The problem facing Cynthia is the so-called *Diffie–Hellman problem*: Given g, g^{k_A}, $g^{k_B} \in \mathbf{F}_q^*$, find $g^{k_A k_B}$. It is easy to see that anyone who can solve the discrete log problem in $\mathbf{F}_q^*$ can then immediately solve the Diffie–Hellman problem as well. The converse is not known. That is, it is conceivable (though generally thought to be unlikely) that someone could invent a way to solve the Diffie–Hellman problem without being able to find discrete logarithms. In other words, breaking the Diffie–Hellman key exchange has not been *proved* to be equivalent to solving the discrete log problem. But for practical purposes it is probably safe to assume that the Diffie–Hellman key exchange in a group G is secure provided that the discrete logarithm problem in G is intractable. See [14] for the latest partial results supporting the conjectured equivalence of the Diffie–Hellman and discrete log problems.

Several related systems based on the discrete logarithm problem in $\mathbf{F}_q^*$ have also been developed. Perhaps the most important is the Digital Signature Standard (DSS) proposed by the U.S. government as a public key signature method that could play the same role in standardizing digital signatures that DES played in standardizing the commercial use of private-key cryptography.

6 Primality Testing

In the implementation of several of the most popular cryptosystems (RSA, Diffie–Hellman in the multiplicative group of a prime field, etc.) one has to find large prime numbers, usually of 100 to 200 digits. Suppose that we generate a random 100-digit odd number n. By the Prime Number Theorem, n has approximately a 1 out of 115 chance of being prime. We then apply a *test of primality* to n. If n turns out to be composite, we choose another random odd number (or perhaps simply replace n by $n + 2$). If n passes the primality test, then we are happy, and choose $p = n$ to be our prime.

Depending on how demanding we are, the term "primality test" means one of three things:

(1) a sequence of easily verified tests of compositeness, each of which is a congruence that is likely (but not certain) to fail if n is composite; thus, if n satisfies all of these congruences, it is a *probable prime*;

(2) an efficient algorithm that tells us with certainty if n is prime;
(3) an efficient algorithm that not only guarantees the primality of n but also gives us a *certificate of primality*, i.e., a list of number theoretic relations that can all be true only if n is prime and that can be verified very rapidly.

We will discuss each type of primality test in turn.

6.1 Compositeness Tests

The simplest such test is *trial division*. Suppose, for example, that we divide a large random odd number n by all odd primes less than 100. If n is composite, then it is easy to see that there is less than a 25% probability that this trial division will fail to reveal a factor of n. However, trial division is practical only as a preliminary step to eliminate trivial composites.

A more promising compositeness test is based on Fermat's Little Theorem, according to which $a^{n-1} \equiv 1 \pmod{n}$ for any a prime to n whenever n is prime. For any a, recall that a^{n-1} can be computed modulo n in $O(\log^3 n)$ bit operations, i.e., very rapidly. If a is chosen at random, then most composite numbers n will fail the Fermat test. However, this test has a serious drawback in that there exist composite numbers n — called *Carmichael numbers* — for which $a^{n-1} \equiv 1 \pmod{n}$ for *all* a prime to n. Carmichael numbers tend to be rather rare (the smallest is $561 = 3 \cdot 11 \cdot 17$); however, it was recently proved that there are infinitely many Carmichael numbers (see [7]).

There is a refined Fermat test due to Miller [75] that does not suffer from this defect and can be performed as rapidly as the a^{n-1} test. Namely, the Miller test has the feature that *no* composite odd number n passes it for more than 25% of the choices of residue class a prime to n. (For most composite n, the chance of passing the Miller test is actually much less than 25%.) The Miller test proceeds as follows. Write $n - 1 = 2^s t$, where t is odd. For any given a prime to n, first compute a^t modulo n. If this is equal to ± 1, then n passes the test. If not, then repeatedly take the square modulo n, obtaining the residue of $a^{2^j t}$, $j = 1, 2, \ldots, s-1$. If any of these residues is -1 modulo n, then n passes the test. If not, then n fails the test. (Notice that a prime number n would pass the test, because: (i) $a^{2^s t} = a^{n-1} \equiv 1 \pmod{n}$, if n is prime; and (ii) -1 is the only square root of 1 modulo n other than 1 itself, if n is prime.)

Applying the Miller test with 10 different random values of a gives — for all practical purposes — an infallible test that n is prime. In fact, n is almost certain to be prime if it passes the Miller test for just the four values $a = 2, 3, 5, 7$; it turns out that there is only one odd composite number less than $25 \cdot 10^9$ that passes the Miller test for all four of these values of a (see [88]).

6.2 Proofs of Primality

If the generalized Riemann Hypothesis is true, then any composite number n must fail the Miller test for some a in the interval $1 < a < 2\log^2 n$ (see [75, 10]). Thus, applying the Miller test for all of these values of a — a procedure which takes only polynomial time, namely $O(\log^5 n)$ bit operations — will (under the generalized Riemann Hypothesis) give a proof either that n is prime or that it is composite.

In the early 1980s, Adleman, Pomerance, and Rumely [4] developed an elegant and efficient deterministic algorithm that gives a proof of primality using Gauss sums. Cohen and Lenstra [24] then improved upon this algorithm, by using Jacobi sums in place of Gauss sums. The idea was to perform a series of Fermat–like tests with Jacobi sums, working in the ring of integers of the cyclotomic field generated by the n-th roots of unity. The Jacobi sum test requires $(\log n)^{O(\log\log\log n)}$ bit operations. Strictly speaking, the algorithm is not quite polynomial time; and the existence of an unconditional polynomial time deterministic proof of primality is still an open question. However, in practice the Jacobi sum algorithm is very fast in the 200-digit range. This shows that whether or not a given algorithm is polynomial time, while of great theoretical importance, is not always what determines its practical utility. For a detailed description of the algorithm and a discussion of implementation, see [23, 24].

6.3 Certificates of Primality

Given a probable prime n, we now want not only to prove primality, but also to do it in such a way that the validity of our proof can be checked in far less time than it took to produce the proof. At present, the most efficient way to do this (first proposed in [32]) uses elliptic curves $Y^2 = X^3 + aX + b$, where the coefficients a and b and the coordinates (x, y) of points on the curve are integers considered modulo n. (We shall need some basic facts about elliptic curves over finite fields; for an introduction to this subject, see [53], [100], or [43].)

The point of departure is an analogue of the following special case of a theorem of Pocklington [81]: *If $q > \sqrt{n}$ is a prime divisor of $n - 1$ and if there exists an integer a satisfying $a^{n-1} \equiv 1 \pmod{n}$ and* g.c.d.$(a^{(n-1)/q} - 1, n) = 1$, *then n is prime.* This result is easy to prove, by observing that if p is any prime divisor of n, then the assumptions imply that the element $a^{(n-1)/q}$ has exact order q modulo p; but then $q|p - 1$, and so $p > q > \sqrt{n}$.

Pocklington's theorem can be of practical use in proving primality of certain special primes n, but it cannot be applied to prove primality of an arbitrary prime n (because $n - 1$ must be factored, and it must be divisible by a prime $> \sqrt{n}$). On the other hand, the elliptic curve analogue of Pocklington's theorem does in fact lead to a primality proof that is generally applicable. This analogue is as follows: *Let E be an elliptic curve defined over the ring $\mathbf{Z}/n\mathbf{Z}$, and let m be an integer having a prime divisor $q > (n^{1/4} + 1)^2$. If there exists a point P on E such that $m \cdot P = O$ and $(m/q) \cdot P$ is defined and different from O (where O denotes the identity point of E), then n is prime.* We make some remarks about how this statement is converted to a primality algorithm.

1. The integer chosen as m has the property that, if n is prime, then m is the order of the group E. Both the group operation on E and the computation of m are performed under the supposition that n is prime; if the corresponding operations break down because of compositeness of n, then at the same time a nontrivial factor of n will be revealed. In practice, of course, this will never happen, since one goes to the trouble of proving primality only for a probable prime n (e.g., n has already passed the Miller test for $a = 2, 3, 5, 7$).

2. If n is prime, then Hasse's theorem says that the order of the group E is between $n+1-2\sqrt{n}$ and $n+1+2\sqrt{n}$. If n is not prime and if $p|n$ is a prime $< \sqrt{n}$, then the order of E mod p (the group of $\mathbf{F}_p$-points of E) is $\leq p+1+2\sqrt{p} < (n^{1/4}+1)^2$.
3. Multiplying a point on E mod n by a large integer such as m or m/q can be done very efficiently, using the "repeated doubling" method that is analogous to the "repeated squaring" method for exponentiation modulo n.
4. The reason why the elliptic curve analogue of Pocklington — and not the original Pocklington theorem — can be used to test any n is that the number m that is analogous to $n-1$ varies as we vary the elliptic curve E. Instead of being stuck with $n-1$, we thus have flexibility in arriving at a number m that is divisible by $q > (n^{1/4}+1)^2$.
5. Because of the difficulty of factoring m, in practice one looks for an m that is a small multiple of a prime q, i.e., q will actually be much greater than $(n^{1/4}+1)^2$.
6. The most time-consuming part of the algorithm is finding a curve E with suitable m. Once we find E and m (and also a point P with the desired property), then it takes much less time to check that with this choice the conditions of the theorem are satisfied, and so n is prime.
7. The point P is selected at random. Once E is found with the desired property (namely, its order m is divisible by a large prime q), almost any point P on E has order dividing m but not m/q.
8. In practice, q, like n itself, is known only to be a *probable prime*. After completing the algorithm for n, one has a proof of primality contingent upon q being prime. One then applies the procedure recursively to prove primality of q.
9. To find a suitable E, two methods have been proposed, respectively by Goldwasser–Kilian and by A.O.L. Atkin: (1) random selection of the coefficients of the defining equation, and then computation of $m = \#E$ using Schoof's algorithm [96] (or one of the recent improvements — see the discussion below); (2) selection of a complex multiplication field first, and then m, followed by the construction of E. See [78, 9] for more details.

Whichever method is used, the final *certificate of primality* of n will be a sequence of 4-tuples (E_i, m_i, q_i, P_i), the first of which satisfies the conditions of the elliptic Pocklington theorem for the probable prime n, the second of which satisfies the conditions for the probable prime q_1 dividing m_1, the third of which satisfies the conditions for the probable prime q_2 dividing m_2, and so on. Given such a certificate, one can verify its correctness — and hence the primality of n — in polynomial time.

It should be mentioned that it has not been proved that one can always find an elliptic curve modulo n for which m is a small multiple of a prime. The difficulty is that little is known about primes in short intervals. One can avoid this problem by using genus-two hyperelliptic curves rather than elliptic curves (see [3]), but at the cost of having a much more complicated algorithm. The hyperelliptic curve primality test is of theoretical but not practical interest.

We conclude this section by remarking that in cryptographic applications one does not usually need a polynomial time certificate of primality. Thus, one normally requires only that n pass the Miller test for several values of a, or (if one wants to be completely certain of primality) that it pass the Jacobi sum test.

7 Factoring

Since the invention in 1978 of the RSA cryptosystem, the security of which depends upon the assumed difficulty of factoring large integers, more work has been devoted to the factoring problem than to any other problem in computational number theory. Despite some major advances, RSA seems still to be secure if the modulus n is the product of two primes of about 100 digits.

All of the subexponential algorithms to factor general integers (i.e., integers not of any special form) are probabilistic rather than deterministic. Most of them (but not all) are based upon the following simple observation. Suppose that n is an odd number having r distinct prime factors. Then every square in $(\mathbf{Z}/n\mathbf{Z})^*$ has 2^r square roots modulo n. Thus, if we can produce a congruence of the form $x^2 \equiv y^2 \pmod{n}$, where x and y are obtained independently of one another, then there is a $1 - 2^{-(r-1)}$ probability that $x \not\equiv \pm y \pmod{n}$. It is easy to see that as soon as that happens, we need only compute g.c.d. $(x + y, n)$ to obtain a nontrivial factor of n. Even in the "worst" case $r = 2$ (as in RSA), we have a 50% chance of factoring n any time we obtain a congruence $x^2 \equiv y^2 \pmod{n}$. If this congruence does not lead to a factorization of n, then we simply repeat the procedure with different random choices, and we again have a 50% chance of success. The chance is only about 1 in 1000 that we will fail to factor n within 10 repetitions of the algorithm.

7.1 Quadratic Sieve

One of the basic factoring algorithms of this type is called the *quadratic sieve*. Consider the quadratic polynomial $f(X) = (X + [\sqrt{n}])^2 - n \in \mathbf{Z}[X]$. This polynomial has two obvious properties: (1) $f(x) \equiv (x + [\sqrt{n}])^2 \pmod{n}$ for any $x \in \mathbf{Z}$; and (2) $f(x)$ is significantly smaller than n in absolute value (of order roughly $\sqrt{n}$) if x is small in absolute value. Suppose that we evaluate $f(x_i)$ at a sequence of fairly small integers $x_1, x_2, \ldots,$ and find that for some subsequence $x_{i_1}, x_{i_2}, \ldots$ the product of the $f(x_{i_l})$ happens to be a perfect square. Then all we have to do is set x equal to the least positive residue modulo n of $\prod_l (x_{i_l} + [\sqrt{n}])$, and set y equal to the least positive residue modulo n of $\sqrt{\prod_l f(x_{i_l})}$; then we have $x^2 \equiv y^2 \pmod{n}$. Note that x and y were formed in quite different ways, so there is no greater-than-random likelihood that $x \equiv \pm y \pmod{n}$. That is, we obtain a congruence of the desired form.

The hard part is to select the x_{i_l} so that $\prod_l f(x_{i_l})$ is a perfect square. To do this we select a set of primes — called a *factor base* — consisting of $p_1 = 2$ and the first $r - 1$ odd primes $p_2, p_3, \ldots, p_r$ for which $\left(\frac{n}{p_j}\right) = 1$ (i.e., n is a quadratic residue modulo p_j). The choice of r will be made later in a certain optimal way. If x_i is small, then there is a good chance that $f(x_i)$ is not divisible by a prime greater than p_r, in which case it is a product of the primes in the factor base. (Note that if $p | f(x_i)$, then n is a square modulo p; so $f(x_i)$ cannot be divisible by any prime p for which $\left(\frac{n}{p}\right) = -1$. This is why we took only primes for which n is a quadratic residue.) Thus, suppose we have a sequence x_i of s

different values of x such that

$$f(x_i) = \prod_{j=1}^{r} p_j^{a_{ij}}, \qquad i = 1, \ldots, s.$$

If s is large enough, then it should be possible to find a subset $\{x_{i_l}\}_l \subset \{x_i\}_i$ such that

$$\prod_l f(x_{i_l}) = \prod_{l,j} p_j^{a_{i_l j}}$$

is a perfect square, i.e., $\sum_l a_{i_l j}$ is an even number for each $j = 1, \ldots, r$. This method of factoring seems to have first appeared in print in [56].

A systematic way to find the required subset $\{x_{i_l}\}$ of $x_1, \ldots, x_s$ is due to Brillhart and Morrison [17]. One simply regards the last condition — that the sum of the exponents $a_{i_l j}$ be an even number— as an equation over the field $\mathbf{F}_2$, and use linear algebra. More precisely, let $u_1, \ldots, u_s$ be unknowns taking the value 0 (if the corresponding x_i is not in the subsequence $\{x_{i_l}\}$) or 1 (if the corresponding x_i is in the subsequence). Then the condition we want is that

$$\sum_{i=1}^{s} a_{ij} u_i \equiv 0 \pmod 2 \qquad \text{for} \qquad j = 1, \ldots, r.$$

If s is somewhat greater than r, then this system has a nontrivial solution — in fact, several solutions. Each solution gives a different subset $\{x_{i_l}\}$, and hence a different congruence of the form $x^2 \equiv y^2 \pmod n$. As explained above, at least one of these congruences is almost certain to give us a nontrivial factor of n.

It is important to choose r wisely. If r is too small, then it will take us a very long time to find even a single x_i for which $f(x_i)$ can be written as a product of primes in the factor base. On the other hand, if r is too large, then our $r \times s$ system of linear congruences modulo 2 will take us too long to solve (and we might also have difficulty storing such a large system). One can use a heuristic argument to show that r should be of order of magnitude roughly equal to $\exp(\frac{1}{2}\sqrt{\log n \log\log n})$ (where log denotes natural log). In the case of the largest numbers factored by this method (for which $\log n \approx 200$), r was chosen to be several tens of thousands.

In 1981, Pomerance developed an important technique — called *sieving* — for speeding up the process of finding the large sequence $\{x_i\}_{i=1,\ldots,s}$ that is needed in the above algorithm. We briefly describe one version of this technique. The idea is to find many of the x_i at once. Consider a long interval of T consecutive values of x. In each of T positions corresponding to the x's, we store an approximation to $\log f(x)$. Next, for each odd prime p in our factor base, we find the two solutions mod p of the congruence $f(x) \equiv 0 \pmod p$. (This amounts to finding the two square roots of n modulo p.) We now go to the interval of T consecutive values of x, and run through the two arithmetic progressions with difference p corresponding to the two solutions of $f(x) \equiv 0 \pmod p$. For each such value of x, we subtract an approximation to $\log p$ from the stored value. (The stored value is equal to the approximation to $\log f(x)$ minus the sum of the approximations to $\log p_j$ for each of the earlier p_j for which x was in one of the two corresponding arithmetic progressions, i.e., for which $p_j | f(x)$.) After completing this procedure for all of the p_j in the factor base, we look once more through the interval of T consecutive values of x. The locations where the stored number is close to zero will be the ones for which $f(x)$ is a product of the p_j.

It turns out that the expected running time for the quadratic sieve factoring method is of the form $\exp\big((1+o(1))\sqrt{\log n \log\log n}\big)$. For a discussion of running time estimates for factoring algorithms — some of which have been rigorously proved, and others of which are based only on heuristic arguments — see [84]. For a more precise discussion of sieving in the above setting and also in the so-called *multiple polynomial* version, see [87].

7.2 Elliptic Curve Factorization

A turning point in the application to cryptography of supposedly "pure" algebraic number theory was the discovery by H. W. Lenstra, Jr. of an ingenious method of using elliptic curves to factor integers. We now outline Lenstra's algorithm.

Let n be the large odd integer that we want to factor, and let p be an as yet unknown prime factor of n. Let $E_{a,b}$ denote the elliptic curve $Y^2 = X^3 + aX + b$ considered modulo n. Strictly speaking, $E_{a,b}$ modulo n is not a group; however, it is a group when considered modulo any prime $p|n$. As in the case of elliptic curve primality testing, we use the usual formulas for the group operation on an elliptic curve, working always modulo n, i.e., we "pretend" that n is prime. In the steps described below, if we ever find ourselves unable to apply those formulas (because of a denominator that is not prime to n and hence not invertible modulo n), we are almost certain to obtain a nontrivial factor of n, and we are done. (Namely, we take the g.c.d. of n with the troublesome denominator.)

Let r be a positive integer chosen in a certain optimal way, let $p_1, \ldots, p_r$ be the first r primes, and let $m = m(r)$ denote $\prod_{j=1}^{r} p_j^{\alpha_j}$, where $p_j^{\alpha_j}$ is the highest power of p_j that is less than $(n^{1/4}+1)^2$.

We start by choosing a random pair consisting of the coefficient a of our elliptic curve and a point $P = (x_0, y_0)$ in the xy-plane modulo n. We want P to lie on the curve $E_{a,b} : Y^2 = X^3 + aX + b$, so the coefficient b is determined by P and a: $b = y_0^2 - x_0^3 - ax_0$. (We should check that $4a^3 + 27b^2$ is prime to n, so that our elliptic curve is not degenerate.) We next compute (or rather, try to compute) the point mP on $E_{a,b}$. If we are able to compute the multiple mP (i.e., if all of the denominators we encounter in the addition and doubling formulas are prime to n), then we have *not* succeeded in factoring n, and we must make another random choice of P and a.

Eventually we are lucky, in that our curve $E_{a,b}$ has the following property: for some prime $p|n$, the order of $E_{a,b}$ modulo p (i.e., the order of the group of $\mathbf{F}_p$-points) is p_r-smooth (this means that $\#E_{a,b}(\mathbf{F}_p)$ has no prime factors greater than p_r). If we suppose that $p < \sqrt{n}$, this means that $\#E_{a,b}(\mathbf{F}_p)$ divides m, because by Hasse's theorem this order is $< (n^{1/4}+1)^2$. In that case the point mP considered modulo p is the identity, i.e., the point at infinity. This situation will snow up in our computations in the form of a denominator divisible by p. At the same time, it is unlikely that the same denominator is divisible by n. Thus, in computing mP we will encounter a denominator whose g.c.d. with n gives a nontrivial factor (most likely the prime factor p).

It remains for us to discuss what the probability is that the randomly chosen $E_{a,b}$ has the desired property. According to Hasse's theorem, the order of $E_{a,b}(\mathbf{F}_p)$ is in the interval

$[p+1-2\sqrt{p}, p+1+2\sqrt{p}]$. Moreover, it follows from work of Deuring and Schoof that as a and b vary, most of these orders fall in the interval $(p-\sqrt{p}, p+\sqrt{p})$, where they are roughly uniformly distributed (see [62] for deta''s). Thus, the chance that $\#E_{a,b}(\mathbf{F}_p)$ is p_r-smooth is about the same as the chance that a random integer in the interval $(p-\sqrt{p}, p+\sqrt{p})$ is p_r-smooth.

Note that r must be chosen in an optimal way. If it is too small, then we might never find an elliptic curve whose order modulo p is p_r-smooth. On the other hand, if r is too big, then it will take us too long to compute mP, since $m = m(r)$ grows rather rapidly with r.

At this point Lenstra needs an unproved conjecture to the effect that the standard results in [20] on the distribution of p_r-smooth integers in the interval $(0, x)$ as $x \to \infty$ also apply to the smaller interval $(x-\sqrt{x}, x+\sqrt{x})$ as $x \to \infty$. Assuming this plausible conjecture, one can find an optimal strategy (i.e., an optimal choice of r) so as to find the desired elliptic curve and factor n as rapidly as possible, and one can rigorously analyze the running time. As in the case of the quadratic sieve factoring algorithm, the expected running time turns out to have the form $\exp((1+o(1))\sqrt{\log n \log\log n})$.

In the case of the hardest numbers to factor, in practice the elliptic curve factorization algorithm is not quite as efficient as its main competitor, the quadratic sieve method. For example, in factoring a number n that is the product of two primes of size roughly $\sqrt{n}$, as in RSA, the best versions of the quadratic sieve are somewhat faster [87]. However, the elliptic curve method is the only subexponential general purpose factoring algorithm whose expected running time actually depends on the size of the smallest prime factor p of n. If $p \ll \sqrt{n}$, then the fastest way to factor n is probably by elliptic curves.

Until recently, all of the contenders for the best general purpose factoring algorithm had running time of the form $\exp(O(\sqrt{\log n \log\log n}))$. Some people even thought that this function of n might be a natural lower bound on the running time of such an algorithm. However, in the early 1990s a new method — called the *number field sieve* — was developed that has a heuristic running time that is much better (asymptotically), namely:

$$\exp(O((\log n)^{1/3}(\log\log n)^{2/3})).$$

The number field sieve is similar to the earlier algorithms that attempt to combine congruences so as to obtain a relation of the form $x^2 \equiv y^2 \pmod{n}$. However, one uses a "factor base" in the ring of integers of a suitably chosen algebraic number field. The details are quite complicated, so we shall not describe the algorithm here. See [61] for details.

8 Discrete Logarithm

Recall that the discrete logarithm problem in a finite field $\mathbf{F}_q$ is the problem, given $x \in \mathbf{F}_q^*$, of solving the relation $g^y = x$ for the integer y. Two types of algorithms are known for solving this problem: (1) algorithms that are inefficient unless $q-1$ happens to have no large prime factors; and (2) index calculus algorithms.

An example of the first type is D. Shanks' "baby-step/giant-step" method (see pp. 9, 575–576 of [46]) combined with the Silver–Pohlig–Hellman algorithm [41]. We give a sketch of this algorithm. First, a routine argument based on the Chinese Remainder

Theorem reduces the problem to that of solving an equation of the form $g^y = x$ when g and x are p-th roots of 1 in $\mathbf{F}_q$, where p is a prime divisor of $q - 1$. Now compute the two sets $S_1 = \{xg^i\}_{0 \le i < \sqrt{p}}$ and $S_2 = \{g^{j([\sqrt{p}]+1)}\}_{0 \le j < \sqrt{p}}$, and compare them. When you have a match, i.e., $xg^i = g^{j([\sqrt{p}]+1)}$, you are done — just take $y = j([\sqrt{p}] + 1) - i$. This method is practical if $q - 1$ is a product of small primes, but not if $q - 1$ is divisible by a prime p of 40 digits or more [79].

For a description of another important algorithm of the first type, due to Pollard, see [82] or [68].

When $q-1$ has a large prime divisor, the most efficient algorithms are all of index calculus type. There are two variants of these algorithms: those that apply when $q = p^f$ is a power of a small prime, and those that apply when $q = p$ is prime itself or a very small power of p. In both cases there are algorithms having running time $\exp(O(\sqrt{\log q \log\log q}))$. See [2] for an algorithm that is subexponential in $q = p^f$ as both p and f increase.

Index calculus algorithms are probabilistic rather than deterministic, and in spirit they resemble the "combination of congruences" procedure used to produce a relationship of the form $x^2 \equiv y^2 \pmod{n}$ in factoring algorithms such as the quadratic sieve (see above).

We shall outline the index calculus method in the case when q is prime. For simplicity we shall suppose that g is a generator of $\mathbf{F}_q^*$. The first part of the algorithm is called *precomputation*. This means that it needs to be performed only once for a given field $\mathbf{F}_q$. The second part of the algorithm must be performed for each individual logarithm, i.e., it depends upon the particular x whose discrete logarithm we want.

In the precomputation, let $p_1, \ldots, p_r$ be the first r primes, where r is chosen in a certain optimal way. Then for random positive integers $l < q-1$ consider the least positive residue of g^l. If this integer is divisible by a prime $> p_r$, we move on to another l. But any time that the least positive residue of g^l factors into a product of the $p_1, \ldots, p_r$ we obtain a relation of the form

$$\prod_{j=1}^{r} p_j^{a_{ij}} = g^{l_i} \quad \text{in} \quad \mathbf{F}_q^*.$$

For $a \in \mathbf{F}_q^*$, let $\mathrm{ind}_g a$ denote the discrete logarithm (also called the "index") of a to the base g. Then the last formula implies the following relation of exponents:

$$\sum_{j=1}^{r} a_{ij} \mathrm{ind}_g p_j \equiv l_i \pmod{q-1}.$$

This is a system of linear congruences in the unknowns $\mathrm{ind}_g p_j$, $j = 1, \ldots, r$. Once we have enough independent congruences of this form, we can solve for the discrete logs of the p_j. It should be noted that, although solving a system of congruences is quite routine in principle, there are subtle questions that arise when one tries to decrease the time spent on this part of the algorithm (see [86]). (For a discussion of index calculus implementation in the case of the fields $\mathbf{F}_{2^k}$, see [37].)

Once the precomputation is complete, it is relatively easy to find $y = \mathrm{ind}_g x$. Namely, we choose random l' and compute $xg^{l'}$ until we find a least positive residue of $xg^{l'}$ that has no prime divisor $> p_r$. Writing $xg^{l'}$ as a product of the p_j and using the values of $\text{ind}_g p_j$ from the precomputation, we find the discrete log of $xg^{l'}$. Then we simply subtract l' to get the desired $y = \text{ind}_g x$.

It remains to explain how r should be chosen. If r is too small, it will take too long to find g^{l_i} for which the least positive residue has no prime factor $> p_r$. On the other hand, if r is too large, the linear system in the r unknowns $\text{ind}_g p_j$ will be too hard to solve. An optimal value of r can be chosen using the asymptotic behavior of the function $\psi(X, s) = \#\{s\text{-smooth positive integers} \leq X\}$. (Recall that a positive integer is said to be s-smooth if it has no prime factors $> s$.) The result of this analysis is that, very roughly speaking, $r \sim \exp(\sqrt{\log q \log\log q})$.

Using such an optimal r, one can show that the running time of the algorithm is of the form $\exp(O(\sqrt{\log q \log\log q}))$. Most of this running time is devoted to (1) finding p_r-smooth values of g^{l_i}, and (2) solving the linear system of congruences for $\text{ind}_g p_j$.

Recently, D. Gordon has modified the number field sieve factoring algorithm so as to obtain a discrete logarithm algorithm for $\mathbf{F}_q$. The result is also an index calculus algorithm, but it is much more complicated, because one works in the ring of integers of an algebraic number field. Although the discrete log number field sieve has not yet been extensively tested in practice, it can be shown to have a much better asymptotic running time than earlier algorithms, namely $\exp(O(\log^{1/3} q \log\log^{2/3} q))$.

We conclude our discussion of algorithms for factoring and discrete log by observing that there is thought to be a close relationship between the complexity of factoring an integer n and the complexity of solving the discrete log problem in $\mathbf{F}_q$, where q and n have the same order of magnitude. See [80] and [101] for discussions of this question. Until recently, all of the best algorithms for both problems had time estimates of the form $L(n, 1/2)$, where for $0 \leq \gamma \leq 1$ one defines $L(n, \gamma) = \exp(O(\log^{\gamma} n \log\log^{1-\gamma} n))$. Then with the invention of the number field sieve for factoring [61], the time estimate for factoring was brought down to $L(n, 1/3)$. Soon after, the number field sieve was also applied to the discrete log problem [35, 36], bringing the time estimate for discrete log down to $L(q, 1/3)$ as well.

9 Elliptic Curve Cryptosystems

Soon after Lenstra showed how to use elliptic curves for factoring, Koblitz [47] and Miller [76] independently proposed using the abelian group of an elliptic curve as the basis for cryptosystems such as the Diffie–Hellman key exchange. We now describe how this works and discuss the corresponding discrete logarithm problem.

We first define the discrete log problem in an arbitrary abelian group G. If the group law in G is written additively, and if $P \in G$ denotes a fixed base element, then the *discrete log problem in G to the base P* is the following: Given $X \in G$, determine an integer y such that $X = yP$ in G (or else determine that X is not in the subgroup generated by P).

The Diffie–Hellman key exchange was first proposed for the group $\mathbf{F}_q^*$, but it can just as well be stated in any finite abelian group G. As before, Alice and Bob want to agree upon a large integer to serve as a key for some private-key cryptosystem. Alice and Bob first select a group G, a map from elements of G to the integers, and a base element $P \in G$. All of this is discussed publicly.

Next, Alice and Bob choose random secret integers k_A and k_B, respectively. Alice sends Bob the group element $k_A P$ (but not her secret integer k_A), and Bob sends Alice the element $k_B P$ (but not the integer k_B). The common key is then $k_A k_B P \in G$ (or rather, the integer associated to $k_A k_B P$ under the agreed upon correspondence). Alice determines this key by multiplying her secret integer k_A times the element $k_B P$ received from Bob; Bob determines the key by multiplying k_B times the element $k_A P$ that he received from Alice. An eavesdropper is in the difficult position of having to determine the element $k_A k_B P$ knowing only P, $k_A P$ and $k_B P$, but not k_A or k_B.

Now take $G = E$ to be an elliptic curve defined over a finite field $\mathbf{F}_q$. Let $P \in E$ be a fixed base point. The multiples $k_A P$, $k_B P$, and $k_A k_B P$ are computed using the addition law on the curve and the "repeated doubling" method. The secret key (known by Alice and Bob but no one else) is the point $k_A k_B P$ — or, more precisely, a certain positive integer corresponding to this point, for example, its x-coordinate in the case when $\mathbf{F}_q$ is a prime field.

The main potential advantage of an elliptic curve cryptosystem is that the discrete logarithm problem seems to be harder to solve in the group E than in the group $\mathbf{F}_q^*$. In the latter group, the generally applicable discrete log algorithms that do not require exponential time — all of which can be categorized as "index calculus" algorithms — rely on the fact that one can find a fairly large set of "small" group elements in terms of which any other element can be expressed with small exponents. For example, if $q = p$ is a large prime, then this set (called a "factor basis") might consist of all prime numbers less than a certain bound. On the other hand, if $q = p^f$, where p is small and f is large, and if $\mathbf{F}_q$ is regarded as $\mathbf{F}_p[X]/F(X)$ for a fixed irreducible polynomial $F(X)$ of degree f, then the factor basis might consist of all irreducible polynomials of degree less than a certain bound, regarded modulo $F(X)$. In the case of the group E, however, it does not seem to be possible to find such a factor basis. As argued in [76], the most natural notion of "small" (the reduction modulo p of a point in characteristic zero having small height) does not work.

Until the early 1990s, the only discrete log algorithms known for an elliptic curve were the ones that work in any group, irrespective of any particular structure. These are exponential time algorithms, provided that the order of the group is divisible by a large prime factor. But then Menezes, Okamoto, and Vanstone [71] found a new approach to the discrete log problem on an elliptic curve E defined over $\mathbf{F}_q$. Namely, they used the Weil pairing (see §III.8 of [100]) to imbed the group E into the multiplicative group of some extension field $\mathbf{F}_{q^k}$. This imbedding reduces the discrete log problem on E to the discrete log problem in $\mathbf{F}_{q^k}^*$.

However, in order for the Weil pairing reduction to help, it is essential for the extension degree k to be small. Essentially the only elliptic curves for which k is small are the so-called "supersingular" elliptic curves, the most familiar examples of which are curves of the form $Y^2 = X^3 + aX$ when the characteristic p of $\mathbf{F}_q$ is $\equiv -1 \pmod 4$, and curves of the form $Y^2 = X^3 + b$ when $p \equiv -1 \pmod 3$. The vast majority of elliptic curves, however, are nonsupersingular. For them, it was shown by Balasubramanian and Koblitz [12] that the reduction almost never leads to a subexponential algorithm.

Thus, the first advantage of elliptic curve cryptosystems is that no subexponential algorithm is known that breaks the system, provided that we avoid supersingular curves and also curves whose order has no large prime factor. A second selling point is that, unlike in the case of the groups $\mathbf{F}_q^*$, there is a tremendous variety of elliptic curves from which to choose. To take full advantage of this, we must have an efficient way to determine the order of a randomly chosen elliptic curve E defined over $\mathbf{F}_q$.

The first polynomial time algorithm to compute $\#E$ was discovered by René Schoof. Schoof's algorithm is even deterministic. It is based on the idea of finding the value of $\#E$ modulo l for all primes l less than a certain bound. This is done by examining the action of the "Frobenius" (the p-th power map) on points of order l.

In the original paper [96] the bound for running time was $O(\log^8 q)$, which is polynomial but quite unpleasant. At first it looked like the algorithm was not practical. However, since then many people have worked on speeding up Schoof's algorithm (see [8, 58, 65, 74]). As a result it has become feasible to compute the order of an arbitrary elliptic curve over $\mathbf{F}_q$ if q is a prime power of a few hundred decimal digits (this is several times more digits than one needs in practice). If this order is found to have a large prime factor, then the discrete log problem on it is intractable at our present level of knowledge and technology, and so the curve can be used for a secure cryptosystem.

Several methods of choosing elliptic curves have been suggested. One possibility is to stick with a simple equation, such as $Y^2 = X^3 + aX$ or $Y^2 = X^3 + b$. If the curve is supersingular, then it is trivial to compute $\#E$. For example, if $q \equiv -1 \pmod 4$ for the first equation or $q \equiv -1 \pmod 3$ for the second equation, then $\#E = q + 1$. But in this case the discrete logarithm problem reduces to discrete logs in $\mathbf{F}^*_{q^2}$ [71], so cryptosystems using E have no strong advantage over cryptosystems using the multiplicative group of a finite field. But even if the curve is nonsupersingular, because of the special form of the equation there are algorithms to find $\#E$ that are simpler and faster than Schoof's algorithm and its variants. For example, computing the order of the curve $Y^2 = X^3 + aX$ over $\mathbf{F}_p$, where $p \equiv 1 \pmod 4$, amounts to expressing p as a sum of two squares (see, for example, §II.2 of [53]), and this can easily be done [15].

A second possibility is to start with a fixed elliptic curve E_1 over a small finite field $\mathbf{F}_q$, and then consider the sequence E_j obtained by regarding the same curve over the extension field $\mathbf{F}_{q^j}$. The sequence $N_j = \#E_j$ is determined in a simple way from $N_1 = \#E_1$, using the congruence zeta-function (see, e.g., §II.1 of [53]).

Another possibility is to generate elliptic curves in some random manner over a fixed finite field $\mathbf{F}_q$. One could take $q = p$ to be an arbitrary prime whose size is roughly the same as that of the group E one needs. (By Hasse's theorem, $|\#E - p - 1|$ is bounded by $2\sqrt{p}$.) Alternatively, one could take $q = 2^f$ to be a fixed large power of 2. The choice of an elliptic curve in characteristic 2 has some advantages in practice, because arithmetic can be performed somewhat more efficiently in such fields. It has been suggested that a random elliptic curve over $\mathbf{F}_{2^f}$ would be particularly suitable for a cryptosystem that could be implemented on a "smart card," where the space limitation is severe [73].

In each case, one must search for an elliptic curve whose order has a large prime factor; otherwise, discrete logs can be found quickly by the baby-step/giant-step and Silver–Pohlig–Hellman algorithms.

Finally, it should be mentioned that, besides elliptic curves and the multiplicative groups of finite fields, other groups have also been proposed for Diffie–Hellman type cryptosystems: the jacobians of hyperelliptic curves [49], the class groups of imaginary quadratic number fields [18], and others.

10 Classical Number Theory Problems and Elliptic Curve Analogues that are of Cryptographic Interest

10.1 Diffie-Hellman in $\mathbf{F}_q^*$

When choosing a finite field $\mathbf{F}_q$ for use in a Diffie-Hellman key exchange, it is crucial that the number of elements in the multiplicative group

$$q - 1 = \#\mathbf{F}_q^*$$

be divisible by a very large prime. This is because, if $q - 1$ is "smooth" (not divisible by a large prime), then the cryptosystem is susceptible to the baby-step/giant-step and Silver–Pohlig–Hellman algorithms.

The two most common ways to choose q are (1) to set q equal to a randomly generated prime p of sufficient size, and (2) to set q equal to 2^r. In each situation we might ask what the chances are that $\#\mathbf{F}_q^*$ is a prime number or a very small factor times a prime number. In the first case, the most we can hope for is that $p - 1$ be equal to twice a prime p_1:

$$p = 2p_1 + 1.$$

A prime p_1 such that $2p_1+1$ is prime is called a "Sophie Germain prime," since in the 1820s Germain proved the so-called "first case" of Fermat's Last Theorem for prime exponents p_1 for which $2p_1 + 1$ is prime. This was the first major result on FLT for a large class of exponents, and it attracted a lot of interest to the set of such primes p_1. Thus, we are led to the question: Are there infinitely many Sophie Germain primes; and if so, then for large x what is the probability that both p_1 and $2p_1 + 1$ will be primes for $p_1 \approx x$? This is a difficult unsolved problem of number theory that was of interest to mathematicians for many decades for a reason unrelated to cryptography: people wanted to know whether the first case of Fermat's Last Theorem had been proved for infinitely many primes.

If we choose $q = 2^r$, then it is possible for $\#\mathbf{F}_{2^r}^*$ itself to be prime. This happens if r is prime and $2^r - 1$ is a Mersenne prime. Thus, we are led to the famous Mersenne prime question: Are there infinitely many Mersenne primes; and, if so, what can be said about their frequency of occurrence for large r?

If we want to work in finite fields of $q = p^r$ elements for small $p > 2$, then, again choosing r prime, we can hope for $(q-1)/(p-1) = (p^r - 1)/(p-1)$ to be prime. Again we have a difficult unsolved problem of classical number theory: For fixed p, are there infinitely many r such that $(p^r - 1)/(p - 1)$ is prime? Almost nothing has been proved about these questions. In fact, it is not even known that there are infinitely many primes of the form $(p^r - 1)/(p - 1)$ as p and r both vary over the set of prime numbers.

10.2 Elliptic Curve Cryptosystems

As mentioned before, there are basically three approaches to choosing an elliptic curve for a cryptosystem. In each case one looks for a curve whose order $\#E$ has a very large prime factor, and in each case the question of the likelihood of encountering such a curve leads to some interesting unsolved problems.

10.2.1 Fix a "Global" Elliptic Curve and Vary the Prime

For example, let E be an elliptic curve $Y^2 = f(X) = X^3 + aX + b$ defined over the field $\mathbf{Q}$ of rational numbers. If p is any odd prime not dividing the denominators of the coefficients or the discriminant of $f(X)$, then one can consider the elliptic curve E over $\mathbf{F}_p$ that is obtained by simply reducing the coefficients modulo p. That elliptic curve will always contain as a subgroup the image of the (finite) torsion subgroup E_{tors} of the curve over $\mathbf{Q}$. But in many cases the quotient will have prime order.

Question: *For a fixed curve E over $\mathbf{Q}$, what can be said about the probability as p varies that*

$$\frac{\#E \bmod p}{\#E_{\text{tors}}}$$

is a prime number? Can one prove (for any fixed E) that there are infinitely many p for which this number is prime?

This question is analogous to the Sophie Germain prime question. Namely, if instead of E we take the multiplicative semigroup of nonzero integers, which has torsion subgroup $\{\pm 1\}$, then we get our earlier question: As p varies, what can be said about the probability that $p_1 = (p-1)/2 = \#\mathbf{F}_p^*/\#\{\pm 1\}$ is prime?

Just as the question about Sophie Germain primes is of interest when using a Diffie–Hellman type cryptosystem in the multiplicative group of a prime field $\mathbf{F}_p$, the analogous elliptic curve question given above is of interest when using an elliptic curve cryptosystem.

It is worth noting that the denominator $\#E_{\text{tors}}$ in the elliptic curve question is often 1, and in any case it cannot be much larger than in the Sophie Germain prime question. According to a deep result of Mazur [67], there are at most 16 torsion points on an elliptic curve over $\mathbf{Q}$.

For a discussion of conjectures on the above question for elliptic curves, see [48].

Another natural question in this context is the following. Suppose that P is a point of infinite order in the group of the elliptic curve E over $\mathbf{Q}$. As p varies, what is the probability that the image modulo p of P generates E modulo p? Results on this problem have been obtained by Gupta and Murty [39]. This question is the analogue of a classical problem of E. Artin, who conjectured formulas for the probability that a fixed integer (like 2 or 10) generates $\mathbf{F}_p^*$ as p varies.

10.2.2 Fix an Elliptic Curve over a Small Field $\mathbf{F}_q$ and then consider it over $\mathbf{F}_{q^r}$ As r Varies

Since $E(\mathbf{F}_{q^{r'}})$ is a subgroup of $E(\mathbf{F}_{q^r})$ whenever $r'|r$, large prime factors of $\#E(\mathbf{F}_{q^r})$ are more likely to occur when r is prime than when r is composite. In the case of prime r, the best one can hope for is that

$$\frac{\#(E(\mathbf{F}_{q^r}))}{\#(E(\mathbf{F}_q))} = \left|\frac{\alpha^r - 1}{\alpha - 1}\right|^2$$

is prime, where α is a reciprocal root of the numerator of the zeta-function $Z(E/\mathbf{F}_q; T)$.

Question: *For fixed $E/\mathbf{F}_q$, what is the probability as r varies that the above number is prime? Can one ever prove that there are infinitely many r such that it is prime?*

Virtually nothing is known about this question. It is analogous to the classical Mersenne prime problem, as we see by replacing α by 2.

10.2.3 Fix the Field of Definition $\mathbf{F}_q$ and Vary the Coefficients

According to Hasse's theorem, $\#E$ falls in a rather small interval around $q+1$, namely, $[q+1-2\sqrt{q}, q+1+2\sqrt{q}]$. As E ranges over all elliptic curves defined over $\mathbf{F}_q$, the number $\#E$ is distributed fairly uniformly in this interval, except that the density drops off near the endpoints (see [62]). Thus, the probability that $\#E$ is prime (or has a prime factor greater than some lower bound) is essentially the same as the probability that a random integer in an interval of the form $[q, q+c\sqrt{q}]$ (c a constant) has this property. But unfortunately, at present almost nothing can be proved about the occurrence of primes in such "short" intervals. It is not even known whether there exists a c such that the interval $[q, q+c\sqrt{q}]$ always contains at least one prime as $q \to \infty$.

It is remarkable that practical questions relating to the security and the efficient implementation of cryptosystems often lead to problems in number theory that are intrinsically interesting from a theoretical and aesthetic point of view; and many number-theoretic concepts and results that were once thought to have no conceivable practical application have recently played a role in the development and analysis of new cryptosystems. There is no *a priori* reason why we should have expected this to be the case. Indeed, the famous statement that Eugene Wigner made in reference to mathematical physics seems to apply to the subject of the present article as well:

> . . .the enormous usefulness of mathematics in the natural sciences is something bordering on the mysterious and . . . there is no rational explanation for it. [102]

References

[1] L.M. Adleman, A subexponential algorithm for the discrete logarithm problem with applications to cryptography, *Proc. 20th IEEE Symp. Foundations of Computer Science* (1979), 55–60.

[2] L.M. Adleman and DeMarrais, J., A subexponential algorithm for discrete logarithms over all finite fields, *Math. Comp.* **61** (1993), 1–15.

[3] L.M. Adleman and Huang, M.-D., Primality Testing and Abelian Varieties over Finite Fields, *Lect. Notes Math.* **1512**, Springer-Verlag (1992).

[4] L.M. Adleman, Pomerance, C. and Rumely, R.S., On distinguishing prime numbers from composite numbers, *Annals Math.* **117** (1983), 173–206.

[5] G.B. Agnew, Mullin, R.C., Onyszchuk, I.M. and Vanstone, S.A., An implementation for a fast public-key cryptosystem, *J. Cryptology* **3** (1991), 63–79.

[6] G.B. Agnew, Mullin, R.C. and Vanstone, S.A., An implementation of elliptic curve cryptosystems over $\mathbf{F}_{2^{155}}$, *IEEE Journal on Selected Areas in Communications* **11** (1993), 804–813.

[7] W.R. Alford, Granville, A. and Pomerance, C., There are infinitely many Carmichael numbers, *Annals Math.* **139** (1994), 703–722.

[8] A.O.L. Atkin, The number of points on an elliptic curve modulo a prime, unpublished manuscript (1991).
[9] A.O.L. Atkin and Morain, F., Elliptic curves and primality proving, *Math. Comp.* **61** (1993), 29–68.
[10] E. Bach, *Analytic Methods in the Analysis and Design of Number-Theoretic Algorithms*, MIT Press (1985).
[11] E. Bach, Shallit, J., *Algorithmic Number Theory*, vol. 1, MIT Press (1996).
[12] R. Balasubramanian and Koblitz, N., The improbability that an elliptic curve has subexponential discrete log problem under the Menezes-Okamoto-Vanstone algorithm, *J. Cryptology* **10** (1997).
[13] I.F. Blake and Fuji-Hara, R., Mullin, R.C. and Vanstone, S.A., Computing logarithms in fields of characteristic two, *SIAM J. Algebraic and Discrete Methods* **5** (1984), 276–285.
[14] D. Boneh and Lipton, R., Algorithms for black-box fields and their applications to cryptography, *Advances in Cryptology — Crypto '96*, Springer-Verlag (1996), 283–297.
[15] J. Brillhart, Note on representing a prime as a sum of two squares, *Math. Comp.* **26** (1972), 1011–1013.
[16] J. Brillhart, Lehmer, D.H., Selfridge, J.L. Tuckerman, B. and Wagstaff, S.S. Jr., Factorizations of $b^n \pm 1, b = 2, 3, 5, 6, 7, 10, 11, 12$ Up to High Powers, *Amer. Math. Soc.* (1988).
[17] J. Brillhart and Morrison, M., A method of factoring and the factorization of F_7, *Math. Comp.* **29** (1975), 183–205.
[18] J. Buchmann and Williams, H.C., A key exchange system based on imaginary quadratic fields, *J. Cryptology* **1** (1988), 107–118.
[19] J. Buchmann and Williams, H.C., A key exchange system based on real quadratic fields, *Advances in Cryptology — Crypto '89*, Springer-Verlag (1990), 335–343.
[20] E.R. Canfield, Paul Erdös and Pomerance, C., On a problem of Oppenheim concerning "Factorisatio Numerorum," *J. Number Theory* **17** (1983), 1–28.
[21] L. Charlap, Coley, R. and Robbins, D., Enumeration of rational points on elliptic curves over finite fields, unpublished manuscript (1991).
[22] H. Cohen, *A Course in Computational Algebraic Number Theory*, Springer-Verlag (1993).
[23] H. Cohen and Lenstra, A.K., Implementation of a new primality test, *Math. Comp.* **48** (1987), 103–121.
[24] H. Cohen and Lenstra, H.W. Jr., Primality testing and Jacobi sums, *Math. Comp.* **42** (1984), 297–330.
[25] D. Coppersmith, Fast evaluation of logarithms in fields of characteristic two, *IEEE Trans. Information Theory* **30** (1984), 587–594.
[26] D. Coppersmith, Odlyzko, A.M. and Schroeppel, R., Discrete logarithms in $GF(p)$, *Algorithmica* **1** (1986), 1–15.
[27] T. Denny, Shirokauer, O. and Weber, D., Discrete logarithms: the effectiveness of the index calculus method, in Henri Cohen, ed., *Algorithmic Number Theory, Proc. Second Intern. Symp., ANTS-II*, Springer-Verlag (1996), 337–361.
[28] W. Diffie and Hellman, M., New directions in cryptography, *IEEE Trans. Information Theory* **22** (1976), 644–654.
[29] J.D. Dixon, Factorization and primality testing, *Amer. Math. Monthly* **91** (1984), 333–352.
[30] T. ElGamal, A public key cryptosystem and a signature scheme based on discrete logarithms, *IEEE Trans. Information Theory* **31** (1985), 469–472.
[31] T. ElGamal, A subexponential-time algorithm for computing discrete logarithms over $GF(p^2)$, *IEEE Trans. Information Theory* **31** (1985), 473–481.
[32] S. Goldwasser and Kilian, J., Almost all primes can be quickly certified, *Proc. 18th ACM Symp. Theory of Computing* (1986), 316–329.

[33] S. Goldwasser and Micali, S. Probabilistic encryption and how to play mental poker keeping secret all partial information, *Proc. 14th ACM Symp. Theory of Computing* (1982), 365–377.
[34] S. Goldwasser and Micali, S., Probabilistic encryption, *J. Comput. System Sci.* **28** (1984), 270–299.
[35] D. Gordon, Discrete logarithms in $GF(p)$ using the number field sieve, *SIAM J. Discrete Math.* **6** (1993), 124–138.
[36] D. Gordon, Discrete logarithms in $GF(p^n)$ using the number field sieve, preprint (1995).
[37] D. Gordon and McCurley, K., Massively parallel computation of discrete logarithms, *Advances in Cryptology — Crypto '92*, Springer-Verlag (1993), 312–323.
[38] A. Granville, Primality testing and Carmichael numbers, *Notices Amer. Math. Soc.* **39** (1992), 696–700.
[39] R. Gupta and Murty, M.R., Primitive points on elliptic curves, *Compositio Math.* **58** (1986), 13–44.
[40] R.K. Guy, *Unsolved Problems in Number Theory*, Springer-Verlag (1981).
[41] M.E. Hellman and Pohlig, S., An improved algorithm for computing logarithms over $GF(p)$ and its cryptographic significance, *IEEE Trans. Information Theory* **24** (1978), 106–110.
[42] M.E. Hellman and Reyneri, J.M., Fast computation of discrete logarithms in $GF(q)$, *Advances in Cryptology — Crypto '82*, Plenum Press (1983), 3–13.
[43] D. Husemöller, *Elliptic Curves*, Springer-Verlag (1987).
[44] K. Ireland and Rosen, M.I., *A Classical Introduction to Modern Number Theory*, 2nd ed., Springer-Verlag (1990).
[45] D.E. Knuth, *The Art of Computer Programming. vol. 2*, 2nd ed., Addison-Wesley (1981).
[46] D.E. Knuth, *The Art of Computer Programming. vol. 3*, Addison-Wesley (1973).
[47] N. Koblitz, Elliptic curve cryptosystems, *Math. Comp.* **48** (1987), 203–209.
[48] N. Koblitz, Primality of the number of points on an elliptic curve over a finite field, *Pacific J. Math.* **131** (1988), 157–165.
[49] N. Koblitz, Hyperelliptic cryptosystems, *J. Cryptology* **1** (1989), 139–150.
[50] N. Koblitz, Constructing elliptic curve cryptosystems in characteristic 2, *Advances in Cryptology — Crypto '90*, Springer-Verlag (1991), 156–167.
[51] N. Koblitz, Elliptic curve implementation of zero-knowledge blobs, *J. Cryptology* **4** (1991), 207–213.
[52] N. Koblitz, CM-curves with good cryptographic properties, *Advances in Cryptology — Crypto '91*, Springer-Verlag (1992), 279–287.
[53] N. Koblitz, *Introduction to Elliptic Curves and Modular Forms*, 2nd ed., Springer-Verlag (1993).
[54] N. Koblitz, *A Course in Number Theory and Cryptography*, 2nd ed., Springer-Verlag (1994).
[55] N. Koblitz, *Algebraic Aspects of Cryptography*, Springer-Verlag (1998).
[56] M. Kraitchik, *Théorie des nombres*, vol. 1, Gauthier-Villars (1922).
[57] B. LaMacchia and Odlyzko, A.M., Computation of discrete logarithms in prime fields, *Designs, Codes and Cryptography* **1** (1991), 47–62.
[58] F. Lehmann, Maurer, M. Müller, V. and Shoup, V., Counting the number of points on elliptic curves over finite fields of characteristic greater than three, *Algorithmic Number Theory, Lect. Notes Comp. Sci.* **877** (1994), Springer-Verlag, 60–70.
[59] A.K. Lenstra, Primality testing, *Cryptology and Computational Number Theory, Proc. Symp. Appl. Math.* **42** (1990), 13–25.
[60] A.K. Lenstra and Lenstra, H.W. Jr., Algorithms in number theory, *Handbook of Theoretical Computer Science*, Elsevier (1990), 673–715.
[61] A.K. Lenstra and Lenstra, H.W. Jr., *The Development of the Number Field Sieve*, Springer-Verlag (1993).

[62] H.W. Lenstra, Jr., Factoring integers with elliptic curves, *Annals Math.* **126** (1987), 649–673.

[63] H.W. Lenstra, Jr., Pila, J. and Pomerance, C., A hyperelliptic smoothness test. I, *Philos. Trans. Roy. Soc. London* **345** (1993), 397–408.

[64] H.W. Lenstra, Jr., and Pomerance, C., A rigorous time bound for factoring integers, *J. Amer. Math. Soc.* **5** (1992), 483–516.

[65] R. Lercier and Morain, F., Counting the number of points on elliptic curves over finite fields: strategies and performances, *Advances in Cryptology — Eurocrypt '95*, Springer-Verlag (1995), 79–94.

[66] R. Lidl and Niederreiter, H., *Introduction to Finite Fields and Their Applications*, Cambridge Univ. Press (1986).

[67] B. Mazur, Modular curves and the Eisenstein ideal, *Inst. Hautes Études Sci. Publ. Math.* **47** (1977), 33–186.

[68] K. McCurley, The discrete logarithm problem, *Cryptology and Computational Number Theory, Proc. Symp. Appl. Math.* **42** (1990), 49–74.

[69] K. McCurley, Odds and ends from cryptology and computational number theory, *Cryptology and Computational Number Theory, Proc. Symp. Appl. Math.* **42** (1990), 145–166.

[70] A. Menezes, Elliptic Curve Public Key Cryptosystems, *Kluwer Acad. Publ.* (1993).

[71] A. Menezes, Okamoto, T. and Vanstone, S.A., Reducing elliptic curve logarithms to logarithms in a finite field, *IEEE Trans. Information Theory* **39** (1993), 1639–1646.

[72] A. Menezes, van Oorschot, P. and Vanstone, S.A., *Handbook of Applied Cryptography*, CRC Press (1996).

[73] A. Menezes and Vanstone, S.A., Elliptic curve cryptosystems and their implementation, *J. Cryptology* **6** (1993), 209–224.

[74] A. Menezes, Vanstone, S.A. and Zuccherato, R.J., Counting points on elliptic curves over $\mathbf{F}_{2^m}$, *Math. Comp.* **60** (1993), 407–420.

[75] G.L. Miller, Riemann's hypothesis and tests for primality, *J. Comput. System Sci.* **13** (1976), 300–317.

[76] V. Miller, Uses of elliptic curves in cryptography, *Advances in Cryptology — Crypto '85*, Springer-Verlag (1986), 417–426.

[77] P.L. Montgomery, Speeding the Pollard and elliptic curve methods of factorization, *Math. Comp.* **48** (1987), 243–264.

[78] F. Morain, Implementation of the Goldwasser–Kilian–Atkin primality testing algorithm, INRIA report 911 (1988).

[79] A.M. Odlyzko, Discrete logarithms and their cryptographic significance, *Advances in Cryptology — Eurocrypt '84*, Springer-Verlag (1985), 224–314.

[80] A.M. Odlyzko, The future of integer factorization, *CryptoBytes*, **1** (No. 2) (1995), 5–12.

[81] H.C. Pocklington, The determination of the prime and composite nature of large numbers by Fermat's theorem, *Proc. Cambridge Philos. Soc.* **18** (1914–1916), 29–30.

[82] J. Pollard, Monte Carlo methods for index computation mod p, *Math. Comp.* **32** (1978), 918–924.

[83] C. Pomerance, Recent developments in primality testing, *Math. Intelligencer* **3** (1981), 97–105.

[84] C. Pomerance, Analysis and comparison of some integer factoring algorithms, in H. W. Lenstra, Jr. and R. Tijdeman, ed., *Computational Methods in Number Theory, Math. Centrum, Amsterdam*, **154** (1982), 89–139.

[85] C. Pomerance, The quadratic sieve factoring algorithm, *Advances in Cryptology —— Eurocrypt '84*, Springer–Verlag (1985), 169–182.

[86] C. Pomerance, Fast, rigorous factorization and discrete logarithm algorithms, *Discrete Algorithms and Complexity* (1987), 119–143.

[87] C. Pomerance, Factoring, *Cryptology and Computational Number Theory, Proc. Symp. Appl. Math.* **42** (1990), 27–47.

[88] C. Pomerance, Selfridge, J.L. and Wagstaff, S.S. Jr., The pseudoprimes to $25 \cdot 10^9$, *Math. Comp.* **35** (1980), 1003–1026.

[89] G. Purdy, A high-security log-in procedure, *Comm. of the ACM* **17** (1974), 442–445.

[90] M.O. Rabin, Probabilistic algorithms for testing primality, *J. Number Theory* **12** (1980), 128–138.

[91] H. Riesel, *Prime Numbers and Computer Methods for Factorization*, Birkhäuser (1985).

[92] R. Rivest, Cryptography, in *Handbook of Theoretical Computer Science*, vol. A, Elsevier (1990), 717–755.

[93] R. Rivest, Shamir, A. and Adleman, L., A method for obtaining digital signatures and public–key cryptosystems, *Communications of the ACM* **21** (1978), 120–126.

[94] K. Rosen, *Elementary Number Theory and Its Applications*, 3rd ed., Addison-Wesley (1993).

[95] R.S. Rumely, Recent advances in primality testing, *Notices Amer. Math. Soc.* **30** (1983), 475–477.

[96] R. Schoof, Elliptic curves over finite fields and the computation of square roots mod p, *Math. Comp.* **44** (1985), 483–494.

[97] E. Seah and Williams, H.C., Some primes of the form $(a^n - 1)/(a - 1)$, *Math. Comp.* **33** (1979), 1337–1342.

[98] D. Shanks, Five number-theoretic algorithms, *Congressus Numerantium* **7** (1972), 51–70.

[99] D. Shanks, *Solved and Unsolved Problems in Number Theory*, 3rd ed., Chelsea (1985).

[100] J. Silverman, *The Arithmetic of Elliptic Curves*, Springer-Verlag (1986).

[101] P. van Oorschot, A comparison of practical public-key cryptosystems based on integer factorization and discrete logarithms, in G. Simmons, ed., *Contemporary Cryptology: The Science of Information Integrity*, IEEE Press (1992), 289–322.

[102] E.P. Wigner, The unreasonable effectiveness of mathematics in the natural sciences, *Comm. Pure Appl. Math.* **13** (1960), 1–14.

[103] M.V. Wilkes, *Time-Sharing Computer Systems*, Elsevier (1968).

Department of Mathematics
Box 354350
University of Washington
Seattle, WA 98195
USA
E-Mail: koblitz@math.washington.edu

Recent Developments in the Mean Square Theory of the Riemann Zeta and Other Zeta-Functions

Kohji Matsumoto

The purpose of the present article is to survey some mean value results obtained recently in zeta-function theory. We do not mention other important aspects of the theory of zeta-functions, such as the distribution of zeros, value-distribution, and applications to number theory. Some of them are probably treated in the articles of Professor Apostol and Professor Ramachandra in the present volume.

Even in the mean value theory, we do not discuss many important recent topics. Those include: Recent progress in the theory of large values and fractional moments made by Heath-Brown [55] and the Indian school (Ramachandra, Balasubramanian, Sankaranarayanan and others, see Ramachandra [172]); mean values taken at the zeros or at the points near the zeros (Gonek [30] [31], Fujii [23]–[26] and others); the mean square of the product of the zeta-function and a Dirichlet polynomial (see Conrey-Ghosh-Gonek [19] and the papers quoted there). All of these three topics are closely connected with the distribution of zeros of zeta-functions, hence the full account of them would require too many pages. We will only discuss the theory of Titchmarsh series very briefly in Section 7. In the fourth power moment theory there have been remarkable developments which may be characterized by the use of the spectral theory of Maass wave forms. We mention this theory occasionally, but only in connection with the mean square problems. For the full details of this theory, see Chapters 4 and 5 of Ivić [68], Motohashi's book [155], and Jutila's series of papers.

In the present article we only discuss the mean square theory of zeta-functions. This is a rather restricted topic, but still it is impossible to mention all the relevant results because the recent progress in this area is very big. The main tools appearing in this article are the approximate functional equations and Atkinson methods, emphasis are laid on the latter. The readers will find, however, that these two tools are not irrelevant (see Sections 4 and 6). Efforts are made to explain the mutual connections among various methods and results.

In Section 1, we summarize the results on the mean square $I_\sigma(T)$ of the Riemann zeta-function, obtained by applying various approximate functional equations. In Sections 3 and 5, $I_\sigma(T)$ is studied from the viewpoint of the method of Atkinson. The background of Atkinson's method is the divisor problem, which is mentioned in Sections 2 and 6. Then, after a brief discussion on some short interval results in Section 7, we proceed to survey the results on more general zeta and L-functions. Sections 8, 9, 10 and 11 are devoted, respectively, to the mean square theory of Dedekind zeta-functions, L-functions attached to cusp forms, Dirichlet L-functions, and Hurwitz zeta and other related zeta-functions.

Throughout this article, $s = \sigma + it$ is a complex variable, $\zeta(s)$ the Riemann zeta-function, $\Gamma(s)$ the gamma-function, γ the Euler constant, $\phi(n)$ the Euler function, $d(n)$ the number of positive divisors of n, and $\sigma_a(n) = \sum_{0<d|n} d^a$. When x tends to infinity, $f(x) \sim g(x)$ means $\lim_{x\to\infty} f(x)/g(x) = 1$, $f(x) = O(g(x))$ or $f(x) \ll g(x)$ means $|f(x)| \leq Cg(x)$ with a certain $C > 0$, $f(x) = \Omega_+(g(x))$ (resp. $f(x) = \Omega_-(g(x))$) means that $f(x_n) > Cg(x_n)$ (resp. $f(x_n) < -Cg(x_n)$) holds for infinitely many x_n such that $x_n \to \infty$, with a certain $C > 0$, and $f(x) = \Omega(g(x))$ means that $|f(x)| = \Omega_+(g(x))$. The letter ε denotes an arbitrarily small positive number, $C, C_1, C_2, \cdots$ denote certain constants, which are not necessarily the same at each occurrence. The references are by no means complete.

The author expresses his gratitude to Professors Martin N.Huxley, Aleksandar Ivić, Matti Jutila, Shigeru Kanemitsu, Masanori Katsurada, Isao Kiuchi, Shin-ya Koyama, Antanas Laurinčikas, K. Ramachandra, Vivek V. Rane, Yoshio Tanigawa and Kai-Man Tsang for valuable comments and information. He is also indebted to Miss Yumiko Ichihara for her laborious work of typesetting this long article.

1 The Approximate Functional Equations

A classical problem in the mean value theory of $\zeta(s)$ is to search for the asymptotic formula of the mean square

$$I_\sigma(T) = \int_0^T |\zeta(\sigma + it)|^2 dt,$$

where $T \geq 2$. (If $\sigma = 1$, we replace the interval of integration by $[1, T]$.) In view of the functional equation $\zeta(s) = \chi(s)\zeta(1-s)$, where

$$\chi(s) = 2(2\pi)^{s-1} \sin\left(\frac{1}{2}\pi s\right)\Gamma(1-s),$$

we may restrict our consideration to the case $\sigma \geq 1/2$. When $\sigma > 1$, the asymptotic formula

$$I_\sigma(T) \sim \zeta(2\sigma)T \tag{1.1}$$

is an easy consequence of the definition of $\zeta(s)$. It was proved by Landau [127, §228, p. 816] and Schnee [180] that (1.1) holds for any $\sigma > 1/2$. To prove this fact, the simple approximate formula

$$\zeta(s) = \sum_{n\leq\xi} n^{-s} - \frac{\xi^{1-s}}{1-s} + O(\xi^{-\sigma}) \qquad (|t| \leq \pi\xi) \tag{1.2}$$

is enough (see Titchmarsh [190, Theorem 7.2]). The most difficult case $\sigma = 1/2$ was settled in 1918 by Hardy-Littlewood [47], who proved

$$I_{\frac{1}{2}}(T) \sim T \log T, \tag{1.3}$$

by using the Mellin transform. Five years later, Hardy-Littlewood [49] gave an alternative proof of (1.3). It is based on the approximate functional equation

$$\zeta(s) = \sum_{n\le\xi} n^{-s} + \chi(s)\sum_{n\le\eta} n^{s-1} + R_1(s;\xi,\eta), \tag{1.4}$$

which is a refinement of (1.2). Here ξ, η are positive, $2\pi\xi\eta = t$, and

$$R_1(s;\xi,\eta) = O(\xi^{-\sigma} + \eta^{\sigma-1}t^{\frac{1}{2}-\sigma}). \tag{1.5}$$

The formula (1.4) first appeared in Hardy-Littlewood [48], with a slightly weaker error estimate than (1.5), and the main instrument of the proof in [48] is the Poisson summation formula

$$\sum_{a\le n\le b}{}' f(n) = \int_a^b f(u)du + 2\sum_{n=1}^{\infty}\int_a^b f(u)\cos(2\pi nu)du \tag{1.6}$$

(for $f \in C^1[a,b]$; the symbol $\sum'$ indicates that $\frac{1}{2}f(a)$ and $\frac{1}{2}f(b)$ are to be taken instead of $f(a)$ and $f(b)$, respectively). In [49], Hardy-Littlewood presented an alternative complex-analytic proof of (1.4) and (1.5).

The formula (1.4) and its relatives are really useful, and dominated the next sixty years of the mean value theory. Littlewood [134] announced that

$$I_{\frac{1}{2}}(T) = T\log T - (1 + \log 2\pi - 2\gamma)T + E(T) \tag{1.7}$$

with $E(T) = O(T^{3/4+\varepsilon})$ (actually the term 2γ is missing in [134]). Ingham [63] improved it to $E(T) = O(T^{1/2}\log T)$. Further improvement was done by Titchmarsh [188], who proved

$$E(T) = O(T^{\frac{5}{12}}\log^2 T). \tag{1.8}$$

Titchmarsh succeeded because he could use the Riemann-Siegel formula, proved by Siegel [182], which gives the very precise asymptotic expansion of $R_1(s;\sqrt{t/2\pi},\sqrt{t/2\pi})$. (See Titchmarsh [190, Theorem 4.16].)

Hardy-Littlewood [49] also studied the fourth power moment

$$I_{2,\sigma}(T) = \int_0^T |\zeta(\sigma+it)|^4 dt,$$

and they showed

$$I_{2,\sigma}(T) \sim \frac{\zeta^4(2\sigma)}{\zeta(4\sigma)}T \quad \left(\frac{1}{2} < \sigma < 1\right) \tag{1.9}$$

and

$$I_{2,\frac{1}{2}}(T) = O(T\log^4 T). \tag{1.10}$$

Ingham [63] improved (1.10) to

$$I_{2,\frac{1}{2}}(T) = (2\pi^2)^{-1} T \log^4 T + O(T \log^3 T), \tag{1.11}$$

by using the approximate functional equation of $\zeta^2(s)$ due to Hardy-Littlewood [50], that is

$$\zeta^2(s) = \sum_{n\le x} \frac{d(n)}{n^s} + \chi^2(s) \sum_{n \le y} \frac{d(n)}{n^{1-s}} + R_2(s; x, y) \tag{1.12}$$

for positive x, y with $4\pi^2 xy = t^2$, where

$$R_2(s; x, y) = O\left(x^{\frac{1}{2}-\sigma} \left(\frac{x+y}{t} \right)^{\frac{1}{4}} \log t \right). \tag{1.13}$$

In [189], Titchmarsh gave a different proof of (1.12) with

$$R_2(s; x, y) = O(x^{\frac{1}{2}-\sigma} \log t). \tag{1.14}$$

Recall that the original proof of (1.4) by Hardy-Littlewood is based on (1.6). Hardy-Littlewood deduced

$$\zeta(s) = \sum_{n \le x} n^{-s} - \frac{x^{1-s}}{1-s} + 2 \sum_{n=1}^{\infty} \int_x^{\infty} u^{-s} \cos(2\pi n u) du + O(x^{-\sigma}) \tag{1.15}$$

from (1.6) in the first stage of their argument. As an analogue of (1.15), Titchmarsh [189] proved

$$\begin{aligned} \zeta^2(s) = & \sum_{n \le x} \frac{d(n)}{n^s} - x^{-s} \sum_{n \le x} d(n) + \frac{2s - s^2}{(s-1)^2} x^{1-s} + \frac{s}{s-1} x^{1-s}(2\gamma + \log x) + \frac{1}{4} x^{-s} \\ & - 2^{4s} \pi^{2s-2} s \sum_{n=1}^{\infty} \frac{d(n)}{n^{1-s}} \int_{4\pi\sqrt{nx}}^{\infty} \frac{K_1(u) + \frac{\pi}{2} Y_1(u)}{u^{2s}} du, \end{aligned} \tag{1.16}$$

where K_1 and Y_1 are Bessel functions. This is the basis of Titchmarsh's proof of (1.14).

A climax of applications of approximate functional equations to the mean value problems came in the late 1970s, with the works of Balasubramanian, Good and Heath-Brown. A very careful analysis based on the Riemann-Siegel formula enabled Balasubramanian [4] to obtain the explicit formula

$$\begin{aligned} E(T) = & 2 \sum_{\substack{m,n \le K \\ m \ne n}} \sum \frac{\sin(T \log(\frac{n}{m}))}{(mn)^{\frac{1}{2}} \log(\frac{n}{m})} \\ & + 2 \sum_{\substack{m,n \le K \\ m \ne n}} \sum \frac{\sin(2\theta_1 - T \log mn)}{(mn)^{\frac{1}{2}} (2\theta_1' - \log mn)} + O(\log^2 T), \end{aligned} \tag{1.17}$$

where $\theta_1 = \theta_1(T) = \frac{1}{2}T\log(T/2\pi) - \frac{1}{2}T - \frac{1}{8}\pi$, θ_1' is the derivative of θ_1, and $K = [(T/2\pi)^{1/2}]$. Then, applying his own idea of multiple integration process to (1.17), Balasubramanian obtained the estimate of the form

$$E(T) = O(T^{\alpha+\varepsilon}) \tag{1.18}$$

with a certain $\alpha < 1/3$. (In [4], the value $\alpha = 27/82$ was given.) Good [34] proved an explicit formula of $E(T)$ similar to (1.17) but with certain smoothing factors, and from this formula he [35] proved

$$E(T) = \Omega(T^{\frac{1}{4}}). \tag{1.19}$$

Heath-Brown's work [53] gives an improvement on (1.11). He proved a new type of approximate functional equation, from which he deduced

$$I_{2,\frac{1}{2}}(T) = T\sum_{j=0}^{4} a_j \log^j T + E_2(T),$$

where a_j's are constants, $a_4 = (2\pi^2)^{-1}$, and

$$E_2(T) = O(T^{\frac{7}{8}+\varepsilon}). \tag{1.20}$$

Heath-Brown's paper also includes an alternative proof of (1.18).

Inspired by these papers, strong interests in the mean value problems revived. Really big progress has been made since 1980, which is the main theme of the present article. But first, we will discuss a closely related problem concerning the behaviour in mean of the divisor function in the next section.

2 The Dirichlet Divisor Problem

The title of this section means the problem of evaluating the error term $\Delta(x)$ defined by

$$\sum_{n\leq x}{}' d(n) = x\log x + (2\gamma - 1)x + \frac{1}{4} + \Delta(x) \tag{2.1}$$

for $x \geq 2$, where $\sum'$ indicates that the last term is to be halved if x is an integer. As can be observed by comparing (2.1) with (1.7), there is a strong analogy between $\Delta(x)$ and $E(T)$. Usually the study of $\Delta(x)$ is easier than that of $E(T)$, hence the results on $\Delta(x)$ are quite suggestive of guessing the behaviour of $E(T)$. Here we quote several known facts on $\Delta(x)$.

Dirichlet himself proved $\Delta(x) = O(x^{1/2})$, and Voronoï [195] improved it to obtain

$$\Delta(x) = O(x^{\frac{1}{3}}\log x). \tag{2.2}$$

The explicit formula

$$\begin{aligned}\Delta(x) &= -\frac{2}{\pi}x^{\frac{1}{2}}\sum_{n=1}^{\infty}\frac{d(n)}{n^{\frac{1}{2}}}\left\{K_1(4\pi\sqrt{nx})+\frac{\pi}{2}Y_1(4\pi\sqrt{nx})\right\}\\ &= \frac{x^{\frac{1}{4}}}{\pi\sqrt{2}}\sum_{n=1}^{\infty}d(n)n^{-\frac{3}{4}}\cos\left(4\pi\sqrt{nx}-\frac{\pi}{4}\right)\\ &\quad -\frac{3}{32\sqrt{2}\pi^2}x^{-\frac{1}{4}}\sum_{n=1}^{\infty}d(n)n^{-\frac{5}{4}}\sin\left(4\pi\sqrt{nx}-\frac{\pi}{4}\right)+O(x^{-\frac{3}{4}}) \qquad (2.3)\end{aligned}$$

is due to Voronoï [196]. Sometimes the truncated form

$$\Delta(x)=\frac{x^{\frac{1}{4}}}{\pi\sqrt{2}}\sum_{n\le N}d(n)n^{-\frac{3}{4}}\cos\left(4\pi\sqrt{nx}-\frac{\pi}{4}\right)+O(x^{\varepsilon}+x^{\frac{1}{2}+\varepsilon}N^{-\frac{1}{2}}) \qquad (2.4)$$

is useful. For instance, the formula

$$\int_2^X \Delta^2(x)dx=\frac{\zeta^4(\frac{3}{2})}{6\pi^2\zeta(3)}X^{\frac{3}{2}}+\delta(X) \qquad (2.5)$$

with $\delta(X)=O(X^{5/4+\varepsilon})$, due originally to Cramér [21], can be proved by substituting (2.4) into the left-hand side and squaring them out. Tong [191] obtained the improved estimate $\delta(X)=O(X\log^5 X)$, and an alternative simple proof was given by Meurman [142]. The best estimate at present is

$$\delta(X)=O(X\log^4 X) \qquad (2.6)$$

due to Preissmann [164].

As for the real order of $\delta(X)$, the author (1992) conjectured (cf.[115]) that $\delta(X)\sim CX(\log X)^B$ with a certain $B>0$. Lau-Tsang [128] proved

$$\int_2^X \delta(x)dx=-\frac{1}{8\pi^2}X^2\log^2 X+C_1X^2\log X+O(X^2), \qquad (2.7)$$

which implies $\delta(X)=\Omega_-(X\log^2 X)$, and conjectured

$$\delta(X)=-\frac{1}{4\pi^2}X\log^2 X+C_2X\log X+O(X). \qquad (2.8)$$

Moreover, Tsang [193] proved that (2.8) is valid for almost all X in a certain mean value sense. A generalization of the result of Lau-Tsang [128] was recently obtained by Furuya [27].

Another mean value formula for $\Delta(x)$ is

$$\begin{aligned}\int_2^X \Delta(x)dx &= \frac{1}{2\sqrt{2}\pi^2}X^{\frac{3}{4}}\sum_{n=1}^{\infty}d(n)n^{-\frac{5}{4}}\sin\left(4\pi\sqrt{nx}-\frac{\pi}{4}\right)\\ &\quad +\frac{15}{64\sqrt{2}\pi^3}X^{\frac{1}{4}}\sum_{n=1}^{\infty}d(n)n^{-\frac{7}{4}}\cos\left(4\pi\sqrt{nx}-\frac{\pi}{4}\right)+O(1), \qquad (2.9)\end{aligned}$$

which is due to Voronoï. Recently the mean value of the above quantity was studied in detail by Furuya-Tanigawa [28]. In several references (2.9) was quoted incorrectly. Note that sometimes $\Delta(x)$ is defined by (2.1) without the term 1/4; then the term $\frac{1}{4}X$ should be added on the right-hand side of (2.9).

The formula (2.5) includes the fact $\Delta(x) = \Omega(x^{1/4})$, and furthermore, it is known that

$$\Delta(x) = \Omega_{-}\{x^{\frac{1}{4}} \exp(C(\log\log x)^{\frac{1}{4}}(\log\log\log x)^{-\frac{3}{4}})\} \tag{2.10}$$

(Corrádi-Kátai [20]) and

$$\Delta(x) = \Omega_{+}\{(x \log x)^{\frac{1}{4}}(\log\log x)^{\frac{3+\log 4}{4}} \exp(-C\sqrt{\log\log\log x})\} \tag{2.11}$$

(Hafner [45]). In view of these Ω-results, it is quite plausible that

$$\Delta(x) = O(x^{\frac{1}{4}+\varepsilon}). \tag{2.12}$$

This is indeed a classical conjecture, but is believed to be extremely difficult. At present, the best known upper-bound is

$$\Delta(x) = O(x^{\frac{23}{73}+\varepsilon}) \tag{2.13}$$

due to Huxley [59]. This is just a small improvement on (2.2), but such a kind of improvement requires quite hard analysis on exponential sums. That is, we should use the theory of exponent pairs, created by van der Corput, and refined by many authors. For instance, Kolesnik [120] proved $\Delta(x) = O(x^{35/108+\varepsilon})$ by using his own elaborated version of the theory of exponent pairs, and later he [121] improved the exponent to $139/429 + \varepsilon$.

Bombieri-Iwaniec [13] [14] invented a new method of treating exponential sums, which gives an essentially new exponent pairs (see Huxley-Watt [62]). Combining this method with the expression

$$\Delta(x) = -2 \sum_{n \le \sqrt{x}} \psi\left(\frac{x}{n}\right) + O(1), \tag{2.14}$$

where $\psi(x) = x - [x] - \frac{1}{2}$, Iwaniec-Mozzochi [80] obtained the estimate $\Delta(x) = O(x^{7/22+\varepsilon})$. Huxley achieved to prove (2.13) by a further refinement of the method of Bombieri-Iwaniec. For the details of the theory of exponent pairs, the readers are referred to Graham-Kolesnik [39] or Huxley [61].

3 The Atkinson Formula and the Recent Results on $E(T)$

As we mentioned in the previous section, there is an analogy between $\Delta(x)$ and $E(T)$. Therefore it is natural to search for a formula analogous to Voronoï's (2.3) or (2.4). This was carried out in 1949 by Atkinson [3]. To state his result, we prepare several notations.

Let $X \asymp T$ (i.e. $T \ll X \ll T$), $\operatorname{arsinh} x = \log(x + \sqrt{1+x^2})$, and define

$$
\begin{aligned}
e(T,n) &= \left(1+\frac{\pi n}{2T}\right)^{-\frac{1}{4}}\left(\frac{2T}{\pi n}\right)^{-\frac{1}{2}}\left(\operatorname{arsinh}\sqrt{\frac{\pi n}{2T}}\right)^{-1},\\
f(T,n) &= 2T\operatorname{arsinh}\sqrt{\frac{\pi n}{2T}} + (\pi^2 n^2 + 2\pi n T)^{\frac{1}{2}} - \frac{\pi}{4},\\
g(T,n) &= T\log\left(\frac{T}{2\pi n}\right) - T + \frac{\pi}{4},\\
B(T,\xi) &= \frac{T}{2\pi} + \frac{1}{2}\xi^2 - \xi\left(\frac{T}{2\pi}+\frac{1}{4}\xi^2\right)^{\frac{1}{2}},\\
\sum\nolimits_{1,\sigma}(T,X) &= \sqrt{2}\left(\frac{T}{2\pi}\right)^{\frac{3}{4}-\sigma}\sum_{n\le X}(-1)^n\sigma_{1-2\sigma}(n)n^{\sigma-\frac{5}{4}}e(T,n)\cos(f(T,n))
\end{aligned}
$$

and

$$
\begin{aligned}
\sum\nolimits_{2,\sigma}(T,X) &= 2\left(\frac{T}{2\pi}\right)^{\frac{1}{2}-\sigma}\sum_{n\le B(T,\sqrt{X})}\sigma_{1-2\sigma}(n)n^{\sigma-1}\\
&\quad\times\left(\log\frac{T}{2\pi n}\right)^{-1}\cos(g(T,n)).
\end{aligned}
$$

Then Atkinson's explicit formula can be stated as

$$
E(T) = \sum\nolimits_{1,\frac{1}{2}}(T,X) - \sum\nolimits_{2,\frac{1}{2}}(T,X) + O(\log^2 T). \tag{3.1}
$$

The starting point of Atkinson's proof of (3.1) is the product $\zeta(u)\zeta(v)$, where u and v are independent complex variables. At first assume $\operatorname{Re} u > 1$, $\operatorname{Re} v > 1$. Then

$$
\zeta(u)\zeta(v) = \sum_{m=1}^{\infty}\sum_{n=1}^{\infty} m^{-u}n^{-v}.
$$

Divide this double sum into three parts according to the conditions $m = n$, $m > n$ and $m < n$ (Atkinson's dissection). The part corresponding to $m = n$ is clearly $\zeta(u+v)$. By using the Poisson summation formula (1.6), Atkinson showed an integral expression of the remaining parts, which enables the analytic continuation. Then he transformed this expression by applying Voronoï's formula (2.3), and evaluated the resulting integrals by his own saddle-point lemma, which gives an asymptotic formula for the integral of the type

$$
\int_a^b g(x)\exp(2\pi i(f(x)+kx))dx \tag{3.2}
$$

with real k and certain functions $f(x)$ and $g(x)$. The details of the proof are rather long and complicated.

When Atkinson published (3.1), no one noticed its usefulness. After the disregard during about thirty years, Heath-Brown first gave attention to Atkinson's paper. In [51], Heath-Brown proved

$$
\int_2^T E^2(t)dt = \frac{2\zeta^4(\frac{3}{2})}{3(2\pi)^{\frac{1}{2}}\zeta(3)}T^{\frac{3}{2}} + F(T) \tag{3.3}
$$

with

$$F(T) = O(T^{\frac{5}{4}} \log^2 T), \tag{3.4}$$

as an application of (3.1). This is the analogy of (2.5), and implies (1.19) of Good. Heath-Brown's another paper [52] deduced the estimate

$$\int_0^T \left|\zeta\left(\frac{1}{2}+it\right)\right|^{12} dt = O(T^2 \log^{17} T) \tag{3.5}$$

from a certain estimate of the mean square of $|\zeta(1/2+it)|$ in short intervals. Atkinson's formula (3.1) was used to prove the latter estimate. (See also Chapter 7 of Ivić [66] for a different proof.)

These works of Heath-Brown showed the fruitfulness of Atkinson's formula (3.1), but it was Jutila [81I] who noticed the real value lying in Atkinson's method. He sketched in [81I] how easily can (1.18) be obtained from (3.1). Following Jutila's idea, and combining with Kolesnik's technique [120], Ivić described a proof of

$$E(T) = O(T^{\frac{35}{108}+\varepsilon}) \tag{3.6}$$

in Section 15.5 of [66].

The main theme of Jutila's aforementioned paper [81I] is the analogy between $E(T)$ and a modification of $\Delta(x)$, that is

$$\Delta^*(x) = -\Delta(x) + 2\Delta(2x) - \frac{1}{2}\Delta(4x).$$

A consequence of his analysis is the hypothetical bound

$$E(T) = O(T^{\frac{5}{16}+\varepsilon}) \tag{3.7}$$

under the assumption of the conjecture (2.12). The exponent 5/16 in (3.7) was later improved to 3/10 by Jutila [81II]. In [82], Jutila proved

$$\int_2^T \left(E(t) - 2\pi\Delta^*\left(\frac{t}{2\pi}\right)\right)^2 dt = O(T^{\frac{4}{3}} \log^3 T). \tag{3.8}$$

The transformation method for Dirichlet polynomials was created by Jutila [83]. The basic tool of this method is Voronoï's summation formula

$$\begin{aligned}\sum_{a\le n\le b}{}' d(n)f(n) &= \int_a^b f(u)(\log u + 2\gamma)du \\ &+ \sum_{n=1}^{\infty} d(n) \int_a^b f(u)(4K_0(4\pi\sqrt{nu}) - 2\pi Y_0(4\pi\sqrt{nu}))du,\end{aligned} \tag{3.9}$$

valid for $f \in C^2[a,b]$, where K_0 and Y_0 are Bessel functions. Jutila's idea is to transform the Dirichlet polynomial

$$S(M_1, M_2; t) = \sum_{M_1\le m\le M_2} d(m)m^{-\frac{1}{2}-it} \tag{3.10}$$

by applying (3.9), and then use a lemma of Atkinson's type to evaluate the resulting expressions. One of his results is an explicit formula for $|\zeta(1/2+it)|^2$, whose shape is similar to Atkinson's formula. Several new ideas are included in his argument. One of them is the device of multiplying the original polynomial (3.10) by trivial factors $e^{2\pi irm}$, where r is an integer. Another novelty is the multiple-averaged version of Atkinson's saddle-point lemma, which gives an asymptotic formula for the integral of the form

$$U^{-J}\int_0^U du_1\cdots\int_0^U du_J\int_{a+u_1+\cdots+u_J}^{b-u_1-\cdots-u_J} g(x)\exp(2\pi i(f(x)+kx))dx \tag{3.11}$$

instead of (3.2). This point was fully developed in Jutila [87]. We will encounter the transformation method again in Sections 7 and 9.

By using the above averaged saddle-point lemma of Jutila, Meurman [142] improved (3.4) to

$$F(T)=O(T\log^5 T). \tag{3.12}$$

In fact, Meurman proved an averaged version of Atkinson's formula, which can be stated as

$$E(T)=\sum\nolimits_{1,\frac{1}{2}}^{*}(T)-\sum\nolimits_{2,\frac{1}{2}}^{*}(T)+\pi+O(T^{-\frac{1}{4}}\log T), \tag{3.13}$$

where $\sum_{j,\frac{1}{2}}^{*}(T)$ is a certain weighted sum similar to $\sum_{j,\frac{1}{2}}(T,X)$ $(j=1,2)$. The deduction of (3.12) from (3.13) is basically analogous to the argument of Heath-Brown [51].

The estimate (3.12) was proved also by Motohashi [149IV] [150] independently. From his asymptotic formula for $R_2(s;t/2\pi)$ (see the next section), Motohashi deduced another version of Atkinson's formula, and from which he obtained (3.12). Motohashi's argument includes an alternative proof of the original formula (3.1) with a slightly better error term $O(\log T)$. Note that another different proof of (3.1) was obtained by Jutila [92]. His argument is based on the Laplace transform of $|\zeta(1/2+it)|^2$, and does not appeal to Atkinson's dissection device.

Inspired by Preissmann's proof of (2.6) (which is an application of the inequality (8.12) below due to Montgomery-Vaughan), Preissmann himself [165] and Ivić [68, (2.100)] independently of each other proved

$$F(T)=O(T\log^4 T), \tag{3.14}$$

which is the best at present. There is a conjecture of the author that $F(T)\sim CT(\log T)^B$ would hold with a certain $B>0$. Probably $B=2$.

Higher power moments of $E(t)$ were first studied by Ivić [64], who proved

$$\int_2^T |E(t)|^\alpha dt = O(T^{1+\frac{\alpha}{4}+\varepsilon}) \qquad \left(0\le\alpha\le\frac{35}{4}\right) \tag{3.15}$$

and

$$\int_2^T |E(t)|^\alpha dt = O(T^{\frac{38+35\alpha}{108}+\varepsilon}) \qquad \left(\alpha\ge\frac{35}{4}\right). \tag{3.16}$$

Heath-Brown [56] proved the existence of the limit

$$\lim_{T\to\infty} T^{-1-\frac{\alpha}{4}} \int_2^T |E(t)|^\alpha dt \tag{3.17}$$

for $0 \le \alpha < 28/3$. Tsang [192] obtained

$$\int_2^T E(t)^k dt = C(k)T^{1+\frac{k}{4}} + O(T^{1+\frac{k}{4}-\delta}) \tag{3.18}$$

for $k = 3$ or 4, where $C(k)$ is an explicitly written positive constant, and $\delta > 0$. Recently Ivić [73] showed, using (3.8), that one can take $\delta = 1/14$ for $k = 3$ and $\delta = 1/23$ for $k = 4$.

In view of the Ω-result (1.19), it is plausible that

$$E(T) = O(T^{\frac{1}{4}+\varepsilon}), \tag{3.19}$$

as an analogue of (2.12). The above results on higher power moments can also be regarded as supporting facts of this conjecture. The best known upper-bound is, however, still far from this conjecture. Heath-Brown and Huxley [57] applied the methods of Bombieri, Iwaniec and Mozzochi (mentioned in Section 2) and some lemmas proved in [53] to obtain $E(T) = O(T^{7/22}(\log T)^{111/22})$, which is better than (3.6), and this was further improved to

$$E(T) = O(T^{\frac{72}{227}}(\log T)^{\frac{679}{227}}) \tag{3.20}$$

by Huxley [60].

The Ω-result (1.19) was refined by Hafner-Ivić [46], who showed, analogously to (2.10) and (2.11), that

$$E(T) = \Omega_-\{T^{\frac{1}{4}} \exp(C(\log\log T)^{\frac{1}{4}}(\log\log\log T)^{-\frac{3}{4}})\} \tag{3.21}$$

and

$$E(T) = \Omega_+\{(T\log T)^{\frac{1}{4}}(\log\log T)^{\frac{3+\log 4}{4}} \exp(-C\sqrt{\log\log\log T})\}. \tag{3.22}$$

The local behaviour of sign-changes of $E(T)$ was studied by Ivić [67], Ivić-te Riele [78], and (independently) Heath-Brown and Tsang [58]. Ivić [67] showed that there exist positive constants C_1 and C_2, such that every interval $[T, T+C_1\sqrt{T}]$ (for $T \ge T_0$) contains numbers t_1, t_2 for which

$$E(t_1) > C_2 t_1^{1/4}, \qquad E(t_2) < -C_2 t_2^{1/4}$$

hold. This result is also included in Heath-Brown and Tsang [58].

From Atkinson's formula Hafner-Ivić [46] deduced, as an analogue of (2.9), that

$$\begin{aligned}\int_2^T E(t)dt &= \pi T + \frac{1}{2}\left(\frac{2T}{\pi}\right)^{\frac{3}{4}} \sum_{n=1}^{\infty}(-1)^n \frac{d(n)}{n^{\frac{5}{4}}} \sin\left(2\sqrt{2\pi nT} - \frac{\pi}{4}\right) \\ &\quad + O(T^{\frac{2}{3}}\log T).\end{aligned} \tag{3.23}$$

This implies that the function $E(t)$ has the mean value π. Hence it is natural to consider the zeros of the function $E(t)-\pi$, which we denote by t_n ($2 \le t_1 < t_2 < \cdots$). Then the above mentioned result implies that

$$t_{n+1} - t_n \ll t_n^{1/2}. \tag{3.24}$$

Let $\kappa = \inf\{c \ge 0\ ;\ t_{n+1} - t_n \ll t_n^c\}$. Then (3.24) implies that $\kappa \le 1/2$. Ivić-te Riele [78] studied $\{t_n\}$ both theoretically and numerically, and proposed the conjecture that $\kappa = 1/4$. This conjecture is very strong because it would lead to (3.19) (see Theorem 1 of [78]). However, this conjecture was disproved by Heath-Brown and Tsang [58]. They showed that for any $\delta > 0$ and any $T \ge T_0(\delta)$, there are at least $C_1\delta T^{1/2}\log^5 T$ disjoint subintervals of length $C_2\delta T^{1/2}(\log T)^{-5}$ in $[T, 2T]$ such that

$$|E(t)| > (B_0 - \delta)t^{\frac{1}{4}}, \qquad B_0 = \frac{\zeta^2(\frac{3}{2})}{2(2\pi)^{\frac{1}{4}}\zeta(3)^{\frac{1}{2}}}$$

whenever t lies in any of these subintervals. In particular $E(t)$ does not change sign in any of these subintervals. Therefore the local behaviour of $E(t)$ is much more mysterious than was expected by Ivić and te Riele.

4 The Remainder Term in the Approximate Functional Equation for $\zeta^2(s)$

In the middle of 1980s, new light was shed on the remainder term $R_2(s; x, y)$ in (1.12). Jutila [83] pointed out that (1.12) with (1.14) can be deduced from the Voronoï summation formula (3.9); this should be compared with the fact, mentioned in Section 1, that (1.4) can be deduced from the Poisson summation formula (1.6). The details are presented in Ivić [65] [66] (see Section 4.2 of [66]).

In the special case $x = y = t/2\pi$ ("the symmetric case"), Motohashi [148I] showed that the estimate (1.14) for $R_2(s; t/2\pi, t/2\pi)$ (which we abbreviate as $R_2(s; t/2\pi)$) follows from the estimate (1.5) for $R_1(s; \sqrt{t/2\pi}, \sqrt{t/2\pi})$ by the Dirichlet device. However, we know much more precise information on $R_1(s; \sqrt{t/2\pi}, \sqrt{t/2\pi})$, that is the Riemann-Siegel formula. What can we obtain if we combine Motohashi's argument with the Riemann-Siegel formula? This idea was pursued by Motohashi himself, and he [148II] [150] [153] obtained a very precise asymptotic formula of $R_2(s; t/2\pi)$. His result includes

$$\begin{aligned}\chi(1-s)R_2\Big(s\,;\,\frac{t}{2\pi}\Big) &= \Big(\frac{t}{2\pi}\Big)^{-\frac{1}{4}}\sum_{n=1}^{\infty} d(n)h(n)n^{-\frac{1}{4}}\sin\Big(2\sqrt{2\pi t n}+\frac{\pi}{4}\Big)\\ &\quad + O(t^{-\frac{1}{2}}\log t),\end{aligned} \tag{4.1}$$

where

$$\begin{aligned}h(n) &= \Big(\frac{2}{\pi}\Big)^{\frac{1}{2}}\int_0^\infty (u+n\pi)^{-\frac{1}{2}}\cos\Big(u+\frac{\pi}{4}\Big)du\\ &= -\frac{1}{\pi}n^{-\frac{1}{2}} + O(n^{-\frac{3}{2}}).\end{aligned}$$

Actually Motohashi's formula is more precise and complicated; the error term in (4.1) is replaced by some more explicit terms and a smaller error $O(t^{-1}\log t)$. A simple consequence of (4.1) and (2.3) is the relation

$$\chi(1-s)R_2\left(s\,;\,\frac{t}{2\pi}\right)=-\sqrt{2}\left(\frac{t}{2\pi}\right)^{-\frac{1}{2}}\Delta\left(\frac{t}{2\pi}\right)+O(t^{-\frac{1}{4}}),\tag{4.2}$$

which was announced in [148I]. A formula of the same type was already given long before in Taylor's posthumous article [186], but Motohashi [153] pointed out that Taylor's argument was incorrect.

Jutila [85] gave an alternative proof of (4.2). His starting point is Titchmarsh's explicit formula (1.16). His idea is to smooth the right-hand side of (1.16) by using multiple integration, and then apply his own saddle-point lemma mentioned in Section 3. So far Jutila's method cannot give a proof of Motohashi's precise formula ((4.1) and more). An advantage of Jutila's approach is that it can be applied to many other Dirichlet series, satisfying a certain functional equation. In [85], Jutila presented analogous results on

$$\varphi(s,F)=\sum_{n=1}^{\infty}a(n)n^{-s},\tag{4.3}$$

where $a(n)$'s are the Fourier coefficients of a holomorphic cusp form $F(z)=\sum_{n=1}^{\infty}a(n)\exp(2\pi inz)$ of weight κ (an even integer) for the full modular group $SL(2,\mathbf{Z})$. There are many analogous properties shared by $\zeta^2(s)$ and $\varphi(s,F)$, but a big difference is that $\zeta^2(s)$ has the good square-root function (i.e. $\zeta(s)$), while $\varphi(s,F)$ does not. This is why Motohashi's approach cannot be applied to $\varphi(s,F)$. Recently, Guthmann [42] [43] [44] has developed another unified approach to the remainder terms in the approximate functional equations for $\zeta^2(s)$ and $\varphi(s,F)$.

The formula (4.2) tells that a strong analogy between $R_2(s\,;\,t/2\pi)$ and $\Delta(t/2\pi)$ should exist (cf. Ivić [69]). Kiuchi-Matsumoto [114] proved, as an analogue of (2.5), that

$$\int_2^T\left|R_2\left(\frac{1}{2}+it\,;\,\frac{t}{2\pi}\right)\right|^2dt=\sqrt{2\pi}C_0T^{\frac{1}{2}}+K(T),\tag{4.4}$$

where

$$C_0=\sum_{n=1}^{\infty}d^2(n)h^2(n)n^{-\frac{1}{2}}\tag{4.5}$$

and $K(T)=O(T^{\frac{1}{4}}\log T)$. The proof is based on (4.1). Using a more precise form of Motohashi's formula, Kiuchi [111] gave the improved bound $K(T)=O(\log^5 T)$, and suggested the conjecture

$$K(T)\sim C\log^3 T.\tag{4.6}$$

The hitherto best upper-bound is

$$K(T)=O(\log^4 T)\tag{4.7}$$

due to Kiuchi [113II]. On the other hand, Ivić [72] proved

$$\int_2^T K(t)dt = C_1 T \log^3 T + C_2 T \log^2 T + O(T \log T) \tag{4.8}$$

with $C_1 < 0$, analogously to (2.7). This implies

$$K(T) = \Omega_-(\log^3 T), \tag{4.9}$$

which supports the conjecture (4.6).

Motohashi's formula has been obtained in the symmetric case $x = y = t/2\pi$. How is the non-symmetric case? Motohashi [148III] [150] proved a formula when $x = \alpha t/2\pi$, $y = t/2\pi\alpha$ with a rational number α, but the result is not so precise as in the symmetric case. Jutila [85] considered the general situation, and in some non-symmetric cases his bound is better than (1.14). See also Jutila [86]. Mean-value results in the non-symmetric case were discussed by Kiuchi [112].

5 The Mean Square of $\zeta(s)$ in the Critical Strip

Now we return to the problem of evaluating $I_\sigma(T)$, and discuss the case $1/2 < \sigma \le 1$. After the classical result (1.1) of Landau and Schnee, the development in this direction had been very slow (cf. Ingham [63] and (8.112) of Ivić [66]). In 1989, the author [135] published the analogue of Atkinson's formula in the strip $1/2 < \sigma < 3/4$. It is stated as

$$E_\sigma(T) = \sum\nolimits_{1,\sigma}(T, X) - \sum\nolimits_{2,\sigma}(T, X) + O(\log T), \tag{5.1}$$

where $E_\sigma(T)$ is defined, for $1/2 < \sigma < 1$, by

$$I_\sigma(T) = \zeta(2\sigma)T + (2\pi)^{2\sigma-1}\frac{\zeta(2-2\sigma)}{2-2\sigma}T^{2-2\sigma} + E_\sigma(T). \tag{5.2}$$

It can be easily seen that $E_\sigma(T) \to E(T)$ as $\sigma \to 1/2 + 0$. In the same paper [135], as applications of (5.1), the author showed that

$$E_\sigma(T) = O(T^{\frac{1}{1+4\sigma}} \log^2 T) \qquad \left(\frac{1}{2} < \sigma < \frac{3}{4}\right) \tag{5.3}$$

(the O-constant may depend on σ) and

$$\int_2^T E_\sigma(t)^2 dt = A_1(\sigma)T^{\frac{5}{2}-2\sigma} + F_\sigma(T) \qquad \left(\frac{1}{2} < \sigma < \frac{3}{4}\right) \tag{5.4}$$

with $F_\sigma(T) = O(T^{7/4-\sigma} \log T)$, where

$$A_1(\sigma) = \frac{2}{5-4\sigma}(2\pi)^{2\sigma-\frac{3}{2}}\frac{\zeta(\frac{3}{2})^2}{\zeta(3)}\zeta\left(\frac{5}{2}-2\sigma\right)\zeta\left(\frac{1}{2}+2\sigma\right).$$

Independently of [135], Laurinčikas [129I] obtained the analogue of Atkinson's formula near the critical line, that is an explicit formula for the error term in the asymptotic formula of the integral

$$\int_0^T |\zeta(\sigma_T+it)|^2 dt,$$

where $\sigma_T = 1/2 + l_T^{-1}$, $0 < l_T \ll \log T$, and l_T tends to infinity as $T \to \infty$. See also Laurinčikas [129II] [131] [132] [133]; in [132], he proved that (5.3) (with a slightly different log-factor) holds uniformly in σ.

An asymptotic formula over short intervals was obtained by Sankaranarayanan-Srinivas [179] by a quite different method. They proved

$$\frac{1}{H}\int_T^{T+H} |\zeta(\sigma+it)|^2 dt = \zeta(2\sigma) + O\left(\exp\left(-\frac{C_1(\log T)^{2-2\sigma}}{\log\log T}\right)\right)$$

for $\exp((\log T)^{2-2\sigma}) \le H \le T$ and $(1/2) + C_2(\log\log T)^{-1} \le \sigma \le 1 - C_3$ under the assumption of the Riemann hypothesis. It should be noted that their method can be applied to much more general Dirichlet series.

The basic tool of the author [135] is Oppenheim's Voronoï-type formula [161] for the error term $\Delta_{1-2\sigma}(x)$ defined by

$$\sum_{n\le x}{}' \sigma_{1-2\sigma}(n) = \zeta(2\sigma)x + \frac{\zeta(2-2\sigma)}{2-2\sigma}x^{2-2\sigma} - \frac{1}{2}\zeta(2\sigma-1) + \Delta_{1-2\sigma}(x). \tag{5.5}$$

The series in Oppenheim's formula is convergent only for $\sigma < 3/4$, which is the reason why the restriction $1/2 < \sigma < 3/4$ exists. It was pointed out in [135] that the coefficient $A_1(\sigma)$ tends to infinity when $\sigma \to 3/4 - 0$, which suggests some singular situation occurring at $\sigma = 3/4$. Now we know that the behaviour of $E_\sigma(T)$ in fact transposes at $\sigma = 3/4$, which can be well observed by the following refinement of (5.4):

$$\int_2^T E_\sigma(t)^2 dt = \begin{cases} A_1(\sigma)T^{\frac{5}{2}-2\sigma} + O(T) & (\frac{1}{2} < \sigma < \frac{3}{4}) \\ A_0 T\log T + O(T) & (\sigma = \frac{3}{4}) \\ O(T) & (\frac{3}{4} < \sigma < 1), \end{cases} \tag{5.6}$$

where $A_0 = \zeta^2(3/2)\zeta(2)/\zeta(3)$. These are due to Matsumoto-Meurman [140II] ($1/2 < \sigma < 3/4$), Lam [125] ($\sigma = 3/4$), and Matsumoto-Meurman [140III] ($3/4 < \sigma < 1$), respectively. (In [140III], the formula for $\sigma = 3/4$ was given with a slightly weaker error term $O(T(\log T)^{1/2})$.)

To prove the result for $1/2 < \sigma < 3/4$, that is $F_\sigma(T) = O(T)$, Matsumoto-Meurman [140II] gave a new averaged version (somewhat similar to (3.13)) of Atkinson-type formula, which is proved by combining the methods of Meurman [142] and Preissmann [165] with some additional new idea. In the same paper [140II], the conjecture

$$F_\sigma(T) \sim 4\pi^2\zeta(2\sigma-1)^2 T \qquad \left(\frac{1}{2} < \sigma < \frac{3}{4}\right) \tag{5.7}$$

was proposed. There are several heuristic arguments which may suggest (5.7) (see [115] [136]). The reason presented in [140II] is the fact that $E_\sigma(T)$ has the mean value $-2\pi\zeta(2\sigma-1)$. This fact was discovered independently by Ivić [68]; he proved

$$\int_2^T E_\sigma(t)dt = B(\sigma)T + O(T^{\frac{5}{4}-\sigma}) \quad \left(\frac{1}{2} < \sigma < \frac{3}{4}\right) \tag{5.8}$$

((3.39) of [68]). The expression of $B(\sigma)$ given in [68] is complicated, but it is actually equal to $-2\pi\zeta(2\sigma-1)$ (see Appendix of Matsumoto-Meurman [140II]). The above (5.8) is a direct consequence of

$$\begin{aligned}\int_2^T E_\sigma(t)dt &= B(\sigma)T + 2^{\sigma-\frac{3}{4}}\left(\frac{T}{\pi}\right)^{\frac{5}{4}-\sigma}\sum_{n=1}^\infty(-1)^n\sigma_{1-2\sigma}(n)n^{\sigma-\frac{7}{4}} \\ &\quad \times\sin\left(\sqrt{8\pi nT}-\frac{\pi}{4}\right) + O(T^{1-\frac{2}{3}\sigma}\log T)\end{aligned} \tag{5.9}$$

((3.30) of Ivić [68]), which is the analogue of (3.23). We mention here that it might be better to define the "real" error term in (5.2) (resp. (1.7)) as $E_\sigma(T)+2\pi\zeta(2\sigma-1)$ (resp. $E(T)-\pi$). The constant $-2\pi\zeta(2\sigma-1)$ (resp. π) corresponds to $-(1/2)\zeta(2\sigma-1)$ in (5.5) (resp. 1/4 in (2.1)).

Matsumoto-Meurman [140III] proved that the formula (5.1) is valid for all σ satisfying $1/2<\sigma<1$. When $\sigma\geq 3/4$ the Voronoï-type formula for the Riesz mean of $\sigma_{1-2\sigma}(n)$ is applied in [140III], because Oppenheim's series is divergent. It is again a certain averaged version of Atkinson-type formula from which the case $3/4<\sigma<1$ of (5.6) was deduced in [140III]. (Here we note that in the statement of Lemma 4 of [140III], σ should be deleted. The author would like to thank Dr. Hideki Nakaya who pointed out this mistake.)

As an extension of (5.7), the author proposed the conjecture that the error terms $O(T)$ in (5.6) could be replaced by $A_2(\sigma)T+o(T)$ for $1/2<\sigma<1$, with a certain constant $A_2(\sigma)$ (see [137]). A refined version is:

Conjecture 1 The error terms $O(T)$ in (5.6) could be replaced by

$$A_2(\sigma)T + O(T^{2-2\sigma}(\log T)^C) \tag{5.10}$$

for $1/2<\sigma\leq 3/4$, where $C\geq 0$, and by

$$A_2(\sigma)T + A_3(\sigma)T^{\frac{5}{2}-2\sigma} + O(T^{2-2\sigma}(\log T)^C) \tag{5.11}$$

with a certain $A_3(\sigma)$ for $3/4<\sigma<1$.

The reason of the error estimates $O(T^{2-2\sigma}(\log T)^C)$ is the result (6.6) mentioned in the next section. The author proposed (5.10) first in correspondence, which is mentioned in Ivić-Kiuchi [74]. The conjecture (5.11) first appeared in [115] (though the term $A_3(\sigma)T^{5/2-2\sigma}$ is missing there). Even the weaker form of the above conjecture is still open.

The formula (5.4) obviously implies $E_\sigma(T)=\Omega(T^{3/4-\sigma})$ for $1/2<\sigma<3/4$. Ivić [68] improved this to $\Omega_\pm(T^{3/4-\sigma})$ with some information about local sign-changes of $E_\sigma(T)$.

The best known Ω-results at present are

$$E_\sigma(T) = \Omega_-\{T^{\frac{3}{4}-\sigma}\exp(C(\log\log T)^{\sigma-\frac{1}{4}}(\log\log\log T)^{\sigma-\frac{5}{4}})\} \qquad \left(\frac{1}{2}<\sigma<\frac{3}{4}\right) \tag{5.12}$$

(Ivić-Matsumoto [75]), exactly corresponding to (3.21), and

$$E_\sigma(T) = \Omega_+(T^{\frac{3}{4}-\sigma}(\log T)^{\sigma-\frac{1}{4}}) \qquad \left(\frac{1}{2}<\sigma<\frac{3}{4}\right) \tag{5.13}$$

(Matsumoto-Meurman [140III]). It is much more difficult to obtain any Ω-result in the strip $3/4 \le \sigma < 1$. The only known result is

$$E_{\frac{3}{4}}(T) = \Omega((\log T)^{\frac{1}{2}}), \tag{5.14}$$

a direct consequence of the case $\sigma = 3/4$ of (5.6).

What is the real order of $E_\sigma(T)$? In view of (5.6) and the above Ω-results, we may formulate the conjecture

$$E_\sigma(T) \ll \begin{cases} T^{\frac{3}{4}-\sigma+\varepsilon} & (\frac{1}{2}<\sigma<\frac{3}{4}) \\ T^{\varepsilon} & (\frac{3}{4}\le\sigma<1). \end{cases} \tag{5.15}$$

In Ivić-Matsumoto [75] this conjecture is stated, and also it is pointed out that if we assume the very strong conjecture that $(\varepsilon, 1/2+\varepsilon)$ would be an exponent pair for any $\varepsilon > 0$, then (5.15) would follow.

The critical behaviour of $E_\sigma(T)$ at $\sigma = 3/4$ is again clear in (5.15); it might suggest some unexpected properties of $\zeta(s)$. In connection with this observation, an interesting discussion concerning the Lindelöf hypothesis is given in Ivić [71]. See also the final section of [136].

The proof of the conjecture (5.15) seems to be out of reach now. As for the upper bound of $E_\sigma(T)$, Motohashi (unpublished) proved that (5.3) holds for any σ satisfying $1/2 < \sigma < 1$. His idea, inspired by his own work [151] on the fourth power mean of $\zeta(s)$, is to use the weighted integral

$$\frac{1}{\Delta\sqrt{\pi}}\int_{-\infty}^{\infty}|\zeta(\sigma+i(T+t))|^2 e^{-(\frac{t}{\Delta})^2}dt \qquad (\Delta > 0). \tag{5.16}$$

Ivić [68] combined Motohashi's idea with the theory of exponent pairs, and obtained various improved upper bounds of $E_\sigma(T)$. This direction was further studied by Ivić-Matsumoto [75] and Kačėnas [94] [95]; for instance, we have

$$E_\sigma(T) = O(T^{\frac{2(1-\sigma)}{3}}(\log T)^{\frac{2}{9}}) \qquad \left(\frac{1}{2}<\sigma<1\right)$$

(Ivić-Matsumoto [75]) and

$$E_\sigma(T) = O(T^{\frac{72}{227}-\frac{1}{2}\delta+\varepsilon}) \qquad \left(\frac{1}{2}<\sigma<\frac{51}{100}\right)$$

with $\delta = \sigma - 1/2$ (Kačėnas [95]). The latter is uniform in σ, and exactly corresponds to Huxley's bound (3.20).

We conclude this section with mentioning the case $\sigma = 1$. No analogue of Atkinson's formula is known in this case. Starting from the simple approximate formula (1.2), Balasubramanian-Ivić-Ramachandra [7] proved the asymptotic formula

$$I_1(T) = \zeta(2)T - \pi \log T + \widetilde{E_1}(T) \tag{5.17}$$

with

$$\widetilde{E_1}(T) = O((\log T)^{\frac{2}{3}}(\log\log T)^{\frac{1}{3}}). \tag{5.18}$$

The connection between $E_\sigma(T)$ and $\widetilde{E_1}(T)$ is given by

$$\lim_{\sigma\to 1-0}\left\{\zeta(2\sigma)T + (2\pi)^{2\sigma-1}\frac{\zeta(2-2\sigma)}{2-2\sigma}(T^{2-2\sigma}-1)\right\} = \zeta(2)T - \pi\log T$$

(Ivić [70]). In [7], they also proved the mean value results

$$\begin{aligned}\int_2^T \widetilde{E_1}(t)dt &= O(T),\\ \int_2^T \widetilde{E_1}(t)^2dt &= O(T(\log\log T)^4)\end{aligned}$$

and conjectured that the latter integral would be asymptotically equal to CT.

To show the estimate (5.18), the method of I.M. Vinogradov and Korobov, based on the deep theory of I.M. Vinogradov on the estimation of exponential sums, is applied. In fact, it is noted in [7] that from (5.18) one can deduce the estimate $\zeta(1+it) = O((\log t)^{2/3}(\log\log t)^{1/3})$, which is very close to the sharpest known bound $\zeta(1+it) = O((\log t)^{2/3})$, obtained by the Vinogradov-Korobov theory (see Chapter 6 of Ivić [66]).

6 Mean Values of $\Delta_{1-2\sigma}(x)$ and $R_2(\sigma + it; t/2\pi)$

In Section 3 we explained that a guiding principle of the study of $E(T)$ is to pursue the analogy with $\Delta(x)$. Similarly, it is useful to study the behaviour of $\Delta_{1-2\sigma}(x)$, defined by (5.5), which is the object analogous to $E_\sigma(T)$.

We already mentioned in Section 5 that the Voronoï-type formula for $\Delta_{1-2\sigma}(x)$ due to Oppenheim [161] was used in the proof of (5.1). By using the truncated Voronoï-type formula, Kiuchi [109] proved that

$$\int_2^X \Delta_{1-2\sigma}(x)^2dx = B_1(\sigma)X^{\frac{5}{2}-2\sigma} + O(X^{\frac{7}{4}-\sigma+\varepsilon}) \qquad \left(\frac{1}{2} < \sigma < \frac{3}{4}\right) \tag{6.1}$$

with

$$B_1(\sigma) = \frac{\zeta(\frac{3}{2})^2}{2\pi^2(5-4\sigma)\zeta(3)}\zeta\left(\frac{5}{2}-2\sigma\right)\zeta\left(\frac{1}{2}+2\sigma\right).$$

It was already mentioned by Cramér [22] that the left-hand side of (6.1) is asymptotically equal to $B_1(\sigma)X^{5/2-2\sigma}$ for $1/2 < \sigma < 3/4$. Meurman [143] refined (6.1) to obtain

$$\int_2^X \Delta_{1-2\sigma}(x)^2 dx = \begin{cases} B_1(\sigma)X^{\frac{5}{2}-2\sigma} + O(X) & (\frac{1}{2} < \sigma < \frac{3}{4}) \\ B_0 X \log X + O(X) & (\sigma = \frac{3}{4}) \\ O(X) & (\frac{3}{4} < \sigma < 1) \end{cases} \tag{6.2}$$

with $B_0 = \zeta^2(3/2)/24\zeta(3)$. The formula (6.2) gives the complete analogue of (5.6). Hence the following analogue of Conjecture 1 can be formulated.

Conjecture 2 The error term $O(X)$ in (6.2) could be replaced by

$$B_2(\sigma)X + O(X^{2-2\sigma}(\log X)^C) \tag{6.3}$$

for $1/2 < \sigma \le 3/4$, and by

$$B_2(\sigma)X + B_3(\sigma)X^{\frac{5}{2}-2\sigma} + O(X^{2-2\sigma}(\log X)^C) \tag{6.4}$$

for $3/4 < \sigma < 1$, with certain $B_2(\sigma)$, $B_3(\sigma)$ and $C \ge 0$.

Meurman first proposed (6.3) in correspondence, while (6.4) appeared in Kiuchi-Matsumoto [115], though the term $B_3(\sigma)X^{5/2-2\sigma}$ is missing there.

In Section 4, we discussed the analogy between $\Delta(x)$ and $R_2(1/2+it; t/2\pi)$. We can find that there also exists an analogy between $\Delta_{1-2\sigma}(x)$ and $R_2(\sigma+it; t/2\pi)$ for $1/2 < \sigma < 1$. Kiuchi [113I] proved that

$$\int_2^T |R_2(\sigma + it; t/2\pi)|^2 dt = \begin{cases} C_1(\sigma)T^{\frac{3}{2}-2\sigma} + O(1) & (\frac{1}{2} < \sigma < \frac{3}{4}) \\ \pi C_0 \log T + O(1) & (\sigma = \frac{3}{4}) \\ O(1) & (\frac{3}{4} < \sigma \le 1), \end{cases} \tag{6.5}$$

where C_0 is defined by (4.5) and $C_1(\sigma) = (2\pi)^{2\sigma-\frac{1}{2}}C_0/(3-4\sigma)$. This precisely corresponds to (5.6) and (6.2).

A remarkable fact is that we can go further in this case. Now it is known that the terms $O(1)$ in (6.5) can be replaced by

$$\begin{cases} C_2(\sigma) + O(T^{1-2\sigma}(\log T)^4) & (\frac{1}{2} < \sigma \le \frac{3}{4}) \\ C_2(\sigma) + C_1(\sigma)T^{\frac{3}{2}-2\sigma} + O(T^{1-2\sigma}(\log T)^4) & (\frac{3}{4} < \sigma \le 1) \end{cases} \tag{6.6}$$

with a certain constant $C_2(\sigma)$. The author [137] showed (6.6) in the case of $1/2 < \sigma \le 3/4$, and in the same paper the weaker result with the error estimate $O(T^{1/4-\sigma})$ was given for $3/4 < \sigma \le 1$. The result of the form (6.6) for $3/4 < \sigma \le 1$ is due to Kiuchi [113II]. The above (6.6) implies that the facts corresponding to Conjectures 1 and 2 are indeed true for $R_2(\sigma + it; t/2\pi)$.

Higher moments of $R_2(\sigma+it; t/2\pi)$ have also been discussed. The results analogous to (3.15)–(3.18) for $R_2(1/2+it; t/2\pi)$ were obtained by Kiuchi [110] and Ivić [69]. The k-th power moment of $R_2(\sigma+it;t/2\pi)$, where k is a positive even integer and $0\le\sigma\le 1$, was studied by Kiuchi-Matsumoto [115]. Their results especially imply that the transposing line for the k-th power moment is $\sigma=1/4+1/k$, unconditionally for k =2,4,6 and 8, and under a certain plausible assumption for any even k.

In the case $3/4<\sigma<1$, the bound $O(X)$ in (6.2) is not the best known result. Already in 1932, Chowla [18] proved the asymptotic formula

$$\int_2^X \Delta_{1-2\sigma}(x)^2dx=\frac{1}{2\pi^2}\left\{\sum_{n=1}^{\infty}\left(\frac{\sigma_{2-2\sigma}(n)}{n}\right)^2\right\}X+O(X^{\frac{5}{2}-2\sigma}\log X)$$
$$\left(\frac{3}{4}<\sigma<1\right), \tag{6.7}$$

which gives the partial solution of the case $3/4<\sigma<1$ of Conjecture 2. Recently, Yanagisawa [198] rediscovered (6.7) and also obtained more general results. The basic tool of both Chowla and Yanagisawa is a generalization of (2.14), that is

$$\Delta_{1-2\sigma}(x)=-G_{1-2\sigma}(x)-x^{1-2\sigma}G_{2\sigma-1}(x)+O(x^{\frac{1}{2}-\sigma}), \tag{6.8}$$

where

$$G_a(x)=\sum_{n\le\sqrt{x}}n^a\psi\left(\frac{x}{n}\right).$$

(As for (6.8), see Kanemitsu [98].) An asymptotic formula for the mean square of $\Delta_{-1}(x)$ was given by Walfisz [197].

On the other hand, as an analogue of (2.7), Lam-Tsang [126] proved

$$\int_2^X \delta_\sigma(x)dx=C(\sigma)X^2+O(X^{2+\frac{(1-2\sigma)(3-4\sigma)}{2(3-2\sigma)}}\log X)\qquad\left(\frac{1}{2}<\sigma<\frac{3}{4}\right), \tag{6.9}$$

where

$$\delta_\sigma(X)=\int_2^X\Delta_{1-2\sigma}(x)^2dx-B_1(\sigma)X^{\frac{5}{2}-2\sigma}$$

and

$$C(\sigma)=-\frac{\zeta(2\sigma)^2\zeta(3-4\sigma)}{12(2\pi)^{3-4\sigma}\zeta(4\sigma)}\Gamma(3-4\sigma)\sin(2\pi\sigma).$$

This result clearly implies the fact

$$\delta_\sigma(X)=\Omega_-(X)\qquad\left(\frac{1}{2}<\sigma<\frac{3}{4}\right), \tag{6.10}$$

which may be regarded as a support for the case $1/2<\sigma<3/4$ of Conjecture 2. The order of $\delta_\sigma(X)$ for $1/2<\sigma<3/4$ is completely determined by (6.2) and (6.10).

It is an interesting problem to prove the analogue of Lau-Tsang's (2.7) or Lam-Tsang's (6.9) for $F(T)$ or $F_\sigma(T)$. Another attractive problem is to search the analogue of the method of Chowla and Yanagisawa for the function $E_\sigma(T)$ in the case $3/4<\sigma<1$; or at least, to find the analogue of (6.8) for $E_\sigma(T)$. The last type of problem was sometimes mentioned by S. Kanemitsu in correspondence and oral communication.

7 Some Mean Value Results in Short Intervals

We mentioned in Section 1 that Good [35] showed $E(T) = \Omega(T^{1/4})$. He actually proved an asymptotic formula for the integral

$$\int_0^T (E(t+U) - E(t))^2 dt \qquad (1 \le U \ll T^{\frac{1}{2}}),$$

and the Ω-result is its corollary. The samdoermula is also used in the proof of Heath-Brown and Tsang [58] mentioned in Section 3.

Next, Jutila [84] studied a similar problem, but for short intervals, by using Atkinson's formula. His result is

$$\begin{aligned}\int_l^{T+H} &(E(t+U) - E(t))^2 dt \\ &= \frac{l}{\sqrt{2\pi}} \sum_{n \le \frac{1}{U}} d(n)^2 n^{-\frac{3}{2}} \int_T^{T+H} t^{\frac{1}{2}} \left| \exp\left(i \left(\frac{2\pi n}{t} \right)^{\frac{1}{2}} U \right) - 1 \right|^2 dt \\ &\quad + O(T^{1+\varepsilon}) + O(HU^{\frac{1}{2}} T^{\varepsilon})\end{aligned} \tag{7.1}$$

for $T \ge 2$, $1 \le U \ll T^{1/2} \ll H \le T$. Note that the right-hand side can be estimated as

$$\ll (HU + T)T^{\varepsilon}. \tag{7.2}$$

Jutila did not give the details of the proof; instead, he described the proof (based on the truncated Voronoï formula (2.4)) of the corresponding formula for $\Delta(x)$, that is

$$\begin{aligned}\int_X^{X+H} &(\Delta(x+U) - \Delta(x))^2 dx \\ &= \frac{1}{4\pi^2} \sum_{n \le \frac{X}{2U}} d(n)^2 n^{-\frac{3}{2}} \int_X^{X+H} x^{\frac{1}{2}} \left| \exp\left(2\pi i \left(\frac{n}{x} \right)^{\frac{1}{2}} U \right) - 1 \right|^2 dx \\ &\quad + O(X^{1+\varepsilon}) + O(HU^{\frac{1}{2}} X^{\varepsilon})\end{aligned} \tag{7.3}$$

for $X \ge 2$, $1 \le U \ll X^{1/2} \ll H \le X$. Moreover, Jutila raised the problem of extending (7.1) and (7.3) to higher power moments. In particular, he pointed out that if one could prove

$$\int_2^T (E(t+U) - E(t))^4 dt = O(T^{1+\varepsilon} U^2),$$

then the very important conjectural bound

$$\int_0^T \left| \zeta\left(\frac{1}{2} + it \right) \right|^6 dt = O(T^{1+\varepsilon})$$

would follow.

A formula for $\Delta_{1-2\sigma}(x)$ $(1/2 < \sigma < 1)$, analogous to (7.3), was recently obtained by Kiuchi-Tanigawa [116]; they actually treated a more general quantity which involves exponential factors. Yanagisawa [199] studied the same problem by the method similar to his another work [198] mentioned in the preceding section. The analogy of (7.1) for $E_\sigma(T)$ $(1/2 < \sigma < 1)$ was given by Kiuchi-Tanigawa [117]. In [118], they studied the same type of short interval mean square of $R_2(\sigma + it; \alpha t/2\pi, t/2\pi\alpha)$ for rational α.

In [88I], Jutila proved the estimate

$$\int_T^{T+H} (E(t+U) - E(t))^2 dt = O((HU + T^{\frac{2}{3}} U^{\frac{4}{3}}) T^{\varepsilon}) \tag{7.4}$$

for $1 \leq H, U \leq T$, which improves (7.2) when $U \ll T^{1/4}$. Jutila noted that (7.4) implies the estimate

$$\int_T^{T+T^{\frac{2}{3}}} \left|\zeta\left(\frac{1}{2} + it\right)\right|^4 dt = O(T^{\frac{2}{3}+\varepsilon}), \tag{7.5}$$

due originally to Iwaniec [79]. In fact, since

$$\int_t^{t+U} \left|\zeta\left(\frac{1}{2} + iu\right)\right|^2 du$$

can be approximated by $E(t+U) - E(t)$, (7.5) easily follows from (7.4) with $H = T^{2/3}$, $U = T^{\varepsilon}$, by applying Lemma 7.1 of Ivić [66].

Iwaniec's proof [79] of (7.5) was really epoch-making, because it was the first successful application of Kuznetsov's trace formula [122] to the mean value theory of zeta-functions, and was followed by many important works of Zavorotnyi, Motohashi, Ivić, Jutila and others in the fourth power moment theory. The full account of this theory is out of the scope of this article, but here we should mention Jutila's alternative proof [89] of (7.5). The basic idea of Jutila is to transform a certain relevant exponential sum by using (3.9), hence it is under the same philosophy as [83] [87]. The remarkable feature of Jutila's proof is that it only uses classical means, without the fancy tools of spectral theory. In Juula's proof, a lemma due to Bombieri-Iwaniec [13] plays an important role. This lemma contains the arithmetic essence of (7.5), which is included in Kuznetsov's formula (or Kloosterman's sum) in Iwaniec's original proof.

Extending the above idea, Jutila [88I, II] studied the integral of the type

$$J = \sum_{r=1}^{R} \int_0^V \left| \sum_{M \leq m \leq M'} d(m) g(m, v, y_r) \exp(2\pi i f(m, v, y_r)) \right|^2 dv, \tag{7.6}$$

where M is a large positive number, $M < M' \leq 2M$, $V > 0$, the functions f and g satisfy certain regularity conditions, and y_r $(1 \leq r \leq R)$ runs over a well-spaced set of numbers lying in [0,1]. Jutila [88I] proved a certain upper-bound of J, and from which he deduced (7.4) as well as its analogue for $\Delta(x)$. A further development of this method, with applications to Dirichlet L-functions, can be found in Jutila [90].

Lastly in this section we mention briefly the theory of Titchmarsh series developed by Ramachandra and his colleagues. Here we do not give the definition of general Titchmarsh series. They are elements of a certain class of Dirichlet series, including $\zeta(s)^k$ (for any positive integer k) as an example. In [169I], Ramachandra raised a conjecture on the lower bound of the mean square of Titchmarsh series over short intervals. Ramachandra (partly with Balasubramanian) wrote many papers ([9], [168]–[170]) on this topic, and finally, Balasubramanian-Ramachandra [12] (and Ramachandra [171]) solved completely the conjecture in a more precise form. This solution especially implies

$$\frac{1}{H}\int_T^{T+H}\left|\zeta\left(\frac{1}{2}+it\right)\right|^{2k}dt \geq C_k(\log H)^{k^2}+O\left(\frac{\log\log T}{H}(\log H)^{k^2}\right) + O((\log H)^{k^2-1}) \tag{7.7}$$

for $\log\log T \ll H \leq T$, where

$$C_k = \frac{1}{\Gamma(k^2+1)}\prod_p\left\{(1-p^{-1})^{k^2}\sum_{m=0}^{\infty}\left(\frac{\Gamma(k+m)}{m!\Gamma(k)}\right)^2 p^{-m}\right\}.$$

Ramachandra [171] includes an interesting lower bound of the mean square of $\zeta(1+it)$ over short intervals. Upper and lower bounds of the mean value

$$\frac{1}{H}\int_T^{T+H}\left|\frac{d^l}{ds^l}\zeta(s)^{2k}\right|_{s=\frac{1}{2}+it} dt$$

were studied by Ramachandra [168].

In the present article we do not discuss the full details of the theory of Titchmarsh series. This theory includes the treatment of the mean value of $|\zeta(s)^{2k}|$ with non-integral complex values of k, various Ω-results, sign-change theorems on $\arg\zeta(s)$, and generalizations etc. The readers are referred to Ramachandra's lecrure note [172].

8 Several General Principles and the Mean Square of Dedekind Zeta-functions

In the previous sections we discussed mainly the mean square of the Riemann zeta-function and related problems. A very important direction of research is to generalize the obtained results to various other zeta-functions. From this viewpoint, it is useful to find some general principles to obtain mean value results. One of them is the following classical theorem of Carlson [15], which is a generalization of (1.1) due to Landau and Schnee. Let

$$f(s) = \sum_{n=1}^{\infty} a_n n^{-s}$$

be a Dirichlet series, convergent in a certain half-plane. Assume that $f(s)$ can be continued to a holomorphic (or with possible poles included in a fixed compact set) function of finite order in the region $\sigma \geq \alpha + \varepsilon > \alpha$. Moreover suppose that

$$\int_{-T}^{T} |f(\sigma + it)|^2 dt = O(T) \tag{8.1}$$

for $\sigma \geq \sigma_0 > \alpha$. Then Carlson's theorem asserts that

$$\int_{-T}^{T} |f(\sigma + it)|^2 dt \sim 2\left(\sum_{n=1}^{\infty} |a_n|^2 n^{-2\sigma}\right) T \tag{8.2}$$

for $\sigma > \sigma_0$. (The part of the range of integration near the poles is omitted.) Potter [163I, II] studied this matter further. Potter's results are especially useful when $f(s)$ can be continued to the whole plane and satisfies a certain functional equation.

Carlson's theorem can be applied to the mean square of the Dedekind zeta-function $\zeta_K(s)$ attached to an algebraic number field K, and the result is

$$\int_{1}^{T} |\zeta_K(\sigma + it)|^2 dt \sim \left(\sum_{n=1}^{\infty} a_K(n)^2 n^{-2\sigma}\right) T \qquad (\sigma > 1 - l^{-1}),$$

where $l = [K : \mathbf{Q}] \geq 2$ and $a_K(n)$ is the number of integral ideals in K with norm n. But this is not the best known result. Chandrasekharan-Narasimhan [16] developed a general theory of approximate functional equations, and the following is a consequence of their theory:

$$\int_{1}^{T} |\zeta_K(\sigma + it)|^2 dt = \left(\sum_{n=1}^{\infty} a_K(n)^2 n^{-2\sigma}\right) T + O(T^{\frac{l-l\sigma+1}{2}} (\log T)^{\frac{l}{2}}) \tag{8.3}$$

if $\sigma > 1 - l^{-1}$, and

$$\int_{1}^{T} |\zeta_K(\sigma + it)|^2 dt = O(T^{l(1-\sigma)} (\log T)^l) \tag{8.4}$$

if $1/2 \leq \sigma \leq 1 - l^{-1}$. When $l = 2$, that is the case that K is a quadratic field, (8.3) gives the asymptotic formula for $\sigma > 1/2$. In the case of $\sigma = 1/2$, (8.4) gives the upper-bound $O(T \log^2 T)$. At the end of their paper [16], Chandrasekharan-Narasimhan conjectured

$$\int_{1}^{T} \left|\zeta_K\left(\frac{1}{2} + it\right)\right|^2 dt \sim C_2 T \log^2 T \tag{8.5}$$

for any real quadratic field K, with a certain constant C_2.

Let D be the discriminant of a quadratic field K, χ_D the Dirichlet character defined as the Kronecker symbol $\chi_D(n) = (\frac{D}{n})$, and $L(s, \chi_D)$ the corresponding Dirichlet L-function. It is well-known that $\zeta_K(s) = \zeta(s) L(s, \chi_D)$, hence the mean square of $\zeta_K(1/2 + it)$ is a generalization of the fourth power moment of $\zeta(1/2 + it)$. As for the latter problem, Titchmarsh [187] proved

$$\int_{0}^{\infty} \left|\zeta\left(\frac{1}{2} + it\right)\right|^4 e^{-\delta t} dt \sim \frac{1}{2\pi^2 \delta} \log^4 \frac{1}{\delta} \tag{8.6}$$

as $\delta \to 0$. This immediately implies

$$\int_0^T \left|\zeta\left(\frac{1}{2}+it\right)\right|^4 dt \sim \frac{1}{2\pi^2} T \log^4 T, \tag{8.7}$$

because there is the general principle that if $f(t) \geq 0$ for all t and

$$\int_0^\infty f(t)e^{-\delta t} dt \sim \frac{1}{\delta} \log^m \frac{1}{\delta} \tag{8.8}$$

as $\delta \to 0$, then

$$\int_0^T f(t)dt \sim T \log^m T \tag{8.9}$$

(see Section 7.12 of Titchmarsh [190]). This principle is, in a sense, a kind of Tauberian theorem. Following the idea of Titchmarsh, Motohashi [147] proved that

$$\int_0^\infty \left|\zeta_K\left(\frac{1}{2}+it\right)\right|^2 e^{-\delta t} dt = C_2 \frac{1}{\delta} \log^2 \frac{1}{\delta} + O\left(\frac{1}{\delta} \log \frac{1}{\delta}\right), \tag{8.10}$$

thereby established the conjecture (8.5). He found that

$$C_2 = \frac{6}{\pi^2} L^2(1, \chi_D) \prod_{p|D} \left(1+\frac{1}{p}\right)^{-1}.$$

Another useful general principle is the mean value theorem for Dirichlet polynomials. For any complex numbers $a_1, \ldots, a_N$, we have

$$\int_0^T \left|\sum_{n\leq N} a_n n^{it}\right|^2 dt = T \sum_{n\leq N} |a_n|^2 + O\left(\sum_{n\leq N} n|a_n|^2\right). \tag{8.11}$$

(This remains valid for $N = \infty$, if the series on the right-hand side converge.) The formula (8.11) is due to Montgomery-Vaughan [146], and the key of their proof is the following generalization of Hilbert's inequality: Let $\lambda_1, \ldots, \lambda_R$ be distinct real numbers and $\delta_n = \min_{m\neq n} |\lambda_m - \lambda_n|$. Then, for any complex numbers $a_1, \ldots, a_R$, we have

$$\left|\sum\sum_{m\neq n} \frac{a_m \bar{a}_n}{\lambda_m - \lambda_n}\right| \leq \frac{3\pi}{2} \sum_n |a_n|^2 \delta_n^{-1}. \tag{8.12}$$

This inequality has a close connection with the theory of large sieve inequalities; see Montgomery [144] [145].

Using (8.11), Ramachandra [166] gave a simple proof of Ingham's (1.11). Applying the same idea to $\zeta_K(1/2+it)$, it is possible to prove

$$\int_0^T \left|\zeta_K\left(\frac{1}{2}+it\right)\right|^2 dt = C_2 T \log^2 T + O(T \log T)$$

for a quadratic field K.

A further refinement was done by Müller [156], who generalized Heath-Brown's proof of (1.20). His result is that

$$\int_0^T \left|\zeta\left(\frac{1}{2}+it\right)L\left(\frac{1}{2}+it,\chi\right)\right|^2 dt = \widetilde{C_2}T\log^2 T + C_1 T\log T + C_0 T$$
$$+ O(q^{\frac{35}{16}+\varepsilon}T^{\frac{7}{8}+\varepsilon}), \tag{8.13}$$

where χ is a primitive Dirichlet character (mod $q \geq 2$), and $\widetilde{C_2}$ is similar to C_2, just replacing $L^2(1,\chi_D)$ by $|L(1,\chi)|^2$.

Another possible way is to generalize the recent spectral-theoretic developments of the fourth power moment theory of $\zeta(s)$. See Motohashi [154], in which a certain explicit formula is given.

9 L-functions Attached to Cusp Forms

In the preceding section we discussed the analogy between $\zeta_K(s)$ for quadratic fields and $\zeta^2(s)$. Another class of Dirichlet series, which may be regarded as an analogue of $\zeta^2(s)$, is L-functions $\varphi(s,F)$ attached to holomorphic cusp forms, defined by (4.3). The function $\varphi(s,F)$ is convergent absolutely for $\sigma > (\kappa+1)/2$, and can be continued to an entire function. The critical strip is $(\kappa-1)/2 \leq \sigma \leq (\kappa+1)/2$. In this section we survey the results on the mean square of $\varphi(s,F)$. Some of the quoted papers actually study more general cases (e.g. congruence subgroups), but here we restrict ourselves to the case of the full modular group for simplicity. Also we assume that $F(z)$ is a normalized eigenform (i.e. a simultaneous eigenfunction of Hecke operators with $a(1)=1$).

The connection between $F(z)$ and $\varphi(s,F)$ was established by Hecke in 1936–37. Just a few years later, the mean square of $\varphi(s,F)$ was already studied by Potter [163I, II]. Let

$$I_\sigma(T,F) = \int_0^T |\varphi(\sigma+it,F)|^2 dt.$$

As a consequence of his general theorem, Potter [163I] proved

$$I_\sigma(T,F) \sim \left(\sum_{n=1}^{\infty} a^2(n)n^{-2\sigma}\right)T \quad \left(\sigma > \frac{\kappa}{2}\right), \tag{9.1}$$

and then he [163II] proved that $I_{\kappa/2}(T,F) = O(T\log T)$.

In the middle of 1970s, Good began deeper investigations of $I_\sigma(T,F)$. First, Good [32] applied Titchmarsh's idea of using the Tauberian principle (8.8)–(8.9) to the present case, and obtained the asymptotic formula

$$I_{\kappa/2}(T,F) \sim 2\kappa A_0 T\log T, \tag{9.2}$$

where

$$A_0 = \frac{12(4\pi)^{\kappa-1}}{\Gamma(\kappa+1)}\int\int_{\mathcal{F}} |F(x+iy)|^2 y^{\kappa-2} dx dy,$$

the integral being taken over a fundamental domain $\mathcal{F}$ of $SL(2, \mathbf{Z})$. The constant A_0 appears in Rankin's celebrated formula

$$\sum_{n\le x} a^2(n) = A_0 x^\kappa + O(x^{\kappa-\frac{2}{5}}),$$

which is essentially used in Good's proof of (9.2).

The next paper [33] of Good gives a certain approximate functional equation for $\varphi(s, F)$. Let $\omega : [0, \infty) \to \mathbf{R}$ be a (fixed) C^∞-function such that $\omega(\rho) = 1$ if $0 \le \rho \le 1/2$ and $\omega(\rho) = 0$ if $\rho \ge 2$. Define $\omega_0(\rho) = 1 - \omega(1/\rho)$. Then, a useful form of Good's formula can be stated as

$$\begin{aligned}\varphi(s, F) &= \sum_{j=0}^{l} \gamma_j(s, |t|^{-1}) \sum_{n=1}^{\infty} a(n) n^{-s} \omega^{(j)}\left(\frac{n}{y_1}\right)\left(-\frac{n}{y_1}\right)^j \\ &\quad + (-1)^{\frac{\kappa}{2}} (2\pi)^{2s-\kappa} \frac{\Gamma(\kappa - s)}{\Gamma(s)} \sum_{j=0}^{l} \gamma_j(\kappa - s, |t|^{-1}) \\ &\quad \times \sum_{n=1}^{\infty} a(n) n^{s-\kappa} \omega_0^{(j)}\left(\frac{n}{y_2}\right)\left(-\frac{n}{y_2}\right)^j \\ &\quad + O(\|\omega^{(l+1)}\|_1 y_1^{\frac{\kappa+1}{2}-\sigma} |t|^{-\frac{l}{2}}) + O(\|\omega_0^{(l+1)}\|_1 y_1^{\frac{\kappa}{2}-\sigma} y_2^{\frac{1}{2}} |t|^{-\frac{l}{2}})\end{aligned} \tag{9.3}$$

for $l \ge (\kappa + 1)/2$ and $4\pi^2 y_1 y_2 = t^2$, where $\|\cdot\|_1$ means the L^1-norm and $\gamma_j(s, |t|^{-1})$ is a quantity defined by a certain integral. Note that $\gamma_0(s, |t|^{-1}) \equiv 1$. It is possible to deduce the approximate functional equation of classical type (like (1.4)) from (9.3), but the above form is more effective in applications. Using (9.3), Good [33] proved

$$I_\sigma(T, F) = \begin{cases} 2\kappa A_0 T \log T + O(T) & (\sigma = \frac{\kappa}{2}) \\ (\sum_{n=1}^{\infty} a^2(n) n^{-2\sigma}) T + O(T^{\kappa+1-2\sigma}) & (\frac{\kappa}{2} < \sigma < \frac{\kappa+1}{2}) \\ \left(\sum_{n=1}^{\infty} a^2(n) n^{-2\sigma}\right) T + O(\log^2 T) & (\sigma = \frac{\kappa+1}{2}). \end{cases} \tag{9.4}$$

The formula (9.3) and its relatives are fundamental in Good's theory. In [34], Good proved a formula of the same type for $\zeta(s)$, and from which he [34] [35] deduced several new facts on $E(T)$ mentioned in Sections 1 and 7. A discrete mean square of $\varphi(s, F)$ was studied by Good [36], as an application of (9.3). As for $I_\sigma(T, F)$, the next step of Good's research is [37], in which he proved

$$I_{\kappa/2}(T, F) = 2\kappa A_0 T \log T + A_1 T + E(T, F) \tag{9.5}$$

where A_1 is a constant, with

$$E(T, F) = O(T^{\frac{5}{6}} (\log T)^{\frac{13}{6}}). \tag{9.6}$$

Moreover he showed that a certain non-vanishing assumption would lead to

$$E(T, F) = \Omega(T^{\frac{1}{2}}). \tag{9.7}$$

To prove these results, Good applied (9.3) to the integral

$$J_F(T,U) = \int_{-\infty}^{\infty} \left|\varphi\left(\frac{\kappa}{2}+it, F\right)\right|^2 \Psi_U\left(\frac{t}{T}\right) dt$$

with a certain weight function Ψ_U, and obtained an explicit formula for $J_F(T,U)$, in which a sum involving the factor $a(l)a(l+n)(l+n/2)^{-\kappa}$ appears. To analyze this sum, Good studied the behaviour of the Dirichlet series

$$\sum_{l=1}^{\infty} a(l)a(l+n)\left(l+\frac{n}{2}\right)^{-s} \tag{9.8}$$

by using the spectral theory. He found an explicit expression of $J_F(T,U)$ written in terms of non-analytic Poincaré series and Fourier coefficients of Maass wave forms, from which he derived the above results (9.5)–(9.7).

In [38], Good developed his theory further, and improved (9.6) to

$$E(T,F) = O(T^{\frac{2}{3}}(\log T)^C) \tag{9.9}$$

with $C = 2/3$. The bound

$$\varphi\left(\frac{\kappa}{2}+it, F\right) = O(|t|^{\frac{1}{3}}(\log|t|)^{\frac{5}{6}}) \qquad (|t| \geq 2) \tag{9.10}$$

is an immediate consequence of (9.9).

Recently, Kamiya [96] generalized (9.3) to the case of

$$\varphi(s,F,\chi) = \sum_{n=1}^{\infty} \frac{\chi(n)a(n)}{n^s} \tag{9.11}$$

with a Dirichlet character χ mod q, and used it to obtain

$$\sum_{\chi \bmod q}{}^{*} \int_{-T}^{T} |\varphi(\sigma+it,F,\chi)|^2 dt \ll \phi(q)T\log(qT) \tag{9.12}$$

if $q \ll T$, uniformly for $\kappa/2 - 1/\log(qT) \leq \sigma \leq \kappa/2 + 1/\log(qT)$ and q. Here, $\sum^*$ denotes the summation running over primitive characters. This is the result corresponding to Montgomery's estimate [144] for the fourth power moment of $L(s,\chi)$.

Another type of mean square of $\varphi(s,F,\chi)$ (including the non-holomorphic case) was discussed in Stefanicki [185] by a different method. See also Kamiya [97].

Jutila's consistent principle is to develop the theory which may treat the both cases $\zeta^2(s)$ and $\varphi(s,F)$ simultaneously. In Chapter 4 of his lecture note [87], Jutila gave a proof of $\varphi(\kappa/2+it,F) = O(|t|^{1/3+\varepsilon})$, slightly weaker than Good's (9.10), in such a unified way. The estimate

$$\int_0^T \left|\varphi\left(\frac{\kappa}{2}+it, F\right)\right|^6 dt = O(T^{2+\varepsilon}), \tag{9.13}$$

an analogue of (3.5), was also proved in the same chapter. Jutila proved those results by means of his transformation method, hence the arguments are of elementary nature. At the end of [87] Jutila proposed the problem of showing (7.5) and the corresponding estimate

$$\int_T^{T+T^{\frac{2}{3}}} \left|\varphi\left(\frac{\kappa}{2}+it, F\right)\right|^2 dt = O(T^{\frac{2}{3}+\varepsilon}), \tag{9.14}$$

an obvious corollary of (9.9), in a unified way. This problem was solved by Jutila himself in [88I], mentioned in Section 7. Hence [88I] includes a proof of (9.14) by classical means. An alternative proof of (9.13) is given in [88II].

We sometimes mentioned the recent spectral-theoretic approach to the fourth power moment of $\zeta(s)$. In view of the analogy between $\zeta^2(s)$ and $\varphi(s, F)$, one may expect, suggested by (9.9), that (1.20) could be improved to

$$E_2(T) = O(T^{\frac{2}{3}+\varepsilon}). \tag{9.15}$$

This was first achieved by Zavorotnyi [201] by using Kuznetsov's convolution formula [124]. Ivić-Motohashi [77] gave an alternative proof, with replacing T^ε by a log-power. In the latter proof, Motohashi's explicit formula [151] for the weighted fourth power mean of $|\zeta(1/2+it)|$ is essentially used. It is worth while noting that the basic idea of [151] is an extension of Atkinson's dissection argument to the fourth power situation.

We already discussed the close connection between $E(T)$ and the Dirichlet divisor problem. Similarly, $E_2(T)$ is closely related to the additive divisor problem, as was first noticed by Atkinson [2]. The additive divisor problem is the problem of evaluating the sum $\sum_{n\leq x} d(n)d(n+r)$, and has a long and rich history. The associated zeta-function is

$$\sum_{n=1}^{\infty} d(n)d(n+r)n^{-s}, \tag{9.16}$$

whose explicit spectral-theoretic expression was obtained by A.I. Vinogradov-Takhtadzhyan [194]. One may notice the similarity between (9.8) and (9.16), both of which were handled by spectral theory. Inspired by those works of Good and Vinogradov-Takhtadzhyan, and also inspired by the classical works of Titchmarsh [187] and Atkinson [2] on the Laplace transform of $|\zeta(1/2+it)|^4$, Jutila [91] developed a new unified approach to $E_2(T)$ and $E(T, F)$. He obtained spectral-theoretic explicit formulas for both $E_2(T)$ and $E(T, F)$, which, in the case of $E_2(T)$, has the same flavour as Motohashi's explicit formula [151]. As a consequence, Jutila reproved (9.9) (with the factor T^ε) and (9.15).

Another approach was given by Motohashi [152]. He sketched the way how to modify the argument in [151] to obtain an explicit formula for the weighted mean square of $\varphi(s, F)$, and to deduce from which the estimate (9.9) as well as the mean square estimate

$$\int_0^T E^2(t, F)dt = O(T^2(\log T)^C). \tag{9.17}$$

The latter is the analogue of

$$\int_0^T E_2^2(t)dt = O(T^2(\log T)^C) \tag{9.18}$$

due to Ivić-Motohashi [76]. Jutila [93] pursued his approach via Laplace transforms further, and proved (9.17) and (9.18) in a unified way.

An important advantage of Jutila's method is that it may also treat the non-holomorphic case. The results corresponding to (9.9) and (9.17) (with replacing $(\log T)^C$ by T^ε) for L-functions attached to Maass wave forms are proved in [91] [93]. The former is an improvement of Kuznetsov's result [123], which gives the exponent $6/7+\varepsilon$ in the error term. See also Müller [157] for another approach to the non-holomorphic case.

10 Dirichlet L-functions

Now we return to the $GL(1)$-situation, and in this section we discuss various mean square formulas for Dirichlet L-functions. Let χ be a Dirichlet character mod q, and $L(s,\chi)$ the corresponding Dirichlet L-function. A natural extension of $I_\sigma(T)$ is the mean value

$$I_\sigma(T,q)=\frac{1}{\phi(q)}\sum_{\chi \bmod q}\int_0^T |L(\sigma+it,\chi)|^2 dt.$$

Serious research on this quantity was started by Ramachandra and his school in 1970s. It was mentioned by Ramachandra [167] that he had obtained the asymptotic formula

$$I_{\frac{1}{2}}(T,q)=\frac{\phi(q)}{q}T\log(qT)+O(T(\log(qT))^\varepsilon).$$

It is easy to see that

$$L(s,\chi)=q^{-s}\sum_{a=1}^{q}\chi(a)\zeta\left(s,\frac{a}{q}\right), \tag{10.1}$$

where $\zeta(s,\alpha)$ is the Hurwitz zeta-function defined by the analytic continuation of the Dirichlet series $\sum_{n=0}^\infty (n+\alpha)^{-s}$. Hence we can reduce the problem of evaluating $I_\sigma(T,q)$ to the study of the mean square of $\zeta(s,\alpha)$. The approximate functional equation of $\zeta(s,\alpha)$, corresponding to (1.4), can be stated as

$$\begin{aligned}\zeta(s,\alpha) &= \sum_{0\le n\le\xi}(n+\alpha)^{-s}+\left(\frac{2\pi}{t}\right)^{\sigma-\frac{1}{2}+it} e^{i(\frac{\pi}{4}+t)}\sum_{1\le n\le\eta}\frac{e^{-2\pi i n\alpha}}{n^{1-s}} \\ &\quad + O(\xi^{-\sigma}\log t), \end{aligned}\tag{10.2}$$

valid for $1\le\xi\le\eta$, $2\pi\xi\eta=t$, and $0<\sigma<1$. Rane [173] proved a more precise approximate formula of the Riemann-Siegel type, and used it to prove

$$\int_0^T\left|\zeta\left(\frac{1}{2}+it,\alpha\right)\right|^2 dt = T\log T+C(\alpha)T-\frac{1}{\alpha}+O(\alpha^{-\frac{1}{2}}T^{\frac{1}{2}}\log T), \tag{10.3}$$

where $C(\alpha)$ is a constant depending on α. From (10.1) and (10.3) Rane [173] proved

$$I_{\frac{1}{2}}(T,q)=\frac{\phi(q)}{q}T\left(\log\frac{qT}{2\pi}+2\gamma-1+\sum_{p|q}\frac{\log p}{p-1}\right)+E(T,q) \tag{10.4}$$

with

$$E(T,q) \ll \frac{\phi(q)}{q}(T^{\frac{1}{2}}\log T + \log q). \tag{10.5}$$

Balasubramanian-Ramachandra [8] gave a simpler proof of (10.4) and (10.5). A key lemma in their argument is a short interval mean square estimate of $\zeta(1/2+it,\alpha)$, which is proved by the idea of Ramachandra [166].

The next step was due to Narlikar [160], who improved (10.5) to

$$E(T,q) \ll \frac{\phi(q)}{q}T^{5/12}(\log T)^2.$$

This is based on her refinement [159] of (10.3). Unfortunately Narlikar's argument includes an error which leads to the existence of extra terms of the order $T^{1/2}$ in her statements, which are to be deleted. Zhan [202] mentioned that it is possible to correct this error and justify Narlikar's argument. Zhan [202] himself adopted a different way; he proved the approximate formula of the type of Heath-Brown [53] for $\zeta(1/2+it,\alpha)$, and using it he gave further improvements on the results mentioned above. In particular he obtained $E(T,q) = O(T^{\alpha+\varepsilon})$ with a certain $\alpha < 1/3$. Other variants of approximate functional equation for $\zeta(s,\alpha)$ were recently given by Rane [177] [178]. See also Balasubramanian-Ramachandra [10].

Meurman [141] proved a generalization of Atkinson's formula to $E(T,q)$, and from which he deduced

$$E(T,q) \ll \begin{cases} \phi(q)^{-1}((qT)^{\frac{1}{3}+\varepsilon} + q^{1+\varepsilon}) & (q \ll T) \\ \phi(q)^{-1}((qT)^{\frac{1}{2}+\varepsilon} + qT^{-1}) & (q \gg T). \end{cases} \tag{10.6}$$

Laurinčikas [130] proved an analogue of Meurman's formula near the critical line, while the analogue for fixed σ, $1/2 < \sigma < 1$, was obtained by Nakaya [158]. Meurman's paper [141] includes the short interval estimate

$$\sum_{\chi \bmod q} \int_T^{T+H} \left|L\left(\frac{1}{2}+it,\chi\right)\right|^2 dt \ll (qH + (qT)^{\frac{1}{3}})(q(T+H))^{\varepsilon}, \tag{10.7}$$

for $H \gg 1$, as a corollary. An alternative proof of (10.7) was obtained by Balasubramanian-Ramachandra [11], in which further improvements by using the theory of exponent pairs were also discussed. It is to be noted that their argument includes, as a special case, a simple proof of the estimate

$$\int_T^{T+T^{1/3}} \left|\zeta\left(\frac{1}{2}+it\right)\right|^2 dt \ll T^{1/3+\varepsilon}.$$

The study of the mean square of individual L-functions

$$\int_0^T |L(\sigma+it,\chi)|^2 dt$$

is more difficult. See the recent article [108] of Katsurada and the author, in which the history of this problem is sketched.

Another version of mean square of L-functions is

$$U(s,q)=\frac{1}{\phi(q)}\sum_{\chi \bmod q}|L(s,\chi)|^2.$$

In the case of $s=1/2+it$, $t\geq 2$, Gallagher [29] proved

$$U\left(\frac{1}{2}+it,q\right)\ll\frac{1}{\phi(q)}(q+t)\log(qt). \tag{10.8}$$

Improved upper-bounds were obtained by Meurman [141] and Rane [176]. Balasubramanian [6] gave an asymptotic formula with the main term $q^{-1}\phi(q)\log(qt)$, which was further refined by W. Zhang [205] [214] and Yu [200]. In [214], it is shown that

$$\begin{aligned}U\left(\frac{1}{2}+it,q\right) &= \frac{\phi(q)}{q}\left\{\log\left(\frac{qt}{2\pi}\right)+2\gamma+\sum_{p|q}\frac{\log p}{p-1}\right\}\\ &\quad+O\left(\frac{1}{\phi(q)}qt^{-1}+\frac{1}{\phi(q)}(qt)^{\frac{1}{2}}\exp\left(\frac{\log(qt)}{\log\log(qt)}\right)\right). \end{aligned}\tag{10.9}$$

In most of the above works, the problem is reduced by (10.1) to the mean square of Hurwitz zeta-functions. And for the latter problem, for example in [214], it is shown that

$$\sum_{a=1}^{q}\left|\zeta\left(\frac{1}{2}+it,\frac{a}{q}\right)\right|^2=q\left\{\log\left(\frac{qt}{2\pi}\right)+2\gamma\right\}+O(qt^{-1}+(qt)^{\frac{1}{2}}\log t) \tag{10.10}$$

(a special case of Lemma 7 of [214]). W. Zhang's papers [206] [215] are devoted to the study of $\sum_{\chi \bmod q}{}^{*}|L(1/2+it,\chi)|^2$, while in [204] he obtained an asymptotic formula for

$$\sum_{q\leq Q}\frac{q}{\phi(q)}\sum_{\chi \bmod q}\left|L\left(\frac{1}{2}+it,\chi\right)\right|^2.$$

On the other hand, Motohashi [149I] applied Atkinson's dissection argument (see Section 3) to the study of $U(s,q)$, and obtained an asymptotic formula for $U(1/2+it,p)$ for any prime p and fixed t. Motohashi's idea was further developed by Katsurada-Matsumoto [103], who proved the following formula:

$$\begin{aligned}U\left(\frac{1}{2}+it,q\right) &= \frac{\phi(q)}{q}\left\{\log\frac{q}{2\pi}+2\gamma+\sum_{p|q}\frac{\log p}{p-1}+\mathrm{Re}\frac{\Gamma'}{\Gamma}\left(\frac{1}{2}+it\right)\right\}\\ &\quad+\frac{2}{q}\sum_{k|q}\mu\left(\frac{q}{k}\right)T\left(\frac{1}{2}+it;k\right), \end{aligned}\tag{10.11}$$

where $\mu(\cdot)$ denotes the Möbius function and $T(1/2+it;\, k)$ satisfies the asymptotic formula

$$\begin{aligned} T\left(\frac{1}{2}+it; k\right) &= \operatorname{Re}\left\{\sum_{n=0}^{N-1}\binom{-\frac{1}{2}+it}{n}k^{\frac{1}{2}+it-n}\zeta\left(\frac{1}{2}+it-n\right)\zeta\left(\frac{1}{2}-it+n\right)\right\} \\ &\quad + O(k^{1-N}t^{2N}) \end{aligned} \tag{10.12}$$

for any positive integer N, where the O-constant depends only on N. The quantity $T(1/2+it;\, 1)$ can be written down in a closed form, while (10.12) gives the asymptotic expansion of $T(1/2+it;\, k)$ with respect to k if $k > 1$. Hence, if $q = p^m$ is a prime power, then from (10.11) we can deduce the asymptotic expansion of $U(1/2+it,\, p^m)$ with respect to p. The special case $t = 0$ of (10.11) was first obtained by Heath-Brown [54] by a different method, but the coefficients of the expansion are not explicitly given there. J. Zhang-Xing [203] gave an alternative proof of (10.11) by using Hurwitz zeta-functions. Another different proof of (10.11) was recently obtained by Katsurada [101]. Rane [175] also gave a similar expansion, but the coefficients are not explicit.

The mean square of $(d/ds)L(s,\chi)$ was considered by W. Zhang [211] [212] and Chen [17]. Katsurada [99III] generalized the method of [103] [99II] to study the case of $(d^k/ds^k)L(s,\chi)$ for any positive integer k.

In [103], the asymptotic expansion formula was proved not only for $\sigma = 1/2$, but for any σ satisfying $0 < \sigma < N+1$, as was pointed out in Katsurada-Matsumoto [106]. The region was further extended by Katsurada [99II] [101]. The formula (10.11) can be derived from this general formula as the limit case $\sigma \to 1/2$. Another important limit case is $\sigma = 1$, and in this way we can deduce a precise formula for

$$V(q) = \sum_{\substack{\chi \bmod q \\ \chi \neq \chi_0}} |L(1,\chi)|^2,$$

where χ_0 is the principal character mod q. Evaluation of $V(q)$ is a classical problem, and in the case that $q = p$ is a prime, the formula

$$V(p) = \zeta(2)p + O(\log^2 p)$$

goes back to Paley [162] and Selberg [181]. This was refined by Slavutskiĭ [183] [184] and then W. Zhang [209] [210]; the result given in [210] is that

$$V(p) = \zeta(2)p - \log^2 p + C + O\left(\frac{1}{\log p}\right).$$

Following the above mentioned method, Katsurada-Matsumoto [106] proved the asymptotic expansion

$$\begin{aligned} V(p) &= \zeta(2)p - \log^2 p + (\gamma^2 - 2\gamma_1 - 3\zeta(2)) - (\gamma^2 - 2\gamma_1 - 2\zeta(2))\frac{1}{p} \\ &\quad + 2\left(1-\frac{1}{p}\right)\left\{\sum_{n=1}^{N-1}(-1)^n\zeta(1-n)\zeta(1+n)p^{-n} + O(p^{-N})\right\} \end{aligned} \tag{10.13}$$

for any positive integer N, where γ_1 is defined by the following Laurent expansion of $\zeta(s)$ at $s=1$:

$$\zeta(s)=\frac{1}{s-1}+\sum_{j=0}^{\infty}\gamma_j(s-1)^j,\qquad \gamma_0=\gamma. \tag{10.14}$$

The formula (10.13) gives the satisfactory answer to the problem of evaluating $V(p)$. Moreover, in [106], a generalization of (10.13) to any composite q is proved. Some explicit expressions of $U(m,q)$, where $m(\neq 1)$ is an integer, are also obtained in [106].

11 Hurwitz Zeta and Other Related Zeta-functions

We already mentioned some mean value results on $\zeta(s,\alpha)$ in the preceding section (see (10.3) and (10.10)). In this final section we mainly discuss the approach by Atkinson's method, due to the recent papers of Katsurada and the author.

First, it is easily seen that a simple modification of the method developed in Katsurada-Matsumoto [103] can be applied to the discrete mean square $\sum_{a=1}^{q}|\zeta(s,a/q)|^2$. The result is an asymptotic expansion formula similar to (10.11) and (10.12), proved in Katsurada-Matsumoto [104]. This should be compared with (10.10).

A more interesting problem is to evaluate the integral

$$H(s)=\int_0^1|\zeta_1(s,\alpha)|^2d\alpha,$$

where $\zeta_1(s,\alpha)=\zeta(s,\alpha)-\alpha^{-s}$. In the case of $s=1/2+it$, $t\geq 2$, this problem was first considered by Koksma-Lekkerkerker [119], who showed $H(1/2+it)=O(\log t)$. This result was used in Gallagher's proof of (10.8). Balasubramanian [5] proved the asymptotic formula

$$H\left(\frac{1}{2}+it\right)=\log t+O(\log\log t),$$

and further refinements were done by Rane [174], Sitaramachandrarao (unpublished), and W. Zhang [207] [213], by using the approximate functional equation (10.2). Zhang [213] arrived at the result

$$H\left(\frac{1}{2}+it\right)=\log\left(\frac{t}{2\pi}\right)+\gamma+O(t^{-\frac{7}{36}}(\log t)^{\frac{25}{18}}),$$

and conjectured that the error estimate could be improved to $O(t^{-1/4})$. Ramachandra independently expressed the same opinion. This conjecture was solved by Andersson [1] and Zhang himself [217], independently of each other, in the following unexpected form:

$$H\left(\frac{1}{2}+it\right)=\log\left(\frac{t}{2\pi}\right)+\gamma-2\mathrm{Re}\frac{\zeta(\frac{1}{2}+it)}{\frac{1}{2}+it}+O(t^{-1}). \tag{11.1}$$

Shortly after their works, Katsurada-Matsumoto [105] [107I] obtained the following asymptotic expansion. For any integer $K \geq 0$, it holds that

$$\begin{aligned} H\left(\frac{1}{2}+it\right) &= \mathrm{Re}\frac{\Gamma'}{\Gamma}\left(\frac{1}{2}+it\right)+\gamma-\log 2\pi - 2\mathrm{Re}\frac{\zeta(\frac{1}{2}+it)-1}{\frac{1}{2}+it} \\ &\quad - 2\mathrm{Re}\sum_{k=1}^{K}\frac{(-1)^{k-1}(k-1)!}{(\frac{3}{2}-k+it)(\frac{5}{2}-k+it)\dots(\frac{1}{2}+it)}\sum_{l=1}^{\infty} l^{-k}(l+1)^{-\frac{3}{2}+k-it} \\ &\quad + O(t^{-K-1}). \end{aligned} \tag{11.2}$$

The starting point of the proof is Atkinson's dissection device (cf. Section 3). For $\mathrm{Re}\, u > 1$, $\mathrm{Re}\, v > 1$, we have

$$\zeta(u,\alpha)\zeta(v,\alpha) = \zeta(u+v,\alpha) + f(u,v;\alpha) + f(v,u;\alpha),$$

where

$$f(u,v;\alpha) = \sum_{m=0}^{\infty}(m+\alpha)^{-u}\sum_{n=1}^{\infty}(m+n+\alpha)^{-v}. \tag{11.3}$$

By the argument similar to [149I] [103], we can prove a contour-integral expression of $f(u,v;\alpha)$, which gives the analytic continuation. Analyzing this expression further, Katsurada-Matsumoto [107I] obtained the following formula, which is fundamental in their theory. Let N be a positive integer, $(s)_n = \Gamma(s+n)/\Gamma(s)$ the Pochhammer symbol, and define

$$S_N(u,v) = \sum_{n=0}^{N-1}\frac{(u)_n}{(1-v)_{n+1}}(\zeta(u+n)-1)$$

and

$$T_N(u,v) = \frac{(u)_N}{(1-v)_N}\sum_{l=1}^{\infty} l^{1-u-v}\int_l^{\infty}\beta^{u+v-2}(1+\beta)^{-u-N}d\beta.$$

Then it holds that

$$\begin{aligned} \int_0^1 \zeta_1(u,\alpha)\zeta_1(v,\alpha)d\alpha &= \frac{1}{u+v-1} \\ &\quad + \Gamma(u+v-1)\zeta(u+v-1)\left(\frac{\Gamma(1-v)}{\Gamma(u)}+\frac{\Gamma(1-u)}{\Gamma(v)}\right) \\ &\quad - S_N(u,v) - S_N(v,u) - T_N(u,v) - T_N(v,u) \end{aligned} \tag{11.4}$$

for $-N+1 < \mathrm{Re}\, u < N+1$, $-N+1 < \mathrm{Re}\, v < N+1$ and $(u,v) \notin E$, where E is the set of (u,v) at which some factor in (11.4) has a singularity. We can derive a formula for $(u,v) \in E$ as a limit case. For instance, (11.2) follows easily from the case $N = 1$ of

(11.4), by integrating $T_1(u,v)$ and $T_1(v,u)$ by parts K-times and taking $u \to 1/2+it$ and $v \to 1/2-it$. Explicit expressions of $H(1+it)$, and of $H(m)$ for any integer $m(\neq 1)$, can also be deduced from (11.4) (see [107I] and [107II], respectively). Letting $N \to \infty$, and then $u \to 1/2+it$ and $v \to 1/2-it$ in (11.4), we obtain

$$H\left(\frac{1}{2}+it\right) = \mathrm{Re}\frac{\Gamma'}{\Gamma}\left(\frac{1}{2}+it\right) + \gamma - \log 2\pi - 2\mathrm{Re}\sum_{n=0}^{\infty}\frac{\zeta(\frac{1}{2}+n+it)-1}{\frac{1}{2}+n+it}, \quad (11.5)$$

due originally to Andersson [1]. (The special case $t=0$ is included also in W. Zhang [217].)

Katsurada [100] presented an alternative proof of (11.4). His proof uses the Mellin-Barnes type of integrals and properties of hypergeometric functions, hence it is under the same principle as his [101]. Actually Katsurada's paper [100] treats a more general situation, that is the mean square of Lerch zeta-functions defined by the analytic continuation of $\sum_{n=0}^{\infty} e^{2\pi i\lambda n}(n+\alpha)^{-s}$, where $\alpha > 0$ and λ is real. He obtained various asymptotic expansions, which includes a refinement of W. Zhang's former result [216] [218].

Next we consider the derivative case

$$H_k(s) = \int_0^1 \left|\frac{d^k}{ds^k}\zeta_1(s,\alpha)\right|^2 d\alpha.$$

The case $k=1$ was studied by W. Zhang [208] and Guo [40] [41], and it is shown that

$$\begin{aligned} H_1\left(\frac{1}{2}+it\right) &= \frac{1}{3}\log^3\left(\frac{t}{2\pi}\right) + \gamma\log^2\left(\frac{t}{2\pi}\right) \\ &\quad + 2\gamma_1\log\left(\frac{t}{2\pi}\right) + 2\gamma_2 + O(t^{-1}\log^2 t) \end{aligned} \quad (11.6)$$

in [40] [41]. On the other hand, as was first noticed by Katsurada [99III], the method based on Atkinson's dissection device is suitable to study the mean square of higher derivatives. The idea is, roughly speaking, to differentiate (11.4) k-times with respect to both u and v and analyze the resulting expression carefully. The result is that

$$\begin{aligned} H_k\left(\frac{1}{2}+it\right) &= \frac{1}{2k+1}\log^{2k+1}\left(\frac{t}{2\pi}\right) + \sum_{j=0}^{2k}\frac{(2k)!}{(2k-j)!}\gamma_j\log^{2k-j}\left(\frac{t}{2\pi}\right) \\ &\quad - 2\mathrm{Re}\left(\frac{k!\zeta^{(k)}(\frac{1}{2}+it)}{(\frac{1}{2}+it)^{k+1}}\right) + O(t^{-2}(\log t)^{2k}) \end{aligned} \quad (11.7)$$

for any $k \geq 1$, where γ_j is defined by (10.14). The case $k=1$ gives a refinement of (11.6). The formula (11.7) was announced in [107II], and the detailed proof is described in [107III].

Finally we mention the author's work [139] (already announced in [138]) on the double zeta-function

$$\zeta_2(s;\alpha,w) = \sum_{m=0}^{\infty}\sum_{n=0}^{\infty}(\alpha+m+nw)^{-s} \quad (11.8)$$

of Barnes, where $\alpha > 0$, $w > 0$ be parameters. Inspired by the similarity of (11.3) and (11.8), the author introduced the generalized double zeta-function

$$\tilde{\zeta}_2(u, v; \alpha, w) = \sum_{m=0}^{\infty}(\alpha + m)^{-u}\sum_{n=1}^{\infty}(\alpha + m + nw)^{-v},$$

and applied the method similar to that in [107I] to $\tilde{\zeta}_2(u, v; \alpha, w)$. Then we put $u = 0, v = s$ to obtain the asymptotic expansion of $\zeta_2(s; \alpha, w)$ with respect to w. Certain asymptotic expansions for double gamma-functions, and for the value at $s = 1$ of Hecke L-functions of real quadratic fields, were also obtained in [139]. Recently, Katsurada [102] introduced another generalization

$$\sum_{m=0}^{\infty}(m + \alpha)^{-u}\sum_{n=0}^{\infty}(m + n + \alpha + \beta)^{-v},$$

where α, β are positive, and studied its properties by using the Mellin-Barnes type of integrals. He actually considered a more general series involving exponential factors.

The results mentioned in the last two sections show that Atkinson's method is indeed useful in a much wider area than was expected before.

The readers probably find that the recent developments in the mean square theory are really impressive. However, the mean square theory has been by no means exhausted; there remain many unsolved problems and uncultivated areas. It will still be one of the main streams in zeta-function theory, and fascinating new methods and results will surely appear in the coming century.

References

[1] J. Andersson, *Mean value properties of the Hurwitz zeta-function*, Math. Scand. **71** (1992), 295–300.

[2] F.V. Atkinson, *The mean value of the zeta-function on the critical line*, Proc. London Math. Soc. (2) **47** (1942), 174–200.

[3] ——, *The mean-value of the Riemann zeta function*, Acta Math. **81** (1949), 353–376.

[4] R. Balasubramanian, *An improvement on a theorem of Titchmarsh on the mean square of* $|\zeta(1/2 + it)|$, Proc. London Math. Soc. (3) **36** (1978), 540–576.

[5] ——, *A note on Hurwitz's zeta-function*, Ann. Acad. Sci. Fenn. Ser. AI Math. **4** (1979), 41–44.

[6] ——, *A note on Dirichlet's L-functions*, Acta Arith. **38** (1980), 273–283.

[7] R. Balasubramanian, Ivić, A. and Ramachandra, K., *The mean square of the Riemann zeta-function on the line* $\sigma = 1$, L'Enseignement Math. **38** (1992), 13–25.

[8] R. Balasubramanian and Ramachandra, K., *A hybrid version of a theorem of Ingham*, in Number Theory, K.Alladi (ed.), LNM **1122**, Springer (1985), 38–46.

[9] ——, *Progress towards a conjecture on the mean value of Titchmarsh series III*, Acta Arith. **45** (1986), 309–318.

[10] ——, *On an analytic continuation of* $\zeta(s)$, Indian J. Pure Appl. Math. **18** (1987), 790–793.

[11] ——, *An alternative approach to a theorem of Tom Meurman*, Acta Arith. **55** (1990), 351–364.

[12] R. Balasubramanian and Ramachandra, K., *Proof of some conjectures on the mean-value of Titchmarsh series I*, Hardy-Ramanujan J. **13** (1990), 1–20; ——*II*, ibid. **14** (1991), 1–20; ——*III*, Proc. Indian Acad. Sci. (Math. Sci.) **102** (1992), 83–91.

[13] E. Bombieri and Iwaniec, H., *On the order of* $\zeta(1/2+it)$, Ann. Scuola Norm. Sup. Pisa (4) **13** (1986), 449–472.

[14] ——, *Some mean-value theorems for exponential sums*, ibid. 473–486.

[15] F. Carlson, *Contributions à la théorie des séries de Dirichlet I*, Ark. Mat. Astr. Fysik **16** (1922), no. 18, 19 pp.

[16] K. Chandrasekharan and Narasimhan, R., *The approximate functional equation for a class of zeta-functions*, Math. Ann. **152** (1963), 30–64.

[17] Zhiming Chen, *On the Dirichlet L-functions*, Pure Appl. Math. (Xi'an) **10** (1994), 70–74 (in Chinese).

[18] S. Chowla, *Contributions to the analytic theory of numbers*, Math. Z. **35** (1932), 279–299.

[19] J.B. Conrey, Ghosh, A. and Gonek, S.M., *Mean values of the Riemann zeta-function with application to the distribution of zeros*, in Number Theory, Trace Formulas and Discrete Groups, K.E. Aubert et al. (eds.), Academic Press (1989), 185–199.

[20] K. Corrádi and Kátai, I., *Egy megjegyzés K.S. Gangadharan "Two classical lattice point problems" címü dolgozatához*, MTA IIIOstály Közlemenyei **17** (1967), 89–97.

[21] H.Cramér, *Über zwei Sätze von Herrn G.H. Hardy*, Math. Z. **15** (1922), 201–210.

[22] ——, *Contributions to the analytic theory of numbers*, in Proc. 5th Scand. Math. Congress, Helsingfors (1922), 266–272.

[23] A. Fujii, *Uniform distribution of the zeros of the Riemann zeta function and the mean value theorems*, in Number Theory, vol. I, Elementary and Analytic, K. Györy and G. Halász (eds.), Colloq. Math. Soc. János Bolyai **51**, North-Holland (1990), 141–161.

[24] ——, *Uniform distribution of the zeros of the Riemann zeta function and the mean value theorems of Dirichlet L-functions II*, in Analytic Number Theory, K. Nagasaka and E. Fouvry (eds.), LNM **1434**, Springer (1990), 103–125.

[25] ——, *Some observations concerning the distribution of the zeros of the zeta functions I*, in Zeta Functions in Geometry, N. Kurokawa and T. Sunada (eds.), Adv. Stud. Pure Math. **21**, Kinokuniya (1992), 237–280; ——*II*, Comment. Math. Univ. St. Pauli **40** (1991), 125–231; ——*III*, Proc. Japan Acad. **68A** (1992), 105–110.

[26] ——, *On a mean value theorem in the theory of the Riemann zeta function*, Comment. Math. Univ. St. Pauli **44** (1995), 59–67.

[27] J. Furuya, *Mean square of an error term related to a certain exponential sum involving the divisor function*, in Number Theory and its Applications, K. Györy and S. Kanemitsu (eds.), Kluwer, to appear.

[28] J. Furuya and Tanigawa, Y., *Estimation of a certain function related to the Dirichlet divisor problem*, in Analytic and Probabilistic Methods in Number Theory, A. Laurinčikas et al. (eds.), New Trends in Probab. and Statist. **4**, VSP/TEV (1997), 171–189.

[29] P.X. Gallagher, *Local mean value and density estimates for Dirichlet L-functions*, Indag. Math. **37** (1975), 259–264.

[30] S.M. Gonek, *Mean values of the Riemann zeta-function and its derivatives*, Invent. Math. **75** (1984), 123–141.

[31] ——, *A formula of Landau and mean values of* $\zeta(s)$, in Topics in Analytic Number Theory, S.W. Graham and J.D. Vaaler (eds.), Univ. of Texas Press (1985), 92–97.

[32] A. Good, *Ein Mittelwertsatz für Dirichletreihen, die Modulformen assoziiert sind*, Comment. Math. Helv. **49** (1974), 35–47.

[33] ——, *Approximative Funktionalgleichungen und Mittelwertsätze für Dirichletreihen, die Spitzenformen assoziiert sind*, ibid. **50** (1975), 327–361.

[34] A. Good, *Ueber das quadratische Mittel der Riemannschen Zetafunktion auf der kritischen Linie*, ibid. **52** (1977), 35–48.

[35] ——, *Ein Ω-Resultat für das quadratische Mittel der Riemannschen Zetafunktion auf der kritischen Linie*, Invent. Math. **41** (1977), 233–251.

[36] ——, *Diskrete Mittel für einige Zetafunktionen*, J. Reine Angew. Math. **303/304** (1978), 51–73.

[37] ——, *Beitraege zur Theorie der Dirichletreihen, die Spitzenformen zugeordnet sind*, J. Number Theory **13** (1981), 18–65.

[38] ——, *The square mean of Dirichlet series associated with cusp forms*, Mathematika **29** (1982), 278–295.

[39] S.W. Graham and Kolesnik, G., *Van der Corput's Method of Exponential Sums*, London Math. Soc. LN **126**, Cambridge Univ. Press (1991).

[40] Jinbao Guo, *On the mean value formula of the derivative of Hurwitz zeta-function*, J. Yanan Univ. **13** (1994), 45–51, 65 (in Chinese).

[41] ——, *A class of new mean value formulas for the derivative of Hurwitz zeta-function*, J. Math. Res. Exposition **16** (1996), 549–553 (in Chinese).

[42] A. Guthmann, *The Riemann-Siegel integral formula for Dirichlet series associated to cusp forms*, in Proceedings of the Conference on Analytic and Elementary Number Theory, Vienna, July 1996, W.G. Nowak and J. Schoissengeier (eds.), Univ. Wien and Univ. Bodenkultur, pp. 53–69.

[43] ——, *Die Riemann-Siegel-Integralformel für die Mellintransformation von Spitzenformen*, Arch. Math. **69** (1997), 391–402.

[44] ——, *New integral representations for the square of the Riemann zeta-function*, Acta Arith. **82** (1997), 309–330.

[45] J.L. Hafner, *New omega theorems for two classical lattice point problems*, Invent. Math. **63** (1981), 181–186.

[46] J.L. Hafner and Ivić, A., *On the mean-square of the Riemann zeta-function on the critical line*, J. Number Theory **32** (1989), 151–191.

[47] G.H. Hardy and Littlewood, J.E., *Contributions to the theory of the Riemann zeta-function and the theory of the distribution of primes*, Acta Math. **41** (1918), 119–196.

[48] ——, *The zeros of Riemann's zeta-function on the critical line*, Math. Z. **10** (1921), 283–317.

[49] ——, *The approximate functional equation in the theory of the zeta-function, with applications to the divisor-problems of Dirichlet and Piltz*, Proc. London Math. Soc. (2) **21** (1923), 39–74.

[50] ——, *The approximate functional equations for $\zeta(s)$ and $\zeta^2(s)$*, ibid. (2) **29** (1929), 81–97.

[51] D.R. Heath-Brown, *The mean value theorem for the Riemann zeta-function*, Mathematika **25** (1978), 177–184.

[52] ——, *The twelfth power moment of the Riemann-function*, Quart. J. Math. Oxford (2) **29** (1978), 443–462.

[53] ——, *The fourth power moment of the Riemann zeta-function*, Proc. London Math. Soc. (3) **38** (1979), 385–422.

[54] ——, *An asymptotic series for the mean value of Dirichlet L-functions*, Comment. Math. Helv. **56** (1981), 148–161.

[55] ——, *Fractional moments of the Riemann zeta-function*, J. London Math. Soc. (2) **24** (1981), 65–78; ——*II*, Quart. J. Math. Oxford (2) **44** (1993), 185–197.

[56] ——, *The distribution and moments of the error term in the Dirichlet divisor problem*, Acta Arith. **60** (1992), 389–415.

[57] D.R. Heath-Brown and Huxley, M.N., *Exponential sums with a difference*, Proc. London Math. Soc. (3) **61** (1990), 227–250.

[58] D.R. Heath-Brown and Tsang, K.-M., *Sign changes of $E(T)$, $\Delta(X)$ and $P(X)$*, J. Number Theory **49** (1994), 73–83.

[59] M.N. Huxley, *Exponential sums and lattice points*, Proc. London Math. Soc. (3) **66** (1993), 279–301; Corrigenda, (3) **68** (1994), 264.

[60] ——, *A note on exponential sums with a difference*, Bull. London Math. Soc. **26** (1994), 325–327.

[61] ——, *Area, Lattice Points and Exponential Sums*, Oxford (1996).

[62] M.N. Huxley and Watt, N., *Exponential sums and the Riemann zeta function*, Proc. London Math. Soc. (3) **57** (1988), 1–24.

[63] A.E. Ingham, *Mean-value theorems in the theory of the Riemann zeta-function*, ibid. (2) **27** (1928), 273–300.

[64] A. Ivić, *Large values of the error term in the divisor problem*, Invent. Math. **71** (1983), 513–520.

[65] ——, *Topics in Recent Zeta Function Theory*, Publ. Math. d'Orsay, Univ. de Paris-Sud (1983).

[66] ——, *The Riemann Zeta-Function*, Wiley (1985).

[67] ——, *Large values of certain number-theoretic error terms*, Acta Arith. **56** (1990), 135–159.

[68] ——, *Lectures on Mean Values of the Riemann Zeta Function*, LN **82**, Tata Inst. Fund. Res., Springer (1991).

[69] ——, *Power moments of the error term in the approximate functional equation for $\zeta^2(s)$*, Acta Arith. **65** (1993), 137–145.

[70] ——, *La valeur moyenne de la fonction zeta de Riemann*, in Séminaire de Théorie des Nombres, Paris 1990–91, S. David (ed.), Progress in Math. **108**, Birkhäuser (1993), 115–125.

[71] ——, *Some problems on mean values of the Riemann zeta-function*, J. Théorie des Nombres de Bordeaux **8** (1996), 101–123.

[72] ——, *On the mean square of the error term in the approximate functional equation for $\zeta^2(s)$*, Arch. Math. **68** (1997), 468–476.

[73] ——, *On some problems involving the mean square of $|\zeta(1/2+it)|$*, Bull. CXVI Acad. Serb. Sci. Arts, Sci. Math. **23** (1998), 71–76.

[74] A. Ivić and Kiuchi, I., *On some integrals involving the Riemann zeta-function in the critical strip*, Univ. Beograd. Publ. Elektrotehn. Fak. Ser. Mat. **5** (1994), 19–28.

[75] A. Ivić and Matsumoto, K. *On the error term in the mean square formula for the Riemann zeta-function in the critical strip*, Monatsh. Math. **121** (1996), 213–229.

[76] A. Ivić and Motohashi, Y., *The mean square of the error term for the fourth power moment of the zeta-function*, Proc. London Math. Soc. (3) **69** (1994), 309–329.

[77] ——, *On the fourth power moment of the Riemann zeta-function*, J. Number Theory **51** (1995), 16–45.

[78] A. Ivić and te Riele, H.J.J., *On the zeros of the error term for the mean square of $|\zeta(1/2+it)|$*, Math. Comput. **56** (1991), 303–328.

[79] H. Iwaniec, *Fourier coefficients of cusp forms and the Riemann zeta-function*, Séminaire de Théorie des Nombres, Univ. de Bordeaux (1979/80), no.18, 36 pp.

[80] H. Iwaniec and Mozzochi, C.J., *On the divisor and circle problems*, J. Number Theory **29** (1988), 60–93.

[81] M. Jutila, *Riemann's zeta-function and the divisor problem*, Ark. Mat. **21** (1983), 75–96; ——*II*, ibid. **31** (1993), 61–70.

[82] ——, *On a formula of Atkinson*, in Topics in Classical Number Theory, vol. I, G. Halász (ed.), Colloq. Math. Soc. János Bolyai **34**, North-Holland (1984), 807–823.

[83] ——, *Transformation formulae for Dirichlet polynomials*, J. Number Theory **18** (1984), 135–156.

[84] M. Jutila, *On the divisor problem for short intervals*, Ann. Univ. Turkuensis Ser. AI **186** (1984), 23–30.

[85] ———, *On the approximate functional equation for $\zeta^2(s)$ and other Dirichlet series*, Quart. J. Math. Oxford (2) **37** (1986), 193–209.

[86] ———, *Remarks on the approximate functional equation for $\zeta^2(s)$*, in The Very Knowledge of Coding, Univ. of Turku (1987), 88–98.

[87] ———, *Lectures on a Method in the Theory of Exponential Sums*, LN **80**, Tata Inst. Fund. Res., Springer (1987).

[88] ———, *Mean value estimates for exponential sums*, in Number Theory, Ulm 1987, H.P. Schlickewei and E. Wirsing (eds.), LNM **1380**, Springer (1989), 120–136; ——*II*, Arch. Math. **55** (1990), 267–274.

[89] ———, *The fourth power moment of the Riemann zeta-function over a short interval*, in Number Theory, K. Györy and G. Halász (eds.), Colloq. Math. Soc. János Bolyai **51**, North-Holland (1990), 221–244.

[90] ———, *Mean value estimates for exponential sums with applications to L-functions*, Acta Arith. **57** (1991), 93–114.

[91] ———, *The fourth moment of Riemann's zeta-function and the additive divisor problem*, in Analytic Number Theory, Proc. Conf. in Honor of H. Halberstam, vol. 2, B.C.Berndt et al. (eds.), Progress in Math. **139**, Birkhäuser (1996), 517–536.

[92] ———, *Atkinson's formula revisited*, in Voronoï's Impact on Modern Science, Book 1, P. Engel and H. Syta (eds.), Inst. Math., Nat. Acad. Sci. Ukraine (1998), 137–154.

[93] ———, *Mean values of Dirichlet series via Laplace transforms*, in Analytic Number Theory, Y. Motohashi (ed.), London Math. Soc. LN **247**, Cambridge Univ. Press (1997), 169–207.

[94] A. Kačėnas, *The asymptotic behaviour of the second power moment of the Riemann zeta-function in the critical strip*, Liet. Mat. Rink. **35** (1995), 315–331 (in Russian)=Lithuanian Math. J. **35** (1995), 249–261.

[95] ———, *A note on the mean square of $\zeta(s)$ in the critical strip*, in Analytic and Probabilistic Methods in Number Theory, A.Laurinčikas et al. (eds.), New Trends in Probab. and Statist. **4**, VSP/TEV (1997), 107–117.

[96] Y. Kamiya, *Zero density estimates of L-functions associated with cusp forms*, Acta Arith. **85** (1998), 209–227.

[97] ———, *A note on the non-vanishing problem for L-functions associated with cusp forms*, Comment. Math. Univ. St. Pauli, to appear.

[98] S. Kanemitsu, *Omega theorems for divisor functions*, Tokyo J. Math. **7** (1984), 399–419.

[99] M. Katsurada, *Asymptotic expansions of the mean values of Dirichlet L-functions II*, in Analytic Number Theory and Related Topics, K. Nagasaka (ed.), World Scientific (1993), 61–71; ——*III*, Manuscripta Math. **83** (1994), 425–442.

[100] ———, *An application of Mellin-Barnes' type integrals to the mean square of Lerch zeta-functions*, Collect. Math. **48** (1997), 137–153.

[101] ———, *An application of Mellin-Barnes type integrals to the mean square of L-functions*, Liet. Mat. Rink. **38** (1998), 98–112.

[102] ———, *Power series and asymptotic series associated with the Lerch zeta-function*, Proc. Japan Acad. **74A** (1998), 167–170.

[103] M. Katsurada and Matsumoto, K., *Asymptotic expansions of the mean values of Dirichlet L-functions*, Math. Z. **208** (1991), 23–39.

[104] ———, *Discrete mean values of Hurwitz zeta-functions*, Proc. Japan Acad. **69A** (1993), 164–169.

[105] ———, *Explicit formulas and asymptotic expansions for certain mean square of Hurwitz zeta-functions*, ibid., 303–307.

[106] M. Katsurada and Matsumoto, K., *The mean values of Dirichlet L-functions at integer points and class numbers of cyclotomic fields*, Nagoya Math. J. **134** (1994), 151–172.

[107] ——, *Explicit formulas and asymptotic expansions for certain mean square of Hurwitz zeta-functions I*, Math. Scand. **78** (1996), 161–177; ——*II*, in Analytic and Probabilistic Methods in Number Theory, A. Laurinčikas et al. (eds.), New Trends in Probab. and Statist. **4**, VSP/TEV (1997), 119–134; ——*III*, preprint.

[108] ——, *A weighted integral approach to the mean square of Dirichlet L-functions*, in Number Theory and its Applications, K. Győry and S. Kanemitsu (eds.), Kluwer, to appear.

[109] I. Kiuchi, *On an exponential sum involving the arithmetic function $\sigma_a(n)$*, Math. J. Okayama Univ. **29** (1987), 193–205.

[110] ——, *Power moments of the error term for the approximate functional equation of the Riemann zeta-function*, Publ. Inst. Math. (Beograd) **52** (66) (1992), 10–12.

[111] ——, *An improvement on the mean value formula for the approximate functional equation of the square of the Riemann zeta-function*, J. Number Theory **45** (1993), 312–319.

[112] ——, *Mean value results for the non-symmetric form of the approximate functional equation of the Riemann zeta-function*, Tokyo J. Math. **17** (1994), 191–200.

[113] ——, *The mean value formula for the approximate functional equation of $\zeta^2(s)$ in the critical strip*, Arch. Math. **64** (1995), 316–322; ——*II*, ibid. **67** (1996), 126–133.

[114] I. Kiuchi and Matsumoto, K., *Mean value results for the approximate functional equation of the square of the Riemann zeta-function*, Acta Arith. **61** (1992), 337–345.

[115] ——, *The resemblance of the behaviour of the remainder terms $E_\sigma(t)$, $\Delta_{1-2\sigma}(x)$ and $R(\sigma+it)$*, in Sieve Methods, Exponential Sums, and their Applications in Number Theory, G.R.H. Greaves et al. (eds.), London Math. Soc. LN **237**, Cambridge Univ. Press (1997), 255–273.

[116] I. Kiuchi and Tanigawa, Y., *The mean value theorem of the divisor problem for short intervals*, Arch. Math. **71** (1998), 445–453.

[117] ——, *The mean value theorem of the Riemann zeta-function in the critical strip for short intervals*, in Number Theory and its Applications, K. Győry and S. Kanemitsu (eds.), Kluwer, to appear.

[118] ——, *The mean value theorem of the approximate functional equation of $\zeta^2(s)$ for short intervals*, preprint.

[119] J.F. Koksma and Lekkerkerker, C.G., *A mean value theorem for $\zeta(s,w)$*, Indag. Math. **14** (1952), 446–452.

[120] G. Kolesnik, *On the order of $\zeta(1/2+it)$ and $\Delta(R)$*, Pacific J. Math. **98** (1982), 107–122.

[121] ——, *On the method of exponent pairs*, Acta Arith. **45** (1985), 115–143.

[122] N.V. Kuznetsov, *Petersson's conjecture for cusp forms of weight zero and Linnik's conjecture. Sums of Kloosterman sums*, Mat. Sbornik **111** (153) (1980), 335–383 (in Russian)=Math. USSR Sbornik **39** (1981), 299–342.

[123] ——, *Mean value of the Hecke series associated with the cusp forms of weight zero*, Zap. Nauchn. Sem. LOMI **109** (1981), 93–130 (in Russian)=J. Soviet Math. **24** (1984), 215–238.

[124] ——, *Convolution of the Fourier coefficients of the Eisenstein-Maass series*, Zap. Nauchn. Sem. LOMI **129** (1983), 43–85 (in Russian)=J. Soviet Math. **29** (1985), 1131–1159.

[125] K.-Y. Lam, *Some results on the mean values of certain error terms in analytic number theory*, M. Phil. Thesis, Univ. of Hong Kong (1997).

[126] K.-Y. Lam and Tsang, K.-M., *The mean square of the error term in a generalization of the Dirichlet divisor problem*, in Analytic Number Theory, Y. Motohashi (ed.), London Math. Soc. LN **247**, Cambridge Univ. Press (1997), 209–225.

[127] E. Landau, *Handbuch der Lehre von der Verteilung der Primzahlen*, Teubner (1909).

[128] Y.-K. Lau and Tsang, K.-M., *Mean square of the remainder term in the Dirichlet divisor problem*, J. Théorie des Nombres de Bordeaux **7** (1995), 75–92.

[129] A. Laurinčikas, *The Atkinson formula near the critical line*, in Analytic and Probabilistic Methods in Number Theory, E. Manstavičius and F. Schweiger (eds.), New Trends in Probab. and Statist. **2**, VSP/TEV (1992), 335–354; ——*II*, Liet. Mat. Rink. **33** (1993), 302–313 (in Russian)=Lithuanian Math. J. **33** (1993), 234–242.

[130] ——, *The Atkinson formula for L-functions near the critical line*, Liet. Mat. Rink. **33** (1993), 435–454 (in Russian)=Lithuanian Math. J. **33** (1993), 337–351.

[131] ——, *On the moments of the Riemann zeta-function near the critical line*, Liet. Mat. Rink. **35** (1995), 332–359 (in Russian)=Lithuanian Math. J. **35** (1995), 262–283.

[132] ——, *A uniform estimate of the error term in the mean square of the Riemann zeta-function*, Liet. Mat. Rink. **35** (1995), 508–517 (in Russian)=Lithuanian Math. J. **35** (1995), 403–410.

[133] ——, *On the mean square of the Riemann zeta-function*, Liet. Mat. Rink. **36** (1996), 354–364 (in Russian)=Lithuanian Math. J. **36** (1996), 282–290.

[134] J.E. Littlewood, *Researches in the theory of the Riemann ζ-function*, Proc. London Math. Soc. (2) **20** (1922), xxii–xxviii.

[135] K. Matsumoto, *The mean square of the Riemann zeta-function in the critical strip*, Japanese J. Math. **15** (1989), 1–13.

[136] ——, *On the function $E_\sigma(T)$*, Sûrikaiseki Kenkyûsho Kôkyûroku **886** (1994), 10–28.

[137] ——, *On the bounded term in the mean square formula for the approximate functional equation of $\zeta^2(s)$*, Arch. Math. **64** (1995), 323–332.

[138] ——, *Asymptotic series for double zeta and double gamma functions of Barnes*, Sûrikaiseki Kenkyûsho Kôkyûroku **958** (1996), 162–165.

[139] ——, *Asymptotic series for double zeta, double gamma, and Hecke L-functions*, Math. Proc. Cambridge Phil. Soc. **123** (1998), 385–405.

[140] K. Matsumoto and Meurman, T., *The mean square of the Riemann zeta-function in the critical strip II*, Acta Arith. **68** (1994), 369–382; ——*III*, ibid. **64** (1993), 357–382.

[141] T. Meurman, *A generalization of Atkinson's formula to L-functions*, Acta Arith. **47** (1986), 351–370.

[142] ——, *On the mean square of the Riemann zeta-function*, Quart. J. Math. Oxford (2) **38** (1987), 337–343.

[143] ——, *The mean square of the error term in a generalization of Dirichlet's divisor problem*, Acta Arith. **74** (1996), 351–364.

[144] H.L. Montgomery, *Topics in Multiplicative Number Theory*, LNM **227**, Springer (1971).

[145] ——, *The analytic principle of the large sieve*, Bull. Amer. Math. Soc. **84** (1978), 547–567.

[146] H.L. Montgomery and Vaughan, R.C., *Hilbert's inequality*, J. London Math. Soc. (2) **8** (1974), 73–82.

[147] Y. Motohashi, *A note on the mean value of the Dedekind zeta-function of the quadratic field*, Math. Ann. **188** (1970), 123–127.

[148] ——, *A note on the approximate functional equation for $\zeta^2(s)$*, Proc. Japan Acad. **59A** (1983), 393–396; ——*II*, ibid., 469–472; ——*III*, ibid. **62A** (1986), 410–412.

[149] ——, *A note on the mean value of the zeta and L-functions I*, Proc. Japan Acad. **61A** (1985), 222–224; ——*IV*, ibid. **62A** (1986), 311–313.

[150] ——, *Lectures on the Riemann-Siegel Formula*, Ulam Seminar, Colorado Univ. (1987).

[151] ——, *An explicit formula for the fourth power mean of the Riemann zeta-function*, Acta Math. **170** (1993), 181–220.

[152] ——, *The mean square of Hecke L-series attached to holomorphic cusp-forms*, Sûrikaiseki Kenkyûsho Kôkyûroku **886** (1994), 214–227.

[153] Y. Motohashi, *An asymptotic expansion of the square of the Riemann zeta-function*, in Sieve Methods, Exponential Sums, and their Applications in Number Theory, G.R.H. Greaves et al. (eds.), London Math. Soc. LN **237**, Cambridge Univ. Press (1997), 293–307.

[154] ——, *The mean square of Dedekind zeta-functions of quadratic number fields*, ibid., 309–324.

[155] ——, *Spectral Theory of the Riemann Zeta-Function*, Cambridge Univ. Press (1997).

[156] W. Müller, *The mean square of the Dedekind zeta-function in quadratic number fields*, Math. Proc. Cambridge Phil. Soc. **106** (1989), 403–417.

[157] ——, *The mean square of Dirichlet series associated with automorphic forms*, Monatsh. Math. **113** (1992), 121–159.

[158] H. Nakaya, *The mean square of the Dirichlet L-function in the critical strip*, preprint.

[159] M.J. Narlikar, *On the mean square value of Hurwitz zeta function*, Proc. Indian Acad. Sci. (Math. Sci.) **90** (1981), 195–212.

[160] ——, *Hybrid mean square of L-functions*, Hardy-Ramanujan J. **9** (1986), 11–16.

[161] A. Oppenheim, *Some identities in the theory of numbers*, Proc. London Math. Soc. (2) **26** (1927), 295–350.

[162] R.E.A.C. Paley, *On the k-analogues of some theorems in the theory of the Riemann ζ-function*, ibid. (2) **32** (1931), 273–311.

[163] H.S.A. Potter, *The mean values of certain Dirichlet series I*, ibid. (2) **46** (1940), 467–478; ——*II*, ibid. (2) **47** (1940), 1–19.

[164] E. Preissmann, *Sur la moyenne quadratique du terme de reste du problème du cercle*, C. R. Acad. Sci. Paris **306** (1988), 151–154.

[165] ——, *Sur la moyenne de la fonction zêta*, in Analytic Number Theory and Related Topics, K. Nagasaka (ed.), World Scientific (1993), 119–125.

[166] K. Ramachandra, *Application of a theorem of Montgomery and Vaughan to the zeta-function*, J. London Math. Soc. (2) **10** (1975), 482–486.

[167] ——, *Some remarks on a theorem of Montgomery and Vaughan*, J. Number Theory **11** (1979), 465–471.

[168] ——, *Some remarks on the mean value of the Riemann zeta-function and other Dirichlet series I*, Hardy-Ramanujan J. **1** (1978), 1–15; ——*II*, ibid. **3** (1980), 1–24; ——*III*, Ann. Acad. Sci. Fenn. Ser. AI Math. **5** (1980), 145–158; ——*IV*, J. Indian Math. Soc. **60** (1994), 107–122.

[169] ——, *Progress towards a conjecture on the mean-value of Titchmarsh series*, in Recent Progress in Analytic Number Theory, vol. 1, H. Halberstam and C. Hooley (eds.), Academic Press (1981), 303–318; ——*II*, Hardy-Ramanujan J. **4** (1981), 1–12.

[170] ——, *Mean-value of the Riemann zeta-function and other remarks I*, in Topics in Classical Number Theory, vol. II, G. Halász (ed.), Colloq. Math. Soc. János Bolyai **34**, North-Holland (1984), 1317–1347; ——*II*, Trudy Mat. Inst. Steklov. **163** (1984), 200–204 = Proc. Steklov Inst. Math. **163** (1985), 233–237; ——*III*, Hardy-Ramanujan J. **6** (1983), 1–21.

[171] ——, *Proof of some conjectures on the mean-value of Titchmarsh series with applications to Titchmarsh's phenomenon*, Hardy-Ramanujan J. **13** (1990), 21–27.

[172] ——, *Lectures on the Mean-Value and Omega-Theorems for the Riemann Zeta-Function*, LN **85**, Tata Inst. Fund. Res., Narosa Publ. House (1995).

[173] V.V. Rane, *On the mean square value of Dirichlet L-series*, J. London Math. Soc. (2) **21** (1980), 203–215.

[174] ——, *On Hurwitz zeta-function*, Math. Ann. **264** (1983), 147–151.

[175] ——, *Dirichlet L-function and power series for Hurwitz zeta function*, Proc. Indian Acad. Sci. (Math. Sci.) **103** (1993), 27–39.

[176] ——, *A footnote to mean square value of Dirichlet L-series*, ibid. **105** (1993), 127–133.

[177] V.V. Rane, *A new approximate functional equation for Hurwitz zeta function for rational parameter*, ibid. **107** (1997), 377–385.

[178] ——, *Approximate functional equation and Fourier series expression for Hurwitz zeta function*, ibid., to appear.

[179] A. Sankaranarayanan and Srinivas, K., *Mean-value theorem of the Riemann zeta-function over short intervals*, J. Number Theory **45** (1993), 320–326.

[180] W. Schnee, *Über Mittelwertsformeln in der Theorie der Dirichlet'schen Reihen*, Sitzungsberichte der Math.-Naturw. Kl. Kaiserl. Akad. Wiss. Wien **118** (1909), 1439–1522.

[181] A. Selberg, *Contributions to the theory of Dirichlet's L-functions*, Skr. Norsk. Vid.-Akad. Oslo I. Mat.-Naturv. Kl. (1946), no. 3, 1–62.

[182] C.L. Siegel, *Über Riemanns Nachlass zur analytischen Zahlentheorie*, Quell. Stud. Gesch. Math. Astr. Phys. **2** (1932), 45–80.

[183] I.Sh. Slavutskiĭ, *Mean value of $\mathcal{L}$-functions and the ideal class number of a cyclotomic field*, in Algebraic Systems with One Action and Relation, Leningrad. Gos. Ped. Inst. (1985), 122–129 (in Russian).

[184] ——, *Mean value of L-functions and the class number of a cyclotomic field*, Zap. Nauchn. Sem. LOMI **154** (1986), 136–143 (in Russian)=J. Soviet Math. **43** (1988), 2596–2601.

[185] T. Stefanicki, *Non-vanishing of L-functions attached to automorphic representations of $GL(2)$ over* **Q**, J. Reine Angew. Math. **474** (1996), 1–24.

[186] P.R. Taylor, *On the Riemann zeta function*, Quart. J. Math. Oxford **16** (1945), 1–21.

[187] E.C. Titchmarsh, *The mean-value of the zeta-function on the critical line*, Proc. London Math. Soc. (2) **27** (1928), 137–150.

[188] ——, *On van der Corput's method and the zeta-function of Riemann (V)*, Quart. J. Math. Oxford **5** (1934), 195–210.

[189] ——, *The approximate functional equation for $\zeta^2(s)$*, ibid. **9** (1938), 109–114.

[190] ——, *The Theory of the Riemann Zeta-Function*, Oxford (1951).

[191] K.-C. Tong, *On divisor problems III*, Acta Math. Sinica **6** (1956), 515–541 (in Chinese).

[192] K.-M. Tsang, *Higher-power moments of $\Delta(x)$, $E(t)$ and $P(x)$*, Proc. London Math. Soc. (3) **65** (1992), 65–84.

[193] ——, *Mean square of the remainder term in the Dirichlet divisor problem II*, Acta Arith. **71** (1995), 279–299.

[194] A.I. Vinogradov and Takhtadzhyan, L.A., *The zeta function of the additive divisor problem and spectral decomposition of the automorphic Laplacian*, Zap. Nauchn. Sem. LOMI **134** (1984), 84–116 (in Russian)=J. Soviet Math. **36** (1987), 57–78.

[195] G.F. Voronoï, *Sur un problème du calcul des fonctions asymptotiques*, J. Reine Angew. Math. **126** (1903), 241–282.

[196] ——, *Sur une fonction transcendante et ses applications à la sommation de quelques séries*, Ann. Sci. École Norm. Sup. (3) **21** (1904), 207–268, 459–534.

[197] A. Walfisz, *Teilerprobleme. Zweite Abhandlung*, Math. Z. **34** (1931), 448–472.

[198] N. Yanagisawa, *An asymptotic formula for a certain mean value in a divisor problem*, J. Number Theory **73** (1998), 339–358.

[199] ——, *On the mean square in the divisor problem for short intervals*, preprint.

[200] Kongting Yu, *On the mean square value formula of L-functions*, J. Math. Res. Exposition **12** (1992), 413–420 (in Chinese).

[201] N.I. Zavorotnyi, *On the fourth moment of the Riemann zeta-function*, in Automorphic Functions and Number Theory I, Collected Sci. Works, Vladivostok (1989), 69–125 (in Russian).

[202] Tao Zhan, *On the mean square of Dirichlet L-functions*, Acta Math. Sinica **8** (1992), 204–224.

[203] Jiankang Zhang and Xiaolong Xing, *An asymptotic series for the mean value of Dirichlet L-functions*, Pure Appl. Math. (Xi'an) **11** (1995), 13–21 (in Chinese).

[204] Wenpeng Zhang, *On Dirichlet L-functions*, Acta Math. Sinica **32** (1989), 824–833 (in Chinese).
[205] ——, *Mean square value formulas for L-functions*, Chinese Ann. Math. **11A** (1990), 121–127 (in Chinese).
[206] ——, *The mean square value of the Dirichlet L-functions*, Adv. in Math. (China) **19** (1990), 321–333 (in Chinese).
[207] ——, *On the Hurwitz zeta-function*, Northeastern Math. J. **6** (1990), 261–267.
[208] ——, *The Hurwitz zeta-function*, Acta Math. Sinica **33** (1990), 160–171 (in Chinese).
[209] ——, *On the mean value of L-function*, J. Math. Res. Exposition **10** (1990), 355–360 (in Chinese).
[210] ——, *On an elementary result of L-functions*, Adv. in Math. (China) **19** (1990), 478–487 (in Chinese).
[211] ——, *On the Dirichlet L-functions*, Acta Math. Sinica **7** (1991), 103–118.
[212] ——, *On the mean value formula of Dirichlet L-functions (II)*, Sci. in China **34** (1991), 660–675.
[213] ——, *On the Hurwitz zeta-function*, Illinois J. Math. **35** (1991), 569–576.
[214] ——, *On the mean square value of Dirichlet's L-functions*, Compositio Math. **84** (1992), 59–69.
[215] ——, *On the mean value formula of Dirichlet L-functions (I)*, Acta Math. Sinica **36** (1993), 245–253 (in Chinese).
[216] ——, *On the mean square value formula of Lerch zeta-function*, Adv. in Math. (China) **22** (1993), 367–369.
[217] ——, *On the mean square value of the Hurwitz zeta-function*, Illinois J. Math. **38** (1994), 71–78.
[218] ——, *On the mean square value formula of Lerch zeta-function*, Chinese Ann. Math. **16A** (1995), 338–343 (in Chinese).

Graduate School of Mathematics
Nagoya University, Chikusa-ku
Nagoya 464–8602
Japan
E-Mail: kohjimat@math.nagoya-u.ac.jp

Algebraic Curves Over Finite Fields with many Rational Points and their Applications

Harald Niederreiter and Chaoping Xing

Algebraic curves over finite fields with many rational points have received a lot of attention in recent years. We present a survey of this subject covering both the case of fixed genus and the asymptotic theory. A strong impetus in the asymptotic theory has come from a thorough exploitation of the method of infinite class field towers. On the other hand, we show by a counterexample that Perret's conjecture on infinite class field towers is wrong, and so Perret's method of infinite ramified class field towers breaks down. In the last two sections of the paper we discuss applications of algebraic curves over finite fields with many rational points to coding theory and to the construction of low-discrepancy sequences.

Key Words. Algebraic curves over finite fields, rational points, global function fields, rational places, algebraic coding theory, low-discrepancy sequences.

1 Introduction

Let q be an arbitrary prime power and let $\mathbb{F}_q$ denote the finite field of order q. By an *algebraic curve over* $\mathbb{F}_q$ we always mean a smooth, projective, absolutely irreducible algebraic curve defined over $\mathbb{F}_q$. If C is such a curve, then we write $g(C)$ for the genus of C. A point of C is called $\mathbb{F}_q$*-rational* if it has homogeneous coordinates which all belong to $\mathbb{F}_q$. Let $N(C)$ denote the number of $\mathbb{F}_q$-rational points of C. The following definition is basic for this paper.

Definition 1 For any prime power q and any integer $g \geq 0$ put

$$N_q(g) = \max N(C),$$

where the maximum is extended over all algebraic curves C over $\mathbb{F}_q$ with $g(C) = g$.

The calculation of $N_q(g)$ is a very difficult problem in algebraic geometry and exact values are currently known only in some isolated cases (see Section 2). Mostly, we have to be satisfied with lower and upper bounds for $N_q(g)$. A well-known general upper bound for $N_q(g)$ is the Weil-Serre bound

$$N_q(g) \leq q + 1 + g\lfloor 2q^{1/2} \rfloor, \tag{1}$$

where $\lfloor u \rfloor$ denotes the greatest integer not exceeding the real number u. Further information on upper bounds is provided in Section 2. In an informal way, we say that an algebraic curve C over $\mathbb{F}_q$ of genus g has many rational points if $N(C)$ is reasonably close to $N_q(g)$ or to a known upper bound for $N_q(g)$.

It will often be convenient to replace the language of algebraic curves over finite fields by the equivalent language of global function fields, i.e., of algebraic function fields with finite constant fields. For an algebraic curve C over $\mathbb{F}_q$, the field K of $\mathbb{F}_q$-rational functions on C is a global function field with full constant field $\mathbb{F}_q$, that is, with $\mathbb{F}_q$ algebraically closed in K. We use the notation $K/\mathbb{F}_q$ if we want to emphasize the fact that $\mathbb{F}_q$ is the full constant field of K. With every global function field $K/\mathbb{F}_q$ we can also associate an algebraic curve C over $\mathbb{F}_q$. By a *rational place* of K we mean a place of K of degree 1. We write $g(K)$ for the genus of K and $N(K)$ for the number of rational places of K. In the correspondence between the curve C and its function field K, the closed points of C can be identified with the places of K and the $\mathbb{F}_q$-rational points of C can be identified with the rational places of K, so that $N(C) = N(K)$. Furthermore, we have $g(C) = g(K)$. In particular, we can interpret the quantity $N_q(g)$ in Definition 1 as the maximum number of rational places that a global function field $K/\mathbb{F}_q$ of genus g can have. In analogy with a corresponding way of speaking for algebraic curves, we say that a global function field $K/\mathbb{F}_q$ of genus g has many rational places if $N(K)$ is reasonably close to $N_q(g)$ or to a known upper bound for $N_q(g)$.

In Section 2 we review some known results on $N_q(g)$ and in Section 3 we discuss the asymptotic behavior of $N_q(g)$ as $g \to \infty$. The conjecture of Perret [19] on infinite class field towers, which arose in the asymptotic theory of $N_q(g)$, is disproved by a counterexample in Section 4. As a consequence, Perret's method of infinite ramified class field towers breaks down. A survey of constructions of algebraic curves over finite fields with many rational points is given in Section 5. Applications of such curves to algebraic coding theory and to the construction of low-discrepancy sequences are discussed in Sections 6 and 7, respectively. Tables of lower and upper bounds for $N_q(g)$ and of an interesting quantity from the theory of low-discrepancy sequences are included.

2 Results on $N_q(g)$

General formulas for $N_q(g)$ are known only for small values of the genus g. It is trivial that $N_q(0) = q + 1$. A formula for $N_q(1)$ is essentially due to Deuring [1]; see also Waterhouse [30]. We have $N_q(1) = q + 1 + \lfloor 2q^{1/2} \rfloor$, except in the case where $q = p^e$ with a prime p dividing $\lfloor 2q^{1/2} \rfloor$ and an odd integer $e \geq 3$, in which case $N_q(1) = q + \lfloor 2q^{1/2} \rfloor$. Serre [20], [21], [22] determined the values of $N_q(2)$. Write again $q = p^e$ with a prime p and an integer $e \geq 1$. If e is even and $q \neq 4, 9$, then $N_q(2) = q + 1 + 4q^{1/2}$, whereas $N_4(2) = 10$ and $N_9(2) = 20$. We call q *special* if either p divides $\lfloor 2q^{1/2} \rfloor$ or q is of the form m^2+1, m^2+m+1, or m^2+m+2 for some integer m. If e is odd and q is not special, then $N_q(2) = q + 1 + 2\lfloor 2q^{1/2} \rfloor$. If e is odd and q is special, then $N_q(2) = q + 2\lfloor 2q^{1/2} \rfloor$ or $q + 2\lfloor 2q^{1/2} \rfloor - 1$, depending on whether the fractional part $\{2q^{1/2}\}$ is greater than $(\sqrt{5} - 1)/2$ or not.

An algebraic curve C over $\mathbb{F}_q$ of genus g is called *maximal* if $N(C) = q + 1 + 2gq^{1/2}$. Clearly, maximal curves can exist only if $g = 0$ or if q is a square. If a maximal curve over $\mathbb{F}_q$ of genus g exists, then $N_q(g) = q+1+2gq^{1/2}$, that is, we have equality in the Weil-Serre bound (1). A well-known family of maximal curves is given by the Hermitian curves. If

$q = r^2$ is a square, then the *Hermitian curve* $\mathcal{H}$ over $\mathbb{F}_q$ is defined by $x^{r+1}+y^{r+1}+z^{r+1} = 0$. This curve satisfies $g(\mathcal{H}) = r(r-1)/2$ and $N(\mathcal{H}) = r^3+1$. Furthermore, for any maximal curve C over $\mathbb{F}_q$ with $q = r^2$ we have $g(C) \leq r(r-1)/2$. For recent work on maximal curves we refer to Fuhrmann and Torres [2] and Xing [31].

If the genus g is sufficiently large relative to q, then the Weil-Serre bound (1) can be improved by using Weil's explicit formula for $N(C)$ in terms of the eigenvalues of the Frobenius of the algebraic curve C over $\mathbb{F}_q$. A general refinement of (1) was established by Serre [20]. Let

$$f(\theta) = 1 + 2\sum_{n=1}^{m} c_n \cos n\theta$$

be a trigonometric polynomial with coefficients $c_n \geq 0$ that are not all 0 and with $f(\theta) \geq 0$ for all real θ. Then

$$N_q(g) \leq \frac{g}{b(q^{-1/2})} + \frac{b(q^{1/2})}{b(q^{-1/2})} + 1, \tag{2}$$

where $b(t) = \sum_{n=1}^{m} c_n t^n$. Another proof of this bound can be found in [23, Section V.3]. The quality of the bound (2) depends on the choice of the trigonometric polynomial f, and the question of this choice was investigated by Oesterlé (see Serre [22]).

At the end of the paper we provide two tables of bounds for $N_q(g)$. Table 1 is for $q = 2, 3, 4, 8, 9, 16, 27$ and $1 \leq g \leq 50$ and for $q = 5$ and $1 \leq g \leq 22$. Table 2 extends Table 1 for the important case $q = 2$ to the range $51 \leq g \leq 95$. In each entry of the tables, the first number is a lower bound for $N_q(g)$ and the second is an upper bound for $N_q(g)$. If only one number is given, then this is the exact value of $N_q(g)$. A program for calculating upper bounds for $N_q(g)$, which is based on (2) and the trigonometric polynomials of Oesterlé, was kindly supplied to us by Jean-Pierre Serre. Tables 1 and 2 are reprinted from [18, Table 3] and [15, Table 2], respectively.

3 Asymptotic Theory

The asymptotic theory of algebraic curves over finite fields with many rational points is concerned with the behavior of $N_q(g)$ for a fixed prime power q and $g \to \infty$. The basic quantity here is given by the following definition.

Definition 2 For any prime power q put

$$A(q) = \limsup_{g\to\infty} \frac{N_q(g)}{g},$$

where g runs through positive values.

It follows from (1) that $A(q) \leq \lfloor 2q^{1/2} \rfloor$. Ihara [4] obtained some improvements on this bound and showed also that $A(q) \geq q^{1/2} - 1$ if q is a square. Vlădut and Drinfeld [29] proved the currently best general upper bound

$$A(q) \leq q^{1/2} - 1 \quad \text{for all} \quad q. \tag{3}$$

This result can be derived quite easily from (2), as demonstrated in [23, Theorem V.3.6]. As a consequence, we have $A(q) = q^{1/2} - 1$ if q is a square.

In the case where q is not a square, no exact values of $A(q)$ are known, but lower bounds are available which complement the general upper bound (3). Serre [20], [22] showed by methods of class field theory that $A(q) \geq c \log q$ with an absolute constant $c > 0$, and an effective version and an alternative proof of this result were recently given by Niederreiter and Xing [16]. Better lower bounds are known for special nonsquares q. For instance, Zink [35] proved that

$$A(p^3) \geq \frac{2(p^2-1)}{p+2} \quad \text{for all primes } p.$$

More generally, for composite nonsquares q the following bounds in Theorem 1 below were recently established by Niederreiter and Xing [13], [15]. It is convenient to state these results as bounds for $A(q^m)$. For a real number u we write $\lceil u \rceil$ for the least integer $\geq u$.

Theorem 1 *Let q be a prime power and $m \geq 3$ an odd integer. Then*

$$A(q^m) \geq \frac{2q+2}{\lceil 2(2q+3)^{1/2} \rceil + 1} \quad \textit{if } q \textit{ is odd},$$

$$A(q^m) \geq \frac{q+1}{\lceil 2(2q+2)^{1/2} \rceil + 2} \quad \textit{if } q \geq 4 \textit{ is even}.$$

If $q = p^e$ with a prime p and an odd integer $e \geq 3$, then Theorem 1 yields a lower bound for $A(q)$ of the order of magnitude $q^{1/(2m)}$, where m is the least prime factor of e. Further refinements of the bounds in Theorem 1 can be found in Niederreiter and Xing [16]. The following result of Niederreiter and Xing [13] yields the currently best lower bounds on $A(q)$ for $q = 2, 3, 5$.

Theorem 2 $A(2) \geq \frac{81}{317} = 0.2555\ldots$, $A(3) \geq \frac{62}{163} = 0.3803\ldots$, $A(5) \geq \frac{2}{3}$.

4 A Counterexample to Perret's Conjecture

Perret [19] described a method of obtaining lower bounds for $A(q)$ which is based on infinite ramified class field towers. However, this method depends on a conjecture which would provide a sufficient condition for the infinitude of certain ramified class field towers. In this section we show by a counterexample that this conjecture is wrong. Before we can state Perret's conjecture, we need further notation and terminology.

Let $K/\mathbb{F}_q$ be a global function field, let D be a positive divisor of K, and let S be a nonempty set of rational places of K with S disjoint from the support supp(D) of D. For a given prime l, the *(l, D, S)-class field* K_1 of K is the maximal abelian extension of K (in a fixed separable closure of K) with a Galois group of exponent 1 or l such that the global conductor of K_1/K divides D and all places in S split completely in K_1/K. If

$$D = \sum_{P \in \text{supp}(D)} m_P P$$

with positive integers m_P, then we define the positive divisor D_1 of K_1 by

$$D_1 = \sum_{P \in \text{supp}(D)} \sum_{Q|P} m_P Q.$$

We let S_1 be the set of all places of K_1 lying over those in S. Now we iterate the construction above by letting K_2 be the (l, D_1, S_1)-class field of K_1, and so on. In this way we obtain the tower

$$K = K_0 \subseteq K_1 \subseteq K_2 \subseteq \dots,$$

which is called the *(l, D, S)-class field tower* of K. This tower is called *infinite* if $K_n \neq K_{n+1}$ for all $n \geq 0$. The following result provides a lower bound for the quantity $A(q)$ in Definition 2.

Lemma 1 *Let $K/\mathbb{F}_q$ be a global function field with $g(K) > 1$, let p be the characteristic of $\mathbb{F}_q$, and let D be a positive divisor of K whose support consists of a single place. If the (p, D, S)-class field tower of K is infinite for some nonempty set S of rational places of K with S disjoint from supp(D), then*

$$A(q) \geq \frac{2|S|}{2g(K) + \deg(D) - 2}.$$

Proof: The degree of the different $\text{Diff}(K_n/K)$ of the extension K_n/K satisfies

$$\deg(\text{Diff}(K_n/K)) \leq ([K_n : K] - 1)\deg(D)$$

by [19, Théorème 1 and Proposition 4]. Therefore the Hurwitz genus formula yields

$$\begin{aligned} 2g(K_n) - 2 &= [K_n : K](2g(K) - 2) + \deg(\text{Diff}(K_n/K)) \\ &\leq [K_n : K](2g(K) + \deg(D) - 2) - \deg(D). \end{aligned}$$

Hence we get

$$\begin{aligned} A(q) &\geq \limsup_{n\to\infty} \frac{2N(K_n)}{2g(K_n)} \\ &\geq \lim_{n\to\infty} \frac{2|S|[K_n : K]}{[K_n : K](2g(K) + \deg(D) - 2) - \deg(D) + 2} \\ &= \frac{2|S|}{2g(K) + \deg(D) - 2}. \end{aligned}$$

□

Let $K/\mathbb{F}_q$, D, S, and l be as at the beginning of this section. For an abelian group G we write $d_l G$ for the l-rank of G. The following conjecture was stated in [19, Conjecture 1].

Perret's Conjecture. *If*

$$d_l G + |S| \leq \frac{1}{4}(d_l G)^2$$

with $G = \text{Gal}(K_1/K)$, then the (l, D, S)-class field tower of K is infinite.

Our construction of a counterexample to Perret's conjecture is based on the theory of narrow ray class fields; see [3, Section 16], [8, Section 2] for convenient summaries of this theory. Let $K/\mathbb{F}_q$ be a global function field with $N(K) \geq 1$ and distinguish a rational place ∞ of K. Let A be the ring of elements of K that are regular outside ∞. The *Hilbert class field* H_A is the maximal unramified abelian extension of K (in a fixed separable closure of K) in which ∞ splits completely. Let M be a nonzero integral ideal of A and $\Lambda(M)$ the corresponding M-torsion module stemming from the action of a sign-normalized Drinfeld A-module of rank 1 defined over H_A. Then $E_M = H_A(\Lambda(M))$ is the *narrow ray class field over K with modulus M*.

Now we take $q = 2$ and we consider specifically a global function field $K/\mathbb{F}_2$ with $g(K) = 6$ and $N(K) = 10$; see [6, Example 6] for an explicit construction of such a function field. Distinguish a rational place ∞ of K and let the ring A be as above. By [34, Lemma 8] or by inspection, there exists a place P of K of degree 18. Now let E_M be the narrow ray class field over K with modulus $M = P^2$. Note that the place ∞ splits completely in E_M/K by the theory of narrow ray class extensions. Furthermore, we have

$$d_2\mathrm{Gal}(E_M/K) \geq d_2\mathrm{Gal}(E_M/H_A) = d_2(A/M)^* = 18,$$

where the last identity follows from the proof of [8, Theorem 3]. Since a subgroup of $\mathrm{Gal}(E_M/K)$ generated by 9 Artin symbols has 2-rank at most 9, it follows that there exists a subfield L of E_M/K such that all 10 rational places of K split completely in L/K and $\mathrm{Gal}(L/K) \simeq (\mathbf{Z}/2\mathbf{Z})^9$. Next we need the following result.

Lemma 2 *With the notation above, the global conductor of L/K divides $2P$.*

Proof: By the theory of narrow ray class extensions, P is the only possible ramified place in L/K. Thus, by [19, Proposition 1] it remains to show that $|G_2| = 1$, where G_i is the i-th ramification group of P in L/K. Let d be the different exponent and e the ramification index of P in L/K and let a be the least integer $k \geq 0$ such that $|G_i| = 1$ for all $i \geq k$. From [15, Lemmas 1 and 2] we deduce that

$$c := \frac{d+a}{e} \leq 2.$$

Note that $e = 2^r$ with $0 \leq r \leq 9$. If $r = 0$, then P is unramified in L/K, and so already $|G_0| = 1$. If $r \geq 1$, then $|G_0| = |G_1| = 2^r$ by ramification theory. If we had $|G_2| > 1$, then the Hilbert different formula shows that

$$d > (2^r - 1) + (2^r - 1) = 2^{r+1} - 2.$$

Also $a \geq 3$, and so

$$c = \frac{d+a}{e} > \frac{2^{r+1}+1}{2^r} > 2,$$

which is a contradiction. □

Altogether, it follows that if S is the set of all 10 rational places of K and if K_1 is the $(2, 2P, S)$-class field of K, then $L \subseteq K_1$. This implies that

$$d_2\mathrm{Gal}(K_1/K) \geq d_2\mathrm{Gal}(L/K) = 9,$$

and so the condition in Perret's conjecture is satisfied. Hence the validity of this conjecture would imply that the $(2, 2P, S)$-class field tower of K is infinite. But then Lemma 1 would yield

$$A(2) \geq \frac{2|S|}{2g(K) + \deg(2P) - 2} = \frac{10}{23} > \sqrt{2} - 1,$$

and this contradicts the bound (3).

This counterexample demonstrates that Perret's conjecture is not valid in general. Consequently, Conjecture 1' in [19] is also wrong and the lower bounds on $A(q)$ in [19, Section III] remain unproved.

5 Constructions of Curves with many Rational Points

In this section we present a brief survey of constructions of algebraic curves over finite fields with many rational points. Since earlier surveys are available in [11], [26], we concentrate on recent construction techniques. It will again be convenient to adopt the equivalent viewpoint of global function fields.

For practical applications of global function fields with many rational places it is often desirable to have the function fields available in explicit form, i.e., in terms of generators over the rational function field and defining equations. Explicit constructions are usually based on special extensions such as Artin-Schreier extensions and Kummer extensions or on cyclotomic function fields. Note that a cyclotomic function field is the special case of a narrow ray class field (see Section 4) in which the base field $K/\mathbb{F}_q$ is the rational function field $\mathbb{F}_q(x)$. A discussion of such explicit constructions and some examples are given in [11, Section 3]. Very recent work on explicit constructions was done by van der Geer and van der Vlugt [27], [28].

A method that has recently been used to obtain global function fields with many rational places is based on Hilbert class fields (see Section 4 for the definition). We refer e.g. to the paper [7] of the authors. It is not so much the Hilbert class field H_A itself that is suitable here, but one has to construct certain subextensions of H_A/K in which one forces rational places of K to split completely. The method works particularly well if the base field K has a relatively small genus and a relatively large divisor class number.

Powerful techniques for the construction of global function fields with many rational places can be based on the theory of narrow ray class extensions. Again, it is not so much the narrow ray class extension itself that is used, but rather certain specially constructed subfields thereof. These methods go back to the papers [8], [33], [34] of the authors. A general method of this type for the case where the constant field is not a prime field was recently introduced in [17]. This method contains earlier constructions, such as the first two constructions in [8], as special cases. The following Theorem 3 summarizes the method.

For a base field $F/\mathbb{F}_q$ we write $F_r = \mathbb{F}_{q^r} \cdot F$ for the constant field extension of F with full constant field $\mathbb{F}_{q^r}$. Furthermore, $h(K)$ denotes the divisor class number of a global function field K.

Theorem 3 *Let $F/\mathbb{F}_q$ be a global function field of genus $g(F)$. For an integer $r \geq 2$ let t be a positive divisor of $q^r - 1$ and put $s = \gcd(q-1, t)$. Suppose that F has a place of degree d with $\gcd(d, r) = 1$ and that $N(F) \geq 1 + \varepsilon_d$, where $\varepsilon_d = 1$ if $d = 1$ and $\varepsilon_d = 0$ if $d \geq 2$. Then for every integer $n \geq 1$ there exists a global function field $K_n/\mathbb{F}_{q^r}$ with*

$$2g(K_n) - 2 = \frac{s(q^r-1)h(F_r)}{t(q-1)h(F)}$$

$$\left(\frac{(q-1)(q^{dr}-1)}{(q^d-1)(q^r-1)} q^{d(r-1)(n-1)}(2g(F) + dn - 2)\right.$$

$$\left. - \frac{d(q-1)(q^{dr}-1)(q^{d(r-1)(n-1)}-1)}{(q^d-1)(q^r-1)(q^{d(r-1)}-1)} - d\right)$$

$$+ \left(\frac{s(q^r-1)}{t(q-1)} - 1\right)\frac{h(F_r)}{h(F)} d + \left(\frac{s(q^r-1)}{t(q-1)} - 1\right)\frac{(q-1)(q^{dr}-1)h(F_r)}{(q^d-1)(q^r-1)h(F)}$$

$$q^{d(r-1)(n-1)}$$

and

$$N(K_n) \geq \frac{s(q^{dr}-1)h(F_r)}{t(q^d-1)h(F)} q^{d(r-1)(n-1)}(N(F) - 1 - \varepsilon_d)$$

$$+ \frac{(q-1)(q^{dr}-1)h(F_r)}{(q^d-1)(q^r-1)h(F)} q^{d(r-1)(n-1)} + \frac{h(F_r)}{h(F)}\varepsilon_d.$$

Theorem 3 has produced a large number of new examples of global function fields $K/\mathbb{F}_{q^r}$ with many rational places, particularly for the cases $q^r = 4, 8, 9, 16, 27$ (see [17]). Further special constructions of global function fields $K/\mathbb{F}_q$ with many rational places using narrow ray class extensions can be found in [34] for $q = 2$, in [10] for $q = 3$, in [8] for $q = 4$, in [12] for $q = 5$, and in [14] for $q = 8, 16$.

6 Applications to Algebraic Coding Theory

Lower bounds on the quantity $A(q)$ in Definition 2, such as those mentioned in Section 3, have important applications to algebraic coding theory. Due to the celebrated work of V.D. Goppa, algebraic curves over $\mathbb{F}_q$ with many rational points can be used to construct good linear codes over $\mathbb{F}_q$. For values of q for which $A(q)$ is larger than a known comparison function, Goppa's construction of algebraic-geometry codes leads to improvements on the classical Gilbert-Varshamov bound for the existence of good linear codes over $\mathbb{F}_q$. We refer to the books [23] and [24] for background on coding theory and algebraic-geometry codes.

For a linear code C over $\mathbb{F}_q$ we denote by $n(C)$, $k(C)$, and $d(C)$ the length, the dimension, and the minimum distance of C, respectively. Let U_q^{lin} be the set of ordered pairs $(\delta, R) \in [0, 1]^2$ for which there exists an infinite sequence $C_1, C_2, \ldots$ of linear codes over $\mathbb{F}_q$ with $n(C_i) \to \infty$ and

$$\delta = \lim_{i\to\infty} \frac{d(C_i)}{n(C_i)}, \quad R = \lim_{i\to\infty} \frac{k(C_i)}{n(C_i)}.$$

According to [24, Section 1.3.1], there exists a continuous function α_q^{lin} on $[0, 1]$ such that

$$U_q^{\text{lin}} = \{(\delta, R) : 0 \le R \le \alpha_q^{\text{lin}}(\delta), 0 \le \delta \le 1\},$$

where it is known that $\alpha_q^{\text{lin}}(0) = 1$, $\alpha_q^{\text{lin}}(\delta) = 0$ for $\delta \in [(q-1)/q, 1]$, and α_q^{lin} decreases on the interval $[0, (q-1)/q]$. The function α_q^{lin} is unknown, but it is an important issue in algebraic coding theory to obtain good lower bounds for α_q^{lin} on the interval $(0, (q-1)/q)$. The Gilbert-Varshamov bound says that

$$\alpha_q^{\text{lin}}(\delta) \ge R_{GV}(q, \delta) := 1 - H_q(\delta) \quad \text{for all } \delta \in \left(0, \frac{q-1}{q}\right),$$

where H_q is the q-ary entropy function

$$H_q(\delta) = \delta \log_q(q-1) - \delta \log_q \delta - (1-\delta)\log_q(1-\delta),$$

with $\log_q$ denoting the logarithm to the base q. Algebraic-geometry codes lead to the bound

$$\alpha_q^{\text{lin}}(\delta) \ge R_{AG}(q, \delta) := 1 - \frac{1}{A(q)} - \delta \quad \text{for all } \delta \in [0, 1].$$

A seminal result of Tsfasman, Vlădut, and Zink [25] shows that $R_{AG}(q, \delta) > R_{GV}(q, \delta)$ if q is a sufficiently large square and δ belongs to a suitable subinterval of $[0, 1]$. The proof of this theorem is based on the fact that $A(q) = q^{1/2} - 1$ if q is a square (see Section 3). It was open until recently whether an analogous result holds for nonsquares q. The following theorem of the authors [13], which is a consequence of Theorem 1 in Section 3, settles this problem for sufficiently large composite nonsquares q.

Theorem 4 *Let $m \ge 3$ be an odd integer and let r be a prime power with $r \ge 100m^3$ for odd r and $r \ge 576m^3$ for even r. Then there exists an open interval $(\delta_1, \delta_2) \subseteq (0, 1)$ containing $(r^m - 1)/(2r^m - 1)$ such that*

$$R_{AG}(r^m, \delta) > R_{GV}(r^m, \delta) \quad \textit{for all } \delta \in (\delta_1, \delta_2).$$

7 Applications to Low-Discrepancy Sequences

Low-discrepancy sequences are sequences of points in an s-dimensional unit cube $[0, 1]^s$ that are distributed very uniformly. Such sequences were studied in the last few decades by number theorists and they are used, for instance, in quasi-Monte Carlo methods for numerical integration over $[0, 1]^s$ (see the book [5] for a general background). The most powerful constructions of low-discrepancy sequences yield so-called digital (t, s)-sequences constructed over $\mathbb{F}_q$; for the precise definition we refer to the recent survey article [9]. Here s is, as above, the dimension in which the sequence lives and the integer $t \geq 0$ is the quality parameter of the sequence. The value of t should be as small as possible for a good low-discrepancy sequence.

Recent work of the authors has shown that global function fields over $\mathbb{F}_q$ with many rational places can be used to obtain the currently best digital (t, s)-sequences constructed over $\mathbb{F}_q$. The most general construction is that in [32] and it has the following ingredients. For a given q and a given dimension $s \geq 1$, let $K/\mathbb{F}_q$ be a global function field containing at least one rational place P_∞ and let D be a positive divisor of K with $\deg(D) = 2g(K)$ and P_∞ not in the support of D. If $P_1, \ldots, P_s$ are s distinct places of K with $P_i \neq P_\infty$ for $1 \leq i \leq s$, then we obtain a digital (t, s)-sequence constructed over $\mathbb{F}_q$ with

$$t = g(K) + \sum_{i=1}^{s} (\deg(P_i) - 1).$$

If at least one of the places P_i, say P_1, is rational, then we can simply take $D = 2g(K)P_1$. In the special case where $N(K) \geq s + 1$, we can choose all P_i, $1 \leq i \leq s$, to be rational places, and then by optimizing K we get

$$t = V_q(s) := \min\{g \geq 0 : N_q(g) \geq s + 1\}, \tag{4}$$

where $N_q(g)$ is as in Definition 1.

For any q and any dimension $s \geq 1$ let $d_q(s)$ be the least value of t such that there exists a digital (t, s)-sequence constructed over $\mathbb{F}_q$. In Table 3 below, which is reprinted from [18, Table 5], we tabulate upper bounds for $d_q(s)$ for $q = 2, 3, 5$ and $1 \leq s \leq 50$. Most of the bounds in Table 3 are obtained from the trivial inequality $d_q(s) \leq V_q(s)$ implied by (4) and from upper bounds for $V_q(s)$ that can be read off immediately from Tables 1 and 2.

Table 1 Bounds for $N_q(g)$

$g\backslash q$	2	3	4	5	8	9	16	27
1	5	7	9	10	14	16	25	38
2	6	8	10	12	18	20	33	48
3	7	10	14	16	24	28	38	56
4	8	12	15	18	25–29	30	45–47	64–68
5	9	12–14	17–18	20–22	29–32	32–36	49–55	55–78
6	10	14–15	20	21–25	33–36	35–40	65	76–88
7	10	16–17	21–22	22–27	33–39	39–43	63–70	64–98
8	11	15–18	21–24	22–29	34–43	38–47	61–76	92–108
9	12	19	26	26–32	45–47	40–51	72–81	82–118
10	13	19–21	27–28	27–34	38–50	54–55	81–87	82–128

(Table 1 Contd.)

$g\backslash q$	2	3	4	5	8	9	16	27
11	14	20–22	26–30	32–36	48–54	55–59	80–92	96–138
12	14–15	22–24	28–31	30–38	49–57	55–63	68–97	109–148
13	15	24–25	33	36–40	50–61	60–66	97–103	136–156
14	15–16	24–26	32–35	39–43	65	56–70	97–108	84–164
15	17	28	33–37	32–45	54–68	64–74	98–113	82–171
16	16–18	27–29	36–38	40–47	56–71	74–78	93–118	81–178
17	17–18	24–30	40	42–49	61–74	56–82	96–124	128–185
18	18–19	26–31	41–42	32–51	65–77	46–85	93–129	94–192
19	20	27–32	36–43	41–54	58–80	84–88	121–134	126–199
20	19–21	30–34	36–45	30–56	68–83	48–91	121–140	133–207
21	21	32–35	41–47	48–58	72–86	82–95	129–145	163–214
22	21–22	28–36	40–48	51–60	66–89	78–98	129–150	112–221
23	22–23	26–37	40–50		68–92	92–101	126–155	114–228
24	20–23	28–38	42–52		66–95	91–104	129–161	166–235
25	24	36–40	51–53		66–97	64–108	144–166	196–242
26	24–25	36–41	55		72–100	110–111	150–171	108–249
27	22–25	39–42	49–56		96–103	60–114	128–176	114–256
28	24–26	37–43	44–58		97–106	105–117	136–181	108–263
29	25–27	42–44	49–60		97–109	104–120	130–187	114–270
30	24–27	34–46	53–61		80–112	60–123	144–192	117–277
31	27–28	40–47	60–63		72–115	84–127	150–197	114–284
32	26–29	38–48	52–65		72–118	81–130	132–202	126–291
33	28–29	37–49	65–66		92–121	78–133	153–207	220–298
34	27–30	44–50	57–68		80–124	111–136	156–213	135–305
35	28–31	47–51	58–69		106–127	84–139	144–218	126–312
36	30–31	46–52	64–71		105–130	110–142	185–223	244–319
37	28–32	48–54	66–72		121–132	120–145	208–228	162–326
38	28–33	36–55	56–74		129–135	105–149	152–233	144–333
39	33	46–56	65–75		117–138	84–152	160–239	271–340
40	32–34	54–57	75–77		100–141	90–155	162–244	244–346
41	32–35	50–58	65–78		112–144	84–158	216–249	153–353
42	30–35	39–59	66–80		129–147	90–161	162–254	280–360
43	33–36	55–60	72–81		100–150	120–164	226–259	196–367
44	32–37	42–61	68–83		129–153	90–167	162–264	153–374
45	32–37	48–62	80–84		144–156	112–170	242–268	171–381
46	34–38	55–63	81–86		129–158	138–173	243–273	162–388
47	36–38	47–65	68–87		120–161	154–177	176–277	174–395
48	34–39	55–66	77–89		126–164	163–180	184–282	244–402
49	36–40	63–67	81–90		130–167	168–183	192–286	268–409
50	40	56–68	91–92		130–170	182–186	200–291	180–416

Table 2 Bounds for $N_2(g)$

g	51	52	53	54	55	56	57	58	59
$N_2(g)$	36–41	34–42	40–42	42–43	36–43	38–44	40–45	40–45	40–46
g	60	61	62	63	64	65	66	67	68
$N_2(g)$	40–47	40–47	44–48	42–48	42–49	48–50	42–50	44–51	45–51
g	69	70	71	72	73	74	75	76	77
$N_2(g)$	49–52	46–53	44–53	48–54	48–54	48–55	48–56	50–56	52–57
g	78	79	80	81	82	83	84	85	86
$N_2(g)$	48–57	52–58	56–59	48–59	53–60	52–60	57–61	52–62	56–62
g	87	88	89	90	91	92	93	94	95
$N_2(g)$	56–63	56–63	56–64	56–65	54–65	60–66	56–66	56–67	65–68

Table 3 Upper bounds for $d_q(s)$

$q\backslash s$	1	2	3	4	5	6	7	8	9	10	11	12	13
2	0	0	1	1	2	3	4	5	6	8	9	10	11
3	0	0	0	1	1	1	2	3	3	4	4	5	6
5	0	0	0	0	0	1	1	1	1	2	2	3	3
$q\backslash s$	14	15	16	17	18	19	20	21	22	23	24	25	26
2	13	15	15	18	19	19	21	23	25	25	29	31	31
3	7	7	9	9	9	11	12	12	13	13	15	15	15
5	3	3	4	4	5	5	6	7	8	9	9	9	10
$q\backslash s$	27	28	29	30	31	32	33	34	35	36	37	38	
2	33	36	36	39	39	39	45	47	47	50	50	50	
3	15	20	20	21	21	25	25	25	25	26	27	27	
5	11	11	11	11	11	13	13	13	13	14	14	14	
$q\backslash s$	39	40	41	42	43	44	45	46	47	48	49	50	
2	50	54	54	59	62	65	65	65	65	69	75	77	
3	29	29	29	33	34	35	35	35	37	40	40	40	
5	16	17	17	18	21	21	21	21	21	22	22	22	

References

[1] M. Deuring, Die Typen der Multiplikatorenringe elliptischer Funktionenkörper, *Abh. Math. Sem. Univ. Hamburg* **14** (1941), 197–272.

[2] R. Fuhrmann and Torres, F., The genus of curves over finite fields with many rational points, *Manuscripta Math.* **89** (1996), 103–106.

[3] D.R. Hayes, A brief introduction to Drinfeld modules, *The Arithmetic of Function Fields* (D. Goss, D.R. Hayes and M.I. Rosen, eds.), pp. 1–32, W. de Gruyter, Berlin, 1992.

[4] Y. Ihara, Some remarks on the number of rational points of algebraic curves over finite fields, *J. Fac. Sci. Univ. Tokyo Sect. IA Math.* **28** (1981), 721–724.

[5] H. Niederreiter, *Random Number Generation and Quasi-Monte Carlo Methods*, SIAM, Philadelphia, 1992.

[6] H. Niederreiter and Xing, C.P., Explicit global function fields over the binary field with many rational places, *Acta Arith.* **75** (1996), 383–396.

[7] H. Niederreiter and Xing, C.P., Cyclotomic function fields, Hilbert class fields, and global function fields with many rational places, *Acta Arith.* **79** (1997), 59–76.

[8] H. Niederreiter and Xing, C.P., Drinfeld modules of rank 1 and algebraic curves with many rational points. II, *Acta Arith.* **81** (1997), 81–100.

[9] H. Niederreiter and Xing, C.P., The algebraic-geometry approach to low-discrepancy sequences, *Monte Carlo and Quasi-Monte Carlo Methods 1996* (H. Niederreiter *et al.*, eds.), Lecture Notes in Statistics, vol. **127**, pp. 139–160, Springer, New York, 1997.

[10] H. Niederreiter and Xing, C.P., Global function fields with many rational places over the ternary field, *Acta Arith.* **83** (1998), 65–86.

[11] H. Niederreiter and Xing, C.P., Algebraic curves over finite fields with many rational points, *Proc. Number Theory: Diophantine, Computational and Algebraic Aspects* (K. Györy *et al.*, eds.), pp. 423–443, W. de Gruyter, Berlin, 1998.

[12] H. Niederreiter and Xing, C.P., Global function fields with many rational places over the quinary field, *Demonstratio Math.* **30** (1997), 919–930.

[13] H. Niederreiter and Xing, C.P., Towers of global function fields with asymptotically many rational places and an improvement on the Gilbert-Varshamov bound, *Math. Nachr.* **195** (1998), 171–186.

[14] H. Niederreiter and Xing, C.P., Algebraic curves with many rational points over finite fields of characteristic 2, *Number Theory in Progress* (K. Györy *et al.*, eds.), pp. 359–380, W. de Gruyter, Berlin, 1999.

[15] H. Niederreiter and Xing, C.P., Global function fields with many rational places and their applications, *Finite Fields: Theory, Applications, and Algorithms* (R.C. Mullin and G.L. Mullen, eds.), Contemporary Math., vol. **225**, pp. 87–111, American Math. Society, Providence, R.I., 1999.

[16] H. Niederreiter and Xing, C.P., Curve sequences with asymptotically many rational points, *Proc. AMS Summer Research Conf.* (Seattle, 1997), to appear.

[17] H. Niederreiter and Xing, C.P., A general method of constructing global function fields with many rational places, *Algorithmic Number Theory* (J.P. Buhler, ed.), Lecture Notes in Computer Science, vol. **1423**, pp. 555–566, Springer, Berlin, 1998.

[18] H. Niederreiter and Xing, C.P., Nets, (t, s)-sequences, and algebraic geometry, *Random and Quasi-Random Point Sets* (P. Hellekalek and G. Larcher, eds.), Lecture Notes in Statistics, vol. **138**, pp. 267–302, Springer, New York, 1998.

[19] M. Perret, Tours ramifiées infinies de corps de classes, *J. Number Theory* **38** (1991), 300–322.

[20] J.-P. Serre, Sur le nombre des points rationnels d'une courbe algébrique sur un corps fini, *C.R. Acad. Sci. Paris Sér. I Math.* **296** (1983), 397–402.

[21] J.-P. Serre, Nombres de points des courbes algébriques sur $\mathbb{F}_q$, *Sém. Théorie des Nombres 1982–1983*, Exp. 22, Université de Bordeaux I, Talence, 1983.

[22] J.-P. Serre, *Rational Points on Curves over Finite Fields*, Lecture Notes, Harvard University, 1985.

[23] H. Stichtenoth, *Algebraic Function Fields and Codes*, Springer, Berlin, 1993.

[24] M.A. Tsfasman and Vlădut, S.G., *Algebraic-Geometric Codes*, Kluwer, Dordrecht, 1991.

[25] M.A. Tsfasman, Vlădut, S.G. and Zink, T., Modular curves, Shimura curves, and Goppa codes, better than Varshamov-Gilbert bound, *Math. Nachr.* **109** (1982), 21–28.

[26] G. van der Geer and van der Vlugt, M., How to construct curves over finite fields with many points, *Arithmetic Geometry* (F. Catanese, ed.), pp. 169–189, Cambridge University Press, Cambridge, 1997.

[27] G. van der Geer and van der Vlugt, M., Generalized Reed-Muller codes and curves with many points, preprint, 1997.

[28] G. van der Geer and van der Vlugt, M., Constructing curves over finite fields with many points by solving linear equations, preprint, 1997.

[29] S.G. Vlădut and Drinfeld, V.G., Number of points of an algebraic curve, *Functional Analysis Appl.* **17** (1983), 53–54.

[30] W.C. Waterhouse, Abelian varieties over finite fields, *Ann. Sci. Ecole Norm. Sup.* (4) **2** (1969), 521–560.

[31] C.P. Xing, Maximal function fields and function fields with many rational places over finite fields of characteristic 2, preprint, 1997.

[32] C.P. Xing and Niederreiter, H., A construction of low-discrepancy sequences using global function fields, *Acta Arith.* **73** (1995), 87–102.

[33] C.P. Xing and Niederreiter, H., Modules de Drinfeld et courbes algébriques ayant beaucoup de points rationnels, *C.R. Acad. Sci. Paris Sér. I Math.* **322** (1996), 651–654.

[34] C.P. Xing and Niederreiter, H., Drinfeld modules of rank 1 and algebraic curves with many rational points, *Monatsh. Math.* **127** (1999), 219–241.

[35] T. Zink, Degeneration of Shimura surfaces and a problem in coding theory, *Fundamentals of Computation Theory* (L. Budach, ed.), Lecture Notes in Computer Science, vol. **199**, pp. 503–511, Springer, Berlin, 1985.

Institute of Discrete Mathematics
Austrian Academy of Sciences
Sonnenfelsgasse 19
A-1010 Vienna
Austria
E-Mail: niederreiter@oeaw.ac.at

Department of Mathematics
National University of Singapore
2 Science Drive 2
Singapore 117543
E-Mail: xingcp@math.nus.edu.sg

A Report on Artin's Holomorphy Conjecture

Dipendra Prasad and C S Yogananda

1 Introduction

The purpose of this paper is to present a report on the current status of Artin's holomorphy conjecture. For a fascinating account of how Artin was led to defining his L-series and his 'reciprocity law' see [19].

2 Preliminaries

2.1 Definition and Properties of Artin's L-series

We begin with the definition of Artin's L-series. Let L/K be a finite normal extension of number fields with Galois group G. For a prime $\mathfrak{p}$ of K and $\mathfrak{q}$ a prime in L above $\mathfrak{p}$ let $G_{\mathfrak{q}}$ denote the decomposition subgroup and $I_{\mathfrak{q}}$ the inertia subgroup of G corresponding to $\mathfrak{q}$:

$$G_{\mathfrak{q}} = \{g \in G | g(\mathfrak{q}) = \mathfrak{q}\}; \quad I_{\mathfrak{q}} = \{g \in G_{\mathfrak{q}} | g(x) \equiv x \pmod{\mathfrak{q}}\}.$$

Let $\sigma_{\mathfrak{q}}$ denote the canonical generator, the Frobenius at $\mathfrak{q}$, of the cyclic group $G_{\mathfrak{q}}/I_{\mathfrak{q}}$. Note that if $\mathfrak{q}'$ is another prime in L above $\mathfrak{p}$ then the corresponding decomposition group, inertia group and Frobenius are conjugates in G of $G_{\mathfrak{q}}$, $I_{\mathfrak{q}}$ and $\sigma_{\mathfrak{q}}$. Let $\mathbf{N}$ denote the norm from K to $\mathbf{Q}$. If a group G acts on a vector space $\mathbf{V}$ and H is a subgroup of G then let $\mathbf{V}^H$ denote the subspace of H-fixed vectors of $\mathbf{V}$, i.e., $\mathbf{V}^H = \{v \in \mathbf{V} | h(v) = v, \forall h \in H\}$.

Definition: Suppose $\rho : G \to \mathrm{GL}(\mathbf{V})$ is a representation of G on an n-dimensional complex vector space $\mathbf{V}$. Following Artin one can associate an L-series to ρ: for $\Re(s) > 1$ put

$$L(s, \rho) = \prod_{\mathfrak{p}} \frac{1}{\det(1 - \rho|_{\mathbf{V}^{I_{\mathfrak{q}}}}(\sigma_{\mathfrak{q}})\mathbf{N}\mathfrak{p}^{-s})},$$

the product ranging over all finite primes of K.

The product above converges for $\Re(s) > 1$ and hence defines an analytic function in that region. Since determinant remains invariant under conjugation, the local factors do not depend upon the choice of the prime $\mathfrak{q}$ in L above $\mathfrak{p}$. Further, since two isomorphic representations have the same determinant, $L(s, \rho)$ depends only on the character χ of the representation ρ so that we may as well write $L(s, \chi)$ instead of $L(s, \rho)$. In fact, Artin gave the following explicit expression of $L(s, \rho)$ which is written purely in terms of the character of a representation:

$$\log L(s, \chi) = \sum_{\mathfrak{p}} \sum_{n=1}^{\infty} \frac{\chi(\sigma_{\mathfrak{q}}^n)}{n \cdot \mathbf{N}\mathfrak{p}^{ns}} \quad \text{where} \quad \chi(\sigma_{\mathfrak{q}}^n) = \frac{1}{|I_{\mathfrak{q}}|} \sum_{\tau \in \sigma_{\mathfrak{q}}^n I_{\mathfrak{q}}} \chi(\tau).$$

Examples:

1. Suppose χ_0 is the trivial character of G. Then the Artin L-series $L(s, \chi_0)$ is nothing but the Dedekind zeta function $\zeta_K(s)$ of K.
2. Suppose χ_R is the regular representation of G. Then the Artin L-series $L(s, \chi_R)$ is the Dedekind zeta function $\zeta_L(s)$ of L. More generally, if H is a subgroup of G with M the subfield of L fixed by H and χ_{R_H} is the character of the left-regular representation of G on functions on G/H then $L(s, \chi_{R_H})$ is the Dedekind zeta function $\zeta_M(s)$ of M.

Properties:

1. If ρ_1, ρ_2 are two representations with characters χ_1, χ_2 then

$$L(s, \chi_1 + \chi_2) = L(s, \chi_1)L(s, \chi_2).$$

   This enables one to define an Artin L-series for virtual characters of G. For example, (with $L/M/K$ as above) $\zeta_L(s)/\zeta_M(s)$ is the Artin L-series associated to the character $\chi_{R_H} - \chi_0$ of G.
2. If H is a subgroup of G and χ a character of H then

$$L(s, \text{Ind}\,\chi) = L(s, \chi)$$

   where Ind χ is the character of G induced from the character χ of H.
3. If H is a normal subgroup of G with $\lambda : G \to G/H$, the canonical map and χ a character of the quotient group G/H then

$$L(s, \text{Infl}\,\chi) = L(s, \chi)$$

   where Infl χ is the character $\lambda \circ \chi$ of G.

2.2 Artin's Conjecture

Conjecture 2.1 If ρ does not contain the trivial representation then $L(s, \rho)$ has an analytic continuation as a holomorphic function to the whole of the complex plane.

A consequence of this conjecture is that for any finite extension of number fields M/K, $\zeta_K(s)$ *divides* $\zeta_M(s)$, i.e., $\zeta_M(s)/\zeta_K(s)$, which is the Artin L-series of the character $\chi_{R_H} - \chi_0$, should be a holomorphic function on $\mathbf{C}$. (Here L is a Galois extension of K containing M with Galois group G and H is the subgroup of G fixing M.)

This statement is known as *Dedekind's conjecture*. In fact, Artin seems to have been led to his L-series while trying to prove this statement which was proved for pure cubic fields by Dedekind in 1873. Artin himself was able to prove such a result when M/K is an Icosahedral extension (and also for some intermediate extensions therein), [1]. In the case of a normal extension L/K Dedekind's conjecture was proved without assuming Artin's conjecture by R. Brauer, [5] and independently by Aramata, (see 3.2 below). In the direction of Dedekind's conjecture in the non-normal case Uchida, [22], and van der Waal, [23], (independently) have proved the following.

Theorem 2.1 (Uchida, van der Waal). *Let M/K be an extension of number fields and $\tilde{M}$ a normal closure of M. Suppose that* $\mathrm{Gal}(\tilde{M}/K)$ *is solvable. Then $\zeta_M(s)/\zeta_K(s)$ is entire.*

3 Meromorphicity

3.1 Artin's Reciprocity Law

In [2] Artin had conjectured that if G is abelian then $L(s, \rho)$ is nothing but a L-series associated to certain characters introduced and studied by E. Hecke in 1917, [12]; Hecke had proved analytic continuation and functional equation for these abelian L-series.

Hecke characters are idele class characters $\chi : C_K \to \mathbf{C}^*$ while Artin's L-series are associated to (characters of) representations of $\mathrm{Gal}(L/K)$. Thus, in order to identify his L-series with Hecke L-series for L/K abelian, Artin needed an identification of the abelian Galois group $\mathrm{Gal}(L/K)$ with a quotient of the idele class group such that for a prime $\mathfrak{p}$ of K and $\mathfrak{q}$ a prime in L above $\mathfrak{p}$ the Frobenius at $\sigma_{\mathfrak{q}}$ corresponds to a uniformising parameter at $\mathfrak{p}$ in C_K. In 1927 Artin proved this 'reciprocity law' in [3] thereby proving his conjecture in the case of abelian G.

3.2 Representations of Finite Groups and Meromorphicity

In 1930, [4] Artin proved the following general result about representations of finite groups.

Theorem 3.1 (Artin). *Every character of a finite group G is a linear combination with rational coefficients of induced characters from cyclic subgroups.*

This shows that an integral power of $L(s, \rho)$ is a product of Hecke L-series and hence is meromorphic. In the case of the regular representation of G, ρ_R, Brauer was able to prove the following, [5], pp. 244.

Lemma 3.1 (Brauer). *The character $\chi_R - \chi_0$ of G can be expressed as a linear combination, with* positive *rational coefficients with denominator $|G|$, of characters of G which are induced by nontrivial irreducible characters of cyclic subgroups.*

This implies that some integral power of $\zeta_L(s)/\zeta_K(s)$, which is the Artin L-series for the character $\chi_R - \chi_0$ of G, is a product of abelian (Hecke) L-series corresponding to nontrivial characters. Since $\zeta_L(s)/\zeta_K(s)$ is meromorphic and Artin's conjecture is known to be true for abelian extensions we have the

Theorem 3.2 (Aramata-Brauer). *The quotient $\zeta_L(s)/\zeta_K(s)$ is entire.*

In [6] R. Brauer was able to improve on Artin's theorem; he proved the following theorem.

Theorem 3.3 (Brauer). *Every character of a finite group G is a linear combination with integer coefficients of characters induced from one dimensional characters of subgroups.*

This showed that $L(s, \rho)$ itself is meromorphic and proved Artin's conjecture for those representations which are *positive* integral linear combinations of representations which are induced from one-dimensional representations, and hence also for those representations which are positive rational linear combinations of representations which are induced from one-dimensional representations.

In [17] Stark proved the following result. The proof is so simple and elegant that we have decided to include it.

Theorem 3.4 (Stark). *If $Ord_{s=s_0}(\zeta_L(s)) \leq 1$ then for any character χ of the Galois group G of L over K, $L(s, \chi)$ is analytic at the point $s = s_0$.*

Proof: Let χ be an irreducible representation of the Galois group G of L over K, and ψ an irreducible representation of a subgroup H of G. (The subgroup H will be defined later.) Let n_χ and n_ψ denote the order of zeros (or poles, counted with negative multiplicity) of $L(s_0, \chi)$ and $L(s_0, \psi)$ respectively. Define generalised characters,

$$\begin{aligned} \theta_G &= \sum_\chi n_\chi \chi \\ \theta_H &= \sum_\psi n_\psi \psi. \end{aligned}$$

We will prove that θ_G is an irreducible character of dimension 1, and therefore $n_\chi \geq 0$, proving the theorem. For this purpose, let

$$\mathrm{Ind}_H^G \psi = \sum_\chi a(\psi, \chi)\chi.$$

Since $L(s, \psi) = L(s, \mathrm{Ind}_H^G \psi)$,

$$n_\psi = \sum_\chi a(\psi, \chi) n_\chi.$$

By Frobenius reciprocity,

$$\chi|_H = \sum_\psi a(\psi, \chi)\psi.$$

Therefore,

$$\theta_G|_H = \sum_{\chi, \psi} n_\chi a(\psi, \chi)\psi = \sum n_\psi \psi = \theta_H \qquad (*).$$

Now let $g \in G$, and H the cyclic subgroup of G generated by g. Since H is an abelian group, Artin's conjecture is true for representations of H. Factorisation of the zeta function of L as the product of the L-functions attached to characters of H and the fact that $Ord_{s_0}(\zeta_L(s)) \leq 1$ implies that $\theta_H = \psi$ for some linear character of H. From the equation (*), this means that $\theta_G(g)$ is a root of unity for all $g \in G$. By Schur orthogonality,

$$1 = \frac{1}{|G|} \sum |\theta_G(g)|^2 = \sum n_\chi^2.$$

Therefore, exactly one n_χ is non-zero. Now take $H = \{1\}$ in (*). This gives $\theta_G(1) = 1$, implying $n_\chi \cdot \chi(1) = 1$, which implies $n_\chi = 1$ and $\chi(1) = 1$.

Using the ideas in [17], Foote and Kumar Murty, [11], were able to reprove Artin's conjecture for some characters, already covered by Brauer's theorem, *without* writing the characters as positive linear combinations of monomial characters. They do this by comparing the order of zero or pole of the $L(s, \chi)$ at a fixed point, as χ varies over the irreducible characters of G, with the order of the Dedekind zeta function $\zeta_L(s)$ at the same point. They also proved the following □

Theorem 3.5 (Foote-Kumar Murty). *Let L/K be a Galois extension of number fields with soluble Galois group G and let $p_1 < p_2 < \cdots < p_n$ be the distinct prime divisors of $|G|$. If $\zeta_L(s)$ has a zero of order r at $s = s_0$, where $r \leq p_2 - 2$, then $L(s, \chi)$ is analytic at s_0 for all irreducible characters χ of G.*

In [14], Michler has explicitly described the relation between the orders of zeros of Dedekind zeta functions and the orders of zeros or poles of Artin L-functions at a point in $\mathbf{C} - \{1\}$ when the Galois group of the extension is S_n, the symmetric group of degree n. Suppose $\mathrm{Gal}(L/K) = S_n$. Denote the set of all partitions λ of n by $\mathcal{P}(n)$. The irreducible characters χ_μ of S_n are parametrised by the partitions $\mu \in \mathcal{P}(n)$. For each $\lambda \in \mathcal{P}(n)$ there is a unique Young subgroup Y_λ of S_n; let $L_\lambda = L^{Y_\lambda}$ be its fixed field. The main result of [14] asserts that for each point s_0 in $\mathbf{C} - \{1\}$ and each partition $\mu \in \mathcal{P}(n)$, the order $n_\mu(s_0)$ of zero or pole of the Artin L-series $L(s, \chi_\mu)$ is uniquely determined by the orders of zeros $r_\lambda(s_0)$ of the Dedekind zeta functions $\zeta_{L_\lambda}(s)$ of the fixed fields L_λ for $\lambda \in \mathcal{P}(n)$.

3.3 Functional Equation

It follows from Brauer's theorem (theorem 3.3) that Artin L-functions can be written as a quotient of product of Hecke L-functions. Since Hecke L-functions have meromorphic continuation to all of the complex plane, and a functional equation, so do Artin L-functions. In this section we state the functional equation satisfied by an Artin L-function. Before we can do that, we must recall the definition of Artin conductor and gamma factors associated to representations of the Galois group.

For every representation V of the Galois group $\mathrm{Gal}(\overline{\mathbf{Q}}/K)$, there is an integral ideal of K, denoted by $f(V)$ and called the Artin conductor of the representation V, which is divisible by exactly those primes in K which are ramified in the representation V. The Artin conductor has the property that $f(V_1 \oplus V_2) = f(V_1)f(V_2)$ for any two representations V_1 and V_2 of $\mathrm{Gal}(\overline{\mathbf{Q}}/K)$. For one dimensional representations of $\mathrm{Gal}(\overline{\mathbf{Q}}/K)$, the Artin conductor is the conductor of the corresponding Grossencharacter. There is a formula for Artin conductor involving the higher ramification groups for which we refer to the book of

Tate [20]. Here we just mention that if the representation V of $\mathrm{Gal}(\overline{\mathbf{Q}}/K)$ is induced from a representation χ of $\mathrm{Gal}(\overline{\mathbf{Q}}/M)$, where M is an extension of K, then one has

$$f(V) = D(M/K)^{\chi(1)}\,\mathrm{Norm}\,(f(\chi))$$

where $D(M/K)$ is the discriminant of M over K, and $f(\chi)$ is the Artin conductor of the character χ.

Suppose that the representation V of $\mathrm{Gal}(\overline{\mathbf{Q}}/K)$ factors through the Galois group of a field extension L of K. Let w be a place of L over an Archimedean place v of K. This defines an element in G of order 1 or 2 which in the representation V has let's say n_v^+, $+1$ eigenspaces, and n_v^-, -1 eigenspaces. If v is a real place, define a gamma factor

$$L_v(s, V) = [\pi^{-s/2}\Gamma(s/2)]^{n_v^+}[\pi^{-(s+1)/2}\Gamma((s+1)/2)]^{n_v^-}.$$

If v is a complex place, define $L_v(s, V) = [2\cdot(2\pi)^{-s}\Gamma(s)]^n$. With this notation, the *completed Artin L-function*

$$\Lambda(s,\chi) = \{|d_K|^{\chi(1)}N_{K/\mathbf{Q}}(f(\chi))\}^{s/2}\prod_{v|\infty}L_v(s,\chi)\cdot L(s,\chi)$$

satisfies the functional equation:

$$\Lambda(1-s,\chi) = W(\chi)\Lambda(s,\overline{\chi})$$

where $W(\chi)$ is a non-zero complex number of absolute value 1 called the *Artin root-number*.

4 Two-dimensional Representations

4.1 Dihedral, Tetrahedral and Octahedral Representations

If ρ is a 2-dimensional Galois representation over a number field k then ρ determines a faithful representation of $\mathrm{Gal}(K/k)$ where K is the finite normal extension of k corresponding to the kernel of the representation and therefore ρ corresponds to a finite subgroup of $GL(2,\mathbf{C})$. The finite subgroups of $GL(2,\mathbf{C})$ have been classified by F. Klein: their image in $PGL(2,\mathbf{C})$ is one of the following.

1. Cyclic.
2. Dihedral.
3. Tetrahedral (isomorphic to A_4).
4. Octahedral (isomorphic to S_4).
5. Icosahedral (isomorphic to A_5).

Thus any finite subgroup of $GL(2,\mathbf{C})$ is a cyclic central extension of one of the above projective groups. The inverse images of the subgroups A_4, S_4, A_5 of $SO(3,\mathbf{R})$ in $SU(2)$ which is the 2 fold cover of $SO(3,\mathbf{R})$:

$$0\longrightarrow \mathbf{Z}/2\longrightarrow SU(2)\longrightarrow SO(3,\mathbf{R})\longrightarrow 1,$$

are $\tilde{A}_4 = SL_2(\mathbf{Z}/3)$, $\tilde{S}_4 = GL_2(\mathbf{Z}/3)$, and $\tilde{A}_5 = SL_2(\mathbf{Z}/5)$. The group $\tilde{A}_4$ has representations of dimensions 1,1,1,3,2,2,2, of which the last 3 are non-trivial on $\mathbf{Z}/2$. The group $\tilde{S}_4$ has representations of dimensions 1,1,3,3,2,2,2,4, of which the last 3 are non-trivial on $\mathbf{Z}/2$. The group $\tilde{A}_5$ has representations of dimensions 1,3,3,5,4,2,2,4,6, of which the last 4 are non-trivial on $\mathbf{Z}/2$.

Artin's conjecture in the cyclic and dihedral case follows from Artin reciprocity and Hecke's proof that abelian L-series are entire (section 3.1). Langlands proved Artin's conjecture for tetrahedral and Tunnell for octahedral representations. Thus the only case where Artin's conjecture is not known over $\mathbf{Q}$ is the case of the icosahedral representations (section 5).

Langlands has in fact generalised Artin's conjecture to ask whether the L-function arising from an irreducible n-dimensional Galois representation of a number field k is in fact the L-function of a cusp form on $GL(n,k)$; because then by known properties of the L-function of automorphic representations of $GL(n,k)$, it will in particular follow that an Artin L-function is entire if $n > 1$. This generalisation of Artin's conjecture which asks if an Artin L-function is the L-function of an automorphic L-function is called *strong Artin conjecture*. Using the theory of base change, Langlands was able to prove such a statement for 2 dimensional tetrahedral representations, which was then extended to Octahedral case by Tunnell.

Theorem 4.1 (Langlands-Tunnell). *A two dimensional representation of* $\mathrm{Gal}(\overline{\mathbf{Q}}/k)$ *with values inside a finite solvable subgroup of* $GL(2,\mathbf{C})$ *comes from an automorphic representation of* $GL(2,\mathbf{A}_k)$.

It follows from the 'converse' theory developed by Weil and Langlands that Artin's conjecture for 2 dimensional Galois representations of $\mathrm{Gal}(\overline{\mathbf{Q}}/\mathbf{Q})$ implies that there is a natural one-to-one correspondence between equivalence classes of odd two-dimensional Galois representations of $\mathrm{Gal}(\overline{\mathbf{Q}}/\mathbf{Q})$ and cusp forms of weight 1 in such a way that the L-series associated (by Artin) to a Galois representation agrees with the L-series associated (by Hecke) to the corresponding cusp form. This conjectural correspondence is to be viewed as a natural generalisation of the equivalence, coming from class field theory, of characters of the Galois group (of an abelian extension) and Hecke characters (Artin's reciprocity). These correspondences are special cases of the Langlands' programme, see [15]. Here is the theorem of Weil and Langlands.

Theorem 4.2 (Langlands-Weil). *Let* ρ *be an odd 2-dimensional Galois representation over* $\mathbf{Q}$ *with conductor* N *and determinant* ϵ. *If* $L(s,\rho\otimes\lambda)$ *is an entire function for all twists* $\rho\otimes\lambda$ *of* ρ *by one-dimensional representations* λ, *then there is a cusp form* f *of weight* 1 *on* $\Gamma_1(N)$ *with nebentypus character* ϵ *such that* $L_f(s) = L(s,\rho)$.

The following theorem due to Deligne and Serre constructs Galois representations associated to cusp forms of weight 1.

Theorem 4.3 (Deligne-Serre). *If* f *is a newform of type* $(1,\epsilon,N)$ *then there is an odd two-dimensional Galois representation* $\rho : \mathrm{Gal}(\overline{\mathbf{Q}}/\mathbf{Q}) \to GL_2(\mathbf{C})$ *with conductor* N *and determinant* ϵ *such that* $L_f(s) = L(s,\rho)$.

4.2 Two Dimensional Representations of Prime Conductor

In this section which has been completely taken from the paper of Serre [16], we summarise the classification of two dimensional representations ρ of $\mathrm{Gal}(\overline{\mathbf{Q}}/\mathbf{Q})$ of prime conductor. We will say that ρ is dihedral, or A_4, S_4, A_5 depending on the image of ρ in $PGL(2, \mathbf{C})$. We will denote the determinant of ρ by ϵ.

Dihedral case: Two dimensional dihedral representations ρ of $\mathrm{Gal}(\overline{\mathbf{Q}}/\mathbf{Q})$ of prime conductor p arises only for $p \equiv 3 \bmod 4$ and is then induced from a non-trivial unramified character of $K = \mathbf{Q}(\sqrt{-p})$. Such Galois representations indeed correspond to a cusp form on $\Gamma_1(p)$ as is well-known from Hecke theory.

Non-dihedral case: Suppose that ρ is a two dimensional non-dihedral representation of $\mathrm{Gal}(\overline{\mathbf{Q}}/\mathbf{Q})$ of prime conductor p. Then,

(1) $p \not\equiv 1 \bmod 8$.
(2) If $p \equiv 5 \bmod 8$, ρ is of type S_4, and the character ϵ of order 4 and conductor p.
(3) If $p \equiv 3 \bmod 4$, ρ is of type S_4 or A_5, and ϵ is the Legendre symbol $n \to (\frac{n}{p})$.

Conversely, start with a Galois extension L of $\mathbf{Q}$ and a prime number p. Consider the following 3 cases.

(i) $\mathrm{Gal}(L/\mathbf{Q}) \cong S_4$ and $p \equiv 5 \bmod 8$.
(ii) $\mathrm{Gal}(L/\mathbf{Q}) \cong S_4$ and $p \equiv 3 \bmod 4$.
(iii) $\mathrm{Gal}(L/\mathbf{Q}) \cong A_5$ and $p \equiv 3 \bmod 4$.

An embedding of $\mathrm{Gal}(L/\mathbf{Q})$ in $PGL(2, \mathbf{C})$ determines a projective representation $\tilde{\rho}_L$ of $\mathrm{Gal}(\overline{\mathbf{Q}}/\mathbf{Q})$. There exists a lifting of $\tilde{\rho}_L$ to $GL(2, \mathbf{C})$ with prime conductor p, if and only if one has the following in the 3 respective cases above:

(i) L is the normal closure of a non-real quartic field of discriminant p^3.
(ii) L is the normal closure of a quartic field of discriminant $-p$.
(iii) L is the normal closure of a non-real quintic field of discriminant p^2.

4.3 Tate's Example

Till 1976 Artin's conjecture was known to be true only for representations which are *positive* rational linear combinations of one-dimensional representations in which cases some integral power of the corresponding Artin L-functions are product of Hecke L-functions. In 1976, in [19] Tate reported on the construction of a tetrahedral representation of conductor $N = 133$ thereby inaugurating the experimental verification of Artin's conjecture. They were able to prove the existence of the corresponding newform of weight 1 and level 133 predicted by the Langlands' conjecture ("by relatively easy hand computation"). By the theorem of Deligne-Serre (theorem 4.3 above), this produced the first example of an Artin L-series which is known to be holomorphic in spite of the fact that no power of it is a product of abelian L-series.

The strategy used by Tate is the following: Given a 2-dimensional representation ρ of the appropriate kind, compute the coefficients a_n of the Artin L-series $L(s, \rho)$ for $n \leq A$, say.

Then look for a modular form of weight 1 of level 133 with Fourier coefficients a_n for $n \leq A$. If A is sufficiently large, for instance $A \geq (N/12)\prod_{p|N}(1+p^{-1})$, then this form is uniquely determined, if it exists. Now invoke the theorem of Deligne-Serre to get a representation ρ_1 corresponding to this form for which the Artin's conjecture is true.

4.4 Dimension of Spaces of New Forms of Weight 1

Let $S_1^{\text{new}}(N,\epsilon)$ denote the complex vector space of newforms of weight 1, level N and character ϵ, and let $d(N,\epsilon)$ be the number of equivalence classes of 2-dimensional, complex, irreducible, continuous, odd Galois representations with conductor N and determinant ϵ. The theorem of Deligne-Serre implies $\dim S_1^{\text{new}}(N,\epsilon) \leq d(N,\epsilon)$. Artin's conjecture is the statement that

$$\dim S_1^{\text{new}}(N,\epsilon) = d(N,\epsilon).$$

Thus one way to verify Artin's conjecture in a particular case is by computing the numbers $\dim S_1^{\text{new}}(N,\epsilon)$ and $d(N,\epsilon)$. But there is no known method of computing the dimension of the space of forms of weight 1. Also, there is no general way of determining the number of equivalence classes of 2-dimensional, complex, irreducible, continuous, odd Galois representations with conductor N and determinant ϵ.

Suppose $N = q$, a prime with $q \equiv 3 \pmod 4$. Hecke proved that

$$f_\chi = \sum_{\mathfrak{a}} \chi(\mathfrak{a}) e^{2\pi i \mathrm{N}(\mathfrak{a})z}$$

where χ is an arbitrary non-trivial character on the ideal class group of $\mathbf{Q}(\sqrt{-q})$ and $\mathfrak{a}$ runs over all non-zero integral ideals in $\mathbf{Q}(\sqrt{-q})$, is a cusp form of weight 1, level q and character $(\frac{\cdot}{q})$. Since $q \equiv 3 \bmod 4$, the class number of $\mathbf{Q}(\sqrt{-q})$ is odd. It is easy to see that $f_\chi = f_{\chi'}$ if and only if $\chi = \chi'$, or $\chi = \chi'^{-1}$. It follows that if the class number of $\mathbf{Q}(\sqrt{-q})$ is h, then there are at least $(h-1)/2$ linearly independent cusp forms of weight 1, level q and character $(\frac{\cdot}{q})$. Therefore Siegel's theorem estimating the order of the class group implies the ineffective lower bound

$$\dim S_1\left(q, \left(\frac{\cdot}{q}\right)\right) >>_\epsilon q^{1/2-\epsilon}$$

for all $\epsilon > 0$. But there are also forms which are not of the Hecke type. However, forms of non-Hecke type are rare and it is conjectured that

$$\dim S_1\left(q, \left(\frac{\cdot}{q}\right)\right) = \frac{1}{2}(h-1) + O_\epsilon(q^\epsilon).$$

In [9] W. Duke has proved that

$$\dim S_1\left(q, \left(\frac{\cdot}{q}\right)\right) << q^{11/12}\log^4 q,$$

with an absolute implied constant.

5 Icosahedral Representations

This section is devoted to the last unsolved case of the Artin's conjecture for 2-dimensional representations, the case of the icosahedral representation. Following the method of Tate, J. Buhler in 1978 constructed for the first time a 2-dimensional icosahedral representation satisfying Artin's conjecture, [7] (see 5.1). About 15 years later, in 1993 Frey and his students discovered seven more examples, [10]. All these examples are for icosahedral representations over **Q**.

5.1 Verification of Artin's Conjecture

The aim of this section is to outline the method followed by J. Buhler [7] who gave the first example of an icosahedral representation for which Artin's conjecture is true. The construction of these examples involve extensive computation on a computer.

The starting point is the construction of icosahedral representations with low conductors; the lowest conductor found was $800 = 2^5 5^2$. This was done by sieving through a large number of quintic polynomials, eliminating first those whose discriminants were not squares, then reducible polynomials and then eliminating those whose Galois groups were proper subgroups of A_5. For each surviving quintic the ring of integers is computed to determine the behavior of ramified primes and the minimal conductor. The quintic which gives rise to the icosahedral representation of conductor 800 is:

$$F(x) = x^5 + 10x^3 - 10x^2 + 35x - 18.$$

Denoting by ρ the icosahedral representation of conductor 800 associated to $F(x)$, the next step is the calculation of the $L(s, \rho) = \sum a_n n^{-s}$. This calculation involves computations in a sextic extension of **Q**. As a sample we reproduce a few values of a_n; note that a_n is 0 if n is not prime to 10. Let i denote a fixed fourth root of 1 and j the positive root of $x^2 - x - 1 = 0$.

n	$\longrightarrow$	1	3	7	9	11	13	17	19
a_n	$\longrightarrow$	1	$-i$	$-ij$	-1	0	j	0	$i - ij$

A cusp form of weight 1 is said to be dihedral if the corresponding representation (via Deligne-Serre) is an irreducible properly induced representation and that the form is tetrahedral (resp. octahedral, icosahedral) if the representation is tetrahedral (resp. octahedral, icosahedral).

Consider the space V of modular forms of type $(1, \epsilon, 800)$ (i.e., of weight 1, character ϵ and level dividing 800) for a carefully chosen Dirichlet character ϵ. The choice of ϵ is such that in V:

- there are two non-cuspidal eigenforms of level 100, denoted by g_1, g_2; and
- there is only one dihedral form of level 100, g_3 (note that g_3 is a cusp form).

Each of these forms can 'pushed-up' to level 800 by the Atkin-Lehner operators, B_d, $d = 1, 2, 4, 8$ ($g|B_d(z) = g(dz)$). Let

$$g_{i,d} = g_i | B_d.$$

If g is a modular form of type $(1, \epsilon, N)$ let $\overline{g}$ denote the 'complex conjugate' of g; $\overline{g}$ is of type $(1, \overline{\epsilon}, N)$ and the Fourier coefficients of $\overline{g}$ are the complex conjugates of the Fourier coefficients of g.

For z in the upper half plane, let $f(z) = \sum_{n\geq 1} a_n e^{2\pi i n z}$ where a_n are as above ($L(s, \rho) = \sum a_n n^{-s}$). The aim is to show that this f coincides with the q-expansion of a cusp form of weight 1 up to a required number of terms. This is done as follows.

By searching in the space of cusp forms of weight 2 and level 800 (which has dimension 97) Buhler discovered the

Fact: For each $i = 1, 2, 3$ and $d = 1, 2, 4, 8$ there is a modular form $h_{i,d}$ of weight 2 and level 800 (and trivial character) such that

$$f\overline{g}_{i,d} \equiv h_{i,d} \pmod{q^{360}}.$$

Here $(\text{mod } q^M)$ means that the first M terms of the two power series agree.

From this it follows that

$$h_{i,d}\overline{g}_{j,d'} \equiv h_{j,d'}\overline{g}_{i,d}$$

for all i, j, d, d'. On each side of this congruence is a modular form of weight 3, level 800 and a certain character. Such modular forms are sections of a bundle on $X_0(800)$ whose degree is 344. Therefore the above congruence is actually an equality and we have that

$$f' = \frac{h_{i,d}}{\overline{g}_{i,d}}$$

is independent of the choice of i, d. It turns out that f' is a cusp form of type $(1, \epsilon, 800)$. It is to prove that it is so that one uses the representation of f' in so many ways as above by showing that the forms $g_{i,d}$ have no common zero in the upper half-plane.

That f' is not of the dihedral type is proved by looking at the eigenvalue of f' for the Hecke operator T_3; and that it is not of the tetrahedral or octahedral type follows from:

- There are no cyclic extensions of $\mathbf{Q}$ of degree 3 unramified outside 2 and 5 and hence there are no A_4 extensions of $\mathbf{Q}$ unramified outside 2 and 5.
- There are exactly three S_4 extensions of $\mathbf{Q}$ unramified outside 2 and 5 and the corresponding representations have conductors not dividing 800.

Therefore it follows that there is an icosahedral form in V! If ρ_1 denotes the corresponding representation (via Deligne-Serre) then ρ_1 is an icosahedral representation which satisfies the Artin's conjecture. It is not known if ρ_1 is same as the representation ρ that we started with. Buhler ([7], pp. 90) proves that if f' is an eigenform for T_{11} as well then it is so.

5.2 Recent Developments

Like the Deligne-Serre theorem which was stated as theorem 4.3 which constructs Galois representations associated to modular forms of weight 1 such that the L-function associated to the Galois representation and the modular form is the same, there is a more classical theorem due to Deligne which constructs ℓ-adic Galois representations for modular forms of any weight $k \geq 2$. These Galois representations can be reduced modulo the maximal ideal in an appropriate ring of ℓ-adic integers to construct Galois representations with values in $GL(2)$ of a finite field. There has been much activity in recent years motivated by a conjecture of Serre which says that any irreducible two dimensional Galois representation of $\mathrm{Gal}(\overline{\mathbf{Q}}/\mathbf{Q})$ with values in a finite field which is irreducible and is *odd* in the sense that the determinant of complex conjugation is -1, comes via this construction for some modular form.

A representation of $\mathrm{Gal}(\overline{\mathbf{Q}}/\mathbf{Q})$ with values in $GL(2, \mathbf{C})$ in fact lies in $GL(2, \mathcal{O})$ where $\mathcal{O}$ is the ring of integers of a number field. By going modulo prime ideals in $\mathcal{O}$, an Artin representation therefore gives rise to representations of $\mathrm{Gal}(\overline{\mathbf{Q}}/\mathbf{Q})$ with values in $GL(2)$ of a finite field. Kevin Buzzard and Richard Taylor have in [8] proved that if this representation is modular in the sense that it comes from a modular form via Deligne's construction, then under a mild hypothesis, Artin's conjecture is true for the original representation.

We note the following result, cf. theorem 1.2 in [18], which relates elliptic curves to mod 5 Galois representations from which the modularity of certain representations of $\mathrm{Gal}(\overline{\mathbf{Q}}/\mathbf{Q})$ with values in $GL_2(\mathbf{Z}/5)$ follows from the work of Wiles, Taylor-Wiles, and Diamond which proves modularity of certain elliptic curves over $\mathbf{Q}$.

Theorem 5.1 *If* $\overline{\rho} : \mathrm{Gal}(\overline{\mathbf{Q}}/\mathbf{Q}) \to GL_2(\mathbf{Z}/5)$ *is any representation whose determinant is the cyclotomic character* mod 5, *then there exists an elliptic curve over* $\mathbf{Q}$ *which realises* $\overline{\rho}$ *on its 5 torsion points.*

However, for the purposes of Artin's conjecture, modularity of a representation of $\mathrm{Gal}(\overline{\mathbf{Q}}/\mathbf{Q})$ with values in $GL_2(\mathbf{Z}/5)$ is not adequate as it is easy to see that the reduction modulo a prime ideal of an Artin representation of A_5-type taking values in $GL_2(\mathcal{O})$ cannot lie in $GL_2(\mathbf{Z}/5)$ (with determinant the cyclotomic character).

In conclusion, one can say that there is now a general approach to Artin's conjecture for 2 dimensional representations of $\mathrm{Gal}(\overline{\mathbf{Q}}/\mathbf{Q})$ which in the absence of knowing enough mod p representations of $\mathrm{Gal}(\overline{\mathbf{Q}}/\mathbf{Q})$ to be modular, has not yet been successful in completing Artin's conjecture for 2 dimensional representations of $\mathrm{Gal}(\overline{\mathbf{Q}}/\mathbf{Q})$.

In a recent work, C. Khare has proved that Serre's conjecture implies Artin's conjecture for 2 dimensional odd representations of $\mathrm{Gal}(\overline{\mathbf{Q}}/\mathbf{Q})$, cf. [13].

6 Stark's Conjecture

In this section we state an important conjecture of Stark which does not depend on Artin's conjecture but uses Artin L-functions. There is a lot of importance attached to Stark's conjecture especially as it is a contribution to Hilbert's 12th problem about generation of classfields of number fields in terms of explicit values of transcendental functions just as was done for $\mathbf{Q}$ and quadratic imaginary fields by attaching values of exponential and elliptic functions respectively.

If L is a number field with r_1 real embeddings and r_2 pairs of complex embeddings, then it is known that the zeta function of L has a zero of order $r_1 + r_2 - 1$ at the origin, and the leading term of the Taylor expansion of $\zeta_L(s)$ at $s = 0$ begins as

$$\zeta_L(s) = -\frac{hR}{e} \cdot s^{r_1+r_2-1} + \cdots,$$

where h is the class number of L, R the regulator, e the number of roots of unity in L.

Stark has defined an analogue of the regulator of a number field for all Artin representations, now called Stark regulator and has conjectured that just like the Dedekind zeta function, the leading term of the Taylor expansion of $L(s, \chi)$ at $s = 0$ divided by the Stark regulator is an algebraic number belonging to the cyclotomic field in which the values of the character of χ lie. We refer to the book of Tate [20] for an exposition of Stark's conjecture which remains open. We note that from proposition 3.4 of [20], the order of vanishing of $L(s, \chi)$ at $s = 0$ is $r(\chi)$ which is

$$r(\chi) = \sum_{v \in \infty} (\dim V^{G_w}) - \dim V^G,$$

where V is the vector space on which the representation χ is defined, and G_w denotes the decomposition group at a place w of L over an infinite place v of K. It follows from this formula for $r(\chi)$ that if L is abelian over K and χ is a faithful character, then $r(\chi) = 1$ if and only if the decomposition group is trivial at exactly one infinite place of K. In this case, the Stark regulator is the determinant of a 1×1 matrix, and unwinding the definitions, it follows from Stark's conjecture that if L is an abelian extension of a number field K, and χ is a character of the Galois group of L over K, then

$$L'(0, \chi) = -\frac{1}{e} \sum_{\sigma} \chi(\sigma) \log |\epsilon^\sigma|,$$

for a suitable element ϵ in L, where e denotes the number of roots of unity in L; the absolute value is taken with respect to a fixed place of L which lies over the unique Archimedean place of K where the decomposition group is trivial. The element ϵ of L is called the Stark unit and has the property that the extension of L obtained by attaching the e-th root of ϵ is abelian over K. If χ is a faithful character of $\mathrm{Gal}(L/K)$, ϵ generates L over K. This case of Stark's conjecture dealing with abelian extensions is also open except when $K = \mathbf{Q}$, or an imaginary quadratic field. The case $K = \mathbf{Q}$ being well-known, and the imaginary quadratic case due to Stark where the Stark unit is an *elliptic unit.*

Acknowledgement

The authors would like to thank C. Khare for some helpful remarks.

References

[1] E. Artin, Über die Zetafunktionen gewisser algebraischer Zahlkörper, *Math. Ann.* **89** (1923), 147–156.

[2] E. Artin, Über eine neue Art von L-reihen, *Hamb. Abh.* (1923), 89–108.

[3] E. Artin, Beweis des allgemeinen Reziprozitätsgesetzes, *Hamb. Abh.* **5** (1927), 353–363.

[4] E. Artin, Zur Theorie der L-Reihen mit allgemeinen Gruppencharakteren, *Hamb. Abh.* **8** (1930), 292–306.

[5] R. Brauer, On the zeta functions of algebraic number fields, *Amer. J. Math.* **69** (1947), 243–250.

[6] R. Brauer, On Artin's L-series with general group characters, *Ann. of Math.* **48** (1947), 502–514.

[7] J. Buhler, Icosahedral Galois Representations, *Lecture Notes in Mathematics* **654** Springer-Verlag, 1978.

[8] K. Buzzard and Taylor, R., Companion forms and weight one forms, *preprint*, 1997.

[9] W. Duke, The dimension of the space of cusp forms of weight one, *International Math. Res. Notices*, No. 2, (1995), 99–109.

[10] G. Frey, On Artin's Conjecture for Odd 2-Dimensional Representations, *Lecture Notes in Mathematics* 1585 Springer-Verlag, 1994.

[11] R. Foote and Murty, V.K., Zeros and poles of Artin L-series, *Math. Proc. Camb. Phil. Soc.* (1989), 105, 5, 5–11.

[12] E. Hecke, Über eine neue Anwendung dur Zetafunktion auf die Arithmetik der Zahlkörper, *Göttinger Nachrichten* (1917), 90–95.

[13] C. Khare, Remarks on mod p forms of weight 1, *International Math. Research Notices* (1997), No. 3, 127–133.

[14] G.O. Michler, On Artin L-series of irreducible characters of the symmetric group S_n, *Algebra and Number Theory*, Ed.: G. Frey and J. Ritter, Walter de Gruyter (1994), 153–163.

[15] M. Ram Murty, A motivated introduction to the Langlands' Programme, *Advances in Number Theory*, Ed.: F. Gouvea and N. Yui, Clarendon Press, Oxford (1993), 37–66.

[16] J.-P. Serre, Modular forms of weight one and Galois representations, in: *Algebraic Number Theory*, edited by Frohlich, Academic Press, 193–268.

[17] H. Stark, Some effective cases of the Brauer-Siegel theorem, *Invent. Math.* **23** (1974), 135–152.

[18] N.I. Shephered-Barron and Taylor, R., Mod 2 and Mod 5 icosahedral representations, *Journal of the AMS*, volume **10** (1997), 283–298.

[19] J. Tate, The general reciprocity law, *Mathematical developments arising from Hilbert Problems*, volume **27** (AMSPSPM series), 1976. Ed.: F.E. Browder, 311–322.

[20] J. Tate, Les conjetures de Stark sur les Fonctions L d'Artin en $s = 0$, *Progress in Mathematics* vol. **47** Birkhauser, 1984.

[21] R. Taylor, Icosahedral Galois Representations, Olga Taussky-Todd: in memoriam. *Pacific Journal of Mathematics* (1997), 337–347.

[22] K. Uchida, On Artin's L-functions, *Tohoku Math. J.* **27** (1975), 75–81.

[23] R.W. van der Waal, On a conjecture of Dedekind on zeta functions, *Indag. Math.* **37** (1975), 83–86.

Mehta Research Institute of
Mathematics & Math. Physics
Chhatnag Road, Jhusi
Allahabad 211 019
E-Mail: dprasad@mri.ernet.in

Olympaid Cell (NBHM)
Department of Mathematics
Indian Institute of Science
Bangalore 560 012
E-Mail: yoga@math.iisc.ernet.in

Siegel's Main Theorem for Quadratic Forms

S. Raghavan

§ 1

A classical question in the Theory of Numbers is one of expressing a positive integer as a sum of squares of integers. The qualitative aspects of this problem require at times no more than rudimentary congruence considerations e.g. a prime number leaving remainder 3 on division by 4 cannot be a sum of two squares of integers; however, in general, subtle arguments are called for. Fermat's Principle of Descent needs to come into play for a proof of the (Euler-Fermat-) Lagrange theorem that every positive integer is a sum of four squares of integers. Skillful use of elliptic theta functions was made by Jacobi to obtain a quantitative refinement of that assertion, viz. according as n is an odd or even natural number, the number of ways of expressing n as a sum of four squares of integers is $8\sigma^*(n)$ or $24\sigma^*(n)$, where $\sigma^*(t)$ for any natural number t is the sum of all the odd natural numbers dividing t; Jacobi's famous identity linking θ_3^4 with other theta constants θ_2, θ_4 and their derivatives is an analytic encapsulation of all these formulae for varying n. An analytic formulation of similar nature arises also as a *special case* of the *Siegel formula* (extended suitably to cover the boundary case of quaternary quadratic forms as well) which connects theta series associated with quadratic forms to Eisenstein series: for complex z with positive imaginary part,

$$\left(\sum_{n\in\mathbb{Z}} \exp(\pi i\, n^2 z)\right)^4 = 1 + \sum_{q\in\mathbb{N}} \sum_{\substack{p\in\mathbb{Z} \\ (p,q)=1,\ p+q\equiv 1 \pmod 2}} (p-qz)^{-2}$$

where the sum over p, q giving the Eisenstein series on the right hand side is only conditionally convergent and can be realized from an absolutely convergent Eisenstein series by analytic continuation via Hecke's Grenzprozess (The inner sum over p is over all integers which are coprime to q and are of opposite parity to q).

§ 2

More generally, let us consider a positive definite quadratic form $f(x_1, \ldots, x_m) := \sum_{1\leq i,j\leq m} s_{ij}x_ix_j$, in m variables $x_1, \ldots, x_m$ with the associated symmetric positive definite integral (coefficient) matrix $S := (s_{ij})$ and then the number $r(f; t) = r(S; t)$ of ways of representing an integer t as $f(a_1, \ldots, a_m)$ with integers $a_1, \ldots, a_m$ (instead of representing

t merely as the sum $a_1^2 + \cdots + a_m^2$ of m squares of integers $a_1, \ldots, a_m$). We know from Minknowski's reduction theory that there are only finitely many integral positive definite quadratic forms, say $f_1 = f, f_2, \ldots, f_h$ such that any integral positive definite form g which is equivalent to f over the ring $\mathbb{Z}_p$ of p-adic integers for every prime p is equivalent to one of $f_1, \ldots, f_h$ over $\mathbb{Z}$ while no two distinct ones among $f_1, \ldots, f_h$ are equivalent over $\mathbb{Z}$. In other words, the *genus* of the positive definite integral quadratic form f splits into h different *classes* each containing precisely one of $f_1, \ldots, f_h$. (We recall here that for two quadratic forms g_1, g_2 with coefficients in a commutative ring R with unit element, g_1 is said to *represent* g_2 over R if there exists a linear transformation of the variables with coefficients from R taking g_1 to g_2 and moreover g_1 is called *equivalent to* g_2 over R (or said to be in the same R equivalence class as g_2) if g_1 and g_2 represent each other mutually over R. The number $r(f; t)$ above is thus the number of representations of the quadratic form ty^2 by $f(x_1, \ldots, x_m)$ over $\mathbb{Z}$). Analogously, let for a power p^s of any given prime number $p, r(f, t; p^s)$ denote the number of representations of t by f over the quotient ring $\mathbb{Z}/p^s\mathbb{Z}$. Then $d_p(f, t)$, the *p-adic density of representations* of t by f is defined as $c_m \lim_{s\to\infty} p^{-s(m-1)} r(f, t; p^s)$ with $c_m := 2$ or 1 according as $m = 1$ or $m > 1$; this p-adic density $d_p(f, t)$ is clearly non-negative and indeed it is a rational number which equals 0 precisely when f fails to represent t over the p-adic ring $\mathbb{Z}_p$. The infinite product $\prod_p d_p(f, t)$ extended over all the prime numbers p converges and is equal to 0 exactly when f fails to represent t over $\mathbb{Z}_p$ for at least one prime i.e. when at least one $d_p(f, t)$ is 0. The *real density* $d_\infty(f, t)$ for measuring the representability of t by f over the field $\mathbb{R}$ of real numbers is defined as $\lim_U \text{ vol } (f^{-1}(U)/\text{vol } U)$, where the limit is taken over measurable neighbourhoods U of t shrinking to $\{t\}$ with vol($\bullet$) denoting Lebesgue measure in the respective spaces; it is known [17] that $d_\infty(f, t) = \pi^{m/2} t^{(m-2)/2} / \{\Gamma(m/2) (\det f)^{1/2}\}$, where Γ is Euler's gamma function and $\det f$ is just the determinant of f. Finally let, for $1 \le i \le h$, e_i denote the number of linear transformations over $\mathbb{Z}$ which preserve the form. We can now state *Siegel's main theorem for positive definite integral quadratic forms* f (in this special case, for $t \in \mathbb{N}$):

$$\left\{\sum_{1\le i\le h} r(f_i, t)/e_i\right\} \Big/ \left\{\sum_{1\le i\le h} 1/e_i\right\} = \frac{1}{1+\delta_{m,2}} d_\infty(f, t) \prod_p d_p(f, t) \qquad (*)$$

with $\delta_{m,2}$ denoting the Kronecker delta function.

For f to represent t over $\mathbb{Z}$, it is clearly necessary that f represents t over $\mathbb{R}$ and over the p-adic ring $\mathbb{Z}_p$ for every prime p. But even all the latter infinitely many conditions put together can not ensure that f represents t over $\mathbb{Z}$ although by the Hasse principle, the representation of t by f over $\mathbb{Q}$ is then assured. Siegel's main theorem guarantees in such an event that at least one of $f_1, \ldots, f_h$ represents t over $\mathbb{Z}$ (even if f does not do so) and is thus a far more subtle refinement and a quantitative one too. The remarkable string of papers ([17], [18], [19]) by Siegel deal with the general situation of quadratic forms f representing quadratic forms g (instead of numbers t), again not necessarily over the ring $\mathbb{Z}$ but over the ring of algebraic integers in a totally real algebraic number field and furthermore with f not having to be a definite quadratic form.

§ 3

For $m > 4$, Siegel reformulated the main theorem stated in (*) as an *analytic identity* between theta series associated with $f_1, \ldots, f_h$ and Eisenstein series:

$$\left\{\sum_{1\le j\le h} \theta(f_j, z)/e_j\right\} \Big/ \left\{\sum_{1\le j\le h} 1/e_j\right\} = 1 + \sum_{\substack{(a,b)=1\\ b>0}}{}' H(f, b, a)(bz-a)^{-m/2} \qquad (**)$$

where, for complex z with positive imaginary part, the theta-series

$$\theta(f_j, z) := \sum_{a_1,\ldots,a_m \in \mathbb{Z}} \exp(\pi i z\ f_j(a_1, \ldots, a_m)),$$

$$H(f, b, a) := (\surd -1/b)^{m/2}(\det S)^{-1/2} \sum_{a_1,\ldots,a_m \bmod b} \exp(\pi i a\ f(a_1, \ldots, a_m)/b)$$

are generalized Gaussian sums and the summation over a, b in the Eisenstein series is carried out over all coprime integer pairs a, b with $b > 0$ and the accent requires ab to be even whenever f represents some odd integer over $\mathbb{Z}$. The Eisenstein series on the right hand side of the analytic identity (**) converges absolutely for $m > 4$. For $m = 4$, one invokes Hecke's limiting process, to suitably define the Eisenstein series via analytic continuation. The significance of the identity (**) can be fully realized only when we notice that it is analytically impossible perhaps to distinguish between the various theta series $\theta(f_i, \mathbb{Z})$ individually in view of their exhibiting the same kind of behaviour as the variable z approaches the 'rational points' a/b or infinity (so that for $i \neq j$, $\theta(f_i, z) - \theta(f_j, z)$ tends to 0 as z tends to a/b or ∞); thus the individual theta series seems to resist being expressible in terms of explicit objects like Eisenstein series. It is on the other hand remarkable that the arithmetical main theorem correctly points to the right weighted mean of theta series $\theta(f_i, z)$ being expressible as an Eisenstein series!

§ 4

For considering the problem of representing forms g in n variables by the given form f (instead of merely representing numbers by f) over $\mathbb{Z}$, the complex variable z above is to be replaced by a point Z of the Siegel upper half plane $\mathbb{H}_n := \{Z = {}^tZ \in \mathfrak{M}_n(\mathbb{Z}) | \frac{1}{2\sqrt{-1}}(Z - \overline{Z})$ is positive definite$\}$ and the associated theta series $\theta(S, Z) := \sum_G \exp(\pi i tr({}^tGSGZ))$, where now G runs over all (m, n) integral matrices and tr denotes matrix trace. The general term in the relevant Eisenstein series that features in an analogue of (**) takes the form $H(S, C, D)(\det(CZ + D))^{-m/2}$, where $H(S, C, D)$ are generalized Gauss sums and the summation in the Eisenstein series is over all n-rowed coprime symmetric pairs (C, D) i.e. with (n, n) integral matrices C, D satisfying the conditions that (i) $C^tD = D^tC$ and (ii) GC, GD integral for any (n, n) rational matrix imply that G is integral. Such n-rowed coprime symmetric pairs C, D make up precisely the last n rows of elements

$M = \begin{pmatrix} A & B \\ C & D \end{pmatrix}$ of the Siegel modular $\Gamma_n := \left\{ M = \begin{pmatrix} A & B \\ C & D \end{pmatrix} \in \mathfrak{M}_{2n}(\mathbb{Z}) \middle| A, B, C, D \in \mathfrak{M}_n(\mathbb{Z}), A^t B = B^t A, C^t D = D^t C \text{ and } A^t D - B^t C = E_n\right.$, the n-rowed identity matrix in $\left.\mathfrak{M}_n(\mathbb{C})\right\}$. The Siegel modular group Γ_n acts on $\mathbb{H}_n$ via the modular transformations $Z \to (AZ + B)(CZ + D)^{-1}$ for $M = \begin{pmatrix} A & B \\ C & D \end{pmatrix} \in \Gamma_n$. Under modular transformations, the behaviour of the Eisenstein series on the right hand side of an analogue of (**) is quite similar to that of the h theta series $\theta(S_i, Z)$. For $n = 1$, as remarked earlier, $\theta(S_i, z) - \theta(S_j, z)$ is a so-called cusp form for an appropriate subgroup of the elliptic modular group Γ_1; this is no longer true, in general, for $n > 1$. For $n = 1$ this phenomenon however leads immediately to an asymptotic formula for $r(S; t)$ with the main term in the asymptotic formula being given by the Fourier coefficient corresponding to the index t in the Fourier expansion of the Eisenstein series on the right hand side of the identity (*) above.

§ 5

In the proof of Siegel's main theorem (*) and more general formulations covering the representation of forms by f or the case of f being indefinite as well or when $\mathbb{Z}$ is replaced by rings of algebraic integers, a crucial step eventually is to show that a certain number $\rho(S)$ equals 2 where, for S positive,

$$\rho(S) := 2(\det S)^{(m+1)/2} \left(\sum_{1 \le i \le h} 1/e_i \right)^{-1} \prod_{1 \le j \le m} \pi^{-j/2}\Gamma(j/2)$$
$$\lim_{q \to \infty} 2^{\omega(q)} q^{(m+1)/2} / e_q(S)$$

where the limit is taken over natural numbers q tending to infinity, through the sequence of factorials $n!$, $\omega(q)$ is the number of prime factors of q and $e_q(S)$ is the number of linear transformations preserving the quadratic form f associated with S over the quotient ring $\mathbb{Z}/q\mathbb{Z}$. A group-theoretic interpretation of this definition of $\rho(S)$ by bringing in the orthogonal group G of f (defined over $\mathbb{Q}$) and choosing the rings $\mathbb{Z}$, $\mathbb{Z}_p$ for all primes p and $\mathbb{R}$ for the base domain, makes the above crucial step equivalent to proving that $\rho(S)$ is just the (Weil-) Tamagawa number $\tau(G)$ for the orthogonal group G corresponding to f; it was actually shown by Tamagawa that $\tau(G) = 2$. By considering the 'adelic zeta function' attached to orthogonal groups of quadratic forms f and its residues at poles (aptly generalizing Siegel's zeta functions associated with f when indefinite too), Weil proved in [25], for orthogonal groups G of (non-degenerate) quadratic forms in $m \geq 3$ variables, by induction on m, the "Siegel-Tamagawa theorem that $\tau(G) = 2$" which "by purely formal calculations" might be seen to be equivalent to "Siegel's main theorem" (on representation by quadratic forms); Weil's generalization of the Siegel formula for classical groups [27] yields again that $\tau(G)$ is the same for all $m \geq 4$.

§ 6

A nice and surprising proof for Siegel's main theorem for representation of integral quadratic forms (or equivalently (n, n) symmetric matrices T) by unimodular positive definite even quadratic forms $f(x_1, \dots, x_m)$ for $m > 2n+2$ has been given by Andrianov ([1], [4, Ch. IV, §6]), by explicit determination of the effect of Hecke operators on theta series attached to such f and invoking properties of Eisenstein series. If S is the integral m-rowed symmetric (m, m) matrix associated with any such given form f, then S has even positive integers as diagonal entries and is of determinant 1; it is well-known that m is then necessarily a multiple of 8. Let $S_1 (= S), S_2, \dots, S_h$ be the matrices corresponding to the representatives $f_1, f_2, \dots, f_h$ of the h different ($\mathbb{Z}$-equivalence) classes in the genus of f. Let $r(S_i, T)$ be the number of integral (m, n) matrices G such that ${}^tGS_iG = T$. Then with notation as in §3, we have for $Z \in \mathbb{H}_n$ and any $S_i (1 \leq i \leq h), \theta(S_i, Z) = \sum_T r(S_i, T) \exp(\pi i \, tr(TZ))$, where the summation in the Fourier expansion on the right hand side is over all (n, n) symmetric non-negative definite even integral matrices T and tr denotes matrix trace as before. For any $M = \begin{pmatrix} A & B \\ C & D \end{pmatrix} \in \Gamma_n$, we have $\theta(S_i; Z)\big|M :=$ $\theta(S_i, (AZ + B)(CZ + D)^{-1}) \det(CZ + D)^{-m/2}$ equals $\theta(S_i, Z)$ and thus $\theta(S_i, Z)$ is a Siegel modular form of degree n and weight $m/2$. For any given prime number p, the double coset $\Gamma_n \begin{pmatrix} E_n & 0 \\ 0 & pE_n \end{pmatrix} \Gamma_n$ of $(2n, 2n)$ matrices is a finite union of left cosets $\Gamma_n N_j$ with $N_j = \begin{pmatrix} A_j & B_j \\ C_j & D_j \end{pmatrix}$. The Hecke operator $T(p)$ acting on Siegel modular forms $f(Z)$ of degree n and weight $m/2$ is given by

$$f(Z)\big|T(p) = p^{(mn-n^2-n)/2} \sum_j f(A_j Z + B_j)(C_j Z + D_j)^{-1} \times (\det C_j Z + D_j)^{-m/2}.$$

Andrianov determined explicitly the effect of the Hecke operator $T(p)$ on $\theta(S_i; Z)$ and showed that the genus invariant $F(S, Z) := \sum_{1 \leq i \leq h} e_i^{-1} \theta(S_i; Z) / \sum_{1 \leq i \leq h} 1/e_i$ which is the weighted mean of $\theta(S_i; Z)$ generalizing the left hand side of the identity (**) in §3 is actually an eigen form of the Hecke operator $T(p)$ for every prime p; the constant term in the Fourier expansion of $F(S, Z)$ is clearly equal to 1. On the other hand, it is known that any such Siegel modular form (i.e. of degree n and weight $m/2$) which has constant term 1 and is further an eigen form of an infinite number of Hecke operators $T(p)$ has to coincide, for $m/2 > n + 1$, with the Eisenstein series $E_{m/2}(Z) = \sum_{C,D} \det(CZ + D)^{-m/2}$, the summation being over a complete set of n-rowed coprime symmetric pairs (C, D) such that no two distinct ones among them, say $(C_1, D_1), (C_2, D_2)$ are related by the condition $C_1 = UC_2, D_1 = UD_2$ with some integral (n, n) matrix U of determinant ± 1. The condition $m/2 > n + 1$ ensures the absolute convergence of this series. Its Fourier coefficients are well-known from Siegel's fundamental paper [20]; comparison of Fourier coefficients on both sides of the relation $F(S, Z) = E_{m/2}(Z)$ yields Siegel's main theorem on representation of (n, n) positive definite integral matrices T by an even (m, m) positive definite integral S of determinant 1 under the condition $m > 2n + 2$. These restrictions on S may seem to be rather severe since the (arithmetical) main theorem of Siegel is true without such 'stringent' conditions; still the function theoretic proof above for Siegel's main theorem even for a special case of positive definite S does represent a spin off for Arithmetic, especially when, as remarked earlier, it seems impossible to distinguish between the various $\theta(S_i, Z)$ analytically. Let us also recall that, in the other direction, Siegel's main theorem for quadratic forms and the resultant analytic identity (linking theta series with totally disparate objects such as Eisenstein series) was the starting point for Siegel's full fledged analytic theory of modular functions of degree n. Siegel modular functions of degree n have the same significance for algebraic function fields of genus n as the usual (elliptic) modular functions for function fields of genus 1; they represent the contribution from Arithmetic towards the study of an important problem of the theory of algebraic functions. Of course, for the development of the theory of authomorphic forms on semi simple Lie groups, the theory of Siegel modular forms has been the driving force and the touchstone, as well.

§ 7

A Siegel modular form of degree n and weight k is a holomorphic (complex-valued) function f on $\mathbb{H}_n$ such that for every $M = \begin{pmatrix} A & B \\ C & D \end{pmatrix} \in \Gamma_n$, $f((AZ + B)(CZ + D)^{-1}) \det (CZ + D)^{-k} = f(Z)$ and further f is regular at infinity for $n = 1$. Given a Siegel modular form f of degree n and weight k, the Siegel operator ϕ associates a Siegel modular form ϕf of degree $n - 1$ and weight k by the prescription $(\phi f)(Z_1) := \lim_{t\to\infty} f\begin{pmatrix} Z_1 & 0 \\ 0 & it \end{pmatrix}$ for $Z_1 \in \mathbb{H}_{n-1}$. Cusp forms of degree n are those belonging to the kernel of ϕ. If k is *large enough* in relation to n, the Siegel operator is surjective as shown by Maass [12] with the use of his Poincaré series and by Klingen [10] through his Eisenstein series 'lifting' cusp forms g of degree $j \le n$ to Siegel modular forms of degree n. The latter Eisenstein series of degree n are series $G(Z, g)$ of the form $\sum g(\pi(AZ + B)(CZ + D)^{-1}) \det (CZ + D)^{-k}$ with g being any given cusp form of degree $j \le n$, $\pi(W)$ represents the principal (j, j) minor for $W \in \mathbb{H}_n$ and the summation over $M = \begin{pmatrix} A & B \\ C & D \end{pmatrix}$ runs through representatives of Γ_n modulo an appropriate subgroup. This series converges absolutely under the condition that $k > n + j + 1$ and is mapped to the cusp form g of degree j under the iterated operator ϕ^{n-j}. For $k \le 2n$, one needs to insert in the general summand a Hecke convergence factor $\left(\frac{\det(\mathrm{Im}(\pi(M\langle Z\rangle)))}{\det(\mathrm{Im}(M\langle Z\rangle))}\right)^{-s}$ with $M\langle Z\rangle := (AZ + B)(CZ + D)^{-1}$, $\mathrm{Im}(W) := \frac{1}{2\sqrt{-1}}(W - \bar{W})$ and a complex parameter s, while real part of s being large clearly ensures absolute convergence of the series; thus the analytic continuation of the Eisenstein series with convergence factors as a function of s has to be studied. For $n = 1$, the first such (vector-valued) Dirichlet series in s arising from Eisenstein series (with Hecke convergence factors) associated to Jacobi theta constants $\theta_1, \theta_2, \theta_3$ or theta series attached to even quadratic forms (of given signature) were thoroughly investigated by Siegel [22, 23] as regards their continuation as a function of s and their functional equation involving s. For general n, the analytic continuation and functional equation for similar Eisenstein series with suitable Hecke convergence factors are subsumed in the general framework occurring in Langlands' theory of Eisenstein series on semi simple Lie groups [11].

It is not clear, on the face of it, if, even for a simple looking Eisenstein series $E_k^{(n)}(Z,s)$ with Hecke convergence factors defined by $E_k^{(n)}(Z,s) := (\det \operatorname{Im}(Z))^s \sum_{C,D} \det (CZ+D)^{-k} |\det (CZ+D)|^{-2s}$ convergent absolutely for $k > n+1$ and complex s with $Re(s) \geq 0$, we have, in the 'boundary case' of $k = n+1 \in 2\mathbb{N}$, a holomorphic function of Z as limit when s tends to 0 say, through positive real values. For $n = 1$ and $k = 2$, we know Hecke [7] that the limit as s tends to 0 exists but is not holomorphic as a function of Z; the first example for higher degrees n, of the Hecke limiting process yielding a non-zero holomorphic modular form of degree n is for the case $n = 3$ and $k = 4$ and may be found in [13]. For general $n \geq 1$ with the 'boundary case' $k = n+1 \in 2\mathbb{N}$, we know from the comprehensive researches of Weissauer [28] and results obtained independently by Shimura [16], when precisely the limiting process yields, holomorphic modular forms; Weissauer shows in particular that Hecke's Grenzprozess yields non-zero holomorphic modular forms for (even) $k > (n+3)/2$ or if $4|k$ with $k \leq (n+1)/2$. Weissauer [28] also determines the obstruction to ϕ-lifting i.e. lifting a cusp form g of degree j and weight k to a holomorphic limit of series resembling Eisenstein series $G(Z,g)$ with appropriate Hecke convergence factors in the case of 'boundary weights' k. The study of Eisenstein series and results on ϕ-lifting are moreover applied in [28] to obtain, in particular, theorems on representation of Siegel modular forms as linear combinations of theta series $\theta(S,Z)$ as above, generalizing Bŏcherer's notable results [2]. Actually Bŏcherer proved that every Siegel modular form (respectively cusp form) of degree n and weight k with $4|k$ and $k > 2n$ is a linear combination of theta series (respectively with spherical harmonics) associated to even unimodular positive definite quadratic forms, by first determining the Fourier expansion of Klingen's Eisenstein series $G(Z,g)$ and then applying important results of Garrett [6], difficult work of Andrianov [1] on Euler products and Siegel's main theorem for quadratic forms. The 'basis problem' for elliptic modular forms (not necessarily of level 1) is one of expressing them as linear combinations of theta series associated with quadratic forms and it is Waldspurger's noteworthy paper [24] that made use of Siegel's main theorem to tackle the 'basis problem', for the first time. "At the other end of the spectrum", when the weight k of Siegel modular forms of degree n is much smaller in relation to n (say, k does not satisfy the condition $k > 2n$ but actually $k < n/2$), we have precisely the singular Siegel modular forms which can also be characterized by having all their Fourier coefficients indexed by (n,n) positive symmetric matrices T (in the Fourier expansion) equal to 0; we know ([4, Ch. IV, §5], [14]) that every singular Siegel modular form of degree n (for Γ_n) is a linear combination of theta series associated with even unimodular positive definite quadratic forms.

§ 8

In two powerful papers ([26], [27]), Weil used the fascinating setting of adelic analysis to present the analytic formulation of Siegel's main theorem for quadratic forms (see also [21]) as an identity between two 'invariant tempered distributions' - the 'theta distribution' and the 'Eisenstein distribution' defined in the context of not just orthogonal groups but more generally classical groups arising from algebras with involution. A vital step in his proof is a very general Poisson formula involving transforms F_Φ^* of Schwartz-Bruhat functions Φ on locally compact abelian groups G and the Fourier transform F_Φ of F_Φ^* and the recognition of the measures 'making up' F_Φ. Suitable specialization of Φ and G leads one to recover Siegel's analytic formulation of his main theorem. Using tools from analysis and deep algebraic geometry such as the Hironaka resolution of singularities in the case of hypersurfaces arising from forms f in n variables and of degree $m \geq 2$, Igusa [9] obtained under appropriate restrictions (e.g. $n > 2m \geq 4$ and the variety defined by $f = 0$ is irreducible and normal) a Poisson formula generalizing that of Weil to the case of forms of higher degree. A local-global theorem of Birch [2] generalizing, to forms f of higher degree, Davenport's results on cubic forms and concerning the existence of rational points on hypersurfaces defined by f, results as an application of Igusa's Poisson formula (see [16] for a nice survey on this).

§ 9

Siegel's main theorem for positive definite quadratic forms f as stated in (*) of §2 is valid without restriction on m while the analytic formulation in (**) of §3 valid for $m > 4$ can be extended to cover the boundary case $m = 4$ as well by an appropriate method viz. Hecke's limiting process to overcome the problem of the Eisenstein series not converging absolutely, for $m = 4$. On the other hand, when f is indefinite, the numbers $r(f, t)$ are no longer finite in general and have to be replaced by 'measures of representation' which are essentially volumes of relevant fundamental domains for discrete subgroups of the orthogonal group of f. Siegel [21] proved an analogue of (*) for indefinite f with $m \geq 4$ and similarly of (**) except when f is a quaternary integral quadratic form with square determinant and represents 0 over $\mathbb{Z}$ non-trivially; the left hand side of this analogue of (**) is given as an integral of a theta series over a fundamental domain for the group of ($\mathbb{Z}-$) automorphisms of f while the right hand side is still an Eisenstein series. Siegel's generalization of all these results of his including those covering representations over algebraic number fields are to be found in [21] while Weil's Siegel formula with its several distinctive features surprisingly unifies the discussion of the definite and indefinite quadratic forms (also with rings of algebraic integers in place of $\mathbb{Z}$) under conditions (e.g. $m > 4$) ensuring absolute convergence of the Eisenstein series. For indefinite ternary or binary quadratic forms, the analogue of Siegel's main theorem fails to be valid [21]; distinct classes in a given genus of binary indefinite forms represent integrally essentially distinct sets of prime numbers.

§ 10

In [15], the Siegel-Weil formula in a range of cases not covered by [27] is established for representation of quadratic forms in n variables by 'anisotropic' (i.e. not representing 0 non-trivially over $\mathbb{Q}$) quadratic forms in m variables over $\mathbb{Q}$ with even integral $m \geq n$ or with all $m \geq 1 = n$. The problem here is far more than just taking care of non-absolute convergence of Eisenstein series (via an analogue of Hecke's procedure)! The proof makes use of Hecke intertwining maps and singular automorphic forms in the sense of Howe [8].

References

[1] A.N. Andrianov, The multiplicative arithmetic of Siegel modular forms, *Russian Math. Surveys* **34** (1979), 75–148.

[2] B.J. Birch, Forms in many variables, *Proc. Roy. Soc. Ser. A* **265** (1961/62), 245–263.

[3] S. Böcherer, Über die Fourier-Jacobi-Entwicklung Siegelscher Eisensteinreihen, *Math Zeit.* **183** (1983), 21–46, . . .II, *Math Zeit.* **189** (1985), 81–110.

[4] E. Freitag, Siegelsche Modulfunktionen, Grundlehren Math. Wissenschaeften, 254, Springer-Verlag, 1983.

[5] P.B. Garrett, Pullbacks of Eisenstein series, Proc. Taniguchi Symposium on Automorphic Forms in several variables, Birkhauser, 1984.

[6] S.J. Haris, Numbertheoretical developments arising from the Siegel formula, *Bull. A.M.S.* **2** (1980), 417–433

[7] E. Hecke, Theorie der Eisensteinschen Reihen höherer Stufe und ihre Anwendung auf Funktionentheorie und Arithmetik, *Abhand. Math. Semin. Hamburg Universität* **5** (1927), 199–224; also Gesamm. Abhand. 461–486.

[8] R. Howe, On a notion of rank for unitary representations of the classical groups (Preprint).

[9] J.-I. Igusa, Criteria for the validity of a certain Poisson Formula, *Algebraic Number Theory, Kyoto* (1976), 43–65.

[10] H. Klingen, Zum Darstellungssatz für Siegelsche Modulformen, *Math. Zeit.* **102** (1967), 30–43.

[11] R.P. Langlands, On the Functional Equation Satisfied by Eisenstein Series, *Springer Lecture Notes in Math.* **544** (1976).

[12] H. Maass, Siegel's Modular Forms and Dirichlet Series, *Springer Lecture Notes in Math.* **216** (1971).

[13] S. Raghavan, On an Eisenstein series of degree 3, *J. Indian Math. Soc.* **39** (1975), 103–120.

[14] S. Raghavan, Singular modular forms of degree s, C.P. Ramanujam - A Tribute, T.I.F.R. Studies in Math. No. 8, (1978), 263–272.

[15] S. Rallis, L-functions and the Oscillator Representation, *Springer Lecture Notes in Math.* **1245** (1987).

[16] G. Shimura, On Eisenstein series, *Duke Math. J.* **50** (1983), 417–476.

[17] C.L. Siegel, Über die analytische Theorie der quadratischen Formen, *Ann. of Math.* **36** (1935), 527–606; Gesamm. Abhand. Bd I, 326–405.

[18] ———, Über die analytische Theorie der quadratischen formen II, *Ann. of Math.* **37** (1936), 230–263; Gesamm. Abhand. Bd I, 410–443.

[19] ———, Über die analytische Theorie der quadratischen formen III, *Ann. of Math.* **38** (1937), 212–291; Gesamm. Abhand. Bd I, 469–548.

[20] ———, Einführung in die Theorie der Modulfunktionen n-ten Grades, *Math. Annalen* **116** (1939), 617–657; Gesamm. Abhand. Bd II, 97–137.

[21] ———, Indefinite quadratische Foremen und Funktionentheorie I, II, *Math. Annalen* **124** (1951), 17–54, **124** (1952), 364–387; Gesamm. Abhand. Bd III, 105–142, 154–177.

[22] ———, Die Funktionalgleichungen einiger Dirichletscher Reihen, *Math. Zeit.* **63** (1956), 363–373; Gesamm. Abhand. Bd III, 228–238.

[23] ———, A generalization of the Epstein zeta function, *J. Indian Math. Soc.* **20** (1956), 1–10; Gesamm. Abhand. Bd II, 239–248.

[24] J.L. Waldspurger, Engendrement par des séries theta de certains espaces de formes modulaires. *Invent. Math.* **50** (1979), 135–168.

[25] A. Weil, Adèles and Algebraic Groups, Princeton, *Inst. for Adv. Study* 1961.

[26] ———, Sur certains groupes d'opérateurs unitaires, *Acta Math.* **111** (1964), 143–211.

[27] A. Weil, Sur la formule de Siegel dans la théorie des groupes classiques, *Acta Math.* **113** (1965), 1–87.

[28] R. Weissauer, Stabile Modulformen und Eistensteinreihen, *Springer Lecture Notes in Math.* **1219** (1986).

A1, Ashok Ganapathim
25, Fourth Main Road
R. A. Puram
Chennai 600 028

Pfister's Work on Sums of Squares

A.R. Rajwade

Historically the theory of quadratic forms was regarded as a topic in number theory. However, Witt's paper "Theorie der quadratischen Formen in beliebigen Körpern" of 1937[15] opened up a new chapter in the theory: that of combining the number theoretic aspect with the algebraic development, by the creation of the famous Witt ring.

Then triggered off by Cassel's paper "On the representation of rational functions as sums of squares" of 1964[4], Albrecht Pfister, about 1966, come up with his celebrated structure theorems, giving birth to a purely algebraic theory of quadratic forms. Special cases of arithmetical aspect of Pfister's theory are his beautiful results about sums of squares and Pfister forms.

Our object here is to give a brief exposition of Pfister's work on sums of squares and related topics, one of the most beautiful and self contained set of results in any field (pun intended).

So let K be a field. We make the following

Definition 1. The smallest integer $s = s(K)$ for which the equation $-1 = a_1^2 + a_2^2 + \cdots + a_s^2$ $(a_j \in K)$ is solvable, is called the *Stufe* (often referred to as *level*) of K. If the equation has no solution, we put $s = \infty$ and call K *formally real*.

In 1932, Van der Waerden had posed the problem of enquiring which numbers can occur as Stufe. For example 3 can never occur as Stufe. Indeed if $-1 = x^2+y^2+z^2 (x, y, z \in K)$, then $0 = 1 + x^2 + y^2 + z^2$. Multiplying by $1 + x^2$ gives

$$\begin{aligned} 0 &= (1+x^2)^2 + (1+x^2)(y^2+z^2) && \text{(a)} \\ &= (1+x^2)^2 + (y+xz)^2 + (z-xy)^2 && \text{(b)} \end{aligned}$$

Now $1 + x^2 \neq 0$, otherwise Stufe $K = 1$. Hence (b) gives

$$-1 = \left(\frac{y+xz}{1+x^2}\right)^2 + \left(\frac{z-xy}{1+x^2}\right)^2$$

showing $s(K) \neq 3$.

It may similarly be shown that no odd number can be the Stufe of any field, but can 6 or 10 or 12 be the Stufe of a suitable field? We shall have to experiment with various fields and then make a conjecture. The rationals Q and the reals R, being formally real, are of no use, the complexes C provide a trivial example: $-1 = \iota^2$ giving $s(C) = 1$. So let us look at the imaginary quadratic fields. Indeed we have the following

Theorem 1. *Let $D > 0$ be a square free integer; then the Stufe $s(K)$ of $K = Q\left(\sqrt{-D}\right)$ is*

$$\begin{cases} 1 & \text{if } D = 1, \\ 2 & \text{if } D \neq 8b+7, \\ 4 & \text{if } D = 8b+7. \end{cases}$$

If $D < 0$, then K is of course formally real.

Proof: [12]. Writing $D = a^2+b^2+c^2+d^2$, $a, b, c, d, \in Z$, we see that $0 = (\sqrt{-D})^2 + a^2 + b^2 + c^2 + d^2$, giving $s(K) \leq 4$. Now $s(K) = 1$ if and only if $\sqrt{-1} \in K$ and this happens only in the case $D = 1$. If $D \not\equiv 7 \pmod 8$, then D is a sum of three squares and so $0 = (\sqrt{-D})^2 + a^2 + b^2 + c^2$, giving $s(K) \leq 3$. But we have already seen that s cannot be 3, hence $s(K) = 2$, the case $s = 1$ being fully cleared.

Finally let $D \equiv 7 \pmod 8$. If $s(K)$ were < 4, then it would be equal to 2, i.e.

$$-1 = (a_1 + b_1\sqrt{-D})^2 + (a_2 + b_2\sqrt{-D})^2, a_1, b_1, a_2, b_2, \in Q.$$

Here, without loss of generality, we may suppose that $b_1 \neq 0$. Equating reals and imaginaries, we get the following two equations:

$$a_1^2 + a_2^2 - D(b_1^2 + b_2^2) = -1$$
$$a_1b_1 + a_2b_2 = 0$$

These imply $D = (\frac{a_2}{b_1})^2 + (\frac{b_1}{b_1^2+b_2^2})^2 + b(\frac{b_2}{b_1^2+b_2^2})^2$. Thus D is a sum of 3 rational squares which is a contradiction since $D \equiv 7 \pmod 8$. Thus $s(K)$ not be less than 4 as required. □

Let us next look at all the finite fields. We have the following easy

Theorem 2. *Let F_q be the finite field of $q = p^\alpha$ elements; then*

$$s(F_q) = \left\{ \begin{array}{ll} 1 & \textit{if either } p = 2 \textit{ or } p \equiv 1(4), \text{ or } p \equiv 3(4), 2|\alpha, \\ 2 & \textit{otherwise i.e. if } p \equiv 3(4), 2 \not| \alpha. \end{array} \right\}$$

Proof: First let $p = 2$. Then $-1 = 1 = 1^2$ giving $s(F_2^\alpha) = 1$

Next if $p \equiv 1(4)$, then $(\frac{-1}{p}) = 1$ i.e. $-1 = x^2$ is solvable in $F_p \subset F_{p^\alpha}$ (for all α). So $s(F_{p^\alpha}) = 1$.

Let now $p \equiv 3(4)$. Let $A = \{-1 - X^2 | X = 1, 2, \ldots, \frac{p-1}{2}, 0\}$ and B $= \{Y^2 | Y = 1, 2, \ldots, \frac{p-1}{2}, 0\}$, A, B, $\subset F_p$. Then $|A| = |B| = \frac{p+1}{2}$. Hence by the pigeon hole principle, there exist $X_0, Y_0 \in F_p$ such that $Y_0^2 = -1 - X_0^2$, i.e. $-1 = X_0^2 + Y_0^2$ in F_p. But -1 is not a square in F_p since $p \equiv 3(4)$. It follows that $s(F_p) = 2$ if $p \equiv 3(4)$.

Now $F_p^2 = F_p(\sqrt{-1})$ and here $-1 = (\sqrt{-1})^2$; so $s(F_{p^2}) = 1$. But $F_{p^\alpha} \supset F_{p^2}$ if $2|\alpha$, so F_{p^α} has Stufe 1 if $2|\alpha$.

If $s(F_{p^\alpha}) = 1$ even for $2 \not| \alpha$, then $-1 = X^2$ is solvable in $F_{p^\alpha}(2 \not| \alpha)$, so $F_{p^\alpha} \supset F_p(X) = F_p(\sqrt{-1}) \cong F_{p^2}$ which is false since $2 \not| \alpha$. Hence $s(F_{p^\alpha}) = 2$ if $2 \not| \alpha$. □

That more or less exhausts all the easy fields using elementary methods. Even allowing the Hasse-Minkowski theorem all those algebraic number fields K for which $s(K)$ exists finite can be dealt with in a single go. The exact result is the following:

Theorem A. *Let $K = Q(\alpha)$ be an algebraic number field with $[K{:}Q] = n$ finite. Then $s(K)$ exists finite iff K is totally complex (i.e. all the zeros of irr (α, Q) are non-real) and then $s(K) \leq 4$ i.e. equals $1, 2$ or 4, because we have seen that Stufe can not be 3 and in fact*

$$s(K) = \begin{cases} 1 & \text{iff } i \in K (i = \sqrt{-1}) \\ 4 & \text{iff } i \notin K \text{ and for all primes } y|2, \text{ the local degree } [K_y : Q_2] \text{ is odd.} \end{cases}$$

For a proof see [13], p 261.

We see that experimentation is not easy and so it was all the more surprising, when Pfister proved the following beautiful

Theorem 3. *For any field K, the Stufe $s(K)$, if finite, is always a power of 2. Conversely, every power of 2 is the Stufe of some field K.*

We shall give a proof of this result in the sequel. In showing that 3 cannot occur as Stufe, the transition from equation (a) to (b) (see before theorem 1) is the crucial step in the process. We have, more generally, the curious looking identity.

$$(X_1^2 + X_2^2)(Y_1^2 + Y_2^2) = (X_1Y_1 - X_2Y_2)^2 + (X_1Y_2 + X_2Y_1)^2 \tag{1}$$

which tells its that *a product of two sums of two squares is itself a sum of two squares.* Known to the Greeks, (1) is equivalent to the statement, The norm of the product of two complex numbers Z_1, Z_2, is the product of their norms:

$$|Z_1Z_2|^2 = |Z_1|^2|Z_2|^2 \tag{1'}$$

for writing $Z_1 = X_1 + iX_2, Z_2 = Y_1 + iY_2$, we see that $Z_1.Z_2 = (X_1Y_1 - X_2Y_2) + i(X_1Y_2 + Y_2X_2)$ and so (1) and (1′) are the same.

This identity (1) enables us to prove another curious result:

For any field K, the set $G_2(K) = \{a \in K^* | a = x^2 + y^2, x, y, \in K\}$ is a multiplicative group.

For, the closure property is the identity (1) while if

$$a = x^2 + y^2 \in G_2(K), \text{ then } \frac{1}{a} = \frac{a}{a^2} = \frac{x^2 + y^2}{a^2} = \left(\frac{x}{a}\right)^2 + \left(\frac{y}{a}\right)^2 \in G_2(K).$$

The following striking identity was already known to Euler in 1770 and he used it to prove Lagrange's theorem that every positive integer is a sum of four squares.

$$(X_1^2 + X_2^2 + X_3^2 + X_4^2)(Y_1^2 + Y_2^2 + Y_3^2 + Y_4^2) = (Z_1^2 + Z_2^2 + Z_3^2 + Z_4^2) \tag{2}$$

where

$$Z_1 = X_1Y_1 - X_2Y_2 - X_3Y_3 - X_4Y_4$$

$$Z_2 = X_1Y_2 + X_2Y_1 + X_3Y_4 - X_4Y_3$$

$$Z_3 = X_1Y_3 + X_3Y_1 - X_2Y_4 + X_4Y_2$$

$$Z_4 = X_1Y_4 + X_4Y_1 + X_2Y_3 - X_3Y_2$$

The discovery of quaternions by William Hamilton, in 1843, brought out the real significance of the identity (2) in as much as (2) is simply the fact that the norm of a product of two quaternions is equal to the product of their norms.

Almost immediately after Hamilton's discovery of the quaternions, Arthur Cayley, in 1845 discovered the octonions (the Cayley numbers) which give rise to the incredible looking identity

$$(X_1^2 + \cdots + X_8^2)(Y_1^2 + \cdots + Y_8^2) = Z_1^2 + \cdots + Z_8^2 \tag{3}$$

where

$$\begin{aligned}
Z_1 &= X_1Y_1 - X_2Y_2 - X_3Y_3 - X_4Y_4 - X_5Y_5 - X_6Y_6 - X_7Y_7 - X_8Y_8,\\
Z_2 &= X_1Y_2 + X_2Y_1 + X_3Y_4 - X_4Y_3 + X_5Y_6 - X_6Y_5 - X_7Y_8 + X_8Y_7,\\
Z_3 &= X_1Y_3 + X_3Y_1 - X_2Y_4 + X_4Y_2 + X_5Y_7 - X_7Y_5 + X_6Y_8 - X_8Y_6,\\
Z_4 &= X_1Y_4 + X_4Y_1 + X_2Y_3 - X_3Y_2 + X_5Y_8 - X_8Y_5 - X_6Y_7 + X_7Y_6,\\
Z_5 &= X_1Y_5 + X_5Y_1 - X_2Y_6 + X_6Y_2 - X_3Y_7 + X_7Y_3 - X_4Y_8 + X_8Y_4,\\
Z_6 &= X_1Y_6 + X_6Y_1 + X_2Y_5 - X_5Y_2 - X_3Y_8 + X_8Y_3 + X_4Y_7 - X_7Y_4,\\
Z_7 &= X_1Y_7 + X_7Y_1 + X_2Y_8 - X_8Y_2 + X_3Y_5 - X_5Y_3 - X_4Y_6 + X_6Y_4,\\
Z_8 &= X_1Y_8 + X_8Y_1 - X_2Y_7 + X_7Y_2 + X_3Y_6 - X_6Y_3 + X_4Y_5 - X_5Y_4.
\end{aligned}$$

Although the identity emerges most naturally from Cayley numbers, it was discovered nearly a quarter of a century earlier by C.F. Degan (1822) with minor sign differences.

Degan stated (erroneously of course) that there is a like formula for 2^n squares. For the case of 16 squares, he gave the literal parts of the 16 bilinear functions $Z_1, Z_2, \ldots, Z_{16}$ but left most of the signs undetermined, saying that the only difficulty is the prolixity of the ambiguities of signs.

Degan was also aware of the 2– and 4– variable Pfister forms $X_1^2 - aX_2^2$ and $X_1^2 + aX_2^2 + b(X_3^2 + aX_4^2)$ both of which satisfy identities similar to (1) and (2).

As before, if we define

$$G_4 = \{a \in K^* | a = x_1^2 + \cdots + x_4^2, x_j \in K\}$$

and G_8 similarly, then it follows from (2) and (3) respectively that G_4 and G_8 are groups under multiplication, so that we have the chain of inclusions

$$K^{*^2} \subseteq G_2 \subseteq G_4 \subseteq G_8 \subseteq K^*.$$

A great many unsuccessful attempts followed Degan's discovery of (3), to extend formulae (1), (2) and (3) to a similar 16 term identity, and many workers, realizing the impossibility of such an extension, tried giving convincing arguments to prove the impossibility. Hamilton's and Cayley's discoveries had reduced the problem to the determination of the so-called normed algebras over the real numbers **R**; the four known ones being **R** (of dimension 1), the complex numbers **C** (of dimension 2), the quaternions **H** (of dimension 4) and the octonions **O** (of dimension 8). It is an astonishing observation how the axioms of the ordered field **R** gradually drop off as we move up these higher dimensional hypercomplex systems: **C** is, no doubt a field, commutative and associative (under multiplication) and a division ring, but the order property is lost. **H** is only an associative division ring; thus commutativity, and order are both lost. Finally **O** is not even associative-it is merely a division ring; thus commutativity, associativity and order are all lost.

The half century following the discovery of these quaternions and octonions then saw many attempts to find a 16-dimensional hypercomplex system over the reals and several erroneous affirmations were given. Finally in 1898, Hurwitz [6] gave a decisive solution to the problem about the dimensionality of all possible normed algebras over **R** and so also about the possible values of n for which there is an identity of the type (3) with n terms. More precisely we have the following.

Theorem 4 (Hurwitz-1898). *Let K be a field with char $K \neq 2$. The only values of n for which there is an identity of the type*

$$(X_1^2 + \cdots + X_n^2)(Y_1^2 + \cdots + Y_n^2) = Z_1^2 + \cdots + Z_n^2 \tag{4}$$

where the Z_k are bilinear functions of the X_i and the Y_j, coefficients in K are $n = 1, 2, 4, 8$.

Actually Hurwitz proved this only over **C** but his proof generalizes to any field K with char $K \neq 2$. We give here a proof given by Dickson in his beautiful expository paper [5] of 1919. A proof using normed algebras can be found in A.A. Albert's *Studies in Modern Algebra* [2].

Proof: (Dickson). The idea is to convert (4) into a system of matrix equations. The bilinearity condition on the Z_k can be written as

$$\begin{pmatrix} Z_1 \\ Z_2 \\ \vdots \\ Z_n \end{pmatrix} = \begin{pmatrix} a_{11} & a_{12} & \cdots & a_{1n} \\ a_{21} & a_{22} & \cdots & a_{2n} \\ \cdots & \cdots & \cdots & \cdots \\ a_{n1} & a_{n2} & \cdots & a_{nn} \end{pmatrix} \begin{pmatrix} Y_1 \\ Y_2 \\ \vdots \\ Y_n \end{pmatrix} = AY,$$

where the a_{ij} are linear functions of $X_1, X_2, \ldots, X_n$. Then (4) becomes

$$(X_1^2 + \cdots + X_n^2)(Y_1, \ldots, Y_n) \begin{pmatrix} Y_1 \\ Y_2 \\ \vdots \\ Y_n \end{pmatrix} I_n = (Z_1, \ldots, Z_n) \begin{pmatrix} Z_1 \\ \vdots \\ Z_n \end{pmatrix} = Y'A'AY,$$

i.e.

$$(Y_1, Y_2, \ldots, Y_n)[(X_1^2 + X_2^2 + \cdots + X_n^2)I_n - A'A] \begin{pmatrix} Y_1 \\ Y_2 \\ \vdots \\ Y_n \end{pmatrix} = 0,$$

and since this is true for all $Y_1, Y_2, \ldots, Y_n$, it follows that

$$A'A = (X_1^2 + \cdots + X_n^2)I_n \tag{a}$$

Now

$$A = \begin{pmatrix} a_{11} & a_{12} & \cdots \\ a_{21} & a_{22} & \cdots \\ \cdots & \cdots & \cdots \end{pmatrix}$$

$$= \begin{pmatrix} b_{11}^{(11)}X_1 + b_{11}^{(12)}X_2 + \cdots, & b_{12}^{(11)}X_1 + b_{12}^{(12)}X_2 + \cdots \\ b_{21}^{(11)}X_1 + b_{21}^{(12)}X_2 + \cdots, & b_{22}^{(11)}X_1 + b_{22}^{(12)}X_2 + \cdots \\ \cdots & \cdots \end{pmatrix}$$

$$= A_1X_1 + A_2X_2 + \cdots + A_nX_n \quad \text{say}$$

By (a),

$$(A_1'X_1 + A_1'X_2 + \cdots + A_n'X_n)(A_1X_1 + A_2X_2 + \cdots + A_nX_n)$$
$$= (X_1^2 + X_2^2 + \cdots + X_n^2)I_n.$$

Since this is true for all X_j, we have

(1) $A_j'A_j = I_n (j = 1, 2, \ldots, n)$, hence also $A_jA_j' = I_n$.
(2) $A_j'A_k + A_k'A_j = 0, 1 \leq j, k \leq n, j \neq k$.

Conversely, the existence of such a system implies that (4) holds with Z_k bilinear in the X_i and the Y_j. Note also that if $n = 1$, (2) is vacuous, and (1) can be trivially satisfied so we may suppose $n > 1$.

Now let $B_i = A_n'A_i (i = 1, 2, \ldots, n-1)$. The B's are easily seen to satisfy

(1) $B_i'B_i = I_n$
(2) $B_i' + B_i = 0 \quad (i, j = 1, 2, \ldots, n-1)$
(3) $B_i'B_j + B_j'B_i = 0 \quad (i \neq j)$

Hence we have

$$\left.\begin{array}{ll} \text{(i)} & B_i' = -B_i (i = 1, 2, \ldots, n-1) \text{ i.e. the } B_i \\ & \quad \text{are skew-symmetric matrices} \\ \text{(ii)} & B_i^2 = -I (i = 1, 2, \ldots, n-1) \\ \text{(iii)} & B_iB_j = -B_jB_i, \ i, j = 1, 2, \ldots, n-1 \quad i \neq j. \end{array}\right\} \tag{b}$$

It follows that $|B_i| = |B_i'| = |-B_i| = (-1)^n|B_i|$, and since $|B_i| \neq 0$ we must have n even. Hence

Proposition 1. *There is no identity of the type* (4) *if* $n (> 1)$ *is odd.*

In future, therefore, we suppose n to be even. Now consider the following set $\mathcal{G}$ of $n \times n$ matrices:

$$\{I, B_{i_1}, B_{i_1}B_{i_2}, B_{i_1}B_{i_2}B_{i_3}, \ldots, B_{i_1}B_{i_2}, \ldots, B_{i_{n-2}}$$
$$\text{and } B_1B_2, \ldots, B_{n-1} (i_1 < n, i_1 < i_2 < n, \ldots)\}.$$

Here B_{i_1} takes $n-1$ values viz. $B_1, B_2, \ldots, B_{n-1}$, while $B_{i_1}B_{i_2}$ takes $\binom{n-1}{2}$ values viz. $B_1B_2, B_1B_3, \ldots$ etc. So altogether there are $1+\binom{n-1}{1}+\cdots+\binom{n-1}{n-1}=2^{n-1}$ elements in the set $\mathcal{G}$. Let $G=B_{i_1}B_{i_2},\ldots,B_{i_r}\in\mathcal{G}$. Then we have

Lemma 1. *G is symmetric if $r\equiv 0$ or 3 modulo 4, and skew-symmetric if $r\equiv 1$ or 2 modulo 4.*

Proof:

$$\begin{aligned} G' &= B'_{i_r},\ldots,B'_{i_1}=(-1)^r B_{i_r},\ldots,B_{i_1} \\ &= (-1)^r(-1)^{r-1}B_{i_1}(B_{i_r},\ldots,B_{i_2}) \end{aligned}$$

by (iii) of (b) to commute B_{i_1} successively with $B_{i_2},\ldots,B_{i_r}$,

$$\begin{aligned} &= (-1)^r(-1)^{r-1}(-1)^{r-2}B_{i_1}B_{i_2}(B_{i_r},\ldots,B_{i_3}) \\ &\qquad \text{and so on} \\ &= (-1)^r(-1)^{r-1}\ldots(-1)^2(-1)B_{i_1}B_{i_2},\ldots,B_{i_r} \\ &= (-1)^{1+2+\cdots+r}G \\ &= (-1)^{r(r+1)/2}G \\ &= \begin{cases} G & \text{if } r\equiv 0,3,(4) \\ -G & \text{if } r\equiv 1,2,(4) \end{cases} \end{aligned}$$

□

Lemma 2. *Let $M\in\mathcal{G}$. Then the set $M\mathcal{G}=\{MG|G\in\mathcal{G}\}$ is simply a permutation of $\mathcal{G}$ with each term prefixed with either $+1$ or -1.*

Proof: The result is clear if the multiplier M is B_1, since then the product will contain or lack B_1 according as the multiplicand of $\mathcal{G}$ lacks or contains B_1 (use again (b)).

If the multiplier is B_2, we first replace $B_1B_2,\ldots$, wherever it appears, by $B_2B_1,\ldots$, and see that the former argument applies.

After thus proving our statement when the multiplier is any B_i, we see that it holds when the multiplier is any product of the B's. □

An Example: $n=4$.

$$\mathcal{G}=\{I, B_1, B_2, B_3, B_1B_2, B_2B_3, B_1B_3, B_1B_2B_3\}.$$

Then

$$\begin{aligned} B_3\mathcal{G} &= \{B_3, B_3B_1, B_3B_2, B_3^2, B_3B_1B_2, B_3B_2B_3, B_3B_1B_3, B_3B_1B_2B_3\}. \\ &= \{B_3, -B_1B_3, -B_2B_3, -I, B_1B_2B_3, B_2, B_1, -B_1B_2\} \\ &= \{-I, B_1, B_2, B_3, -B_1B_2, -B_2B_3, -B_1B_3, B_1B_2B_3\}. \end{aligned}$$

Our aim is now the following.

Proposition 2. *At least half of the elements of $\mathcal{G}$ are linearly independent.*

With this in view, we look for any linear relations that can exist amongst the elements of $\mathcal{G}$.

Definition: A relation $\lambda_1 G_1 + \lambda_2 G_2 + \cdots + \lambda_s G_s = 0, G_j \in \mathcal{G}, \lambda_j \in \mathbf{R}$, or $R = 0$ for short, is called *irreducible* if it is not possible to express R as $R_1 + R_2$, where $R_1 = 0$, $R_2 = 0$ represent two linear relations that hold between the subsets R_1 and R_2 of R with $R_1 \cap R_2 = \emptyset$, i.e. there are no matrices common to R_1 and R_2.

We have the following.

Lemma 3. *An irreducible relation $R = 0$ cannot involve both symmetric and skew-symmetric matrices.*

Proof: Let M_1 be the subset of all symmetric matrices in R and M_2 the set of all skew-symmetric matrices in R. Then $M_1 + M_2 = 0$, i.e. $M_1 = -M_2$. Hence $M_1 = M_1' = -M_2' = M_2$, i.e. $M_1 = M_2$. It follows that $M_1 = 0$, $M_2 = 0$ which contradicts the irreducibility of $R = 0$. □

Now let $R = 0$ be any irreducible relation between the matrices of $\mathcal{G}$. By multiplying R by a suitable λG ($\lambda \in \mathbf{R}, G \in \mathcal{G}$) we get a new relation $T = 0$, one term of which is I and all the remaining terms are products of matrices of $\mathcal{G}$ by real constants. For suppose μG ($\mu \in \mathbf{R}, G \in \mathcal{G}$) is a term in R which we wish should become I in the relation $T = 0$. One just multiplies $R = 0$ by $\pm\, \mu^{-1} G^{-1}$ and notes that one of $\pm\, G^{-1} \in \mathcal{G}$.

For example if $4B_2B_3$ is one term of R, then on multiplying $R = 0$ by

$$-\frac{1}{4}(B_2B_3)^{-1} = -\frac{1}{4}B_3^{-1}B_2^{-1} = -\frac{1}{4}(-B_3)(-B_2) = -\frac{1}{4}B_3B_2 = \frac{1}{4}B_2B_3,$$

we get what is required.

This new relation $T = 0$ is also irreducible, for if $T = 0$ were to split as $T_1 = 0$, $T_2 = 0$, then since $T = \lambda G R$ we have $\lambda^{-1}G^{-1}T = R$ and so $R = 0$ splits as $\lambda^{-1}G^{-1}T_1 = 0$, $\lambda^{-1}G^{-1}T_2 = 0$, which gives a contradiction.

Hence we may suppose that $T = 0$ looks like

$$I = \sum c_{i_1 i_2 i_3} B_{i_1} B_{i_2} B_{i_3} + \sum d_{i_1 i_2 i_3 i_4} B_{i_1} B_{i_2} B_{i_3} B_{i_4} + \cdots \tag{$*$}$$

where by Lemma 3, each of the matrices $B_{i_1}B_{i_2}B_{i_3}$, $B_{i_1}B_{i_2}B_{i_3}B_{i_4}$, etc. is symmetric since I is symmetric. That is why no singleton B_i nor any of the products B_iB_j of two B's can be involved in $(*)$ since B_i and B_iB_j are skew-symmetric by Lemma 1.

Now multiply $(*)$ throughout on the right by B_i to obtain an irreducible relation which then involves only skew-symmetric matrices since one term (on the left side) is the skew-symmetric matrix B_i. But by Lemma 1, $B_{i_1}B_{i_2}B_{i_3}B_{i_4}$ is symmetric. So all the c_j are 0 if only i is distinct from $i_1 i_2 i_3$. Since i may have any value $\leq n - 1$, we see that each c is 0 unless $n - 1 = 3$ for then i cannot be chosen different from i_1, i_2, i_3.

Next we show that all the d's are 0 too; for multiply $(*)$ by B_{i_4} and it becomes

$$B_{i_4} = \sum d_{i_1 i_2 i_3 i_4} B_{i4} B_{i_1} B_{i_2} B_{i_3} B_{i_4} + \cdots$$

But $B_{i_4} B_{i_1} B_{i_2} B_{i_3} B_{i_4} = (-1)^3 B_{i_1} B_{i_2} B_{i_3} B_{i_4}^2 = B_{i_1} B_{i_2} B_{i_3}$. So $(*)$ becomes

$$B_{i_4} = \sum d_{i_1 i_2 i_3 i_4} B_{i_1} B_{i_2} B_{i_3} + \cdots.$$

Here $B_{i_1}B_{i_2}B_{i_3}$ is symmetric, while B_{i_4} is skew-symmetric (by Lemma 1). It follows that all the d's are 0 too.

The method used in proving $c = 0$ applies when the number r of factors in $B_{i_1}B_{i_2}, \ldots, B_{i_r}$ is $\equiv 3(4)$ and $r < n - 1$. Similarly the method used in proving $d = 0$ also applies when $r \equiv 0(4)$.

Hence if our relation exists, it has the form

$$I = kB_1B_2, \ldots, B_{n-1}$$

the right hand term being the only survivor. Now I is symmetric so $B_1B_2, \ldots, B_{n-1}$ is symmetric i.e. $n - 1 \equiv 0$ or $3(4)$, but n is even so $n - 1 \equiv 3(4)$ i.e. $n \equiv 0(4)$

We have thus proved the following.

If an irreducible relation between the elements of $\mathcal{G}$ does exist, then $n \equiv 0(4)$.

Now square this relation to get

$$\begin{aligned} I &= k^2B_1B_2, \ldots, B_{n-1}B_1B_2, \ldots, B_{n-1} \\ &= k^2(-1)^{n-2}B_2 \ldots B_{n-1}B_2 \ldots B_{n-1} \\ &= \ldots\ldots\ldots\ldots\ldots \\ &= k^2(-1)^{\frac{1}{2}(n-1)n}I. \end{aligned}$$

Since $n \equiv 0(4)$, we see that $k^2 = 1$ i.e. $k = \pm 1$. Hence we have the following.

Lemma 4. *If $n \equiv 2(4)$ then the 2^{n-1} matrices of $\mathcal{G}$ are linearly independent, while for $n \equiv 0(4)$, they are either linearly independent or are connected by the relations which arise from the relation $I = \pm B_1B_2, \ldots, B_{n-1}$ through multiplication by the various elements of $\mathcal{G}$, but are connected by no further irreducible linear relations.*

Example: let $n = 4$. Then

$$\mathcal{G} = \{I, B_1, B_2, B_3, B_1B_2, B_2B_3, B_1B_3, B_1B_2B_3\}$$

and these eight matrices are either linearly independent or are connected by the following four irreducible linear relations and no others:

$$I = \pm B_1B_2B_3, \; B_1 = \mp B_2B_3, \; B_2 = \pm B_1B_3, \; B_3 = \mp B_1B_2.$$

These express $B_1B_2B_3$, B_2B_3, B_1B_3, B_1B_2 linearly in terms of I, B_1, B_2, B_3; so that these latter matrices are, in any case, linearly independent.

Now consider all the irreducible linear relations that exist between the element of $\mathcal{G}$. As we have seen, they are all of the type

$$G \cdot I = \pm G \cdot B_1B_2, \ldots, B_{n-1}(G \in \mathcal{G})$$

and no others. Now reduce the right side of this using (b). Then one of G or the reduced right side obviously contains fewer than half of the B's while the other contains more than half of the B's.

Thus these irreducible linear relations merely serve to express the products containing more than half of the B's in terms of those with less than half of the B's.

So in every case (i.e. irrespective of whether $n \equiv 0$ or 2 (mod 4)) the 2^{n-2} matrices of $\mathcal{G}$, which are products of less than $\frac{n-1}{2}$ B's, are linearly independent. Hence for all values of n (necessarily even) if there is to be an identity of the type (4), the 2^{n-2} matrices of $\mathcal{G}$ consisting of the product of at most $\frac{n-2}{2}$ B's are linearly independent.

This completes the proof of Proposition 2. □

We can now give a proof of our main result.

The elements of $\mathcal{G}$ are all $n \times n$ matrices and the maximum number of linearly independent $n \times n$ matrices is n^2 since they form, over the reals, a vector space of dimension n^2. Hence by the proposition we get

$$2^{n-2} \leq n^2.$$

This is satisfied if $n \leq 8$ but fails if $n = 10$. Now if it fails for $n = m$, then it fails for $n = m + 1$ for we have

$$\begin{aligned} 2^{m+1-2} &= 2 \cdot 2^{m-2} > 2 \cdot m^2 \text{ (since the relation fails for } m) \\ &> (m+1)^2 \text{ if } m \geq 3. \end{aligned}$$

It follows that if an identity of the type (4) exists, then $n \leq 8$ (and n is even). For $n = 2, 4, 8$ we already have the required type of identities. It remains to dispose off the case $n = 6$.

Suppose an identity exists for $n = 6$. Then since $6 \equiv 2(4)$, we see that

the 2^5 matrices of $\mathcal{G}$ are linearly independent. (i)

Of these 32 matrices, 16 are skew-symmetric by Lemma 1, viz. the ones that are products of 1, 2 or 5 B's. But

between any 16 skew-symmetric 6×6 matrices there exists a linear relation. (ii)

This is because the 15 matrices

$$\begin{pmatrix} 0 & 1 & 0 & \cdots \\ -1 & 0 & 0 & \cdots \\ 0 & 0 & 0 & \cdots \\ \cdots & \cdots & \cdots & \cdots \end{pmatrix}, \begin{pmatrix} 0 & 0 & 1 & \cdots \\ 0 & 0 & 0 & \cdots \\ -1 & 0 & 0 & \cdots \\ \cdots & \cdots & \cdots & \cdots \end{pmatrix}, \ldots$$

with a 1 in the one place above the main diagonal, -1 in the corresponding place below, and 0's elsewhere, form a basis for the subspace of all 6×6 skew-symmetric matrices and so this subspace has dimension 15. This proves (ii).

(i) and (ii) above are contradictory. Hence no identity of type (4) can exist for $n = 6$.

That at last completes the proof of Hurwitz's theorem. □

Remark: The proof works for any field K of characteristic $\neq 2$.

Although the impossibility of the identity (4) for n $\neq 1, 2, 4, 8$ has been proved, it was under the stringent restriction that the Z_k are bilinear polynomials in the X_i and the Y_j. One could look into the possibility of the existance of other values of n for which (4) holds, if we relax this bilinear condition and allow the Z_k to be more general polynomials in the X_i and the Y_j. However, in 1966, Frank Adams [1] showed that when n is not 1, 2, 4, 8, there are no identities of the type (4) even if the Z_k are allowed to be any bi-skew, continuous functions of the X_i and the Y_j (where a mapping $f : K^r \times K^s \to K^n$ is called bi-skew if

$$f(-\underline{x}, \underline{y}) = f(\underline{x}, -\underline{y}) = -f(x, y) \text{ for all } \underline{x} \in K^r, \underline{y} \in K^s).$$

It was thus totally unexpected when in 1965, Albrecht Pfister [10] proved the following remarkable.

Theorem 5. *Let K be a field and let $n = 2^m$ be a power of 2. Then there are identities.*

$$(X_1^2 + \cdots\cdots + X_n^2)(Y_1^2 + \cdots\cdots + Y_n^2) = Z_1^2 + \cdots\cdots + Z_n^2 \tag{5}$$

where the Z_k are linear functions of the Y_j with coefficients in $K(X_1, \ldots, X_n)$: $Z_k = \sum_j T_{kj} \cdot Y_j$ with $T_{kj} \in K(X_1, \ldots, X_n)$.

Conversely suppose n is not a power of 2. Then there is a field K such that there is no identity (5) with the $Z_k \in K(X_1, \ldots, X_n, \quad Y_1, \ldots, Y_n)$. Here the Z_k are not even demanded to be linear in the Y_j.

We shall now give proofs of theorems 3 and 5. In the process we shall get other results which are interesting in their own right.

The proof cf the first part of theorem 5 requires no elaborate algebraic machinery and is, indeed, remarkably simple. We dispose of it first.

Proof of the first part of theorem 5: We use induction on m. We know that (5) holds for $m = 1, 2, 3$ (see (1), (2), (3)). Suppose it holds for m. Write $T = (T_{ij})$ so that

$$\begin{pmatrix} Z_1 \\ Z_2 \\ \vdots \\ Z_n \end{pmatrix} = T \begin{pmatrix} Y_1 \\ Y_2 \\ \vdots \\ Y_n \end{pmatrix} \tag{i}$$

Then (5) can be written as

$$\begin{aligned}(X_1^2 + \cdots + X_n^2)(Y_1^2 + \cdots + Y_n^2) &= (X_1^2 + \cdots + X_n^2)(Y_1, \ldots, Y_n)\begin{pmatrix} Y_1 \\ \vdots \\ Y_n \end{pmatrix} \\ &= (Z_1^2 + \cdots + Z_n^2) = (Z_1, \ldots, Z_n)\begin{pmatrix} Z_1 \\ \vdots \\ Z_n \end{pmatrix} \\ &= (Y_1, \ldots, Y_n)T'T\begin{pmatrix} Y_1 \\ \vdots \\ Y_n \end{pmatrix}, \text{ by (i), i.e.}\end{aligned}$$

$$(X_1^2 + \cdots + X_n^2)(Y_1, \ldots, Y_n)I_n\begin{pmatrix} Y_1 \\ \vdots \\ Y_n \end{pmatrix} - (Y_1, \ldots, Y_n)T'T\begin{pmatrix} Y_1 \\ \vdots \\ Y_n \end{pmatrix} = 0$$

or

$$(Y_1, \ldots, Y_n)\{(X_1^2 + \cdots + X_n^2)I_n - T'T\}\begin{pmatrix} Y_1 \\ \vdots \\ Y_n \end{pmatrix} = 0.$$

Since this is true for all $Y_1, \ldots, Y_n$, we must have

$$TT' = (X_1^2 + \cdots + X_n^2)I_n$$

and so also $T'T = (X_1^2 + \cdots + X_n^2)I_n$, T being orthogonal.

We now prove (5) for $2^{m+1} = 2n$. Write

$$(X_1, \ldots, X_{2n}) = (\mathbf{X}^{(1)}, \mathbf{X}^{(2)})$$

where $\mathbf{X}^{(1)} = (X_1, \ldots, X_n)$ and $\mathbf{X}^{(2)} = (X_{n+1}, \ldots, X_{2n})$. By the induction hypothesis there exist two matrices $T^{(1)}, T^{(2)}$ say, corresponding to $\mathbf{X}^{(1)}, \mathbf{X}^{(2)}$ respectively such that

$$\left.\begin{array}{l} (X_1^2 + \cdots + X_n^2)I_n = \mathbf{X}^{(1)}\mathbf{X}^{(1)'}I_n = T^{(1)}T^{(1)'} = T^{(1)'}T^{(1)} \\ \text{and} \\ (X_{n+1}^2 + \cdots + X_{2n}^2)I_n = \mathbf{X}^{(2)}\mathbf{X}^{(2)'}I_n = T^{(2)}T^{(2)'} = T^{(2)'}T^{(2)} \end{array}\right\} \quad \text{(ii)}$$

and we wish to show that there exists a matrix T say, such that

$$T'T = (X_1^2 + \cdots + X_n^2 + X_{n+1}^2 + \cdots + X_{2n}^2)I_{2n}. \quad \text{(iii)}$$

Try $T = \begin{pmatrix} T^{(1)} & T^{(2)} \\ T^{(2)} & X \end{pmatrix}$-a partitional matrix, where X will be determined by (iii). We have

$$T'T = \begin{pmatrix} T^{(1)'} & T^{(2)'} \\ T^{(2)'} & X' \end{pmatrix}\begin{pmatrix} T^{(1)} & T^{(2)} \\ T^{(2)} & X \end{pmatrix},$$

and using block multiplication of matrices this equals

$$\begin{pmatrix} T^{(1)'}T^{(1)} + T^{(2)'}T^{(2)} & T^{(1)'}T^{(2)} + T^{(2)'}X \\ T^{(2)'}T^{(1)} + X'T^{(2)} & T^{(2)'}T^{(2)} + X'X \end{pmatrix}$$

$$= \begin{pmatrix} (X_1^2 + \cdots + X_n^2 + X_{n+1}^2 + \cdots + X_{2n}^2)I_n & A \\ B & C \end{pmatrix}$$

say; we want to choose X so that $A = B = 0$ and

$$C = (X_1^2 + \cdots + X_n^2 + X_{n+1}^2 + \cdots + X_{2n}^2)I_n.$$

To make $A = 0$ we have to have $X = -T^{(2)'^{-1}} T^{(1)'} T^{(2)}$. This automatically makes $B = 0$ (just check). Now it seems too much to expect C to be what we want. But we have

$$\begin{aligned} C &= T^{(2)'} T^{(2)} + T^{(2)'} T^{(1)} T^{(2)'^{-1'}} T^{(2)'^{-1}} T^{(1)'} T^{(2)} \\ &= (X_{n+1}^2 + \cdots + X_{2n}^2) I_n + (X_{n+1}^2 + \cdots + X_{2n}^2)^{-1} T^{(2)'} T^{(1)} T^{(1)'} T^{(2)} \\ &= (X_{n+1}^2 + \cdots + X_{2n}^2) I_n + (X_{n+1}^2 + \cdots + X_{2n}^2)^{-1} \\ &\quad (X_1^2 + \cdots + X_n^2) T^{(2)'} T^{(2)} \\ &= (X_{n+1}^2 + \cdots + X_{2n}^2) I_n + (X_1^2 + \cdots + X_n^2) I_n \\ &= (X_1^2 + \cdots + X_n^2 + X_{n+1}^2 + \cdots + X_{2n}^2) I_n. \end{aligned}$$

This completes the proof of the first part of Theorem 5. □

We now come to a result of Cassels [4], which, in a way, was the starting point of this whole business and which is an indispensible tool in our further developments.

Cassels' Lemma (1964). *Let $f(X) \in K(X)$ be a polynomial with coefficients in K. If $f(X)$ is a sum of n squares of elements of the field $K(X)$, then it is a sum of n squares of elements of the ring $K[X]$.*

Note: What is new in this enunciation is that the same number n of squares suffice in $K[X]$; without this condition, the result had been proved by Artin [3].

Proof: There are three trivial cases of the lemma which we dispose of first.

(i) $n = 1$. Then $f(X) = (p(X)|q(X))^2$, *so* $q(X)\big|p(X)$.

(ii) char $K = 2$. Then $a^2 + b^2 = (a+b)^2$ and so if

$$f(X) = \gamma_1^2(X) + \cdots + \gamma_n^2(X)$$

then combining two squares at a time into one, $f(X)$ reduces to a single square, i.e. we land up in case (i).

(iii) -1 is a sum of $n-1$ squares of elements in K.

Say $-1 = b_1^2 + \cdots + b_{n-1}^2$. Then for any $f(X)$, we have

$$\begin{aligned} f(X) &= \left(\frac{f+1}{2}\right) - \left(\frac{f-1}{2}\right)^2 \\ &= \left(\frac{f+1}{2}\right)^2 + \left(b_1 \frac{(f-1)}{2}\right)^2 + \cdots + \left(b_{n-1} \frac{(f-1)}{2}\right)^2 \end{aligned}$$

a sum of n squares of elements of $K[X]$.

So now let us suppose none of these three cases holds and let

$$f(X) = (p_1(X)/q_1(X))^2 + \cdots + (p_n(X)/q_n(X))^2.$$

Dropping the X from now on and clearing the denominators, this gives

$$fZ^2 = Y_1^2 + \cdots + Y_n^2, \quad Z, Y_1, \ldots, Y_n \in K[X], Z \neq 0.$$

Thus the equation

$$fZ^2 = Y_1^2 + \cdots + \cdots + Y_n^2 \tag{a}$$

has a solution $(Z, Y_1, \ldots, Y_n)$ with $Z \neq 0$ and we have to show that there exists a solution of (a) with $Z \in K (Z \neq 0)$, i.e. with degree of Z (in X) $= 0$. Now since (a) has a solution with $Z \neq 0$, so there is a solution, call it $(\zeta, \eta_1, \ldots, \eta_n)$, with $\zeta \neq 0$ for which deg ζ is as small as possible:

$$f\zeta^2 = \eta_1^2 + \cdots + \eta_n^2. \tag{b}$$

We shall show that this degree is 0 i.e. that $\zeta \in K$, by showing that if not, then there exists a solution, say $(\zeta^*, \eta_1^*, \ldots, \eta_n^*)$ with $\zeta^* \neq 0$ and deg $\zeta^* <$ deg ζ.

So suppose deg $\zeta > 0$. By the division al rithm in $K[X]$, we can write, for $j = 1, 2, \ldots, n$,

$$\eta_j = \lambda_j \zeta + \gamma_j$$

where either $\gamma_j = 0$ or deg $\gamma_j <$ deg ζ. i.e.

$$\eta_j/\zeta = \lambda_j + \gamma_j/\zeta = \lambda_j + \Lambda_j, \tag{c}$$

say. Note that not all the γ_j can be zero, otherwise ζ divides all the η_j so that (b) becomes $f = \lambda_1^2 + \cdots + \lambda_n^2$ -a contradiction, since the degree of ζ was least possible.

Now let

$$\zeta^* = \zeta \left\{ \sum_i \lambda_i^2 - f \right\} - 2 \left\{ \sum_i \lambda_i \eta_i - f\zeta \right\}$$

and

$$\eta_j^* = \eta_j \left\{ \sum_i \lambda_i^2 - f \right\} - 2\lambda_j \left\{ \sum \lambda_i \eta_i - f\zeta \right\}.$$

Then visibly, all of $\zeta^*, \eta^*, \ldots, \eta_n^* \in K[X]$. We now claim

(i) that $(\zeta^*, \eta_1^*, \ldots, \eta_n^*)$ is a solution of (a),
(ii) $\zeta^* \neq 0$ and
(iii) deg $\zeta^* <$ deg ζ.

This would then contradict the definition of ζ and so would prove the lemma.

We prove (i) by brute force: we must show that $\sum_j \eta_j^{*^2} - f\zeta^{*^2} = 0$ i.e. that

$$\sum_j \left[\eta_1^2 \left\{ \sum_i \lambda_i^2 - f \right\}^2 + 4\lambda_j^2 \left\{ \sum_i \lambda_i \eta_i - f\zeta \right\}^2 \right.$$
$$\left. -4\lambda_j \eta_j \left\{ \sum_i \lambda_i^2 - f \right\} \left\{ \sum_i \lambda_i \eta_i - f\zeta \right\} \right]$$
$$= f \left[\zeta^2 \left\{ \sum_i \lambda_i^2 - f \right\}^2 + 4 \left\{ \sum_i \lambda_i \eta_i - f\zeta \right\}^2 \right.$$
$$\left. -4\zeta \left\{ \sum_i \lambda_i^2 - f \right\} \left\{ \sum_i \lambda_i \eta_i - f\zeta \right\} \right]$$

Here the first terms from both sides cancel since $\sum \eta_j^2 = f\zeta^2$ and it remains to prove that

$$4\left\{\sum_i \lambda_i\eta_i - f\zeta\right\}\left[\left(\sum \lambda_i\eta_i - f\zeta\right)\sum \lambda_j^2 - \left(\sum_i \lambda_i^2 - f\right)\sum_j \lambda_j\eta_j \right.$$
$$\left. - \left(\sum_i \lambda_i\eta_i - f\zeta\right)f + \left(\sum \lambda_i^2 - f\right) + f\zeta\right] = 0$$

Here the expression in square brackets just cancels out.

To prove (ii) and (iii) we substitute for λ_j from (c) in ζ^*. Then

$$\begin{aligned}\zeta^* &= \zeta\left(\sum_i\left(\frac{\eta_i^2}{\zeta^2} + \Lambda_i^2 - \frac{2\eta_i\Lambda_i}{\zeta}\right) - f\right) - 2\left(\sum_i\left(\frac{\eta_i}{\zeta} - \Lambda_i\right)\eta_i - \eta\zeta\right)\\ &= \zeta(\Lambda_i^2 + \cdots + \Lambda_n^2)\ (\text{using } f\zeta^2 = \eta_i^2 + \cdots + \eta_n^2)\\ &= \zeta\sum_i \gamma_i^2/\zeta^2 = 1/\zeta\sum_i \gamma_i^2.\end{aligned}$$

Here not all the γ_i are zero (as already noted) and so $\sum \gamma_i^2$ is non-zero since otherwise by equating the coefficient of the highest power in X to 0, we find that 0 is a sum of at most n squares of elements of K, which is the third trivial case of the lemma. Thus $\zeta^* \neq 0$, which proves (ii).

Finally $\zeta^* = 1/\zeta\sum_i \gamma_i^2$ giving $\zeta\zeta^* = \sum_i \gamma_i^2$. Equating degrees, we get deg $\zeta+$ deg $\zeta^* = 2\max_i$ (deg γ_i) $<$ 2 deg ζ since deg $\gamma_i <$ deg ζ (for all i). Thus deg $\zeta^* <$ deg ζ, which proves (iii).

This completes the proof of Cassels' lemma. □

Remark: The solution $(\zeta^*, \eta_1^*, \ldots, \eta_n^*)$ does not just come out of the blue. It is the second point of intersection Q of the quadric (a) with the line joining the points $P = (\zeta, \eta_1, \ldots, \eta_n)$ (on the quadric) and $P' = (1, \lambda_1, \ldots, \lambda_n)$ (in space) in the n-dimensional projective space over the field $K(X)$. The simplest way to get this point Q is as follows: a general point of the line PP' is

$$(\theta\zeta + \varphi, \theta\eta_1 + \varphi\lambda_1, \ldots, \theta\eta_n + \varphi\lambda_n)$$

θ/φ being a parameter for various points, $\varphi = 0$ giving the point P. To get Q we substitute this general point in the quadric (a):

$$f(\theta^2\zeta^2 + \varphi^2 + 2\theta\varphi\zeta) = \sum_{j=1}^{n}(\theta^2\eta_j^2 + \varphi^2\lambda_j^2 + 2\theta\varphi\eta_j\lambda_j)$$

i.e. $\theta^2(\zeta^2 f - \sum \eta_j^2) + 2\theta\varphi(f\zeta - \sum \lambda_j\eta_j) + \varphi^2(f - \sum \lambda_j^2) = 0$. But $\zeta^2 f = \sum \eta_j^2$ so this becomes $2\theta\varphi(f\zeta - \sum \lambda_j\eta_j) + \varphi^2(f - \sum \lambda_j^2) = 0$. This has a root $\varphi = 0$ as expected giving the point P. The other root is $\theta/\varphi = -(\sum \lambda_j^2 - f)/2(\sum \lambda_j\eta_j - \zeta f)$, and substituting this in the general point and multiplying by a suitable factor (allowed in a projective space) we get our point Q as required.

We now deduce a few corollaries from this lemma.

Corollary 1. *Let* char $K \neq 2$ *let* $f(X_1, \ldots, X_m) \in K(X_1, \ldots, X_m)$ *be a sum of* n *squares of elements of* $K(X_1, \ldots, X_m)$. *Let* $a_1, a_2, \ldots, a_m \in K$ *be such that* $f(a_1, \ldots, a_m)$ *is defined* (*i.e. the dominator is not* 0). *Then* $f(a_1, \ldots, a_m)$ *is a sum of* n *squares in* K.

Remark: The point is that although $f(X_1, \ldots, X_m)$ is defined at $(a_1, \ldots, a_m)$, it may well happen that the summands $f_j^2(X_1, \ldots, X_m)$ of the right hand side of $f(X_1, \ldots, X_m) = f_1^2 + \cdots + f_n^2$ may not be defined at $(a_1, \ldots, a_m)$, but still according to the corollary, $f(a_1, \ldots, a_m)$ is a sum of n squares in K.

Proof: We use induction on m. For $m = 1$, we have

$$f(X) = g(X)/h(X) = \gamma_1^2(X) + \cdots + \gamma_n^2(X).$$

Then $gh = (\gamma_1 h)^2 + \cdots + (\gamma_n h)^2$. Thus gh, which is in $K[X]$, is a sum of n squares in $K(X)$ and so by Cassels' lemma, it is a sum of n squares in $K[X]$:

$$gh = f_i^2 + \cdots + f_n^2 \ (f_j \in K[X]).$$

Hence $g(X)/h(X) = (\frac{f_1(X)}{h(X)})^2 + \cdots + (\frac{f_n(X)}{h(X)})^2$. Now by hypothesis, $f(a) = g(a)/h(a)$ is defined; i.e. $h(a) \neq 0$, so each $f_j(a)/h(a)$ is defined.

Let now $m > 1$. Let $L = K(X_1, \ldots, X_{m-1})$. Assume the result for $m - 1$ variables and let $g(X_1, \ldots, X_m)/h(X_1, \ldots, X_m)$ be a rational function which is a sum of n squares in $K(X_1, \ldots, X_m)$. Regard g/h as a rational function of X_m belonging to $L(X_m)$. So by the case $m = 1$, we see that $g(X_1, \ldots, X_{m-1}, a_m)/h(X_1, \ldots, X_{m-1}, a_m)$ is a sum of n squares in $L = K(X_1, \ldots, X_{m-1})$. So by the induction hypothesis $g(a_1, \ldots, a_m)/h(a_1, \ldots, a_m)$ is a sum of n squares in K. This completes the proof of the corollary. □

Corollary 2. *Suppose* $n = 2^m$. *Let* G_n *be the set of all non-zero elements of* K *which are sums of* n *squares in* K. *Then* G_n *is a group under multiplication.*

Proof: Let $\alpha\beta \in G_n$ say, $\alpha = \alpha_1^2 + \cdots + \alpha_n^2$, $\beta = \beta_1^2 + \cdots + \beta_n^2$. Then $\alpha^{-1} = \alpha/\alpha^2 = (\alpha_1/\alpha)^2 + \cdots + (\alpha_n/\alpha)^2 \in G_n$ and it remains to prove that $\alpha\beta \in G_n$. Consider the identity

$$(X_1^2 + \cdots + X_n^2)(Y_1^2 + \cdots + Y_n^2) = Z_1^2 + \cdots + Z_n^2$$

which exists since $n = 2^m$. In this let $X_1 \to \alpha_1, \ldots, X_n \to \alpha_n, Y_1 \to \beta_1, \ldots, Y_n \to \beta_n$. Then the left side is well defined and equal to $\alpha\beta$ and so by Corollary 1, the right side is a sum of n squares of elements of K, i.e. $\alpha\beta \in G_n$ as required. □

We see that it is the identity (5) that does the trick.

We can now prove the first part of Theorem 3: $s(K)$ *is always a power of* 2.

Proof of the first part of Theorem 3: Let

$$n = 2^m \leq s(K) < 2^{m+1} \tag{$*$}$$

Then $a_1^2 + \cdots + a_n^2 + a_{n+1}^2 + \cdots + a_s^2 + 1 = 0$ $(a_j \in K)$. Let $A = a_1^2 + \cdots + a_n^2$, $B = a_{n+1}^2 + \cdots + a_s^2 + 1$. Here A, B are both non-zero, otherwise $s(K) < s$. Also A, B, both $\in G_n$ (by adding a suitable number of 0^2's to B if necessary). Then $A + B = 0$ so $A = -B$ i.e. $-1 = B/A \in G_n$ since G_n is a group i.e. $-1 = c_1^2 + \cdots + c_n^2$ giving $s(K) \leq n$. Comparing with $(*)$, we get $s(K) = n = 2^m$.

To prove the remaining parts of Theorems 3 and 5 we need to deduce some more corollaries from Cassels' lemma; see [4].

Corollary 3. *Let* char $K \neq 2$. *A necessary and sufficient condition for $X^2 + d \in K[X]$ to be a sum of n squares in $K(X)$ (and so in $K[X]$ by Cassels' lemma) is that either*

(i) -1 *is a sum of $n-1$ squares in K or*
(ii) d *is a sum of $n-1$ squares in K.*

Proof: If $-1 = b_1^2 + \cdots + b_{n-1}^2$, then for any polynomial $f(X) \in K[X]$, we have

$$\begin{aligned} f &= \left(\frac{f+1}{2}\right)^2 - \left(\frac{f-1}{2}\right)^2 \\ &= \left(\frac{f+1}{2}\right)^2 + \left(\frac{b_1(f-1)}{2}\right)^2 + \cdots + \left(\frac{b_{n-1}(f-1)}{2}\right)^2. \end{aligned}$$

In particular $X^2 + d$ is a sum of n squares.

If d is a sum of $n-1$ squares then visibly $X^2 + d$ is a sum of n squares in $K[X]$.

For the converse, suppose $X^2 + d$ is a sum of n squares in $K[X]$. If (i) holds, well and good; otherwise let $X^2 + d = p_1^2(X) + \cdots + p_n^2(X)$ say. Here we may suppose the $p_j(X)$ to be linear poynomials in X for if not, then equating to 0 the coefficient of the highest power of X gives (i). Then

$$X^2 + d = (a_1X + b_1)^2 + \cdots + (a_nX + b_n)^2 \tag{$*$}$$

Now one of the equations $C = \pm(a_nC + b_n)$ is always solvable in K. For if $a_n \neq 1$ then $C = +(a_nC + b_n)$ is solvable, while if $a_n = 1$ then $C = -(a_nC + b_n)$ is solvable since char $K \neq 2$. Now put $X = C$ in $(*)$:

$$C^2 + d = (a_1C + b_1)^2 + \cdots + (a_{n-1}C + b_{n-1})^2 + (a_nC + b_n)^2.$$

Cancelling C^2 with $(a_nC + b_n)^2$ we see that d is *a* sum of $n-1$ squares in K. This completes the proof. □

Corollary 4. *Let* **R** *be the field of real numbers. Then $X_1^2 + \cdots + X_n^2$ is not a sum of $n-1$ squares of elements in* $\mathbf{R}(X_1, \ldots, X_n)$.

Proof: We use induction on n. For $n = 1$, the result is trivial. So suppose the result is true for $n-1$. Let $K = \mathbf{R}(X_1, \dots, X_{n-1})$, $X_n = X$ and $d = X_1^2 + \dots + X_{n-1}^2$. If $X_1^2 + \dots + X_n^2$ is a sum of $n-1$ squares in $K(X) = \mathbf{R}(X_1, \dots, X_n)$, then by Corollary 3, $d = X_1^2 + \dots + X_{n-1}^2$ is a sum of $n-2$ squares in K, since -1 is clearly not a sum of $n-2$ squares in K-indeed not a sum of squares at all in K, which is formally real. This contradicts the induction hypothesis and completes the proof of Corollary 4. □

We are now in a position to complete the proofs of the remaining parts of Theorems 3 and 5.

Every power of 2 *is the Stufe of some field K.*

Proof: Let $n = 2^m$ and let $K = \mathbf{R}(X_1, \dots, X_{n+1}, Y)$ where $X_1, \dots, X_{n+1}$ are independent transcendentals over $\mathbf{R}$ and Y satisfies the equation

$$Y^2 + X_1^2 + \dots + X_{n+1}^2 = 0 \tag{i}$$

We claim that $s(K) = n = 2^m$. In any case by (i), $s(K) \le n+1$ and so is at most n since $n+1$ cannot be a power of 2 whereas $s(K)$ *is* (except in the trivial case $n = 1$ i.e. $m = 0$).

If $s(K) < n$ then there exist $t_1, \dots, t_n \in K$, not all zero such that

$$t_1^2 + \dots + t_n^2 = 0 \tag{ii}$$

Let $L = \mathbf{R}(X_1, \dots, X_{n+1})$ so that $K = L(Y)$. By (i), Y is algebraic over L of degree 2 and so each element of K is a linear polynomial in Y with coefficients from L. Write $t_j = a_j + Yb_j$, $a_j, b_j \in L$. Then by (ii) we see that

$$\sum a_j^2 + Y^2 \sum b_j^2 = 0$$

and

$$\sum a_j b_j = 0.$$

Here not all the a_j are zero, otherwise $\sum b_j^2 = 0$ and so each $b_j = 0$ since the $b_j \in L = \mathbf{R}(X_1, \dots, X_{n+1})$ is formally real. Then each t_j would be zero which is not true. Similarly not all the b_j are zero. Hence

$$\begin{aligned} -Y^2 &= \sum_{j=1}^{n} a_j^2 / \sum_{j=1}^{n} b_j^2 \in G_n \text{ (by the group property of } G_n) \\ &= c_1^2 + \dots + c_n^2, \text{ say } c_j \in L, \end{aligned}$$

i.e. $X_1^2 + \dots + X_{n+1}^2$ is a sum of n squares in L which contradicts Corollary 4. Thus $s(K)$ is *not* less than n and so equals n. This completes the proof. □

Remark: The proof also works for $\mathbf{Q}(X_1, \dots, X_{n+1}, Y)$.

Finally we prove the remaining part of Theorem 5.

Suppose n is not a power of 2. *Then there is a field K such that there is no identity*

$$(X_1^2 + \dots + X_n^2)(Y_1^2 + \dots + Y_n^2) = z_1^2 + \dots + Z_n^2$$

with $Z_j \in K(X_1, \dots, X_n, Y_1, \dots, Y_n)$.

Proof: Let $2^{m-1} < n < 2^m$. Let K be a field having Stufe $2^m = \nu$, say. Then $a_1^2+\cdots+a_n^2+a_{n+1}^2+\cdots+a_\nu^2+1=0$. Let $A=a_1^2+\cdots+a_n^2$, $B=a_{n+1}^2+\cdots+a_\nu^2+1$; hence $A, B \in G_n$ and if an identity of the above type exists, then G_n is a group (see the proof of Corollary 2). So $-1 = B/A \in G_n$, i.e. $-1 = C_1^2+\cdots+C_n^2$ $(C_j \in K)$ hence $s(K) \le n < \nu$. But K was chosen to have Stufe ν. This gives a contradiction and so completes the proof. □

Remark 1. In our examples of fields with high Stufe both the fields $\mathbf{R}(X_1, \ldots, X_{n+1}, Y)$ and $\mathbf{Q}(X_1, \ldots, X_{n+1}, Y)$ are of high transcendence degree over $\mathbf{R}$ or $\mathbf{Q}$ as the case may be. We have the following

Problem: *Does high Stufe always imply high degree of transcendence? (over* **R** *or* **Q***).*

Let us now go back to the identity (4):

$$(X_1^2+\cdots+X_n^2)(Y_1^2+\cdots+Y_n^2) = (Z_1^2+\cdots+Z_n^2)$$

where the Z_k are bilinear functions of the X_i and the Y_j with coefficients in the field K.

There are three obvious ways of generalizing this identity (one of which we have already looked at in theorem 5). They are

a) Allow the Z_k to be the rational functions of the X_i and the $Y_j : Z_k \in K(X_1, \ldots, X_n, Y_1, \ldots, Y_n)$. Then, as we have seen in theorem 5, such identities can be found for each power $n = 2^m$ $(m = 0, 1, \ldots)$ of 2 and for no other value of n.
b) Consider the (r, s, n) identity

$$(X_1^2+\cdots+X_r^2)(Y_1^2+\cdots+Y_s^2) = Z_1^2+\cdots+Z_n^2 \tag{6}$$

with Z_k bilinear in the X_i and the Y_j, and determine, for given r, s the least value of n for which (6) holds. We could, alternatively look for the maximum value of r, for given s and n, for which (6) holds.

For general values of r, s, n, little is known about (6). However, for $s = n$, Hurwitz and Radon gave a solution of (6) in 1922–3, for the field R of real numbers. Before giving the exact statement of the Hurwitz-Radon theorem we make the following

Definition 2. We say the triple (r, s, n) is *admissible* over K if (6) holds.

Thus (r, s, rs) is trivially admissible over any field K; so that what we want is the most economical n for which (r, s, n) is admissible for a given pair r, s of integers. In view of this we have the

Definition 3. We denote by r^*s (or rather r_k^*s) the least n for which (r, s, n) is admissible$/K$.

We have the trivial bounds.

$$\max(r, s) \le r^*s \le r.s$$

It is not easy to determine r^*s, even for small values or r, s.

Alternatively, as already said above, we could ask, for given s, n the maximum value of r for which (r, s, n) is admissible/K. This is the approach adopted by Hurtwitz and Radon in their treatment of (6) for the field R of real numbers. Simultaneously, Hurwitz solved (6), in this special case, for the field C of complex numbers, published posthumously in 1923. Various authors have since dealt with other fields.

As illustrations of definitions 2 and 3, we have the following:

Examples:

(i) (n, n, n) is admissible over $\mathbf{R}$, indeed over any field K, char $K \neq 2$, iff $n = 1, 2, 4, 8$. Thus is Hurwitz's theorem (Theorem 4).

(ii) $(1, n, n)$, and indeed (r, s, rs), is admissible for all n, r, s over any field K.

(iii) If char$K = 2$, then $r *_K s = 1$ for all r, s for then $a^2 + b^2 = (a + b)^2$.

(iv) $8 * 8 = 8$ for $\max(8, 8) \leq 8 * 8 \leq 8$. Similarly $4 * 4 = 4$ and $2 * 2 = 2$.

(v) *The* 16-*square problem*: Before Hurwitz, studies about the (r, s, n)-identites (6) were exclusively restricted to the polynomial ring $\mathbf{Z}[X_1, \ldots, X_r, Y_1, \ldots, Y_s]$ over $\mathbf{Z}$. One then speaks of the $(r, s, n)_{\mathbf{Z}}$-identities. It has recently been confirmed that $16 *_{\mathbf{Z}} 16 = 32$, thereby completing the solution of the so-called 16-square problem in the integer coefficient (case see [16]). However, the integer $\nu = 16 *_{\mathbf{R}} 16$ is not known to date. Various methods developed by K.Y. Lam and J. Adem narrow down the range of ν to $23 \leq \nu \leq 32$. The values 23, 24 were subsequently ruled out by Lam and Yuzvinski. By going more deeply into the geometry of sums of square formulae and using sophisticated algebraic topology, it has now been established by Lam and Yiu that $29 \leq \nu \leq 32$.

It is trivial to see that $\nu \leq 32$; indeed

$$\left(\sum_1^{16} X_j^2\right)\left(\sum_1^{16} Y_j^2\right) = \left(\sum_1^{8} X_j^2 + \sum_9^{16} X_j^2\right)\left(\sum_1^{8} Y_j^2 + \sum_9^{16} Y_j^2\right)$$

which is a sum of 32 squares, using the 8-square identity four times.

(vi) Amongst small values of r, s, n, even $(10, 11, 25)$ is not known to be admissible or otherwise.

Definition 4. For any positive integer n, define the so-called *Radon function* $\rho(n)$ as follows:

Write $n = 2^m \cdot u (u \text{ odd})$; then

$$\rho(n) = \begin{cases} 2m+1 \\ 2m \\ 2m \\ 2m+2 \end{cases} \text{according as } m \equiv \begin{cases} 0 \\ 1 \\ 2 \\ 3 \end{cases} (\text{modulo } 4).$$

Equivalently write $m = 4a + b, 0 \leq b \leq 3$; then

$$\rho(n) = 8a + 2^b.$$

We now have the following

Theorem B (Radon, Hurwitz-1922, 1923). *The triple (r, n, n) is admissible over the field of real numbers (indeed over any field K, char $K \neq 2$) iff $r \leq \rho(n)$.*

For a proof see [13], p 127 or [14].

(c) Instead of a "product formula" (4) for the form $X_1^2 + \cdots + X_n^2$, look for such a formula for more general quadratic forms $q(X_1, \ldots, X_n)$; i.e. determine other quadratic forms $q(X_1, \ldots, X_n)$, if any, for which

$$q(X_1, \ldots, X_n) \bullet q(Y_1, \ldots, Y_n) = q(Z_1, \ldots, Z_n) \tag{7}$$

where $Z_k \in K(X_1, \ldots, X_n, Y_1, \ldots, Y_n)$

Examples: (i) If $q(X_1, X_2) = X_1^2 + aX_2^2 (a \in K)$, then we have the curious identity (cf. identity (1))

$$(X_1^2 + aX_2^2)(Y_1^2 + aY_2^2) = (X_1Y_1 + aX_2Y_2)^2 + a(X_1Y_2 - aX_2Y_1)^2.$$

(ii) In 4 variables, we have the striking identity (cf. identity(2))

$$\begin{aligned}(X_1^2 + aX_2^2 + bX_3^2 + baX_4^2)&(Y_1^2 + aY_2^2 + bY_3^2 + baY_4^2)\\ &= (X_1Y_1 + aX_2Y_2 + bX_3Y_3 + abX_4Y_4)^2\\ &+ a(-X_1Y_2 + X_2Y_1 - bX_3Y_4 + bX_4Y_3)^2\\ &+ b(-X_1Y_3 + X_3Y_1 + aX_2Y_4 - aX_4Y_2)^2\\ &+ ab(-X_1Y_4 + X_4Y_1 - bX_2Y_3 + bX_3Y_2)^2\end{aligned}$$

Pfister has given a complete solution of (c). He shows that for every power $n = 2^m$ of 2, there is this form in n variables generalizing the forms $X_1^2+aX_2^2$ and $X_1^2+aX_2^2+bX_3^2+abX_4^2$ by an obvious induction, which satisfies a product identity. These are the so called Pfister forms. Further there are no other forms that satisfy a product formula (7). For a proof of these results see [10], [11], [13].

In the identity (5), we have proved that the Z_k may be chosen linear functions of the Y_j with coefficients in $K(X_1, \ldots, X_n)$:

$$Z_k = \sum_{j=1}^{n} T_{kj} Y_j \text{ with } T_{kj} \in K(X_1, \ldots, X_n).$$

For $n = 2, 4,$ and 8, these T_{kj} are linear in the X_i as well. It is natural to enquire how simple we can take the T_{kj} as functions of the X_i. By theorem 4, they can not all be taken linear in the X_i nor indeed polynomials, by Frank Adam's theorem so that some of them at least have to have a denominator; but can we make at least some of them linear forms in the X_i?

Let us first see what we can do with the first term Z_1 and prove the following

Theorem 6. *Let $n = 2^m$ and let $X_1, \ldots, X_n, Y_1, \ldots, Y_n \in K$. Then*

$$(X_1^2 + \cdots + X_n^2)(Y_1^2 + \cdots + Y_n^2) = (X_1Y_1 + \cdots + X_nY_n)^2 + Z_2^2 + \cdots + Z_n^2$$

for some $Z_2, \ldots, Z_n \in K$.

We first prove: Let $n = 2^m$ and let $c = c_1^2 + \cdots + c_n^2$ $(c_j \in K)$. Then there exists an $n \times n$ matrix S with first row $(c_1, \ldots, c_n)$ such that $SS' = S'S = cI_n$.

Proof: First let $c = 0$. If all the $c_j = 0$, we take S to be the zero matrix.

So suppose, say, $c_1 \neq 0$. Let R be the row vector $(c_1, \ldots, c_n)$ and take $S = c_1^{-1}R'R$, which has first row R as required. Further

$$\begin{aligned} SS' &= c_1^{-1}R'Rc_1^{-1}R'R \\ &= c_1^{-2}R'(RR')R \\ &= 0, \end{aligned}$$

since $RR' = c_1^2 + \cdots + c_n^2 = c = 0$. Similarly $S'S = 0$ and the proof is complete. So we may now suppose that $c \neq 0$ and we proceed by induction on m.

Write

$$\begin{aligned} \underline{R} = (c_1, \ldots, c_{2^m}) &= (c_1, \ldots, c_{2^{m-1}}, c_{2^{m-1}+1}, \ldots, c_{2^m}) \\ &= (\underline{R}_1, \underline{R}_2). \end{aligned}$$

Let $a = c_1^2 + \cdots + c_{2^{m-1}}^2$, $b = c_{2^{m-1}+1}^2 + \cdots + c_{2^m}^2$, so that $c = a + b$. Here since $c \neq 0$, so a, b cannot be both zero; say, without loss of generality, that $a \neq 0$. By the induction hypothesis, there exist square matrices S_1, S_2 of size 2^{m-1} such that

$$\begin{aligned} S_1S_1' &= S_1'S_1 = aI_{2^{m-1}} \\ S_2S_2' &= S_2'S_2 = bI_{2^{m-1}}. \end{aligned}$$

Furthermore the first row of S_1 is $(c_1, \ldots, c_{2^{m-1}})$ and that of S_2 is $(c_{2^{m-1}+1}, \ldots, c_2^m)$. Now let

$$S = \begin{pmatrix} S_1 & S_2 \\ -a^{-1}S_1'S_2'S_1 & S_1' \end{pmatrix}.$$

This has first row equal to R as required and an easy matrix computation gives $SS' = S'S = cI_n$, e.g.

$$\begin{aligned} SS' &= \begin{pmatrix} S_1 & S_2 \\ -a^{-1}S_1'S_2'S_1 & S_1' \end{pmatrix} \begin{pmatrix} S_1' & -a^{-1}S_1'S_2S_1 \\ S_2' & S_1 \end{pmatrix} \\ &= \begin{pmatrix} aI_{2^{m-1}} + bI_{2^{m-1}} & -a^{-1}aS_2S_1 + S_2S_1 \\ -a^{-1}S_1'S_2'aI_{2^{m-1}} + S_1'S_2' & a^{-2}S_1'S_2'S_1S_1'S_2S_1 + S_1'S_1 \end{pmatrix} \\ &= \begin{pmatrix} cI_{2^{m-1}} & 0 \\ 0 & bI_{2^{m-1}} + aI_{2^{m-1}} \end{pmatrix} = cI_{2^m}. \end{aligned}$$

□

Proof of Theorem 6: Write

$$\begin{aligned} X &= X_1^2 + \cdots + X_n^2 \\ Y &= Y_1^2 + \cdots + Y_n^2. \end{aligned}$$

Then there exist $n \times n$ matrices U, V such that $UU' = U'U = XI_n$, $VV' = V'V = YI_n$ and

$$U \text{ has 1st row} = (X_1, \ldots, X_n),$$
$$V \text{ has 1st row} = (Y_1, \ldots, Y_n).$$

Then

$$XYI_n = XVV' = V(U'U)V' = (VU')(VU')' = V(U'U)V' = WW', \quad \text{where } W = VU'.$$

This equation says that if $(Z_1, \ldots, Z_n)$ is the first row of W, then $XY = Z_1^2 + \cdots + Z_n^2$. But since $W = VU'$, we have $Z_1 = X_1Y_1 + \cdots + X_nY_n$. □

Theorem 6 enables us to give another proof of the important group property of the set. $G_n = \{a \in K^* | a = \alpha_1^2 + \cdots + \alpha_n^2, \alpha_j \in K\}$, when $n = 2^m$, a power of 2 (see corollary 2 after cassels' lemma). For, theorem 6 implies closure of G_n under multiplication while as before $a^{-1} \in G_n$ whenever $a \in G_n$.

Going back to our enquiry about how simple we can take the Z_k as functions of the X_i, we now state the following striking result of Shapiro:

Theorem 7 (Shapiro-1978). *Let $n = 2^m$ and let K be any field. In the n-square identity* (5)

$$(X_1^2 + \cdots + X_n^2)(Y_1^2 + \cdots + Y_n^2) = Z_1^2 + \cdots + Z_n^2 \qquad (*)$$

with Z^k linear in the Y_j with coefficient in $K(X_1, \ldots, X_n)$, the first r terms $Z_1, \ldots, Z_r$ of the right side of (5) *can also be taken linear in the X_i iff $r \leq \rho(n)$, the Radon function!*

For a proof of this result, see [13], page 183.

Incidentally, we note that in $(*)$ above we can easily arrange a formula where 8 of the Z_k are bilinear (when $n \geq 8$). To do this start with the known (8, 8, 8) bilinear identity and apply the "doubling" process given in Pfister's theorem 5. Indeed write the 8-square identity twice over, once for the variables $X_1, \ldots, X_8$; $Y_1, \ldots, Y_8$; $Z_1, \ldots, Z_8$ and once for $X_9, \ldots, X_{16}$; $Y_9, \ldots, Y_{16}$; $Z_9, \ldots, Z_{16}$. Thus

$$\begin{pmatrix} Z_1 \\ Z_2 \\ \vdots \\ Z_8 \end{pmatrix} = \begin{pmatrix} X_1 & -X_2 & -X_3 & -X_4 & -X_5 & -X_6 & -X_7 & -X_8 \\ X_2 & X_1 & -X_4 & X_3 & -X_6 & X_5 & X_8 & -X_7 \\ \cdots & \cdots & \cdots & \cdots & \cdots & \cdots & \cdots & \cdots \\ X_8 & X_7 & -X_6 & -X_5 & X_4 & X_3 & -X_2 & -X_1 \end{pmatrix} \begin{pmatrix} Y_1 \\ Y_2 \\ \vdots \\ Y_8 \end{pmatrix}$$

and

$$\begin{pmatrix} Z_9 \\ Z_{10} \\ \vdots \\ Z_{16} \end{pmatrix} = \begin{pmatrix} X_9 & -X_{10} & -X_{11} & -X_{12} & -X_{13} & -X_{14} & -X_{15} & -X_{16} \\ X_{10} & X_9 & -X_{12} & X_{11} & -X_{14} & X_{13} & X_{16} & -X_{15} \\ \cdots & \cdots & \cdots & \cdots & \cdots & \cdots & \cdots & \cdots \\ X_{16} & X_{15} & -X_{14} & -X_{13} & X_{12} & X_{11} & -X_{10} & X_9 \end{pmatrix} \begin{pmatrix} Y_9 \\ Y_{10} \\ \vdots \\ Y_{16} \end{pmatrix},$$

(simply read off the identity (3)); say $\underline{Z}_1 = S_1\underline{Y}_1$ and $\underline{Z}_2 = S_2\underline{Y}_2$ for short. Then

$$\begin{pmatrix} \underline{Z}_1 \\ \underline{Z}_2 \end{pmatrix} = \begin{pmatrix} S_1 & S_2 \\ S_2 & -S_2^{t^{-1}} S_1^t S_2 \end{pmatrix} \begin{pmatrix} \underline{Y}_1 \\ \underline{Y}_2 \end{pmatrix}$$

by Pfister's Theorem 5. We see that $Z_1, Z_2, \ldots, Z_8$ are bilinear in the X_i and the Y_j as claimed. The process can be repeated for $32, 64, \ldots,$ variables; the 8 bilinear terms will persist.

But of course even for $n = 16$, Theorem 7 is stronger than the above method as it gives us nine fully bilinear terms.

This problem was posed by Baeza and solved by Shapiro in a letter to Baeza in 1976.

Pfister has other very interesting results about Hilbert's 17th problem in the function fields $\mathbf{R}\,(X, Y)$ and more generally in $\mathbf{R}(X_1, X_2, \ldots, X_n)$. We refer our readers again to [13], [9].

References

[1] J.F. Adams, Vector fields on Sphere, *Annals of Maths* **75** (1962), 603–632.

[2] A.A. Albert (editor), Studies in Modern Algebra, vol. 2, MAA Studies in Maths (1963).

[3] E. Artin, Uber die Zerlegung definiter Funktionen in Quadrate, *Hamb. Abh.* **5** (1927), 100–115.

[4] J.W.S. Cassels, On the representation of rational functions as sums of squares, *Acta Arith.* **9** (1964), 79–82.

[5] L.E. Dickson, On quaternions and their generalizations and the history of the 8-square theorem, *Annals of Maths.* **20** (1919), 155–171.

[6] Adolf Hurwitz, Uber der Komposition der Quadratischen Formen von beliebig vielen Variabeln, Nachrichten von der Königlichen Gessellschaft der Wissenschaften in Göttingen (1898), 309–316; = Math. werke, II 565–571.

[7] A. Pfister, Zur Darstellung von-1 also Summe von Quadraten in einem Körper, JLMS, **40** (1965), 159–165.

[8] A. Pfister, Zur Darstellung definiter Funktionen als Summe Von Quadraten, *Inventiones Math.* **4** (1967), 224–236.

[9] A. Pfister, Hilbert's 17th problem and related problem on definite forms, *Proc. of Symposia in Pure Maths.* **28** (1976), 483–489.

[10] A. Pfister, Multiplicative quadratische Formen, *Arch. Math.* **16** (1965), 363–370.

[11] A. Pfister, Quadratrische Fromen in beliebigen Körpern, *Inventiones Maths.* **1** (1966), 116–132.

[12] A.R. Rajwade, A note on the Stufe of quadratic fields. *Indian J. Pure and App. Maths.* **6** (1975), 725–6.

[13] A.R. Rajwade, Squares, *London Math Soc.* Lecture note series no. 171, (1993).

[14] J. Radon, Lineare Scharen orthogonaler Matrizen, *Abh. Math. Sem Univ. Hamburg* **1** (1922), 1–14.

[15] E. Witt, Theorie der quadratischen Formen in beliebigen Körpern, *J. reine angew. Math.* 17 **6** (1937), 31–34.

[16] Paul Yiu, On the product of 2 sums of 16 squares as a sum of squares on integral bilinear forms, *Quart. J. Maths.* (2) **41** (1990), 463–500.

Centre for Advanced Study in Mathematics
Panjab University
Sector 14
Chandigarh 160 014

Notes on the Prime Number Theorem-I

K. Ramachandra

To Professor A.A. Karatsuba, Yu.V. Nesterenko and A.B. Shidlowsky

§ 1 Introduction and Notation

We begin by stating the Prime Number Theorem in a way somewhat different from the usual. Let p_n denote the n-th prime (viz. $p_1 = 2, p_2 = 3, p_3 = 5, \ldots$).

Then p_n is given by the approximate formula

$$n = \int_2^{p_n} \frac{du}{\log u}$$

Of course this is only an approximation. The Prime Number Theorem in the usual form states that the error here is $O(p_n^{\frac{1}{2}} \log p_n)$ (on Riemann hypothesis namely $\zeta(s) \neq 0$ if $Re(s) > \frac{1}{2}$) and $O(p_n(\exp((\log p_n)^{\frac{3}{5}}(\log\ \log p_n)^{-\frac{1}{5}}))^{-\mu})$ (unconditionally) u.c., where $\mu > 0$ is an absolute constant, which is not very important ($O(\ldots)$ means a term whose absolute value is less than a constant times...). Let $y \geq 2$ and $li\ y = \int_2^y \frac{du}{\log u}$. Then the equation $x = li\ y$ has a unique solution say $y = f(x)$ with $f(0) = 2$. From $x = li\ y$ it follows that $1 = (\log y)^{-1}\frac{dy}{dx}$ and hence $f'(x) = \log\ y$. From these it follows (by inverting the approximation for p_n stated above), that

$$|p_n - f(n)| \leq \begin{cases} Cn^{\frac{1}{2}}(\log n)^{\frac{5}{2}}, & \text{on RH,} \\ CnE^{-D}, & u.c., \end{cases} \tag{1}$$

where C and D are absolute positive constants and $E = \exp((\log n)^{\frac{3}{5}}\ (\log\ \log\ n)^{-\frac{1}{5}})$. (Of course on the assumption $\zeta(s) = O((t^{(1-\sigma)^a}(\log t))^A)(t \geq 2, \frac{1}{2} \leq \sigma \leq 1)$ where a ≥ 1 and $A > 0$ are absolute constants there follow appropriate u.c forms of the error terms in P N T and the above things can be formulated in terms of such error terms also; the function E above corresponds to $a = \frac{3}{2}$). Also a famous work of J.E. LITTLEWOOD states that $\pi(x)$ defined as $\sum_{p \leq x} 1$ satisfies

$$\pi(x) = li\ x + \Omega_{\pm}\left(\frac{x^{\frac{1}{2}}}{\log x} \log\log\log x\right).$$

this means that the upper and lower limits of $\pi(x) - li\ x$ divided by $\frac{x^{\frac{1}{2}}}{\log x}\ \log\log\log x$ as $x \to \infty$ are respectively positive and negative. From this it follows (on putting $x = p_n$ and inverting) that

$$p_n - f(n) = \Omega_{\pm}((n \log n)^{\frac{1}{2}} \log\log\log n)$$

with a similiar meaning (for $\Omega_{\pm}$) as explained just now. Thus on RH we have a very precise result for $p_n - f(n)$ namely both O and $\Omega_{\pm}$ results nearly of the same order of precision. What is usually customary is to state the results for $\sum_{p^n \le x} \log p$, because this admits of a nice formula (called explicit formula) in terms of the zeros of $\zeta(s)$. It should be mentioned that all the results on p_n stated above are equivalent to the corresponding (O and $\Omega_{\pm}$) results on $\sum_{p^n \le x} \log p, \pi(x)$ and so on.

The results of this paper are fairly well-known to experts working in this area. (It is meant for non-experts). The only justification for their publication is that we give a unified (and further generalised) version of the prime number theorem (on the number of primes $\le x$) and Landau's theorem (on the number of numbers $\le x$, which are either squares or sums of two squares) with the Vinogradov's error term. (There are also other results such as Montgomery-Vaughan theorem on square free numbers). The first step (see §2) is that for any generalised Dirichlet series $F(s)$ (defined by $\sum_{n=1}^{\infty} a_n \lambda_n^{-s}, s = \sigma + it$) satisfying some conditions we prove

$$\frac{1}{2\pi i} \int_{C-iT}^{C+iT} F(s) \frac{x^s}{s} ds = \sum_{\lambda_n \le x} a_n + E(T) \quad \left(C = 1 + \frac{1}{\log x}, 2 \le T \le x\right) \tag{1*}$$

where $E(T) = O(xT^{-1} \exp((\log x)^{\epsilon}))$ for every fixed $\epsilon > 0$ and the O-constant depends only on ε. (If f_1 is any complex number depending on some parameter and $f_2 > 0$ then $f_1 = O(f_2)$ will mean that $|f_1 f_2^{-1}|$ is bounded above. Some times we write $O\ldots(f_2)$ to indicate that the bound depends on the constants...). The next step (see §3) is to prove that $F(s)$ is analytic in the rectangle $(\sigma \ge 1 - \lambda(T), C_0 \le |t| \le T)$ for a constant $C_0 > 0$ and a suitable small positive function $\lambda(T)$ and establish the bound

$$F(s) = O(\exp((\log x)^{\epsilon})) \tag{2}$$

on the boundary of this rectangle. The net result is

$$\left.\begin{array}{l} \sum_{\lambda_n \le x} a_n = M(x) + E_0(x, T) \\ \text{with } E_0(x,T) = O(x(\exp((\log x)^{2\epsilon}))(T^{-1} + x^{-\lambda(T)})) \\ \text{where } M(x) = \frac{1}{2\pi i} \int_K F(s) \frac{x^s}{s} ds \end{array}\right\} \tag{3}$$

K being the contour obtained by joining $1-\lambda(T)-iC_0$, $C-iC_o$, $C+iC_0$ and $1-\lambda(T)+iC_o$ in this order by straight line segments. Choosing T by the requirement $T = \exp(\lambda(T) \log x)$ we are led to our main theorem. (Note that $M(x)$ can sometimes be so small as the error term itself, for example when $F(s) = (\zeta(s))^{-10}$)

In our final choice of $F(s)$, we can improve the error estimate in our main theorem by assuming Riemann hypothesis and its generalisations to L-functions involved in $F(s)$. (See Remark 3 below). For unconditional results we have to depend on the deep estimate

$$\left|\zeta(s) - \frac{1}{s-1}\right| \le ((|t| + 10)^{(1-\sigma)^a} \log(|t| + 10))^A \tag{4}$$

where $(\frac{1}{2} \leq \sigma \leq 1) A > 0$ and $a > 1$ are constants. This with $a = \frac{3}{2}$ is a deep estimate (see [E.C.T], [A.I], [A.A.K; S.M.V] and $[K.R, A.S]_1$) due to the Soviet Mathematician I.M.VINOGRADOV. (It is well-known that $a = 1$ is trivial and leads to $\lambda(T)$ with $(\lambda(T))^{-1} = O(\log T)$ and hence to

$$\pi(x) \left(\equiv \sum_{p \leq x} 1 \right) = \int_2^x \frac{du}{\log u} + O(x(\exp(\sqrt{\log x}))^{-\mu}) \tag{5}$$

where $\mu > 0$ is a constant which is unimportant). (The symbol $\equiv$ denotes a definition). But the merit of VINOGARDOV'S estimate is that it leads to the $\lambda(T)$ with $(\lambda(T))^{-1} = O((\log T)^{\frac{2}{3}} (\log \log T)^{\frac{1}{3}})$ (see Remark 1 below) and hence to the O-term

$$O(x(\exp((\log\ x)^{\frac{3}{5}}\ (\log\ \log\ x)^{-\frac{1}{5}}))^{-\mu}) \tag{6}$$

in (5). It must be mentioned that any constant $> \frac{1}{2}$ in place of $\frac{3}{5}$ in (6) is a very very important result from many points of view.

Although the estimate (4) is not known (nor likely to be known for many centuries to come) for $a > \frac{3}{2}$, we can assume (4) for any $a > 1$, and arrive at the O-term

$$O(x(\exp((\log\ x)^{\frac{a}{a+1}}\ (\log\ \log\ x)^{-\frac{a-1}{a+1}}))^{-\mu}) \tag{7}$$

in (5).

Remark 1 The only method of getting the best known unconditional $\lambda(T)$ seems to be via (4). The best $\lambda(T)$ follows from (4) by an easy function-theoretic device due to J. HADAMARD, de-la-VALLEE POUSSIN and E. LANDAU. (For the relation between a and $\lambda(T)$ namely $(\lambda(T))^{-1} = O((\log\ T)^{\frac{1}{a}}\ (\log\ \log\ T)^{1-\frac{1}{a}})$ see the booklet $[K.R]_1$ by K. RAMACHANDRA). Instead of the assumption (4) we can assume a $\lambda(T)$ as a hypothesis and proceed and get the corresponding error estimate in the main result. (A more illuminating method of getting $\lambda(T)$ from (4) is due to H.L. MONTGOMERY and later to K. RAMACHANDRA $[K.R]_5$ where he proves a slight refinement on the result of H.L. MONTGOMERY).

Remark 2 The inequality (4) has been extended suitably (by a few mathematicians notably by T. MITSUI [T.M]) to cover zeta and L-functions of algebraic number fields (with $a = \frac{3}{2}$ and $A > 0$ depending on the L-functions). The corresponding $\lambda(T)$ satisfies $(\lambda(T))^{-1} = O((\log T)^{\frac{1}{a}} (\log\ \log T)^{1-\frac{1}{a}})$ where the O-constant depends on the L-function. This leads to the error estimate (7) in our main result which is fairly general.

Remark 3 If we assume the analogue of Riemann hypothesis for all the L-functions occuring in our final choice of $F(s)$, then we can take $\lambda(T) = \frac{1}{2} - B_0(\log\ \log\ T)^{-1}$, where B_0 is a positive constant depending on $F(s)$. Using some results of J.E. LITTLEWOOD; A. SELBERG; K. RAMACHANDRA and A. SANKARANARAYANAN (for reference see the paper $[K.R; A.S]_2$ by the last two authors) we can obtain the error estimate $O(x^{\frac{1}{2}+\mu(x)})$,

in S_0. Let S_2 be a power product (with non negative integer constants as exponents) of functions which are logarithms (in $\sigma > 1$) of functions in S_0. Let S_3 be a power product (with non negative integer constants as exponents) of all possible derivatives (with bounded order) of functions of S_0. Finally let $f(s) = \sum_{n=1}^{\infty} b_n n^{-s}$ be a Dirichlet series for which $\sum_{n=1}^{\infty} |b_n| n^{-\mu_0} < \infty$ for some constant $\mu_0 < 1$. Then our function $F(s)$ is defined by

$$F(s) = S_1\, S_2\, S_3\, f(s)\ (\sigma > 1) \tag{14}$$

and its analytic continuations in ($\sigma > \mu_0$) along all possible paths in the complex s-plane with certain lines Λ removed. (Λ consists of all lines, parallel to the real axis, which contain a singularity of L-functions in S_1, S_2 and S_3). We select that branch for which $\log\, L(s) \to 0$ as $\sigma \to \infty$.

We now make the following hypothesis and proceed.

Hypothesis: $L(s) \neq 0$ *in* $(\sigma \geq 1 - \lambda(T),\ |t| \leq T)$ where $(\lambda(T))^{-1} = O((\log T)^{\frac{1}{a}}$ $(\log\, \log\, T)^{1-\frac{1}{a}})$ for some constant $a \geq 1$.

Lemma 1 *In* $(\sigma > 1 - \lambda(T),\ |t| \leq T)$, *we have* $L(s) = O(T^d)$ *where d is the degree of the number field, provided* $L(s)$ *is formed with the non-principal character. If* $L(s)$ *is formed with the principal character then* $|L(s) - K_0(s-1)^{-1}| = O(T^d)$, *where* K_0 *is a positive constant.*

Remark: In the main term $M(x)$ defined by (3) the constant C_0 is not very important. Hence we can confine our attention to sufficiently large values of $|t|$ (i.e. to $|t|$ exceeding a large constant at our choice). Moreover plainly we see that it sufficies to consider large values of t(that of $-t$ is covered by complex conjugation). In this case the lemma is well-known.

Lemma 2 (BOREL-CARATHEODORY). *Suppose* $f(z)$ *is analytic in* $|z - z_0| \leq R$ *and on the circle* $z = z_0 + Re^{i\theta}\,(0 \leq \theta \leq 2\pi)$ *we have* $Re\ f(z) \leq U$. *Then in* $|z - z_0| \leq r < R$, *we have*

$$|f(z) - f(z_0)| \leq \frac{2r(U - Ref(z_0))}{R - r}.$$

Proof: See Theorem 1.6.1 of $[K.R]_3$ (This is a well-known Theorem and we have stated it with the notation adopted therein). □

Lemma 3 *We have, in* $(\sigma \geq 1 - \frac{1}{2}\lambda(T),\ C_0 \leq |t| \leq T)$, *the estimate*

$$\log\ L(s) = O((\log\ T)^{20}).$$

Also $\log\ L(s) = O(1)$ *in* $\sigma \geq 1 + \epsilon$, *where the* O*-constant depends on* ϵ.

Proof: The second part is trivial by looking at the (absolutely convergent) Dirichlet series for $\log\ L(s)$. To prove the first part we apply lemma 2 with $f(z) = \log\ L(z)$, $z_0 = 2 + it$, $R = 1 + \lambda(T)$, $r = 1 + \frac{1}{2}\lambda(T)$. By lemma 1 we can take U to be $(\log\ T)^2$. This gives the first part. □

Lemma 4 *For all* $\epsilon (0 < \epsilon < \frac{1}{10})$, *we have in* $(\sigma \geq 1 - \epsilon\lambda(T),\ C_0 \leq |t| \leq T)$ *the estimate*

$$\log\ L(s) = O((\log\ T)^{\epsilon}) \tag{15}$$

where the O*-constant depends only on* ϵ.

Proof: We apply maximum modulus principle. Let s_0 belong to the rectangle in question. Put $s_0 = \sigma_0 + it_0$, $G(s) = e^{(s-s_0)2} X^{s-s_0} \log\ L(s)$. (We can assume that $\sigma_0 \leq 1 + \epsilon$ since otherwise we are through).

Then $|G(s_0)| \leq \max |G(s)|$ taken over the sides of the rectangle

$$\left(|t - t_0| \leq \log\ t_0,\ 1 - \frac{1}{2}\lambda(T) \leq \sigma \leq 1 + \epsilon\right).$$

Note that $e^{(s-s_0)^2} = O(e^{-(t-t_0)^2})$. Lemma follows by a proper choice of X. □

Theorem 2 *Subject to our choice of* $F(s)$ *as stated in the beginning of this section, we have*

$$\sum_{n \leq x} a_n = M(x) + O(x(\exp((\log\ x)^{\frac{a}{a+1}}(\log\ \log\ x)^{-\frac{a-1}{a+1}}))^{-\mu}). \tag{16}$$

Proof: In (3) we have only to replace $\lambda(T)$ by $\epsilon\lambda(T)$ and use Lemma 4 and choose T suitably. □

Remark: We remark that $\log\ \zeta(\sigma + it) = O(\log\ \log\ t)$ in $(\sigma \geq 1,\ t \geq 100)$ and a similar result for $\log\ L(\sigma + it)$. These are not hard to prove.

§ 4 Discussion of $M(x)$

Note that $M(x)$ depends on T. We will now see that

$$M(x) = M_0(x) + O(x(\exp((\log\ x)^{\frac{1}{a+1}}(\log\ \log\ x)^{-\frac{a-1}{a+1}}))^{-\mu})$$

where $M_0(x)$ is a certain contour integral. Since $L(1 + it) \neq 0$ and $\zeta_k(s)(s - 1) \neq 0$ for $s = 1 + it(\zeta_k(s)$ denotes the zeta-function of the number field k), it follows that $\log\ L(s)$ and $\log(\zeta_k(s)(s - 1))$ are regular in $(\sigma \geq 1 - \delta,\ |t| \leq C_0)$ (for some constant $0 < \delta \leq 1$) and are bounded in $(\sigma \geq 1 - \frac{1}{2}\delta,\ |t| \leq C_0)$. Hence if we write

$$H(s) = F(s)(s - 1)^{\alpha_0} \left(\log \frac{1}{s - 1}\right)^{-\beta_0}$$

and choose the complex constant α_0 and the non-negative integer constant β_0, then $H(s)$ is regular in $(\sigma \geq 1 - \frac{1}{2}\delta,\ |t| \leq C_0)$ and is bounded there. Hence we can replace the contour (in $M(x)$) namely the one obtained by joining by straight line segments the points $1 - \lambda(T) - iC_0,\ C - iC_0,\ C + iC_0,\ 1 - \lambda(T) + iC_0$ by the new contour obtained by joining $1 - \lambda(T) - iC_0,\ 1 - \frac{1}{2}\delta - iC_0,\ 1 - \frac{1}{2}\delta - i0,\ 1 - \frac{1}{2}\delta + i0,\ 1 - \frac{1}{2}\delta + iC_0,\ 1 - \lambda(T) + iC_0$.

Here we mean by straight lines except the ones with a bar below where we mean the circle D with centre 1 and radius $\frac{1}{2}\delta$ with the point $1-\frac{1}{2}\delta$ removed. Apart from the circular contour D the contribution of the rest of the contour is $O(x^{1-\lambda(T)}))$ since $H(s)=O(1)$ and

$$(s-1)^{-\alpha_0}\left(\log\frac{1}{s-1}\right)^{\beta_0}=O(1).$$

Thus

$$\left.\begin{array}{l} M(x)=M_0(x)+O(x(\exp((\log x)^{\frac{a}{a+1}}(\log\log x)^{-\frac{a-1}{a+1}}))^{-\mu}) \\ \textit{where} \\ M_0(x)=\frac{1}{2\pi i}\int_D F(s)\frac{x^s}{s}\,ds. \end{array}\right\} \qquad (17)$$

Now $H(s)$ being analytic in the disc $|s-1|\le\delta$, we have

$$\frac{H(s)}{s}=\sum_{n=0}^{\infty}a_n^*(s-1)^n$$

where

$$a_n^*=\frac{1}{2\pi i}\int_{D^+}\frac{H(s)}{(s-1)^{n+1}}\frac{ds}{s}$$

where D^+ is the disc $|s-1|=\delta$ and so $a_n^*=O(\delta^{-n})$. Thus in $|s-1|\le\frac{1}{2}\delta$, we have

$$s^{-1}H(s)=\sum_{n=0}^{N}a_n^*(s-1)^n+R,$$

where

$$R=O\left(\sum_{n=N+1}^{\infty}\delta^{-n}\left(\frac{1}{2}\delta\right)^n\right)=O(2^{-N}).$$

Finally we arrive at the asymptotic expansion

$$M_0(x)=\sum_{0\le n\le N}a_n^*\left(\frac{1}{2\pi i}\int_D(s-1)^{n-\alpha_0}\left(\log\frac{1}{s-1}\right)^{\beta_0}x^s ds\right)$$
$$+O(2^{-N}x(\log x)^{l_0}).$$

where l_0 is a real constant. This can be seen by deforming contour $D(for\ n\le N)$ suitably. Here the integral on the right is

$$\frac{1}{2\pi i}\int_D(s-1)^{n-\alpha_0}\left(\log\frac{1}{s-1}\right)^{\beta_0}x^s ds$$
$$=\frac{1}{2\pi i}\int_{-\infty}^{1+}(s-1)^{n-\alpha_0}\left(\log\frac{1}{s-1}\right)^{\beta_0}x^s ds+O(x^{1-\frac{1}{2}\delta}(\log x)^{l_0}\delta^{-2N}).$$

The integral $-\infty, 1+$ is the line joining $-\infty - i0$ to $1 - i0$ and this to $1 + i0$ by an indentation at $s = 1$ and then $1 + i0$ to $-\infty + i0$. The last integral is (with an obvious notation).

$$\frac{x}{2\pi i}\int_{-\infty}^{0+} s^{n-\alpha_0}\left(\log\frac{1}{s}\right)^{\beta_0} x^s ds.$$

Thus we are led to study (in the first instance)

$$\frac{1}{2\pi i}\int_{-\infty}^{0+} s^{-\gamma_0}x^s ds$$
$$= (\log x)^{\gamma_0-1}\left(\frac{1}{2\pi i}\int_{-\infty}^{0+} w^{-\gamma_0}e^w\,dw\right)$$

(by the substitution $s = w(\log x)^{-1}$), where γ_0 is a complex constant.

Now

$$J(\gamma_0) \equiv \frac{1}{2\pi i}\int_{-\infty}^{0+} w^{-\gamma_0}e^w du$$

is analytic in γ_0 and it is

$$\frac{1}{2\pi i}\int_0^{\infty}(ue^{-i\pi})^{-\gamma_0}e^{-u}du + \frac{1}{2\pi i}\int_0^{\infty}(ue^{i\pi})^{-\gamma_0}e^{-u}du$$

(on splitting the path of integration into $(-\infty, 0)$ and $(0, -\infty)$ respectively),

$$= \frac{1}{2\pi i}(2i\sin(\pi\gamma_0))\,\Gamma(1-\gamma_0) = \frac{1}{\pi}\frac{\sin(\pi\gamma_0)}{\Gamma(\gamma_0)}\Gamma(\gamma_0)\Gamma(1-\gamma_0) = \frac{1}{\Gamma(\gamma_0)},$$

and this gives the expression for $J(\gamma_0)$ in terms of the familiar function $\Gamma(z)$. Also for any complex constant γ_0 we have

$$\frac{1}{2\pi i}\int_{-\infty}^{0+} s^{-\gamma_0}\left(\log\frac{1}{s}\right)^{\beta_0} x^s ds$$
$$= \frac{d^{\beta_0}}{d\,z^{\beta_0}}((\log x)^{z-1}J(z))\Big]_{z=\gamma_0}$$
$$= \frac{d^{\beta_0}}{d\,z^{\beta_0}}\left(\frac{(\log x)^{z-1}}{\Gamma(z)}\right)\Big]_{z=\gamma_0}$$

This completes the discussion of $M(x)$. We have still to check the condition (8). For this observe that the coefficients of the Dirichlet series for $F(s)$ are majorised by those of

$$\left(1 + \sum_{n=2}^{\infty}\frac{d_q(n)(\log n)^q 2^q}{n^s}\right)\left(\sum_{n=1}^{\infty}\frac{|b_n|}{n^s}\right)$$

for some integer constant $q > 0$. Now

$$\sum_{X \le mn \le X+H} |b_n|(d_q(m)(2\log m)^q + 1)$$

$$= O\left(\sum_{X \le mn \le X+H} |b_n| d_q(m)\right)(\log X)^{2q})$$

and the last sum is

$$\sum_{1 \le n \le X} |b_n| \sum_{\frac{X}{n} \le m \le \frac{X+H}{n}} d_q(m) = \sum_1 + \sum_2 \text{ say,}$$

where $\sum_1 = \sum_{1 \le n \le X^\eta} \ldots$ (and) $\sum_2 = \sum_{X^\eta < n \le X} \ldots$ and $0 < \eta < 1$ is at our choice.

Since $d_q(n) = O_{\epsilon,q}(n^\epsilon)$ we have

$$\sum_2 = O\left(\sum_{X^\eta \le n \le X}\left(\frac{H}{n}+1\right)|b_n| X^\epsilon\right)$$

$$= O\left(X^\epsilon H X^{-(1-\mu_0)\eta} \sum_{n=1}^{\infty} \frac{|b_n|}{n^{\mu_0}} + X^\epsilon \sum_{n \le X} |b_n|\right)$$

where the last sum is $\le X^\epsilon \left(\sum_{n \le X} |b_n| n^{-\mu_0}\right) X^{\mu_0} = O(X^{\mu_0+\epsilon})$, and the sum previous to it is $O(HX^{-(1-\mu_0)\eta+\epsilon})$.

To treat $\sum_1$ we use

$$\sum_{n \le x} d_k(n) = x P_{k-1}(\log x) + O(x^{1-\frac{1}{k}}(\log x)^{k-2})$$

where $k > 0$ is an integer constant, $x \ge 2$ is a variable and $P_{k-1}(\log x)$ is a polynominal in $\log x$ of degree $k-1$. (For this result see equation (12.1.4) on page 313 of [E.C.T]). Now on using this result,

$$\sum_1 = O\left(\sum_{n \le X^\eta} |b_n| \left(\frac{H}{n}(\log X)^q\right)\right) = O(H(\log X)^q),$$

(provided $H = x^{1-\delta}$ and δ and η are small constants). Condition (8) now follows. Thus we have the following theorem (by choosing $N = (\log x)^{B^*}$ where $B^*(0 < B^* < 1)$ is an arbitrary positive constant).

Theorem 3 (MAIN THEOREM) *Subject only to the choice of $F(s)$ as in the beginning of §3, we have*

$$\sum_{\lambda_n \le x} a_n = xM^*(x) + O(x(\exp((\log x)^{\frac{a}{a+1}}(\log\log x)^{-\frac{a-1}{a+1}}))^{-\mu}), \tag{18}$$

where $M^*(x)$ has the asymptotic expansion

$$M^*(x) = \sum_{\nu=j}^{\infty} A_\nu (\log\, x)^{-\nu+\gamma_0} \sum_{0\le k\le\beta_0} B_k(\log\log x)^k, \tag{19}$$

where j is an integer constant, $\beta_0 > 0$ is an integer constant, and A_ν and $B_k = B_k(\nu)$ are complex constants and γ_0 is a complex constant.

Remark 1 It must be remembered that Theorem 3 is subject to the hypothesis in §3 made after equation (14). Theorem 3 is unconditional only if $a = \frac{3}{2}$. The hypothesis just refered to with $a = \frac{3}{2}$ is a deep theorem established by the Soviet mathematician I.M. VINOGRADOV. It may take many centuries to prove the hypothesis with some $a > \frac{3}{2}$.

Remark 2 It may happen that many A_ν are zero in the beginning so as to make $xM^*(x)$ as small as the O-term (for example when $F(s) = (\zeta(s))^{-1}$). If F (s) has a pole at $s = 1$ even then the treatment is simpler.

§ 5 An Alternative Method

Consider $F(s) = \sum_{n=1}^{\infty} a_n\lambda_n^{-s}$ where $a_n \geq 0$ and $0 < \lambda_1 < \lambda_2 < \ldots$ and $\lambda_n \to \infty$. Then it is possible to get rid of the condition (8) in the following way. Put $A(x) = \sum_{\lambda_n\le x} a_n$ and $B(x) = \int_0^x A(u)du = \sum_{\lambda_n\le x} a_n(x-\lambda_n)$. It is much simpler to deal with $B(x)$. Suppose that $B(x) = \int_0^x M(u)du + O(x^2\Delta)$ where $0 \le \Delta \le \frac{1}{4}$ with $M(u) = \frac{1}{2\pi i}\int_{K_1} F(s)u^s\frac{ds}{s}$, K_1 being a suitable contour.

Then we have the following Tauberian Lemma.

A Tauberian Lemma: *We have*

$$A(x) = M(x) + O\left(x\Delta^{\frac{1}{2}}\left(1 + \max_{\frac{x}{2}\le\xi\le 2x} |M'(\xi)|\right)\right).$$

Proof: Let $0 \le \delta \le \frac{1}{2}$. Then

$$B(x+\delta x) - B(x) = \int_x^{x+\delta x} A(u)du \ge \delta x A(x).$$

Here the LHS is

$$\delta x\left\{M(x) + O\left(\delta x \max_{x\le\xi\le\frac{3x}{2}} |M'(\xi)|\right)\right\} + O(x^2\Delta).$$

Thus

$$A(x) \le M(x) + O\left(\delta x \max_{x\le\xi\le\frac{3x}{2}} |M'(\xi)|\right) + O(x\Delta\delta^{-1}).$$

Similarly we have $\delta x A(x) \geq B(x) - B(x-\delta x) = \int_{x-\delta x}^{x} M(u)du + O(x^2\Delta)$

$$= \delta x \left\{ M(x) + O\left(\delta x \max_{\frac{x}{2}\leq\xi\leq x} |M'(\xi)|\right)\right\} + O(x^2\Delta) \text{ and so}$$

$$A(x) \geq M(x) + O\left(\delta x \max_{\frac{x}{2}\leq\xi\leq x} |M'(\xi)|\right) + O(x\Delta\delta^{-1}).$$

Combining both we have

$$A(x) = M(x) + O\left(\delta x \max_{\frac{x}{2}\leq\xi\leq 2x} |M'(\xi)|\right) + O(x\Delta\delta^{-1})$$

the choice $\delta = \Delta^{\frac{1}{2}}$ gives the lemma. □

Remark 1 In many cases it is possible to prove (without bringing in condition (8)) that $B(x) = \int_0^x M(u)du + O(x^2\Delta)$ where $\Delta = O((\exp((\log x)^{\frac{3}{5}}(\log\log x)^{-\frac{1}{5}}))^{-\mu})$ and $M'(u) = O((\log(|u|+10))^g)$ where $g > 0$ is an integer constant. Thus for example if $F(s) = -\zeta'(s)(\zeta(s))^{-1}$ we can recover the prime number theorem with the VINOGRADOV error term.

Remark 2 Consider $F(s) = \sum_{n=1}^{\infty}(\varphi(n))^{-s}$, where $\varphi(n)$ is the Euler's totient function. In this case we can apply the alternative method. But we have only to be contented with the best known (unconditional) result

$$\sum_{\varphi(n)\leq x} 1 = d_0 x + O(x(\exp\sqrt{\log x \log\log x})^{-\mu}), \left(d_0 = \pi\left(1 + \frac{1}{p(p-1)}\right)\right),$$

due to P.T. BATEMAN (see [P.T.B]). Even the assumption of Riemann hypothesis that $\zeta(s) \neq 0$ in $\sigma > \frac{1}{2}$ does not seem to result in any improvement over this unconditional result. It may be noted that for this special choice of $F(s)$ we have

$$F(s) = \zeta(s)\pi\left(1 - \frac{1}{p^s} + \frac{1}{(p-1)^s}\right).$$

Thus $F(s)$ and $\zeta(s)$ have the same zeros in $\sigma > o$ (since the infinite product is regular in $\sigma > 0$ and does not vanish there). Note that the infinite product is

$$\exp(O(t^{1-\sigma}\log\log t))) \left(t \geq 100, \frac{1}{2} \leq \sigma \leq 1\right)$$

and Bateman's result follows from this and the trivial estimate $\zeta(s) = O(t^{1-\sigma}\log t)(t \geq 10, \frac{1}{2} \leq \sigma \leq 1)$.

§ 6 Squarefree Numbers and other Problems

We now proceed to explain a method due to H.L. MONTGOMERY and R.C. VAUGHAN (see [H.L.M; R.C.V]). For instance we can apply this method to estimate things like $\sum_{n\leq x}|\mu(n)| - (\frac{6}{\pi^2})x$ (they devised this method in connection with this problem on the assumption of RH (Riemann hypothesis)), and

$$\sum_{mn^2\leq x} \mu(n)d(m) - Ax \log x - Bx \text{ (for suitable constants A, B)}$$

where $\mu(n)$ and $d(n)$ are defined by $(\zeta(s))^{-1} = \sum \mu(n)n^{-s}$ and $(\zeta(s))^2 = \sum_{n=1}^{\infty} d(n)n^{-s}$. (The best known unconditional results are due to A. WALFISZ [A.W]. His method gives $O(x^{\frac{2}{5}+\varepsilon})$, on RH, for the first problem. However the method of A. WALFISZ was discovered first by A. AXER [A.A] who assumed RH and proved that $\sum_{n\leq x}\mu_k(n) = (\zeta(k)^{-1}x + O(x^{(2+\epsilon)/(2k+1)})$ where $\zeta(s)(\zeta(ks))^{-1} = \sum_{n=1}^{\infty}\mu_k(n)n^{-s}, k > 1$ being an integer constant. AXER assumed that $\sum_{n\leq x}\mu(n) = O(x^{\frac{1}{2}+\varepsilon})$ a fact which was shown by J.E. LITTLEWOOD [J.E.L.] to follow from RH). The method of H.L. MONTGOMERY and R.C. VAUGHAN can also be applied to things like $\sum_{mn^2\leq x}\mu(n)r(n)(r(m)$ being defined by $\zeta(s)(1-\frac{1}{3^s}+\frac{1}{5^s}-+\cdots) = \sum_{n=1}^{\infty} r(n)n^{-s})$ and also to $\sum_{mn^2\leq x}\mu(n)d_3(n)(d_3(n)$ being defined by $(\zeta(s))^3 = \sum_{n=1}^{\infty} d_3(n)^{n-s})$. We will remark about some other applications at the end of this section. (It is good to mention here the results $\sum_{n\leq x} d(n) = x \log x + (2\gamma - 1)$ $x + O(x^{\frac{23}{73}}(\log x)^{\frac{461}{146}})$

$$\sum_{n\leq x} r(n) = \pi x + O(x^{\frac{23}{73}}(\log x)^{\frac{315}{146}})$$

both due to M.N. HUXLEY [M.N.H] and the result $\sum_{n=1} d_3(n) = xP_2(\log x) + O(x^{\frac{43}{96}+\epsilon})$ ($P_2(\log x)$ is a polynomial in $\log x$ of degree 2) due to G. KOLESNIK [G.K]). We begin with $\zeta(s)(\zeta(ks))^{-1} = \sum_{n=1}^{\infty}\mu_k(n)n^{-s}$ where $k \geq 2$ is an integer constant $\mu_k(n) = 1$ if n is k power free and zero otherwise. When $k = 2$ this is the characteristic function of square-free numbers. We have

$$\begin{aligned}
\sum_{n\leq x}\mu_k(n) &= \sum_{n^k m\leq x}\mu(n) \\
&= \sum_{n\leq x^{\frac{1}{k}}}\mu(n)\left[\frac{x}{n^k}\right] = \sum_{n=1}^{\infty}\mu(n)\left(\left[\frac{x}{n^k}\right]-\frac{x}{n^k}\right) + x\sum_{n=1}^{\infty}\frac{\mu(n)}{n^k} \\
&= E_0 + \frac{x}{\zeta(k)}, \text{ say.}
\end{aligned}$$

Now $E_0 = \sum_{n\leq Y} + \sum_{n>Y} = E_1 + E_2$ say. Clearly $E_1 = O(Y)$ and

$$E_2 = \sum_{n>Y}\mu(n)\left(\left[\frac{x}{n^k}\right]-\frac{x}{n^k}\right).$$

Note that $[\frac{x}{n^k}]$ and $\frac{x}{n^k}$ are monotonic.

Thus assuming

$$\sum_{n\leq x}\mu(n)=O(x(E_0(x))^{-1}) \text{ where } E_0(x)=\exp(\mu(\log x)^{\frac{3}{5}}(\log\log x)^{-\frac{1}{5}})$$

($\mu>0$ is some absolute constant), we have $E_2=O(x\sum_{U\geq Y}U^{1-k}(E_0(U))^{-1})$, ($U$ runs over numbers $Y2^n;n=0,1,2,\ldots)\leq(E_0(Y))^{-1}x\sum_{U\geq Y}U^{1-k}\leq(E_0(Y))^{-1}xY^{1-k}$.

Hence

$$E_0=O(Y+(E_0(Y))^{-1}xY^{1-k}).$$

Choose Y by

$$Y=x^{\frac{1}{k}}(\exp((200)^{-1}\mu k^{-1}(\log Y)^{\frac{3}{5}}\quad(\log\log Y)^{-\frac{1}{5}}))^{-1}$$

so that

$$\begin{aligned}(E_0(Y))^{-1}xY^{1-k} &= x^{\frac{1}{k}}(\exp((200)^{-1}\mu k^{-1}(\log Y)^{\frac{3}{5}}(\log\log Y)^{-\frac{1}{5}})^{k-1}(E_0(Y))^{-1}\\ &\leq x^{\frac{1}{k}}\left(\exp\left(\frac{1}{2}\mu(\log Y)^{\frac{3}{5}}(\log\log Y)^{-\frac{1}{5}}\right)\right)^{-1}.\end{aligned}$$

Also from the equation defining Y we have (for fixed k, as $x\to\infty$)

$$\log Y\sim\frac{1}{k}\log x$$

and so

$$\begin{aligned}E_0 &= O(Y+xY^{1-k}(E_0(Y))^{-1})\\ &= O(x^{\frac{1}{k}}(\exp(\mu_1k^{-\frac{8}{5}}(\log x)^{\frac{3}{5}}(\log\log x)^{-\frac{1}{5}}))^{-1})\end{aligned}$$

where $\mu_1>0$ is an absolute constant. Thus we have the following theorem.

Theorem 4 (A. WALFISZ). *Let $k\geq 2$ be an integer constant and let $\mu_k(n)=1$ if n not divisible by the k-th power of any prime and zero otherwise. Then for all $x\geq x_0(k)$, we have*

$$\sum_{n\leq x}\mu_k(n)=\frac{x}{\zeta(k)}+O(x^{\frac{1}{k}}(\exp(k^{-\frac{8}{5}}(\log x)^{\frac{3}{5}}(\log\log x)^{-\frac{1}{5}}))^{-\mu_1}).$$

where $\mu_1>0$ is an absolute constant.

H.L. MONTGOMERY and R.C. VAUGHAN assumed RH and treated the term E_2 (more effectively than A. AXER) and arrived at the error estimate $O(x^{(1+\epsilon)/(k+1)})$. (In fact they treated E_1 also more effectively and got even better results on RH). Their method (see [H.L.M; R.C.V]), is embodied in the following lemma.

Lemma: (H.L. MONTGOMERY and R.C. VAUGHAN). *Let Y be bounded above by a fixed positive power of x. Then for every $\epsilon > 0$ and all $x \geq x_0(\epsilon, k)$, we have*

$$E_2 \equiv \sum_{n \geq Y} \mu(n) \left(\left[\frac{x}{n^k} \right] - \frac{x}{n^k} \right) = O(x^{\epsilon} + x^{\frac{1}{2}+\epsilon} Y^{(1-k)/2}).$$

Proof: We have, by Perron's formula (for $20 \leq T \leq x$)

$$\sum_{n \geq Y} \mu(n) \left[\frac{x}{n^k} \right] = \frac{1}{2\pi i} \int_{1+\epsilon-iT}^{1+\epsilon+iT} \zeta(s) \left(\sum_{n \geq Y} \frac{\mu(n)}{n^{ks}} \right) \frac{x^s}{s} ds$$

$$+ \; O\left(x^{2\epsilon} + \frac{x^{1+2\epsilon}}{T} \right).$$

Denoting the integral on the right by I and the sum over n by $\zeta_Y(ks)$, we have

$$I = \frac{1}{2\pi i} \int_{\frac{1}{2}+\epsilon-iT^*}^{\frac{1}{2}+\epsilon+iT^*} \zeta(s) \zeta_Y(ks) \frac{x^s}{s} ds + \sum_{n \leq Y} \mu(n) \left(\frac{x}{n^k} \right) + O\left(\frac{x^{1+2\epsilon}}{T} + x^{2\epsilon} \right)$$

where T^* lies between $\frac{T}{2}$ and $2T$ (since $\zeta_Y(ks) = O(1)$ and $\int_{\frac{1}{2}T}^{T} |\zeta(s)| dt = O(T \log T)$ for $\sigma \geq \frac{1}{2}$). Thus

$$\sum_{n \leq Y} \mu(n) \left(\left[\frac{x}{n^k} \right] - \frac{x}{n^k} \right) = \frac{1}{2\pi i} \int_{\frac{1}{2}+\epsilon-iT^*}^{\frac{1}{2}+\epsilon+iT^*} \zeta(s) \zeta_Y(ks) \frac{x^s}{s} ds + O\left(\frac{x^{1+2\epsilon}}{T} + x^{2\epsilon} \right).$$

We now assume RH and show that on ($\sigma = \frac{1}{2} + \epsilon$, $|t| \leq T$) we have

$$\zeta_Y(ks) = O(Y^{-\frac{k}{2}+\frac{1}{2}+3\epsilon} T^{3\epsilon}).$$

Let $V \geq 10$ and let U run over the numbers $Y2^n (n = 0, 1, 2, \ldots)$. Then by Perron's formula again, we have (with $s = \frac{1}{2} + \epsilon + it$)

$$\frac{1}{2\pi i} \int_{1-iV}^{1+iV} (\zeta(ks + W))^{-1} U^W (2^W - 1) \frac{dW}{W}$$

$$= \sum_{U \leq m < 2U} \mu(m) m^{-ks} + O\left(\frac{U^{1+\epsilon}}{V} \right) + O(U^{-\frac{k}{2}})$$

We use $(\zeta(\frac{1}{2} + \epsilon + it))^{-1} = O_{\epsilon,\delta}((|t| + 2)^{\delta})$ for all $\epsilon > 0, \delta > 0$ (this result is due to J.E. LITTLEWOOD [J.E.L]) and move the line of integration to Re W $= -\frac{k}{2} + \frac{1}{2} - \epsilon$ and take rough estimates. Thus we obtain

$$\sum_{U \leq m < 2U} \mu(m) m^{-ks} = O\left(V^{\epsilon} T^{\epsilon} U^{\epsilon - \frac{k-1}{2}} + \frac{U^{1+\epsilon} T^{\epsilon} V^{\epsilon}}{V} + U^{-\frac{k}{2}} \right).$$

We next choose V to be a large power of U and sum with respect to U. Thus we obtain the lemma (since $\epsilon > 0$ is an arbitrary constant.) □

From the lemma it follows that

$$\sum_{n=1}^{\infty} \mu(n) \left[\frac{x}{n^k}\right] = \frac{x}{\zeta(k)} + O((Y + x^{\frac{1}{2}} Y^{\frac{1}{2}-\frac{k}{2}})(xY)^{\epsilon})$$

choosing $Y = x^{\frac{1}{k+1}}$ we get the MONTGOMERY-VAUGHAN Theorem namely

Theorem 5 (H.L. MONTGOMERY and R.C. VAUGHAN [H.L.M., R.C.V]). *Let $k \geq 2$ be any integer constant. Assume RH. Let $Q_k(x)$ be the number of k-free numbers $n \leq x$. Then*

$$Q_k(x) = \frac{x}{\zeta(k)} + O_{k,\epsilon}(x^{\frac{1}{k+1}+\epsilon})$$

holds for every $\epsilon > 0$.

From the proof it is plain that we can make the following generalisation.

Theorem 6 *Let $G(s) = \sum_{n=1}^{\infty} B_n n^{-s}$ (where B_n is any sequence of complex numbers) converge absolutely in $\sigma > 1$. Suppose that $G(s)$ is regular in $\sigma > \frac{1}{2}$ except for a finite number of poles and that for all $\epsilon > 0$ and $\delta > 0$ we have*

$$\frac{1}{T}\int_T^{2T} \left|G\left(\frac{1}{2}+\epsilon+it\right)\right| dt = O_{\epsilon,\delta}(T^{\delta}).$$

For $x > 0$, let $P(x)$ denote the sum of the residues of $x^s G(s)(\zeta(sk))^{-1}s^{-1}$ at the poles (in $\sigma > \frac{1}{2}$) and that

$$\sum_{n\leq x} B_n - P(x) = O(x^{\alpha})$$

for some positive constant $\alpha < \frac{1}{k}$. Further let $B_n = O_{\epsilon}(Xx^{\epsilon})$ where $X = x^{\frac{1-\alpha}{k+1-2k\alpha}}$ for every $\epsilon > 0$. Then, we have

$$\sum_{m^k n\leq x} \mu(m) B_n = \sum_{m=1}^{\infty} \mu(m) P\left(\frac{x}{m^k}\right) + O_{\epsilon}(Xx^{\epsilon}).$$

for every $\epsilon > 0$.

Remark 1 Instead of $\zeta(ks)$ one can take any finite power product of $\zeta(ks)$ and its derivatives (with integer constants as exponents) which is regular at $s = \frac{1}{k}$. We can do the same thing with $L(ks, \chi)$ provided we assume that $L(s, \chi) \neq 0$ in $\sigma > \frac{1}{2}$). A curious application of the MONTGOMERY-VAUGHAN method gives (of course on RH),

$$\sum_{p\leq x} \log p \sum_{n\leq xp^{-k}} \mu_k(n) = \left((\zeta(k))^{-1} \sum \frac{\log p}{p^k}\right) x + O(x^{\frac{1}{k+1}+\epsilon}),$$

where $k \geq 2$ is any integer constant and $\mu_k(n)$ is one or zero according as n is k-th power free or not. In the last sum above p runs over all primes.

Remark 2 We mention that we can impose $\sum_{n \leq x} B_n = O_\epsilon(x^\epsilon)$. Then the above mean-value condition in $(T, 2T)$ is automatically satisfied (we leave this as an exercise to the reader) and trivially $\alpha = \epsilon$ for every $\epsilon > 0$. Hence in the last statement of the theorem the error term is $O(x^{(1+\epsilon)(1+k)^{-1}})$. If $k = 2$ or 3 we can take $G(s) = \zeta^2(s)$ since $\alpha > \frac{23}{73}$. If $k = 2$ we can take $G(s) = \zeta^3(s)$ since $\alpha > \frac{43}{96}$. However we cannot take $G(s) = \zeta^4(s)$ and prove the error term to be $O(x^{\frac{1}{2}-\delta})$ for some $\delta > 0$. It would be interesting to prove this since

$$\frac{\zeta^4(s)}{\zeta(2s)} = \sum_{n=1}^{\infty} d^2(n) n^{-s},$$

an identity due to S. Ramanujan.

Remark 3 In §1 to §5 we have borrowed freely from the papers [R.B; K.R] and $[K.R]_4$ although we do not refer to them in the text. Theorem 6 and Remarks 1 and 2 above, though new, are essentially due to H.L. MONTGOMERY and R.C. VAUGHAN.

Remark 4 We hope to publish a paper II with the same title where we study the oscillation of the error terms in the prime number theorem in a more general set up.

References

[A.A] A. Axer, Über einige grenzertstaze, S-B. Wiss. Wien IIa **120** (1911), 1253–1298.

[R.B; K.R] R. Balasubramanian and Ramachandra, K., On the number of integers n such that $nd(n) \leq x$, *Acta Arith.* **49** (1988), 313–322.

[P.T.B] P.T. Bateman, The distribution of the values of the Euler function, *Acta Arith.* **21** (1972), 329–345.

[M.N.H] M.N. Huxley, Exponential sums and Lattice points-II, *Proc. London Math. Soc.* **3** (66) (1993), 279–301; Corrigenda ibid **68** (1994), 264.

[A.I] A. Ivic, The Riemann zeta-function, The John Wiley interscience, New York (1985).

[A.A.K; S.M.V] A.A. Karatsuba and Voronin, S.M., The Riemann zeta-function, Expositions in Mathematics, Walter de Gruyter, **5**, N.Y. (1972) (English translation by N. Koblitz).

[G.K] G. Kolesnik, On the estimation of multiple exponential sums, Recent progress in analytic Number Theory, Symp. Durham (1979) Volume I (ed. H. Halberstam and C. Hooley), Academic Press (1981), 231–246.

[J.E.L] J.E. Littlewood, Quelques consequences de l' hypothese que la fonction $\zeta(s)$ de Riemann n'a pas de z'eros dans le demi-plan $Re(s) > \frac{1}{2}$, comptes Rendus *Acad. Sci.* Paris **154** (1912), 263–266.

[T.M] T. Mitsui, On the prime ideal theorem, *J. Math. Soc.* Japan, **20** (1968), 233–247.

[H.L.M; R.C.V] H.L. Montgomery and Vaughan, R.C., The distribution of squarefree numbers, Recent progress in analytic Number Theory, Symp. Durham (1979) Volume I (ed. H. Halberstam and C. Hooley), Academic Press (1981), 247–256.

$[K.R]_1$ K. Ramachandra, Riemann zeta-function, Ramanujan Institute (1979) (A phamplet with 16 pages).

[K.R]$_2$ K. Ramachandra, A Remark on Perron's formula, *J. Indian Math. Soc.* vol. **65** (1998) (to appear).

[K.R]$_3$ K. Ramachandra, On the mean-value and omega theorems for the Riemann Zeta function, LN 85, TIFR, Bombay (1995) (Published for TIFR by Springer-Verlag).

[K.R]$_4$ K. Ramachandra, Some problems of analytic Number Theory-I, *Acta Arith.* vol. **31** (1976), 313–324.

[K.R]$_5$ K. Ramachandra, On the zeros of the Riemann zeta-function and L-series-I, *Acta Arith.* **34** (1978), 211–218.

[K.R; A.S]$_1$ K. Ramachandra and Sankaranarayanan, A., A Remark on Vinogradov's mean-value theorem, *The J. of Analysis* (Chennai) **3** (1995), 111–129.

[K.R; A.S]$_2$ K. Ramachandra and Sankaranarayanan, A., On some theorems of Littlewood and Selberg-III, *Annales Acad. Sci. Fenn.* Ser. AI(Math), **16** (1991), 139–149.

[E.C.T] E.C. Titchmarsh, The theory of the Riemann zeta-function (second edn., Revised and edited by D.R. Heath-Brown), Clarendon Press, Oxford (1986).

[A.W] A. Walfisz, Weylsche Exponentialsummen in der nueren Zahlentheorie, Berlin (1963).

Appendix

We begin by recalling Grim's conjecture (for reference to Grim's paper see the two important papers [A-1] and [A-2] below, on Grim's conjecture by K. RAMACHANDRA, T.N. SHOREY and R. TIJDEMAN). Of course the conjecture is very much open today.

GRIM'S CONJECTURE (G.C): *Let $n \geq 1$ and $n+1, n+2, \ldots, n+r$ be $r(\geq 1)$ consecutive integers with the property that none of them is prime. Then their product is divisible by at least r distinct primes.*

Remark: Note for example that $(n+1)\ldots(2n)$ can be divisible by at most $O(n(\log n)^{-1})$ primes.

Now we come to the point. The object of this appendix is to see that a remark of Professor PAUL ERDÖS (written in a letter to the author) does not get lost to the mathematical community. ERDÖS showed in the letter that G.C. implies that for all sufficiently large positive integers N there is a prime between N^2 and $(N+1)^2$. The following theorem is essentially due to Professor PAUL ERDŐS.

Theorem: *Let $r \geq 2$ and $n \geq 2$ be integers and C any positive constant. Put $R = r - [Cr(\log r)^{-1}]$ and assume that $R \geq 2$. Then the product of $n+1, n+2, \ldots, n+r$ is divisible by at most R distinct primes, provided n exceeds a large positive constant and r exceeds a certain large positive constant times $n^{\frac{1}{2}}(\log n)^{-\frac{1}{2}}$.*

Corollary: *The conjecture G.C. implies that for all $N \geq N_0$, there is a prime between N^2 and $(N+1)^2$. In fact, if p_n denotes the n-th prime, G.C. implies $p_{n+1} - p_n = O(p_n^{\frac{1}{2}}(\log n)^{-\frac{1}{2}})$.*

Remark: The deduction of the corollary is easy and hence left to the reader.

Proof: Assume that the product of $n+1, \dots, n+r$ is divisible by at least $R+1$ distinct primes. Note that the integer.

$$I = \frac{(n+1)\dots(n+r)}{r!}$$

does not exceed

$$(n+r)^r \left\{\sqrt{2\pi} r^{r+\frac{1}{2}} e^{-r}\left(1+O\left(\frac{1}{r}\right)\right)\right\}^{-1}$$

which is $\leq (\frac{n+r}{r})^r C_1^r$ where $C_1 > 0$ is some constant.

Denote by $p_1, \dots, p_j$ all the primes dividing $r!$. Then

$$I p_1 p_2, \dots, p_j \leq \left(\frac{n+r}{r}\right) C_2^r$$

where $C_2 > 0$ is a constant. (Here we have used the Chebyshev's result $\sum_{p\leq x} \log p = O(x)$). Hence (using $1 + x \leq e^x$ for $x > 0$).

We have

$$I p_1, \dots, p_j \leq \left(\frac{n}{r}\right)^r \left(1+\frac{r}{n}\right)^r C_2^r \left(\leq \left(\frac{C_2 n}{r}\right)^r \exp\frac{r^2}{n}\right).$$

Here by our assumption (using Chebyshev's result $\pi(x) = O(\frac{x}{\log x})$ or equivalently the n-th prime exceeds a positive constant times $n \log n$ for $n \geq 2$)

$$\begin{aligned} I p_1, \dots, p_j &\geq p_1, \dots, p_{R+1} \geq \pi_{2\leq k\leq R}(C_3 k \log k) \\ &= C_3^{R_1}(R!)\pi_{2\leq k\leq R} \log k, \end{aligned}$$

where $C_3 > 0$ is a constant. It is not hard to see (since $\log\log k$ for $k \geq 3$ is increasing) that the last expression is $\geq C_4^r r^r (\log r)^r$, where $C_4 > 0$ is some constant. Thus finally we end up with the inequality

$$(C_4 r \log r)^r \leq \left(C_2 n r^{-1} \exp\frac{r}{n}\right)^r$$

i.e.

$$r^2 \log r = O\left(n \exp\frac{r}{n}\right)$$

The choice $r = [C_5 n^{\frac{1}{2}} (\log n)^{-\frac{1}{2}}]$, where $C_5 > 0$ is a large constant leads to a contradiction which proves the theorem. □

Remark: We have also the following stronger conjecture due to Grim: If $n \geq 1$ and $r \geq 1$ are integers and $n+1, n+2, \dots, n+r$ are all composite there exist distinct primes $q_1, \dots, q_r$ such that q_k divides $n+k$ for $k = 1, 2, \dots, r$. But this does not seem to give more information than $p_{n+1} - p_n = O(p_n^{\frac{1}{2}} (\log n)^{-\frac{1}{2}})$.

[A-1] K. Ramachandra, Shorey, T.N. and Tijdeman, R., On grim's conjecture relating to factorisation of a block of consecutive integers-I, J. Reine u. Angew Math. **273** (1975), 109–124.

[A-2] K. Ramachandra, Shorey, T.N. and Tijdeman, R., On Grim's conjecture relating to factorisation of a block of consecutive integers-II, J.Reine u.Angew. Math. **288** (1976), 192–201.

Senior Scientist of the INSA

TIFR Centre
P.B. 1234
IISc. Campus
Bangalore 560 012

National Institute of Advanced Studies
IISc. Campus
Bangalore 560 012

Sums of Squares: An Elementary Method

R.A. Rankin

1 Introduction

If $x_1, x_2, \ldots, x_s$ are integers positive negative or zero such that

$$x_1^2 + x_2^2 + \cdots + x_s^2 = n,$$

then $(x_1, x_2, \ldots, x_s)$ is called a representation of n as a sum of s squares, and the total number of representations is denoted by $R_s(n)$. Two representations $(x_1, x_2, \ldots, x_s)$ and $(y_1, y_2, \ldots, y_s)$ are considered to be different unless

$$x_1 = y_1, x_2 = y_2, \ldots, x_s = y_s.$$

Further, using a notation introduced by J.W.L. Glaisher, we write $R_{\alpha,\beta}(n)$ for the number of representations of n as a sum of squares of which α are odd and β are even, no restriction being placed upon the order of the squares. Observe that $R_s(0) = 1$, and that $R_{\alpha,\beta}(0) = 1$, if $\alpha = 0$, but that otherwise $R_{\alpha,\beta}(0) = 0$.

Mathematicians have been interested in evaluating $R_s(n)$ and $R_{\alpha,\beta}(n)$ since the 17th century and earlier, as a study of Chapters 6–9 of [3] will confirm and various methods have been used. Nowadays the use of modular forms provides the simplest answer. In the case of $R_s(n)$ the problem boils down to expressing the s-th power of the theta function ϑ_3 as an Eisenstein series plus a cusp form, both being holomorphic modular forms of weight $s/2$. The problem is simpler when s is even, as we shall generally assume, mainly because no complicated multiplier systems are then involved. See, for example, §7.4 of [10].

In a series of papers in the old Quarterly Journal (summarised in [4]) Glaisher obtained formulae for $R_{2s}(n)$ and $R_{\alpha,\beta}(n)$ for $s = 1, 2, \ldots, 9$ and $\alpha + \beta = 2s$ by means of elliptic function equations. His method, now regarded as old-fashioned, was essentially based on the consideration of various power series involving ϑ_3 and the Jacobi functions k and k'. Among these there appeared a number of what we now call cusp forms, whose Fourier coefficients had interesting multiplicative, or partly multiplicative, properties; see [8]. Where these cusp forms were absent, or had a zero Fourier coefficient, $R_{2s}(n)$ or $R_{\alpha,\beta}(n)$ was expressed in terms of divisor functions of different types. Among Glaisher's formulae the only ones which do not involve arithmetical functions other than divisor functions are formulae (1)–(24) given below. The divisor functions are defined in equations (25) and (26). Then we have

$$R_2(n) = 4E_0(n), \tag{1}$$

$$R_4(n) = 8\{2 + (-1)^n\}\Delta_1(n), \tag{2}$$

$$R_6(n) = 16E_2'(n) - 4E_2(n), \tag{3}$$

$$R_8(n) = \begin{cases} 16\Delta_3(n) & (n \text{ odd}) \\ \frac{16}{7}\{8\Delta_3'(n) - 15\Delta_3(n)\} & (n \text{ even}) \end{cases} \tag{4}$$

If $N = 2^\alpha N_1$, where $\alpha \geq 0$ and $N_1 \equiv 3 \pmod 4$,

$$R_{10}(N) = \frac{4}{5}\{E_4(N) - 16E_4'(N)\}, \tag{5}$$

$$R_{12}(n) = \frac{24}{31}\{21\Delta_5(2n) + 10\Delta_5'(2n)\}, \tag{6}$$

$$R_{2,0}(8n+2) = 4E_0(4n+1), \tag{7}$$

$$R_{1,1}(4n+1) = 4E_0(4n+1), \tag{8}$$

$$R_{4,0}(8n+4) = 16\Delta_1(2n+1), \tag{9}$$

$$R_{3,1}(4n+3) = 8\Delta_1(4n+3), \tag{10}$$

$$R_{2,2}(4n+2) = 24\Delta_1(2n+1), \tag{11}$$

$$R_{1,3}(4n+1) = 8\Delta_1(4n+1), \tag{12}$$

$$R_{6,0}(8n+6) = -8E_2(4n+3), \tag{13}$$

$$R_{4,2}(4n) = 240E_2'(n), \tag{14}$$

$$R_{3,3}(4n+3) = -20E_2(4n+3), \tag{15}$$

$$R_{2,4}(4n+2) = 60E_2'(2n+1), \tag{16}$$

$$R_{8,0}(8n) = 256\Delta_3'(n), \tag{17}$$

$$R_{4,4}(4n) = 1120\Delta_3'(n). \tag{18}$$

If N and N_1 are as before equation (5), then

$$R_{8,2}(4N) = 576E_4'(N), \tag{19}$$

$$R_{6,4}(2N_1) = 168E_4'(N), \tag{20}$$

$$R_{4,6}(4N) = 2688E_4'(N), \tag{21}$$

$$R_{2,8}(2N_1) = 36E_4'(N_1), \tag{22}$$

$$R_{8,4}(8n) = 3960\Delta_5'(2n), \tag{23}$$

$$R_{4,8}(8n) = 3960\Delta_5'(2n). \tag{24}$$

In formulae (7)–(18), $R_{\alpha,\beta}(m) = 0$ unless m is of the form stated. This is obvious since, if x is odd, then $x^2 \equiv 1 \pmod 8$, while, otherwise, we have $x^2 \equiv 0 \pmod 4$. The functions $\Delta_\nu(n)$, $\Delta_\nu'(n)$, $E_\nu(n)$ and $E_\nu'(n)$ are defined as follows. Let δ be an odd divisor of n and write $\delta\delta' = n$. Also let $\chi(n)$ be the character defined by

$$\chi(n) = 0 \ (n \text{ even}), \quad \chi(n) = (-1)^{(n-1)/2} (n \text{ odd}).$$

Then

$$\Delta_\nu(n) = \sum_{\delta|n} \delta^\nu, \quad \Delta'_\nu(n) = \sum_{\delta|n} (\delta')^\nu, \tag{25}$$

$$E_\nu(n) = \sum_{\delta|n} \chi(\delta)\delta^\nu, \quad E'_\nu(n) = \sum_{\delta|n} \chi(\delta)(\delta')^\nu, \tag{26}$$

Note that formulae for $R_{0,2s}(4n)$ are omitted since $R_{0,2s}(4n) = R_{2s}(n)$.

2 Historical Remarks

Some of these formulae have, over the years, been proved by elementary methods not involving elliptic or modular functions, by mathematicians such as Pepin, Dirichlet and Uspensky. The most complete elementary account so far published is Helmut Bessel's doctoral dissertation submitted to the University of Königsberg [2] in 1929. He bases his work on three formulae of Liouville in articles in the Journal de Mathématiques between the years 1840 and 1850, with the title 'Sur quelques formules générales qui peuvent être utiles dans la théorie des nombres'. See Chapter 11 of [3] or pp. 365–463 of [1]. I give the following as an example:

$$\sum\{f(x+y) - f(x-y)\} = \frac{1}{2}\sum_{\delta|n} \delta'\{f(0) - f(\delta')\},$$

where f is an odd function and the summation is over all solutions of the equation $ax+by = 2n(a, b, x, y \text{ odd})$.

By means of these formulae Bessel proves the majority of the results stated above, excluding formulae (19)–(24). He also makes use of certain cusp form coefficients such as $\sum(x_1^4 - 3x_1^2x_2^2)$, the summation, in this case, being taken over all solutions of the equation $x_1^2 + x_2^2 = n$. This is a multiple of Glaisher's multiplicative function $\chi_4(n)$, which is the Fourier coefficient of a cusp form of weight 5, and arises in the study of 10 squares.

I completed the work of the present article in July 1944 but it was not until October 1947 that I managed to borrow a copy of Bessel's thesis, through Inter Library Loan. I have now, many years later, succeeded in obtaining a photocopy of his thesis through the courtesy of the Mathematics Library of the University of Illinois at Urbana.

In what follows I assume (1) and deduce the remaining formulae from it. A proof of (1) can be found in various treatises on number theory, for example in Chapter 2 of [5] or §6.7 of [7]. An interesting account of the history of the subject will be found in Chapter 9 of [6] and the notes to that chapter. Finally, a good account of Liouville's methods is in Chapter 13 of [11] and in [1].

In my treatment of the subject I do not use any of Liouville's formulae, but base my account on the elementary Theorem 1 below. It may be mentioned that this work has already been used in my paper [9]. Inevitably there are certain similarities between my method and Bessel's, but the two methods are not identical.

3 General Results

All letters, with the exception of π, denote integers; in particular $x, y, a, b, c, \xi, \eta, n, d$ and δ denote non-negative integers, and will usually be positive. We write d for any divisor of n and δ for an odd divisor of n, and put

$$d' = \frac{n}{d}, \quad \delta' = \frac{n}{\delta}. \tag{27}$$

It follows from (25) and (26) that, if $\alpha \geq 0$,

$$\Delta_\nu(2^\alpha n) = \Delta_\nu(n), \quad \Delta'_\nu(2^\alpha n) = 2^{\alpha\nu}\Delta'_\nu(n), \tag{28}$$

$$E_\nu(2^\alpha n) = E_\nu(n), \quad E'_\nu(2^\alpha n) = 2^{\alpha\nu}E'_\nu(n). \tag{29}$$

If n is odd,

$$\Delta'_\nu(n) = \Delta_\nu(n), \quad E'_\nu(n) = \chi(n)E_\nu(n). \tag{30}$$

In particular, by (26) and (30),

$$E_0(4n+3) = E'_0(4n+3) = 0. \tag{31}$$

The number of representations of a number as a sum of s squares may be expressed in terms of the representations by less than s squares by means of the formulae

$$R_{a+b}(n) = \sum_{m=0}^{n} R_a(m)R_b(n-m), \tag{32}$$

and

$$R_{a+b,c+d}(n) = \frac{(a+b+c+d)!a!b!c!d!}{(a+b)!(c+d)!(a+c)!(b+d)!}\sum_{m=0}^{n} R_{a,c}(m)R_{b,d}(n-m). \tag{33}$$

The proof of (32) is obvious. If $R'_{\mu,\nu}(n)$ is the number of representations of m in the form

$$x_1^2 + x_2^2 + \cdots + x_\mu^2 + y_1^2 + y_2^2 + \cdots + y_\nu^2,$$

where $x_1, x_2, \ldots, x_\mu$ are odd, and $y_1, y_2, \ldots, y_\nu$ are even, then clearly,

$$R_{\mu,\nu}(m) = \frac{(\mu+\nu)!}{\mu!\nu!}R'_{\mu,\nu}(m),$$

and (33) follows since

$$R'_{a+b,c+d}(n) = \sum_{m=0}^{n} R'_{a,c}(m)R'_{b,d}(n-m).$$

As an illustration of the use of formulae (32) and (33), we have

$$R_{8,0}(8n) = \sum_{m=0}^{n} R_{4,0}(m)R_{4,0}(n-m) = \sum_{m=1}^{n} R_{4,0}(8m-4)R_{4,0}(8n-8m+4).$$

If we assume the truth of (9) for the moment, we obtain

$$R_{8,0}(8n) = 256 \sum_{m=1}^{n} \Delta_1(2m-1)\Delta_1(2n-2m+1).$$

The right hand side may be written as $256 \sum xy$, where the summation is extended over all solutions a, b, x, y of the equation

$$ax + by = 2n$$

for which $abxy$ is odd. Thus the problem of evaluating $R_{8,0}(8n)$ is reduced to the investigation of the solutions of this equation. This example is typical of the methods used to derive (2)–(24). We denote by $T(n)$ the set of all positive a, b, x, y which satisfy

$$ax + by = n, \quad x, y \text{ odd},$$

and write $T_1(n)$ for the subset of $T(n)$ for which ab is odd. Clearly $T_1(n)$ is empty unless n is even.

Further, we denote by $S(n)$ the set of all positive ξ, η, x, y that satisfy

$$\xi x + \eta y = n, \quad xy \text{ odd}, \quad (\xi, \eta) = 1.$$

We write $T(n)$, $T_1(n)$, or $S(n)$ below the summation sign to indicate a sum carried out over all solutions of $T(n)$, $T_1(n)$, or $S(n)$, respectively. If there are no solutions (e.g. if $n = 1$), such a sum is empty and has the value zero.

Let $f(x, y, a, b)$ be any function of x, y, a, b. Put $d = (a, b), a = d\xi, b = d\eta$. Then we have

$$\sum_{T(n)} f(x, y, a, b) = \sum_{\delta|n} \sum_{S(d')} f(x, y, d\xi, d\eta). \tag{34}$$

Similarly, if n is even,

$$\sum_{T_1(n)} f(x, y, a, b) = \sum_{\delta|n} \sum_{S(d')} f(x, y, \delta\xi, \delta\eta). \tag{35}$$

If

$$f(x, y, ca, cb) = c^{\nu} f(x, y, a, b) \tag{36}$$

for every c, (34) and (35) may be written

$$\sum_{T(n)} f(x, y, a, b) = \sum_{d|n} d^{\nu} \sum_{S(d')} f(x, y, \xi, \eta), \tag{37}$$

$$\sum_{T_1(n)} f(x, y, a, b) = \sum_{\delta|n} \delta^{\nu} \sum_{S(\delta')} f(x, y, \xi, \eta). \tag{38}$$

Equation (38) holds only if n is even.

The preceding formulae show that we need only consider solutions of $S(n)$.

Write

$$\delta_n = \frac{1}{2}\{1 - (-1)^n\}.$$

We assume that $n > 1$, as otherwise $S(n)$ is empty.

Let $C(n)$ be the set of solutions of $S(n)$ for which

$$|\xi - \eta| = \delta_n. \tag{39}$$

When n is odd we write $C(n) = C_1(n) + C_2(n)$, where $C_1(n)$ is the set of solutions of $C(n)$ for which $\xi = \eta - 1$, and $C_2(n)$ is the set for which $\xi = \eta + 1$.

The solutions of $C(n)$ are easy to obtain, and are given in the following three Lemmas.

Lemma 1 *If n is even $C(n)$ consists of the $\frac{1}{2}n$ solutions*

$$x = 2u - 1, \quad y = n - 2u + 1, \quad \xi = \eta = 1, \quad 1 \leq u \leq \frac{1}{2}n.$$

This follows, since $\xi = \eta$ by (39), and therefore $\xi = \eta = 1$, since $(\xi, \eta) = 1$.

Lemma 2 *If n is odd $C_1(n)$ consists of the solutions*

$$\xi = \left[\frac{n}{2t}\right], \quad \eta = \xi + 1, \quad x = 2t - y, \quad y = n - 2\xi t,$$

where t is any integer satisfying

$$0 < t < \frac{1}{2}n. \tag{40}$$

Write $2t = x + y$ for any solution of $C(n)$. Then clearly (40) holds, and the lemma follows since

$$2\xi t < 2\xi t + y = n < 2\xi t + 2t.$$

We have, similarly,

Lemma 3 *If n is odd $C_2(n)$ consists of the $\frac{1}{2}(n-1)$ solutions*

$$\xi = \eta + 1, \quad \eta = \left[\frac{n}{2t}\right], \quad x = n - 2\eta t, \quad y = 2t - x,$$

where t is any integer satisfying (40).

We write $c(n)$ for the number of members of $C(n)$. By Lemmas 1, 2, and 3,

$$c(n) = \begin{cases} \frac{1}{2}n & (n \text{ even}), \\ n - 1 & (n \text{ odd}). \end{cases} \tag{41}$$

We now consider the properties of another subclass of solutions of $S(n)$. Let $C'(n)$ be the set of solutions of $S(n)$ for which $x = y$. Then we have $x(\xi + \eta) = n$. Thus x must be a divisor of n. Let δ be any odd divisor of n. Then, if there is a solution of $S(n)$ with $x = y = \delta$, we must have, by (27),

$$\xi + \eta = \delta'. \tag{42}$$

It follows that $\delta' > 1$, i.e.

$$\delta < n. \tag{43}$$

If (43) is satisfied, there are exactly $\phi(\delta')$ solutions of (42); here ϕ is Euler's function. For, since $(\xi, \eta) = 1$, the only values which ξ can take are the $\phi(\delta')$ numbers less than δ' that are prime to δ'.

Lemma 4 *For any positive integer n*

$$\sum_{\delta|n} \phi(\delta') = \begin{cases} n & (n \text{ odd}), \\ \frac{1}{2}n & (n \text{ even}). \end{cases}$$

The case when n is odd follows from the well known result, which holds for odd and even n:

$$\sum_{\delta|n} \phi(d) = n.$$

Suppose that n is even of the form $2^\alpha m$, where m is odd and $\alpha > 0$. Then

$$\sum_{\delta|n} \phi(\delta') = \sum_{\delta|n} \phi(2^\alpha)\phi(m/d) = 2^{\alpha-1} \sum_{d|m} \phi(d) = 2^{\alpha-1} m = \frac{1}{2}n.$$

Lemma 5 *The solutions of $C'(n)$ are given by*

$$x = y = \delta, \quad \xi = u, \quad \eta = \delta' - u, \quad 0 < u < \delta',$$

where δ is any odd divisor of n which is less than n, $\delta' = n/\delta$, and $(u, \delta') = 1$. The total number of solutions is $c(n)$.

The first part of the lemma has already been proved, so that it remains to prove the second part. The number of solutions of $C'(n)$ is

$$c'(n) = \sum_{\delta|n, \delta<n} \phi(\delta').$$

By Lemma 4, if n is odd,

$$c'(n) = \sum_{\delta|n} \phi(\delta') - 1 = n - 1,$$

and, if n is even,

$$c'(n) = \sum_{\delta|n} \phi(\delta') = \frac{1}{2}n.$$

It follows from (41) that $c'(n) = c(n)$.

We now show that the solutions of $S(n)$ may be divided into sets, or 'chains', each associated with a unique solution (x, y, ξ, η) of $S(n)$.

Theorem 1 *The solutions of $S(n)$ may be divided into $c(n)$ chains of solutions such that each solution belongs to one and only one chain. Each chain consists of a sequence of solutions $(x_m, y_m, \xi_m, \eta_m)(m = 1, 2, \dots, q)$, with the following properties:*

$$|\xi_1 - \eta_1| = \delta_n, \quad x_q = y_q. \tag{44}$$

Also, if $q > 1, 1 \leq m < q$,

$$0 \leq |x_{m+1} - y_{m+1}| < |x_m - y_m|, \tag{45}$$

$$\delta_n \leq |\xi_m - \eta_m| < |\xi_{m+1} - \eta_{m+1}|, \tag{46}$$

$$x_{m+1} + y_{m+1} = |x_m - y_m|, \tag{47}$$

$$\xi_{m+1} - \eta_{m+1} = -(\xi_m + \eta_m)\mathrm{sgn}(x_m - y_m), \tag{48}$$

$$\chi(x_{m+1}) - \chi(y_{m+1}) = -\{\chi(x_m) + \chi(y_m)\}\mathrm{sgn}(x_m - y_m). \tag{49}$$

Proof: Since $n > 1$ there exists at least one solution for which

$$|\xi - \eta| = \delta_n, \quad x = y. \tag{50}$$

Any such solution we set in a chain by itself. Clearly (44) is satisfied. If $n = 2$ or 3, there are no solutions other than those given by (50), so that the theorem holds. □

We suppose, therefore, that $n > 3$. Then there exist solutions for which $|\xi - \eta| > 1$, and also solutions for which $x \neq y$. We consider those two cases separately in what follows.

Suppose that we have a solution with $x \neq y$. Write

$$\epsilon = \mathrm{sgn}(x - y).$$

We can generate a new solution by means of the following transformation,

$$\begin{aligned} x' &= \lambda x - (\lambda + \epsilon)y, \xi' = \lambda\xi + (\lambda - \epsilon)\eta, \\ y' &= -(\lambda - \epsilon)x + \lambda y, \eta' = (\lambda + \epsilon)\xi + \lambda\eta, \end{aligned} \tag{51}$$

where λ is an integer to be chosen. From this we obtain the inverse transformation

$$\begin{aligned} x &= \lambda x' + (\lambda + \epsilon)y', \xi = \lambda\xi' - (\lambda - \epsilon)\eta', \\ y &= (\lambda - \epsilon)x' + \lambda y', \eta = -(\lambda + \epsilon)\xi' + \lambda\eta'. \end{aligned} \tag{52}$$

It is evident from (51) and (52) that $(\xi', \eta') = 1$, and that both x' and y' are odd for all choices of λ. Also $\xi'x' + \eta'y' = n$. It remains to show that we can choose λ so that ξ', η', x' and y' are all positive. For x' and y' to be positive we must have

$$\epsilon y < \lambda(x - y) < \epsilon x,$$

i.e.

$$0 < \frac{\min(x, y)}{|x-y|} < \lambda < \frac{\max(x, y)}{|x-y|}. \tag{53}$$

The two numbers on the left and right of λ in this inequality differ by unity, and neither is an integer, since $|x-y|$ is even and both x and y are odd. Hence there exists one integral value of λ and one only for which x' and y' are positive. By (53), $\lambda \geq 1$, and hence ξ' and η' are positive. It follows that the new solution belongs to $S(n)$. By (51)

$$x' + y' = \epsilon(x-y) = |x-y|, \tag{54}$$

and

$$\xi' - \eta' = -\epsilon(\xi + \eta). \tag{55}$$

Also

$$|x-y| = x' + y' > |x' - y'|, \tag{56}$$

and

$$|\xi' - \eta'| = \xi + \eta > |\xi - \eta|. \tag{57}$$

It follows from (55) that

$$\epsilon = -\text{sgn}(\xi' - \eta').$$

If $x \equiv y$ (mod4),

$$x' \equiv -\epsilon y \pmod 4, \quad y' \equiv \epsilon x \pmod 4,$$

so that

$$\chi(x') - \chi(y') = -\epsilon\{\chi(x) + \chi(y)\}. \tag{58}$$

If $x \equiv -y \pmod 4$, $\chi(x) + \chi(y) = 0$, and

$$x' \equiv (2\lambda + \epsilon)x \pmod 4, \quad y' \equiv -(2\lambda - \epsilon)x \pmod 4,$$

so that

$$\chi(x') = (-1)^\lambda \epsilon \chi(x), \quad \chi(y') = (-1)^\lambda \epsilon \chi(y),$$

and therefore (58) holds in this case too.

If $x' \neq y'$, we can continue this process and arrive at a new solution, and so on. By (56), we shall eventually obtain a solution for which $x'' = y''$.

We can also generate new solutions by proceeding in the opposite direction. Suppose that (x', y', ξ', η') is a solution for which $|\xi' - \eta'| > \delta_n$. Then we obtain a new solution (x, y, ξ, η) by the transformation

$$\begin{aligned} x &= \mu x' + (\mu + \epsilon')y', \; \xi = \mu\xi' - (\mu - \epsilon')\eta', \\ y &= (\mu - \epsilon')x' - \mu y', \; \eta = -(\mu + \epsilon')\xi' + \mu\eta', \end{aligned} \tag{59}$$

where

$$\epsilon' = -\text{sgn}(\xi' - \eta').$$

Clearly x and y are both odd, and if we solve for ξ' and η' in terms of ξ' and η' we can show that $(\xi, \eta) = 1$. Also $\xi x + \eta y = n$. It remains to show that μ can be chosen so that ξ, η, x and y are all positive. For ξ and η to be positive we must have

$$\epsilon'\xi' < \mu(\eta' - \xi') < \epsilon'\eta',$$

i.e.

$$0 < \frac{\min(\xi', \eta')}{|\xi' - \eta'|} < \mu < \frac{\max(\xi', \eta')}{|\xi' - \eta'|}. \tag{60}$$

The two numbers on either side of μ in (60) differ by unity, and neither is an integer since $|\xi' - \eta'| > 1$. This is obvious if n is odd; if n is even ξ' and η' are both odd, so that $\xi' - \eta'$ is even, and hence $|\xi' - \eta'| \geq 2$. Hence there exists one integral value of μ and only one for which ξ and η are both positive. By (10), μ is positive, and hence x and y are also positive, and it follows that the solution (x, y, ξ, η) belongs to $S(n)$. By (59),

$$x - y = -\epsilon'(x' + y'), \tag{61}$$

and

$$\xi + \eta = -\epsilon'(\xi' - \eta') = |\xi' - \eta'|. \tag{62}$$

Hence

$$\epsilon' = \operatorname{sgn}(x - y).$$

Thus (61) and (62) are identical with (54) and (55) with ϵ' in place of ϵ. Formulae (56), (57), and (58) may be deduced in the same manner.

If $|\xi - \eta| \neq \delta_n$ we can continue the process and derive a new solution, and so on. By (57), we shall eventually obtain a solution $(x^*, y^*, \xi^*, \eta^*)$ for which $|\xi^* - \eta^*| = \delta_n$.

Now $\epsilon' = \epsilon$ since each is equal to $-\operatorname{sgn}(\xi' - \eta')$, and on comparing equations (52) and (59) we see that we must have $\lambda = \mu$, since both are uniquely determined. Thus to each solution (x, y, ξ, η) of $S(n)$ we can assign a unique successor, if $x \neq y$, and a unique predecessor if $|\xi - \eta| \neq \delta_n$. Thus we have shown that every solution of $S(n)$ is a member of a unique sequence or chain of solutions, and that the members of any such chain have the properties (44)–(49). Since $|\xi_1 - \eta_1| = \delta_n$ for the first member, and $x_q = y_q$ for the last member of each chain, it follows that each chain corresponds to a unique solution of $C(n)$ and to a unique solution of $C'(n)$. There are therefore $c(n)$ chains. This completes the proof of Theorem 1. It may be remarked that the number q is not necessarily the same for different chains.

Let $\psi_0(u, v)$ and $\psi_1(u, v)$ be any two functions of the integers u and v, such that ψ_0 is an even function of both u and v, and ψ_1 is an even function of u and an odd function of v. Then, by Theorem 1, for two successive members of a chain, indexed by m and $m + 1$,

$$\psi_0(x_m - y_m, \xi_m + \eta_m) = \psi_0(x_{m+1} + y_{m+1}, \xi_{m+1} - \eta_{m+1}) \tag{63}$$

and

$$\begin{aligned}\psi_1(x_m - y_m, \xi_m + \eta_m)\{\chi(x_m) + \chi(y_m)\} \\ = \psi_1(x_{m+1} + y_{m+1}, \xi_{m+1} - \eta_{m+1})\{\chi(x_{m+1}) - \chi(y_{m+1})\}.\end{aligned} \tag{64}$$

Put

$$f_0(x, y, \xi, \eta) = \psi_0(x + y, \xi - \eta) - \psi_0(x - y, \xi + \eta), \tag{65}$$

and

$$\begin{aligned} &f_1(x, y, \xi, \eta) \\ &\quad = \psi_1(x + y, \xi - \eta)\{\chi(x) - \chi(y)\} - \psi_1(x - y, \xi + \eta)\{\chi(x) + \chi(y)\}. \end{aligned} \tag{66}$$

Then we have, by (63) and (64), for any chain of more than one member,

$$\sum_{m=1}^{q} f_0(x_m, y_m, \xi_m, \eta_m) = \psi_0(x_1 + y_1, \xi_1 - \eta_1) - \psi_0(0, \xi_q + \eta_q), \tag{67}$$

and

$$\begin{aligned} &\sum_{m=1}^{q} f_1(x_m, y_m, \xi_m, \eta_m) \\ &\quad = \psi_1(x_1 + y_1, \xi_1 - \eta_1)\{\chi(x_1) - \chi(y_1)\} - 2\psi_1(0, \xi_q - \eta_q)\chi(x_q). \end{aligned} \tag{68}$$

Formulae (67) and (68) also hold for a chain containing only one solution. It may be noted that it is possible to prove results of the same type for functions $\psi(u, v)$ that are odd functions of u, but such results have no application since the sums over $S(n)$ of the corresponding functions f vanish.

Theorem 2 *If $f_0(x, y, \xi, \eta)$ and $f_1(x, y, \xi, \eta)$ are defined as in* (65) *and* (66), *we have (i) if n is odd,*

$$\sum_{S(n)} f_0(x, y, \xi, \eta) = 2 \sum_{t=1}^{(n-1)/2} \psi_0(2t, 1) - \sum_{\delta | n} \phi(\delta)\psi_0(0, \delta) + \psi_0(0, 1), \tag{69}$$

$$\begin{aligned} \sum_{S(n)} f_1(x, y, \xi, \eta) = 4\chi(n) &\sum_{u=1}^{\frac{1}{4}\{n-2+\chi(n)\}} \psi_1(4u, 1) \\ &- 2\sum_{\delta | n} \chi(\delta)\psi_1(0, \delta')\phi(\delta') + 2\chi(n)\psi_1(0, 1), \end{aligned} \tag{70}$$

and (ii) if n is even

$$\sum_{S(n)} f_0(x, y, \xi, \eta) = \frac{1}{2}\psi_0(n, 0) - \sum_{\delta | n} \phi(\delta')\psi_0(0, \delta), \tag{71}$$

$$\sum_{S(n)} f_1(x, y, \xi, \eta) = -2\sum_{\delta | n} \chi(\delta)\phi(\delta')\psi_1(0, \delta'). \tag{72}$$

By (67)

$$\sum_{S(n)} f_0(x, y, \xi, \eta) = \sum_{C(n)} \psi_0(x+y, \xi-\eta) - \sum_{C'(n)} \psi_0(0, \xi+\eta),$$

and (69) and (71) follow from Lemmas 1, 2, 3 and 5. Also, by (68),

$$\sum_{S(n)} f_1(x, y, \xi, \eta) = \sum_{C(n)} \psi_1(x+y, \xi-\eta)\{\chi(x) - \chi(y)\} - 2\sum_{C'(n)} \psi_1(0, \xi+\eta)\chi(n).$$

By Lemma 5,

$$2\sum_{C(n)} \psi_1(0, \xi+\eta)\chi(n) = 2\sum_{\delta|n, \delta<n} \chi(\delta)\phi(\delta')\psi_1(0, \delta'). \tag{73}$$

To evaluate

$$\sum_{C(n)} \psi_1(x+y, \xi-\eta)\{\chi(x) - \chi(y)\}$$

we observe that $\chi(x) - \chi(y) = 0$ unless $x + y \equiv 0 \pmod 4$, so that it is only necessary to consider values of x and y which satisfy this congruence.

Suppose that n is odd. Then

$$n = \xi(x+y) + (\eta - \xi)y \equiv (\eta - \xi)y \pmod 4,$$

and therefore

$$\chi(x) - \chi(y) = -2\chi(y) = 2\chi(n)\chi(\xi - n).$$

Thus

$$\sum_{C(n)} \psi_1(x+y, \xi-\eta)\{\chi(x) - \chi(y)\} = 2\chi(n) \sum_{1\le u<n/4} \{\psi_1(4u, 1) - \psi_1(4u, -1)\}$$

$$= 4\chi(n) \sum_{u=1}^{\frac{1}{4}\{n-2+\chi(n)\}} \psi_1(4u, 1). \tag{74}$$

Finally, if n is even,

$$\sum_{C(n)} \psi_1(x+y, \xi-\eta)\{\chi(x) - \chi(y)\} = \psi_1(n, 0) \sum_{x+y=n} \{\chi(x) - \chi(y)\} = 0, \tag{75}$$

by Lemma 1.

Equations (70) and (72) now follow from (73), (74) and (75).

4 Two Squares

We assume (1). As stated in §2, it may be proved by elementary methods. Equations (7) and (8) are particular cases of (1).

5 Four Squares

Take

$$\psi_0(u, v) = \frac{1}{2}(1 - \cos \pi u/2),$$

so that, by (65),

$$\begin{aligned} f(x, y) &= f_0(x, y, \xi, \eta) \\ &= -\frac{1}{2}\left\{\cos\frac{\pi}{2}(x+y) - \cos\frac{\pi}{2}(x-y)\right\} = \sin\frac{\pi}{2}x \sin\frac{\pi}{2}y = \chi(x)\chi(y). \end{aligned}$$

By Theorem 2, if n is odd,

$$\sum_{S(n)} f(x, y) = \sum_{t=1}^{(n-1)/2} \{1 - (-1)^t\} = \frac{1}{2}\{n - \chi(n)\}.$$

If n is even,

$$\sum_{S(n)} f(x, y) = \frac{1}{4}n\left(1 - \cos\frac{\pi}{2}n\right) = \frac{1}{4}n\{1 - (-1)^{n/2}\}.$$

Hence, by (34),

$$\sum_{T(2n+1)} f(x, y) = \frac{1}{2}\sum_{\delta|2n+1} \{\delta - \chi(\delta)\} = \frac{1}{2}\{\Delta_1(2n+1) - E_0(2n+1)\}, \tag{76}$$

and

$$\begin{aligned} \sum_{T(2n)} f(x, y) &= \sum_{d|2n}\sum_{S(d)} f(x, y) = \frac{1}{2}\sum_{\delta|n}\{\delta - \chi(\delta)\} + \frac{1}{2}\sum_{d|n} d\{1 - (-1)^d\} \\ &= \frac{3}{2}\Delta_1(n) - \frac{1}{2}E_0(n). \end{aligned} \tag{77}$$

It follows from (76) and (77) that

$$\sum_{T(n)} f(x, y) = \frac{1}{2}\{(2 + (-1)^n)\Delta_1(n) - E_0(n)\}. \tag{78}$$

Also, by (35),

$$\sum_{T_1(2n)} f(x, y) = \frac{1}{2}\sum_{\delta|n} \delta'\{1 - (-1)^{\delta'}\} = \frac{1}{2}\{1 - (-1)^n\}\Delta_1(n). \tag{79}$$

Now, by (32) and (78),

$$\begin{aligned} R_4(n) &= \sum_{m=1}^{n-1} R_2(m)R_2(n-m) + 2R_2(n) = 16\sum_{m=1}^{n-1} E_0(m)E_0(n-m) + 8E_0(n) \\ &= 16\sum_{T(n)} f(x, y) + 8E_0(n) = 8\{2 + (-1)^n\}\Delta_1(n). \end{aligned}$$

Also, by (7), (31), (33) and (79),

$$\begin{aligned} R_{4,0}(8n+4) &= \sum_{m=0}^{n} R_{2,0}(4m+2)R_{2,0}(8n-4m+2) \\ &= 16\sum_{m=0}^{2n} E_0(2m+1)E_0(4n-2m+1) \\ &= 16\sum_{T_1(4n+2)} f(x,y) = 16\Delta_1(2n+1), \end{aligned}$$

and, by (7), (8), (29), (31), (32) and (76),

$$\begin{aligned} R_{3,1}(4n+3) &= 2\sum_{m=0}^{2n} R_{2,0}(2m+2)R_{1,1}(4n-2m+1) \\ &= 32\sum_{m=0}^{2n} E_0(2m+2)E_0(4n-2m+1) \\ &= 16\sum_{\mu=1}^{4n+2} E_0(\mu)E_0(4n+3-\mu) \\ &= 16\sum_{T(4n+3)} f(x,y) = 8\Delta_1(4n+3). \end{aligned}$$

In a similar manner it can be shown that

$$R_{2,2}(4n+2) = 24\sum_{T_1(4n+2)} f(x,y) = 24\Delta_1(2n+1),$$

and (by using the fact that $R_{0,2}(4m) = R_2(m)$)

$$R_{1,3}(4n+1) = 16\sum_{T(4n+1)} f(x,y) + 2R_{1,1}(4n+1) = 8\Delta_1(4n+1).$$

6 Six Squares

Take $\psi_1(u,v) = \frac{1}{2}v$. Then, by (66),

$$\begin{aligned} f(x,y,\xi,\eta) &= -\frac{1}{2}(\xi-\eta)\{\chi(x)-\chi(y)\} + \frac{1}{2}(\xi+\eta)\{\chi(x)+\chi(y)\} \\ &= \xi\chi(y) + \eta\chi(x), \end{aligned} \tag{80}$$

and, by Theorem 2,

$$\sum_{S(n)} f_1(x,y,\xi,\eta) = \sum_{\delta|n} \delta'\phi(\delta')\chi(\delta),$$

when n is even; and therefore, since (36) is satisfied with $\nu = 1$, we have, by (38) and Lemma 4,

$$\begin{aligned}
\sum_{T_1(2n)} f_1(x,y,a,b) &= \sum_{\delta|2n} \delta \sum_{S(2n/\delta)} f_1(x,y,\xi,\eta) \\
&= \sum_{\delta|n} \delta \sum_{\delta_1|\frac{n}{\delta}} \frac{2n}{\delta\delta_1}\phi\left(\frac{2n}{\delta\delta_1}\right)\chi(\delta_1) \\
&= 2\sum_{\delta_1|n} \frac{n}{\delta_1}\chi(\delta_1) \sum_{\delta|n/\delta_1} \phi\left(\frac{2n}{\delta\delta_1}\right) \\
&= 2\sum_{\delta_1|n} \chi(\delta_1)\left(\frac{n}{\delta_1}\right)^2 = 2E_2'(n). \qquad (81)
\end{aligned}$$

Write

$$U_0(n) = \sum_{ax+by=n,2|ax} x\chi(y), \qquad (82)$$

$$U_1(n) = \sum_{ax+by=n,2\nmid ax} x\chi(y), \qquad (83)$$

and

$$U(n) = U_0(n) + U_1(n). \qquad (84)$$

By (80), (81) and (82),

$$\begin{aligned}
U_1(2n) = \sum_{T_1(2n)} x\chi(y) &= \sum_{T_1(2n)} a\chi(y) = \frac{1}{2}\sum_{T_1(2n)} \{a\chi(y) + b\chi(x)\} \\
&= \frac{1}{2}\sum_{T_1(2n)} f_1(x,y,a,b) \\
&= E_2'(n), \qquad (85)
\end{aligned}$$

since a and b are odd as well as x and y, for members of $T_1(2n)$.

With the help of (85) we can evaluate $U_0(n)$ and $U_1(n)$ in the general case. For, by (1), (2), (7)–(12) and (33),

$$\begin{aligned}
R_{4,2}(4n) &= \frac{5}{2}\sum_{m=0}^{n-1} R_{2,2}(4m+2)R_{2,0}(4n-4m-2) \\
&= 240\sum_{m=0}^{n-1} \Delta_1(2m+1)E_0(2n-2m-2) \\
&= 240U_1(2n), \qquad (86)
\end{aligned}$$

$$\begin{aligned}
R_{4,2}(8n+4) &= 15\sum_{m=0}^{n} R_{4,0}(8m+4)R_{0,2}(8n-8m) \\
&= 960U_1(2n+1) + 240\Delta_1(2n+1) \qquad (87)
\end{aligned}$$

$$\begin{aligned} R_{2,4}(4n+2) &= \frac{5}{2}\sum_{m=0}^{n} R_{2,2}(4m+2)R_{0,2}(4n-4m) \\ &= 240U_1(2n+1)+60\Delta_1(2n+1), \end{aligned} \tag{88}$$

$$\begin{aligned} R_{2,4}(8n+2) &= 15\sum_{m=0}^{n} R_{2,0}(8m+2)R_{0,4}(8n-8m) \\ &= 1440\sum_{m=0}^{n-1} E_0(4m+1)\Delta_1(4n-4m)+60E_0(4n+1) \\ &= 1440\sum_{m=0}^{2n-1} E_0(2n+1)\Delta_1(4n-2m)+60E_0(4n+1) \\ &= 1440U_0(4n+1)+60E_0(4n+1), \end{aligned} \tag{89}$$

$$\begin{aligned} R_{2,4}(8n+6) &= 15\sum_{m=0}^{n} R_{2,0}(8m+2)R_{0,4}(8n-8m+4) \\ &= 480\sum_{m=0}^{n} E_0(4m+1)\Delta_1(4n-4m+2) \\ &= 480\sum_{m=0}^{2n} E_0(2n+1)\Delta_1(4n-2m+2) \\ &= 480U_0(4n+3), \end{aligned} \tag{90}$$

$$\begin{aligned} R_{6,0}(8n+6) &= \sum_{m=0}^{n} R_{4,0}(8m+4)R_{2,0}(8n-8m+2) \\ &= 64\sum_{m=0}^{n} E_0(4n-4m+1)\Delta_1(4m+2) \\ &= 64\sum_{m=0}^{2n} E_0(4n-2m+1)\Delta_1(2m+2) \\ &= 64U_0(4n+3), \end{aligned} \tag{91}$$

$$\begin{aligned} R_{3,3}(4n+3) &= \frac{5}{2}\sum_{m=0}^{n}\{R_{3,1}(4m+3)R_{0,2}(4n-4m) \\ &\quad + R_{1,3}(4m+1)R_{2,0}(4n-4m+2)\} \\ &= 80\sum_{m=0}^{2n} \Delta_1(2m+1)E_0(4n-2m+2)+20\Delta_1(4n+3) \\ &= 80U_1(4n+3)+20\Delta_1(4n+3), \end{aligned} \tag{92}$$

and

$$\begin{aligned} R_6(n) &= \sum_{m=0}^{n} R_4(m)R_2(n-m) \\ &= 96U(n) - 64U_1(n) + 4E_0(n) + 8\{2+(-1)^n\}\Delta_1(n). \end{aligned} \tag{93}$$

It follows from (85), (86) and (87) that

$$\begin{aligned} U_1(2n+1) &= \frac{1}{4}\{U_1(4n+2) - \Delta_1(2n+1)\} \\ &= \frac{1}{4}\{E_2'(2n+1) - \Delta_1(2n+1)\}. \end{aligned} \tag{94}$$

This formula may be combined with (85) to give

$$U_1(n) = \frac{1}{4}\left\{E_2'(n) - \frac{1}{2}(1-(-1)^n)\Delta_1(n)\right\}. \tag{95}$$

By (88), (89) and (90),

$$\begin{aligned} 24U_0(4n+1) &= 4U_1(4n+1) - E_0(4n+1) + \Delta_1(4n+1) \\ &= E_2'(4n+1) - E_0(4n+1), \end{aligned}$$

and

$$24U_0(4n+3) = 12U_1(4n+3) + 3\Delta_1(4n+3) = 3E_2'(4n+3),$$

which combine to give, by (30) and (31),

$$U_0(2n+1) = \frac{1}{24}\{2E_2'(2n+1) - E_2(2n+1) - E_0(2n+1)\}. \tag{96}$$

From (94) and (96) it follows that

$$U(n) = \frac{1}{24}\{8E_2'(n) - E_2(n) - E_0(n) - 6\Delta_1(n)\}, \tag{97}$$

if n is odd, and we shall show that this holds also when n is even. For we have, if $n = 2^\alpha m$ where m is odd and α is positive,

$$\begin{aligned} U(n) &= \sum_{\beta=1}^{\alpha} U_1(2^\beta m) + U(m) \\ &= \sum_{\beta=1}^{\alpha} E_2'(2^\beta m) + U(m) \\ &= \frac{1}{3}(2^{2\alpha}-1)E_2'(m) + \frac{1}{24}\{8E_2'(m) - E_2(m) - E_0(m) - 6\Delta_1(m)\} \\ &= \frac{1}{3}E_2'(n) - \frac{1}{24}\{E_2(n) + E_0(n) + 6\Delta_1(n)\}, \end{aligned}$$

by (28) and (29), and this is the same as (97). It follows from (83) and (95) that

$$\begin{aligned} U_0(2n) &= U(2n) - U_1(2n) \\ &= \frac{1}{24}\{8E_2'(n) - E_2(n) - E_0(n) - 6\Delta_1(n)\} = U(n). \end{aligned}$$

This may also be deduced immediately from (82), (83) and (84).

Formulae (3), (13), (14), (15) and (16) now follow from (86), (88), (91), (92) and (93), with the help of (85), (94), (95), (96) and (97).

7 Eight Squares

Take $\psi_0(u, v) = \frac{1}{4}u^2$. Then, by (65),

$$f_0(x, y, \xi, \eta) = f(x, y) = xy.$$

It follows from Theorem 2 that

$$\sum_{S(n)} f(x, y) = 2 \sum_{t=1}^{(n-1)/2} t^2 = \frac{1}{12}n(n^2 - 1),$$

when n is odd, and

$$\sum_{S(n)} f(x, y) = \frac{1}{8}n^3$$

when n is even.

Therefore, if n is odd, by (34),

$$\sum_{T(n)} f(x, y) = \frac{1}{12}\sum_{\delta|n} \delta'(\delta'^2 - 1) = \frac{1}{12}\{\Delta_3(n) - \Delta_1(n)\}. \tag{98}$$

Suppose that $n = 2^\alpha m$, where m is odd and α is positive. Then

$$\begin{aligned} \sum_{T(n)} f(x, y) &= \sum_{\delta|n}\left\{\frac{1}{12}\delta(\delta^2 - 1) + \frac{1}{8}\delta^3(2^3 + 2^6 + \cdots + 2^{3\alpha})\right\} \\ &= \sum_{\delta|n}\left\{\frac{1}{12}\delta(\delta^3 - 1) + \frac{1}{7}(2^{3\alpha} - 1)\delta^3\right\} \\ &= \frac{1}{84}\{12\Delta_3'(n) - 5\Delta_3(n) - 7\Delta_1(n)\}, \end{aligned} \tag{99}$$

and, by (98) this holds also when n is odd. Also, by (35),

$$\sum_{T_1(n)} f(x, y) = \Delta_3'(n). \tag{100}$$

We now apply these results. We have

$$R_8(n) = 2R_4(n) + \sum_{m=1}^{n-1} R_4(m)R_4(n-m). \tag{101}$$

If n is odd one of the numbers $m, n-m$ is odd and the other is even, so that, by (2),

$$\begin{aligned} R_8(n) &= 16\Delta_1(n) + 192 \sum_{m=1}^{n-1} \Delta_1(m)\Delta_1(n-m) \\ &= 16\Delta_1(n) + 192 \sum_{T(n)} f(x, y) \\ &= 16\Delta_3(n). \end{aligned}$$

Also, by (2), (28), (99), (100), and (101),

$$\begin{aligned} R_8(2n) &= 48\Delta_1(n) + 64 \sum_{m=1}^{n-1} \Delta_1(2m-1)\Delta_1(2n-2m+1) \\ &\quad + 576 \sum_{m=1}^{n-1} \Delta_1(2m)\Delta_1(2n-2m) \\ &= 48\Delta_1(n) + 64 \sum_{T_1(2n)} f(x, y) + 576 \sum_{T(n)} f(x, y) \\ &= \frac{16}{7}\{8\Delta_3'(2n) - 15\Delta_3(2n)\}. \end{aligned}$$

Formula (4) follows from the last two results.

By (9) and (100),

$$\begin{aligned} R_{8,0}(8n) &= \sum_{m=0}^{n-1} R_{4,0}(8m+4)R_{4,0}(8n-8m-4) \\ &= 256 \sum_{m=0}^{n-1} \Delta_1(2m+1)\Delta_1(2n-2m-1) \\ &= 256 \sum_{T_1(2n)} f(x, y) = 256\Delta_3'(n), \end{aligned}$$

which is formula (17). By (11) and (33),

$$\begin{aligned} R_{4,4}(4n) &= \frac{35}{18}\sum_{m=0}^{n-1} R_{2,2}(4m+2)R_{2,2}(4n-4m-2) \\ &= 1120\sum_{m=0}^{n-1} \Delta_1(2m+1)\Delta_1(2n-2m-1) \\ &= 1120\sum_{T_1(2n)} f(x,y) = 1120\Delta_3'(n), \end{aligned}$$

which is (18).

We conclude by proving a formula which we shall require in the next section. By (33), (10) and (12),

$$\begin{aligned} R_{6,2}(4n+2) &= \frac{7}{4}\sum_{m=0}^{n-1} R_{3,1}(4m+3)R_{3,1}(4n-4m-1) \\ &= 112\sum_{m=0}^{n-1} \Delta_1(4m+3)\Delta_1(4n-4m-1), \end{aligned}$$

and

$$\begin{aligned} R_{2,6}(4n+2) &= \frac{7}{4}\sum_{m=0}^{n-1} R_{1,3}(4m+1)R_{1,3}(4n-4m+1) \\ &= 112\sum_{m=0}^{n-1} \Delta_1(4m+1)\Delta_1(4n-4m+1). \end{aligned}$$

Hence, by (100),

$$\begin{aligned} R_{6,2}(4n+2)+R_{2,6}(4n+2) &= 112\sum_{m=0}^{2n} \Delta_1(2m+1)\Delta_1(4n-2m+1) \\ &= 112\sum_{T_1(4n+2)} f(x,y) \\ &= 112\Delta_3(2n+1). \end{aligned} \tag{102}$$

8 Ten Squares

Take $\psi_1(u,v) = -\frac{1}{4}v^3$, so that, by (66),

$$f_1(x,y,a,b) = \frac{1}{2}\{a^3\chi(y)+b^3\chi(a)\} + \frac{3}{2}ab\,\{a\chi(x)+b\chi(y)\}. \tag{103}$$

It follows from Theorem 2 and (38) that

$$\sum_{S(2n)} f_1(x, y, \xi, \eta) = 4 \sum_{\delta | n} \chi(\delta)\phi\left(\frac{2n}{\delta}\right)\left(\frac{n}{\delta}\right)^3,$$

and

$$\begin{aligned}
\sum_{T_1(2n)} f_1(x, y, a, b) &= \sum_{\delta_1 | n} \delta_1^3 \sum_{S(2n/\delta_1)} f_1(x, y, \xi, \eta) \\
&= 4 \sum_{\delta_1 | n} \delta_1^3 \sum_{\delta | n/\delta_1} \chi(\delta)\phi\left(\frac{2n}{\delta\delta_1}\right)\left(\frac{n}{\delta\delta_1}\right)^3 \\
&= 4 \sum_{\delta | n} \left(\frac{n}{\delta}\right)^3 \chi(\delta) \sum_{\delta_1 | n/\delta} \phi\left(\frac{2n}{\delta\delta_1}\right) \\
&= 4 \sum_{\delta | n} (\delta')^4 \chi(\delta) \\
&= 4E_4'(n). \qquad (104)
\end{aligned}$$

On the other hand, by (103),

$$\begin{aligned}
\sum_{T_1(n)} f_1(x, y, a, b) &= \sum_{m=0}^{n-1} \{\Delta_3(2m+1)E_0(2n-2m-1) \\
&\quad + 3E_2'(2m+1)\Delta_1(2n-2m-1)\}. \qquad (105)
\end{aligned}$$

Put $N = 2^\alpha N_1$, where $\alpha \geq 0$ and $N_1 \equiv 3 \pmod 4$. Write

$$V_1(N) = \sum E_2(4m+1)\Delta_1(N-4m-1), \qquad (106)$$

$$V_3(N) = \sum E_2(4m+3)\Delta_1(N-4m-3), \qquad (107)$$

$$V_0(N) = \sum E_2(2m)\Delta_1(N-2m), \qquad (108)$$

$$V_0'(N) = \sum E_2'(2m)\Delta_1(N-2m), \qquad (109)$$

$$W(N) = \sum \Delta_3(2m+1)E_0(2N-2m-1), \qquad (110)$$

where in each case the summation is extended over all m for which the arguments of the functions Δ_1, Δ_3, E_0, E_2 and E_2' are positive.

It follows from (104), (105), (106), (107) and (110) that

$$W(N) + 3\{V_1(2N) - V_3(2N)\} = 4E_4'(N). \qquad (111)$$

By (33), (102) and (110),

$$\begin{aligned}
3R_{4,6}(4N) + 14R_{8,2}(4N) &= \frac{45}{2}\sum_{m=0}^{N-1}\{R_{6,2}(4m+2) + R_{2,6}(4m+2)\} \\
&\quad \times R_{2,0}(4N-4m-2) = 10080W(N). \qquad (112)
\end{aligned}$$

Also, by (2), (3) and (32),

$$\begin{aligned} R_{10}(N) &= \sum_{m=1}^{N-1} R_6(m)R_4(N-m) + R_6(N) + R_4(N) \\ &= 32\sum_{m=1}^{N-1}\{4E_2'(m) - E_2(m)\}\{2+(-1)^{N-m}\}\Delta_1(N-m) + R_6(N) + R_4(N). \end{aligned}$$

Hence we have, if $\alpha = 0$,

$$\begin{aligned} R_{10}(N) &= 288V_1(N) - 480V_3(N) + 128V_0'(N) - 32V_0(N) \\ &\quad + 8\Delta_1(N) + 20E_2'(N), \end{aligned} \tag{113}$$

and, if $\alpha > 0$,

$$\begin{aligned} R_{10}(N) &= 96V_1(N) - 160V_3(N) + 384V_0'(N) - 96V_0(N) \\ &\quad + 24\Delta_1(N) + 16E_2'(N) - 4E_2(N). \end{aligned} \tag{114}$$

The following results may be proved in a similar manner by means of (33), and (1)–(4), (7)–(8). The group of four numbers after each formula indicates the values of (a, b, c, d) used in applying (33).

$$R_{6,4}(2N_1) = 3360V_0'(N_1) \quad (4,2,2,2), \tag{115}$$

$$= 13440V_1(N_1) \quad (4,0,2,4), \tag{116}$$

$$= -40320V_3(N_1) + 1680E_2'(N_1) \quad (6,0,0,4), \tag{117}$$

$$R_{2,8}(2N_1) = 1440\{V_1(N_1) - 3V_3(N_1)\} + 180E_2'(N_1)(2,4,0,4), \tag{118}$$

$$R_{8,2}(4N) = -1440V_3(2N) \quad (6,0,2,2), \tag{119}$$

$$= 11520V_0'(N)(\alpha = 0) \quad (4,0,4,2), \tag{120}$$

$$= 11520\{V_1(N) - V_3(N)\}(\alpha > 0) \quad (4,0,4,2), \tag{121}$$

$$R_{4,6}(4N) = 3360\{V_1(2N) - V_3(2N)\} \quad (2,4,2,2), \tag{122}$$

$$R_{4,6}(N) = 13440\{4V_0'(N) - V_1(N)\} + 3360\Delta_1(N)(\alpha = 0) \quad (4,0,0,6), \tag{123}$$

$$\begin{aligned} &= 26880\{V_0'(N) + 3V_1(N) - 3V_3(N)\} + 3360E_2'(N) \\ &\quad (\alpha = 0)(4,2,0,4), \end{aligned} \tag{124}$$

$$\begin{aligned} &= 26880\{3V_0'(N) + V_1(N) - V_3(N)\} + 3360E_2'(N) \\ &\quad (\alpha > 0)(4,2,0,4). \end{aligned} \tag{125}$$

By (112), (119) and (122),

$$V_1(2N) - 3V_3(2N) = W(N), \tag{126}$$

and therefore, by (111),

$$2V_1(2N) - 3V_3(2N) = 2E_4'(N). \tag{127}$$

We consider first the case $\alpha = 0$. It follows from (115), (116) and (117) that

$$V_1(N_1) = \frac{1}{4}V_0'(N_1), \quad V_3(N_1) = -\frac{1}{24}\{2V_0'(N_1) - E_2'(N_1)\}, \tag{128}$$

and, from (122), (123), (124) and (128), that

$$V_0(N_1) = \frac{1}{4}\Delta_1(N_1), \quad V_1(2N_1) - V_3(2N_1) = 16V_0'(N_1). \tag{129}$$

Thus we have, in addition to (115) and (120),

$$\begin{aligned} R_{4,6}(4N_1) = 53760V_0'(N_1), R_{2,8}(2N_1) &= 728V_0'(N_1), R_{10}(N_1) \\ &= 240V_0'(N_1), \end{aligned} \tag{130}$$

and therefore, by (112),

$$W(N_1) = 32V_0(N_1). \tag{131}$$

It follows from (111) that

$$V_0'(N_1) = \frac{1}{20}E_4'(N_1). \tag{132}$$

Thus we have, by (127), (128), (129), (131) and (132),

$$V_1(N_1) = \frac{1}{80}E_4'(N_1), V_3(N_1) = -\frac{1}{240}\{E_4'(N_1) - 10E_2'(N_1)\}, \tag{133}$$

$$W_1(N_1) = \frac{8}{5}E_2'(N_1), V_1(2N_1) = \frac{1}{41}E_4'(2N_1), \tag{134}$$

$$V_3(2N_1) = -\frac{1}{40}E_4'(2N_1). \tag{135}$$

The functions $V_0(N_1)$ and $V_0'(N_1)$ are given by (129) and (132).

We now suppose that $\alpha > 0$. By (119) and (121),

$$V_3(2N) + 8V_1(N) - 8V_3(N) = 0. \tag{136}$$

By (127) with $\frac{1}{2}N$ in place of N,

$$2V_1(N) - 3V_3(N) = \frac{1}{8}E_4'(N). \tag{137}$$

Eliminating $V_1(N)$ between (136) and (137), we obtain

$$V_3(2N) + 4V_3(N) + \frac{1}{2}E_4'(N) = 0,$$

which may be set in the following form:

$$V_3(2N) + \frac{1}{40}E_4'(2N) = -4\left\{V_3(N) + \frac{1}{40}E_4'(N)\right\}. \tag{138}$$

It follows from (135) by repeated application of (138) that

$$V_3(N) = -\frac{1}{40}E_4'(N). \tag{139}$$

Hence, by (126) and (137),

$$V_1(N) = \frac{1}{40}E_4'(N), \; W(N) = \frac{8}{5}E_4'(N). \tag{140}$$

By (122), (125), (139) and (140),

$$V_0'(N) = \frac{1}{60}E_4'(N) - \frac{1}{24}E_2'(N). \tag{141}$$

It remains to evaluate $V_0(N)$ when N is even. By (108), (139) and (140),

$$\begin{aligned} V_0(2N) &= \sum_{m=1}^{N-1} E_2(2m)\Delta_1(2N-2m) = \sum_{m=1}^{N-1} E_2(m)\Delta_1(N-m) \\ &= V_0(N) + V_1(N) + V_3(N) = V_0(N). \end{aligned} \tag{142}$$

Similarly, by (129) and (133),

$$\begin{aligned} V_0(2N_1) &= V_0(N_1) + V_1(N_1) + V_3(N_1) \\ &= \frac{1}{4}\Delta_1(N_1) - \frac{1}{120}E_4(N_1) - \frac{1}{24}E_2(N_1). \end{aligned} \tag{143}$$

It follows from (143) by repeated application of (142) that

$$V_0(N) = \frac{1}{4}\Delta_1(N) - \frac{1}{120}E_4(N_1) - \frac{1}{24}E_2(N). \tag{144}$$

Formulae (5), (19) and (21) for $\alpha > 0$ now follow from (114), (119), (122), (139), (140), (141) and (144).

In the preceding analysis it has been assumed that $N_1 \equiv 1 \pmod 4$. It might be thought that formulae for the case $N_1 \equiv 3 \pmod 4$ could be obtained in a similar manner. This, however, is not the case. For if we set up the equations corresponding to equations (113)–(125) we find that we cannot eliminate the five functions (106)–(110) with the help of (111) and (112) and obtain formulae for the number of representations $R_{2\alpha,10-2\alpha}$. The reason for this is, as is well known, that in the general case it is necessary to introduce a new function which is not expressible in terms of divisor functions. This new function is Glaisher's function $\chi_4(n)$ which is defined by

$$\chi_4(n) = \sum (a+ib)^4,$$

where the summation is carried out over all representations of n as a^2+b^2. When n is not representable as a sum of two squares, e.g. when $n \equiv 3 \pmod 4$, $\chi_4(n)$ vanishes.

9 Twelve Squares

Take $\psi_0(u, v) = \frac{1}{16}u^4$, so that, by (65),

$$f(x, y) = f_0(x, y) = \frac{1}{2}(xy^3 + x^3y).$$

By Theorem 2, if n is odd,

$$\begin{aligned}\sum_{S(n)} f(x, y) &= 2\sum_{t=1}^{(n-1)/2} t^4 \\ &= \frac{1}{240}n(n^2-1)(3n^2-7), \end{aligned} \tag{145}$$

and, if n is even,

$$\sum_{S(n)} f(x, y) = \frac{1}{32}n^5. \tag{146}$$

Write

$$Z(2n) = \sum_{m=1}^{2n-1} \Delta_1(m)\Delta_3(2n-m), \tag{147}$$

$$Z'(n) = \sum_{m=1}^{n-1} \Delta_1(m)\Delta_3'(n-m), \tag{148}$$

$$Z_1(2n) = \sum_{m=1}^{n} \Delta_1(2m-1)\Delta_3(2n-2m+1), \tag{149}$$

It follows from these definitions and from (34), (35), (145) and (146) that

$$\begin{aligned}Z_1(2n) &= \sum_{T_1(2n)} f(x, y) = \sum_{\delta|2n}\sum_{S(2n)/\delta} f(x, y) \\ &= \frac{1}{32}\Delta_5'(2n), \end{aligned} \tag{150}$$

and if $n = 2^\alpha m$, where m is odd and $\alpha \geq 0$,

$$\begin{aligned}Z(2n) &= \sum_{T_1(2n)} f(x, y) = \sum_{d|2n}\sum_{S(d)} f(x, y) = \sum_{\delta|n}\sum_{\beta=0}^{\alpha+1}\sum_{S(2^\beta\delta)} f(x, y) \\ &= \sum_{\delta|n}\left\{\frac{1}{240}\delta(\delta^3-1)(3\delta^2-7) + \delta^5(1+2^5+\cdots+2^{5\alpha})\right\} \\ &= \tfrac{1}{240}\{3\Delta_5(2n) - 10\Delta_3(2n) + 7\Delta_1(2n)\} \\ &\quad + \tfrac{1}{31}\{\Delta_5'(2n) - \Delta_5(2n)\}. \end{aligned} \tag{151}$$

We have, by (2), (4) and (32)

$$\begin{aligned}
R_{12}(2n) &= R_4(2n) + R_8(2n) + \sum_{m=1}^{2n-1} R_4(m)R_8(2n-m) \\
&= R_4(2n) + R_8(2n) + 128Z_1(2n) \\
&\quad + \tfrac{384}{7}\sum_{m=1}^{n-1} \Delta_1(2m)\{8\Delta_3'(2n-2m) - 15\Delta_3(2n-2m)\} \\
&= 24\Delta_1(2n) + \tfrac{16}{7}\{8\Delta_3'(2n) - 15\Delta_3(2n)\} \\
&\quad + 512Z_1(2n) + \tfrac{384}{7}\{8Z'(2n) - 15Z(2n)\}. \qquad (152)
\end{aligned}$$

Similarly, by (33) and the formulae already proved, we obtain

$$\begin{aligned}
R_{4,8}(8n) &= 126720Z_1(2n), \quad (4,0,0,8) \qquad (153) \\
R_{8,4}(8n) &= 126720Z_1(2n), \quad (4,0,4,4) \qquad (154) \\
&= 3041280Z'(n) + 126720\Delta_3'(n). \quad (8,0,0,4) \qquad (155)
\end{aligned}$$

It follows from (150), (153) and (154) that

$$R_{4,8}(8n) = R_{8,4}(8n) = 3960\Delta_5'(2n). \qquad (156)$$

Finally, by (155) and (156),

$$Z'(n) = \frac{1}{24}\{\Delta_5'(n) - \Delta_3'(n)\}, \qquad (157)$$

and (6) follows from (150), (151), (152) and (157).

10 Concluding Remarks

If we examine the formulae that have been proved we notice certain relations connecting different types of representations, such as

$$5R_{6,0}(8n+6) = 2R_{3,3}(4n+3)$$

and

$$3R_{4,0}(8n+4) = 2R_{0,4}(8n+4).$$

These are particular cases of the following three theorems.

Theorem 3 *Let α, β, a, b and k be non-negative integers such that*

$$2k \equiv \alpha \pmod 4, \quad k \equiv a \pmod 4, \quad 0 \le k < 4,$$

and suppose that

$$R_{\alpha,\beta}(8n+2k) = AR_{a,b}(4n+k)$$

for all positive $4n+k$ *and fixed* A. *Then*

$$R_{\alpha+2m,\beta}(8n+2k+2m) = A_m R_{a+m,b+m}(4n+k+m) \tag{158}$$

for $m \geq 0$, *where*

$$A_m = 2^m A \frac{(\alpha+\beta+2m)!\alpha!(a+b)!(a+m)!(b+m)!}{(\alpha+\beta)!(\alpha+2m)!(a+b+2m)!a!}.$$

Theorem 4 *If* α, β, a, b *and* k *are non-negative integers such that*

$$k \equiv \alpha \equiv a \pmod 4, \quad 0 \leq k < 8,$$

and if

$$R_{\alpha,\beta}(8n+k) = B R_{a,b}(8n+k)$$

for all positive $8n+k$ *and fixed* B, *then*

$$R_{\alpha+2m,\beta}(8n+k+2m) = B_m R_{a+2m,b}(8n+k+2m)$$

for $m \geq 0$, *where*

$$B_m = B \frac{(\alpha+\beta+2m)!\alpha!(a+b)!(a+2m)!}{(\alpha+\beta)!(\alpha+2m)!(a+b+2m)!a!}.$$

Theorem 5 *If* α, β, a, b *and* k *are non-negative integers such that*

$$k \equiv \alpha \equiv a \pmod 4, \quad a \leq k < 4,$$

and if

$$R_{\alpha,\beta}(4n+k) = C R_{a,b}(4n+k)$$

for all positive $4n+k$ *and fixed* C, *then*

$$R_{\alpha+m,\beta+m}(4n+k+m) = C_m R_{a+m,b+m}(4n+k+m),$$

for $m \geq 0$, *where*

$$C_m = C \frac{(\alpha+\beta+2m)!\alpha!\beta!(a+b)!(a+m)!(b+m)!}{(\alpha+\beta)!(\alpha+m)!(\beta+m)!(a+b+2m)!a!b!}.$$

In all the applications of these theorems $a+b=\alpha+\beta$ so that

$$\begin{aligned} A_m &= 2^m A \frac{\alpha!(a+m)!(b+m)!}{(a+2m)!a!b!}, \\ B_m &= B \frac{\alpha!(a+2m)!}{(\alpha+2m)!a!}, \\ C_m &= C \frac{\alpha!\beta!(a+m)!(b+m)!}{(\alpha+m)!(\beta+m)!a!b!}. \end{aligned}$$

We may prove Theorem 3 by induction as follows. Assume that (158) holds for a certain value of m. Write

$$\begin{aligned}\alpha' &= \alpha+2m, a'=a+m, b'=b+m, k'=k+m, n'=n+[k/4],\\ D_m &= \frac{(\alpha'+\beta+2)(\alpha'+\beta+1)}{(\alpha'+2)(\alpha'+1)}, E_m=\frac{(a'+b'+1)(a'+b'+2)}{2(a'+1)(b'+1)}.\end{aligned}$$

Then, by (33),

$$R_{\alpha'+2,\beta}(8n+2k'+2)=D_m\sum_{\mu=0}^{n'}R_{\alpha',\beta}(8n-8\mu+2k')R_{2,0}(8\mu+2),$$

and

$$R_{a'+1,b'+1}(4n+k'+1)=E_m\sum_{\mu=0}^{n'}R_{a',b'}(4n-4\mu+k')R_{1,1}(4\mu+1).$$

But

$$R_{2,0}(8\mu+2)=R_{0,1}(4\mu+1),$$

and it follows by induction, since $A_mD_m=A_{m+1}E_m$, and since (158) holds for $m=0$.

Theorems 4 and 5 may be proved similarly.

Theorem 6 *As particular cases of Theorem* 3 *we have the following:*

(i) *If* $r>0$,

$$R_{2r,0}(8n+2r)=\frac{2^r(r!)^2}{(2r)!}R_{r,r}(4n+r). \tag{159}$$

(ii) *If* $r\geq 0$,

$$R_{2r,2}(8n+2r)=\frac{2^{r-1}r!(r+2)!}{(2r)!}R_{r,r+2}(4n+r).$$

Formula (159) was proved by Glaisher by elliptic function theory.

Theorem 7 *As particular cases of Theorem* 4 *we have the following:*

(i) *If* $r>1$,

$$R_{2r-4,4}=\frac{(2r)!}{2.4!(2r-4)!}R_{2r,0}(8n+2r).$$

(ii) *If* $r\geq 0$,

$$R_{2r+4,4}(8n+2r+4)=\frac{8!(2r)!}{4!(2r+4)!}R_{2r,8}(8n+2r+4).$$

References

[1] P. Bachman, *Niedere Zahlentheorie*, Leipzig, 1910, vol. 2.

[2] H. Bessel, *Elementar-arithmetische Untersuchungen über die Darstellung einer natürlichen Zahl als Summe von Quadratzahlen*, Inaugural-Dissertation, Königsberg, 1929.

[3] L.E. Dickson, *History of the theory of numbers*, New York, 1934.

[4] J.W.L. Glaisher, On the number of representations of a number as a sum of $2r$ squares, *Proc. London Math. Soc.* (2) **5** (1907), 479–490.

[5] E. Grosswald, *Representations of integers as sums of squares*, Springer, Heidelberg, 1985.

[6] G.H. Hardy, *Ramanujan*, Cambridge, 1940.

[7] L.K. Hua, *Introduction to number theory*, Springer, Heidelberg, 1982.

[8] R.A. Rankin, Hecke operators on congruence subgroups of the modular group, *Math. Ann.* **168** (1964), 40–58.

[9] R.A. Rankin, Elementary proofs of relations between Eisenstein series, *Proc. Roy. Soc. Edinburgh*, **76A** (1976), 107–117.

[10] R.A. Rankin, *Modular forms and functions*, Cambridge, 1977.

[11] J.V. Uspensky and Heaslet, M.A., *Elementary number theory*, New York, 1939, Chapter XIII.

Department of Mathematics
University of Glasgow
Glasgow G12 8QT
Scotland
E-Mail: rar@math.gla.ac.uk

Solution of the Basic Problems of Discrete Geometry on the Plane

S.S. Ryshkov, R.G. Barykinskii, Y.V. Kucherinenko

1 Motivations, Literature and Basic Notions

§1.1 Discrete Geometry and Geometry of Numbers

1^0. There are two geometric disciplines which have the sufficient large intersection. Now we mean the discrete geometry and the geometry of numbers. It is accepted to refer various problems about dispositions of points and figures in a space to the discrete geometry. The same problems are accepted in the geometry of numbers, when they are somehow connected with point lattices in the n-dimensional euclidean space $\mathbb{E}^n$, and also some other problems about lattices are accepted as well.

The majority of successes in the basic problems of the discrete geometry for $n > 2$ is reached, where they were managed to be reduced to the problems of the geometry of numbers, and solved it there.

2^0. To the basic problems of the discrete geometry are usually referred:

1) Problem about the densest packing of equal balls in the Euclidean space.
2) Problem about the thinnest covering of the Euclidean space by equal balls.
3) Problem about enumeration of all parallelohedra.
   Nevertheless, it is often forgotten, that the following problem had been standing for many years.
4) Problem about enumeration of all groups of Euclidean space motions, which have a compact fundamental region.
   It is explained by the fact, that only this problem of the discrete geometry is reduced completely at any n (by the famous Schönfliess-Bieberbach theorem) to the problem of the geometry of numbers, and for $n \leq 4$ it is solved with concepts of geometry of numbers and algebra.
   As the theorem of Schönfliess-Bieberbach has numerous good detailed accounts, especially for $n = 2$, see for example [1], we shall not be concerned with problem (4).
   It seems to us that another simple problem, which is solved for all n, that should be returned to the basic problems of discrete geometry is the following:
5) Problem about characteristic criteria of lattices, which mark them out among other point systems.

3^0. The case $n = 2$ is sharply distinguished from the common case by the fact, that the problems (1), (2), (3) are solved immediately within the framework of the discrete geometry.

In this case all theories and concepts (the theory of point systems, the theory of L- and DV-tilings, the theory of lattices, the theory of parallelohedra etc.), which accompany solutions of the mentioned problems, are easier and more obvious, than for $n > 2$. These circumstances have induced us to write this paper.

4^0. We shall note in inference, that the paper is a summary of a terminal course of lectures, which was delivered by the first author at the mechanical-mathematical department of the Moscow State Lomonosov University. Thus, it is "text-book" and survey of results at the same time. When delivering this course, many references including the following books and papers were used, the authors recommend them to the reader.

[1] D. Hilbert, and Cohn-Vossen, S., "Anschauliche Geometrie" // Berlin, 1932.

[2] L. Fejes Tóth, "Lagerungen in der Ebene auf der Kugel und im Raum" // Springer-Verlag, Berlin · Göttingen · Heidelberg, 1953.

[3] B.N. Delone, "The geometry of positive quadratic forms" // *Uspehi Mat. Nauk*, v. **3** (1937), p. 16–62; v. **4** (1938), p. 102–164. (In Russian.)

[4] S.S. Ryshkov, "Theory of point lattices" // (In Russian, under preparation.)

[5] P.M. Gruber, and Lekkerkerker, C.G., "Geometry of numbers" // *North-Holland Math. Library*, v. **37**. *North-Holland*, Amsterdam · New York · Oxford · Tokyo.

5^0. After the numbers of theorems we shall sometimes specify, as for example:

$n.n$ – analogical theorem is correct (quite right) at any n, the given proof is also spreaded on any n,

$n.2$ – analogical theorem is correct (quite right) at any n, but the given proof is not spreaded on any n,

2.2 – theorem is correct only at $n = 2$.

Besides we systematically use the following lables:

$\triangleright$ – the beginning of a proof,

$\bullet$ – the end of a proof,

$\triangleright\bullet$ – a proof is obvious,

iff – if and only if.

6^0. We are very grateful to all mathematicians, on whose results the paper is based.

Finally, our gratitudes go to E.P. Nikiforova who helped us in the translation of the paper into English and to Z.D. Lomakina who carefully read the Russian version of the paper.

The first author is supported by RFFI, grant No. 97-01-00266.

§1.2 Dispositions. Local Finiteness

1^0. Throughout the paper $\mathbb{E}^2$ denotes the euclidean plane with an orthonormal system of coordinates (O, ξ_1, ξ_2), which has the origin in some fixed point $O \in \mathbb{E}^2$.

The word "*figure*" everywhere means a bounded Jordan-measurable subset of the plane. The word "convex set" means a convex set of the plane, containing at least one interior point.

Definition: An arbitrary denumerable system $\mathfrak{T} = \{T_1, T_2, \ldots\}$ with sets $T_1, T_2, \ldots \subset \mathbb{E}^2$ is called a *disposition* (of these sets).

Generally speaking, some of elements (sets) of a disposition or even all of them can be pairwise congruent, and some of them can intersect or even, for example, coincide with each other.

Further, we shall consider dispositions of bounded sets only, without any special reserve.

One of examples of dispositions is any arbitrary (denumerable) point system Σ (see below), i.e. points can be considered as elements of a disposition.

A disposition of sets, in which any two of them don't intersect (i.e., if $T_i \cap T_j = \emptyset$, for $i, j = 1, 2, \ldots$ and $i \neq j$), is called an *exact (disjunctive) packing* (of these sets).

A disposition $\mathfrak{T} = \{T_1, T_2, \ldots\}$, for which

$$\mathbb{E}^2 = \bigcup_{i=1}^{\infty} T_i$$

is called a *covering* of the plane $\mathbb{E}^2$ (by these sets).

A disposition $\mathfrak{T} = \{T_1, T_2, \ldots\}$, being both a covering and an exact packing, is called an *exact (disjunctive) tiling* of the plane $\mathbb{E}^2$ (by or with the sets). The sets T_i are called *tiles*.

A disposition $\mathfrak{T} = \{T_1, T_2, \ldots\}$, in which any two elements don't intersect on interior points (i.e., if $\mathrm{int}T_i \cap \mathrm{int}T_j = \emptyset$, for $i, j = 1, 2, \ldots$ and $i \neq j$), is called *(classical) packing* (of these sets).

A disposition $\mathfrak{T} = \{T_1, T_2, \ldots\}$ of closed sets, being both a covering, and a (classical) packing, is called *(classical) tiling* of the plane $\mathbb{E}^2$ (by or with the tiles T_i).

When reviewing a classical tiling below, we shall not stress the fact, that all its tiles are closed sets.

Further, if it is clear what kinds of tilings or packings we are speaking about: exact or classical, or if some statement is right for both of them, we shall use only the words "tiling" or "packing". Besides, it is natural, that we shall not mention, that we consider tilings of the (whole) plane.

Definition: A disposition $\mathfrak{T} \subset \mathbb{E}^2$ is called *regular*, if there exists a group of plane motions, which is acting transitively on this disposition.

Definition: A disposition $\mathfrak{T} \subset \mathbb{E}^2$ is called *translationally regular*, if there exists a group of parallel translations of the plane $\mathbb{E}^2$ which is acting transitively on this disposition.

As an example, let's consider an "infinite" sheet of checked paper. The family of closed squares of this sheet form a regular classical tiling and the family of open squares form a regular packing. The families of edges and knots are regular dispositions. The dispositions of squares and knots are also translationally regular. At the same time the disposition of edges is regular, but not translationally regular, because the group transitively acting on this disposition necessarily contains the turn of 90°.

2^0. A disposition $\mathfrak{T} = \{T_1, T_2, \ldots\}$ is called *locally finite*, if any point $A \in \mathbb{E}^2$ has a neighbourhood intersecting only finite number of elements (sets) of the disposition $\mathfrak{T}$.

Lemma 1 n.n. *For a disposition $\mathfrak{T}$ to be locally finite, it is necessary and sufficient, that any bounded set $M \subset \mathbb{E}^2$ has nonempty intersection only with a finite number of elements of the disposition $\mathfrak{T}$.*

▷ The sufficiency is trivial. Let's prove the necessity. Let there be given a bounded set M. If its closure intersects only a finite number of elements of the disposition $\mathfrak{T}$, then it is also intersects only a finite number of elements of $\mathfrak{T}$. Therefore, we can consider the set M closed. Let's construct a neighbourhood of each point of the set M, which has nonempty intersection only with a finite number of elements of the disposition $\mathfrak{T}$. The obtained system of the neighbourhoods is an open covering of the compact set M. We choose a finite subcovering from this covering. It is obvious, that the set M intersectes only a part of those elements of $\mathfrak{T}$, which intersect the union of all elements of the constructed subcovering.•

If the topological theorem used in the proof is unknown to the reader, he can include the set M in a sufficient large square and take its points' neighbourhoods homothetical to this square. Further, to reach the purpose readers can use several times the Heine-Borel lemma.

Lemma 2 n.n. *Let for some packing $\mathfrak{T}$ there exist numbers $r > 0$ and $R > 0$, such that for any set $T_i \in \mathfrak{T}$ there is a closed comprehending disk of radius R and an open inserted disk of radius r, then the packing $\mathfrak{T}$ is locally finite.*

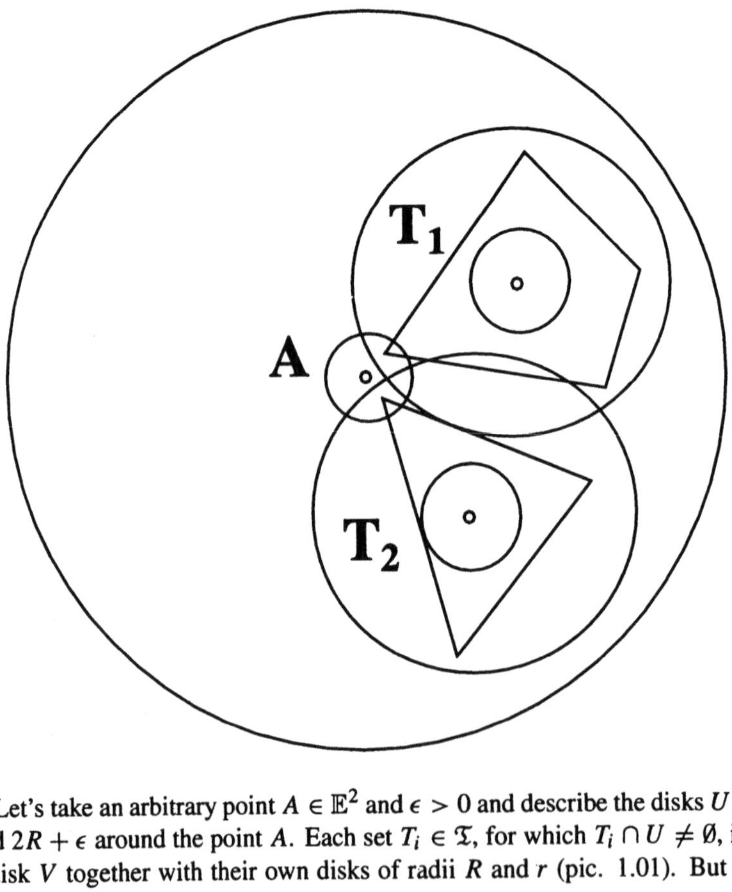

pic. 1.01

▷ Let's take an arbitrary point $A \in \mathbb{E}^2$ and $\epsilon > 0$ and describe the disks U and V of radii ϵ and $2R + \epsilon$ around the point A. Each set $T_i \in \mathfrak{T}$, for which $T_i \cap U \neq \emptyset$, is contained in the disk V together with their own disks of radii R and r (pic. 1.01). But being inserted in the sets $T_i \in \mathfrak{T}$ the disks of radii r are not pairwise intersected. Therefore, there are, obviously, not more, than $(\frac{2R+\epsilon}{r})^2$ of them in the disk V.•

3^0. **Theorem 1.1 n.n.** *Each locally finite tiling with bounded convex sets is a tiling with convex polygons.*

▹ As all sets of the tiling are convex, that for any two sets, having not less then one point in the intersection, there exists a straight line, containing their intersection (pic. 1.02). It means that each two elements of the tiling are either not intersected, or have a common point, or have a common segment. From local finiteness we infer that every set of the tiling intersects with a finite number of other sets.•

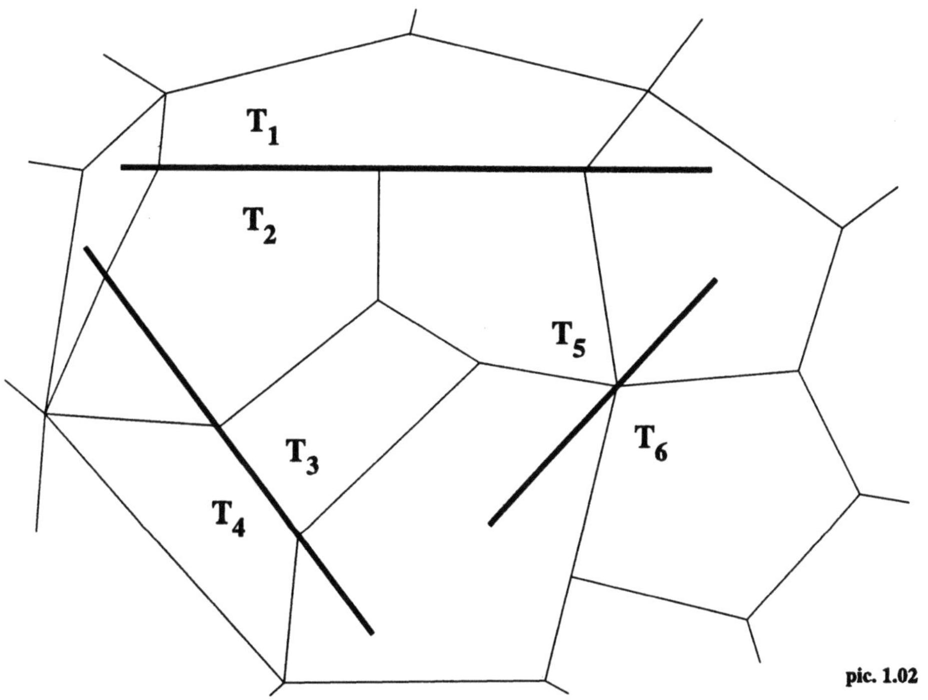

pic. 1.02

Let's introduce some notions, which we will need further in different measure.

A packing of convex polygons, in which each two polygons are either not intersected, or have a common vertex or a common edge, is called a *face-to-face packing*.

A *face-to-face tiling* is such a face-to-face packing, which is also a covering.

Let's consider a face-to-face tiling with convex polygons, then *faces* of the tiling are the polygons themselves, their vertices and edges. A *facet* of a convex polygon is its edge (1-dimensional face).

A tiling with convex polygons is *facet-to-facet* if any facet of a tile is also a facet of another tile.

Lemma 3 n.2. *A locally finite tiling with convex polygons is face-to-face iff it is facet-to-facet.* ▹•

A face-to-face tiling with convex polygons is *primitive*, if in each vertex of the tiling only three edges converge.

§1.3 Point Systems. Delone Systems

1^0. We'll call by a *point system* $\Sigma \subset \mathbb{E}^2$ an arbitrary denumerable set of points (an arbitrary disposition of single-point sets).

Definition: A point system $\Sigma \subset \mathbb{E}^2$ is called *a* $\langle 1, 0\rangle$*-system*, if for some $r > 0$ it has the following "r-property":

In each open disk $U \subset \mathbb{E}^2$ of radius r, not more than one point, belonging to the system Σ, will be discovered see pic. 1.03. The least upper bound of the set of such numbers r will be denoted by r_Σ.

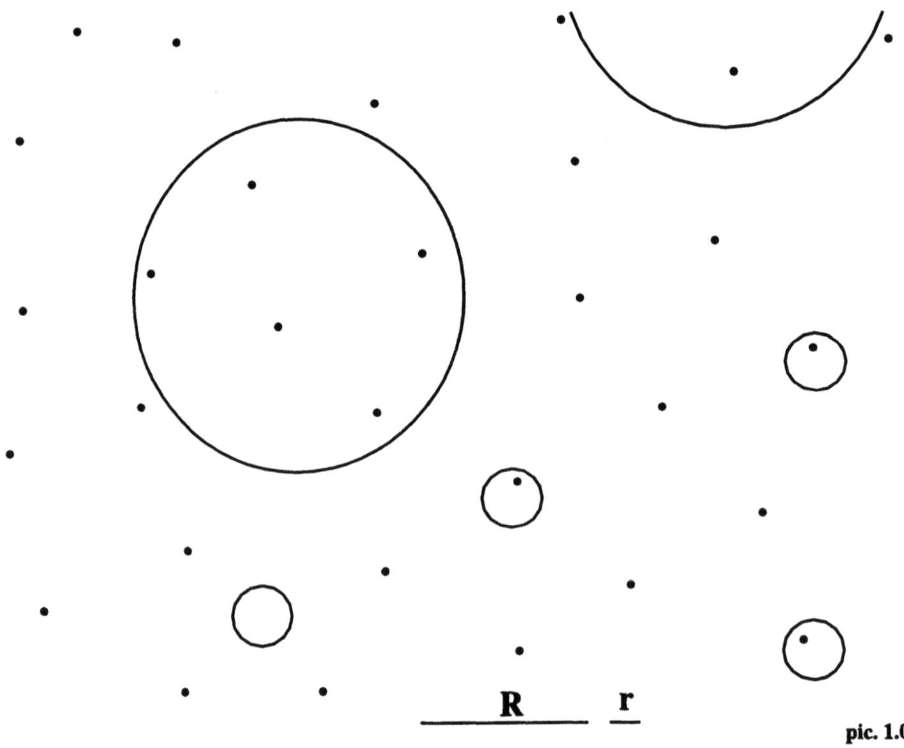

pic. 1.03

It is obvious, that each $\langle 1, 0\rangle$-system is locally finite, i.e. is a locally finite disposition of points.

Definition: A point system $\Sigma \subset \mathbb{E}^2$ is called *a* $\langle 0, 1\rangle$*-system*, if for some $R > 0$ it has the following "R-property":

In each closed disk $U \subset \mathbb{E}^2$ of radius R, at least one point, belonging to the system Σ will be discovered see pic. 1.03. The greatest lower bound of the set of such numbers R will be denoted by R_Σ.

Definition: Point system $\Sigma \subset \mathbb{E}^2$ is called *a* $\langle 1,1\rangle$*-system* or a *Delone system*, if it is both a $\langle 1,0\rangle$-system and a $\langle 0,1\rangle$-system (pic. 1.03).

Lemma 4 n.n. *The affine image of any* $\langle 1,1\rangle$*-system* ($\langle 1,0\rangle$-, $\langle 0,1\rangle$*-system) is again a* $\langle 1,1\rangle$*-system* ($\langle 1,0\rangle$-, $\langle 0,1\rangle$*-system).* ▷•

Usually (with the replacement of the radius r to $r/2$ in the definition of a $\langle 1,1\rangle$-system) a $\langle 1,1\rangle$-system is called an (r, R)-system or a *uniform discrete* system. Sometimes a $\langle 1,1\rangle$-system is called an (r, R)-system, meaning that it has the r-property and the R-property exactly just for given $r > 0$ and $R > 0$. The constants r_Σ, R_Σ are called the *basic constants* of Σ.

Lemma 5 n.n. *Each* (r, R)*-system* Σ *is an* (r_Σ, R_Σ)*-system, but it is not an* (r', R')*-system at any* $r' > r_\Sigma$ *and (or)* $R' < R_\Sigma$. ▷•

Examples: Let's consider on the Cartesian plane xOy three systems $\mathfrak{T}_1$, $\mathfrak{T}_2$, $\mathfrak{T}_3$ of points, which have the following coordinates (x_i, y_j), where $x_i = S_i$, $y_j = S_j : i, j \in \mathbb{Z}$. Let, firstly, $S_0 = 0$, $S_i = S_{i-1} + i$, $S_{-i} = -S_i$ with $i = 1, 2, \ldots$ (pic. 1.04). It is obvious, that the obtained system $\mathfrak{T}_1$ is a $\langle 1,0\rangle$-system (for any $r \leq \frac{\sqrt{2}}{2}$) but it is not a $\langle 0,1\rangle$-system.

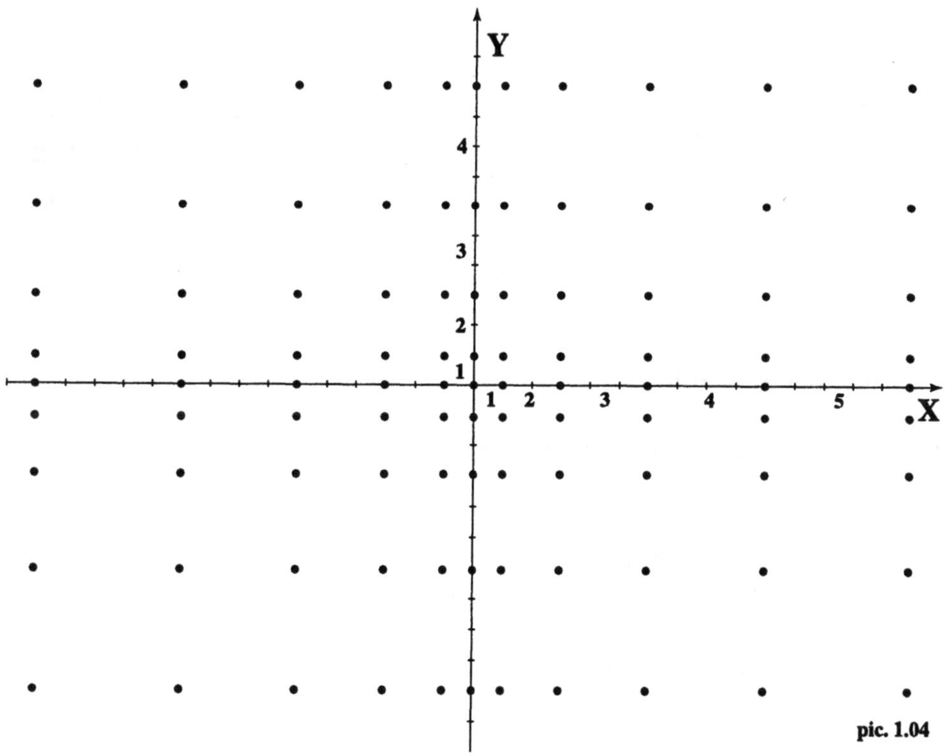

pic. 1.04

Let, secondly, $S_0 = 0$, $S_i = S_{i-1} + \frac{1}{i}$, $S_{-i} = -S_i$ with $i = 1, 2, \ldots$ (pic. 1.05). In this case the point system $\mathfrak{T}_2$ (its local finitness follows from divergence of the harmonic series) is a $\langle 0, 1\rangle$-system (for any $R \geq \frac{\sqrt{2}}{2}$) but it is not a $\langle 1, 0\rangle$-system.

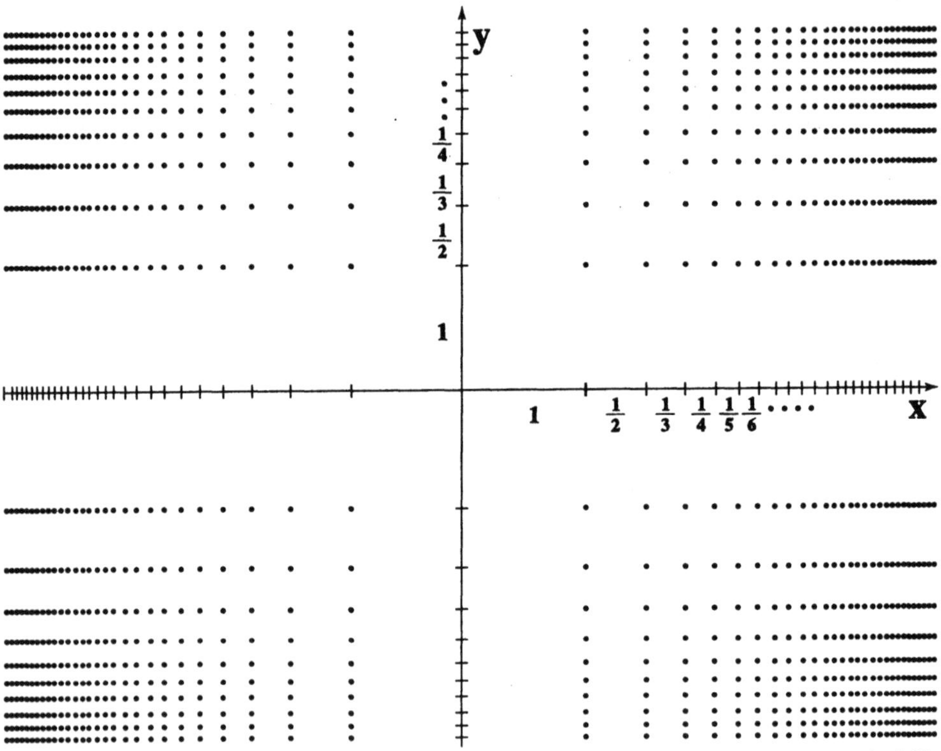

pic. 1.05

Let the system $\mathfrak{T}_3$ coincides with the system $\mathfrak{T}_1$ in the right half-plane and with the system $\mathfrak{T}_2$ in the left one. Then, it is easy to see, the system $\mathfrak{T}_3$ is neither a $\langle 0, 1\rangle$-, nor a $\langle 1, 0\rangle$-system.

2 L-tilings and DV-tilings

§2.1 *L-tilings. Empty disk ("la sphère vide"). Necessary and sufficient criterion*

1^0. Let us be given a $\langle 1, 1\rangle$-system $\Sigma \subset \mathbb{E}^2$.

Definition: Every convex figure $L \subset \mathbb{E}^2$ is called an *L-figure,* of the $\langle 1, 1\rangle$-system $\Sigma \subset \mathbb{E}^2$, if L is the convex hull of a subset $\Sigma' \subset \Sigma$, satisfying the following conditions:

1) $\Sigma' \subset L^\circ$, where L° – some circle.
2) Σ' uniquely determines the circle L°.
3) $L^* \cap \Sigma = \Sigma'$, where L^* – the disk, bounded by the circle L°, and the interior of the disk L^* is empty from points of the system Σ.

The circle L° is called an *L-circle* of the system Σ and the disk L^* is called an *L-disk* of Σ. Further, through $\{L\}_\Sigma$, $\{L^\circ\}_\Sigma$ and $\{L^*\}_\Sigma$, accordingly, the families of all L-figures, L-circles and L-discs of the system Σ will be denoted. (pic. 2.01.)

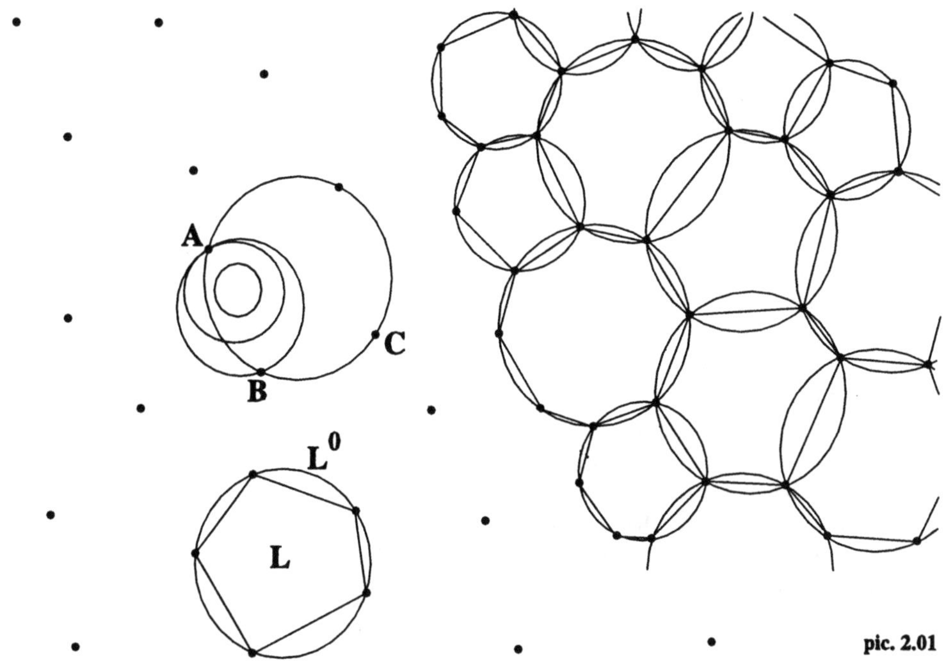

pic. 2.01

Lemma 1 n.n. *Every L-figure, of an arbitrary $\langle 1, 1\rangle$-system is a convex bounded polygon.*

⊳ The set Σ' is finite as a bounded subset of the given $\langle 1, 1\rangle$-system. As the set Σ' uniquely determines the circle and, thus, has rank 2, the lemma is proved.•

Lemma 2 n.n. ("Empty disk.") *Every $\langle 1, 1\rangle$-system has at least one L-polygon.*

⊳ Let us be given a $\langle 1, 1\rangle$-system $\Sigma \subset \mathbb{E}^2$. Let's consider "la sphère vide", that is a disk, containing neither inside, nor on the boundary points of the system Σ (pic. 2.01). Let's move by any mode this disk, simultaneously increasing its radius. At some moment our disk will meet, generally speaking, one point of the system Σ; let it be the point A. Then we shall move our disk in such a way incessantly increasing its radius, that the point A will remain on the boundary of the disk. At some moment our disk will meet one more point of Σ; let it be the point B. Then we shall continue to increase radius of the disk in such a way, that the points A and B will remain on the boundary of the disk. At some moment our disk will meet one more point of Σ; let it be the point C. Every point from the mentioned ones exists, that on account of the fact, that any disk of radius R_Σ is to contain points of the system Σ. Thus, the system of the points which have appeared on the boundary of this disk determines it completely. The obtained disk, obviously, is an L-disk of Σ, the convex hull of the points of Σ, lying on its boundary, is an L-figure of the system Σ.•

Lemma 3 n.n. *For every $\langle 1, 1\rangle$-system Σ, the sets $\{L\}_\Sigma$, $\{L^\circ\}_\Sigma$ and $\{L^*\}_\Sigma$ are denumerable. That means, that these sets are dispositions.* ▷•

2^0. **Theorem 2.1 n.n.** *The family $\{L\}_\Sigma$ of all L-polygons of an arbitrary $\langle 1, 1\rangle$-system $\Sigma \subset \mathbb{E}^2$ is a locally finite face-to-face tiling of the plane.*

▷ Let us be given a $\langle 1, 1\rangle$-system Σ and all its L-polygons. Let's divide the proof into few parts.

1) We shall remark, that every bounded set intersects only with a finite number of L-polygons. Really, let $M \subset \mathbb{E}^2$ be such a set, d – its diameter and $A \in M$. The point A can belong only to those L-polygons, whose vertices are away from it by not more than $2R_\Sigma$, thus the diameter of the set of vertices of those L-polygons, which can intersect the set M, doesn't exceed $d + 4R_\Sigma$. The finitness of the number of L-polygons, with such a set of the vertices, is obvious.
2) Let's show, that the set $\{L\}_\Sigma$ is a face-to-face packing.
Really, let two different L-polygons L_1 and L_2 have common points. Hence, it follows immediately, that appropriate for them L-disks L_1^* and L_2^* are different, but their intersection is not empty. Let's consider a chordal straight line $\mathcal{P}$ of these disks. This straight line divides each of the disks into two parts – "caps", it will be more convenient to us to consider both the caps of every disk closed. One of two caps of the disk L_1^* lies completely in the disk L_2^* – we shall call it the interior cap and another one the exterior cap (with respect to the disk L_2^*). The according names shall be given to the caps of the disk L_2^*. It is obvious, that points of the system Σ can be located only, firstly, outside of the intersection of the disks $L_1^* \cap L_2^*$ – we don't need now such points and, secondly, in the intersection of the L-circles L_1° and L_2° (there are vertices of the polygons L_1 and L_2, pic. 2.02). The vertices of the polygon L_1 (L_2) lies on the boundary of the exterior cap of the disk L_1^* (L_2^*). Hence, it follows that the intersection $L_1 \cap L_2$ lies in the intersection of the exterior caps, i.e. the polygons L_1 and L_2 adjoin each other either on a vertex or on an edge.
3) Let's consider now any polygon $L_1 \subset \{L\}_\Sigma$ and one of its edges F. Let's also consider the straight line $\mathcal{P}$, which contains F. The center O of the disk L_1^* lies on the perpendicular to $\mathcal{P}$ in the point O' – the middle of F. The straight line $\mathcal{P}$ divides the disk L_1^* into two caps (see above), one of which contains points of the system Σ only in the ends of the edge F – the points A and B of intersection L_1° with $\mathcal{P}$. We shall put on the straight line OO' the point O_t on the distance t from the point O in the leg of the empty cap (pic. 2.03). Further, we shall construct the disk U_t with the center at the point O_t, intersecting the straight line $\mathcal{P}$ in the points A and B. For small t the disk U_t contains only the points A and B. For large values t, in the disk U_t are contained new points of Σ, we shall designate through t' the least of such t. It is obvious, that the disk $U_{t'}$ contains new points only on the boundary. From here implies, that the disk $U_{t'}$ is an L-circle, and the edge F together with the new

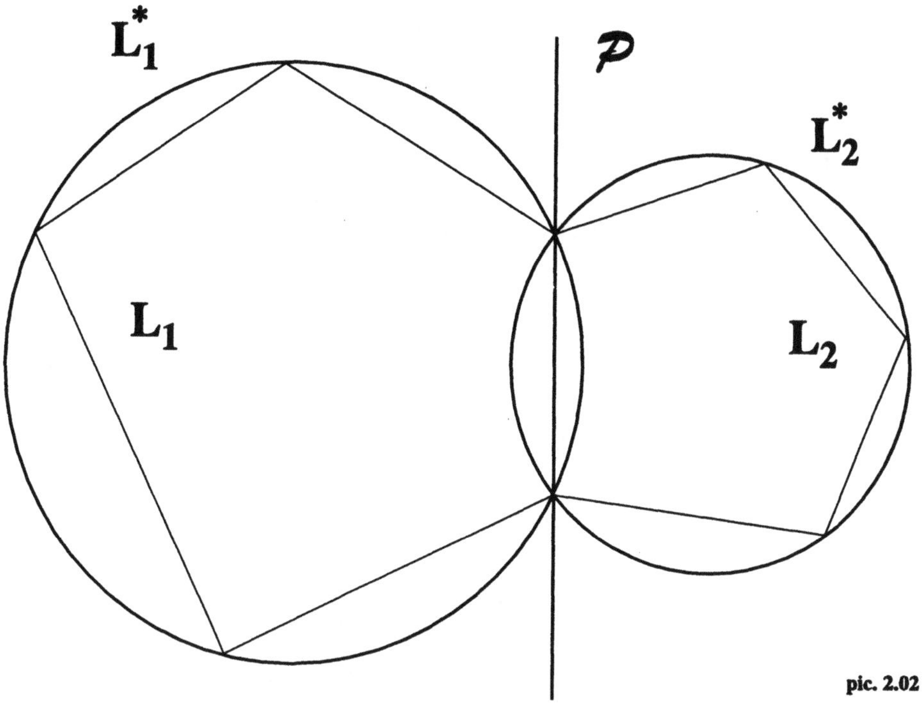

pic. 2.02

points determine the L-polygon $L_2 \in \{L\}_\Sigma$, adjacent with the polygon L_1 on F. So, for any L-polygon we have found other L-polygon, adjacent with initial on an initial beforehand given edge.

4) It remains for us to show, that each point $A \in \mathbb{E}^2$ lies at least in one polygon of the set $\{L\}_\Sigma$.

Let's consider an arbitrary point $A \in \mathbb{E}^2$, an arbitrary polygon $L \in L_\Sigma$, a point $O \in \text{int}L$ and the segment $[O, A]$. The segment $[O, A]$ is intersected with a finite number of L-polygons, because the disposition L_Σ is locally finite (pic. 2.04). Without loss of generality, we may choose the point $O' \in \text{int}L$, that the half-interval $[O', A)$ will intersect L-polygons only on their interior points and interior points of their edges (**"Lemma about shashlik", n.n.**) Then, applying a finite number of times the result of item (3), it will be possible to find the L-polygon, which contains the point A.•

Let $\Sigma \subset \mathbb{E}^2$ be an arbitrary $\langle 1, 1\rangle$-system. The above theorem allows us to introduce the following definition:

Definition: The tiling $\{L\}_\Sigma$ is called the *L-tiling* or the *Delone tiling* of the $\langle 1, 1\rangle$-system Σ (for Σ) (pic. 2.01).

Corollary: n.n. (from theorem 2.1.) *For any* $\langle 1, 1\rangle$*-system* Σ *the tiling* $\{L\}_\Sigma$ *is unique.* ▷•

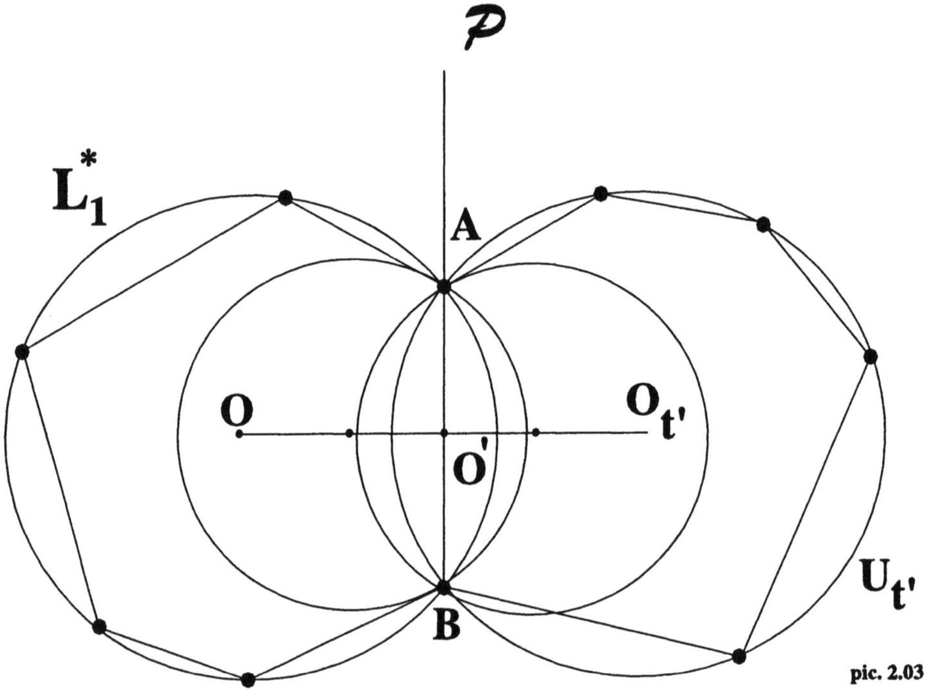

pic. 2.03

We shall note, that for any $\langle 1, 1\rangle$-system Σ the *star* of any *vertex* of the tiling $\{L\}_\Sigma$, i.e. the set of all its tiles, incident to this vertex, is finite. It follows from the local finitness of the L-tilings for such systems. (The *star* of any *edge* of the tiling $\{L\}_\Sigma$, i.e. the set of all its tiles, incident to this edge, consists of two tiles.)

We now make another one simple, but important statement.

Theorem 2.2 *Let some $\langle 1, 1\rangle$-system Σ be translated into itself by some motion (of first or second sort), then the tiling L_Σ is translated into itself by this motion too.* ▹•

3^0. ***L*-tiling. (Necessary and sufficient criterion.)** Let us be given some $\langle 1, 1\rangle$-system $\Sigma \subset \mathbb{E}^2$ and a locally finite face-to-face tiling $\mathfrak{T}$ with convex polygons.

We shall say, that the tiling $\mathfrak{T}$ is *compatible* with the $\langle 1, 1\rangle$-system Σ, if every point of Σ is a vertex of $\mathfrak{T}$, and every vertex of $\mathfrak{T}$ is a point of Σ.

Theorem 2.3 n.n. *Let $\mathfrak{T}$ be a tiling compatible with some $\langle 1, 1\rangle$-system Σ, than the tiling $\mathfrak{T}$ is the L-tiling of Σ, iff:*

1) *Every polygon of the tiling $\mathfrak{T}$ is a polygon, inscribable in a circle.*
2) *For any pair of polygons of the tiling $\mathfrak{T}$, adjacent on an edge, any of their vertices, which don't belong to this edge, lie outside of the disk, which is circumscribed around the other one.*

▹ The necessity of the specified conditions follows from the definition.

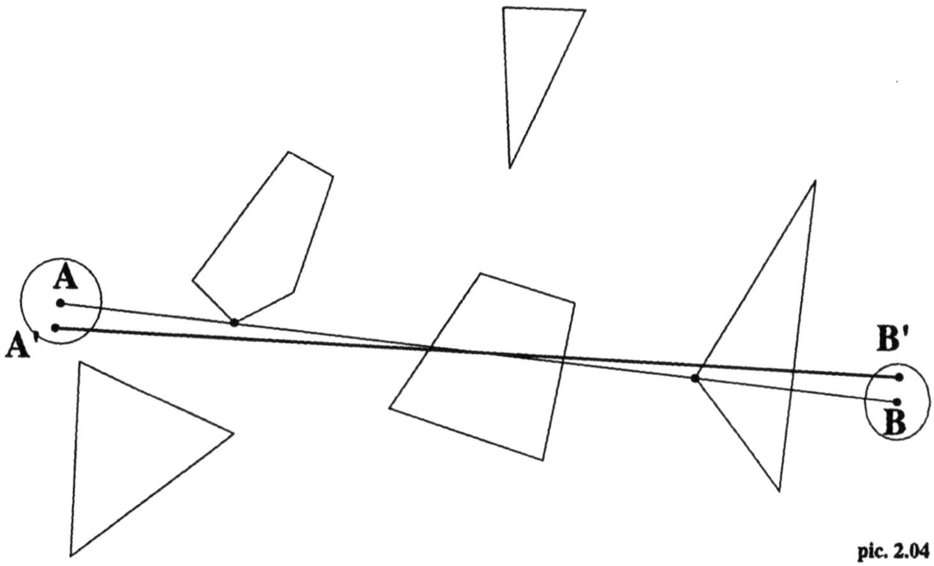

pic. 2.04

Let's prove the sufficiency. Let T_1 be an arbitrary polygon of the tiling $\mathfrak{T}$ and T_1^* be its disk, circumscribed around it. It is obvious, that for the proof of the theorem it is enough to prove the emptyness of the disk T_1^* from points of Σ, which are not vertices of T_1.

Let's consider an arbitrary point $A \in \Sigma$, which is not a vertex of T_1. As in the item (3) of the proof of theorem 2.1., let's choose a point $O \in \text{int}T_1$ and consider the polygons, intersected with the half-interval $[O, A)$. For more convenient description of the situation, let's consider, that the segment $[O, A]$ is horizontal and that the sequence of the polygons goes from left to right. Let these polygons be numbered in sequence $T_1, T_2, \dots, T_m$ so, that $A \in T_m$. Let T_i^*, where $i = 1, 2, \dots, m$, be the disk, circumscribed around the polygon T_i, and let $\mathcal{P}_j$, where $j = 1, 2, \dots, m-1$, be the straight line, containing the intersection $T_j \cap T_{j+1}$ (pic. 2.05).

By our construction of the segment $[O, A]$, the point A lies more to the right than all $\mathcal{P}_j$. Let's consider the polygons T_{m-1} and T_m. By the condition (2), the point A does not belong to the disk T_{m-1}^*. The interior cap of the disk T_{m-2}^* relatively to the disk T_{m-1}^* is contained inside T_{m-1}^*, that means that the cap doesn't contain A. And as the point A lies more to the right than the straight line $\mathcal{P}_{m-2}$, it is not contained in the disk T_{m-2}^* (pic. 2.05). On carrying out sequentially such a reasoning for the remaining polygons $T_i, i = m-3, m-4, \dots, 1$, we shall have, that the point A lies outside T_1^*. That means (once again), that whatever polygon $T_1 \in \mathfrak{T}$ is given, any point $A \in \Sigma$, not being its vertex, cannot lie either inside or on the boundary of T_1^*.•

4^0. **Example 1** Let's consider a tiling $\mathfrak{T}_\Delta$ by regular triangles (with an edge a). Let Σ_1 and Σ_2 be the point systems consisting accordingly of the vertices of $\mathfrak{T}_\Delta$ and of weight centers of $\mathfrak{T}_\Delta$'s triangles (pic. 2.06). It is obvious, that the L-tiling of Σ_1 is $\mathfrak{T}_\Delta$. (Discover L-tiling of Σ_2.)

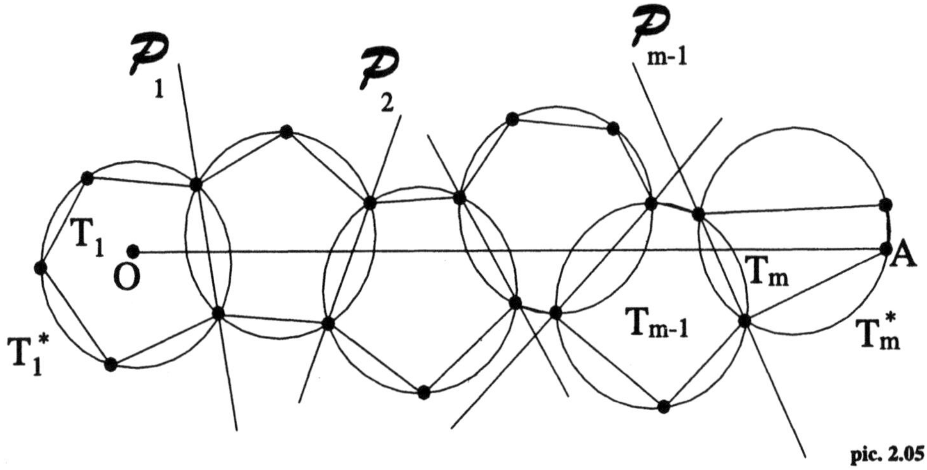

pic. 2.05

Example 2. Let's consider the system Σ, consisting of points of the lattice $\mathbb{Z}^2$ and all points $(x, y) = (k + 1/4, l + 1/2)$, where k and l are integers. The system Σ is a $\langle 1, 1\rangle$-system with basic constants $r_\Sigma = \sqrt{10}/4$ and $R_\Sigma = 5/8$.

Let's consider the triangles (pic. 2.07):

$$\begin{aligned} T_1 &= \{(-3/4, 1/2), (0, 1), (1/4, 1/2)\}, \\ T_2 &= \{(-3/4, 1/2), (0, 0), (1/4, 1/2)\}, \\ T_3 &= \{(0, 0), (1, 0), (1/4, 1/2)\}, \\ T_4 &= \{(0, 1), (1, 1), (1/4, 1/2)\}. \end{aligned}$$

It is easy to see, that the set $\mathfrak{T}$ of all triangles, each of them being parallel congruent to one of four constructed, will form face-to-face tiling of the plane, compatible with the system Σ.

Let's check, that $\mathfrak{T}$ is the L-tiling of Σ. The first condition of the sufficient criterion is trivial. For checking-up of the second condition we shall write out the equations of the circles, circumscribed around our four triangles:

$$\begin{aligned} T_1^o &= \{(X + 1/4)^2 + (Y - 9/16)^2 = 65/256\}, \\ T_2^o &= \{(X + 1/4)^2 + (Y - 7/16)^2 = 65/256\}, \\ T_3^o &= \{(X - 1/2)^2 + (Y - 1/16)^2 = 65/256\}, \\ T_4^o &= \{(X - 1/2)^2 + (Y - 15/16)^2 = 65/256\}. \end{aligned}$$

Vertices of triangles, adjacent with researched ones, have, correspondingly, such coordinates:

$$\begin{aligned} &T_1{:}\{(-1, 1), (1, 1), (0, 0)\}, \\ &T_2{:}\{(-1, 0), (0, 1), (1, 0)\}, \end{aligned}$$

pic. 2.06

T_3:{(−3/4, 1/2), (5/4, 1/2), (1/4, −1/2)},
T_4:{(−3/4, 1/2), (5/4, 1/2), (1/4, 3/2)}.

Substituting these coordinates in the equations of circles T_1^o, T_2^o, T_3^o and T_4^o, we can convince that the condition (2) is carried out and $\mathfrak{T}$ is an L-tiling.

5^0 **(n.n)**. The content of this item is not used further, therefore all information, which is given in it, is given without proofs.

Let $\epsilon > 0$. Then a $\langle 1, 1\rangle$-system Σ' is called *ϵ-close* to a system Σ, if points of the system Σ' is possible to be numbered in such a way, that for any natural i the following inequality is carried out $\rho(A_i, A_i') < \epsilon$, where $A_i \in \Sigma$ and $A_i' \in \Sigma'$.

Theorem 2.4 *For an arbitrary $\langle 1, 1\rangle$-system Σ and any number $\epsilon > 0$ there exists ϵ-close to it a $\langle 1, 1\rangle$-system Σ' such that the tiling $L_{\Sigma'}$ is simplicial.*

The statement "converse" to this theorem is incorrect. That means that there exists such a Delone system Σ with the simplicial tiling L_Σ, that for any number $\epsilon > 0$, there exists close to it a Delone system Σ' which L-tiling is not simplicial.

(We recommend to the reader to construct such example independently.)

Theorem 2.5 *For such an arbitrary system Σ of Delone, where the tiling L_Σ is simplicial, and for any compact set $\mathbf{F} \subset \mathbb{E}^2$ there exists such a number $\epsilon > 0$, that any Delone system Σ', distinguished from the system Σ only on ϵ-neighbourhoods of the compact set $\mathbf{F}$, ϵ-close to Σ, has the simplicial L-tiling.*

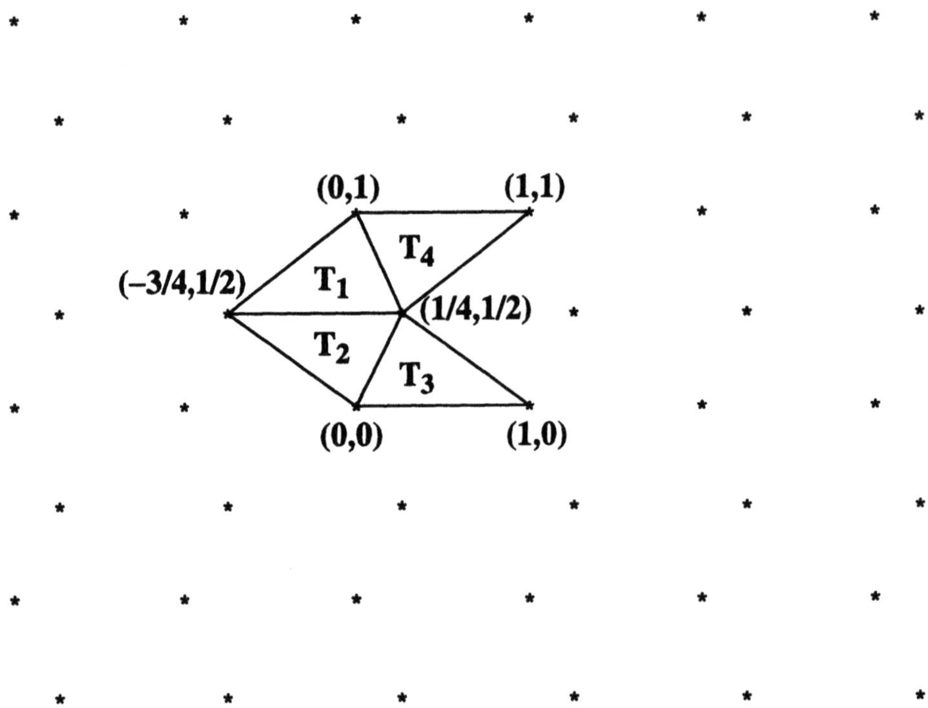

pic. 2.07

§2.2 DV-tiling

1^0. **Definition: 2** Let us be given a $\langle 1, 1\rangle$-system $\Sigma \subset \mathbb{E}^2$. *A Dirichlet-Voronoii domain (DV-domain)* of Σ with center in a point $A \in \Sigma$ is the set DV_A with the property: if $X \in DV_A$, then $\rho(A, X) \leq \rho(B, X)$ for all points $B \in \Sigma$.

The family of all DV-domains of Σ with centers in all points of Σ, we shall denote through $\{DV\}_\Sigma$ (pic. 2.08).

Theorem 2.6 n.n. *For any $\langle 1, 1\rangle$-system, i.e. a system of Delone, $\Sigma \subset \mathbb{E}^2$ and for any point $A \in \Sigma$ the domain DV_A is a closed polygon, which has such diameter and number of edges, which don't exceed those magnitudes, which depending on the basic constants of the system Σ.*

▷ Let us be given a $\langle 1, 1\rangle$-system Σ and a point $A \in \Sigma$. Any point $X \in \mathbb{E}^2$, where $\rho(A, X) > R_\Sigma$, cannot belong to the set DV_A, as in its neighbourhood of radius R_Σ there exists a point $B \in \Sigma$, obviously, different from the point A.

Further, the DV_A-domain is the intersection of closed half-planes, bounded by straight lines, which are orthogonal to the middles of segments $[A, B]$, where $B \in \Sigma - \{A\}$ (pic. 2.08). Let's take notes of the fact, that the half-planes, for which $[A, B] > 2R_\Sigma$, can't take part in formation of the DV_A-domain, as the additions of these half-planes are away from the DV_A-domain on a positive distance. Thus, not more than $[(2R_\Sigma + r_\Sigma)/r_\Sigma]^2$ of half-planes can take part in formation of the DV_A-domain.●

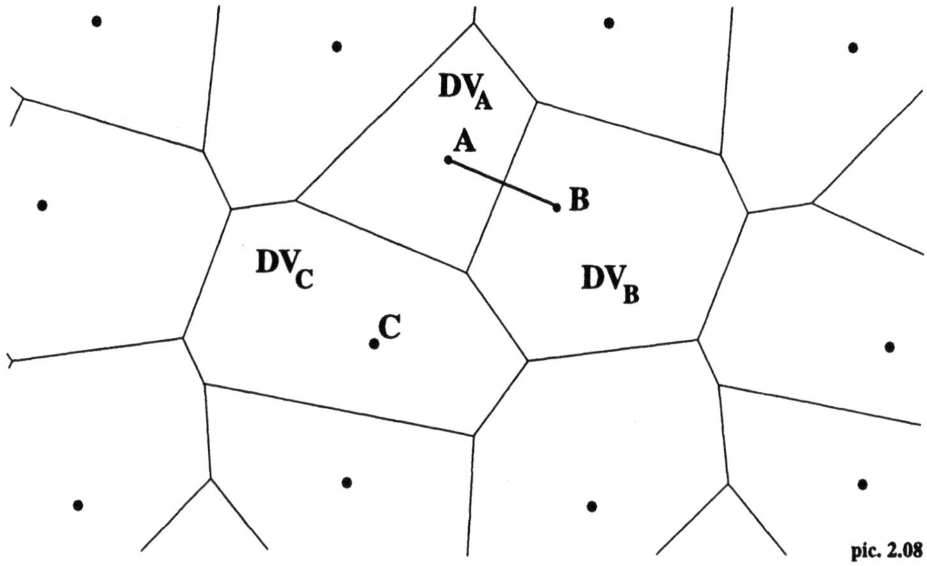

pic. 2.08

Theorem 2.7 *For any convex polygon $M \subset \mathbb{E}^2$ there exists such a $\langle 1, 1\rangle$-system Σ, that $M \in \{DV\}_\Sigma$.*

▷ Let's choose an arbitrary point $A \in \operatorname{int} M$ and reflect it in all edges of the polygon M. Let A' be one of such reflected points, F be the corresponding edge and $\mathcal{P}$ be the carrying straight line of that edge. All points of the plain $\mathbb{E}^2$, which are closer to the point A, than to the point A', together with the straight line $\mathcal{P}$ form one of the half-planes, which determine the polygon M. Thus, the point A and all its mirror images give a $\langle 1, 0\rangle$-system, for which the polygon M satisfies to all conditions of the DV_A-domain's definition.

Let's construct now a $\langle 1, 1\rangle$-system, necessary to us. Let the diameter of M be equal to d. Let's take an arbitrary $\langle 1, 1\rangle$-system, throw out from it all points, which hitting in the disk of radius $3d$ and with center in A and we'll add instead of them the $\langle 1, 0\rangle$-system constructed above.•

Theorem 2.8 n.n. *The family $\{DV\}_\Sigma$ of all DV-domains of an arbitrary $\langle 1, 1\rangle$-system Σ is a locally finite face-to-face tiling. (This tiling is named the Dirichlet-Voronoii tiling or the DV-tiling.)*

▷ The fact, that $\{DV\}_\Sigma$ is a classical tiling, is obvious. Also, it is obvious, any bounded set intersects only a finite number of DV-domains of Σ.

Let now for some $< 1, 1 >$-system Σ the tiling $\{DV\}_\Sigma$ is not face-to-face. Then, there is a domain $DV_A \subset \{DV\}_\Sigma$ and its edge F, such that on interior points of the edge F there adjoin to the DV_A-domain, at least, two DV-domains DV_B and DV_C. But then the segments $[A, B]$ and $[A, C]$, not being segments of the same straight line, are simultaneously orthogonal to the straight line, which contains the edge F, that is impossible.•

We'll mark, that the star of any vertex and any edge of $\{DV\}_\Sigma$ for any $\langle 1, 1\rangle$-system is finite (pic. 2.08). It follows from the local finitness of DV-tilings for such systems.

Theorem 2.9 *Let some $\langle 1, 1\rangle$-system be translated into itself by some motion (of first or second sort), then the DV-tiling of the $\langle 1, 1\rangle$-system will be translated into itself by the same motion.* ▷•

2^0. **Example 1** Let's consider $\mathfrak{T}_\Delta$ (pic. 2.06), and systems Σ_1 and Σ_2 as well. (See §2, item 4^0). It is obvious, that the DV-tiling of Σ_2 is the tiling $\mathfrak{T}_\Delta$. (Discover the L-tiling of Σ_1.)

Example 2 Let's consider the Delone system Υ, constructed in item 3^0, §2 (pic. 2.07). It is obvious, that there exist DV-domains of two and only two kinds: with integer and with fractional centers. Let's construct these domain. The domain $DV_{(0,0)}$ has vertices $(-1/4, 7/16)$, $(1/2, 1/16)$, $(1/2, -1/16)$, $(-1/4, -7/16)$, $(-1/2, -1/16)$, $(-1/2, 1/16)$.

The domain $DV_{(1/4,1/2)}$ has vertices $(1/2, 15/16)$, $(3/4, 9/16)$, $(3/4, 7/16)$, $(1/2, 1/16)$, $(-1/4, 7/16)$, $(-1/4, 9/16)$.

§2.3 Duality of L-tilings and DV-tilings

1^0. Let us be given a face-to-face tiling $\mathfrak{T} = \{T_1, T_2, \ldots\}$ of the plane $\mathbb{E}^2$. We shall denote all faces of the tiling by letters, with upper indexes 0, 1, 2 (dimensionalities of the faces), meaning dimensionality 2 for tiles. Let, further, the letters r and q (may be with indexes) satisfy to the following relation: $r + q = 2$.

Definition: Two face-to-face tilings $\mathfrak{T}$ and $\mathfrak{M}$ of the plane $\mathbb{E}^2$ are called *combinatorially-metric dual*, if there exists the one-to-one correspondence between faces of the tilings $\mathfrak{T}$ and $\mathfrak{M}$, with the following conditions:

1) An arbitrary r-dimensional face of the tiling $\mathfrak{T}$ ($\mathfrak{M}$) corresponds to some q-dimensional face of the tiling $\mathfrak{M}$ ($\mathfrak{T}$).
2) At the specified correspondence the inclusion $T^r \subset T^{r'}$ ($M^q \subset M^{q'}$) entails the inclusion $M^q \subset M^{q'}$ ($T^r \subset T^{r'}$).
3) Each of the two corresponding faces of the tilings $\mathfrak{T}$ and $\mathfrak{M}$ are mutually orthogonal.

(If only two of the first conditions are realized then the face-to-face tilings $\mathfrak{T}$ and $\mathfrak{M}$ are called *combinatorially dual.*)

2^0. **Theorem 2.10 n.2.** *For any $\langle 1, 1\rangle$-system $\Sigma \subset \mathbb{E}^2$ the tilings $\{DV\}_\Sigma$ and $\{L\}_\Sigma$ are combinatorially-metric dual.*

▷ To begin with we'll note, that each vertex of the $\{L\}_\Sigma$-tiling, being a point of Σ, is the centre of some polygon of the $\{DV\}_\Sigma$-tiling. On the contrary, the centre of each polygon of $\{DV\}_\Sigma$, being a point of Σ, is a vertex of $\{L\}_\Sigma$. That means, that vertices of $\{L\}_\Sigma$ and polygons of $\{DV\}_\Sigma$ are in one-to-one correspondence.

Centres of disks, circumscribed around $\{L\}_\Sigma$-tiles and only they have this property: the set of points of Σ, laied from them on the minimum distance, has the rank 2. Therefore, the set of these centres coincides with the set of $\{DV\}_\Sigma$-vertices. That means, that $\{L\}_\Sigma$-tiles and $\{DV\}_\Sigma$-vertices are in one-to-one correspondence.

It is obvious, that the constructed correspondences satisfy all conditions of the duality. Thus, there exists a natural duality of the $\{DV\}_\Sigma$- and $\{L\}_\Sigma$-tilings in zero and maximum dimensions.

Let's consider an arbitrary edge $L^1 \subset \{L\}_\Sigma$, with the vertices A and B. This edge is the common edge of two $\{L\}_\Sigma$-tiles, which we shall denote by L_1 and L_2. Denote the centers of the disks L_1^* and L_2^* (see §2) by O_1 and O_2 (pic. 2.09). According to aforesaid, the points O_1 and O_2 are 0-faces (vertices) of some two tiles of $\{DV\}_\Sigma$. Let's show, that there are the tiles DV_A and DV_B, and the segment O_1O_2 is their common face. The distances from any point of the straight line O_1O_2 up to the points A and B are equal, i.e. the tile DV_A lies in the half-plane, containing the point A, relatively O_1O_2. We shall also note, that medial perpendiculars of segments, which connect the point A with vertices of the tiles L_1 and L_2 (not including A and B), intersect the straight line O_1O_2 in the points O_1 and O_2 correspondingly. And medial perpendiculars of segments, which connect the point A with another L-vertex X, intersect the straight line O_1O_2 outside of the segment O_1O_2. In the opposite case $X \in L_1^* \cap L_2^*$ (pic. 2.09). Hence, the segment O_1O_2 is a face of the tile DV_A. Similarly for DV_B.

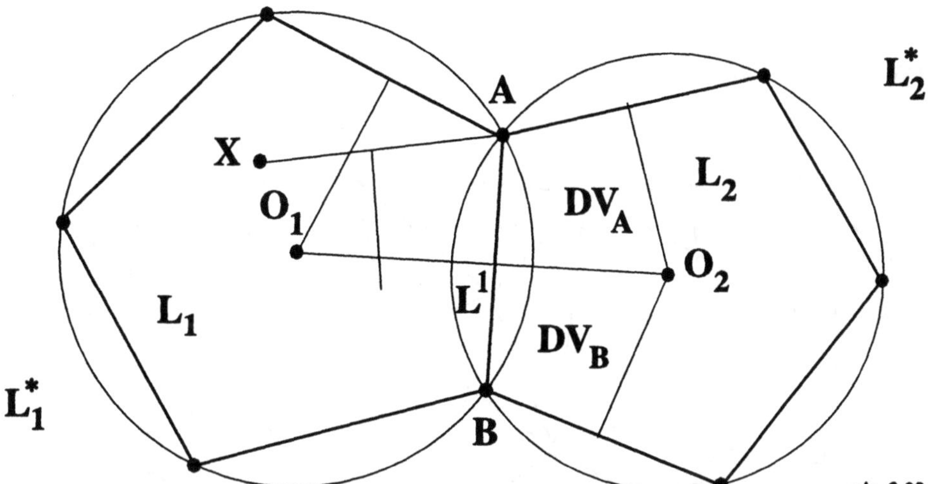

pic. 2.09

Let's correspond to every edge $L^1 \subset \{L\}_\Sigma$ the constructed edge $O_1O_2 \subset \{DV\}_\Sigma$. From the construction of such a correspondence one can see, that for every L-edge (DV-edge) there exists some DV-edge (L-edge), and these edges are orthogonal.

Let's prove, that this correspondence between edges (1-faces) of the L-tiling and edges (1-faces) of the DV-tiling is one-to-one.

Suppose it isn't so. Let to some L-edge there corresponds not one DV-edge, for example, two. But then these edges have the same ends, that reduces to the inconsistency. Let, on the contrary, to some DV-edge there corresponds not one L-edge, for example, two. Hence, these two edges lie on one straight line. That also is incorrect.

Thus, the correspondence between 1-dimensional faces satisfy conditions (1) and (3) of the dualities. The condition (2) is checked immediately.

Thus, the tilings $\{DV\}_\Sigma$ and $\{L\}_\Sigma$ are combinatorially-metric dual.•

3^0. As a corollary of the proof of this theorem, we have the following important fact:

Lemma 8 *There exists one-to-one correspondence between L-edges and pairs of adjacent DV-tiles on the corresponding DV-edges. Every L-edge is orthogonal to the straight line, which contains the corresponding DV-edge, and the L-edge is divided by this straight line in half.* ⊳•

It enables us to introduce the following definition:

Definition: (n,·). Every edge of an L-tiling with vertex at a point A and with attached direction from the point A is called a *contiguous vector* of DV_A.

Theorem 2.11 n.n. *Let us be given a Delone system* Σ, *then: if the tiling* $\{L\}_\Sigma$ *is simplicial, then the tiling* $\{DV\}_\Sigma$ *is primitive. Conversely: if the tiling* $\{DV\}_\Sigma$ *is primitive, then the tiling* $\{L\}_\Sigma$ *is simplicial.* (pic. 2.10.)

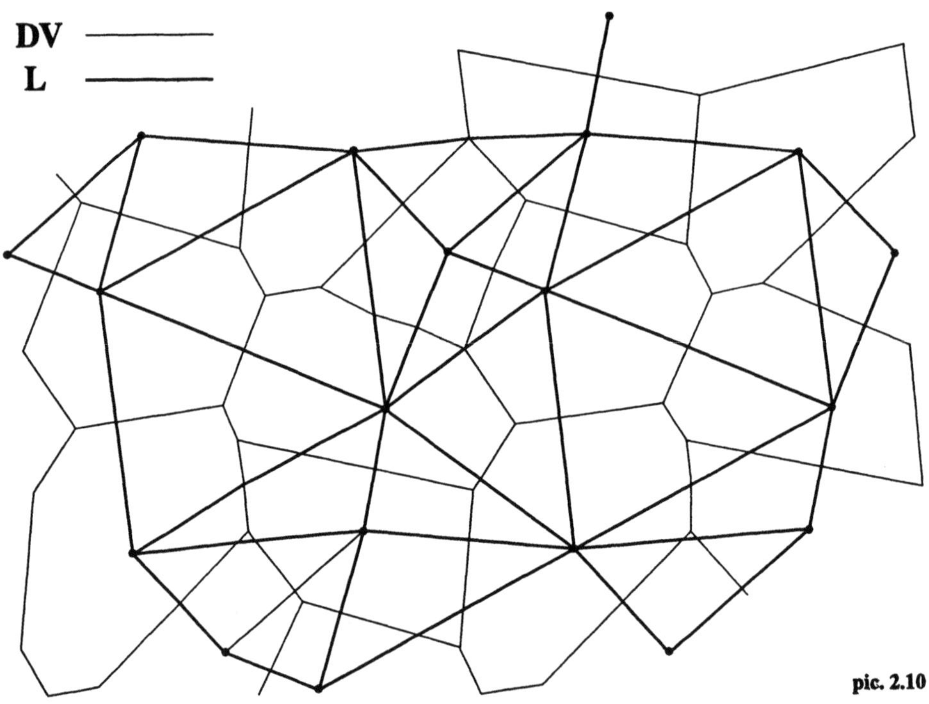

pic. 2.10

⊳ The statement directly follows from the duality of the tilings.•

4^0. See item 5^0 §1.

Theorem 2.12 n.n. *For an arbitrary Delone system Σ and for any $\epsilon > 0$ there exists an ϵ-close to it Delone system Σ' such that the tiling $DV_{\Sigma'}$ is primitive.*

▷ The statement directly follows from theorem 2.4, and also from the duality of tilings. •

Theorem 2.13 n.n. *For an arbitrary Delone system Σ, with the primitive DV-tiling, and for any bounded set $\mathbf{F} \subset \mathbb{E}^2$ there exists $\epsilon > 0$, such that any ϵ-close to Σ Delone system Σ', which differs from the system Σ only on ϵ-neighbourhoods of $\mathbf{F}$, has the primitive DV-tiling.* ▷•

5^0. **Example 1** Let's consider the tiling $\mathfrak{T}_\Delta$ and the systems Σ_1 and Σ_2 (see §2, item 4^0 pic. 2.06). It is obvious, that DV_{Σ_2} is $\mathfrak{T}_\Delta$. (Discover the L-tiling for the system Σ_1.)

Example 2 Let's consider the system of Delone Σ from item 3^0, §2 (pic. 2.07). It is obvious, that there exist DV-domains of two and only two kinds: with integer and with fractional centers. Let's construct these domains, using the correspondence between edges and vertices of the DV- and L-tilings.

The domain $DV_{(0,0)}$ has vertices $(-1/4, 7/16)$, $(1/2, 1/16)$, $(1/2, -1/16)$, $(-1/4, -7/16)$, $(-1/2, -1/16)$, $(-1/2, 1/16)$.

The domain $DV_{(1/4,\,1/2)}$ has vertices $(1/2, 15/16)$, $(3/4, 9/16)$, $(3/4, 7/16)$, $(1/2, 1/16)$, $(-1/4, 7/16)$, $(-1/4, 9/16)$.

3 Lattices

§3.1 Lattices, its Basic Frames

1^0. Let's consider in the plane $\mathbb{E}^2$ an arbitrary frame $\mathcal{E} = \mathcal{E}(O, e_1, e_2)$, i.e. two linearly independent vectors e_1, e_2 with the beginning in the same point $O \in \mathbb{E}^2$, which is convenient to consider coinciding with the origin of the coordinate system (O, ξ_1, ξ_2). See § 1.2.

Definition: The set $\Gamma_{\mathcal{E}}$ of all points, which have integer coordinates in the frame $\mathcal{E}$, i.e. the set of points $g = g_1e_1 + g_2e_2$, where $g_1, g_2 \in \mathbb{Z}$, is called a *point lattice of rank* 2 or a 2-*dimensional point lattice*, or it is called simply a *lattice* (pic. 3.01). (Later we shall denote $\Gamma_{\mathcal{E}}$ through Γ if it will not be important which frame determines the lattice.)

In other way the lattice Γ is the affine image of the integer lattice $\mathbb{Z}^2 = \{(\xi_1, \xi_2) : \xi_1, \xi_2 \in \mathbb{Z}\}$.

Each frame, which determines a lattice Γ, is called a *basic frame* of Γ, and the parallelogram $\Pi(\mathcal{E})$, constructed on any basic frame $\mathcal{E}$ is called a *basic parallelogram* of Γ, we shall denote its area through $S(\mathcal{E})$.

A *vector of a lattice* is any vector equal to a vector which has the beginning and the extremity in points of the lattice, i.e. any vector with integer coordinates in an arbitrary basic frame of the lattice.

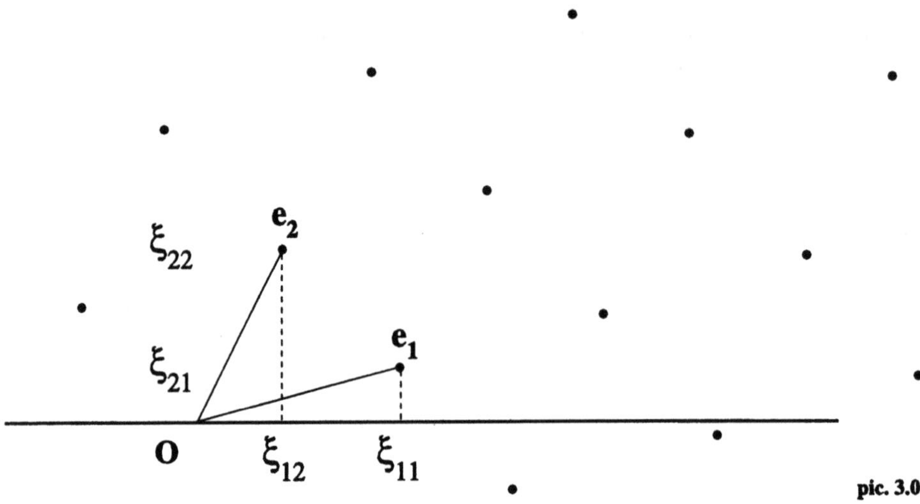

pic. 3.01

We will also consider *lattices of rank* 1, i.e. the set of points $\{ke : k \in \mathbb{Z}\}$, where e is an arbitrary vector with the fixed beginning. We will sometimes call a point a *lattice of rank* 0.

Any subset Γ_1 of a lattice Γ, which is a lattice itself, is called a *sublattice* of Γ.

Theorem 3.1 n.n. *For any lattice $\Gamma_{\mathcal{E}}$ the following three statements hold:*

(1) *Every frame $\mathcal{E}' = (O', e_1, e_2)$, where $O' \in \Gamma$, is basic for* Γ (pic. 3.02).
(2) *The lattice is symmetric with respect to each of its points.*
(3) *The lattice is symmetric with respect to the middle of every segment which joins two of its arbitrary points* (pic. 3.03).

$\triangleright$ Let's prove (3). Let's consider any two points P and Q of the lattice Γ with coordinates (p_1, p_2) and (q_1, q_2) in the frame $\mathcal{E}$. Then the coordinates of the middle R of PQ are equal to $(\frac{p_1+q_1}{2}, \frac{p_2+q_2}{2})$. Let's take now an arbitrary point $X \in \Gamma$ with coordinates (x_1, x_2), then, as one can easily see, the point symmetric to it with respect to R will have the coordinates $(p_1 + q_1 - x_1, p_2 + q_2 - x_2)$, i.e. integer coordinates in the frame $\mathcal{E}$. The statement (3) is proved.

Hence, if $P = Q$, we obtain the proof of statement (2).

The statement (1) is obvious.$\bullet$

Corollary of the statement (1). *Every lattice is a translationly regular disposition of points.*

2^0. **Lemma 1 n.n.** *An integer matrix L is unimodular iff the matrix L^{-1} reciprocal to L is integer.* $\triangleright\bullet$

The set of all integer unimodular (2×2)-matrices is usually denoted by $GL(2, \mathbb{Z})$.

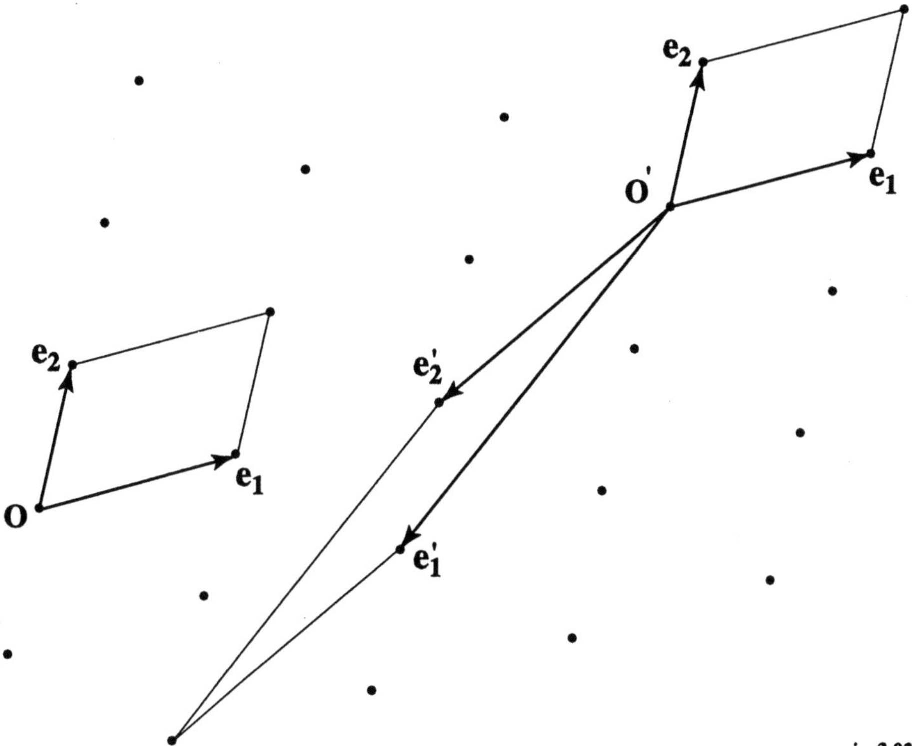

pic. 3.02

Theorem 3.2 n.n. *Let a lattice Γ be determined by a frame $\mathcal{E} = \mathcal{E}(O, e_1, e_2)$. A frame $\mathcal{E}' = (O, e_1', e_2')$, where e_1', e_2' are vectors of Γ, is a basic frame of the lattice Γ iff the matrix of transformation of the frame $\mathcal{E}$ into $\mathcal{E}'$ belongs to $GL(2, \mathbb{Z})$.*

▷ Let the frame $\mathcal{E}'$ be a basic frame of Γ (pic. 3.02). Let's consider the integer matrix L, which rows are the coordinate rows of the vectors e_1', e_2' in the frame $\mathcal{E}$. The rows of the reciprocal matrix L^{-1} are the coordinate rows of the vectors e_1, e_2 in the basic frame $\mathcal{E}'$. That means that L^{-1} is also integer. But then from lemma 1 it follows, that $L \in GL(2, \mathbb{Z})$.

If the matrix of transformation of the frame $\mathcal{E}$ into $\mathcal{E}'$ is integer and unimodolar, it is easy to check up, that each point of Γ has integer coordinates in the frame $\mathcal{E}'$, i.e. the frame $\mathcal{E}'$ is basic one.•

3^0. Let vectors e_1 and e_2 of an arbitrary frame $\mathcal{E} = \mathcal{E}(O, e_1, e_2)$ have coordinates (ξ_{11}, ξ_{21}) and (ξ_{12}, ξ_{22}) accordingly in the coordinate system (O, ξ_1, ξ_2). See pic. 3.01. Then with the frame $\mathcal{E}$ two important matrices are brought in line: the *coordinate matrix* Ξ and the Gramian matrix A. That is

$$\Xi = \left\| \begin{matrix} \xi_{11} & \xi_{12} \\ \xi_{21} & \xi_{22} \end{matrix} \right\|$$

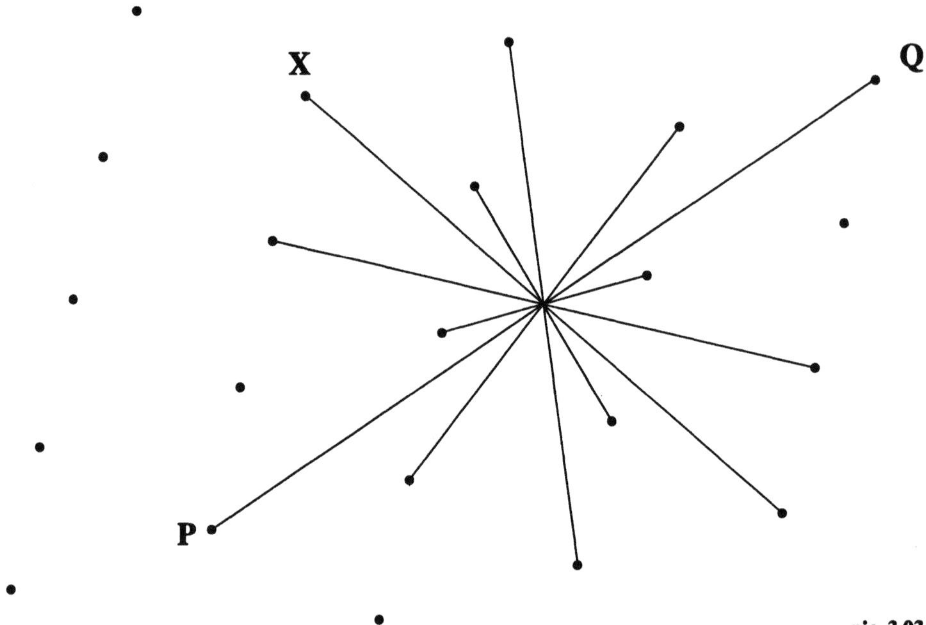

pic. 3.03

and

$$A = \begin{Vmatrix} a_{11} & a_{12} \\ a_{21} & a_{22} \end{Vmatrix} = \begin{Vmatrix} (e_1, e_1) & (e_1, e_2) \\ (e_1, e_2) & (e_2, e_2) \end{Vmatrix}.$$

As one can easily see, equality $A = \Xi' \Xi$ holds for the Gramian matrix $A = \|a_{ij}\|$ of the frame $\mathcal{E}$.

Let's note that $S_{\mathcal{E}} = |det \Xi| = \sqrt{det A}$.

Corollary 1 n.n. *of theorem 3.2. Areas of basic parallelograms of any basic frames are equal.*

⊳ In the used notation we have:

$$S_{\mathcal{E}'} = |det \Xi'| = |det(\Xi L^t)| = |det \Xi||det L| = |det \Xi| = S_{\mathcal{E}}.\bullet$$

4^0. Let $\mathcal{E} = \mathcal{E}(O, e_1, e_2)$ be a frame. The frame $\mathcal{E}^{(-1)} = \mathcal{E}(O, e_1^{(-1)}, e_2^{(-1)})$, which vectors satisfy the correlations $(e_i, e_i^{(-1)}) = 1, (e_i, e_j^{(-1)}) = 0$, with $i \neq j$, is called the *reciprocal frame* to $\mathcal{E}$.

For the frame $\mathcal{E}$ not only the reciprocal frame, but also the *adjoint* frame $\mathcal{E}^* = \mathcal{E}(O, e_1^*, e_2^*)$ is considered, with $e_i^* = (det A)e_i^{(-1)}$, where A is the Gramian matrix of $\mathcal{E}$.

From the equality

$$det A^* = (det A)^2 det A^{-1} = (det A)$$

it follows, that

$$S(\mathcal{E}) = [S(\mathcal{E}^{-1})]^{-1} = S(\mathcal{E}^*).$$

Let's note one of aspects of difference between reciprocal and adjoint frames.

1) Lengths of the vectors $e_1^{(-1)}$, $e_2^{(-1)}$ of the reciprocal frame $\mathcal{E}^{(-1)}$ depend on a standard of length, **(n.n)**.

The following example makes it clear. Let the length of e_1 be equal to 2 cm, and let the angle between e_1 and $e_2^{(-1)}$ be equal to $\frac{\pi}{3}$, then the length of $e_2^{(-1)}$ is equal to 1 cm. But if we primary measure lengths of vectors in millimeters, length of e_1 will not vary, and the length of $e_2^{(-1)}$ will be equal to $\frac{1}{10}$ mm, that is obviously not equal to 1 cm.

2) Lengths of the vectors e_1^*, e_2^* of the adjoint frame $\mathcal{E}^*$ do not depend on a standard of length, **(2.2)**. See pic. 3.04.

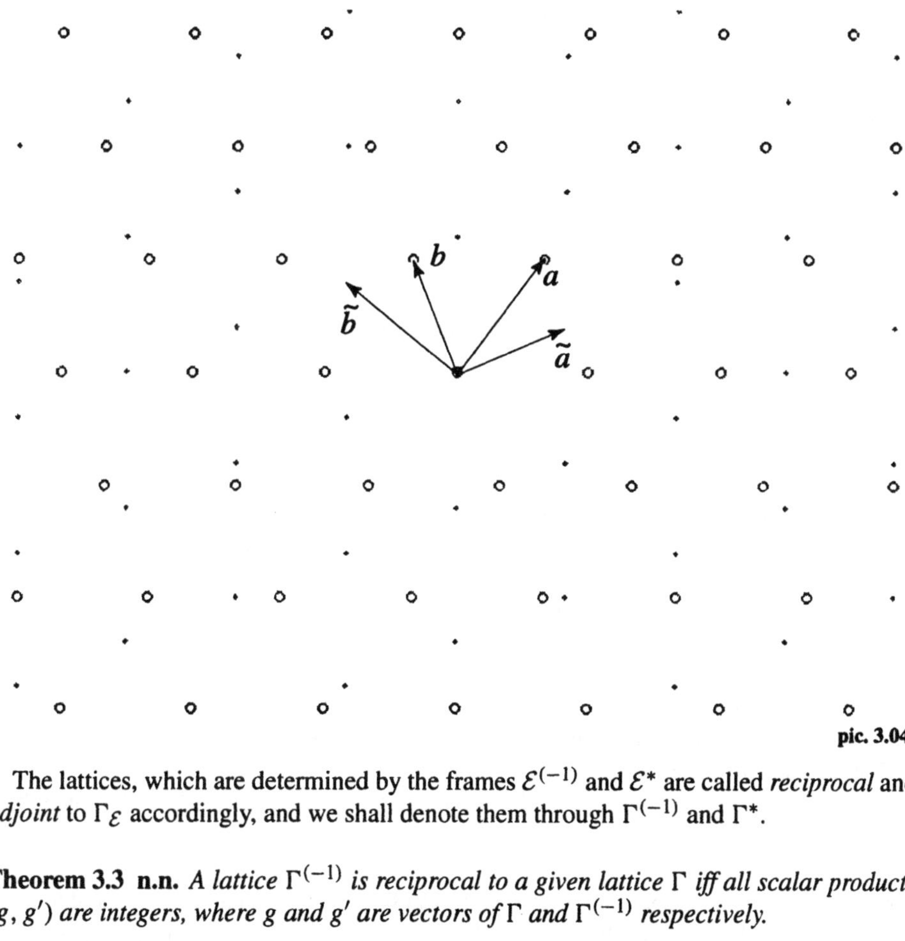

pic. 3.04

The lattices, which are determined by the frames $\mathcal{E}^{(-1)}$ and $\mathcal{E}^*$ are called *reciprocal* and *adjoint* to $\Gamma_{\mathcal{E}}$ accordingly, and we shall denote them through $\Gamma^{(-1)}$ and Γ^*.

Theorem 3.3 n.n. *A lattice $\Gamma^{(-1)}$ is reciprocal to a given lattice Γ iff all scalar products (g, g') are integers, where g and g' are vectors of Γ and $\Gamma^{(-1)}$ respectively.*

This theorem gives the arithmetical interpretation of a reciprocal lattice. As we won't need it further, we are leaving its proof for the reader.

§3.2 Frames, Lattices and Positive Definite Quadratic Forms (PQF)

1^0. Here, as well as everywhere further, we keep all designations of item 3^0 of §1.

Let x_1, x_2 be the coordinates of any point X with respect to a frame $\mathcal{E}$. Then the square of the distance between the origin and X is equal to

$$\begin{aligned} |OX|^2 &= (x_1e_1 + x_2e_2, x_1e_1 + x_2e_2) \\ &= a_{11}x_1^2 + 2a_{12}x_1x_2 + a_{22}x_2^2 = f_{\mathcal{E}}(x_1, x_2) = f_{\mathcal{E}}. \end{aligned}$$

As the square of the distance between distinct points is positive, the quadratic form $f_{\mathcal{E}}$ is a positive definite quadratic form (PQF). The form $f_{\mathcal{E}}$ is called the *metric form* of $\mathcal{E}$. Let's note, that the matrix of $f_{\mathcal{E}}$ is the Gramian matrix of $\mathcal{E}$.

Theorem 3.4 n.n. *Every PQF of two variables determines (correct to motion) a frame of the plane* $\mathbb{E}^2$.

▷ Let's consider PQF $f = f(x_1, x_2)$. It is known from the linear algebra, and for $n = 2$ it is proved trivially, that such a form f can be decomposed by infinite number of modes to a sum of two squares of independent linear forms of variables x_1, x_2. Let us have one of such decompositions:

$$f = (\xi_{11}x_1 + \xi_{12}x_2)^2 + (\xi_{21}x_1 + \xi_{22}x_2)^2$$

Then the form f can be presented in the following manner

$$\begin{aligned} f &= \| \xi_{11}x_1 + \xi_{12}x_2 \quad \xi_{21}x_1 + \xi_{22}x_2 \| \left\| \begin{matrix} \xi_{11}x_1 + \xi_{12}x_2 \\ \xi_{21}x_1 + \xi_{22}x_2 \end{matrix} \right\| \\ &= (x_1e_1 + x_2e_2, x_1e_1 + x_2e_2), \end{aligned}$$

where e_1 and e_2 are linearly independent vectors with coordinates (ξ_{11}, ξ_{21}) and (ξ_{12}, ξ_{22}), i.e. after a choice of the point O, the frame $\mathcal{E} = \mathcal{E}(O, e_1, e_2)$ is determined by the vectors e_1, e_2.

The metric of $\mathcal{E}$ does not depend on a choice of form's decomposition to a sum of squares and a choice of O. Really, lengths of the frame's vectors and their scalar product are simply determined by the coefficients of the form f. Compatibility of any two frames with identical metric by a motion is obvious.•

Let's note that theorem 3.4 makes clear the geometric sense of various decompositions of PQF in a sum of squares.

Let's also note that theorem 3.4 establishes the one-to-one correspondence between PQF and frames given correct to motion.

Definition: PQF, corresponding to the reciprocal frame $\mathcal{E}^{(-1)}$ is called *reciprocal* to PQF $f_{\mathcal{E}}$ and is denoted $f^{(-1)}$, i.e. the quadratic form determined by the reciprocal matrix $A^{(-1)}$ for a given matrix A of PQF f, which is appropriate to a frame $\mathcal{E}$. PQF f^*, corresponding to $\mathcal{E}^*$ (adjoint frame to $\mathcal{E}$) is called *adjoint* to $f_{\mathcal{E}}$.

2^0 It is obvious, that it is possible to determine a lattice Γ both with the help of its basic frame $\mathcal{E}$ and (correct to motion) by PQF $f_{\mathcal{E}}$. Thus, from theorem 3.4 one that there exists a one-to-one correspondence between PQF and congruence classes of lattices. In spite of this to the forms f^{-1} and f^*, i.e. reciprocal and adjoint to f, there correspond congruence classes of the reciprocal $\Gamma_{f^{-1}}$ and adjoint Γ_{f^*} lattices to Γ_f.

Let A and A' be the Gramian matrices of arbitrary basic frames $\mathcal{E}$ and $\mathcal{E}'$ of a lattice Γ. Let $f(X)$ and $f'(X')$, where $X = (x_1, x_2)^t$, $X' = (x_1', x_2')^t$, the metric forms of the frames $\mathcal{E}$ and $\mathcal{E}'$. Let, at last, as in theorem 3.2, L be the matrix of the transformation of the frame $\mathcal{E}$ into $\mathcal{E}'$. Then, obviously, we have $X' = (L^{(-1)})^t X = L^* X$ and

$$A' = LAL^t. \tag{3.1}$$

Let's consider PQF f and f' determined by matrices A and A'.

Definition: PQF f and f' (the matrices A and A') are called *integral unimodular equivalent* ($f \sim f'$, $A \sim A'$), if there exists a matrix $L \in GL(\mathbb{Z}, 2)$ for which (3.1) holds.

Sometimes we will denote the form f' through Lf. Further we shall omit the words "integral unimodular".

As far as we had seen here (see also theorems 3.2 and 3.4), every lattice corresponds to an equivalence class of PQF, there exists a one-to-one correspondence between equivalence classes of PQF and congruence classes of lattices.

There immediately appears the problem of a choice in a lattice of its one definite basic frame. This problem is solved in the so-called theory of reduction to which the following paragraph is devoted.

§3.3 Theories of Reduction. Minimal Vector. Lagrange Reduction

1^0. A quadratic form f (the frame $\mathcal{E}_f$), chosen on the basis of the special requirements ("conditions of reductions") from an equivalence class of forms (frames) (generally speaking, not necessary positive) is called the *reduced* form (frame). An algorithm of such a choice is called an *algorithm of reduction*. As a class of equivalence $\{f\}$ is usually determined by its some representative – a form f_0 (frame $\mathcal{E}_{f_0}$), an algorithm of reduction consists in that fact, that beginning from any form f_0 to discover the reduced form $f \sim f_0$. According to one or another condition of reduction there are known various partial theories of reduction.

Though the requirement of uniqueness of the reduced form (frame) for a given equivalence class is desirable, in used modes of reduction a finite number of the reduced forms (frames) is supposed for some equivalences classes. In such cases with every equivalence class the whole population of the reduced forms (frames) is brought in the correspondence.

The first theory of reduction for PQF of two variables was constructed only in 1768 by Lagrange, we give its geometric variant in 3^0.

Gauss also investigated the problem of reduction, however the problem of reduction for PQF of three variables was solved only by Zeeber in 1831. Just in the review for this Zeeber's paper Gauss introduced the concept of a lattice for the first time.

Historically the first developed algorithm of reduction, suitable for PQF of any number n of variables, was offered by A.N. Korkin and E.I. Zolotarev.

The problem of reduction is one of the primary problems in the theory of quadratic forms, and to the present time besides the mode of A.N. Korkin and E.I. Zolotarev, there are now available some modes of reduction (n-dimensional, though constructed in details only for the first few values of n): Hermite – Minkowski, Zelling – Sharve, Voronoi reductions. The method of Delone reduction is popular in the cristallography. Rather general mode of reduction, enveloping as special cases some of those mentioned above, was offered by Venkov.

2^0. Further, when speaking about vectors of a lattice, we shall consider, that their beginning is the origin O (the origin of a lattice).

Lemma 2 n.n. *In every lattice Γ for any $R > 0$ there exists only a finite number of vectors with the lengths not more than R.*

▷ Let's take a disk D of radius R with center in O and the affine transformation of Γ into the lattice $\mathbb{Z}^2$ (pic. 3.05). The image of the disk D at this transformation is the set bounded by an ellipse, which obviously contains only a finite number of points.•

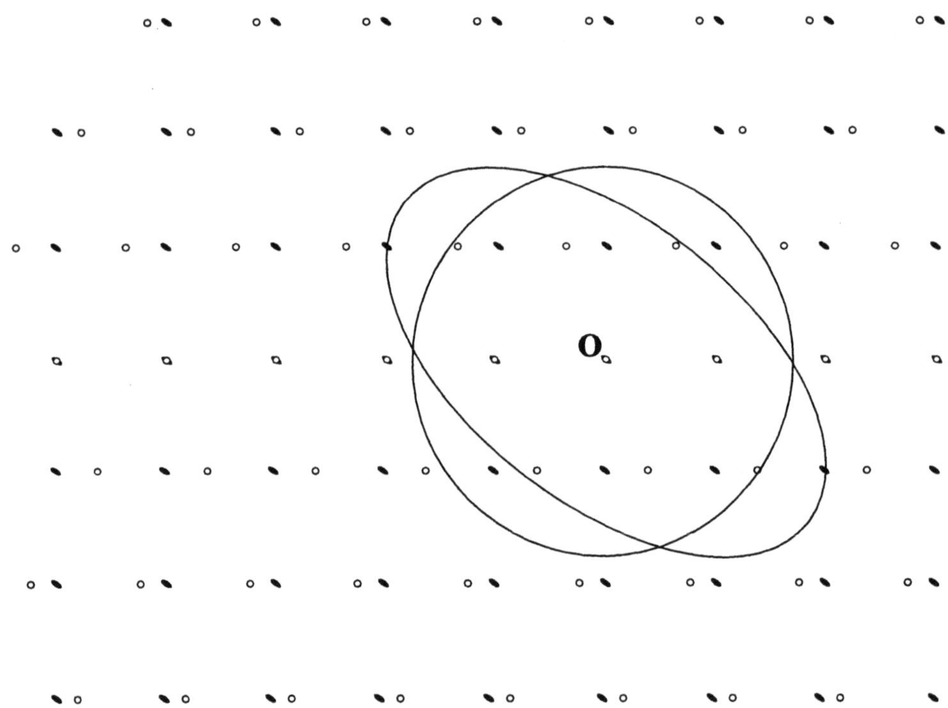

pic. 3.05

Definition: A *minimal* vector of a lattice is its vector which has the minimal positive length.

Lemma 3 *In every lattice $\Gamma \in \mathbb{E}^2$ there exist minimal vectors* (**n.n**), *and their number is not more then 6. This quantity of minimal vectors can be only at the lattice constructed on the regular triangle* (**2.2**).

▷ The existence of minimal vectors, obviously, follows from lemma 2. Let's consider a lattice $\Gamma \in \mathbb{E}^2$ with the beginning in O and also the circle C of radius m, which is equal to the length m of a minimal vector, with the center in O. All vectors with beginnings in O and extremities in the points of the intersection $C \cap \Gamma$ are minimal and their length is equal to m. So, distances between the extremities of any such two vectors not smaller than m, since their residuals is vectors of the lattice. That means, that the number of points in the intersection $C \cap \Gamma$ cannot be more then 6.

The last statement is already obvious.•

3^0. **Definition:** A basic frame $\mathcal{E} = \mathcal{E}(O, e_1, e_2)$ of an arbitrary lattice Γ is called *Lagrange reduced*, if the orthogonal projection of every vector e_i on e_j does not exceed $\frac{|e_j|}{2}$, where $i, j = 1, 2; i \neq j$ and the angle between e_1 and e_2 is not obtuse.

We shall call the set $\Gamma_k^1 = \{ke_2 + le_1\}$, where $k \in \mathbb{Z}$ is fixed and l runs over $\mathbb{Z}$, as k-series of Γ with respect to a given frame $\mathcal{E}$.

Lagrange algorithm: We shall consider a lattice Γ (pic. 3.06) with a basic frame $\mathcal{E} = \mathcal{E}(O, e_1, e_2)$. Let $|e_1| \leq |e_2|$, otherwise we shall change their indexing. Let's choose a vector in the series Γ_1^1 such that the orthogonal projection of it on e_1 has length not more than $\frac{|e_1|}{2}$. If this projection is vectored as the vector e_1, we shall denote the chosen vector through e_2' and if it is not vectored as e_1, we shall denote the opposite vector to chosen through e_2'.

Two cases are possible:

1) $|e_1| > |e_2'|$ and 2) $|e_1| \leq |e_2'|$.

In case 1) we come back to the beginning of the algorithm, by taking for the new vector e_1 – the vector e_2' and for the new vector $e_2 - e_1$.

Lemma 4 2.2. *In case* 2) *the vector* e_1 *is a minimal vector of* Γ, *and* e_2' *is one of (not more, than* 4) *the shortest vectors not collinear to* e_1. *Length of the orthogonal projection of* e_1 *on* e_2' *is not more, than* $\frac{|e_2'|}{2}$.

▷ Let's denote the distance between 0- and 1-series, i.e. between the straight lines, carrying these series Γ_0^1 and Γ_1^1 through h. We shall estimate the magnitude $2h$. We have:

$$|e_2'|^2 - h^2 = \frac{|e_1|^2}{4} < \frac{|e_2'|^2}{4}, \qquad 2h > \sqrt{3}|e_2'| > |e_2'|.$$

From here we have, that there are no vectors shorter, than e_2' in k-serieses, with $|k| > 1$. And in k-serieses, with $|k| = 1$ it is obvious, that there cannot be more then 4 vectors, which have length equil to $|e_2'|$. The last statement of the theorem is already obvious.•

Thus, in case 2), the frame $\mathcal{E}(O, e_1, e_2')$ is the Lagrange reduced frame.

Theorem 3.5 *The algorithm of Lagrange is finite.*

▷ On each next step of the algorithm we take a vector e_1 smaller, than we have taken on the previous step. But there are only a finite number of vectors which have lengths smaller than $|e_1|$.•

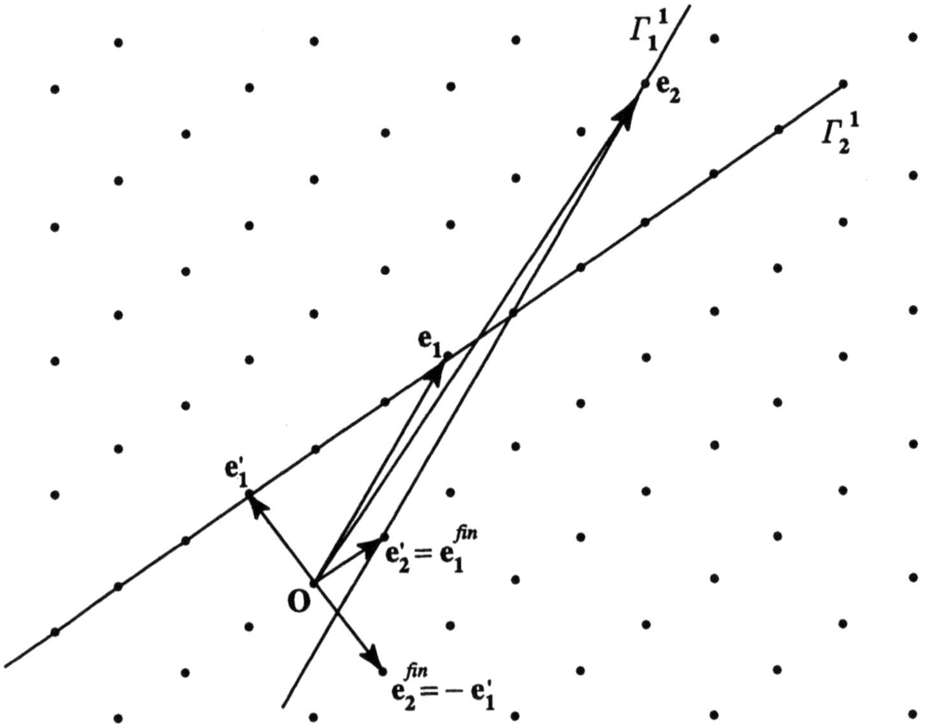

pic. 3.06

Let's note that there can be a few Lagrange reduced frames with the beginning in the origin in a lattice, but number of them is finite and all of them are pairwise congruent. We submit the proof of it and analysis cases to the reader.

Let also note, that the Lagrange algorithm is one of most effective modes of search of minimal vectors of a lattice of rank 2 determined by an arbitrary basic frame.

In inference we shall aduce the definition of two variables Lagrange reduced PQF.

Definition: PQF $f = a_{11}x_1^2 + 2a_{12}x_1x_2 + a_{22}x_2^2$ is called *Lagrange reduced* if $2a_{12} \leq a_{11}, a_{11} \leq a_{22}, a_{12} \geq 0$.

§3.4 "Semiopen" parallelogram. Lemmas of Blichfeldt and Minkowski

1^0. **Definition:** Let us be given a lattice $\Gamma \subset \mathbb{E}^2$, a point $A \in \Gamma$ and a pair $\mathcal{A} = (a_1, a_2)$ of linearly independent vectors of this lattice, with the same beginning in A. A parallelogram $\Pi(A, \mathcal{A})$ determined by the point A and the vectors a_1, a_2 is called a *parallelogram* of Γ.

Rather often, including this paragraph, it will be convenient to consider the parallelogram $\Pi(A, \mathcal{A})$, as a *"semiopen" parallelogram*, i.e. as the following set of points $\{\alpha a_1 + \beta a_2 | 0 \leq \alpha, \beta < 1\}$.

In particular, a "semiopen" basic parallelogram $\Pi(O, \mathcal{E})$ of Γ consists of those and only those points of the plane $\mathbb{E}^2$, which have coordinates (x_1, x_2) in the frame $\mathcal{E}$, satisfying the inequalities $0 \leq x_i < 1$, with $i = 1, 2$.

From the translational regularity of a lattice we infer that it is possible to (disjunctive) divide the whole plane into semiopen parallelograms, i.e. the following theorem takes place:

Theorem 3.6 n.n. *Let us be given a lattice Γ determined by a basic frame $\mathcal{E} = \mathcal{E}(e_1, e_2)$. Then*

$$\mathbb{E}^2 = \bigcup_{A \in \Gamma} \Pi(A, \mathcal{E}).$$

It is the disjunctive tiling by semiopen basic parallelograms $\Pi(A, \mathcal{E})$ (i.e. every point of the plane belongs to one and only one such basic parallelogram).

▷ Let the point $X \in \mathbb{E}^2$ have coordinates (x_1, x_2) in $\mathcal{E}$. Let also $A = ([x_1], [x_2])$, where through [] Antie symbol is denoted. Then X, obviously, belongs to $\Pi(A, \mathcal{E})$ and doesn't belong to any other parallelogram $\Pi(B, \mathcal{E})$, with $B \in \Gamma \setminus \{A\}$.•

We shall denote the tiling obtained in theorem 3.6 through $\mathfrak{T}_{\mathcal{E}}$.

Let's note, that the plane can also be submitted as a tiling $\mathfrak{T}$ with semiopen parallelograms of Γ, constructed on an arbitrary pair (a_1, a_2) of its linearly independent vectors. (For the proof it is sufficient to take the lattice $\Gamma_{\mathcal{A}}$ and the tiling $\mathfrak{T}_{\mathcal{A}}$.)

Lemma 5 n.n. *A semiopen parallelogram $\Pi(O, \mathcal{A})$ of a lattice Γ is a basic parallelogram iff it does not contain any points of the lattice except O.*

▷ The necessity of the condition by virtue of theorem 3.6 is obvious, let's prove the sufficiency of the condition. Let us be given a semiopen parallelogram $\Pi = \Pi(O, \mathcal{A})$, with $\mathcal{A} = \mathcal{A}(a_1, a_2)$, in which there are no points of the lattice, except O. Then we shall take the tiling $\mathfrak{T}_{\mathcal{A}}$ with parallelograms congruent to parallelogram Π. As in the parallelogram Π there are no points of Γ, except one point, and in any other parallelogram of this tiling there are no points of Γ, except its beginning (here we have used translational regularity of the lattices Γ and $\Gamma_{\mathcal{A}}$). That is, in other words, that the whole lattice $\Gamma_{\mathcal{A}}$ consists only of points of Γ. Hence, Π is a basic parallelogram of Γ.•

2^0. Here for the completeness of the account we aduce the famous lemmas of Blichfeldt and Minkowski. They are given without proofs, but one can find them in many references.

Further, we shall consider, that on $\mathbb{E}^2$ there is given a lattice Γ^2 with the area S of a basic parallelogram.

Theorem 3.7 n.n. *(Blichfeldt's lemma). Let M be a set with the area $S(M) > S$* (pic. 3.07). *Then there exists such a parallel transposition $\mathfrak{A}$, that $\#(\mathfrak{A}(M) \cap \Gamma^2) \geq 2$ (i.e. in the set M there will be discovered, at least, one pair of points determine a nonzero vector of Γ^2).*

Blichfeldt is known, perhaps, for this beautiful lemma. His main achievements belong to the theory of packing. Here a really great result belongs to him: he has found the densest lattice packings of equal balls at $n = 6, 7, 8$. Blichfeldt had published the theorem completely already in 1935 and since then nothing for $n > 8$ is known.

Theorem 3.8 n.n. *(Lemma of Minkowski on a convex body). Let $M \subset \mathbb{E}^2$ be a central symmetric convex figure, with the centre at one of points of Γ^2* (pic. 3.08). *Then, if $S(M) > 4S$, M contains, at least, one more point of the lattice.*

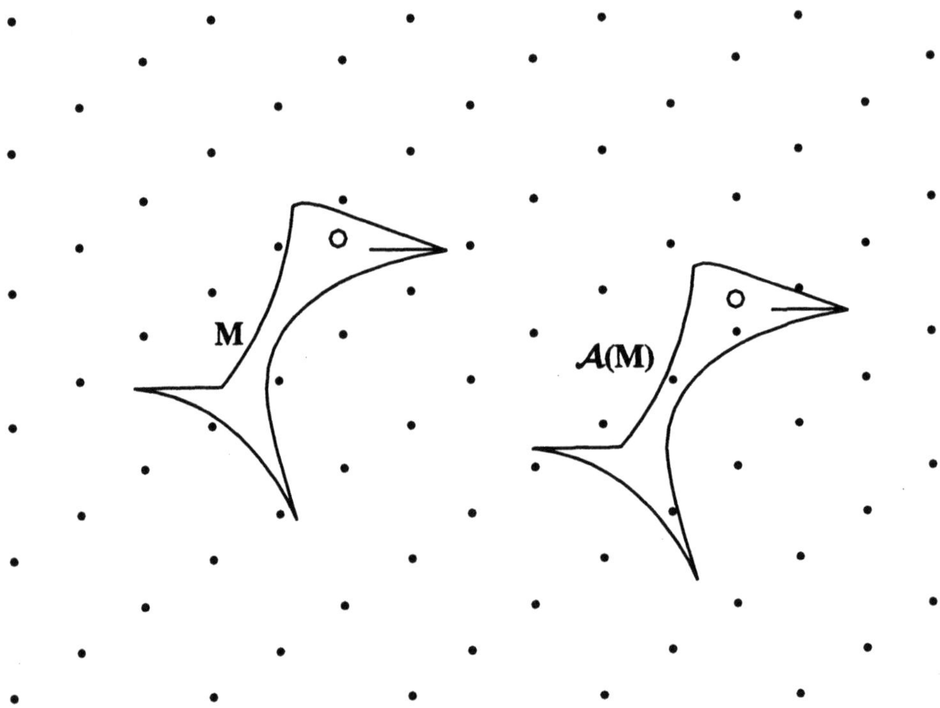

pic. 3.07

§3.5 Characteristic Criteria of Lattices among Point Systems

1^0. From the definition of a lattice it follows that a lattice is a $\langle 1, 1\rangle$-system. The following theorem gives the condition, at which an arbitrary $\langle 1, 1\rangle$-system becomes a lattice, .

Theorem 3.9 n.n. *A translationaly regular* $\langle 1, 1\rangle$*-system is a lattice.*

▷ We shall, at first, prove the theorem in case of a straight line $\mathbb{E}^1$. According to the R-property on any segment with length $2R_\Sigma$ there will be discovered two points of Σ, at which neighbourhoods of radii r_Σ there is no other points of the system according to the r-property.

Let there be no other points of the system between chosen ones. Then the system is a lattice with a basic vector, drawn from the first point to the second. Let there be other points of the system between them. Then according to the r-property there is a finite number of them and the distance between any two of them does not exceed $2r_\Sigma$. Let's take two nearest points and everything will be reduced to the previous case.

Let's consider now the Delone system Σ on the plane. Let's take any two points of the system and draw the straight line l through them. The intersection $\Sigma \cap l$ is an lattice of rank 1. Let's choose the origin as one of points of this lattice. From the R-property it follows, that there exists a point $A \in \Sigma$ not lieing on l (pic. 3.09). Let the parallelogram, constructed on the vector at the origin terminating at A and on the basic vector of the one-dimensional

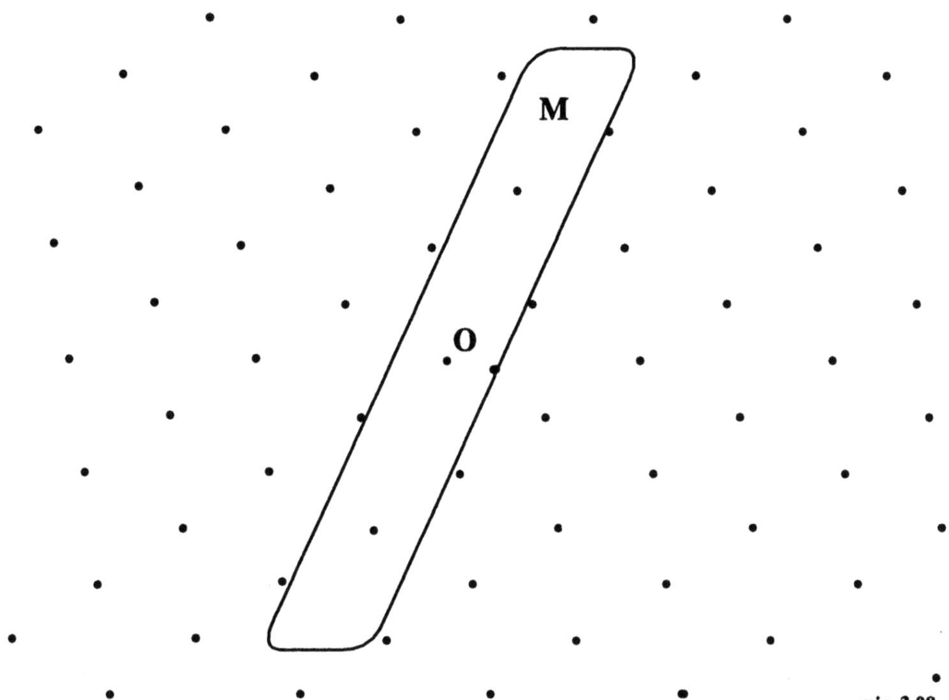

pic. 3.08

lattice, does not contain other points of Σ. Then Σ is obviously a lattice. Let now in our parallelogram there are other points of the system. From the r-property infer that there is a finite number of them. Then we shall choose instead of A another point, nearest to the straight line, among them, we obtain a new parallelogram and come to the first case.•

Each of the two following theorems is an immediate corollary of theorem 3.9.

Theorem 3.10 n.n. *Every translationly regular* $\langle 1, 0\rangle$*-system* $\Sigma \subset \mathbb{E}^2$ *is a lattice of rank* m *with* $m \leq 2$.

Theorem 3.11 n.n. *Every translationly regular* $\langle 1, 1\rangle$*-system (i.e. every translationly regular system of Delone)* $\Sigma \subset \mathbb{E}^2$ *is a lattice of rank* 2.

In inference of this paragraph we shall prove the theorem as kind of inverse to theorems 3.10 and 3.11, which finally establishes the connection between $(\langle 1, 1\rangle)$-systems and lattices.

Theorem 3.12 n.n. *Every lattice* Γ *of rank* m *with* $m \leq 2$ *is a translationly regular* $\langle 1, 0\rangle$*-system. It is a Delone system iff* $m = 2$.

▷ The corollary to theorem 3.1 actually affirms an existence of the number r from the definition of a $\langle 1, 0\rangle$-system for a lattice. Translational regularity of a lattice is also affirmed by the corollary. Thus the first part is proved.

The last statement is obvious.•

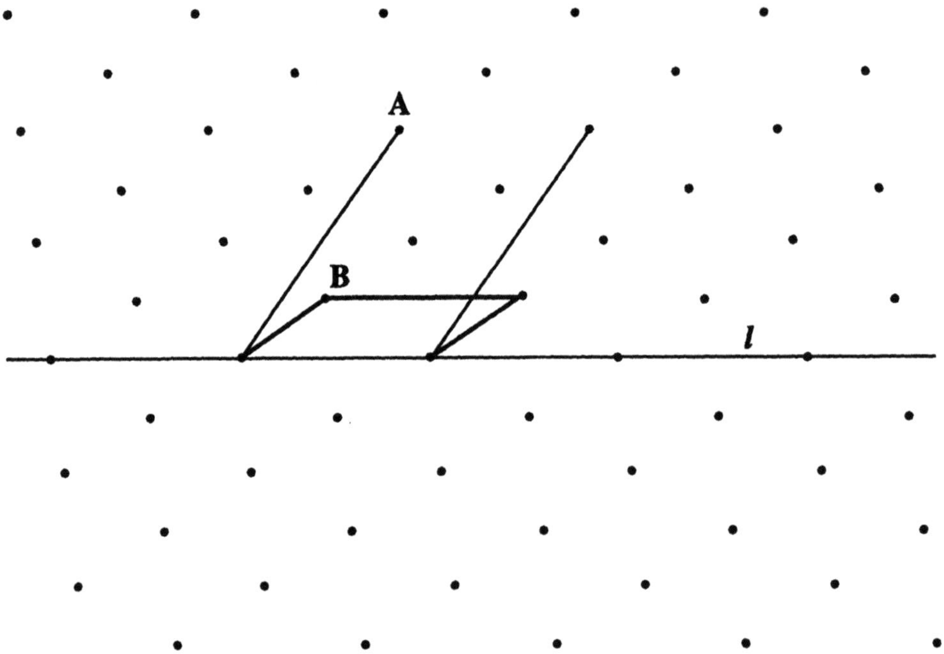

pic. 3.09

4 Parallelohedra

§4.1 DV-tilings and L-tilings of Lattices

Lemma 1 2.2. *No L-polygons of a lattice has obtuse angles.*

▷ Let M be an L-polygon of a given lattice Γ and let it has the obtuse angle between the edges AB and BC, see pic. 4.01. The point D symmetric to B with respect to the middle of AC belongs to the lattice Γ (theorem 3.1) and lies inside the L-disk, whose boundary is determined by the points A, B and C. Hence M is not an L-polygon.•

Lemma 2 *No L-polygons of a lattice can have more than four vertices.*

▷ Any convex polygon, which has more than four vertices, has at least one obtuse angle.•

Theorem 4.1 2.2. *Every L-polygon of a lattice is either an acute triangle or a rectangle.* ▷•

Theorem 4.2 *All L-tilings of lattices of rank 2 are divided in two types of affine equivalence:*

1) *General* (pic. 4.02). *In this type the star of every vertex of an L-tiling consists of* 6 *sequentially cyclically contiguous to each other triangles. And every two adjacent*

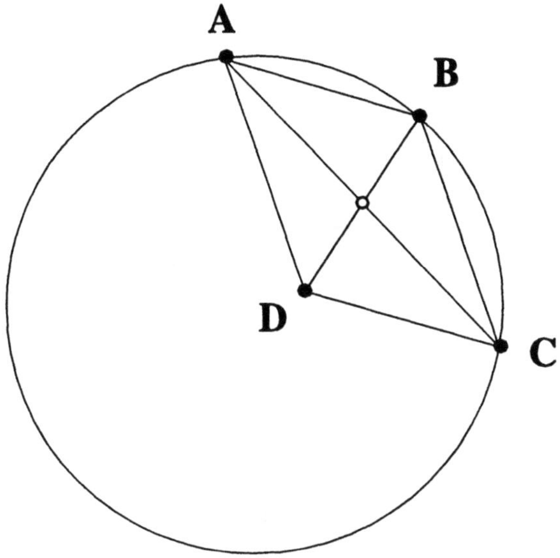

pic. 4.01

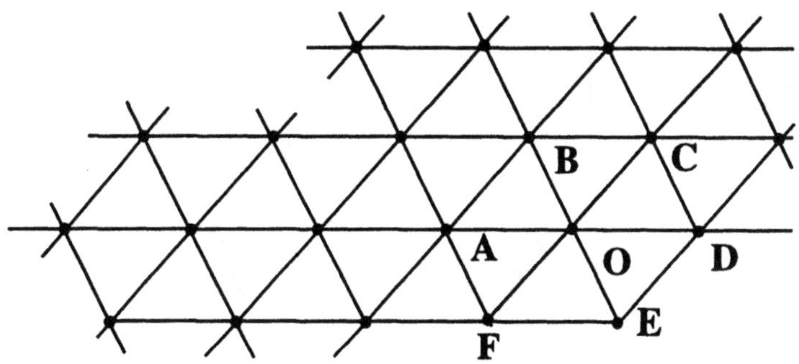

pic. 4.02

triangles are symmetric to each other with respect to the middle of their common edge.

2) *Special* (pic. 4.03). *In this type the star of every vertex of an L-tiling consists of* 4 *sequentially cyclically contiguous to each other rectangles. And every two adjacent rectangles are symmetric to each other with respect to the middle of their common edge.*

▷ Suppose in the L-tiling of some lattice Γ, there can be found at least one triangle. From the translational regularity of a lattice and theorem 3.1 it follows that we can suppose that it is the triangle OAB, see pic. 4.04. From lemma 1 (and theorem 3.1) we infer, that the triangles OAB' and $OA'B$ symmetric to OAB with respect to the middle of OA and OB accordingly are L-triangles. On calculating the sum of the angles we can see, that the points O, A' and B' lie on the same straight line. Therefore, the construction of the L-star of the point O is finished, by the triangles $OB'B''$, $OA''B''$, $OA'A''$ symmetric to the triangles $OA'B$, OAB, OAB' accordingly, with respect to O. From the translational regularity

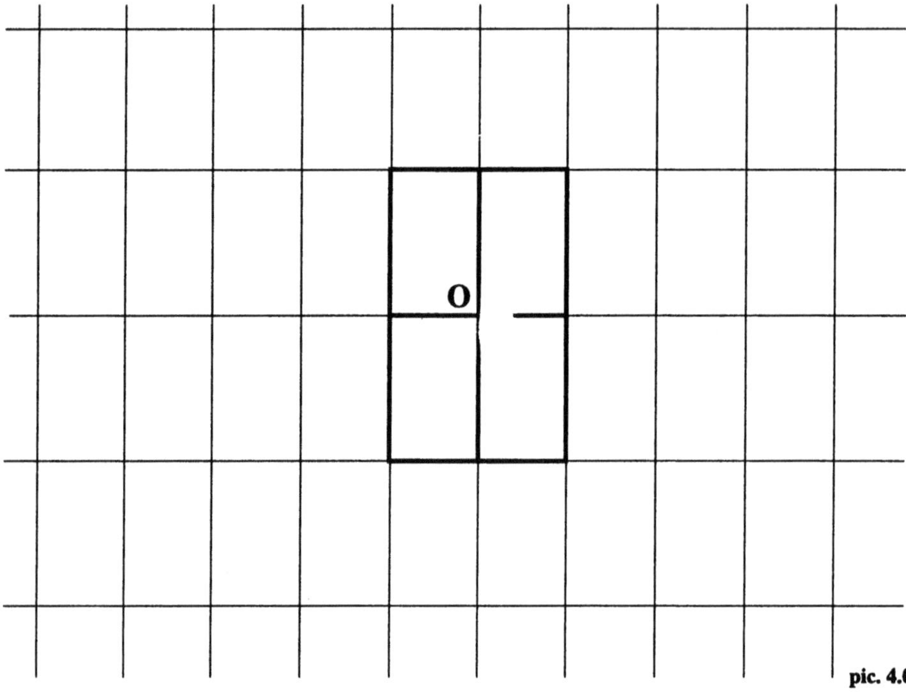

pic. 4.03

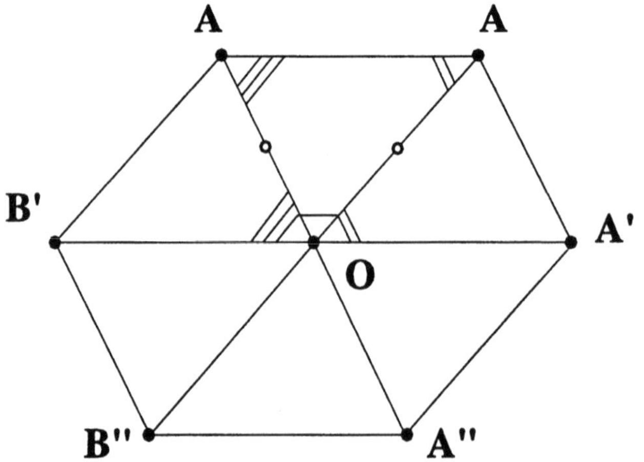

pic. 4.04

and theorem 3.1 it follows, that the L-star of each point of Γ is constructed. These stars, obviously, satisfy to the description of the general type given in the theorem.

Let in the L-tiling of some lattice Γ there are no triangles. Then all L-polygons in this tiling are rectangles. The fact, that the tiling satisfies to the description of the special type given in the theorem, is easy to check, see pic. 4.05.•

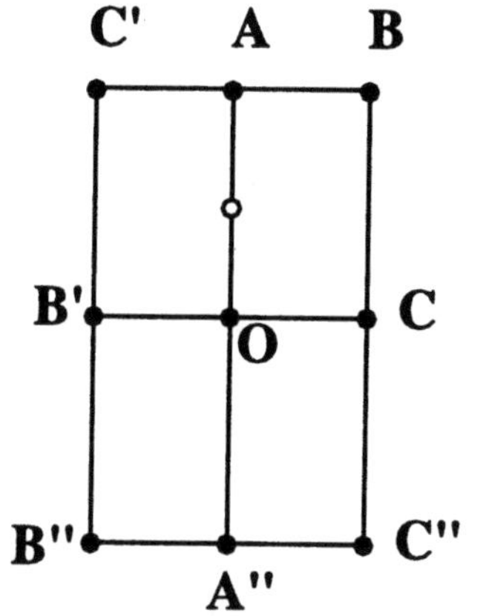

pic. 4.05

Definition: A lattice is called *general*, if its L-tiling has the general type, and it is called *special*, if its L-tiling has the special type.

Theorem 4.3 2.2. *A DV-domain of a lattice is either a rectangle (for special lattices) or is a centrally symmetric inscribed hexagon (for general lattices).*

Conversely. For every rectangle (every centrally symmetric inscribed hexagon) there exists a lattice, for which this figure is a DV-domain.

$\triangleright$ We shall use the duality of DV- and L-tilings. Let's remind, that all DV_O-vertices, with $O \in \Gamma$, are the centres of all circles circumscribed around the polygons of the L-star of the point O.

In the general case we have a hexagon. It is central symmetric, as an L-star is centrally symmetric. It is inscribed, as all triangles of an L-star are pairwise equal. The special case is absolutely trivial.

Conversely. We shall consider an arbitrary convex central symmetric inscribed hexagon **M**, with the vertices U, V, W, X, Y, Z and with the center O (pic. 4.06). Let's denote through A, B, C, D, E, F the points symmetric to O with respect to ZU, UV, VW, WX, XY, YZ.

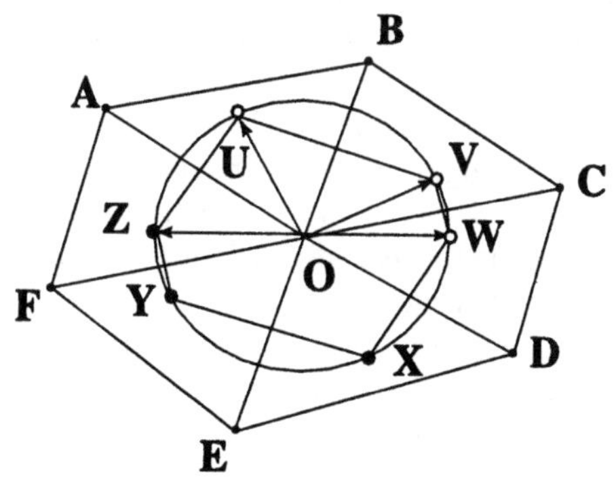

pic. 4.06

Let's note, see pic. 4.06, that $\vec{OA} = \vec{OZ} + \vec{OU}$, $\vec{OB} = \vec{OU} + \vec{OV}$, $\vec{OC} = \vec{OV} + \vec{OW}$ and $\vec{OZ} + \vec{OW} = \vec{0}$, because **M** is a central symmetric hexagon. Hence, $\vec{OB} = \vec{OA} + \vec{OC}$. From here and again from the central symmetry of **M** it follows, that the points A, B, C, D, E, F have the integer coordinates with respect to the frame $\{\vec{OA}, \vec{OC}\}$, i.e. they belong to the lattice Γ, which have the frame $\{\vec{OA}, \vec{OC}\}$ as a basic frame. We shall show, that the triangles AOB, BOC, COD, DOE, EOF and FOA form the L-star of Γ. By transferring this L-star on all vectors of Γ, we shall obtain the triangulation $\mathcal{T}$ of the plane, obviously, compatible with the lattice Γ.

Let's note that each convex central symmetric inscribed polygon, except a rectangle, has all angles obtuse. We leave the proof of this elementary fact to the reader.

Taking into account this note, we have that all angles of the triangles AOB, BOC, COD, DOE, EOF and FOA, as supplementary to the angles of **M**, are acute. From this follows the realization of the necessary and sufficient criterion of L-tiling for the triangulation $\mathcal{T}$. Really to each edge of an acute triangle of $\mathcal{T}$ there adjoins the central symmetric to it triangle of $\mathcal{T}$. It is obvious, that in the disk circumscribed around one of these triangles the third vertex of the other does not get into it.

Thus $\mathcal{T}$ is the L-tiling of Γ. According to the definitions, **M** is the DV-polygon of Γ with the center in the point O.

The case of the special type is trivial.•

§4.2 Enumeration of all parallelohedra on the Plane

Definition: A convex polygon **P** is called a *parallelohedron* iff there exists a tiling $\mathfrak{P}$ in which every tile is parallel congruent to **P**.

If the tiling $\mathfrak{P}$ is face-to-face, then **P** is called *normal*, else *abnormal*. If the tiling $\mathfrak{P}$ is primitive, then **P** is called *primitive*, else *non-primitive*.

Theorem 4.4 2.2. *Each parallelohedron, which is abnormal, is a parallelogram. Each parallelogram can appear as a normal parallelohedron and as an abnormal parallelohedron.*

▹ As the tiling $\mathfrak{P}$ is not face-to-face, there will be discovered an edge AB of a parallelohedron $\mathbf{P_0} \in \mathfrak{P}$, which contains a point C – vertex of some parallelohedra $\mathbf{P}_1, \mathbf{P}_2 \in \mathfrak{P}$, which adjoin to the edge AB on the edges DC and CE accordingly (pic. 4.07a).

For convenience we shall consider, that the segment AB is horizontal, $\mathbf{P}_0$ is located above AB, and $\mathbf{P}_1, \mathbf{P}_2$ are below AB, as it is shown on pic. 4.07a. As to the straight line AB there adjoin the parallelohedron $\mathbf{P}_0$ above and $\mathbf{P}_1, \mathbf{P}_2$ below, each parallelohedron of $\mathcal{P}$ has the "upper" and "lower" horizontal edges (while it is not known, that they have equal length). The parallelohedron $\mathbf{P}_2$ is parallel congruent to $\mathbf{P}_1$. Therefore, the (upper) edge CE of the parallelohedron $\mathbf{P}_2$ is equal to the (upper) edge DC of $\mathbf{P}_1$. Analogically the upper edge of any parallelohedron of our tiling is equal to $DC = CE$.

The edge CF of the parallelohedron $\mathbf{P}_2$, adjoining to the vertex C and different from DC coincides with the edge CF^* of $\mathbf{P}_1$, adjoining to C and different from CE. Firstly they lay on the same straight line, otherwise we can choose such a point X, not belonging to the parallelohedra $\mathbf{P}_1$ and $\mathbf{P}_2$, laying below AB and rather close to the point C, that any parallelohedron of $\mathfrak{P}$ covering it will have common interior points with the polygons $\mathbf{P}_1$

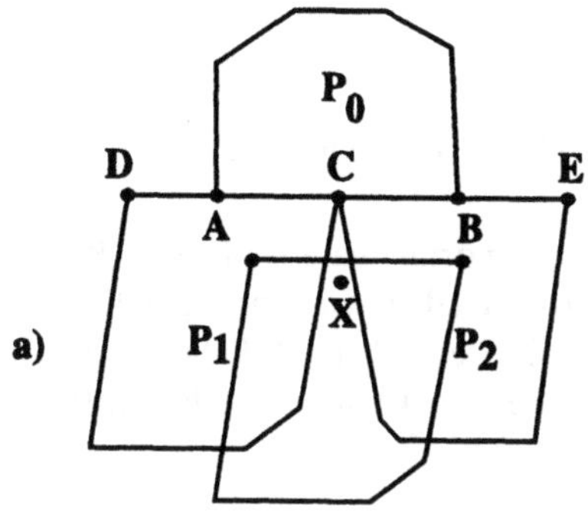

pic. 4.07a

and (or) $\mathbf{P}_2$, see pic. 4.07b. From the same origin it follows that $F = F^*$. From this it follows, that the edges of the parallelohedra $\mathbf{P}_1$ and $\mathbf{P}_2$ adjoining to the vertices D, C and E are parallel and equal to each other.

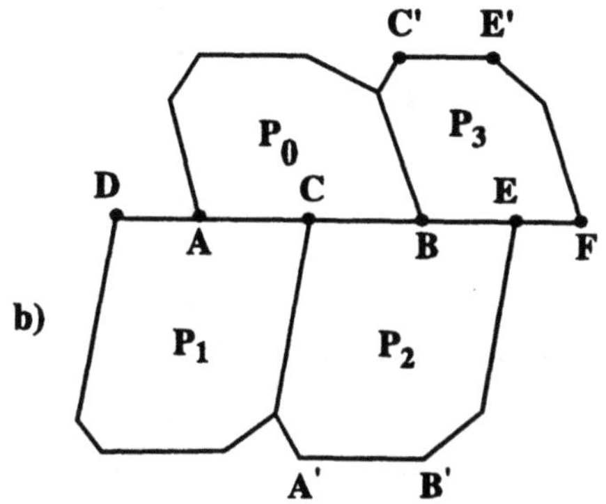

pic. 4.07b

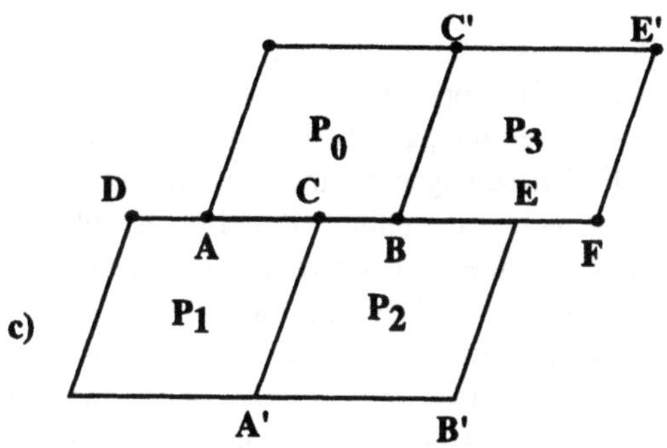

pic. 4.07c

Further, it is obvious, that the lower edges of $\mathbf{P}_1$ and $\mathbf{P}_2$ lie on the same straight line. From here and from the reasoning twice used above (pic. 4.07c) we infer that the point F must lay on the mentioned straight line.

The second statement of the theorem is obvious.•

Theorem 4.5 2.2. *Each normal parallelohedron, which is not a parallelogram, is a centrally symmetric hexagon.*

Each centrally symmetric hexagon is a normal parallelohedron.

⊳ Let's consider an arbitrary normal parallelohedron $\mathbf{P}_0$ and the tiling $\mathfrak{P}$ formed by it. At first, we shall note, that to each edge of a normal parallelohedron there exists, obviously, the edge equal and parallel to it. Let AB and $A'B'$ be the such edges of $\mathbf{P}_0$, pic. 4.08. From the convexity of the parallelohedron we infer that the edges BC and $B'D'$ of $\mathbf{P}_0$ should lie outside of the parallelogram $ABB'A'$.

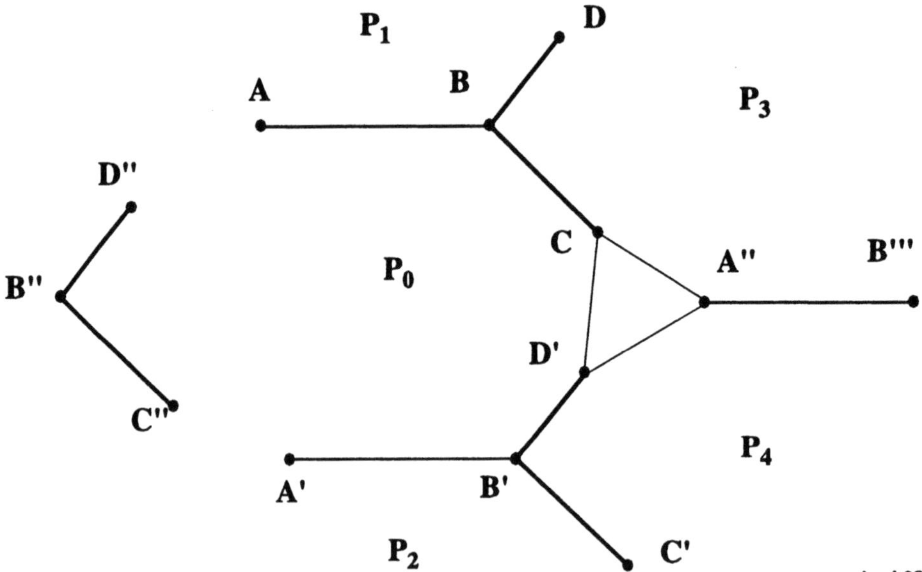

pic. 4.08

The parallelohedron $\mathbf{P}_1$ adjoins to $\mathbf{P}_0$ on the edge AB. Thus there exists one more edge BD, which has B as their end. The appropriate edges $B'D'$ and $B'C'$ are exist also for the point B'. There can be discovered the pair of edges of $\mathbf{P}_0$ accordingly equal and parallel to the edges BC and $B'D'$. These edges, by the convexity of $\mathbf{P}_0$, cannot belong to the sequence of edges of $\mathbf{P}_0$ connecting the points D' and C and which is not contains AB. Let's show, that they should have the common vertex B''. Really, the parallelohedron $\mathbf{P}_3$ contiguous to $\mathbf{P}_0$ on the edge BC, should have the pair of edges, equal and parallel to BD. These two edges should be placed inside the angle DBC, but by the convexity of $\mathbf{P}_3$ they cannot to be placed both strictly inside DBC. Hence, one of them should coincide with BD. Thus there exists the point $B'' \in \mathbf{P}_0$ corresponding to the point $B \in \mathbf{P}_3$.

Secondly, the edges BC and BD ($B'C'$ and $B'D'$) are edges of the same parallelohedron $\mathbf{P}_3$ ($\mathbf{P}_4$), which are adjacent in the point B (B').

Let's consider such a parallel translation $\mathcal{A}$, that $\mathcal{A}(\mathbf{P}_1) = \mathbf{P}_0$ and $\mathcal{A}(\mathbf{P}_0) = \mathbf{P}_2$. It is obvious, that $\mathcal{A}(\mathbf{P}_3) = \mathbf{P}_4$. Thus, the parallelohedra $\mathbf{P}_3$ and $\mathbf{P}_4$ have the common edge $A'''B'''$ equal and parallel to the edge AB.

The triangle $A'''D'C$ is strictly situated in the band between the straight lines AB and $A'B'$, that's why an entire parallelohedron of $\mathfrak{P}$ cannot be contained in it. Hence, each point of $A'''D'C$ belongs, at least, to one of the parallelohedra $\mathbf{P}_0$, $\mathbf{P}_3$ and $\mathbf{P}_4$. But the letter, if all three points A''', D' and C do not coincide among themselves, contradicts to the convexity of the polygons $\mathbf{P}_0, \mathbf{P}_3$ and $\mathbf{P}_4$.

As the points D' and C coincide, the points A and D'' also coinside and also C'' with A'. So the parallelohedron $\mathbf{P}_0$ is a hexagon.

The second statement of the theorem is obvious.•

§4.3 *The Main Theorems about parallelohedra*

After the completed now enumeration of all parallelohedra on the plane, in the two-dimensional case, there become obvious or almost obvious many theorems, which are fair (but are difficult or very difficult to prove) for parallelohedra of any dimensionality.

Theorem of Minkowski n.2. *Every parallelohedron is central symmetric.*

▷ See theorems 4.4 and 4.5.•

Theorem of Venkov n.2. *Every parallelohedron, which is abnormal, can appear as a normal parallelohedron.*

▷ See theorem 4.4.•

Theorem of Voronoi n.2. *Every primitive parallelohedron is affine equivalent to the DV-domain of some lattice.*

▷ According to theorems 4.4 and 4.5 each primitive two-dimensional parallelohedron is a central symmetric (convex) hexagon. We need to prove, that every hexagon of this kind is affine equivalent to inscribed one.

At first, we shall prove, by the convenient method for us, the rather known fact, that it is possible to describe an ellipse around each hexagon of this kind. Let's take four of the six vertices of our hexagon (pic. 4.09), which form the parallelogram $ABDE$ and let's transform this parallelogram into a rectangle, whose edges will be called "vertical" and "horizontal". Our rectangle determines the set of ellipses, which limiting figures are horizontal and vertical bands determined by the edges of the rectangle. The images of the two remaining vertices of the hexagon lie by its convexity in one of these bands and select one ellipse from the constructed set of them. Now the auxiliary statement is obvious.

On transforming this ellipse in a circle, we prove the theorem.•

Note: The similar theorem is proved by Zhitomirskii also for the large class of non-primitive parallelohedra. And though two-dimensional non-primitive parallelohedra do not belong to this class, it is obvious, that each of them is affine equivalent to the DV-domain of some lattice.

The following theorem is a kind of the theorem of uniqueness to the theorem of Voronoii as to the theorem of existence.

Theorem MRS (Michel, Ryshkov, Senechal) n.2. *Let a parallelohedron be affine equivalent to the DV-domain D of some lattice. Then, if this domain D is not direct product*

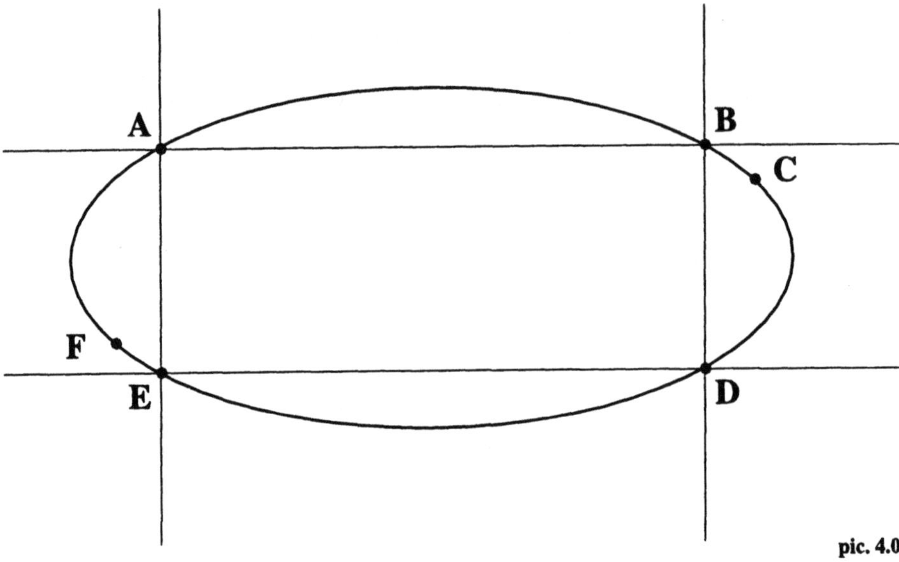

pic. 4.09

of two (or several) DV-domains of lattices of smaller rank, it, correct to a transformation similitude, is unique.

▷ At first, we shall note, that a rectangle is the direct product of two one-dimensional DV-domains (segments) and consequently the theorem is not formulated for it. The appropriate hypothesis is not correct, as there exists the continuum of hyperbolic turns, which transform a rectangle into rectangles with different ratioes of edges' lengthes. So only the case of six-vertixes parallelohedra is to be considered.

So, let us be given a central symmetric (convex) hexagon **M**, for which there exist two affine maps $\mathcal{A}$ and $\mathcal{B}$ in inscribed hexagons. We can, carring out a homothetic transformation if it is necessary, consider that these hexagons are inscribed in the same circle C. Let's carry out the transformation $\mathcal{A}\mathcal{B}^{-1}$. Thus one of inscribed hexagons will transform into another one, and hence, the circle, in which they are inscribed, will transform into itself. But the unique kind of the affine transformation of a circle into itself is a motion.•

5 Density

§5.1 Ray Sets. Mesuring Figures

1^0. **Definition.** A set $M \subset \mathbb{E}^2$ is called a *ray set* with respect to a point $O \in M$ (*centre*), if for any point $A \in M$ the interval OA is contained in M (pic. 5.01).

We shall suppose everywhere that, as soon as we are given a ray set, we are given its centre.

We shall call a set $M \subset \mathbb{E}^2$ a *strongly ray set* with respect to $O \in M$, if there exists such a neighborhood (*radiance neighborhood*) **U** of O, that M is a ray set respectively any point $O' \in \mathbf{U}$.

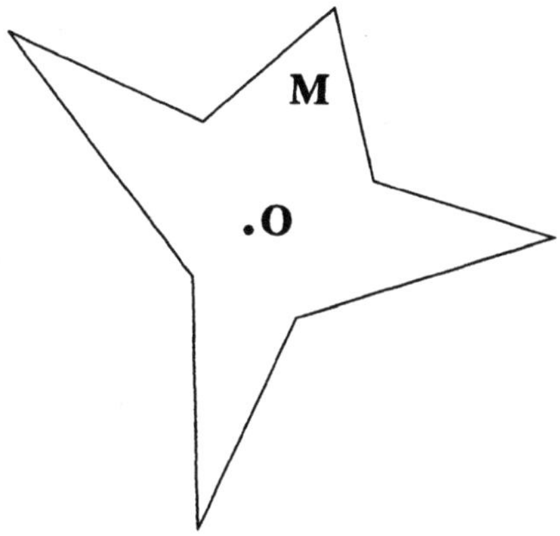

pic. 5.01

Further, we shall consider ray sets of two kinds: bounded closed polygons not necessary convex (with centres, which do not lie on the straight lines containing its edges) and bounded closed convex figures (with centres, which do not lie on its boundary). No other ray sets are considered in this article.

It is obvious, that all ray sets, which we consider, are strongly ray.

We shall denote the set homothetical to M with respect to O through rM, where $r > 0$ is the coefficient of the homothety.

Theorem 5.1 *Let us be given a ray set M with a center O. For an arbitrary number $t > 0$ there exists such $k_t > 0$, that for all $k > k_t$ $\inf \rho(X, Y_1)$ does not depend on k, where $X \in fr(kM)$ and $Y_1 \in fr((k+t)M)$.*

▷ At first we shall note, that $\inf \rho(X, Y_1)$ is the infimum of a continuous function on a compact set. Let $\inf \rho(X, Y_1)$ be reached on points $A \in fr(kM)$ and $B_1 \in fr((k+t)M)$.

We shall proof the theorem (1) for convex figures and (2) for ray polygons separately.

(1) Let M be a (bounded closed) convex figure. We shall note, firstly, that the whole segment AB_1 (except A) lies, obviously, outside of the figure kM. Secondly, the supporting straight line l' to $(k+t)M$ in the point B_1 is unique and it is orthogonal to AB_1, otherwise, see pic. 5.02, by moving the point B_1 we could have received a smaller distance. Thirdly, we shall note that among supporting straight lines to kM at the point A there is the straight line l orthogonal to AB_1, otherwise, see pic. 5.03, kM would intersect l and we could have find in the figure kM a point closer to B_1, than A.

Let's consider now the point $A_1 \in (k+t)M$ corresponding to A by the homothety. There exists the supporting straight line l_1 to $(k+t)M$ in it which is parallel to l. As the segment AA_1 lies outside of kM, the figures kM and $(k+t)M$ are disposed outside of the strip between l_1 and l'. Therefore l_1 and l' coincide. Thus $\rho(A, A_1) = \rho(l, l_1)$ and this distance, obviously, does not depend on k and is linearly on t.

(2) Here we shall distinguish "convex" and "concave" angles of a given polygon M, see pic. 5.04, 5.05. Let EF and E_1F_1 and also FG and F_1G_1 be pairs of edges of polygons

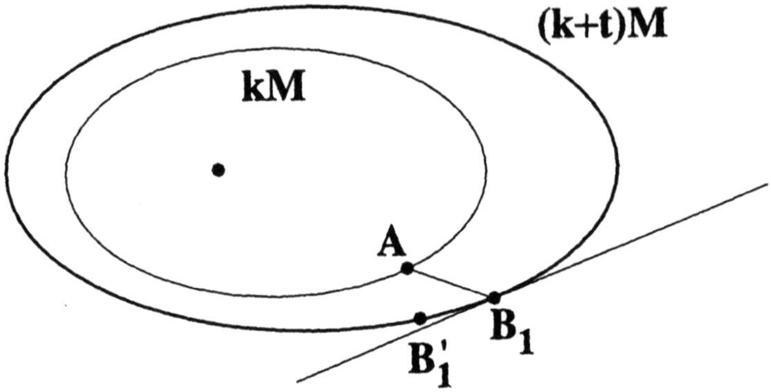

pic. 5.02

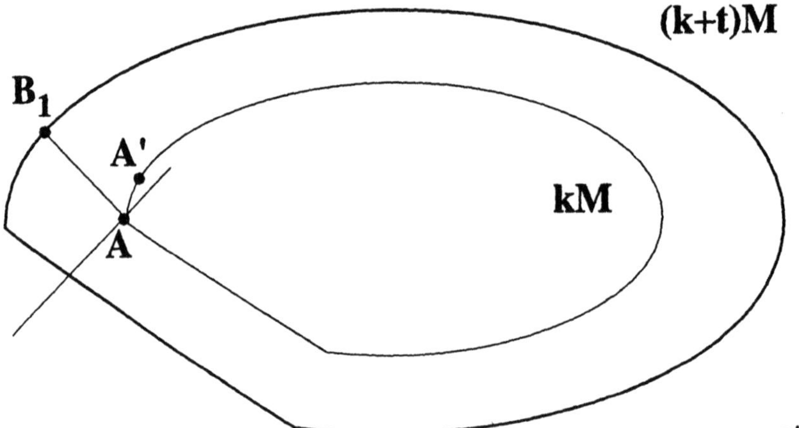

pic. 5.03

kM and $(k+t)M$ which are homothetic to each other. The straight lines EF, E_1F_1, FG and F_1G_1 we shall denote through l, l_1, m and m_1 respectively. If k increase then lengthes of the mentioned four edges increase too (the points E and G and also E_1 and G_1 move off, accordingly, from the points F and F_1). At the same time, the distances between l and l_1 (m and m_1) do not change. From above it is clear, that for each $t > 0$ there exists such $k_t > 0$, that at all $k > k_t$, the perpendiculars from F on l_1 and m_1 will be intersect the segments E_1F_1 and F_1G_1, if the angle EFG is convex (pic. 5.04), or the perpendiculars droping from F_1 on l and m will intersect the segments EF and FG, if the angle EFG is concave (pic. 5.05). At the expense of increase of k_t, because the number of M's vertices is finite, it is possible to achieve, that the described situation would be observed in all angles of kM. Further, we suppose that the number k_t is already chosen large.

Now, it is already clear, that $\rho(A, B_1) = \rho(l', l'')$, where l' and l'' are the straight lines, which carry a pair of homothetical edges of the polygons kM and $(k+t)M$, and this distance, obviously, does not depend on k and is linearly on t.•

Further, we shall denote inf $\rho(A, B_1)$ (not depending on k) at a given t through $d(t)$.

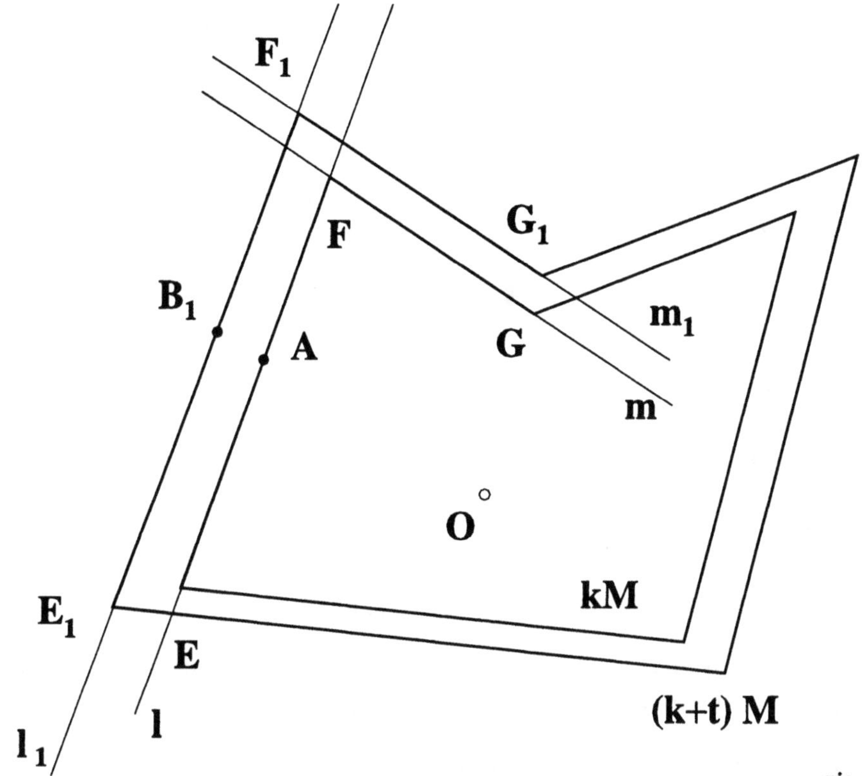

pic. 5.04

Theorem 5.2 *$d(t)$ is a linear homogeneous function of t. That is $d(t) = \lambda t$, where $\lambda > 0$ is some constant which depends on a ray set M and on its centre.*

⊳ See the proof of theorem 5.1.

Let M be a ray set with the centre in a point O and M' be a set parallel congruent (parallel) to it with the centre in O' corresponding to O. We shall denote the set homotetical to $M(M')$ with respect to $O(O')$ through $rM(rM')$, where $r > 0$ is the coefficient of the homothety.

Theorem 5.3 *Let $d > 0$ be the distance from O up to O'. Then at rather large $k > 0$ for every $d > 0$ there exists $t > 0$ that does not depend on k, such that*

$$(k - t)M \subset kM' \subset (k + t)M.$$

⊳ Let's take rather large $k > 0$ and λ from theorem 5.2 and put $t = 2\rho/\lambda$. Then from theorems 5.1 and 5.2 we infer that each point of the set kM is removed from each point of $\mathbb{E}^2 \setminus (k + t)M$ by not less than on 2ρ. At the same time each point of kM' is away from the point of kM, according to it, on the distance not greater, than ρ. From here it follows immediately, that $kM' \subset (k + t)M$.

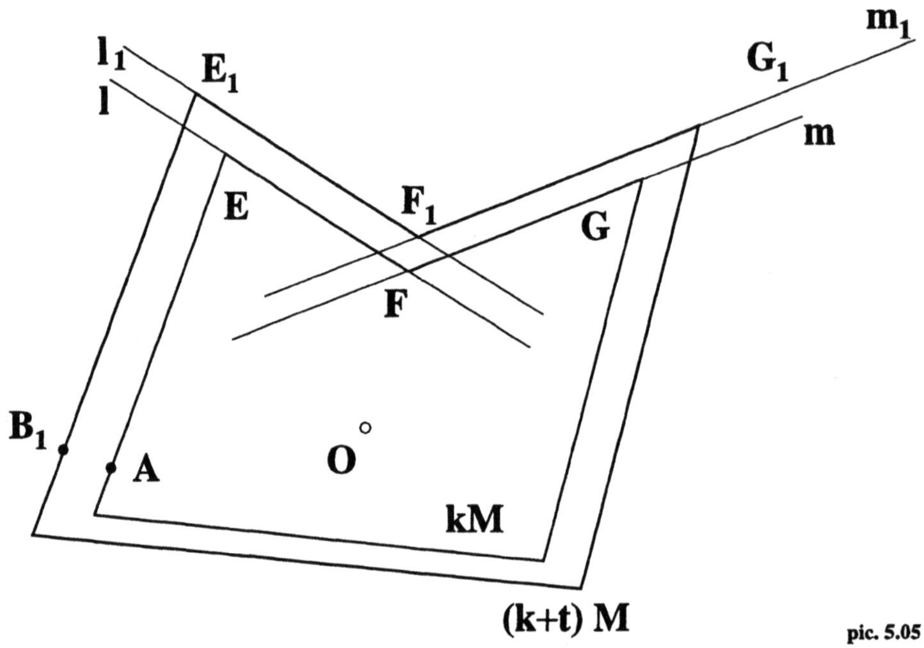

pic. 5.05

We can obtain the inclusion $(k-t)M \subset kM'$ either by the similar reasonings or by replacing M on M', M' on M and k on $k-t$ in the previous inclusion.

§5.2 Density, Its Independence on Begining and Boundary Effects

1^0. The word "*functional*" in this section means a numerical function of sets $T \subset \mathbb{E}^2$. Examples can serve such functionals: the functional $\mathcal{F}_1(T)$ identically equal to 1 on all $T \subset \mathbb{E}^2$, functional $\mathcal{F}_d = d(T)$ equal to the diameter of T, functional $\mathcal{F}_S = S(T)$ equal to the area of T.

The domain of definition of some functionals are not all sets $T \subset \mathbb{E}^2$ (for example, every nonmeasurable set isn't contained to the domain of the functional $\mathcal{F}_S$). However, in those places, where we not concretize a functional, we shall consider, that it is given on the considered sets by us.

Further, all considered functionals are finite on bounded sets.

We pay attention, that in all three examples the functional $\mathcal{F}$ is possessed of the following property $\mathcal{F}(T) \geq \mathcal{F}(T')$, with $T \subset T'$. This property is called *monotony* of $\mathcal{F}$.

Definition: Let us be given some locally finite disposition $\mathfrak{T}$ of sets $\{T_i\}$ bounded in total (i.e. there exists a number C that for any T_i, $\mathcal{F}_d(T_i) < C$) and a functional $\mathcal{F}(T)$. Let us also be given a ray set $M \in \mathbb{E}^2$ with centre in O. The *inferior density* of the functional $\mathcal{F}$ with respect to M on the disposition $\mathfrak{T}$ is

$$\underline{\Delta} = \underline{\Delta}(\mathfrak{T}) = \underline{\Delta}(\mathfrak{T}, \mathcal{F}) = \liminf_{r\to\infty} \frac{\sum \mathcal{F}(T_i)}{S(rM)},$$

where the summation is over the sets $T_i \subset rM$.

Definition: Let us be given some locally finite disposition $\mathfrak{T}$ of sets $\{T_i\}$ bounded in total and a functional $\mathcal{F}(T)$. Let us be also given a ray set $M \in \mathbb{E}^2$ (*measuring figure*) with centre O. The *superior density* of the functional $\mathcal{F}$ respectively to M on the disposition $\mathfrak{T}$ is

$$\bar{\Delta} = \bar{\Delta}_M(\mathfrak{T}) = \bar{\Delta}_M(\mathfrak{T}, \mathcal{F}) = \limsup_{r \to \infty} \frac{\sum \mathcal{F}(T_i)}{S(rM)}$$

where the summation is over the sets T_i, that $T_i \cap rM \neq \emptyset$.

Definition: The *density* with respect to a given measuring figure M of a functional $\mathcal{F}$ on a disposition $\mathfrak{T}$ is the number $\Delta_M(\mathfrak{T}, \mathcal{F}) = \bar{\Delta}_M(\mathfrak{T}, \mathcal{F})$, if $\bar{\Delta}_M(\mathfrak{T}, \mathcal{F}) = \underline{\Delta}_M(\mathfrak{T}, \mathcal{F})$.

Examples. For the functional $\mathcal{F}_1(T)$ the (superior, inferior) density is the (superior, inferior) average number of elements of $\mathfrak{T}$ respectively the set M. For the functionals $d(T)$ and $S(T)$ density is the average (respectively to M!) diameter or area of elements of the disposition.

2^0. **Independence on beginning effects.**

Theorem 5.4 n.n. *Let M be a strongly ray set with centre in O and M' be a parallel set to it with centre in O'. Let $\mathfrak{T}$ be an arbitrary locally finite disposition of sets bounded in total and $\mathcal{F}(T)$ be a functional. Then the densities $\bar{\Delta}$, $\underline{\Delta}$ (and Δ, if $\bar{\Delta} = \underline{\Delta}$) of $\mathcal{F}(T)$ with respect to the sets M and M', accordingly, coincide.*

▷ Let $\mathbf{U}_O$ be the radiance neighborhood of the set M. By replacing, if it is necessary, the sets M and M' on homothetical to them ones with centers of homothety, accordingly, in the points O and O', let's assume that $O' \in \mathbf{U}_O$. Let $\rho > 0$ be the distance between O and O'. We shall assume, that the embeddings from theorem 5.4 are carried out for some $t > 0$, since $k_0 > 0$. We shall note now, that the distance between each point of $\mathbf{M}$ and the set $\mathbf{M}'$ is not greater, than ρ, therefore for any $k \geq k_0$ from theorem 3.2 we have:

$$(k - t)M \subset kM' \subset (k + t)M$$

Let's consider sums: $\sum_r \mathcal{F}(T_i)$ and $\sum_r' \mathcal{F}(T_i)$. Further, in the proof of the theorem for $\bar{\Delta}$ ($\underline{\Delta}$) the summations is over the sets T_i which there are in the definition of $\bar{\Delta}$ ($\underline{\Delta}$). The first of these sums for the set M, the second for M'.

From the mentioned above embeddings we have:

$$\sum_{k-t} \mathcal{F}(T_i) < \sum_{k}^{\prime} \mathcal{F}(T_i) < \sum_{k+t} \mathcal{F}(T_i)$$

further, we have:

$$\frac{S((k-t)M)}{S(kM)} \frac{\sum_{k-t} \mathcal{F}(T_i)}{S((k-t)M)} < \frac{\sum_k' \mathcal{F}(T_i)}{S(kM)} < \frac{S((k+t)M)}{S(kM)} \frac{\sum_{k+t} \mathcal{F}(T_i)}{S((k+t)M)}.$$

from here we obtain:

$$\frac{(k-t)^2}{k^2} \frac{\sum_{k-t} \mathcal{F}(T_i)}{S((k-t)M)} < \frac{\sum_k' \mathcal{F}(T_i)}{S(kM)} < \frac{(k+t)^2}{k^2} \frac{\sum_{k+t} \mathcal{F}(T_i)}{S((k+t)M)}.$$

Taking in the case of $\bar{\Delta}$ the superior limit and in the case of $\underline{\Delta}$ the inferior limit, we finish the proof.•

3^0. **Independence on boundary effects.**

Theorem 5.5 n.n. *Let M be a strongly ray set with centre in O. Let also $\mathfrak{T}$ be an arbitrary locally finite disposition of sets bounded in total and $k > 0$. Then*

$$\liminf_{r\to\infty}\frac{\sum S(T_i)}{S(rM)} \quad \textit{and} \quad \limsup_{r\to\infty}\frac{\sum S(T_i)}{S(rM)}$$

do not depend on that, which of T_i intersecting close$[(r+k)M \setminus (r-k)M]$ or which its subsets (besides of all sets T_i wholly lying in $(r-k)M$) are taken into account in the sums. These limits are equal, accordingly, to

$$\liminf_{r\to\infty}\frac{\sum S(T_i)}{S(rM)} \quad \textit{and} \quad \limsup_{r\to\infty}\frac{\sum S(T_i)}{S(rM)},$$

where the summation may be over $T_i \subset (r-k)M$ or over T_i, that $T_i \cap (r+k)M \neq \emptyset$. Through $S(rM)$ in all limits the area of rM is denoted.

▷ We shall denote through $\sum' S(T_i)$ the sum, in which the summation is over $T_i \subset (r-k)M$ or also is over some arbitrary set of such sets T_i, that $T_i \cap (r+k)M \neq \emptyset$ and subsets of such sets.

Let $d(T_i) < d$ for each $T_i \in \mathfrak{T}$, where d is some number. We shall consider such a number t, that (see theorem 5.2) $d(t) > d$.

Let's denote through $\sum S(T_i)$ the sum, in which the summation is over all $T_i \subset (r-k-t)M$, and through $\sum'' S(T_i)$ the sum, in which the summation is over all $T_i \subset (r+k+t)M$. We have:

$$\frac{\sum S(T_i)}{S((r-k-t)M)} \le \frac{\sum' S(T_i)}{S((r)M)} \le \frac{\sum'' S(T_i)}{S((r+k+t)M)}$$

On noticing, that the superior (inferior) limit of the right part is equal to the superior (inferior) limit of the left part, we finish the proof.•

Meaning theorems 5.4 and 5.5, further, we shall use only strongly ray measuring figures, when calculating the density of a functional.

4^0. We shall adduce one example of rather general character. Let's firstly note, that the proved theorems about the independence allow in this example not to care for the sets of a disposition $\mathfrak{T}$, which are near the boundary of kM and for a measuring figure M with correct to a parallel translation. (As for the form of the set M, we shall see, that the following situation is absolutly different.)

We shall consider the plane $\mathbb{E}^2$ determined by the rectangular coordinate system and the following disposition $\mathfrak{T}$ of individual squares. The right half-plane ($x \geq 0$) is divided on such squares with whole vertices and the left half-plane does not contain internal points of any square of $\mathfrak{T}$. Let's consider two strongly ray sets M and M' with common centre in the origin. We shall denote the set M as the rectangular with the vertices $(-1,1), (2,1), (2,-1)$ and $(-1,-1)$. M' we shall denote as the rectangular with the vertices $(-2,1), (1,1), (1,-1)$ and $(-2,-1)$. See pic. 5.06. It is obvious, that with respect to the first set $\Delta(\mathfrak{T}, \mathcal{F}_S) = 2/3$, and with respect to the second one $\Delta(\mathfrak{T}, \mathcal{F}_S) = 1/3$.

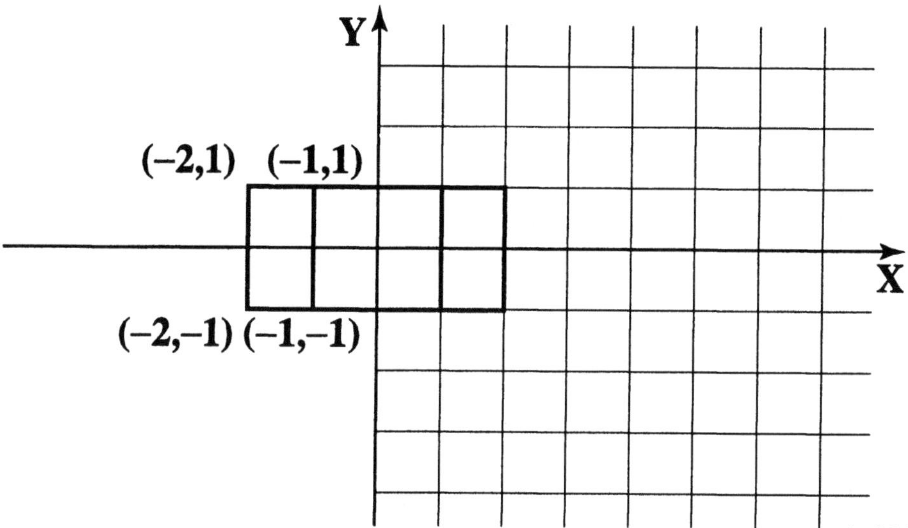

pic. 5.06

So, we have seen, that the density of the same functional on the same disposition, but respectively various measuring figures, may appear various.

Exercise

Construct an example of a dispositon for which density with respect to a square differs from density with respect to a disk (both with centre in the centre of symmetry).

§5.3 Periodic Disposition.

1^0. Here and in §4 we shall prove a number of theorems, allowing, in particular, to concern to the concept of density not so pessimistic, as it could be after the last example from §2.

Let $\mathfrak{T}$ be some locally finite disposition and M be a given measuring figure. As in §2 we shall denote the inferior(superior) density of $\mathfrak{T}$ measured by M through $\underline{\Delta}_M(\mathfrak{T})$ ($\bar{\Delta}_M(\mathfrak{T})$). If the disposition $\mathfrak{T}$ has the density with respect to M, we shall denote it through $\Delta_M(\mathfrak{T})$.

Further, we will need for several times the following design. Let's take on the plane the square q_0 with the centre in the origin O and with the edges parallel to the axes of coordinates, with length equal to 1. Further, we construct the tiling (*quadriliage*) $\mathbb{Q} = q_0 \oplus \mathbb{Z}^2$ of $\mathbb{E}^2$. (Through $X \oplus Y$ the *vectorial sum* or the *sum of Minkowski* is denoted here, i.e. the set of points determined by all vectors $\vec{Ox} + \vec{Oy}$, with $x \in X$, $y \in Y$ and O is an arbitrary chosen point, in this case it is the origin.)

We shall denote the quadriliage homotetical to $\mathbb{Q}$ with the coefficient $\lambda > 0$ through $\lambda\mathbb{Q}$.

2^0. A disposition $\mathfrak{T}$ is called *periodic*, if it can be represented as $\mathfrak{T}^* \oplus \Gamma$, where $\mathfrak{T}^*$ is some finite disposition and Γ is a lattice. See pic. 5.07.

Theorem 5.6 n.n. *For each periodic disposition $\mathfrak{T} = \mathfrak{T}^* \oplus \Gamma$ there exists the density not dependent on a measuring figure.*

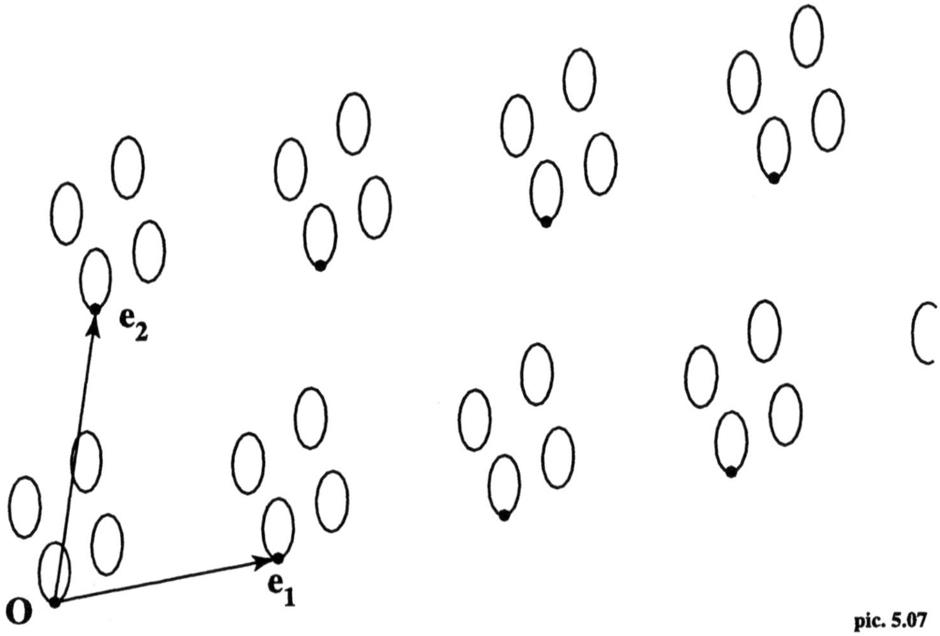

pic. 5.07

▷ By making, if it is necessary, parallel translation, we can consider that the origin of Γ and one of points of some element of $\mathfrak{T}^*$ are placed in the origin O. Let's take an arbitrary basic parallelogram Π_O of the lattice Γ with the vertex in O and denote its diameter and area through d and S. We shall construct the (affine equivalent to the tiling $\mathbb{Q}$) face-to-face tiling $\mathbb{Q}' = \Pi_O \oplus \Gamma$ with parallelograms $\Pi_A = \mathfrak{A}_{OA}(\Pi_O)$, where $A \in \Gamma$. We shall attribute the finite disposition $\mathfrak{A}_{OA}(\mathfrak{T}^*)$ to the parallelogram Π_A.

Let $\mathfrak{T}^* = \{T_1, T_2, \dots, T_k\}$ and d^* be the diameter of the set $T_1 \cup T_2 \cup \dots \cup T_k$. Let us fix a measuring figure M with centre in O.

We shall show, that $\Delta_M(\mathfrak{T}) = \frac{\sum^* S(T_i)}{S}$, where the summation is over all $T_i \in \mathfrak{T}^*$.

Really, let's choose a number $\rho > d + 2d^*$. Then, according to theorem 5.5, in the following expressions

$$\liminf_{r\to\infty} \frac{\sum S(T_i)}{S(rM)} \; and \; \limsup_{r\to\infty} \frac{\sum S(T_i)}{S(rM)}$$

the sum $\sum S(T_i)$ is possible to be considered with all elements of $\mathfrak{T}$, attributed to those parallelograms of $\mathbb{Q}'$ which lie in rM. This sum is equal to $k \sum^* S(T_i)$, where k is the number of such parallelograms. We have

$$\begin{aligned}\liminf_{r\to\infty} \frac{\sum S(T_i)}{S(rM)} &= \liminf_{r\to\infty} \frac{K \sum^* S(T_i)}{kS} \frac{kS}{S(rM)} \\ &= \frac{\sum^* S(T_i)}{S} \liminf_{r\to\infty} \frac{kS}{S(rM)}.\end{aligned}$$

Accordingly to same theorem 5.5, we have $\liminf_{r\to\infty} \frac{kS}{S(rM)} = 1$ and analogically calculating $\limsup_{r\to\infty} \frac{\sum S(T_i)}{S(rM)}$, we obtain

$$\liminf_{r\to\infty} \frac{\sum S(T_i)}{S(rM)} = \limsup_{r\to\infty} \frac{\sum S(T_i)}{S(rM)} = \Delta_M(\mathfrak{T}) = \frac{\sum^* S(T_i)}{S}.$$

But this magnitude does not depend on M.•

As the separate remark we shall note once again the important fact, obtained in the proof of theorem 5.6.

Remark: **n.n.** For each periodic disposition

$$\mathfrak{T} = \{T_1, T_2, \dots, T_k\} \oplus \Gamma$$

the density is equal to $\sum_{i=1}^{i=r} S(T_i)/S(\Gamma)$ for any measuring figure.

§5.4 The Densiest Packing and the Thinnest Covering Densities' Independence on Measuring Figure

1^0. **Packings.** Let $\{T\}$ be some family of sets with diameters not greater than the number $d/2$. In this item through $\mathfrak{T}$ (may be with indexes) will be denoted packings of such sets T_i each of which is congruent (may be by a given set of ways) to one of the sets $T \in \{T\}$.

Theorem 5.7 *For any measuring figure M, any $\epsilon > 0$ and for any packing $\mathfrak{T}_0$ there exists such a periodic packing $\mathfrak{T}_{0M}$, for which the inequality $\bar{\Delta}_M(\mathfrak{T}_0) - \Delta_M(\mathfrak{T}_0 M) \le \epsilon$ is valid.*

▹ We consider the figure M, the number $1 > \epsilon > 0$ and the packing $\mathfrak{T}_0$ given to us. Let's choose such $a > 0$, that the following inequality is valid

$$(a+d)^2 - a^2 < \frac{1}{3}\epsilon a^2. \tag{5.1}$$

Let's consider the quadriliage $a\mathbb{Q}$ and choose (see theorem 5.5) $\rho = a\sqrt{2} + d$ and such r_0, that at all $r > r_0$ the next inequality is valid

$$0 < 1 - \frac{ka^2}{S(rM)} < \frac{1}{3}\epsilon,$$

where k is the number of squares of $\mathbb{Q}$ containing in rM. We shall consider that for all $r > r_0$ the following condition is also valid

$$\bar{\Delta}_M(\mathfrak{T}_0) - \frac{\sum S(T_i)}{S(rM)} < \frac{1}{3}\epsilon,$$

where the sum is taken over all elements of $\mathfrak{T}_0$, intersected with the mentioned squares (theorem 5.5 with $d(t) = \rho$). From here we have

$$\bar{\Delta}_M(\mathfrak{T}_0) - \frac{\sum S(T_i)}{ka^2}\frac{ka^2}{S(rM)} < \frac{1}{3}\epsilon.$$

Further, as $ka^2 < S(rM)$, we obtain

$$\bar{\Delta}_M(\mathfrak{T}_0) - \frac{\sum S(T_i)}{ka^2} < \frac{1}{3}\epsilon.$$

Hence, among our k squares, by the Dirihlet principle, we can find, at least, one square (we shall denote it through Q), for which

$$\bar{\Delta}_M(\mathfrak{T}_0) - \frac{\sum' S(T_i)}{a^2} < \frac{1}{3}\epsilon, \tag{5.2}$$

where the sum $\sum'$ is taken over all elements of $\mathfrak{T}_0$ containing in the square Q', concentric to Q and homothetical to it with the coefficient $(1 + \frac{d}{a})$ (as well over all elements of $\mathfrak{T}_0$, intersected with Q).

Moreover the next inequality is obvious:

$$\frac{\sum' S(T_i)}{a^2} < \frac{(a+d)^2}{a^2} < \left(1 + \frac{1}{3}\epsilon\right).$$

We shall make necessary estimates using this inequality and also the inequalities (5.1) and (5.2):

$$\begin{aligned} \bar{\Delta}_M(\mathfrak{T}_0) - \frac{\sum' S(T_i)}{(a+d)^2} &= \bar{\Delta}_M(\mathfrak{T}_0) - \frac{\sum' S(T_i)}{a^2}\frac{a^2}{(a+d)^2} \\ < \bar{\Delta}_M(\mathfrak{T}_0) - \frac{\sum' S(T_i)}{a^2}\left(1 - \frac{1}{3}\epsilon\right) &< \frac{1}{3}\epsilon + \frac{1}{3}\epsilon\frac{\sum' S(T_i)}{a^2} \\ < \frac{1}{3}\epsilon + \frac{1}{3}\epsilon\left[1 + \frac{1}{3}\epsilon\right] &< \left(\frac{1}{3}\epsilon + \frac{2}{3}\epsilon\right) = \epsilon. \end{aligned} \tag{5.3}$$

Let's construct the packing $\mathfrak{T}_{0M}$. (We shall note, that the family $\{T\}$ may be reduced.) For this purpose we shall take the quadriliage $\mathbb{Q}' = (a+d)\mathbb{Q}$ and replace in it each square with the square $Q\prime$ together with that part of the packing $\mathfrak{T}_0$, which is contained in $Q\prime$. According to theorem 5.1, the packing $\mathfrak{T}_{0M}$ has the density $\Delta_M(\mathfrak{T}_{0M})$ and this density satisfies, by (5.3), the demanded inequality.•

Theorem 5.8 *Whatever measuring figures M and M' were for packings of an aspect $\mathfrak{T}$, the following equalities take place*

$$\sup \underline{\Delta}_M(\mathfrak{T}) = \sup \bar{\Delta}_M(\mathfrak{T}) = \sup \bar{\Delta}_{M'}(\mathfrak{T}) = \sup \underline{\Delta}_{M'}(\mathfrak{T}).$$

(The greatest possible density of packings of an aspect $\mathfrak{T}$ does not depend on a measuring figure.)

▷ Let for the given measuring figure $\sup \bar{\Delta}_M(\mathfrak{T})$ be not attained. Let's choose $\epsilon > 0$ and such sequence of packings $\{\mathfrak{T}_j\}$ with $j = 1, 2, \ldots,$ that $\sup \bar{\Delta}_M(\mathfrak{T}) - \bar{\Delta}_M(\mathfrak{T}_j) < 2^{-j-1}\epsilon$. Replacing in theorem 5.6 ϵ on $2^{-j-1}\epsilon (j = 1, 2, \ldots)$, we shall obtain the sequence of periodic packings $\{\mathfrak{T}_{jM}\}$, for which the density (and by that, the inferior density) satisfies to inequalities

$$\sup \bar{\Delta}_M(\mathfrak{T}) - \Delta_M(\mathfrak{T}_{jM}) < 2^{-j}\epsilon. \tag{5.4}$$

Therefore the first and the third inequalities from the confirmed chain of inequalities are proved.

Further. As all $\mathfrak{T}_{jM}$ are periodic, their densities do not depend on a measuring figure, i.e. for any other measuring figure M' there is the equality $\Delta_{M'}(\mathfrak{T}_{jM}) = \Delta_M(\mathfrak{T}_{jM})$. The same constructions may be carried out for $\sup \bar{\Delta}_{M'}(\mathfrak{T})$ and it is possible to construct the appropriate sequence $\mathfrak{T}_{jM'}$, for which $\Delta_M(\mathfrak{T}_{jM'}) = \Delta_{M'}(\mathfrak{T}_{jM'})$. From here it follows, at once, the second inequality from the confirmed chain.

Let for the given measuring figure $\sup \bar{\Delta}_M(\mathfrak{T})$ be attained for some packing $\mathfrak{T}_0$. The sequence $\mathfrak{T}_{jM}$, which satisfies to inequalities (5.4), in this case is constructed directly on the packing $\mathfrak{T}_0$, and all further reasonings are kept.•

2^0. **Coverings.** Let $\{T\}$ be some family of sets whose diameters are not greater than d. In this item through $\mathfrak{T}$ (may be with indices) will be denoted an arbitrary (locally finite) covering by sets T_i, each of which is congruent (may be by a given set of ways) to one of the sets $T \in \{T\}$.

Lemma 1 *For any measuring figure M $\inf \bar{\Delta}_M(\mathfrak{T})$ is finite.*

▷ Let's pay attention on the fact, that no sets of any locally finite covering and thus no sets $T \in \{T\}$ can have the zero area. Let's consider a set $T \in \{T\}$ such that $S(T) > 0$. As the set T has the positive internal measure, there will be discovered a square q with the edges, parallel to the axes of coordinates. We shall denote the length of the edge through ϵ and consider the quadriliage $\epsilon\mathbb{Q}$. We shall place T in such a way, that the square q has coincided with the square ϵq_0 of $\epsilon\mathbb{Q}$. The periodic packing $T \oplus \epsilon\mathbb{Z}^2$ has the density not greater than

$$S(T)/\epsilon^2 > \inf \bar{\Delta}_M(\mathfrak{T}).\bullet$$

Theorem 5.9 *For arbitrary measuring figure M, $\epsilon > 0$ and for an arbitrary covering $\mathfrak{T}_0$ there exists such a periodic covering $\mathfrak{T}_{0M}$, for which the inequality $\Delta_M(\mathfrak{T}_{0M}) - \underline{\Delta}_M(\mathfrak{T}) \leq \epsilon$ is valid.*

▷ We consider the figure M, the number $1 > \epsilon > 0$ and the covering $\mathfrak{T}_0$ given to us. Let's choose such $a > 0$, that the following inequality is carried out

$$a^2 - (a-d)^2 < \frac{1}{3}\left[\frac{\epsilon}{\underline{\Delta}_M(\mathfrak{T}_0)}\right](a-d)^2 = \alpha(a-d)^2. \tag{5.5}$$

Let's consider the quadriliage $a\mathbb{Q}$ and choose (see theorem 5.5) $\rho = a\sqrt{2}$ and such r_0, that at all $r > r_0$ the next inequality is valid

$$0 < \frac{ka^2}{S(rM)} - 1 < \alpha,$$

where k is the number of squares of a $\mathbb{Q}$ intersecting rM. We shall consider, that for r the following condition is also carried out

$$\frac{\sum S(T_i)}{S(rM)} - \underline{\Delta}_M(\mathfrak{T}_0) < \alpha,$$

where the sum is taken over all elements of $\mathfrak{T}_0$, intersected with the mentioned squares (theorem 5.5). From here we have

$$\frac{\sum S(T_i)}{ka^2}\frac{ka^2}{S(rM)} - \underline{\Delta}_M(\mathfrak{T}_0) < \alpha.$$

Further, as $ka^2 > S(rM)$, see (5.7), we obtain

$$\frac{\sum S(T_i)}{ka^2} - \underline{\Delta}_M(\mathfrak{T}_0) < \alpha.$$

Hence, among our k squares, by the Dirihlet principle, we can find, at least, one square (we shall denote it through Q), for which

$$\frac{\sum' S(T_i)}{a^2} - \underline{\Delta}_M(\mathfrak{T}_0) < \alpha, \tag{5.6}$$

where the sum $\sum'$ is taken over all elements of $\mathfrak{T}_0$ containing the square Q. This elements cover the square Q', concentric to Q and homothetical to it with the coefficient $(1 - \frac{d}{a})$. We shall make necessary estimates using (5.5) and twice (5.6):

$$\begin{aligned}\frac{\sum' S(T_i)}{(a-d)^2} - \underline{\Delta}_M(\mathfrak{T}_0) &= \frac{\sum' S(T_i)}{a^2}\frac{a^2}{(a-d)^2} - \underline{\Delta}_M(\mathfrak{T}_0)\\ &< \frac{\sum' S(T_i)}{a^2}(1+\alpha) - \underline{\Delta}_M(\mathfrak{T}_0) < \alpha + \alpha\frac{\sum' S(T_i)}{a^2}\\ &< \alpha + \alpha\underline{\Delta}_M(\mathfrak{T}_0) + \alpha) < \frac{1}{3}\epsilon(2+\frac{1}{3}\epsilon) < \epsilon.\end{aligned} \tag{5.7}$$

Let's construct the periodic covering $\mathfrak{T}_{0M}$. (We shall note, that the family $\{T\}$ may be reduced.) For this purpose we shall take the quadriliage $\mathbb{Q}' = (a-d)\mathbb{Q}$ and replace in it each square with the square $Q\prime$ together with that part of the covering $\mathfrak{T}_0$, which is contained in $Q\prime$. According to theorem 5.6, the covering $\mathfrak{T}_{0M}$ has the density $\Delta_M(\mathfrak{T}_{0M})$ and this density satisfies, by (5.7), the demanded inequality.•

Theorem 5.10 *Whatever measuring figures M and M' were for coverings of an aspect $\mathfrak{T}$, the following equalities take place*

$$\inf \bar{\Delta}_M(\mathfrak{T}) = \inf \underline{\Delta}_M(\mathfrak{T}) = \inf \underline{\Delta}_{M'}(\mathfrak{T}) = \inf \bar{\Delta}_{M'}(\mathfrak{T}).$$

(The least possible density of coverings of an aspect $\mathfrak{T}$ does not depend on a measuring figure.)

▷ Let for the given measuring figure $\sup \underline{\Delta}_M(\mathfrak{T})$ is not reached. Let's choose $\epsilon > 0$ and such sequence of coverings $\{\mathfrak{T}_j\}$ with $j = 1, 2, \ldots,$ that $\bar{\Delta}_M(\mathfrak{T}) - \inf \underline{\Delta}_M(\mathfrak{T}_j) < 2^{-j-1}\epsilon$. Replacing in theorem 5.4 ϵ on $2^{-j-1}\epsilon (j = 1, 2, \ldots)$, we shall obtain the sequence of periodic coverings $\{\mathfrak{T}_{jM}\}$, for which the density (and by that, the superior density) satisfies to inequalities

$$\Delta_M(\mathfrak{T}_{jM}) - \inf \underline{\Delta}_M(\mathfrak{T}) < 2^{-j}\epsilon. \tag{5.8}$$

Therefore, the first and the third inequalities from the confirmed chain of inequalities are proved.

Further. As all $\mathfrak{T}_{jM}$ are periodic, their densities do not depend on a measuring figure, i.e. for any other measuring figure M', the equality $\Delta_{M'}(\mathfrak{T}_{jM}) = \Delta_M(\mathfrak{T}_{jM})$ holds. The same constructions may be carried out for $\inf \Delta_{M'}(\mathfrak{T})$ and it is possible to construct the appropriate sequence $\mathfrak{T}_{jM'}$, for which $\Delta_M(\mathfrak{T}_{jM'}) = \Delta_{M'}(\mathfrak{T}_{jM'})$. From here it follows, at once, the second inequality from the confirmed chain.

Let for the given measuring figure $\inf \underline{\Delta}_{M'}(\mathfrak{T})$ is reached on some covering $\mathfrak{T}_0$. The sequence $\mathfrak{T}_{jM}$, which satisfies to inequalities (5.8), in this case is constructed directly on the covering $\mathfrak{T}_0$, and all further reasonings are kept.•

6 Packing of Equal Balls. Covering by Equal Balls

§6.1 Results. Elementary Theorems

1^0. Here we adduce the result of two classical problems of the discrete geometry for the case of the plane, using only the concept of L-tilings and simple theorems of the elementary geometry. Use of theorems from §5 facilitates the reception of the result and a little specifies it, generalizing on arbitrary measuring figures.

Theorem 6.1 2.2. *For any packing of equal disks on the plane (and whatever a measuring figure there was) the density (superior density) of a packing cannot be greater, than* $\pi/\sqrt{12} = \pi/2\sqrt{3}$.

Such density, with respect to any measuring figure, is reached on the "hexagonal" lattice packing. See pic. 6.01.

Theorem 6.2 2.2 *For any covering by equal disks of the plane (and whatever a measuring figure there was) the density (inferior density) of a covering cannot be less, than* $2\pi/3\sqrt{3}$.

Such density, with respect to any measuring figure, is reached on the "hexagonal" lattice packing. See pic. 6.02.

In the proofs of theorems 6.1 and 6.2 we shall fix on the plane an arbitrary measuring figure M with centre in O (see §2 of chapter 5) and we shall consider figures of an aspect λM, where λ is a real number.

2^0. The proof will be preceded by a number of definitions and lemmas.

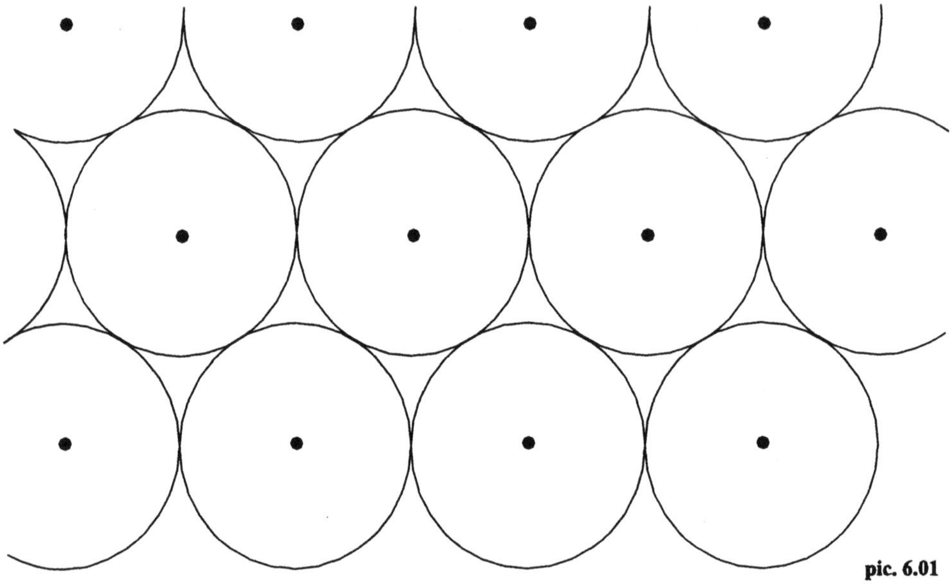

pic. 6.01

Lemma 1 *For any* $\alpha_i, \beta_i > 0$, *with* $i = 1, 2, \ldots, N$, *the following inequalities take place:*

$$\max(\alpha_i/\beta_i) \geq \frac{\alpha_1 + \alpha_2 + \cdots + \alpha_N}{\beta_1 + \beta_2 + \cdots + \beta_N} \geq \min(\alpha_i/\beta_i). \triangleright \bullet$$

Lemma 2 *Among triangles with edges* $a, b, c \geq 2r$ *and angles less than* $120°$, *the least area is equal to* $r^2\sqrt{3}$, *of the regular triangle with the edge* $2r$.

$\triangleright$ Let $\angle ab$ be the greatest of the angles of the considered triangle (see pic. 6.03), then $120° > \angle ab \geq 60°$. We have

$$S = \frac{1}{2} ab \sin\angle ab \geq 2r^2 \sin \angle ab \geq r^2\sqrt{3},$$

where the equalities are carried out, accordingly, only at $a = b = 2r$ and $\angle ab$=60°.$\bullet$

Lemma 3 *Among triangles enclosed in a disk of radius* r *the greatest area, equal to* $3r^2\sqrt{3}/4$, *has the inscribed regular triangle.*

$\triangleright$ Let some triangle ABC be enclosed in a disk of radius r, with the boundary (circle) C, see pic. 6.04. May be, by making a motion of the triangle, we may consider that $A \in C$, and by making, if it is necessary, a homothety with the centre in A (and increasing, by that, the area of the triangle), we may consider that one more its vertex, let it be B, belongs to the circle. If after this the triangle ABC is not inscribed in C, then we shall find the intersection (point C') of the straight line AC and the circle C and obtain the triangle ABC' with greater area and already inscribed in the circle. For this case our theorem is trivial.$\bullet$

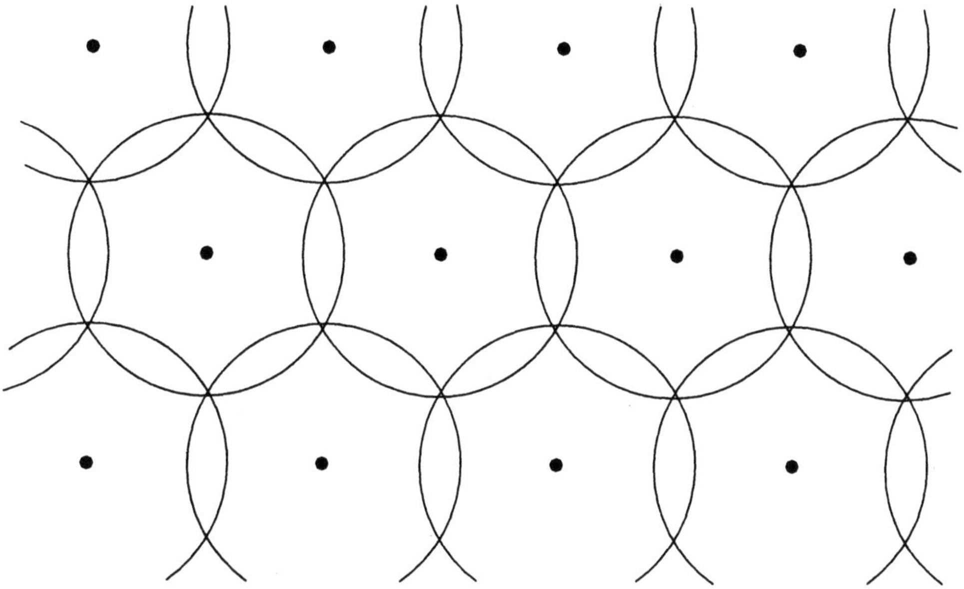

pic. 6.02

§6.2 The Densest Packing of Equal Disks

1^0. **Definition:** A packing $\mathfrak{T} = \{T_1, T_2, \ldots, T_k, \ldots\}$ of pairwise congruent sets in $\mathbb{E}^2$ is called *saturated*, if for any set $T \subset \mathbb{E}^2$ congruent to T_1 there exists such a set $T_k \subset \mathfrak{T}$, that $T \cap T_k \neq \emptyset$.

Below, in §2, we shall consider only packings of pirewise congruent disks and assume, that through $\Sigma(\mathfrak{T})$ is denoted the system of the disks' centres.

Let's note, that for a saturated packing $\mathfrak{T}$ the system $\Sigma(\mathfrak{T})$ is a Delone system.

Lemma 4 *Whatever was an L-polygon in the L-tiling of $\Sigma(\mathfrak{T})$, with an arbitrary saturated packing $\mathfrak{T}$, each its angle is less, than* 120°.

▷ Assume the contrary. Let in a point $M \in \Sigma$ some L-polygon L_1 has an angle $\angle M \geq 120°$. The edges MP and MQ of the polygon L_1 have length not smaller, than $2r$ (you see, Σ is the system of the centres of packing's disks of radius r). From these circumstances it follows that the circle circumscribed around ΔMPQ and thus around the polygon L_1, has the radius $R \geq 2r$. Thus, we have obtained the empty disk of radius $R \geq 2r$, but it is impossible, because $\mathfrak{T}$ is a satureted packing.•

Corollary: *In the conditions of lemma* 4 *only triangles, quadrangles and pentagons can be L-polygons.*

▷ Each inscribed n-angel, with $n > 5$, has, on extreme, one internal angle greater, than 120°.•

We shall note also, that in the conditions of lemma 4 neither length of an edge, nor length of a diagonal of an L-polygon can exceed $4r$.

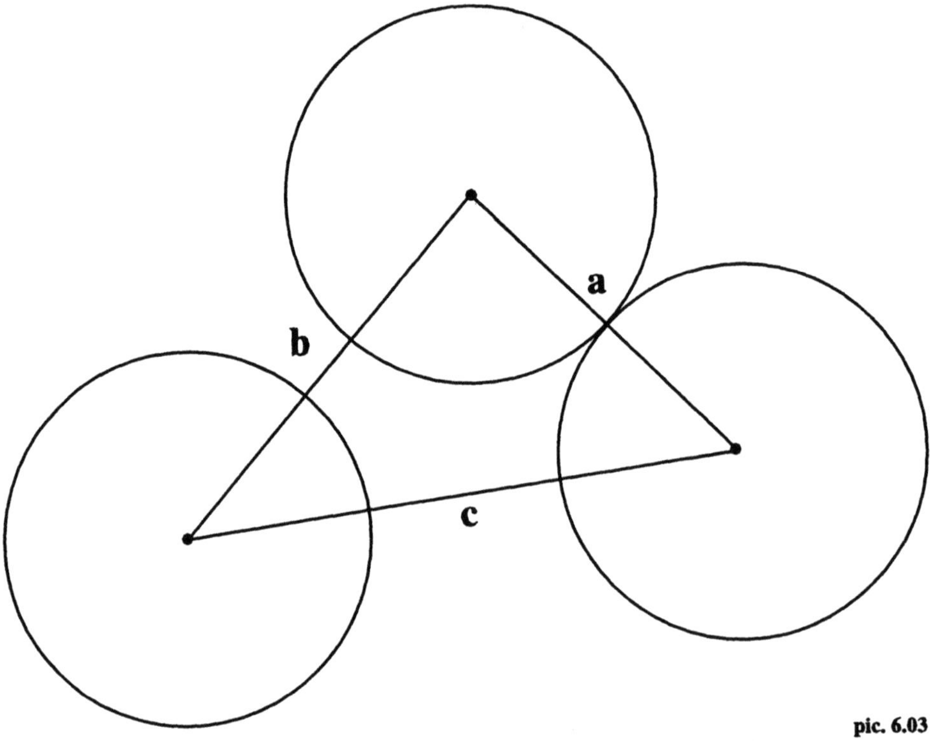

pic. 6.03

Lemma 5 *In the conditions of lemma 4, every L-triangle is intersected only with those disks of the packing whose centres are located in its vertices.*

▷ Each point of Σ is covered together with their neighbourhood only with "its own" disk. Therefore for us it is enough to show that in our conditions the circle of radius r drew from an edge of a triangle does not intersect the edge opposite to it. Let the triangle has the edges a, b, c, the area S and radius of circumscribed circle R. We shall calculate its height h_a. We have $abc = 4RS$. From here, taking into account, that $R < 2r$ and $a, b, c \geq 2r$, we obtain

$$H_a = 2S/a = bc/2R > b/2 \geq r.\bullet$$

For further, for such L-tilings let's divide each quadrangle and pentagon by its diagonals, with one of its ends in any one vertex, on two and three triangles, accordingly. It is obvious, that everything told above about L-triangles is fair, as well for the triangles obtained in such a way.

2^0. **The proof of theorem 6.1**

▷ Let's take an arbitrary packing $\mathfrak{T}$ of disks of radius 1, which have as centres the point system $\Sigma(\mathfrak{T})$. If this packing is not satureted, we shall supplement it with the same disks up to satureted. The density (superior density) of the packing, whatever a measuring figure was, will not decrease because of it, so we can assume at once that the packing $\mathfrak{T}$ is satureted. Hence, we can consider the system $\Sigma(\mathfrak{T})$ as a Delone system.

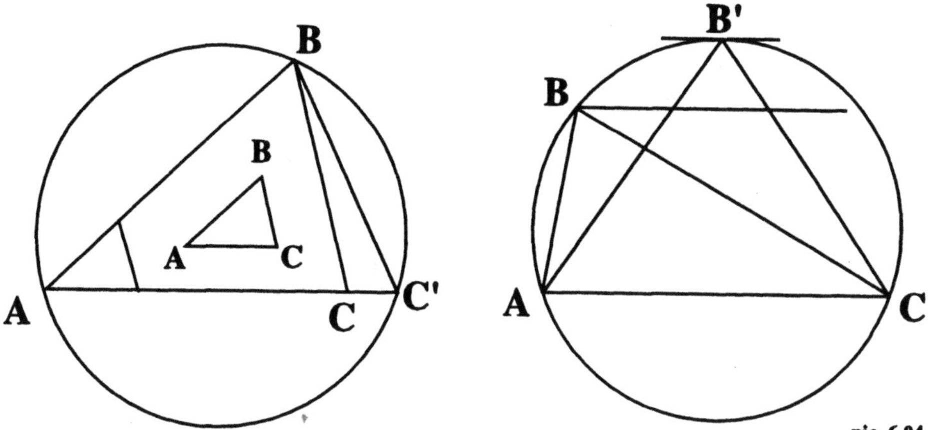

pic. 6.04

Let $\delta_{\lambda M}(\mathfrak{T})$ be the common area of all disks of $\mathfrak{T}$, which have got in the figure λM, and are divided on the common area of all triangles of $\{L\}_{\Sigma(\mathfrak{T})}$, which have the nonempty intersection with λM. We shall denote the area of an L-triangle through S_{tr}, the area of the $\mathfrak{T}$-disks' sectors, which have got in it, through S_{cir} and we shall denote the summation over all L-triangles, which intersected λM, through Σ_L. See pic. 6.05 (for simplicity in this picture the packing is not saturated).

We shall estimate the density of the packing by applying lemmas 1, 5 and 2:

$$\delta_{\lambda M}(\mathfrak{T}) \le \sum_L S_{\text{cir}} \Big/ \sum_L S_{\text{tr}} \le \max(S_{\text{cir}}/S_{\text{tr}})$$
$$\le \frac{\pi}{2} \Big/ \sqrt{3} = \pi/\sqrt{12}.$$

As, according to theorem 5.2, it is possible to neglect boundary effects, we have $\bar{\Delta}_M(\mathfrak{T}) \le \pi/\sqrt{12}$.

The right part of the inequality does not depend on M, so we have

$$\bar{\Delta}(\mathfrak{T}) \le \pi/\sqrt{12}.$$

As the hexagonal packing is periodic, the second statement of the theorem is obvious.•

§6.3 The Thinnest Covering by Equal Disks

1^0. **Lemma 6.** *Let radius of the circle circumscribed around a triangle ABC be not more than r. Let also S_A, S_B, S_C be the sectors of the disks of radius r with the centres in the points A, B, C accordingly, determined by the straight lines AB and AC, BA and BC, CA and CB. Then the triangle ABC is covered by the sectors S_A, S_B, S_C.* See pic 6.06.

▷ For the proof, it is enough to consider three figures formed by perpendiculars, dropping from the centre of the circumscribed circle of ABC to its vertices, and by halves of the edges of ABC.•

2^0. **The proof of theorem 6.2**

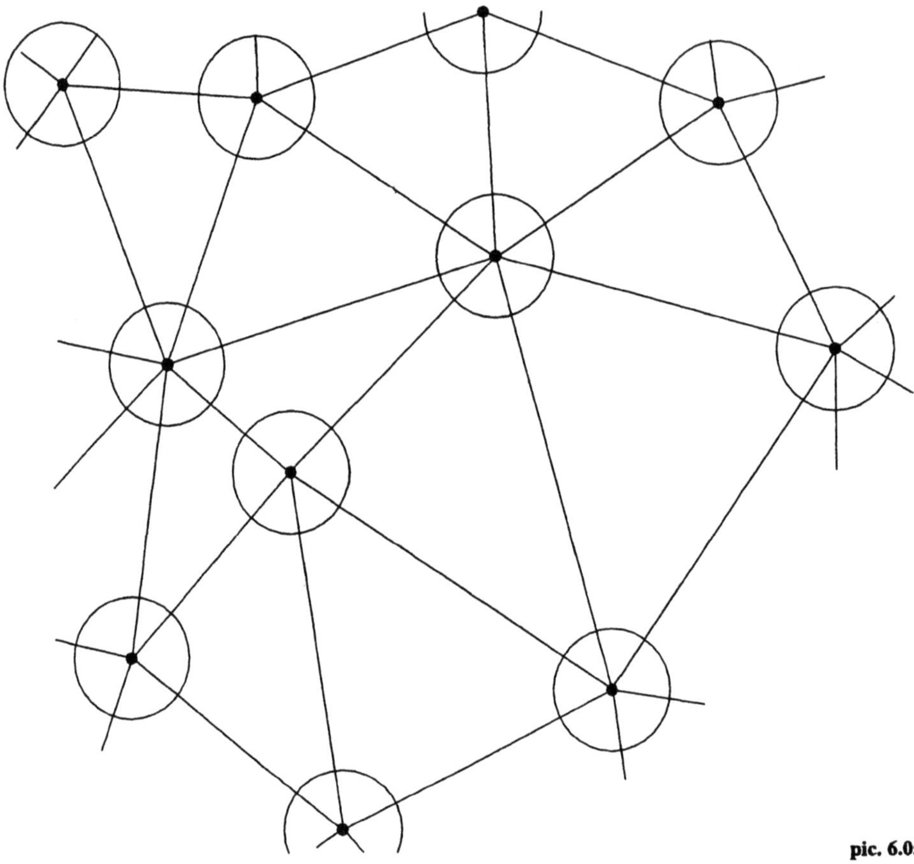

pic. 6.05

▷ Let's take an arbitrary covering $\mathfrak{T}$ of the plane $\mathbb{E}^2$ by disks of radius 1, which have the point system $\Sigma(\mathfrak{T})$ as centres. Thus the system $\Sigma(\mathfrak{T})$ has $R_{\Sigma(\mathfrak{T})} \leq 1$. And it has no limit points, otherwise the covering $\mathfrak{T}$ is not locally finite. We shall estimate the density in a bounded domain of an aspect λM, therefore there will be discovered in it the radius r_Σ^λ of the discreteness for the system $\Sigma(\mathfrak{T})$. By replacing outside of λM the system $\Sigma(\mathfrak{T})$ on an arbitrary Delone system, which has $R_\Sigma \leq 1$, we shall obtain the Delone system $\Sigma = \Sigma_\lambda$. For each system Σ_λ we can construct the L-tiling $\{L\}_\lambda$. If $\lambda < \mu$, the tilings $\{L\}_\lambda$ and $\{L\}_\mu$ do not differ inside $(\lambda - 1)M$. We shall further denote through $\{L\}_\lambda$ only this part of the first tiling . By subdividing, as well as above, each L-polygon on triangles, we shall consider $\{L\}_\lambda$ as a complex, which consists of triangles only.

Let's consider an arbitrary triangle $L_0 \in \{L\}_\lambda$. Let S_{tr} be its area. The sectors which are cut out by the angles of the triangle L_0 from the disks of $\mathfrak{T}$, which have the vertices of L_0 as the centres, according to lemma 6, cover whole this triangle. Let $S_{\mathrm{cir}} = \pi$ be the sum of these sectors' areas. Through Σ_L we shall denote the summation over all such triangles.

Let $\sigma_{\lambda M}(\mathfrak{T})$ be the common area of all disks of $\mathfrak{T}$, which have a nonempty intersection with the figure λM, and are divided on the common area of all triangles of $\{L\}_\lambda$.

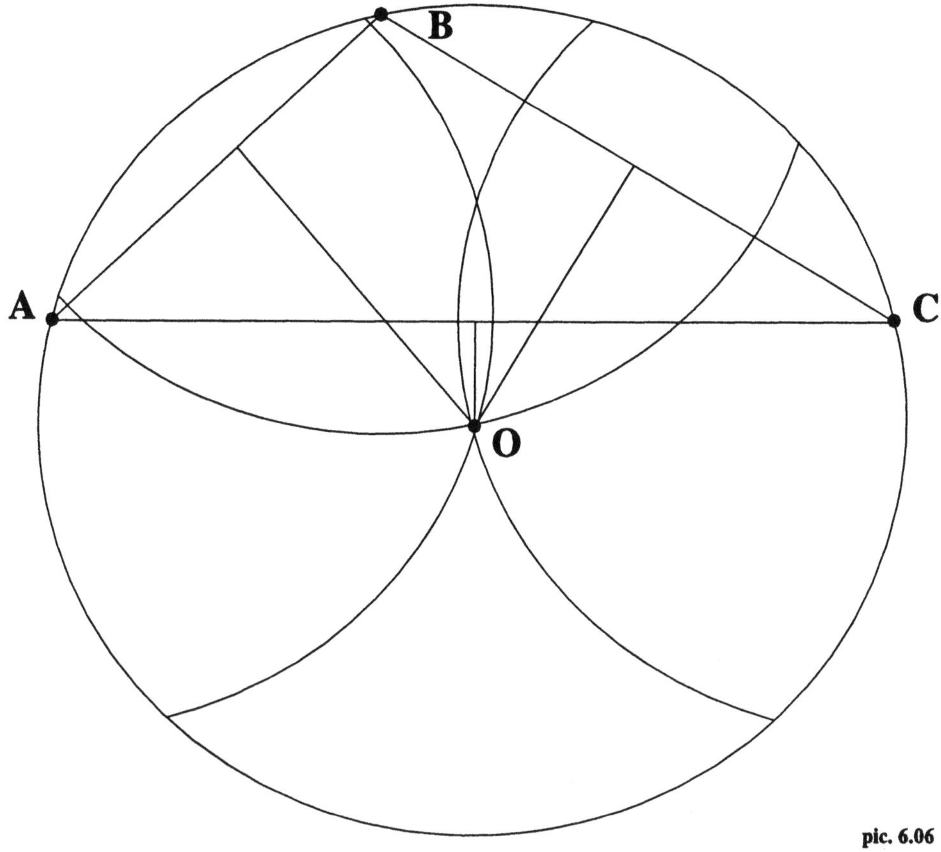

pic. 6.06

We shall estimate the density of the covering, by applying lemmas 1, 5 and 2:

$$\begin{aligned}\sigma_{\lambda M}(\mathfrak{T}) &\geq \sum_L S_{\mathrm{cir}} \Big/ \sum_L S_{\mathrm{tr}} \geq \min(S_{\mathrm{cir}}/S_{\mathrm{tr}}) \\ &\geq \frac{\pi}{2} \Big/ (3\sqrt{3}/4) = 2\pi/3\sqrt{3}.\end{aligned}$$

As, according to theorem 2.2, it is possible to neglect boundary effects, we have $\Delta_M(\mathfrak{T}) \geq 2\pi/3\sqrt{3}$.

The right part of the inequality does not depend on M, therefore we have

$$\Delta(\mathfrak{T}) \geq 2\pi/3\sqrt{3}.$$

As the hexagonal covering is periodic, the second statement of the theorems is obvious.•

Exponential Diophantine Equations Involving Products of Consecutive Integers and Related Equations

T.N. Shorey

Dedicated to Professor Alan Baker on his 60th birthday

0 Introduction

This paper contains an account of the results on the following topics:

1. Squares in products from a block of consecutive integers
2. Equal products of consecutive integers
3. An equation of Goormaghtigh
4. An equation of Nagell-Ljunggren
5. Equal products of integers in arithmetic progressions
6. The greatest prime factor of integers in arithmetic progression
7. Cubes and higher powers in products from a block of consecutive integers
8. Perfect powers in products of integers in arithmetic progression

It is not our intention to give a historical survey. We have included 112 references and we hope that they will provide an access to the related results which are not mentioned in this paper. In Section 1, we have described the method of Erdös and Rigge that a product of two or more consecutive positive integers is never a square. A sketch of developments on the method of Erdös and Rigge is also included. In Section 2, we have explained an extension of Runge's method to exponential diophantine equations by giving a proof that for a given integer $m \geq 2$, there are only finitely many instances when a product of k consecutive positive integers is equal to a product of mk consecutive positive integers. In the next section, the above method is combined with the theory of linear forms in logarithms for applying to an equation of Goormaghtigh. The equation of Goormaghtigh asks for integers whose digits are all equal to one with respect to two distinct bases. In Section 4, we give a survey of the results on the equation of Nagell-Ljunggren asking for perfect powers in integers with all the digits equal to one. It is shown that abc conjecture implies that the equations of Goormaghtigh and Nagell-Ljunggren have only finitely many solutions. We apply the latter equation and its particular case (30) with m even to show that certain numbers considered by Mahler are irrational. Solving (30) with m even is equivalent to solving simultaneously a Nagell-Ljunggren equation and an equation of Catalan. The equation of Catalan is a particular case of the equation of Pillai. Thus we turn to considering the equations of Catalan and Pillai in Section 4. As an extension of the theorem of Schinzel and Tijdeman on hyper-elliptic equations in the theory of exponential diophantine equations, we formulate a conjecture which includes a well-known conjecture of Pillai on his equation. The formulated conjecture follows from generalised *abc* conjecture. The Section 5 is a continuation of Section 2. It gives an account of developments in the extension of Runge's method to exponential diophantine equations on equal products in arithmetic progressions. The Section 7 explains an elementary method of Erdös by proving that there are only finitely many possibilities when a product of two or more consecutive positive integers is a cube or a higher power. We also consider a more general question on an upper bound for the number of integers in a block of consecutive positive integers such that their product is a cube or higher power. The treatment to this more general question is no more elementary. The Section 8 gives analogues of the results of Section 7 for arithmetic progressions. An assumption gcd $(n, d) = 1$ in a result on equation $n(n+d)\cdots(n+(k-1)d) = by^{\ell}$ has been relaxed to $d \nmid n$ in Section 8. Further we show that *abc* conjecture implies that k is bounded by an absolute constant whenever the preceding equation with gcd $(n, d) = 1$ and $l > 3$ holds. The Section 6 is concerned with improvements and extensions of a well-known theorem of Sylvester that a product of k consecutive positive integers $> k$ is divisible by a prime exceeding k. It gives estimates on particular cases of equations considered in Sections 1, 7, 8 and the results of Sections 7, 8 imply some estimates related to the ones considered in Section 6. The paper points out crucial basic tools and it suggests several problems for further investigations. It aims at updating the book of Shorey and Tijdeman [95] on exponential diophantine equations written in 1986. An article with similar intention was already written by Tijdeman [109]. The proofs of several results of this book depend on its Theorems B.3 and B.4 due to van der Poorten on p-adic linear forms in logarithms. The proofs of van der Poorten of these theorems have turned out to be incorrect. But correct proofs have been established by Yu [111], [112]. Hence all the consequences of Theorems B.3 and B.4 in the book are valid. All the constants appearing in this paper are effectively computable. Unless otherwise specified, the results mentioned in this paper are effective. For an integer $\nu > 1$, we write $P(\nu)$ and $\omega(\nu)$ for the greatest prime factor and the number of distinct prime divisors of ν, respectively, and we put $P(0) = P(1) = 1$, $\omega(0) = \omega(1) = 0$.

1 Squares in Products from a Block of Consecutive Integers

Erdös [22] and Rigge [63] proved in 1939 that a product of two or more consecutive positive integers is never a square. Further Erdös and Selfridge [27] proved in 1975 the analogous statement for cubes and higher powers. Therefore a product of two or more consecutive positive integers is never a power. The first contribution in this direction dates back to 1724 when Goldbach, in a letter to D. Bernoulli, showed that a product of three consecutive positive integers is not a square. The theorem of Erdös-Rigge is equivalent to saying that equation

$$n(n+1)\cdots(n+k-1) = y^2 \text{ in integers } n > 0, y > 0, k \geq 2 \tag{1}$$

has no solution. First we give a sketch of the proof that (1) has no solution whenever k exceeds a sufficiently large absolute constant.

If $n \leq k$, we refer to Bertrand's postulate to find a prime p satisfying $n \leq \frac{n+k}{2} \leq p \leq n+k-1$ and dividing the left hand side of (1) to the first power. Therefore (1) implies that

$n > k$. Now we apply a well-known theorem of Sylvester [105] to find a prime q exceeding k and dividing the left hand side of (1). Consequently there is unique i with $0 \le i < k$ such that $q\,|\,(n+i)$. Furthermore we derive from (1) that $n+i$ is divisible by q^2. Hence

$$n + k > n + i \ge q^2 \ge (k+1)^2$$

which implies that

$$n > k^2. \tag{2}$$

We assume that k exceeds a sufficiently large absolute constant so that all our estimates are valid. By (1), we write

$$n + i = a_i x_i^2 \quad \text{for } 0 \le i < k$$

where a_i is square free and $P(a_i) \le k$. The inequality (2) implies that $a_o, a_1, \ldots, a_{k-1}$ are distinct and this is fundamental for the proof. Let $a_i = a_j$ with $i > j$. Then

$$\begin{aligned} k > i - j &= a_j(x_i^2 - x_j^2) \\ &= a_j(x_i - x_j)(x_i + x_j) \\ &\ge 2a_j x_j \ge 2(a_j x_j^2)^{1/2} \ge 2n^{1/2} \end{aligned} \tag{3}$$

which contradicts (2). The proof depends on comparing a lower and an upper bound for

$$\Delta =: a_o, a_1, \ldots, a_{k-1}.$$

Since $a_o, a_1, \ldots, a_{k-1}$ are distinct, it is clear that

$$\Delta \ge k!.$$

This can be sharpened to

$$\Delta \ge \left(\frac{3}{2}\right)^k k!, \tag{4}$$

by using that $a_o, a_1, \ldots, a_{k-1}$ are square free. Now we turn to an upper bound for Δ. If a prime p divides a_i and a_j with $i > j$, then $i \equiv j \pmod{p}$. Therefore, since a_i's are square free, we have

$$\operatorname{ord}_p(\Delta) \le \left[\frac{k}{p}\right] + 1.$$

Thus

$$\Delta \,\Big|\, k! \prod_{p \le k} p.$$

In particular

$$\Delta \le k! \prod_{p \le k} p \le k! 3^k.$$

Our intention is to find an upper bound for Δ less than the lower bound (4). The above upper bound is not sufficient for this purpose and we must sharpen it. Let $q \in \{2, 3\}$ and we write

$$g_q = \operatorname{ord}_q(\Delta), \; h_q = \operatorname{ord}_q(k!).$$

In fact we have

$$\Delta \Big| k! \prod_{p \le k} p \, 2^{g_2-h_2} 3^{g_3-h_3}.$$

Since a_i is square free, a prime divides a_i if and only if it divides $n+i$ to an odd power. Thus one can imagine that g_q is much smaller than h_q. This is the case:

$$g_2 - h_2 \le -\frac{2k}{3} + 2\frac{\log k}{\log 2} + 2$$

and

$$g_3 - h_3 \le -\frac{k}{4} + 2\frac{\log k}{\log 3} + 2.$$

We obtain

$$\Delta \le k! 3^k 2^{-2k/3} 3^{-k/4} 36 k^4. \tag{5}$$

Finally we combine (4) and (5) to obtain

$$(1.04)^k \le \left(\frac{3}{2}\frac{1}{3} 2^{2/3} 3^{1/4}\right)^k \le 36k^4$$

which is not possible if k is sufficiently large. Hence (1) has no solution whenever k exceeds a sufficiently large absolute constant.

We always write b for a positive integer such that $P(b) \le k$. Let $d_1, \dots, d_t$ with $t \ge 2$ be distinct integers in $[0, k)$. We consider

$$(n+d_1)\cdots(n+d_t) = by^2. \tag{6}$$

Equation (6) with $t = k$ and $b = 1$ coincides with (1). It is natural to suppose that the left hand side of (6) is divisible by a prime exceeding k. Then Saradha [67], [68] showed that equation (6) with $t = k$ and $k \ge 3$ has no solution other than the one given by $\binom{50}{3} = 140^2$. This includes a result of Erdös [23] of 1951 on (6) with $t = k$ and $b = k!$ i.e.

$$\binom{n+k}{k} = y^2.$$

The assumption $k \ge 3$ is necessary in the above result of Saradha; equation (6) with $t = k = b = 2$ has infinitely many solutions. Now we consider the question of relaxing the assumption $t = k$. Erdös [25] observed in 1955 that the proof given above allows to show that there exist absolute constants C_1 and C_2 such that (6) with

$$n > k^2, t \ge k - C_1\frac{k}{\log k} \tag{7}$$

implies that $k \le C_2$. The assumption $n > k^2$ is satisfied if the left hand side of (6) is divisible by a prime exceeding k. Saradha [67] showed that equation (6) with $n > k^2$,

$t \geq k - [\frac{(0156)k}{\log k}]$ implies that $k < 870$. It is easy to see that the above proof fails if (7) is replaced by

$$n > k^2, \ t \geq k - \frac{kf(k)}{\log k}$$

with $f(k) \to \infty$ as $k \to \infty$. Therefore we should look for replacing the idea of saving powers of 2 and 3 in the proof given above by a different argument and it was found by Shorey [84] in 1986. For $\epsilon > 0$, Shorey [87] applied in 1987 Baker's theorem on integral solutions of hyper-elliptic equations and Sieve-theoretic arguments to show that (6) with

$$n > k^2, \ t \geq k - (1-\epsilon)k \frac{\log \log k}{\log k} \tag{8}$$

implies that k is bounded by a number depending only on ϵ. This was asked by Erdös [25] in 1955 when he proved the analogous result for higher powers. We shall explain the proof. Before this we continue describing further results on (6). Balasubramanian and Shorey [7] obtained a further relaxation of the assumption (8), namely,

$$n > k^2, \ t \geq \mu_k \tag{9}$$

where

$$\mu_k = k\left(1 - \frac{\log \log k}{\log k} + \frac{\log \log \log k}{\log k} + \frac{\theta_0}{\log k}\right)$$

for some absolute constant θ_0. We fix θ_0 and write

$$F(k) = k(\log k)/\log \log k \ \text{ for } \ k \geq 3.$$

We consider (6) with

$$e^{1-\theta_0+\epsilon}F(k) < n \leq k^2. \tag{10}$$

Then

$$P(n + d_i) \leq k \ \text{ for } 1 \leq i \leq t.$$

For every prime $p \leq k$, we choose an $f(p) \in \{n + d_1, \ldots, n + d_t\}$ such that p does not appear to a higher power in the factorisation of any other $n + d_i$ with $1 \leq i \leq t$. Then, by an argument of Erdös [25], we have

$$n^{t-\pi(k)} \leq \prod^{*}(n + d_i) \leq \prod_{p \leq k} p^{[\frac{k}{p}]+[\frac{k}{p^2}]+\cdots} \leq k^k$$

which, by (10), implies that $t < \mu_k$. Here the product $\prod^*$ is taken after omitting all $f(p)$ with $p \leq k$. The above argument has turned out to be basic and we shall apply it again in Sections 6, 7, 8 where we refer to it as 'a fundamental argument of Erdös'. Further we observe that the assumption (9) can be replaced by

$$n > e^{1-\theta_0+\epsilon}F(k), \ t \geq \mu_k.$$

On the other hand, Balasubramanian and Shorey [7] showed that the preceding assumption is close to the best possible. More precisely, they [7] proved: Let

$$\epsilon > 0, k \geq 3 \text{ and } n \leq e^{-1-\gamma-\theta_0-\epsilon} F(k)$$

where γ denotes the Euler's constant. For $k \geq k_0 = k_0(\epsilon)$, we can find distinct integers $d_1, \ldots, d_t \in [0, k)$ with $t \geq \mu_k$ such that $(n+d_1)\cdots(n+d_t)$ is a square. Now we give a sketch of the proof of this assertion. For a set T of positive integers, we write $\omega(T)$ for the number of distinct prime divisors of all elements of T. Then it is well-known that there exists a subset T' of T with

$$|T'| \geq |T| - \omega(T)$$

such that the product of all elements of T' is a square. We take T for all $n+i$ with $0 \leq i < k$ such that $P(n+i) \leq k$ and we write T_1 for the complement of T in $\{n, n+1, \ldots, n+k-1\}$. It is enough to show that

$$|T| \geq \mu_k + \pi(k)$$

i.e.

$$|T_1| \leq k - \mu_k - \pi(k).$$

It is easy to check that

$$|T_1| \leq \sum_{1 \leq \lambda \leq (n+k-1)/k} \left(\pi\left(\frac{n+k-1}{\lambda}\right) - \pi\left(\frac{n-1}{\lambda}\right)\right).$$

The proof is completed by applying deep results on difference between consecutive primes for obtaining the desired estimate.

As promised, we turn to giving a sketch of the proof that (6) with (8) implies that k is bounded by a number depending only on ϵ. We may assume that k exceeds a sufficiently large number depending only on ϵ. By (6), we have

$$n + d_i = a_i x_i^2 \text{ for } 1 \leq i \leq t \tag{11}$$

where a_i and x_i are positive integers such that $P(a_i) \leq k$ and a_i is square free. We put

$$S = \{a_1, \ldots, a_t\}.$$

As earlier we observe from $n > k^2$ that $|S| = t$ and

$$a_1, \ldots, a_t \leq \prod_{p \leq k} p^{[k/p]+1} \leq (3k)^k.$$

Then there exists a subset S_1 of S such that

$$|S_1| \geq \epsilon k/2 \tag{12}$$

and

$$a_i \leq k(\log k)^{1-(\epsilon/2)} \text{ if } a_i \in S_1. \tag{13}$$

Further we notice from (11), (8) and (13) that

$$x_i \geq k^{1/4} \quad \text{if} \quad a_i \in S_1. \tag{14}$$

We write S_2 for the set of all $a_i \in S_1$ such that $a_i \leq 3k$ and we denote by S_3 the complement of S_2 in S_1. We split the proof in two cases:

Case I $|S_2| \leq \epsilon k/4$

Then we observe from (12) that

$$|S_3| \geq \epsilon k/4. \tag{15}$$

Further we use $a_i > 3k$ for showing as in (3) that the products

$$a_i a_j \text{ with } a_i \in S_3, a_j \in S_3$$

are distinct. This property is restrictive; Erdös [25] showed by Sieve that

$$|S_3| \leq 2G/\log \mathrm{G} \tag{16}$$

where G is the largest element of S_3. By (16) and (13), we have

$$|S_3| \leq k(\log k)^{-\epsilon/4}$$

which contradicts (15).

Case II $|S_2| > \epsilon k/4$

We start with three distinct elements of S_2. By permuting $d_1, \ldots, d_t$, there is no loss of generality in numbering them as a_1, a_2 and a_3. We write from (11) that

$$a_2 x_2^2 = a_1 x_1^2 + (d_2 - d_1)\,,\; a_3 x_3^2 = a_1 x_1^2 + (d_3 - d_1).$$

We put

$$a = \text{ gcd } (a_1, a_2, a_3), \quad b_i = a^{-1} a_i \text{ for } 1 \leq i \leq 3$$

and

$$R = a^{-1}(d_2 - d_1), \; R' = a^{-1}(d_3 - d_1).$$

Then

$$b_2 b_3 (x_2 x_3)^2 = (b_1 x_1^2 + R)(b_1 x_1^2 + R').$$

Thus we have arrived at a hyper-elliptic equation. For applying results on integral solutions of hyper-elliptic equations, we must have control on the magnitude of b_1, b_2 and b_3. This is provided by a Sieving argument due to Erdös. In this case, we can find three distinct elements of S_2 such that b_1, b_2 and b_3 are bounded. Now we apply a theorem of Baker [3] to derive that $\max(x_1, x_2, x_3)$ is bounded which, together with (14), implies that k is bounded. Hence we conclude the assertion that (6) with (8) implies that k is bounded by a number depending only ϵ. The proof of Baker's theorem depends on his theory of linear forms in logarithms. This deep theory finds applications at several places in this paper and we refer to Baker [4] for an account.

2 Equal Products of Consecutive Integers

We consider (1) with the left hand side replaced by two blocks of consecutive integers. More precisely, we consider the equation

$$X(X+1)\cdots(X+K-1)Y(Y+1)\cdots(Y+K'-1)=Z^2,$$

where $K \geq 2$, $K' \geq 2$, $K+K' \geq 5$ and $X \geq Y+K'$. It has been conjectured (see [26]) that this equation has only finitely many solutions in all the integral variables $X > 0, Y > 0$, $Z > 0$, K and K'. This conjecture implies that

$$x(x+1)\cdots(x+k-1)=y(y+1)\cdots(y+k+\ell-1)$$

has only finitely many solutions in integers $x > 0$, $y > 0, k \geq 2$ and $\ell \geq 0$ satisfying $x \geq y+k+\ell$. More generally, for relatively prime positive integers A and B, Erdös [26] conjectured in 1975 that there are only finitely many integers $x > 0$, $y > 0$, $k \geq 3$, $\ell \geq 0$ with $x \geq y+k+\ell$ satisfying

$$Ax(x+1)\cdots(x+k-1)=By\,(y+1)\cdots(y+k+\ell-1). \tag{17}$$

All the solutions of (17) have been determined for several values of (A, B, k, ℓ) and we refer to [91, p. 239] for an account of these results. The first result in this direction is due to Mordell [48] that (17) with $A = B = 1$, $k = 2$, $\ell = 1$ implies that $x = 2, y = 1$ and $x = 14$, $y = 5$. For a given (A, B, k, ℓ), Beukers, Shorey and Tijdeman [11] confirmed the above conjecture of Erdös. For the proof, they showed that the underlying curve is irreducible and it has positive genus. Then the assertion follows from a theorem of Siegel [104] on integral points on curves but it provides no explicit bound for the magnitude of the solutions. In fact they determined all (A, B, k, ℓ) for which the curve has genus one and (17) has only finitely many rational solutions for all other (A, B, k, ℓ) by a theorem of Faltings [28] on Mordell's conjecture. In particular, they [11] proved that (17) with $A = B = 1$ has only finitely many rational solutions. These results are non-effective in the sense that they provide no explicit estimate for the heights of the solutions. If $\ell = 0$, Shorey [82] confirmed the conjecture of Erdös when $x-1$ and $y-1$ are composed of fixed primes. Saradha and Shorey [69] extended this result to all $\ell \geq 0$ and the proof depends on the theory of linear forms in logarithms. Further Saradha and Shorey [69] showed that (17) with $x \geq y+k+\ell$ implies that

$$x-y \geq C_3 x^{2/3}$$

for certain number $C_3 > 0$ depending only on A and B. It is a difficult problem to confirm the above conjecture in general. Even a very particular case $A = B = 1$, $k = 2$ and $y = 1$ of (17) is an open problem. We consider (17) with $A = B = 1$ and $k+\ell$ an integral multiple of k. For an integer $m \geq 2$, we re-write (17) in this case as

$$x(x+1)\cdots(x+k-1)=y(y+1)\cdots(y+mk-1)\ \text{ in integers } x>0, y>0, k\geq 2. \tag{18}$$

MacLeod and Barrodale [40] considered (18) with $m = 2$. If $m = 2$, equation (18) has a solution given by

$$8.9.10 = 6!$$

i.e.

$$x = 8,\ y = 1,\ k = 3. \tag{19}$$

MacLeod and Barrodale [40] observed that (19) is the only solution of (18) with $m = 2$ and $k \leq 5$. Saradha and Shorey [69] showed that (19) is the only solution of (18) with $m = 2$. Further Saradha and Shorey [70] proved that (18) with $m = 3, 4$ has no solution and Mignotte and Shorey [47] verified it for $m = 5, 6$. In general Saradha and Shorey [71] showed that (18) implies that

$$\max(x, y, k) \leq C_4(m) \tag{20}$$

where $C_4(m)$ is a number depending only on m. We have not been able to replace $C_4(m)$ by an absolute constant. It is likely that (18) with $m \geq 3$ has no solution.

Now we give a sketch of the proof of (20). There is no loss of generality in proving (20) for integers $x \geq 0$, $y \geq 0$ and $k \geq 2$ satisfying

$$(x + 1) \cdots (x + k) = (y + 1) \cdots (y + mk). \tag{21}$$

By counting the powers of 2 on both the sides of equation (21), we find that x and y are large as compared with k. In fact

$$x \geq 2^k - k.$$

We write

$$(z + 1) \cdots (z + mk) = \sum_{j=0}^{mk} A_j(m, k) z^{mk-j}. \tag{22}$$

Further we determine rational numbers

$$B_j = B_j(m, k) \text{ for } 1 \leq j \leq m$$

such that

$$(z^m + B_1 z^{m-1} + \cdots + B_m)^k = \sum_{j=0}^{mk} H_j(m, k) z^{mk-j} \tag{23}$$

satisfies

$$H_j(m, k) = A_j(m, k) \text{ for } 0 \leq j \leq m. \tag{24}$$

By (23) and (24),

$$\begin{aligned} kB_1 &= A_1(m, k), \\ kB_2 + \binom{k}{2} B_1^2 &= A_2(m, k), \\ \cdots\cdots\cdots\cdots &\ \cdots\cdots \\ \cdots\cdots\cdots\cdots &\ \cdots\cdots \\ \cdots\cdots\cdots\cdots &\ \cdots\cdots \end{aligned}$$

Therefore $B_1, \ldots, B_m$ are determined recursively.

Let z be a sufficiently large number as compared with k and m. The relations (24) imply that the left hand side of (22) is close to the left hand side of (23). Therefore the k-th root of the left hand side of (22) is close to the k-th root of the left hand side of (23). We use this observation with z replaced by x and y, since x and y are large as compared with k and m. The k-th root of the left hand side of (21) is close to $x + \frac{k+1}{2}$ and the k-th root of the right hand side of (21) is close to $y^m + B_1 y^{m-1} + \cdots + B_m$. Consequently we derive that $x + \frac{k+1}{2}$ is close to $y^m + B_1 y^{m-1} + \cdots + B_m$. In fact, we show that

$$\left| x - \left(y^m + B_1 y^{m-1} + \cdots + B_m - \frac{k+1}{2} \right) \right| < \tau$$

where

$$\tau = (2 \text{ lcm } (\text{den } (B_1), \ldots, \text{ den } (B_m)))^{-1}.$$

Therefore

$$x = y^m + B_1 y^{m-1} + \cdots + B_m - \frac{k+1}{2}.$$

We substitute this relation in (21) to get

$$H_j(m,k) = A_j(m,k) \quad \text{for} \quad 0 \le j < 2m.$$

Thus we have added $m-1$ relations to (24) with which we started. This is the main idea of the proof. Further Balasubramanian (see [71, Appendix]) showed that these $2m-1$ relations can not hold whenever k is sufficiently large. For a different and more general proof of the preceding assertion, see Balasubramanian and Shorey [6]. Thus we may suppose that k is bounded and y is sufficiently large. Then we derive the polynomial identity

$$(X+1)\cdots(X+k) = (Y+1)\cdots(Y+mk),$$
$$X = Y^m + B_1 Y^{m-1} + \cdots + B_m - \frac{k+1}{2}.$$

This is not possible and the assertion (20) follows.

We observe that k is a variable in equation (21). The left hand side of (21) is a polynomial in x of degree k and the right hand side is a polynomial in y of degree mk. Thus the exponents also appear as variables in equation (21). Therefore (21) is an exponential diophantine equation. Thus the above method can be considered as an extension of a method of Runge [64] to some exponential diophantine equations. We shall apply the above method to an equation of Goormaghtigh which we consider in the next section. Let $\delta = 1$ or $\delta = 2$ according as m is odd or even, respectively. It has been conjectured in Saradha and Shorey [71] that $H_{m+\delta}(m,k) \ne A_{m+\delta}(m,k)$. It is easy to see that $H_{m+1}(m,k) = A_{m+1}(m,k)$ if m is even. Glesser [70, Appendix] and Mignotte and Shorey [47] checked the conjecture for $m \le 12$ and $13 \le m \le 20$, respectively. Further Mignotte and Shorey [47] showed that (21) with $H_{m+\delta}(m,k) \ne A_{m+\delta}(m,k)$ implies that

$$y \le 1.026 \times 2^m (m+1+\delta)! k^{m+1+\delta} (2mk)^{m+2+\delta}$$

and

$$y \ge 2^{1/m} 2^{k(1-1/m-(\log_2(mk))/mk)} - \left(\frac{mk+2}{2}\right)$$

where $\log_2$ denotes the logarithm with respect to the base 2. Thus the above estimates are valid for $7 \leq m \leq 20$. We recall that (21) with $2 \leq m \leq 6$ has no solution other than $m = 2,\ x = 7,\ y = 0,\ k = 3$.

3 An Equation of Goormaghtigh

We start this section with the well-known $a\ b\ c$ conjecture: *Let $\epsilon > 0$ and a, b, c be relatively prime positive integers satisfying*

$$a + b = c.$$

Then there exists a number κ depending only on ϵ such that

$$c < \kappa G^{1+\epsilon}$$

where

$$G = \prod_{p|abc} p.$$

Next we recall that Nagell [51] confirmed a conjecture of Ramanujan [60] that the solutions of

$$x^2 + 7 = 2^n \text{ in integers } x > 0, n > 0 \tag{25}$$

are given by $(x, n) = (1, 3), (3, 4), (5, 5), (11, 7), (181, 15)$. We consider an equation of Goormaghtigh

$$\frac{x^m - 1}{x - 1} = \frac{y^n - 1}{y - 1} \text{ in integers } x > 1, y > 1, m > 2, n > 2 \text{ with } x \neq y. \tag{26}$$

It has been conjectured that (26) has only finitely many solutions. Goormaghtigh [30] observed in 1917 that

$$31 = \frac{2^5 - 1}{2 - 1} = \frac{5^3 - 1}{5 - 1},\ 8191 = \frac{2^{13} - 1}{2 - 1} = \frac{90^3 - 1}{90 - 1}. \tag{27}$$

There is no loss of generality in assuming that $x < y$ in (26) and then $m > n$. Further we re-write (26) as

$$(y - 1)x^m - (x - 1)y^n = y - x. \tag{28}$$

We show that $a\ b\ c$ conjecture implies that (26) has only finitely many solutions. By applying $a\ b\ c$ conjecture to (28) after dividing both the sides by $g = \gcd\left((x-1)y^n, y-x\right)$, we have

$$(y - 1)x^m < \kappa_1 G_1^{1+\epsilon}$$

where $\epsilon = 1/100$, κ_1 is an absolute constant and G_1 is the product of all prime divisors of $xy(x - 1)(y - 1)(y - x)/g$. Therefore

$$x^{m-2-2\epsilon} < \kappa_1 y^{2+3\epsilon}$$

and we may suppose that y exceeds a sufficiently large absolute constant. On the other hand, we see from (26) that $y^{n-1} < 2x^{m-1}$ which implies that

$$x^{m-2-2\epsilon} > \frac{1}{2} y^{(n-1)(m-2-2\epsilon)/(m-1)}.$$

Finally we combine the preceding two inequalities to conclude that $n = 3$. I thank A. Granville for pointing out an inaccuracy in the above application of abc conjecture in an earlier draft of this paper. Further N. Saradha observed that abc conjecture implies that (26) with $n = 3$ has only finitely many solutions. For this, we observe that $m \geq 4$ and we re-write (26) with $n = 3$ as

$$(2y + 1)^2 = 4(x^{m-1} + \cdots + x) + 1$$

and we check that the roots of the polynomial $4(X^{m-1} + \cdots + X) + 1$ are simple. Then we may suppose that $m \geq 6$ by a result of Baker [3] on integral solutions of hyper-elliptic equations. Now we re-write (26) as

$$4x^m = (x - 1)(2y + 1)^2 + (3x + 1)$$

and we apply $a\ b\ c$ conjecture to conclude the assertion.

It is not known that (26) with $n = 3$ has only finitely many solutions. Nesterenko and Shorey [53] proved that (26) with $m \equiv 1 \pmod 2$, $m \leq 23$ and $n = 3$ has no solution other than the ones given by (27). Let $n = 3$ and $x = 2$ in (26). Then $y > 2$ and we re-write (26) as

$$(2y + 1)^2 + 7 = 2^{m+2}.$$

Now we apply the result of Nagell stated above in this section to conclude that (26) with $x = 2$ and $n = 3$ implies (27). Thus the numbers 31 and 8191 are given by Ramanujan-Nagell equation (25). Perhaps (26) has no solution other than the ones given by (27). If x and y are fixed, we read the exponents mod 3 to write (28) as Thue equations with fixed coefficients and hence (26) has only finitely many solutions. Balasubramanian and Shorey [5] extended the preceding result by showing that (26) has only finitely many solutions whenever x and y are composed of primes from a given finite set. Further Shorey [83] showed that this is also the case if gcd $(x, y) = 1$ and either x, $y - x$ or y, $y - x$ are composed of primes from a given finite set. For a given m and n, Davenport, Lewis and Schinzel [20] proved that (26) has only finitely many solutions. The proof depends on the well-known theorem of Siegel [104] on integral solutions of polynomial equations in two variables and therefore the proof of Davenport, Lewis and Schinzel does not allow an explicit bound for the magnitude of the solutions. Further for a given m and n with gcd $(m-1, n-1) > 1$, Davenport, Lewis and Schinzel [20] gave an effective version of their result. Now the proof depends on an elementary method of Runge [64]. Further Nesterenko and Shorey [53] showed that (26) with $h = \gcd(m-1, n-1) > 1$ implies that max (x, y, m, n) is bounded by a number depending only on $(m - 1)/h$. This includes the effective version of Davenport, Lewis and Schinzel stated above. The proof depends on an extension of Runge's method to exponential diophantine equations sketched in the last section. We counted the power of 2 on both the sides of (21) to show that k is small as compared with x. Analogous assertion for (26) should be that

m is small as compared with y. This is achieved by the theory of linear forms in logarithms. For a given pair $(r, s) \neq (1, 1)$ of relatively prime positive integers, let $S_{r,s}$ be the set of all pairs $(m, n) = (1+dr, 1+ds)$ with $d = 2, 3, \ldots$. This is an infinite set and the above result states that (26) has only finitely many solutions when the exponents m and n are restricted to the elements of $S_{r,s}$. This is the first result on (26) of the type where the exponents m and n are unbounded and there is no restriction on variables x and y. We observe that (26) asks for integers with all the digits equal to one with respect to two distinct bases. Further we notice that 31 and 8191 are prime numbers such that $\omega(N-1) = 3$ if $N = 31$ and $\omega(N-1) = 5$ if $N = 8191$. Shorey [90] showed that 31 and 8191 are the only prime numbers N such that $\omega(N-1) \leq 5$ and N has all the digits equal to one with respect to two distinct bases.

For relatively prime positive integers A and B, a more general equation than (26), namely,

$$A\frac{x^m - 1}{x - 1} = B\,\frac{y^n - 1}{y - 1} \tag{29}$$

in integers $x > 1, y > 1, m > 2, n > 2$ with $B(x-1) \neq A(y-1)$ has been considered in [5], [83], and [86]. It has been shown in [5] that (29) implies that max (x, y, m, n) is bounded by a number depending only on A, B and the greatest prime factor of xy. Let A, B, X and Y be positive integers satisfying $1 \leq A < X, 1 \leq B < Y$ and $B(X-1) \neq A(Y-1)$. Then it has been proved in [86] that the number of integers whose digits are all equal to A with respect to base X and equal to B with respect to base Y does not exceed 24. That the estimate is independent of A, B, X and Y is an interesting feature of the preceding result.

4 An Equation of Nagell-Ljunggren

We consider the equation

$$\frac{x^m - 1}{x - 1} = y^q \text{ in integers } x > 1, y > 1, m > 2, q \geq 2. \tag{30}$$

This equation asks for perfect powers whose digits are all equal to one with respect to base x. By writing $y^q = (y^{q/p})^p$, there is no loss of generality in assuming that q is a prime number. We observe that $(x, y, m, q) = (3, 11, 5, 2), (7, 20, 4, 2)$ and $(18, 7, 3, 3)$ are solutions of (30). It has been conjectured that (30) has no other solution. Ljunggren [39] confirmed it whenever $q = 2$. Therefore we always suppose that $q > 2$ in (30). Further it follows from the results of Ljunggren [39] and Nagell [50] that (30) implies that $3 \nmid m$, $4 \nmid m$ and if $q = 3$ then $m \equiv 5 \pmod 6$ unless $(x, y, m, q) = (18, 7, 3, 3)$. The proofs of these results are also available in Ribenboim [61].

We show that $a\ b\ c$ conjecture implies that (30) has only finitely many solutions. We re-write (30) as

$$x^m = (x-1)y^q + 1.$$

Now we apply $a\ b\ c$ conjecture with $\epsilon = 1/8$ to the preceding equation for deriving that

$$x^{m-2-2\epsilon} < \kappa_2 y^{1+\epsilon} \tag{31}$$

where κ_2 is an absolute constant. On the other hand, we see from (30) that $y^q < 2x^{m-1}$ which implies that

$$x^{m-2-2\epsilon} > \frac{1}{2} y^{q(m-2-2\epsilon)/(m-1)}.$$

By the results of Ljunggren and Nagell stated above, we may suppose that $m \geq 5$. Therefore, since $\epsilon = 1/8$ and $q \geq 3$, we see that the exponent of y in the preceding inequality is at least $3/2$. This contradicts (31) since y is sufficiently large.

From now onward in this section, we confine to a weaker version of the conjecture that (30) has only finitely many solutions. Shorey and Tijdeman [94] confirmed this conjecture when x is fixed. In particular there are only finitely many perfect powers with all the digits equal to one in their decimal expansions. This settles an old problem and the proof depends on the theory of linear forms in logarithms. Now we apply the above result of Shorey and Tijdeman to show that it suffices to prove the conjecture when m is a prime power. Let $m = P^a m_1$ where $a > 0$ is an integer, $P = P(m)$, gcd $(m_1, P) = 1$ and we re-write (30) as

$$\frac{x^m - 1}{x^{P^a} - 1} \frac{x^{P^a} - 1}{x - 1} = y^q.$$

Let p be a prime dividing each of the factors on the left hand side. Then p divides m_1. Let ν be the least positive integer such that $x^\nu \equiv 1 \pmod{p}$. Then ν divides P^a as well as $p-1$. Thus either $\nu = 1$ or $P \leq \nu < p \leq P$ since $P = P(m)$. Consequently $\nu = 1$ which implies that that $p = P$ divides m_1 contradicting gcd $(m_1, P) = 1$. Therefore we conclude that the factors on the left hand side are relatively prime. Thus the second factor is a q-th power. Hence the assertion follows from the result of Shorey and Tijdeman stated above. The above factorisation on equation (30) appears for the first time in [84, Lemma 7] and it has been very useful. Shorey [84], [85] confirmed the conjecture when $\omega(m) > q - 2$. For the proof of this result, Shorey [84], [85] showed that (30) has only finitely many solutions when either $m \equiv 1 \pmod{q}$ or x is a q-th power. The proofs depend on the theory of linear forms in logarithms and a result of Baker [1] on the approximations of certain algebraic numbers by rationals proved by Thue-Siegel hypergeometric method. The latter result already found application in Shorey and Tijdeman [94] that (30) with $x = 10$ and $q \leq 19$ has no solution. Le [38] proved that (30) has no solution whenever x is a q-th power. Further Le (Acta Arith. 69 (1995), 91–97) claimed that (30) with $m \equiv 1 \pmod{q}$ has no solution but his proof is not correct. A correct proof has been given by Bennett [9]. This is an immediate consequence of his result [9] that for integers $a > 0$ and $n \geq 3$, the equation

$$(a+1)x^n - ay^n = 1$$

has no solution in integers $x > 0$ and $y > 0$ other than $x = y = 1$. A weaker version of the preceding result was given by Bennett and de Weger [10]. We replace the above stated results of Shorey that (30) has only finitely many solutions if x is a q-th power or $m \equiv 1 \pmod{q}$ by the results of Le and Bennett that (30) has no solution if either x is a q-th power or $m \equiv 1 \pmod{q}$ in the proof of Shorey that (30) with $\omega(m) > q - 2$ has only finitely many solutions. Then we conclude that (30) with $\omega(m) > q - 2$ has no solution. Now we show that (30) with $q | m$ implies that m is a power of q. We write $m = q^e m'$

where $e \geq 1$ and gcd $(m', q) = 1$. Further we may suppose that m' is not divisible by a prime $\equiv 1 \pmod{q}$ otherwise the assertion follows as in [84] from the above factorisation on (30) and the result of Bennett . Now we re-write (30) as

$$\frac{x^m - 1}{x^{q^e} - 1} \frac{x^{q^e} - 1}{x - 1} = y^q.$$

Let p be a prime dividing each of the factors on the left hand side. Then p divides m'. Let ν be the least positive integer such that $x^\nu \equiv 1 \pmod{p}$. Then ν divides q^e and $p - 1$. Therefore $\nu = 1$ since $p \not\equiv 1 \pmod{q}$. Now we see that $p = q$ divides m' contradicting gcd $(m', q) = 1$. Hence we derive that the factors on the left hand side are relatively prime. Therefore the first factor on the left hand side is a q-th power. Then we may suppose that $m' = 1$ or $m' = 2$ since $x^{q^e} = (x^{q^{e-1}})^q$ is a q-th power. The latter possibility is ruled out since two consecutive positive integers cannot be q-th powers and the assertion is proved. On the other hand, a very particular case of the conjecture that (30) has only finitely many solutions when m is a power of q is an open problem. An easier question than the conjecture that (30) has only finitely many solutions is to replace $\omega(m) > q - 2$ by $\omega(m) \geq 2$ in the result of Shorey stated above. This is possible if m is divisible by q as derived above.

Now we consider (30) with m even. Then $m = 2n$ where $m > 1$ is odd by Nagell [50] and

$$\frac{x^n - 1}{x - 1} = y_1^q, \quad x^n + 1 = y_2^q$$

where y_1 and y_2 are relatively prime positive integers greater than one. The second is the equation of Catalan. Catalan [17] conjectured that the only solution of this equation is given by $x = 2,\ n = 3,\ y_2 = 3,\ q = 2$. In other words, the conjecture of Catalan states that 9 and 8 are the only perfect powers that differ by one. This remains open but Tijdeman [108] proved that the equation of Catalan has only finitely many solutions in integers in $x > 1,\ y_2 > 1,\ n > 1,\ q > 1$. Therefore we derive that (30) with m even has only finitely many solutions. More generally, Shorey and Tijdeman [94] showed that (30) has only finitely many solutions if m is divisible by a fixed prime. It has not yet been possible to show that (30) has no solution whenever m is even. Mignotte and Roy [46] proved that $x^n + 1 = y_2^q$ implies that

$$\max (n, q) > 10^6, \quad \min (n, q) > 10^5.$$

For positive integers a, b and non-zero integer k, the equation of Catalan is a particular case of the following equation of Pillai:

$$ax^m - by^n = k \quad \text{in integers } x > 1, y > 1, m > 1, n > 1 \text{ with } mn \geq 6.$$

Pillai [56] conjectured that this equation has only finitely many solutions. This conjecture has been confirmed if at least one of the four variables in the above equation is fixed, see Shorey and Tijdeman [95, Chapter 12] which we also refer for a survey on exponential diophantine equations. This equation is known as Pillai's equation.

The height of a rational number, in its reduced form, is the maximum of the absolute values of its numerator and denominator. Let $f(X)$ be a polynomial of degree n with rational coefficients such that it has at least two distinct roots and $f(0) \neq 0$. We write

$$f(X) = a_0X^n + a_1X^{n-1} + \cdots + a_n$$

where $a_0, a_1, \ldots, a_n$ are rational numbers with $a_0 \neq 0$ and $a_n \neq 0$. Let H be the maximum of the heights of a_i with $0 \leq i \leq n$ and L be the number of i with $0 \leq i \leq n$ such that $a_i \neq 0$. Let $m \geq 2$, x and y with $|y| > 1$ be integers satisfying

$$y^m = f(x)$$

which we suppose without reference in this paragraph. By a proper subsum of $y^m - f(x)$, we understand a proper subsum of $y^m - a_0x^n - a_1x^{n-1} - \cdots - a_n$ where the terms a_ix^{n-i} with $a_i = 0$ are ignored. Schinzel and Tijdeman [79] proved that m is bounded by a number depending only on f. We conjecture: *There exists a number C_5 depending only on L and H such that either*

$$m \leq C_5$$

or $y^m - f(x)$ has a proper subsum which vanishes. We refer to the above conjecture as Conjecture 1. For the proof of the conjecture, we may suppose that m exceeds a sufficiently large number depending only on L, H and $y^m - f(x)$ has no proper subsum which vanishes. Then we apply generalised $a\ b\ c$ conjecture (see Darmon and Granville [18, p. 533]) to derive that n is bounded by a number depending only on L and H. Now we see from the theorem of Schinzel and Tijdeman stated above that m is bounded by a number depending only on L and H. Thus Conjecture 1 follows from generalised $a\ b\ c$ conjecture. Further conjecture 1 implies that the exponents in Pillai's equation are bounded and then the conjecture of Pillai follows from the result of Baker [3] on integral solutions of hyper-elliptic equations. Next, we see that Conjecture 1 implies that either $m \leq C_5$ or $|x| \leq 2H^2$. Therefore Conjecture 1 includes the theorem of Schinzel and Tijdeman with $f(0) \neq 0$ stated above. For positive integers $m,\ x,\ \mu,\ \nu$ with $m > 1,\ x > 1,\ \mu > \nu$ and $\lambda = (\mu^m - \nu^m)^2$, we consider conjecture 1 in the following cases:

$$f(X) = (X^2 - \lambda)/4,\ x = \mu^m + \nu^m,\ y = \mu\nu$$

and

$$f(X) = (x-1)(X^{m-1} + \cdots + X) + x,\ y = x.$$

We observe that no proper subsum of $y^m - f(x)$ vanishes in either of the cases. Since $f(x) = (\mu\nu)^m$ in the first case and $f(x) = x^m$ in the second case, we see that the dependance of C_5 on L as well as H is necessary in Conjecture 1. Further it is clear that the assumption that f has at least two distinct roots such that $f(0) \neq 0$ is also necessary in Conjecture 1.

Let us consider the equation

$$\frac{x^m - 1}{x - 1} = y^q + 1 \quad \text{in integers } x > 1, y > 1, m > 2, q \geq 2. \tag{32}$$

By subtracting one on both the sides of (32), we have

$$x\,\frac{x^{m-1}-1}{x-1} = y^q.$$

We see that $m \geq 4$ since a product of two consecutive integers is not a power. Further we observe that the factors on the left hand side are relatively prime. Therefore

$$x = y_3^q, \quad \frac{x^{m-1}-1}{x-1} = y_4^q$$

where $y_3 > 1$ and $y_4 > 1$ are integers. Then Shorey [89] applied his result stated above to conclude that (32) has only finitely many solutions. Further Le [38] concluded that (32) has no solution. This is an immediate consequence of his result that (30) has no solution whenever x is a q-th power. The first result in the direction of the preceding result is due to Inkeri [33] that (30) with $q = 3$ has no solution whenever x is a cube. Saradha and Shorey [74] proved that (30) with $x = z^2$ implies that $z \leq 31$ and $z \notin \{2, 3, 4, 8, 9, 16, 25, 27\}$. The proof depends on an estimate of Laurent, Mignotte and Nesterenko [36] on linear forms in two logarithms, irrationality measures of Baker [1] of certain algebraic numbers, p-adic method of Skolem as given by Le [37] and congruence arguments due to Inkeri [33]. Mignotte [45] has recently improved the above referred estimate on linear forms in two logarithms. By combining this estimate with estimates on solutions of certain Thue-Mahler equations and difficult and elaborate computations, Bugeaud, Mignotte, Roy and Shorey [15] showed that (30) with $x = z^2$ and $z \leq 31$, $z \notin \{2, 3, 4, 8, 9, 16, 25, 27\}$ has no solution. The preceding result was proved, independently and differently, by Bennett [9]. Hence (30) has no solution whenever x is a square. The proof of Bennett depends on his result on the equation $(a+1)x^n - ax^n = 1$ stated above. Further Hirata-Kohno and Shorey [54] proved that for a prime number $\mu \geq 3$, equation (30) with $x = z^\mu$ and $q > 2(\mu - 1)(2\mu - 3)$ implies that max (x, y, m, q) is bounded by a number depending only on μ. If $\mu = 3$, we may suppose that $q \neq 3$ in view of a result of Inkeri stated above. Therefore (30) with $x = z^3$ and $q \neq 5, 7, 11$ has only finitely many solutions. Linear forms in logarithms with α_i's very close to one appear in the proof. If α_i's are very close to one Shorey [80], by integrating the auxiliary function on a large circle, obtained for the first time sharp estimates for linear forms in logarithms and these are close to the best possible in the sense that $\log A_1 \cdots \log A_n$ can be replaced by $\log A$ where $A = \max_{1 \leq j \leq n} A_j$. The proof of Hirata-Kohno and Shorey shows that sharp estimates for these special linear forms in logarithms combine well with the irrationality measures of Baker [1], [2] proved by hypergeometric method. This feature appeared for the first time in Shorey [84]. See also Sections 7, 8 below and Mignotte [44], Bennett and de Weger [10] where linear forms in logarithms with α_i's close to one find applications.

We recall that (30) has only finitely many solutions if x is fixed. A natural p-adic extension of the above result is not known. In other words, we do not know whether (30) has only finitely many solutions if x is composed of primes from a given finite set. Even a proof of weaker assertion that (30) has only finitely many solutions if x is a power of an arbitrary fixed integer is not available. Saradha and Shorey [74] gave an infinite set S_4 containing all integers in the inteval (1,20] other than 11 such that for $h \in S_4$ and integer

$t > 0$, equation (30) with $x = h^t$ implies that max (x, y, m, q) is bounded by an absolute constant. We refer to [74] and [12] for an explicit construction of the set S_4. Now we give an application of this result to show that certain numbers considered by Mahler are irrational. Let $g \geq 2$ and $h \geq 2$ be integers. For any integer $n \geq 1$, we write $n = a_1 h^{r-1} + \cdots + a_r$ for some integers $r > 0$ and $0 \leq a_i < h$ for $1 \leq i \leq r$ with $a_1 \neq 0$. We define $(n)_h = a_1, \ldots, a_r$ i.e. the sequence of digits of n written in h-ary notation. For a sequence $\{n_i\}_{i=1}^{\infty}$ of non-negative integers, we put

$$a_h(g) = 0.(g^{n_1})_h (g^{n_2})_h \cdots.$$

Mahler [41] proved that $a_{10}(g)$ is irrational for $\{n_i\}_{i=1}^{\infty} = \{i-1\}_{i=1}^{\infty}$. It is now known that $a_h(g)$ is irrational for any unbounded sequence $\{n_i\}_{i=1}^{\infty}$ of non-negative integers, see Sander [65]. If an element occurs in a sequence infinitely many times, it is called a limit point of the sequence. Let $\{n_i\}_{i=1}^{\infty}$ be a bounded sequence of non-negative integers. If it has only one limit point, it is ultimately periodic and hence $a_h(g)$ is rational. We always suppose now onward that it has exactly two limit points $N_1 < N_2$ such that $g^{N_2 - N_1} \neq h + 1$ whenever $g^{N_1} < h$ and it is not ultimately periodic. Then the above result of Saradha and Shorey on (30) implies that for integers $g \geq 2$ and $h \in S_4$, if $a_h(g)$ is rational then N_2 is bounded by an absolute constant. In fact Sander [65] claimed more general irrationality result but it depends on an unproved assertion on (30) stated in the beginning of this paragraph. The connection between (30) and irrationality of $a_h(g)$ is due to Sander [65]. More precisely, Sander [65] showed that if $a_h(g)$ is rational, then

$$g^{N_2 - N_1} = \frac{h^{tL} - 1}{h^t - 1}$$

for some integer $L \geq 2$ and t given by $h^{t-1} \leq g^{N_1} < h^t$. Bugeaud, Mignotte and Roy [14] improved the assertion of Saradha and Shorey by showing that $a_h(g)$ with $g \geq 2$ and $h \in S_4$ is irrational unless $g = 1 + h + \cdots + h^{L-1}$ for every $L \geq 2$ if $(N_1, N_2) = (0, 1)$ or $(N_1, N_2, g, h) = (0, 2, 11, 3), (0, 2, 20, 7), (0, 3, 7, 18), (1, 4, 7, 18)$. On the other hand, it is easy to see that $a_h(g)$ is rational in any of the above possibilities. A more general statement of irrationality of $a_h(g)$ is available if g is even and h is odd. In this case, we observe that L appearing above in the result of Sander has to be even. Thus we are led to (30) with m even and it has only finitely many solutions. Hence we conclude that, for integers $g \geq 2$ and $h \geq 2$ such that g is even and h is odd, if $a_h(g)$ is rational then N_2 is bounded by an absolute constant. For bounded sequences with more than two limit points, Sander [65] obtained some partial results and further investigations are desirable. For proving the above result on (30) applied to derive an irrationality criterion, Saradha and Shorey [74] showed that (30) implies that either max (x, y, m, q) is bounded by an absolute constant or there exists a prime p such that $p | x$ and $p \nmid (y-1)$. M. Le (Acta Arith. 69 (1995), 91–97) and L. Yu and M. Le (Acta Arith. 73 (1995), 363–365) claimed stronger version of the preceding result but their proofs are not correct. Correct proofs have recently been given by Bugeaud, Mignotte and Roy [14] who proved that (30) has no solution other than $(x, y, m, q) = (18, 7, 3, 3)$ if every prime divisor of x divides $y - 1$. This implies their result on irrationality of $a_h(g)$ stated above. This also includes a result of Bugeaud and Mignotte [13] that (30) with $2 \leq x \leq 10$ has no solution settling a problem of Inkeri [33].

Thus it is not possible to find a perfect power greater than one with digits identically equal to one in its decimal expansion. Further the result of Bugeaud and Mignotte that (30) with $x = 3$ has no solution finds application in group theory. We remark that p-adic linear forms in logarithms with $\alpha_i's$ p-adically close to one appear in the work of Bugeaud, Mignotte, Roy and sharp estimates of Bugeaud [12] for these p-adic linear forms in logarithms are utilised. The work also involves very heavy computations depending on a result of Bennett that (30) with $m \equiv 1 \pmod{q}$ has no solution.

Now we consider a more general equation than (30). For positive integers A, B and prime number q with $AB > 1$, $\gcd(A, B) = 1$ and q-free A, we consider the equation

$$A\,\frac{x^m - 1}{x - 1} = By^q \text{ in integers } x > 1, y > 1, m > 2, q \geq 2. \tag{33}$$

Obläth [55] showed that (33) with $1 < A < 10$, $B = 1$ and $x = 10$ has no solution. Inkeri [33] determined all solutions of (33) with $1 < A < x \leq 10$ and $B = 1$. Shorey and Tijdeman (see [94], [93]) showed that (33) implies that either $\max(x, y, m, q)$ is bounded by a number depending only on A and B or $m = 2n$ with

$$\frac{x^n - 1}{x - 1} = y_5^q, \quad A(x^n + 1) = By_6^q$$

where $y_5 > 1$ and $y_6 > 1$ are integers satisfying $\gcd(y_5, y_6) = 1$ and $y_5 y_6 = y$. Now it is clear from the above relations that it is easier to deal with (33) than (30). Therefore stronger results on (33) than (30) are available though we have not been able to show that (33) has only finitely many solutions. By combining the preceding result with the known results in the theory of exponential diophantine equation, Shorey [93] observed that (33) implies that $\max(x, y, m, q)$ is bounded by a number depending only on A, B and the greatest prime factor of x. As already mentioned, the preceding assertion remains unproved for (30). For integer $x > 1$ and prime number q, we write $U(A, B, x, q)$ and $U(x, q)$ for the number of solutions of (33) and (30) in integers $y > 1$ and $m > 2$, respectively. Shorey [85] showed that

$$U(x, q) \leq q + C_6$$

where C_6 is an absolute constant. Le [38] sharpened it to

$$U(x, q) \leq q.$$

Shorey [93] showed that

$$U(A, B, x, q) \leq C_7$$

where C_7 is a number depending only on A and B. We refer to Shorey [93] for stronger results on equation (33) when the exponents m run through infinite set satisfying

$$\gcd(m, AB\varphi(AB)) = 1 \tag{34}$$

where φ is the Euler totient function. For example, (33) with $x = z^2$, (34) and (33) with $q = 2$, (34) have no solution.

5 Equal Products of Integers in Arithmetic Progression

In this section we shall continue applications of extension of Runge's method to exponential diophantine equations sketched in Section 2. Further these applications result in developing the method. For positive integers m, d_1 and d_2, we consider

$$x(x+d_1)\cdots(x+(k-1)d_1) = y(y+d_2)\cdots(y+(mk-1)d_2) \text{ in integers } x>0, y>0, k\geq 2. \tag{35}$$

Equation (35) with $d_1 = d_2 = 1$ coincides with (18). Saradha and Shorey [72] showed that (35) with $m \geq 2$ and $d_1 = d_2 = d$ implies that $\max(x, y, k)$ is bounded by a number depending only on m, d and this includes (20) corresponding to the case $d_1 = d_2 = 1$. If $m = 2$, Saradha and Shorey [73] proved that (35) implies that either $\max(x, y, k)$ is bounded by a number depending only on $d_1,\ d_2$ or $k = 2,\ d_1 = 2d_2^2,\ x = y^2 + 3d_2y$. On the other hand, equation (35) with $m = 2$ is satisfied whenever the latter possibility holds. Saradha, Shorey and Tijdeman [77] proved that (35) with $m = 2,\ d_1 = 1$ and $d_2 \leq k+1$ implies that $k \leq 35$ and $d_2 = 2^\ell$ for some integer $\ell \geq 2$ and consequently they showed that $32.33 = 1.6.11.16, 207.208 = 8.13.18.23$ are the only solutions of (35) with $m = 2,\ d_1 = 1$ and $d_2 = 5, 6$. Saradha and Shorey [73] proved that (35) with $m > 2$ implies that k is bounded by a number depending only on m, d_1 and d_2. For a fixed k, they [73] showed that (35) implies that either $\max(x, y)$ is bounded by a number depending only on m, d_1, d_2 or d_1/d_2^m is a product of m distinct positive integers composed of primes not exceeding m. It is clear that the latter possibility can not hold whenever $d_1 = d_2$. Further it is replaced in [73] by $m \geq \alpha(k)$ where

$$\alpha(k) = \begin{cases} 14 & \text{for } 2 \leq k \leq 7 \\ 50 & \text{for } k = 8 \\ \exp(k \log k - (1.25475)k - \log k + 1.56577) & \text{for } k \geq 9. \end{cases}$$

We observe that $\alpha(k) \geq 2568$ for $k \geq 9$. Finally Saradha, Shorey and Tijdeman [75] proved that (35) with $m > 2$ implies that max (x, y) is bounded by a number depending only on d_1, d_2, k and m. In fact, for positive integers d_1, d_2, ℓ and m with $\ell < m, m > 2$ and gcd $(\ell, m) = 1$, Saradha, Shorey and Tijdeman [75] showed that if $x > 0$, $y > 0$ and $k \geq 2$ are integers satisfying

$$x(x+d_1)\cdots(x+(\ell k-1)d_1) = y(y+d_2)\cdots(y+(mk-1)d_2) \tag{36}$$

then max (x, y, k) is bounded by a number depending only on d_1, d_2 and m. Let $f(X)$ be a monic polynomial of degree $\nu > 0$ with rational coefficients. Then we consider

$$f(x)f(x+d_1)\cdots f(x+(\ell k-1)d_1) = f(y)f(y+d_2)\cdots f(y+(mk-1)d_2) \tag{37}$$

in integers x, y and $k \geq 2$ such that $f(x + jd_1) \neq 0$ for $0 \leq j \leq \ell k - 1$. We observe that equation (37) with $f(X) = X$ is (36). Saradha, Shorey and Tijdeman [78] proved that (37) implies that k is bounded by a number depending only on d_1, d_2, m and f. Further, if f is a power of an irreducible polynomial, they [78] proved that (37) implies that max $(|x|, |y|, k)$ is bounded by a number depending only on d_1, d_2, m and f unless $\ell = 1$,

$m = k = 2, d_1 = 2d_2^2$ and $f(X) = (X+r)^\nu$ with integer r, $x+r = (y+r)(y+r+3d_2)$. An extension of this result to all non-constant monic polynomials with rational coefficients remains unproved. Further it is clear from the proof that the preceding result extends to all non-constant monic polynomials with rational coefficients if we show that the underlying curve for (37) is irreducible. A proof for irreducibility in a particular case $f(X) = X$ and $(\ell, m, k) \neq (1, 2, 2)$ was given by Beukers, Shorey and Tijdeman [11]. A particular case $d_1 = d_2 = \ell = 1$ of the result of Saradha, Shorey and Tijdeman on (37) was already proved by Balasubramanian and Shorey [6]. Saradha [66] considered (37) when f is reducible. Let $s_1, \ldots, s_\mu$ with $\mu \geq 2$ be distinct positive integers. Suppose that x, y and $k \geq 1$ are integers satisfying

$$\prod_{i=1}^{\mu} \prod_{j=0}^{(\ell k-1)} (x - s_i + jd_1) = \prod_{i=1}^{\mu} \prod_{j=0}^{(mk-1)} (y - s_i + jd_2)$$

such that the factors on the left hand side as well as the factors on the right hand side are non-zero and pairwise distinct. Then Saradha [66] proved that max $(|x|, |y|, k)$ is bounded by a number depending only on d_1, d_2, ℓ, m, μ and $s_1, \ldots, s_\mu$ whenever

$$m = 2 \text{ or } \mu = 2, 3, 4 \text{ or } d_2 = 1$$

unless

$$m = \mu = 2,\ \ell = k = d_2 = 1,\ x = y^2 + y(1 - s_1 - s_2) + s_1 s_2.$$

Further we observe that the above infinitely many possibilities satisfy the preceding equation.

Finally we turn to considering (35) with $m = 1$:

$$x(x+d_1)\cdots(x+(k-1)d_1) = y(y+d_2)\cdots(y+(k-1)d_2). \tag{38}$$

This equation was suggested by Gabovich [29] in 1966. It is clear that (38) with $k = 2$ has infinitely many solutions. Further Gabovich [29] gave an infinite class of solutions of (38) with $k = 3, 4$. Some infinite classes of solutions of (38) with $k = 5$ were given by Szymiczek [106] and Choudhry [16]. Choudhry [16] also provided an infinite class of solutions of (38) with arbitrary k and unbounded d_1, d_2. Finally we consider equation (38) with fixed d_1 and d_2. If $d_1 = d_2$, we observe that $x = y$. Therefore we suppose that d_1 and d_2 are distinct. Further there is no loss of generality in assuming that $d_1 < d_2$ and gcd $(x, y, d_1, d_2) = 1$. Then Saradha, Shorey and Tijdeman [76] proved that $\max(x, y, k)$ is bounded by a number depending only on d_2 unless

$$x = k+1,\ y = 2,\ d_1 = 1,\ d_2 = 4. \tag{39}$$

In view of the relations due to Makowski [42],

$$(k+1)\cdots(2k) = 2.6...(4k-2) \text{ for } k = 2, 3, \ldots,$$

the possibilities (39) can not be excluded. The proof depends on Prime Number Theorem for arithmetic progressions. On the other hand, it is possible to make the proof independent of Prime Number Theorem for certain values of d_1 and d_2. This led Saradha,

Shorey and Tijdeman [77] to show that all the solutions of (38) with $d_1 = 1$ and $d_2 = 2, 3, 5, 6, 7, 9, 10$ are given by $2.3 = 1.6, 7.8.9 = 4.9.14, 8.9 = 6.12, 5.6 = 3.10, 4.5.6 = 1.8.15, 15.16.17 = 10.17.24, 9.10 = 6.15, 7.8 = 4.14, 24.25 = 20.30$ and $32.33.34 = 24.34.44$. This follows by computations from their result that (38) with $d_1 = 1$ and $d_2 \leq k+1$ implies that $y \equiv 2 \pmod 4$ and $d_2 = 2^\ell$ for some integer $\ell \geq 2$.

6 The Greatest Prime Factor of Integers in Arithmetical Progression

Let us consider equation (6) with $t = k$ and $P(y) \leq k$. Then

$$P =: P(n(n+1)\cdots(n+k-1)) \leq k.$$

This leads us to considering lower bound for the greatest prime factor of the product of k consecutive positive integers. The first result dates back to Sylvester [105] in 1892 that

$$P > k \quad \text{if } n > k.$$

If $n < k^{3/2}$ and k is sufficiently large, it follows from well-known results on difference between consecutive primes that there is prime p satisfying $n \leq p \leq n+k-1$. Therefore we shall consider lower bound for P when $n \geq k^{3/2}$. Then Erdös [24] proved that

$$P > C_8 k \log k \tag{40}$$

where $C_8 > 0$ is an absolute constant. We denote by T the number of integers $n+i$ with $0 \leq i < k$ such that $P(n+i) \leq k$. Then we apply a fundamental argument of Erdös for deriving that

$$(k^{3/2})^{T-\pi(k)} \leq n^{T-\pi(k)} \leq k^k$$

and the assertion (40) follows from Prime Number theory. Ramachandra [57] applied Selberg's Sieve to prove (40) with $C_8 = (1-\epsilon)$ for $\epsilon > 0$ and k exceeding a number depending only on ϵ. Ramachandra [58] also obtained partial results for (40) with $C_8 = (2-\epsilon)$. Combining these results with the method of Roth and Halberstam on difference between consecutive ν-free integers, Tijdeman [107] proved (40) with $C_8 = 2-\epsilon$. Further Ramachandra and Shorey [59] sharpened (40) to

$$P > C_9 k \log k \left(\frac{\log\log\log k}{\log\log\log\log k}\right)^{1/2} \quad \text{for } \log k > e^e \tag{41}$$

where $C_9 > 0$ is an absolute constant. The next improvement follows from the work of Jutila [34] and Shorey [81], namely,

$$P > C_{10} k \log k \, \frac{\log\log k}{\log\log\log k} \quad \text{for } \log k > e \tag{42}$$

where $C_{10} > 0$ is an absolute constant. The best possible estimate with $\log A$ in place of $\log A_1 \cdots \log A_n$ for linear forms in logarithms with α_i's close to one are crucial for the proofs of (41) and (42). Furthermore the proof of (42) depends on estimates for exponential sums. We shall continue with the results on lower bound for P at the end of the next section.

For an integer $d > 1$, we write

$$P_d = P(n(n+d)\cdots(n+(k-1)d)), \ \chi = n+(k-1)d.$$

Improving on a result of Langevin [35], Shorey and Tijdeman [98] gave a stronger version of Sylvester's theorem that

$$P_d > k \ \text{ if } d > 1 \text{ and } (n,d,k) \neq (2,7,3). \tag{43}$$

It is necessary to exclude the possibility $(n, d, k) = (2, 7, 3)$ in the above result since $P(2.9.16) = 3$. The proof depends on estimates from Prime Number theory. Further Shorey and Tijdeman [99], [103] applied the theory of linear forms in logarithms for showing that

$$P_d > C_{11} k \log\log \chi \quad \text{if} \quad \chi > k(\log k)^{\epsilon}$$

where $\epsilon > 0$ and $C_{11} > 0$ is a number depending only on ϵ. It is clear that the assertion is not valid if the assumption $\chi > k(\log k)^{\epsilon}$ is replaced by $\chi > k(\log\log k)^{\delta}$ with $0 < \delta < 1$. An analogue of (42) for P_d with $d > 1$ is not known.

7 Cubes and Higher Powers in Products from a Block of Integers

Let us recall the result of Erdös and Selfridge mentioned in the beginning of this paper. It states that a product of two or more consecutive positive integers is never a cube or a higher power. This is equivalent to saying that equation

$$n(n+1)\cdots(n+k-1) = y^{\ell} \ \text{ in integers } n > 0, y > 0, k \geq 2, \ell > 2 \tag{44}$$

has no solution. The proof of Erdös and Selfridge depends on an earlier method due to Erdös [25] of 1955 that (44) does not hold if k exceeds a sufficiently large absolute constant. Another proof of this assertion was given by Erdös and Siegel (unpublished). Now we explain the method of Erdös by giving a sketch of the proof that (44) does not hold if k is sufficiently large.

Suppose that (44) is satisfied and k exceeds a sufficiently large absolute constant. As in Section 1, we apply Bertrand's postulate and Sylvester's theorem to conclude that $n > k^{\ell}$. Further we write

$$n + i = a_i x_i^{\ell} \ \text{ for } 0 \leq i < k$$

where a_i is ℓ-free and $P(a_i) \leq k$. We see from $n > k^{\ell}$ that the elements of $S_5 = \{a_0, a_1, \ldots, a_{k-1}\}$ are distinct. Then we apply a fundamental argument of Erdös for finding a subset S_6 of S_5 containing at least $k/2$ elements and $a_i \leq 6k$ for $a_i \in S_6$. Further we see from $n > k^{\ell}$ and $\ell \geq 3$ that $a_i a_j$ with $a_i \in S_6, a_j \in S_6$ are distinct. This property, as mentioned in Section 1, implies that $|S_6| \leq 2k/\log k$. This is a contradiction.

As in Section 1, we consider a more general equation analogous to (6) for cubes and higher powers, namely,

$$(n+d_1)\cdots(n+d_t) = by^{\ell} \tag{45}$$

in integers $n > 0,\ y > 0,\ \ell > 2,\ k \geq 2,\ d_1, \ldots, d_t$ and b. Equation (45) with $t = k$ and $b = 1$ is (44). We recall that $d_1, \ldots, d_t$ with $t \geq 2$ are distinct integers in $[0, k)$ and $P(b) \leq k$. We always suppose that the left hand side of (45) is divisible by a prime exceeding k. Then Saradha [67] proved that (45) with $t = k$ and $k \geq 4$ has no solution. This extends a result of Erdös [23] on

$$\binom{n+k}{k} = y^\ell.$$

Further Győry [32] proved that (45) with $t = k$ and $k = 2, 3$ has no solution. A theorem of Wiles [110] on the most famous exponential diophantine equation, namely Fermat's equation, states that

$$x^n + y^n = z^n \text{ in integers } n > 2, x > 0, y > 0, z > 0$$

has no solution. This led to a striking theorem on equation

$$x^n + y^n = 2^\alpha z^n. \tag{46}$$

Győry derived his result from the theorems of Ribet [62] and Darmon and Merel [19] that if $x > 0,\ y > 0,\ z > 0,\ n > 2, \alpha \geq 1$ are integers satisfying x, y, z relatively prime, n prime, $\alpha < n$ and (46), then $x = y = z = 1$.

Erdös [25] observed that his elementary method sketched above yields that for $\epsilon > 0$, equation (45) with $t \geq k - (1-\epsilon)k\, \frac{\log\log k}{\log k}$ implies that k is bounded by a number depending only on ϵ. We put

$$\nu_\ell = \frac{1}{2}\left(1 + \frac{4\ell^2 - 8\ell + 7}{2(\ell-1)(2\ell^2 - 5\ell + 4)}\right).$$

We observe that

$$\nu_3 = \frac{47}{56},\ \nu_4 = \frac{45}{64} \text{ and } \nu_\ell < 2/3 \text{ for } \ell \geq 5.$$

Shorey [84], [87] improved considerably the result of Erdös by proving that (45) with

$$t \geq \nu_\ell k \tag{47}$$

implies that k is bounded by an absolute constant. Further Shorey [84] proved that (45) with

$$t \geq k\ell^{-1/11} + \pi(k) + 2 \tag{48}$$

implies that min (k, ℓ) is bounded by an absolute constant. In fact the exponent $1/11$ in (48) can be replaced by $(1/3) + \epsilon$ for $\epsilon > 0$, see [52]. The assumption (47) has been relaxed by Nesterenko and Shorey. For stating their result, we introduce the following notation

$$\nu'_\ell = \begin{cases} \frac{112\ell^2 - 160\ell + 29}{28\ell^3 - 76\ell + 29} & \text{if } \ell \equiv 1 \pmod 2 \\ \frac{112\ell^2 - 160\ell + 17}{28\ell^3 - 188\ell + 129} & \text{if } \ell \equiv 0 \pmod 2. \end{cases}$$

For $\ell \geq 7$, we observe that $\nu'_\ell > 3/\ell$,

$$\nu'_\ell \leq \begin{cases} \frac{4}{\ell}(1 - \frac{1}{(.875)\ell}) & \text{if } \ell \equiv 1 \pmod 2 \\ \frac{4}{\ell}(1 - \frac{1}{(1.412)\ell}) & \text{if } \ell \equiv 0 \pmod 2 \end{cases}$$

and

$$\nu'_7 \leq .4832,\ \nu'_8 \leq .4556,\ \nu'_9 \leq .3878,\ \nu'_{10} \leq .3664,$$
$$\nu'_{11} \leq .3243,\ \nu'_{12} \leq .3076,\ \nu'_{13} \leq .2787,\ \nu'_{14} \leq .2655.$$

Nesterenko and Shorey [52] proved that (45) with

$$\ell \geq 7,\ t \geq \nu'_\ell k$$

implies that k is bounded by a number depending only on ℓ. The proofs depend on the theory of linear forms in logarithms, irrationality measures of Baker proved by Pade approximations and the method of Roth and Halberstam on difference between consecutive ν-free integers. Here sharp estimates for linear forms in logarithms with α_i's close to 1 are crucial and, as remarked in Section 4, they combine well with the irrationality measures of Baker. We suggest to examine that for $\epsilon > 0$, equation (45) with $t \geq \epsilon k$ implies that k is bounded by a number depending only on ϵ. We have not been able to derive this even from $a\ b\ c$ conjecture.

Let $\epsilon > 0$ and k exceeds a sufficiently large number depending only on ϵ. It has been conjectured that there is a prime dividing $n(n+1)\cdots(n+k-1)$ to the first power. If $n \leq k^{3/2}$, the conjecture is confirmed by the well-known results on difference between consecutive primes. Thus we suppose that $n > k^{3/2}$. Let $\ell \geq 3$ and

$$\nu''_\ell = \begin{cases} \nu_\ell & \text{if } \ell = 3, 4, 5, 6 \\ \nu'_\ell & \text{if } \ell \geq 7. \end{cases}$$

We derive from the above results that there exists a prime p satisfying

$$p > (1 - \nu''_\ell - \epsilon)k \log k$$

and

$$\text{ord}_p(n(n+1)\cdots(n+k-1)) \not\equiv 0 \pmod \ell$$

whenever k exceeds a number depending only on ϵ and ℓ. For a prime p between k and $(1-\nu''_\ell-\epsilon)k \log k$ dividing $n(n+1)\cdots(n+k-1)$, we omit the unique $n+i$ with $0 \leq i < k$ such that p divides $n+i$ and we denote the remaining ones from $n, n+1, \ldots, n+k-1$ by $n+d_1, \ldots, n+d_t$. We observe from Prime Number theory that $t \geq \nu''_\ell k$ and we may suppose that equation (45) is satisfied. If $n \leq k^\ell$, the assertion follows from (42). Thus we may suppose that $n > k^\ell$ and we see from a fundamental argument of Erdös that the number of i with $1 \leq i \leq t$ and $P(n+d_i) \leq k$ is at most $k\ell^{-1} + \pi(k)$. Therefore the left hand side of (45) is divisible by a prime exceeding k. Finally we apply the above results of Shorey and Nesterenko to conclude the proof. A weaker version of the assertion is contained in Shorey and Tijdeman [102]. By a similar reasoning, the above assertion is valid with $p > (1-\epsilon)k \log k$ whenever min (k, ℓ) exceeds a number depending only on ϵ.

8 Perfect Powers in Products of Integers in Arithmetic Progression

Fermat (see [49, p. 21]) proved that there are no four squares in arithmetic progression. As stated in the preceding section, Darmon and Merel [19] proved that

$$x^n + y^n = 2z^n$$

has no solution in positive integers $n > 2$, x, y and z other than the ones given by $x = y = z$. Thus for $\ell > 2$ there are no three ℓ-th powers in arithmetic progression. For earlier results in this direction, see [21], [96] and [100]. Now we consider a more general situation. Let b, d and $k \geq 2$ be positive integers such that $P(b) \leq k$. We consider

$$n(n+d)\cdots(n+(k-1)d) = by^\ell \text{ in integers } n > 0, \\ y > 0, k \geq 2, \ell \geq 2 \text{ with gcd } (n,d) = 1. \tag{49}$$

The first result on (49) is due to Euler (see [49, p. 21]) that a product of four positive integers in arithmetic progression is never a square. A related question is that a product of four positive integers in arithmetic progression can not be a product of two consecutive positive integers. The answer to this question is negative. Here the underlying equation is (35) with $m = 2$, $d_1 = 1$, $k = 2$ which, as pointed out in [77], has infinitely many solutions in positive integers x, y and d_2. Our survey of the case $d = 1$ of (49) is already complete in Section 7 and we assume that $d \geq 2$ in this section. Then we see from (43) that $P(y) \leq k$ if and only if $(n, d, k) = (2, 7, 3)$. Thus the left hand side of (49) is divisible by a prime exceeding k whenever $(n, d, k) \neq (2, 7, 3)$. There is no loss of generality in assuming that ℓ is a prime number in equation (49) and the subsequent equations (50), (51). Erdös conjectured that (49) implies that k is bounded by an absolute constant. We shall show at the end of this section that the conjecture of Erdös with $l > 3$ is a consequence of a b c conjecture. Saradha [67] proved that (49) with $d \leq 6$ and $k \geq 3$ has no solution. If $\ell = 2$, Saradha [68] showed that the assertion is valid even when $d \leq 22$ unless $(n, d, k) = (2, 7, 3)$, $(18, 7, 3)$, $(64, 17, 3)$. Marszalek [43] confirmed the conjecture for fixed d. If $\ell \geq 3$, Shorey [88] proved that (49) implies that k is bounded by a number depending only on the greatest prime factor of d. Further Shorey and Tijdeman [97] proved that (49) implies that k is bounded by a number depending only on ℓ and $\omega(d)$. It is an open problem to bound k by a number depending only on $\omega(d)$ in the preceding result. If $\ell \geq 7$, Shorey [92] showed that k is bounded by a number depending only on n whenever (49) is satisfied. The proof of the former result is elementary whereas the proof of the latter depends on combining sharp estimates from linear forms in logarithms with α_i's close to one and irrationality measures proved by hypergeometric method.

Now we relax the assumption gcd $(n, d) = 1$ in the above stated results on (49). For this, we consider

$$n(n+d)\cdots(n+(k-1)d) = by^\ell \text{ in integers } n > 0, y > 0, k \geq 2, \ell \geq 2 \tag{50}$$

where we do not subject the solutions to restriction gcd $(n, d) = 1$. We combine the result of Saradha stated above with the results of Saradha and Györy of Section 7. We derive that (50) with $d \in \{2, 3, 4, 6\}$ and $k \geq 3$ does not hold unless $k = 3$, $\ell = 2$ and $n = 48d$

if the left hand side of (50) is divisible by a prime exceeding k. Further the assertions of Marszalek and Shorey are also valid for (50) such that its left hand side is divisible by a prime exceeding k. Finally we turn to relaxing the assumption gcd $(n, d) = 1$ in the result of Shorey and Tijdeman. For this, we consider an extension of (45) and (49):

$$(n + d_1 d) \cdots (n + d_t d) = by^\ell \tag{51}$$

in integers $n > 0,\ y > 0,\ \ell \geq 2,\ k \geq 2,\ d_1, \ldots, d_t$ and b such that gcd $(n, d) = 1$ and $P(b) \leq k$. We recall that $d_1, \ldots, d_t$ with $t \geq 2$ are distinct non-negative integers less than k. We suppose that the left hand side of (51) is divisible by a prime exceeding k. Let $\epsilon > 0$ and

$$h(k) = \begin{cases} \log\log k & \text{if } \ell \geq 5 \\ \log\log\log k & \text{if } \ell = 2, 3. \end{cases}$$

Then a consequence of a result of Shorey and Tijdeman [102] states that equation (51) with

$$t \geq k - (1 - \epsilon)k \, \frac{h(k)}{\log k}$$

implies that k is bounded by a number depending only on ℓ and $\omega(d)$. This includes a result of Shorey and Tijdeman stated above on (49). In view of Balasubramanian and Shorey [8, Theorem 2], the assumption that the left hand side of (50) is divisible by a prime exceeding k is necessary.

We apply the preceding result of Shorey and Tijdeman to show that (50) with $d \nmid n$ implies that k is bounded by a number depending only on ℓ and $\omega(d)$. Suppose that (50) with $d \nmid n$ is satisfied and k exceeds a sufficiently large number depending only on ℓ and $\omega(d)$. There is no loss of generality in assuming that every prime divisor of y is greater than k. We put

$$G = \text{gcd}\,(n, d), \ G = G_1 G_2$$

where $P(G_1) \leq k$ and every prime divisor of G_2 exceeds k. Further we set

$$n' = nG^{-1},\ d' = dG^{-1},\ b' = bG_1^{-k}.$$

We observe that gcd $(n', d') = 1$ and $d' \geq 2$ since $d \nmid n$. By dividing both the sides of (50) by G^k, we see that

$$n'(n' + d') \cdots (n' + (k-1)d') = b' y^\ell G_2^{-k}$$

where b' and $y^\ell G_2^{-k}$ are positive integers such that $P(b') \leq k$. For a prime p dividing G_2 and $y^\ell G_2^{-k}$, we observe that $p > k$ and there is precisely one integer $F(p)$ in $[0, k)$ such that p divides $n' + F(p)d'$. We write $\{d_1, \ldots, d_t\}$ for the set obtained by deleting all $F(p)$ with p dividing gcd $(G_2, y^\ell G_2^{-k})$. We observe that

$$t \geq k - \min\,(\omega'(n), \omega'(d)) \geq k - \omega(d) \tag{52}$$

where $\omega'(\nu)$ denotes the number of prime divisors of ν exceeding k and

$$n'(n' + d_1 d') \cdots (n' + d_t d') = b_1 y_1^\ell \tag{53}$$

where b_1 and y_1 are positive integers such that $P(b_1) \leq k$ and all the prime divisors of y_1 are greater than k. We show that the left hand side of (53) is divisible by a prime exceeding k. We prove by contradiction. Let $\epsilon = \frac{1}{4}$ and suppose that the greatest prime factor of the left hand side of (53) does not exceed k. Then we apply a fundamental argument of Erdös for deriving that

$$\prod_{j=0}^{t-\pi(k)-1} (n' + jd') \leq k^k$$

which implies that $n' \leq 4k$ and $d' \leq 8$. Now we apply Prime Number Theorem for arithmetic progressions and (52) to derive that there are at least $(1-\epsilon)d'k/(\varphi(d') \log k)$ prime numbers among $n'+d_1d', \ldots, n'+d_td'$. Therefore there are at least $(1-\epsilon)d'k/(\varphi(d') \log k)$ prime numbers p satisfying $p \equiv n' \pmod{d'}$ and $p \leq k$. This, since $d' \geq 2$, is not possible by Prime Number Theorem for arithmetic progressions. Now we apply the result of Shorey and Tijdeman [102] stated above to (53) for concluding the desired assertion.

For a result analogous to the last assertion of Section 7, see Shorey and Tijdeman [102]. Several estimates on equation (49) have been proved. For example, it is proved in [97], [101] and [92] that (49) implies that

$$d \geq k^{C_{12} \log\log k}, \quad P(d) \geq C_{12}\ell \log k \log\log k$$

and

$$n \geq k^{C_{12} \log\log k} \text{ if } \ell \geq 7$$

where $C_{12} > 0$ is an absolute constant. Finally we apply the above estimate for d to conclude that $a\ b\ c$ conjecture implies the conjecture of Erdös on (49) with $l > 3$. We suppose (49) with $l > 3$ and k exceeding a sufficiently large absolute constant. Then we write

$$n + id = a_i x_i^l \quad \text{for} \quad 0 \leq i < k$$

where $P(a_i) \leq k$ and every prime divisor of x_i exceeds k. By a fundamental argument of Erdös, we find positive integers $f < g < h$ such that $a_f,\ a_g,\ a_h$ do not exceed k^2. We have

$$(g-f)(n+hd) + (h-g)(n+fd) = (h-f)(n+gd)$$

i.e.

$$(g-f)a_h x_h^l + (h-g)a_f x_f^l = (h-f)a_g x_g^l.$$

We observe that $x_f^l < k^3 x_g^l$ which implies that $x_f < kx_g$ since $l > 3$. Similarly $x_h < kx_g$. The greatest common divisor ν of the summands on the left hand side in the above relation do not exceed k^3 and we divide both the sides of the above relation by ν to apply $a\ b\ c$ conjecture. We conclude from $a\ b\ c$ conjecture with $\epsilon = 1/6$ that

$$n + gd = a_g x_g^l \leq k^{18} x_g^{7/2}.$$

Therefore $n + gd \leq k^{60}$ since $l \geq 5$. Finally we conclude from $n + gd > d > k^{C_{12} \log\log k}$ that k is bounded by an absolute constant.

References

[1] A. Baker, *Rational approximations to $\sqrt[3]{2}$ and other algebraic numbers*, Quart. Jour. Math. Oxford Ser. (2) **15** (1964), 375–383.

[2] A. Baker, *Simultaneous rational approximations to certain algebraic numbers*, Proc. Cambridge Philos. Soc. **63** (1967), 693–702.

[3] A. Baker, *Bounds for the solutions of the hyperelliptic equation*, Proc. Cambridge Philos. Soc. **65** (1969), 439–444.

[4] A. Baker, *The theory of linear forms in logarithms*, Transcendence Theory: Advances and Applications, ed. by A. Baker and D.W. Masser, Academic Press, London and New York (1977), 1–27.

[5] R. Balasubramanian and Shorey, T.N., *On the equation $a(x^m-1)/(x-1) = b(y^n-1)/(y-1)$*, Math. Scand. **46** (1980), 177–182.

[6] R. Balasubramanian and Shorey, T.N., *On the equation $f(x+1)\cdots f(x+k) = f(y+1)\cdots f(y+mk)$*, Indag. Math. N.S. **4** (1993), 257–267.

[7] R. Balasubramanian and Shorey, T.N., *Squares in products from a block of consecutive integers*, Acta Arith. **65** (1994), 213–220.

[8] R. Balasubramanian and Shorey, T.N., *Perfect powers in products of terms in an arithmetical progression* (IV), Number Theory, ed. by V. Kumar Murty and Michel Waldschmidt, Contemporary Mathematics **210** (1997), American Mathematical Society.

[9] M. Bennett, *Rational approximation to algebraic numbers of small height: The diophantine equation $|ax^n - by^n| = 1$*, to appear.

[10] M. Bennett and de Weger, B.M.M., *On the diophantine equation $|ax^n - by^n| = 1$*, Math. Comp. **67** (1998), 413–438.

[11] F. Beukers, Shorey, T.N. and Tijdeman, R., *Irreducibility of polynomials and arithmetic progressions with equal products of terms*, Number Theory in Progress, ed. by Györy, Iwaniec, Urbanowicz, Walter de Gruyter, Berlin (1999), 11–26.

[12] Y. Bugeaud, *Linear forms in p-adic logarithms and the Diophantine equation $\frac{x^n-1}{x-1} = y^q$*, Math. Proc. Cambridge Philos. Soc., to appear.

[13] Y. Bugeaud and Mignotte, M., *On integers with identical digits*, Mathematika, to appear.

[14] Y. Bugeaud, Mignotte, M. and Roy, Y., *On the Diophantine equation $\frac{x^n-1}{x-1} = y^q$*, Pacific Jour. Math., to appear.

[15] Y. Bugeaud, Mignotte, M., Roy, Y. and Shorey, T.N., *The equation $\frac{x^n-1}{x-1} = y^q$ has no solution with x square*, Math. Proc. Camb. Philos. Soc., to appear.

[16] A. Choudhry, *On arithmetic progressions of equal lengths and equal products of terms*, Acta Arith., to appear.

[17] E. Catalan, *Note extraite dúne lettre adressée á l'éditeur*, Jour. reine angew. Math. **27** (1844), 192.

[18] H. Darmon and Granville, A., *On the equations $z^m = F(x,y)$ and $Ax^p + By^q = Cz^r$*, Bull. London Math. Soc. **27** (1995), 513–543.

[19] H. Darmon and Merel, L., *Winding quotients and some variants of Fermat's Last Theorem*, Jour. reine angew. Math. **490** (1997), 81–100.

[20] H. Davenport, Lewis, D.J. and Schinzel, A., *Equations of the form $f(x) = g(y)$*, Quart. Jour. Math. Oxford (2) **12** (1961), 304–312.

[21] P. Dénes, *Über die diophantische Gleichung $x^\ell + y^\ell = cz^\ell$*, Acta Math. **88** (1952), 241–251.

[22] P. Erdös, *Note on the product of consecutive integers* (1), Jour. London Math. Soc. **14** (1939), 194–198.

[23] P. Erdös, *On a diophantine equation*, Jour. London Math. Soc. **26** (1951), 176–178.

[24] P. Erdös, *On consecutive integers*, Nieuw Arch. Voor Wiskunde **3** (1951), 124–128.

[25] P. Erdös, *On the product of consecutive integers* III, Indag. Math. **17** (1955), 85–90.

[26] P. Erdös, *Problems and results on number theoretic properties of consecutive integers and related questions*, Proc. Fifth Manitoba Conf. Numerical Math., Utilitas Math. (1975), 25-44.

[27] P. Erdös and Selfridge, J.L., *The product of consecutive integers is never a power*, Illinois Jour. Math. **19** (1975), 292–301.

[28] G. Faltings, *Endlichkeitssätze für abelsche Varietäten über Zahlkörpern*, Invent. Math. **73** (1983), 349–366.

[29] Ya. Gabovich, *On arithmetic progressions with equal products of terms*, Colloq. Math. **15** (1966), 45–48.

[30] R. Goormaghtigh, *L' Intermédiaire des Mathématiciens* **24** (1917), 88.

[31] K. Györy, *On the diophantine equation* $\binom{n}{k} = x^\ell$, Acta Arith. **80** (1997), 289–295.

[32] K.Györy, *On the diophantine equation* $n(n+1)\cdots(n+k-1) = bx^\ell$, Acta Arith. **83** (1998), 87–92.

[33] K. Inkeri, *On the diophantine equation* $a(x^n - 1)/(x - 1) = y^m$, Acta Arith. **21** (1972), 299–311.

[34] M. Jutila, *On numbers with large prime factor* II, Jour. Indian Math. Soc. **38** (1974), 125–130.

[35] M. Langevin, *Plus grand facteur premier d'entiers en progression arithmétique*, Sém. Delange-Pisot-Poitou, 18 e année, 1976/77, No. 3, 6 pp.

[36] M. Laurent, Mignotte, M. and Nesterenko, Y., *Formes linéaires en deux logarithmes et déterminants d'interpolation*, Jour. Number Theory **55** (1995), 285–321.

[37] M.H. Le, *A note on the diophantine equation* $\frac{x^m-1}{x-1} = y^n$, Acta Arith. **64** (1993), 19–28.

[38] M.H. Le, *A note on the diophantine equation* $(x^m - 1)/(x - 1) = y^n + 1$, Math. Proc. Cambridge Philos. Soc. **116** (1994), 385–389.

[39] W. Ljunggren, *Noen Setninger om ubestemte likninger av formen* $(x^n - 1)/(x - 1) = y^q$, Norsk. Mat. Tidsskr. 1. Hefte **25** (1943), 17–20.

[40] R.A. MacLeod and Barrodale, I., *On equal products of consecutive integers*, Canadian Math. Bull. **13** (1970), 255–259.

[41] K. Mahler, *On some irrational decimal fractions*, Jour. Number Theory **13** (1981), 268–269.

[42] A. Makowski, *problèmes P543 et P545*, R1, Colloq. Math. **19** (1968), 179–180.

[43] R. Marszalek, *On the product of consecutive elements of an arithmetic progression*, Monatsh. für. Math. **100** (1985), 215–222.

[44] M. Mignotte, *A note on the equation* $ax^n - by^n = c$, Acta Arith. **75** (1997), 287–295.

[45] M. Mignotte, *A corollary to a Theorem of Laurent - Mignotte - Nesterenko*, Acta Arith., to appear.

[46] M. Mignotte and Roy, Y., *Minorations pour l'équation de Catalan*, C. R. Acad. Sci. Paris **324** (1997), 377–380.

[47] M. Mignotte and Shorey, T.N., *The equations* $(x+1)\cdots(x+k) = (y+1)\cdots(y+mk)$, $m = 5, 6$, Indag. Math. N.S. **7** (1996), 215–225.

[48] L.J. Mordell, *On the integer solutions of* $y(y+1) = x(x+1)(x+2)$, Pacific Jour. **13** (1963), 1347–1351.

[49] L.J. Mordell, *Diophantine equations*, Academic Press, 1969.

[50] T. Nagell, *Note sur l' équation indéterminée* $(x^n - 1)/(x - 1) = y^q$, Norsk. Mat. Tidsskr. **2** (1920), 75–78.

[51] T. Nagell, *The diophantine equation* $x^2 + 7 = 2^n$, Ark. Mat. **4** (1961), 185–188.

[52] Yu.V. Nesterenko and Shorey, T.N., *Perfect powers in products of integers from a block of consecutive integers* (II), Acta Arith. **76** (1996), 191–198.

[53] Yu.V. Nesterenko and Shorey, T.N., *On an equation of Goormaghtigh,* Acta Arith. **83** (1998), 381–389.

[54] Noriko Hirata-Kohno and Shorey, T.N., *On the equation* $(x^m - 1)/(x - 1) = y^q$ *with x power,* Analytic Number Theory, ed. by Y. Motohashi, London Math. Soc. Lecture Note Series **247** (1997), 343–351, Cambridge University Press, Cambridge.

[55] R. Obláth, *Une propriété des puissances parfaites*, Mathesis **65** (1956), 356–364.

[56] S.S. Pillai, *On the equation* $2^x - 3^y = 2^X + 3^Y$, Bull. Calcutta Math. Soc. **37** (1945), 15–20.

[57] K. Ramachandra, *A note on numbers with a large prime factor* II, Jour. Indian Math. Soc. **34** (1970), 39–48.

[58] K. Ramachandra, *A note on numbers with a large prime factor* III, Acta Arith. **19** (1971), 49–62.

[59] K. Ramachandra and Shorey, T.N., *On gaps between numbers with a large prime factor,* Acta Arith. **24** (1973), 99–111.

[60] S. Ramanujan, *Question 464,* Jour. Indian Math. Soc. 5 (1913), 120. Collected papers, Cambridge University Press, 1927, 327.

[61] P. Ribenboim, *Catalan's Conjecture*, Academic Press (1994).

[62] K.A. Ribet, *On the equation* $a^p + 2^\alpha b^p + c^p = 0$, Acta Arith. **79** (1997), 7–16.

[63] O. Rigge, *Über ein diophantisches Problem*, 9th Congress Math. Scand. Helsingfors, 1938, Mercator (1939), 155–160.

[64] C. Runge, *Ueber ganzzahlige Lösungen von Gleichungen zwischen swei Veränderlichen*, Jour. reine angew. Math. **100** (1887), 425–435.

[65] J.W. Sander, *Irrationality Criteria for Mahler's numbers*, Jour. Number Theory **52** (1995), 145–156.

[66] N. Saradha, *On blocks of arithmetic progressions with equal products,* Jour. de Théorie des Nombres de Bordeaux **9** (1997), 183–199.

[67] N. Saradha, *On perfect powers in products with terms from arithmetic progressions*, Acta Arith. **82** (1997), 147–172.

[68] N. Saradha, *Squares in products with terms in an arithmetic progression*, Acta Arith. **86** (1998), 27–43.

[69] N. Saradha and Shorey, T.N., *On the ratio of two blocks of consecutive integers*, Proc. Indian Acad. Sci. (Math. Sci.) **100** (1990), 107–132.

[70] N. Saradha and Shorey, T.N., *The equations* $(x + 1) \cdots (x + k) = (y + 1) \cdots (y + mk)$ *with* $m = 3, 4$, Indag. Math. N.S. **2** (1991), 489–510.

[71] N. Saradha and Shorey, T.N., *On the equation* $(x + 1) \cdots (x + k) = (y + 1) \cdots (y + mk)$, Indag. Math. N.S. **3** (1992), 79–90.

[72] N. Saradha and Shorey, T.N., *On the equation* $x(x+d) \cdots (x+(k-1)d) = y(y+d) \cdots (y+(mk-1)d)$, Indag. Math. N.S. **3** (1992), 237–242.

[73] N. Saradha and Shorey, T.N., *On the equation* $x(x+d_1) \cdots (x+(k-1)d_1) = y(y+d_2) \cdots (y+(mk-1)d_2)$, Proc. Indian Acad. Sci. (Math. Sci.) **104** (1994), 1–12.

[74] N. Saradha and Shorey, T.N., *The equation* $(x^n - 1)/(x - 1) = y^q$ *with x square,* Math. Proc. Cambridge Philos. Soc. **125** (1999), 1–19.

[75] N. Saradha, Shorey, T.N. and Tijdeman, R., *On arithmetic progressions with equal products,* Acta Arith. **68** (1994), 89–100.

[76] N. Saradha, Shorey, T.N. and Tijdeman, R., *On arithmetic progressions of equal lengths with equal products*, Math. Proc. Cambridge Philos. Soc. **117** (1995), 193–201.

[77] N. Saradha, Shorey, T.N. and Tijdeman, R., *On the equation* $x(x + 1) \cdots (x + k - 1) = y(y + d) \cdots (y + (mk - 1)d)$, $m = 1, 2$, Acta Arith. **71** (1995), 181–196.

[78] N. Saradha, Shorey, T.N. and Tijdeman, R., *On values of a polynomial at arithmetic progressions with equal products*, Acta Arith. **72** (1995), 67–76.

[79] A. Schinzel and Tijdeman, R., *On the equation* $y^m = P(x)$, Acta Arith. **31** (1976), 199–204.

[80] T.N. Shorey, *Linear forms in the logarithms of algebraic numbers with small coefficients* I, Jour. Indian Math. Soc. **38** (1974), 271–284.

[81] T.N. Shorey, *On gaps between numbers with a large prime factor* II, Acta Arith. **25** (1974), 365–373.

[82] T.N. Shorey, *On the ratio of values of a polynomial*, Proc. Indian Acad. Sci. (Math. Sci.) **93** (1984), 109–116.

[83] T.N. Shorey, *On the equation* $a(x^m - 1)/(x - 1) = b(y^n - 1)/(y - 1)$ (II), Hardy Ramanujan Jour. **7** (1984), 1–10.

[84] T.N. Shorey, *Perfect powers in values of certain polynomials at integer points*, Math. Proc. Camb. Philos. Soc. **99** (1986), 195–207.

[85] T.N. Shorey, *On the equation* $z^q = (x^n - 1)/(x - 1)$, Indag. Math. **48** (1986), 345–351.

[86] T.N. Shorey, *On the equation* $ax^m - by^n = k$, Indag. Math. **48** (1986), 353–358.

[87] T.N. Shorey, *Perfect powers in products of integers from a block of consecutive integers*, Acta Arith. **49** (1987), 71–79.

[88] T.N. Shorey, *Some exponential Diophantine equations*, New Advances in Transcendence Theory, ed. by A. Baker, Cambridge University Press (1988), 352–365.

[89] T.N. Shorey, *Some exponential diophantine equations* II, Number Theory and Related Topics, ed. by S. Raghavan, Tata Institute of Fundamental Research, Bombay (1988), 217–229.

[90] T.N. Shorey, *Integers with identical digits*, Acta Arith. **53** (1989), 81–99.

[91] T.N. Shorey, *On a conjecture that a product of k consecutive positive integers is never equal to a product of mk consecutive positive integers except for* 8.9.10 = 6! *and related problems*, Number Theory, Paris 1992–3, ed. by S. David, London Math. Soc. Lecture Note Series, Cambridge University Press, **215** (1995), 231–244.

[92] T.N. Shorey, *Perfect powers in products of arithmetical progessions with fixed initial term*, Indag. Math. N.S. **7** (1996), 521–525.

[93] T.N. Shorey, *The equation* $a(x^m - 1)/(x - 1) = by^q$ *with* $ab > 1$, Number Theory in Progress, ed. by Győry, Iwaniec, Urbanowicz, Walter de Gruyter, Berlin (1999), 473–485.

[94] T.N. Shorey and Tijdeman, R., *New applications of Diophantine approximations to Diophantine equations*, Math. Scand. **39** (1976), 5–18.

[95] T.N. Shorey and Tijdeman, R., *Exponential Diophantine equations*, Cambridge Tracts in Mathematics **87** (1986), Cambridge University Press, Cambridge.

[96] T.N. Shorey and Tijdeman, R., *Perfect powers in arithmetical progression*, Jour. Madras University (Section B) **51** (1988), 173–180.

[97] T.N. Shorey and Tijdeman, R., *Perfect powers in products of terms in an arithmetical progression*, Compositio Math. **75** (1990), 307–344.

[98] T.N. Shorey and Tijdeman, R., *On the greatest prime factor of an arithmetical progression*, A Tribute to Paul Erdös, ed. by A. Baker, B. Bollobás and A. Hajnal, Cambridge University Press (1990), 385–389.

[99] T.N. Shorey and Tijdeman, R., *On the greatest prime factor of an arithmetical progression* (II), Acta Arith. **53** (1990), 499–504.

[100] T.N. Shorey and Tijdeman, R., *Perfect powers in arithmetical progression* II, Compositio Math. **82** (1992), 107–117.

[101] T.N. Shorey and Tijdeman, R., *Perfect powers in products of terms in an arithmetical progression* (II), Compositio Math. **82** (1992), 119–136.

[102] T.N. Shorey and Tijdeman, R., *Perfect powers in products of terms in an arithmetical progression* (III), Acta Arith. **61** (1992), 391–398.

[103] T.N. Shorey and Tijdeman, R., *On the greatest prime factor of an arithmetical progression* (III), Diophantine Approximation and Transcendental Numbers, Luminy 1990, edited by Ph. Philippon, Walter de Gruyter, Berlin (1992), 275–280.

[104] C.L. Siegel, *Ueber einige Anwendungen diophantischer Approximationen*, Abh. Preuss. Akad. Wiss. Phys-math. K1. no. 1 (1929), 70 pp.

[105] J.J. Sylvester, *On arithmetical series*, Messenger Math. **21** (1892), 1–19, 87–120.

[106] K. Szymiczek, *Note on arithmetical progressions with equal products of five terms*, Elem. Math. **27** (1972), 11–12.

[107] R. Tijdeman, *On the maximal distance of numbers with a large prime factor*, Jour. London Math. Soc. **5** (1972), 313–320.

[108] R. Tijdeman, *On the equation of Catalan*, Acta Arith. **29** (1976), 197–209.

[109] R. Tijdeman, *Exponential diophantine equations* 1986–1996, Number Theory, Diophantine, Computational and Algebraic Aspects, Proceedings of the International Conference in Eger, Hungary, 1996, ed. by Györy, Pethö and Sös, Walter de Gruyter, Berlin (1998), 523–539.

[110] A. Wiles, *Modular elliptic curves and Fermat's Last Theorem*, Annals of Math. **141** (1995), 443–551.

[111] Kunrui Yu, *Linear forms in p-adic logarithms*, Acta Arith. **53** (1989), 107–186.

[112] Kunrui Yu, *Linear forms in p-adic logarithms* II, Compositio Math. **74** (1990), 15–113.

School of Mathematics
Tata Institute of Fundamental Research
Homi Bhabha Road, Colaba
Mumbai 400 005
E-Mail: shorey@math.tifr.res.in

Algebraic Independence of Transcendental Numbers: A Survey

Michel Waldschmidt

A survey on algebraic independence of transcendental numbers by Gel'fond's method was published by the author in 1984 [84 W1]. Here we cover the recent period 1984/1997.

Introduction

Important progress concerning algebraic independence of transcendental numbers has been achieved recently. Our goal is to introduce some of the most important ones. So many new results have been produced that we need to make a choice: here we wish to concentrate our survey on the description of new results which are related with Gel'fond's method. However we shall also include other topics (like Nesterenko's results on modular functions). Moreover our list of references will cover a broader scope.

Here is a short list of some of the many topics which are not covered here.

- Works involving the method of Siegel and Shidlovskiĭ, including values of hypergeometric functions. One main reference on this topic is Shidlovskiĭ's book [87 Shi2] and [89 Shi2]. Among many contributions to this subject, we mention those of Shidlovskiĭ himself [87 Shi1], [89 Shi1], [89 Shi3], [91 Shi1], [91 Shi2], as well as his work with Yu.V. Nesterenko [96 NS]. Several papers have been produced by other members of Shidlovskiĭ's school, including Yu.V. Nesterenko [88 Ne], V. Kh. Salikhov [84 Sal], [89 Sal], [90 Sal1], [90 Sal2], V.A. Kulagin [87 K], [91 K], [92 K], [96 K], N.I. Lossov [89 Lo], M.A. Cherepnev [90 Chr], [91 Chr], [94 Chr], [95 Chr], A.I. Zhukov [91 Zh] and V.A. Gorelov [92 Go]. Further references are [84 BV], [85 Br], [86 Br], [88 Beu] and [88 BeBrH]. Most of these articles are devoted to the study of algebraic independence of the values of solutions of differential equations, applying Siegel's or Shidlovskiĭ's main theorems.
- The method of Bézivin-Robba [89 BR], [90 BBR], which produces a new proof of the Lindemann-Weierstraß Theorem. We refer to Y. André's papers [96 An] and [98 An2], which explain connections with E-functions as well as with G-functions.
- *Mahler's method.* An excellent reference on this subject is Kumiko Nishioka's Lecture Notes [96 Ni1] where many further references are provided. See also [86 Ni1], [86 Ni2], [86 Ni3], [87 Ni], [88 Bec], [89 BN], [89 Ni], [90 Ni], [90 NN], [90 Wa], [91 Am], [91 Bec], [91 Ni], [92 Bec], [94 Ni], [94 Tö1], [95 Tö1], [95 Tö2], [96 Ni2], [96 Ni3], [96 Ta], [97 Ni] and [97 To].

- Liouville's argument yields not only transcendental numbers, but also algebraically independent numbers. This way of constructing algebraically independent numbers is related with results of transcendence or algebraic independence on values of gap series. Many authors have studied this question (see a survey in [90 W1]), including Zhu Yao Chen [84 Z], [85 Z2], [85 Z3], [87 Z], [88 Z], [89 WZ], [91 Z1], [91 Z2], as well as W.W. Adams [85 Ad], M. Amou [85 Am], [91 Am], P. Bundschuh [88 Bu], [90 Bu], R. Müller [93 Mü] and T. Töpfer [94 Tö2].
- *Transcendence and algebraic independence in finite characteristic.* The recent survey by W.D. Brownawell [98 Br] gives a good description of this rich theory, including the important achievements by Jing Yu [92 Y], [97 Y] as well as the works of A. Thiery [92 Th], W.D. Brownawell [93 Br], [94 BBT], [96 Br], R. Tubbs [96 Tu] and L. Denis [93 Den1], [93 Den2], [95 Den1], [95 Den2], [96 Den], [97 Den]. We also quote the work [93 Mü] by R. Müller which involves non archimedean valued fields.
  Also related to this topic is the paper [88 AMP] by J-P. Allouche, M. Mendès-France and A.J. van der Poorten, dealing with the algebraic independence over $F(X)$ of powers $f^{\lambda_1}, \ldots, f^{\lambda_s}$, where f is a formal power series, with coefficients in a finite field F, which is algebraic over $F(X)$.
- Metric theory and classifications of transcendental numbers.
- Lower bounds for linear forms in logarithms of algebraic numbers.

In this survey we first state a few conjectures (§1). Next we describe the methods of transcendence and algebraic independence (§2); some proofs of algebraic independence results involve a transcendence criterion, other ones rest on an effective version of Hilbert's Nullstellensatz; a more recent approach is via simultaneous Diophantine approximation. Finally we state some of the most important recent progress on this topic, starting with Gel'fond's method (§3) considering results of algebraic independence related to the modular functions (§4) and finishing with Philippon's Diophantine rings (§5).

We would like to point out two important directions related with Gel'fond's method of algebraic independence:

- *Commutative algebraic groups*: the currently available tools and methods are well adapted to this geometrical setting, and a rather strong set of general results are available (even if we are far from being able to establish the main conjectures). A survey of this topic is [96 Ca].
- *Diophantine approximation and algebraic independence.* The connection between these two topics opens the way to a promising field of research. We discuss some of these new advances below (§2.4); a good starting point to get acquainted with this subject is [97 RW2].

1 Conjectures

In this section we collect various conjectures of algebraic independence. We start with Schanuel's one (see [84 W1] for references):

Conjecture 1.1 (Schanuel). Let $x_1, \dots, x_n$ be $\mathbb{Q}$-linearly independent complex numbers. Then the transcendence degree over $\mathbb{Q}$ of the field

$$\mathbb{Q}(x_1, \dots, x_n, e^{x_1}, \dots, e^{x_n})$$

is at least n.

The following problem of A.O. Gel'fond and Th. Schneider (again, see [84 W1]) is a special case of Conjecture 1.1:

Conjecture 1.2 (Gel'fond-Schneider). Let α be a non zero complex algebraic number, $\log \alpha$ a non zero logarithm of α and β an algebraic number of degree $d \geq 2$. For $z \in \mathbb{C}$, define $\alpha^z := \exp(z \log \alpha)$. Then the $d-1$ numbers

$$\alpha^\beta, \alpha^{\beta^2}, \dots, \alpha^{\beta^{d-1}}$$

are algebraically independent.

It is equivalent to say that *for algebraic numbers $\beta_1, \dots, \beta_m$ which are pairwise distinct modulo $\mathbb{Z}$*[1]*, the numbers*

$$\alpha^{\beta_1}, \dots, \alpha^{\beta_m}$$

are linearly independent over $\mathbb{Q}$ (or over $\overline{\mathbb{Q}}$, as you wish). However partial results are known on the question of algebraic independence, but almost nothing is known on the question of linear independence!

In 1949, A.O. Gel'fond solved Conjecture 1.2 in the case $d = 3$. More generally, he proved that for $d \geq 3$, two at least of the $d-1$ numbers in Conjecture 1.2 are algebraically independent. Such a result, which shows that some field has transcendence degree ≥ 2 over $\mathbb{Q}$, is a statement of "small transcendence degree". Showing that the transcendence degree tends to infinity with d (as achieved first by G.V. Chudnovsky) is proving a result of "large transcendence degree". Working with the exponential function, A.O. Gel'fond introduced two sets of $\mathbb{Q}$-linearly independent numbers, $(x_1, \dots, x_d)$ and $(y_1, \dots, y_\ell)$, and proved results of small transcendence degree involving the $d\ell$ values $e^{x_i y_j}$ $(1 \leq i \leq d, 1 \leq j \leq \ell)$.

Available methods of algebraic independence on such values of the exponential function which yield "large transcendence degrees" all require a so-called "technical hypothesis", namely a measure of linear independence for the tuple $(x_1, \dots, x_d)$ as well as for the tuple $(y_1, \dots, y_\ell)$.

Definition: We shall say that a set $\{x_1, \dots, x_n\}$ of $\mathbb{Q}$-linearly independent complex numbers *satisfies a measure of linear independence* if, for any $\epsilon > 0$, there exists a positive number H_0 such that, for any $H \geq H_0$ and n-tuple $(h_1, \dots, h_n)$ of rational integers satisfying $0 < \max\{|h_1|, \dots, |h_n|\} \leq H$, the inequality

$$|h_1 x_1 + \cdots + h_n x_n| \geq \exp\{-H^\epsilon\}$$

holds.

---

[1] This means $\beta_i - \beta_j \notin \mathbb{Z}$ for $i \neq j$.

As a matter of fact, in the original papers, the actual results which we are going to discuss sometimes involve weaker assumptions (see [98 FN] for a more careful discussion). But we shall use this definition for convenience and simplicity.

Schanuel's Conjecture suggests that such a technical hypothesis should not be necessary, and it is a challenge to remove it (see [92 Del], [93 Del], [96 Del1] and [96 Del2]). Unfortunately, so far, only results of small transcendence degree are free of such an assumption.

By the way, we take this opportunity to point out that the generalization of Schanuel's Conjecture which is suggested p. 566 of [84 W1] is too optimistic: it does not hold without a technical hypothesis, as shown by the following result of A. Bijlsma[2]

Lemma 1.3 (Bijlsma). *For any fixed natural number κ, there exist irrational numbers $a, b \in (0,1)$ such that for infinitely many triples (α, β, γ) of rational numbers*

$$\max\{|a-\alpha|, |b-\beta|, |a^b-\gamma|\} < \exp(-(\log H)^\kappa),$$

where H denotes the maximum of the heights of α, β and γ.

Recall that the *usual height* of a polynomial P with rational (and more generally complex) coefficients is the maximum of the absolute values of its coefficients, and the *(usual) height* of an algebraic number is the usual height of its minimal polynomial over $\mathbb{Z}$.

In place of the generalization of Schanuel's Conjecture suggested in [84 W1] section IV.2, we now propose the following statement, whose assumptions include a technical hypothesis:

Conjecture 1.4 Let $x_1, \ldots, x_n$ be $\mathbb{Q}$-linearly independent complex numbers which satisfy a measure of linear independence. Let d be a positive integer. Then there exists a positive number $C = C(x_1, \ldots, x_n, d)$ with the following property: for any integer $H \geq 2$ and any $n+1$ tuple $P_1, \ldots, P_{n+1}$ of polynomials in $\mathbb{Z}[X_1, \ldots, X_n, Y_1, \ldots, Y_n]$ with degrees $\leq d$ and usual heights $\leq H$, which generate an ideal of $\mathbb{Q}[X_1, \ldots, X_n, Y_1, \ldots, Y_n]$ of rank[3] $n+1$, we have

$$\sum_{j=1}^{n+1} |P_j(x_1, \ldots, x_n, e^{x_1}, \ldots, e^{x_n})| \geq H^{-C}.$$

One of the most important special cases of Schanuel's Conjecture deals with the set

$$\mathcal{L} = \{\ell \in \mathbb{C};\ e^\ell \in \overline{\mathbb{Q}}^\times\}.$$

of logarithms of algebraic numbers:

Conjecture 1.5 (Conjecture on algebraic independence of logarithms of algebraic numbers) If $\ell_1, \ldots, \ell_n$ are $\mathbb{Q}$-linearly independent elements of $\mathcal{L}$, then they are algebraically independent.

[2]Bijlsma, Alex – On the simultaneous approximation of a, b and a^b. Compositio Math. **35** (1977), no. 1, 99–111.

[3]The rank of a prime ideal $\mathfrak{P} \subset \mathbb{Q}[T_1, \ldots, T_m]$ is the largest integer $r \geq 0$ such that there exists an increasing chain of prime ideals $(0) = \mathfrak{P}_0 \subset \mathfrak{P}_1 \subset \cdots \subset \mathfrak{P}_r = \mathfrak{P}$.
The rank of an ideal $\mathfrak{J} \subset \mathbb{Q}[T_1, \ldots, T_m]$ is the minimum rank of a prime ideal containing $\mathfrak{J}$.

A formal analog of Conjecture 1.5, which answers a question of Y. Hellegouarch, has been established by D.L. McQuillan [85 Mc]:

Let k be a field of zero characteristic, $P_1, P_2, \ldots$ non constant polynomials in $k[x]$ which are pairwise relatively prime and satisfy $P_i(0) = 1$. Define, for $i \geq 1$,

$$\mathrm{Log} P_i = -\sum_{n=1}^{\infty} \frac{1}{n}(1 - P(x))^n \in k((x)).$$

Then $\mathrm{Log} P_1, \mathrm{Log} P_2, \ldots$ *are algebraically independent over $k(x)$.*

Let us come back to the classical case. From Conjecture 1.4 one deduces the following quantitative version of Conjecture 1.5: *Let $n \geq 2$ and $d \geq 1$ be positive integers and $a_1, \ldots, a_n$ be multiplicatively independent positive integers, i.e. positive elements in $\mathbb{Z}$ such that the only relation $a_1^{b_1} \cdots a_n^{b_n} = 1$ with $(b_1, \ldots, b_n) \in \mathbb{Z}$ arises from the trivial case $b_1 = \cdots = b_n = 0$. Then there exists a constant $C > 0$, depending on n, $a_1, \ldots, a_n$ and also on d, such that, for any integer $H \geq 2$ and any non zero polynomial P in $\mathbb{Z}[X_1, \ldots, X_n]$ of degree $\leq d$ and height $\leq H$,*

$$|P(\log a_1, \ldots, \log a_n)| \geq H^{-C}.$$

Only the case of an homogeneous linear form $P = b_1X_1 + \cdots + b_nX_n$ is known (Theorem of Baker-Fel'dman).

The work of D. Roy [92 R] yields a new approach to Conjecture 1.5, involving matrices whose coefficients are linear combinations, with algebraic coefficients, of elements of $\mathcal{L}$. For such a matrix M, say of size $d \times \ell$, he defines the *structural rank* $r_{\mathrm{str}}(M)$ of M as the rank[4] of the matrix $\widetilde{M}$ obtained from M by replacing with indeterminates a basis of the vector space over $\overline{\mathbb{Q}}$ spanned by the coefficients of M: if

$$M = M_0 + M_1\ell_1 + \cdots + M_k\ell_k,$$

where $1, \ell_1, \ldots, \ell_k$ are $\overline{\mathbb{Q}}$-linearly independent, $\ell_j \in \mathcal{L}\ (1 \leq j \leq k)$ and $M_j \in \mathrm{Mat}_{d\times\ell}(\overline{\mathbb{Q}})$ $(0 \leq j \leq k)$, then

$$\widetilde{M} = M_0 + M_1X_1 + \cdots + M_kX_k \in \mathrm{Mat}_{d\times\ell}(\overline{\mathbb{Q}}(X_1, \ldots, X_k)).$$

It is neither surprising nor difficult to check that if Conjecture 1.5 holds, then the rank of M is equal to $r_{\mathrm{str}}(M)$. What is remarkable is the converse: *if the rank of M is always equal to $r_{\mathrm{str}}(M)$, then Conjecture* 1.5 *is true.*

In [95 R], D. Roy studies the set of points, whose coordinates are logarithms of algebraic numbers, which belong to a given algebraic variety. This work may be viewed as a first step towards the conjecture on algebraic independence of logarithms, since D. Roy remarks in [95 R] that Conjecture 1.5 is also equivalent to the following statement: *For each positive integer n and each algebraic subvariety V of $\mathbb{C}^n$ defined over $\overline{\mathbb{Q}}$, the set $V \cap \mathcal{L}^n$ is contained in the union of all vector subspaces of $\mathbb{C}^n$ defined over $\mathbb{Q}$ and contained in V.*

[4]The rank of a matrix is the rank over the field generated by its entries.

Further related conjectures of algebraic independence dealing with commutative algebraic groups are discussed in [96 W].

Conjectural extensions of Lindemann-Weierstraß Theorem to algebraic groups are considered in [87 P].

A conjectural description of the ideal of relations of algebraic dependence between periods of algebraic varieties which are defined over the field of algebraic numbers has been proposed by A. Grothendieck in 1966 and made more precise by S. Lang the same year. A joint generalization of Grothendieck's and Schanuel's Conjectures is suggested by Y. André [98 An1], who also proposes a transcendence conjecture on uniformization of Shimura's varieties, and discusses a non abelian variant of Grothendieck's Conjecture (see [97 Pr]), due to C. Simpson [90 Si], concerning the Riemann-Hilbert correspondence. Moreover in [98 An1] one can find comments on links between these conjectures and several topics, especially the Conjectures of Rohrlich and Lang on the algebraic dependence relations between values of the Gamma function, the results of Nesterenko on the modular functions (see §4 below), and the results of P. Cohen, H. Shiga and J. Wolfart, who extend in higher dimension Schneider's Theorem on the modular invariant j.

Notice that the conjectures in [86 W] are much less ambitious (and therefore should be easier to reach) than the general ones of Y. André in [98 An1]: they are more closely connected with currently available methods.

A conjecture in the "folklore" is that the zeroes of the Riemann zeta function (say their imaginary parts, assuming it > 0) are algebraically independent. As suggested by J-P. Serre, one might be tempted to consider also

- The eigenvalues of the zeroes of the hyperbolic Laplacian in the upper half plane modulo $\mathrm{SL}_2(\mathbb{Z})$ (i.e. to study the algebraic independence of the zeroes of Selberg zeta function).
- The eigenvalues of the Hecke operators acting on the corresponding eigenfunctions (Maass forms).

2 Methods

2.1 Transcendence and Algebraic Independence

Given functions $f_1, \dots, f_d$ and points $u_1, \dots, u_\ell$, a central problem is to give, under suitable assumptions, a lower bound for the transcendence degree of the field generated over $\mathbb{Q}$ by the $d\ell$ numbers $f_i(u_j)$ $(1 \le i \le d,\ 1 \le j \le \ell)$. Here is an example with $d = 2$, $\ell = 1$ and $f_1(z) = z$: consider a given function f and an algebraic number α such that $f(\alpha)$ is also algebraic; one wishes to deduce that α belongs to a restricted set. If $f(z) = e^z$, the restricted set is just $\{0\}$; if $f(z) = e^{i\pi z}$, the restricted set is $\mathbb{Q}$.

In general, one deals with a finite set of numbers $\{\theta_1, \dots, \theta_n\}$. In the last example one would take $n = 2$, $\theta_1 = \alpha$, $\theta_2 = f(\alpha)$

The basic methods of transcendence usually involve the construction of some number γ in the field $K = \mathbb{Q}(\theta_1, \dots, \theta_n)$. This number γ typically is either the value of an auxiliary function, or the determinant of an interpolation matrix.

To start with, assume that all numbers $\theta_1, \dots, \theta_n$ are algebraic. Therefore the number γ also is algebraic. Three main steps in the proof are:

- *Zero estimate*[5]: This number γ does not vanish.
- *Arithmetic estimate:* Lower bound for $|\gamma|$.
- *Analytic estimate:* Upper bound for $|\gamma|$.

We are interested here in the lower bound, which is a Liouville type argument: the size inequality, or the product formula, always rest on the fact that a non zero rational integer has absolute value at least 1.

Such a method yields the transcendence of values of f at points which do not belong to the restricted set. In general, it enables one to prove that some fields $\mathbb{Q}(\theta_1, \dots, \theta_n)$ have transcendence degree ≥ 1 over $\mathbb{Q}$. Variants of this argument yield *measures of simultaneous approximation by algebraic numbers:* a quantitative refinement of the transcendence result is a lower bound for $\max_{1\leq i\leq n} |\theta_i - \gamma_i|$, when $\gamma_1, \dots, \gamma_n$ are algebraic numbers (such a lower bound depends explicitly on the degrees and heights of the algebraic numbers $\gamma_1, \dots, \gamma_n$).

Here is an example of a measure of simultaneous approximation by algebraic numbers (see [97 RW2]). We define the *absolute logarithmic height* of an algebraic number γ by

$$\mathrm{h}(\gamma) = \frac{1}{d}\log a_0 + \frac{1}{d}\sum_{i=1}^{d} \log\max\{1, |\gamma_i|\}$$

when the minimal polynomial $a_0X^d + \cdots + a_d$ of γ over $\mathbb{Z}$ (with $a_0 > 0$) decomposes over $\mathbb{C}$ as

$$a_0\prod_{i=1}^{d}(X - \gamma_i).$$

Proposition 2.1 *Let $\beta_1, \dots, \beta_n$ be $\mathbb{Q}$-linearly independent algebraic numbers. There exists a positive constant $C = C(\beta_1, \dots, \beta_n)$ such that, if $\gamma_1, \dots, \gamma_n$ are algebraic numbers satisfying*

$$[\mathbb{Q}(\gamma_1, \dots, \gamma_n) : \mathbb{Q}] \leq D \quad \textit{and} \quad \max_{1\leq i\leq n} \mathrm{h}(\gamma_j) \leq h$$

with $h \geq e$, then

$$\begin{aligned}|e^{\beta_1} - \gamma_1| &+ \cdots + |e^{\beta_n} - \gamma_n| \\ &\geq \exp\{-CD^{1+(1/n)}h(\log h + D\log D)(\log h + \log D)^{-1}\}.\end{aligned}$$

In the case $n = 1$, it is not difficult to deduce from a measure of algebraic approximation for $\theta = \theta_1$ a *transcendence measure* of θ, i.e. a lower bound for $|P(\theta)|$, when $P \in \mathbb{Z}[X]$ is a non zero polynomial (this lower bound is explicit in terms of the degrees and heights of P). Conversely, from a transcendence measure one easily deduces a measure of approximation.

Now assume one wants to prove that some field $K = \mathbb{Q}(\theta_1, \dots, \theta_n)$ has transcendence degree ≥ 2 (result of small transcendence degree). Assume not: if θ denotes any transcendence basis, then the field K is a finite extension of $\mathbb{Q}(\theta)$. Using the same scheme of proof

---

[5] We have the disposition of strong general zero estimates for commutative algebraic groups; see for instance Bertrand, Daniel–Lemmes de zéros et nombres transcendants. *Séminaire Bourbaki*, vol. 1985/86. Astérisque No. **145–146** (1987), 3, 21–44.
In the proofs of the results stated in section 3 a result of P. Philippon plays a fundamental role. However, so far, apart from the setting of algebraic groups, the situation is not as good.

as above, one constructs a number γ in K, and one meets a difficulty for the lower bound. One solution, suggested by S. Lang in 1966 (in his book on transcendental numbers), is to assume a transcendence measure for θ; such a transcendence measure needs to be rather sharp (the field $\mathbb{Q}(\theta)$ should have "finite transcendence type"). This suggests an inductive procedure for algebraic independence. However, so far, results obtained by this way are rather weak, partly because the Diophantine estimates which can be proved are far from the best possible ones. Also this method is not universal, since some numbers like Liouville ones will not satisfy the desired measure.

Another much more efficient solution arises from Gel'fond's work in 1949: the transcendence construction yields not only one element $\gamma \in K$, but a sequence γ_N of such elements. Instead of asking for a lower bound for each single γ_N, Gel'fond succeeded to prove that not all elements γ_N in this sequence can be too small.

One should point out that it is often possible to complete transcendence proofs (at least when they involve only functions of a single variable) without a sharp zero estimate, but for algebraic independence (as well as for quantitative results), one really needs precise zero estimates.

2.2 *Criteria for Algebraic Independence and Measures*

The proof of A.O. Gel'fond's criterion involved only elementary properties of resultants of two polynomials in a single variable. It enables one to prove that some fields have transcendence degree ≥ 2 over $\mathbb{Q}$. In few cases it shows that two numbers are algebraically independent (one such example is A.O. Gel'fond's solution of Conjecture 1.2 for $d = 3$). Several authors contributed to refine it (see [84 W1]). Criteria for algebraic independence ("large transcendence degree") have been worked out by G.V. Chudnovsky (see Chap. 4 of [84 Chu]), E. Reyssat, R. Dvornicich, Yu.V. Nesterenko, P. Philippon, M. Waldschmidt and Zhu Yao Chen (see [84 W1]), as well as Zhu Yao Chen ([85 Z1], [89 Z1], [89 Z2], [90 Z]). One main tool, due to G.V. Chudnovsky, is the semi-resultant. These statements were sufficient to yield lower bounds for the transcendence degree of fields generated by values of the exponential function for instance, but these lower bounds were not sharp. The first sharp estimate has been announced by P. Philippon in [84 P] and proved in [86 P]. Here is Corollary 0.6 of [86 P] (the main result of [86 P] also splits degree and height; further statements and references are given in [98 NP]).

Theorem 2.2 (Philippon's Criterion for Algebraic Independence). *Let n be a positive integer, C a sufficiently large real number, $(\theta_1, \dots, \theta_n)$ an element of $\mathbb{C}^n$ and η a positive real number. Assume that for all sufficiently large N, there exist a positive integer $m = m(N) \geq 1$ and polynomials $Q_1, \dots, Q_m$ in $\mathbb{Z}[X_1, \dots, X_n]$ with*

$$\max_{1\leq j\leq m} \deg Q_j \leq N, \qquad \max_{1\leq j\leq m} \mathrm{H}(\mathrm{Q_j}) \leq e^N$$

and

$$0 < \max_{1\leq j\leq m} |Q_j(\theta_1, \dots, \theta_n)| \leq e^{-CN^\eta},$$

such that the set of common zeros of the polynomials $Q_1, \dots, Q_m$ in the ball

$$\left\{ z \in \mathbb{C}^n \, ; \, \max_{1 \le i \le n} |z_i - \theta_i| \le e^{-3CN^\eta} \right\}$$

is finite. Then the transcendence degree over $\mathbb{Q}$ of the field $\mathbb{Q}(\theta_1, \dots, \theta_n)$ is $> \eta - 1$.

This statement allows one to give lower bounds for the transcendence degree of some fields. Quantitative refinements of such criteria have been studied by several authors: the goal is now to produce a result of Diophantine approximation, which includes the lower bound for the transcendence degree. When the conclusion of the proof of algebraic independence is that some numbers $\theta_1, \dots, \theta_n$ are algebraically independent (an example is the Lindemann-Weierstraß Theorem, another one is A.O. Gel'fond's result on α^β and α^{β^2}; a few other examples are due to G.V. Chudnovsky), the most natural quantitative refinement is a lower bound for $|P(\theta_1, \dots, \theta_n)|$, when P is a non zero polynomial with rational (or algebraic) coefficients. This is called a *measure of algebraic independence*. As shown by P. Philippon [85 P] and Yu.V. Nesterenko [85 Ne1], similar measures can also be given when the conclusion is only that the transcendence degree of the field $\mathbb{Q}(\theta_1, \dots, \theta_n)$ over $\mathbb{Q}$ is greater than some number k which is $< n$. Another quantitative refinement is to give a lower bound for $\max_{1 \le i \le n} |\theta_i - \gamma_i|$ when $\gamma_1, \dots, \gamma_n$ belong to a field of transcendence degree $\le k$; this is called a *measure of simultaneous approximation for* $\theta_1, \dots, \theta_n$. An example with $k = 0$ is Proposition 2.1. In [85 P], P. Philippon compares both measures; he produces measures of algebraic independence and deduces measures of simultaneous approximations for various transcendental numbers.

In [88 Ab] and [89 Ab], M. Ably proves a criterion for measures of simultaneous approximation which yields sharper estimates than [85 P] for numbers of the form $e^{x_i y_j}$, α^{β^j}, and also for G.V. Chudnovsky's result on η/ω and π/ω. This later example is inspired by G. Philibert's result in [88 Ph] (see Theorem 3.1 below).

In [92 Jb], E.M. Jabbouri improves Philippon's results [85 P] and [86 P] and produces a criterion which yields effective measures of algebraic independence, splitting degree and height.

In order to prove general results for one parameter subgroups of commutative algebraic groups, M. Ably in [92 Ab2] combines the previous criteria for measures of algebraic independence [92 Jb] and measures of simultaneous approximation [89 Ab] into a single one. Finally C. Jadot [96 Jd] refines the previous criteria of algebraic independence due to P. Philippon and E.M. Jabbouri, so that he also includes criteria for linear independence due to Yu. V. Nesterenko, P. Bundschuh and T. Töpfer. Again Jadot's results are effective (i.e. his criterion can be used to produce measures of linear independence as well as measures of algebraic independence).

The reference [98 NP] contains an extensive exposition of the basic tools from commutative algebra and elimination theory, including *Chow forms*, which have been introduced by Yu.V. Nesterenko in this context. Following a lecture of D. Roy, we only give here the definition of *the Chow form associated with an algebraic variety $V \subset \mathbb{P}_m(\overline{\mathbb{Q}})$ of dimension k defined over the field $\mathbb{Q}$ of rational numbers.* Let

$$U_i = U_0^{(i)} X_0 + \cdots + U_m^{(i)} X_m \qquad (0 \le i \le k)$$

be generic linear forms. There exists an irreducible polynomial

$$F \in \mathbb{Q}[\underline{U}^{(0)}, \dots, \underline{U}^{(k)}],$$

where $\underline{U}^{(i)}$ stands for the m-tuple $(U_0^{(i)}, \dots, U_m^{(i)})$, such that, for any linear forms $L_0, \dots, L_k$ in $\overline{\mathbb{Q}}[X_0, \dots, X_m]$, say

$$L_i = u_0^{(i)} X_0 + \cdots + u_m^{(i)} X_m \qquad (0 \leq i \leq k)$$

the condition

$$F(\underline{u}^{(0)}, \dots, \underline{u}^{(k)}) = 0$$

holds if and only if

$$L_0(\underline{x}) = \cdots = L_k(\underline{x}) = 0$$

have a solution $\underline{x}$ in V. This form F, which is unique up to a multiple in $\mathbb{Q}^\times$, is the *Chow form of* V.

2.3 Hilbert Nullstellensatz

The first results on large transcendence degree for values of elliptic functions are due to D.W. Masser and G. Wüstholz in the early 80's. They proved elliptic analog of Chudnovsky's result on the exponential function for elliptic functions with algebraic invariants without complex multiplication. In order to provide a replacement for the elimination technique of G.V. Chudnovsky by means of semi-resultants, they prove (and use) an explicit lower bound for the degree in Hilbert's Nullstellensatz. They need also tools from commutative algebra in order to refine their zero estimate. Another tool is an effective elliptic version of a Theorem of Kolchin on subgroups of products of algebraic groups.

The bounds of D.W. Masser and G. Wüstholz for the degree in Hilbert's Nullstellensatz were not sharp. W.D. Brownawell's succeeded to get essentially best possible estimates, by means of the powerful tools of Yu.V. Nesterenko and P. Philippon in elimination theory, as well as deep results from the theory of several complex variables of C. Berenstein and A. Yger. However, the method of Masser-Wüstholz then produces weaker results than Philippon's one in [86 P] (already in 1983 P. Philippon had much stronger results for large transcendence degree than those obtained by D.W. Masser and G. Wüstholz).

However an alternative approach to this question has been worked out by W.D. Brownawell in [89 Br]. He establishes sharp lower bounds for the maximum absolute values of integral polynomials having no common zero within a ball of given radius centered at a fixed point in $\mathbb{C}^n$. This allows him to relax the technical hypothesis: in place of a measure of linear independence, he only needs to assume that some lower bound holds for infinitely many values of the parameter. In [89 B], he works out the case of algebraic independence of values of the exponential function, or of elliptic functions without complex multiplication. The case of complex multiplication is considered in his joint paper with R. Tubbs [89 BT]. These two papers [89 B] and [89 BT] also give sharp effective quantitative measures of algebraic independence. Further references on this topic are [91 Tu] and [96 Den2].

2.4 Simultaneous Diophantine Approximation and Algebraic Independence

A new link between simultaneous approximation and algebraic independence arises from the work of D. Roy, M. Waldschmidt, M. Laurent, P. Philippon and G. Diaz (see [95 RW], [97 Di 2], [97 P], [97 RW1], [97 RW2], [98 LR2], [98 P1], [98 P2] and [98 P3]).

The basic idea is as follows. A complex transcendental number θ can be approximated by algebraic numbers. For instance Dirichlet's box principle enables one to produce a non zero polynomial $P \in \mathbb{Z}[X]$ such that $|P(\theta)|$ is small; then one considers the root of P which is closest to θ. This last step involves some technical complications, but E. Wirsing[6] showed in 1961 how to overcome them. Denote by $\mathrm{M}(\gamma)$ *Mahler's measure* of the algebraic number γ, which is defined by

$$\mathrm{M}(\gamma) = e^{d\mathrm{h}(\gamma)}$$

where $d = [\mathbb{Q}(\gamma) : \mathbb{Q}]$ is the degree of γ.

Theorem 2.3 (Wirsing). *For any integer $D \geq 2$ there exist infinitely many algebraic numbers γ which satisfy*

$$[\mathbf{Q}(\gamma) : \mathbf{Q}] \leq D, \qquad \text{and} \qquad |\theta - \gamma| \leq \mathrm{M}(\gamma)^{-D/4}.$$

Wirsing's arguments are related with Gel'fond's proof of his transcendence criterion. Apart from a result by H. Davenport and W.M. Schmidt (1968), the constant in Wirsing's result has not yet been substantially improved, but many useful recent variants have recently been produced. In [97 RW1] and [97 RW2], one considers more generally the simultaneous approximation of complex numbers by algebraic numbers. Given $\theta_1, \ldots, \theta_n$ in $\mathbb{C}$, one shows the existence of algebraic numbers $\gamma_1, \ldots, \gamma_n$ such that $\max |\theta_i - \gamma_i|$ is bounded from above in terms of the degree $\mathbb{Q}(\gamma_1, \ldots, \gamma_n)$, as well as the heights of γ_i's. The transcendence degree over $\mathbb{Q}$ of the field $\mathbb{Q}(\theta_1, \ldots, \theta_n)$ is the invariant which should, at least conjecturally, control the situation.

For instance, assume that the transcendence degree over $\mathbb{Q}$ of the field $K = \mathbb{Q}(\theta_1, \ldots, \theta_n)$ is 1. It is not too hard to check that an algebraic approximation of a single transcendental element in K will enable one to produce "simultaneous" algebraic approximations to $\theta_1, \ldots, \theta_n$. In other terms, if $\theta_1, \ldots, \theta_n$ have the property that they cannot be simultaneously approximated by algebraic numbers, then they cannot lie in a field of transcendence degree 1, and two at least of them are algebraically independent. This is how a link between simultaneous Diophantine approximation and algebraic independence arises.

Usually, "simultaneous" approximation (for instance of real numbers by rational ones) means that one has a good control of a common denominator of the rational approximations; more generally, "simultaneous" approximation of complex numbers by algebraic ones usually means that one has a good control of the absolute logarithmic height of the projective point whose coordinates are the algebraic approximations. Here, we do not pay too much attention to the constants, but we are interested with algebraic numbers of large degree, and we are concerned with bounds for the degree of the number field generated by the algebraic approximations.

---

[6] Wirsing, Eduard – Approximation mit algebraischen Zahlen beschränkten Grades. J. reine angew. Math. **206** (1960), 67–77.

The following statement is Conjecture 1.7 of [97 RW2].

Conjecture 2.4 Let $a \geq 1, b \geq 1$ be real numbers, t a non negative integer and $\theta_1, \ldots, \theta_n$ complex numbers. Let $\varphi : \mathbb{N} \times \mathbb{R}_+ \to \mathbb{R}_+ \cup \{\infty\}$ be a mapping with the following property: there exist a positive integer D_0 together with a real number $h_0 \geq 1$ such that, for any integer $D \geq D_0$, any real number $h \geq h_0$ and any n-tuple $(\gamma_1, \ldots, \gamma_n)$ of algebraic numbers satisfying

$$[\mathbb{Q}(\gamma_1, \ldots, \gamma_n) : \mathbb{Q}] \leq D \quad \text{and} \quad \max_{1 \leq i \leq n} \mathrm{h}(\gamma_i) \leq h,$$

we have

$$\max_{1 \leq i \leq n} |\theta_i - \gamma_i| \geq \exp\{-\varphi(D, h)\}.$$

Let $(D_\nu)_{\nu \geq 1}$ be an increasing[7] sequence of positive integers and $(h_\nu)_{\nu \geq 1}$ an increasing sequence of positive real numbers with $D_\nu + h_\nu \to \infty$. Assume

$$D_{\nu+1} \leq a D_\nu, \qquad h_{\nu+1} \leq b h_\nu, \qquad (\nu \geq 1)$$

and

$$\limsup_{\nu \to \infty} \frac{1}{D_\nu^{1+(1/t)} h_\nu} \varphi(D_\nu, h_\nu) = 0$$

Then the transcendence degree over $\mathbb{Q}$ of the field $\mathbb{Q}(\theta_1, \ldots, \theta_n)$ is $> t$.

So far, only the case $n = 1$ is solved, by M. Laurent and D. Roy [98 LR2].

Lindemann-Weierstraß Theorem on the algebraic independence of $e^{\beta_1}, \ldots, e^{\beta_n}$ (when $\beta_1, \ldots, \beta_n$ are $\mathbb{Q}$-linearly independent algebraic numbers) follows from Conjecture 2.4 and Proposition 2.1. The special case $n = 2$ (algebraic independence of e^{β_1} and e^{β_2}) follows from Proposition 2.1 together with the special case of Conjecture 2.4 which is proved in [97 RW2].

This approach yields results of algebraic independence (only small transcendence degree, so far) by means of Laurent's interpolation determinants. Previously, the method of interpolation determinants was limited to proofs involving only algebraic numbers: it was efficient only for showing that some fields have transcendence degree ≥ 1, as well as for providing quantitative versions of such results, namely measures of transcendence or of simultaneous algebraic approximation by algebraic numbers. The point is that such measures turn out to yield algebraic independence results!

Another solution to the problem of proving algebraic independence by means of interpolation determinants has been found by M. Laurent and D. Roy: in [98 LR1] they first extend Gel'fond's criterion by including multiplicities; next they show that the interpolation determinant produces a polynomial which, not only is small at some point, but also has small derivatives at that point.

In a forthcoming paper they also include derivatives in Philippon's criterion of algebraic independence. This will produce results of large transcendence degree by means of interpolation determinants. Unfortunately these new tools, which are definitely valuable from

[7] Increasing sequences are sometimes called non decreasing sequences: we do not exclude a constant sequence for instance.

the point of view of the method, do not seem to produce new results which could not be reached by previous arguments, so far.

Notice also that A. Sert's proof in [97 Se] of a measure of algebraic independence for $e^{\beta_1}, \dots, e^{\beta_n}$ (quantitative version of the Lindemann-Weierstraß Theorem) involves interpolation determinants. Another way of replacing the auxiliary function by an interpolation determinant has been worked out by P. Philippon in [97 P], [98 P1] and [98 P2]. One main difference between [97 RW1] and [97 P] is that Philippon considers approximation by algebraic cycles rather than by algebraic numbers. For instance, in the one dimensional case, given a transcendental number θ, instead of considering $|\theta - \gamma|$ for some algebraic number γ, he only needs to get a non zero polynomial $P \in \mathbb{Z}[X]$ such that $|P(\theta)|$ is small; and Dirichlet's box principle is just sufficient!

3 Gel'fond's Method and its Developments

3.1 Small Transcendence Degree

We start with "small transcendence degree". Many very interesting results are due to G.V. Chudnovsky [84 Chu], [84 W1]. We recall the following one: *Let $\mathcal{E}$ be an elliptic curve over $\overline{\mathbb{Q}}$, ω a non zero period of a differential form of first kind and η the corresponding period of a differential form of second kind. Then the two numbers ω/π and η/π are algebraically independent.*

Let ξ be an algebraic number and $\mathcal{E}$ the elliptic curve $y^2 = (1 - x^2)(1 - \xi x^2)$. In [96 An], Y. André states Chudnovsky's result in terms of values of hypergeometric functions, and gives a completely new proof by means of the Gel'fond-Dèbes method involving G-functions. Also he produces p-adic analog.

Here is an extension of G V. Chudnovsky's result to Abelian varieties, due to K.G. Vasilev [96 V]. Let $F \in \overline{\mathbb{Q}}[X, Y]$ be an irreducible polynomial such that the curve $F(x, y) = 0$ has genus $g \geq 1$. Let $\varphi_1, \dots, \varphi_{2g}$ be differentials of first or second kind and $\gamma_1, \dots, \gamma_{2g}$ be closed integration paths on the Riemann surface such that the $2g \times 2g$ matrix

$$\Omega = \left(\int_{\gamma_j} \varphi_i\right)_{1 \leq i, j \leq 2g}$$

is non singular. Then *any $g + 1$ rows of Ω contains at least 2 algebraically independent numbers.* As a consequence, *any subset with $[(n + 1)/2]$ elements of the following set of values of the Beta function*

$$\{B(k/n, 1/2)\,;\ 1 \leq k \leq n - 1,\ k \neq n/2\}$$

contains at least 2 algebraically independent numbers.

The fact that $\Gamma(1/4)$ and $\Gamma(1/3)$ are not Liouville numbers, which was claimed by G.V. Chudnovsky already in 1980 (see also [84 Chu] Th. 3 p. 18), has been proved only recently by P. Philippon in [98 P2]. More generally Philippon produces a sharp measure of algebraic independence of π/ω and η/ω:

Theorem 3.1 *Let $\wp$ be a Weierstraß elliptic function with algebraic invariants g_2, g_3. Let ω be a non zero period, and η be the corresponding quasi-period of the associated Weierstraß* *zeta function. There exists a constant $C > 0$ such that, for any non zero polynomial $P \in \mathbb{Z}[X, Y]$ of degree $\leq D$ and usual height $\leq H$,*

$$|P(\pi/\omega, \eta/\omega)| \geq \exp\{-CD^2(\log H + D\log D)\}.$$

After Chudnovsky's pioneer work, previous weaker estimates had been obtained by G. Philibert in [88 Ph], by M. Ably in [89 Ab] and by E.M. Jabbouri in [92 Jb]. Notice that G. Philibert's measure in [88 P] played an interesting role in connection with the work of P. Philippon related to the algebraic independence of π and e^{π} (see §4 below).

Gel'fond's algebraic independence Theorem (1949) on α^{β} and α^{β^2} has been extensively studied. The measure of algebraic independence he obtained with N.I. Feldman has been improved, especially by G. Diaz in [87 Di], [88 Di], [89 Di1] and [90 Di]; see also [90 She]. Here is the best known result by G. Diaz :

Theorem 3.2 *Let α be a non zero algebraic number, $\log\alpha$ a non zero logarithm of α and β a cubic number. There exists a constant $c > 0$ such that if $P \in \mathbb{Z}[X, Y]$ is a non zero polynomial of degree $\leq D$ and usual height $\leq H$, then*

$$\log\left|P(\alpha^{\beta}, \alpha^{\beta^2})\right| \geq -\exp\{cD(D + \log H)\}.$$

The elliptic analog of Gel'fond's Theorem on α^{β} and α^{β^2}, announced by D.W. Masser and G. Wüstholz in the early 80's, has been proved in [86 MW]. Much sharper results have then been achieved by P. Philippon in 1983. In [86 Tu2], R. Tubbs gives a lower bound for $|P(\wp(\beta u), \wp(\beta^2 u))|$ in the CM case (here, $\wp$ is an elliptic function of Weierstraß with algebraic invariants and complex multiplication, u is a complex number such that $\wp(u)$ is algebraic, P a polynomial with rational integer coefficients and β is an algebraic number of degree 3 over the field of complex multiplications). In [88 Tu], he removes the assumption that $\wp(u)$ is algebraic and produces a lower bound for pairs of relatively prime integral polynomials evaluated at $(\wp(u), \wp(\beta u), \wp(\beta^2 u))|$ (elliptic analog of a result of W.D. Brownawell in 1979). The elliptic analog of Theorem 3.2 has been established by S.O. Shestakov [92 She]. For an updated survey of such questions with new results, see [96 Ca].

General results of small transcendence degree related with one parameter subgroups of commutative algebraic groups are obtained in [85 W]. The main difference with [86 W] is that no technical hypothesis is needed for small transcendence degree. These results do not exhaust the possibilities of the method: some results by G.V. Chudnovsky [84 Chu] are not covered by the statements of [85 W] and many variations are possible (see for instance the papers [86 TY] by M. Toyoda and T. Yasuda and [86 Tu1], [87 Tu1] by R. Tubbs). A more systematic study has been achieved by R. Tubbs in [87 Tu2] and [90 Tu2]. Further results related to Baker's method (see Chap. 2 of [84 Chu]) are due to G. Diaz [92 Di].

Transcendence measures enable one to construct algebraically independent numbers: an early example is due to N.I. Feldman (1964): using a lower bound for $|\beta - \log\alpha|$ when α and β are algebraic numbers, he proved that for some "Liouville" numbers a, the two numbers a and e^a are algebraically independent. Here, "Liouville" number means a complex number a which is very well approximated by rational, or more generally algebraic, numbers.

In the same way, R. Tubbs [90 Tu1] uses a lower bound for elliptic linear forms in logarithms and proves, for algebraic β, the algebraic independence of u and $\wp(\beta u)$ for some Liouville numbers u, and also the algebraic independence of $\wp(u)$ and $\wp(\beta u)$ for some complex numbers u such that $\wp(u)$ is Liouville. In [93 Ca], D. Caveny shows the algebraic independence of a and a^b when a and b are Liouville numbers; with R. Tubbs in [93 CT] they refine Fel'dman's above mentioned result on a and e^a and also discuss the elliptic analog. Using a lower bound for linear forms in logarithms, D. Caveny gets in [94 Ca] the algebraic independence of a and a^β for a Liouville number a and an algebraic β. A similar elliptic situation is considered by M. Ably in [91 Ab].

Usually, each single recent paper deals with either transcendence degree ≥ 2 or else large transcendence degree. One of the few exceptions are [89 Tu] where R. Tubbs produces fields of transcendence degree ≥ 3 generated by values of an elliptic function, and [98 P2].

We conclude this section with a partial result concerning Conjecture 1.5. It is still not proved that the field generated by all the logarithms of algebraic numbers has transcendence degree ≥ 2 over $\mathbb{Q}$: it might just as well be an algebraic extension of $\mathbb{Q}(\pi)$ (but this would be unexpected!). One of the very few known results in this direction arises from [97 RW1] and [97 RW2]:

Theorem 3.3 *Let $\ell_1, \ldots, \ell_n$ be $\mathbb{Q}$-linearly independent elements of $\mathcal{L}$. Assume that the transcendence degree over $\mathbb{Q}$ of the field $\mathbb{Q}(\ell_1, \ldots, \ell_n)$ is 1. Then for each non zero homogeneous polynomial $Q \in \mathbb{Q}[X_1, \ldots, X_n]$ of degree 2 (quadratic form), the number $Q(\ell_1, \ldots, \ell_n)$ does not vanish.*

3.2 Large Transcendence Degree

In [84 W1], we noticed that Philippon's results on the elliptic analog of the Gel'fond-Schneider problem were stronger than the exponential result. Shortly after, P. Philippon succeeded to prove the expected result in the exponential case.

Let t be the transcendence degree over $\mathbb{Q}$ of the field $\mathbb{Q}(\alpha^\beta, \alpha^{\beta^2}, \ldots, \alpha^{\beta^{d-1}})$. After the works of G.V. Chudnovsky, P. Warkentin, P. Philippon, E. Reyssat, R. Endell, W.D. Brownawell and Yu.V. Nesterenko (see [84 W1] as well as [84 Ne1]), the best known lower estimate for t was

$$2^t > \frac{d+1}{2}.$$

In 1984 P. Philippon succeeded to reach the lower bound $t \geq (d-1)/2$, which can be written $t \geq [d/2]$ where [ ] denotes the integral part. In [84 P] he announced the result, together with a sharp criterion for algebraic independence, as well as a quantitative refinement (measure of algebraic independence) which he discussed in [85 P]. Proofs are provided in [86 P]. They involve the tools from commutative algebra introduced earlier by Yu.V. Nesterenko. Algebraic independence measures are given in [85 Ne1] and [86 Ne]. Further references on this topic are [85 P] and [87 Ne]. It should be pointed out that the arguments used in the proofs also work in the context of Mahler's method (see [85 Ne1]).

In 1987, G. Diaz [87 Di] obtained a refinement, which is the best known estimate so far:

Theorem 3.4 (G. Diaz). *Let α be a non zero complex algebraic number,* $\log \alpha$ *a non zero logarithm of α and β an algebraic number of degree $d \geq 2$. Then the transcendence degree t over $\mathbb{Q}$ of the field $\mathbb{Q}(\alpha^{\beta}, \alpha^{\beta^2}, \ldots, \alpha^{\beta^{d-1}})$ satisfies*

$$t > \frac{d-1}{2}.$$

An equivalent formulation is $t \geq [(d+1)/2]$. One main tool in Diaz' proof is Philippon's Criterion (Proposition 2.2). Another important feature of Diaz' proof is that he succeeds to work out an idea, going back to G.V. Chudnovsky, which consists in deriving the coefficients of the auxiliary polynomial with respect to a transcendence basis. The elimination process has been revisited by Yu.V. Nesterenko [89 Ne] who avoids the use of Philippon's criterion.

All above works, including Diaz' results in [89 Di1] and [89 Di2], deal more generally with fields generated over $\mathbb{Q}$ by numbers of the form $e^{x_i y_j}$, where $x_1, \ldots, x_d$ are $\mathbb{Q}$-linearly independent complex numbers, and also $y_1, \ldots, y_\ell$ are $\mathbb{Q}$-linearly independent numbers. Of course, in such a general setup, a technical hypothesis is needed. An extension to large transcendence degree of Diaz' result [92 Di] using Baker's method is due to Chen Gong Liang [93 Chn], [95 Chn].

Several quantitative refinements of the Lindemann-Weierstraß Theorem, i.e. measures of algebraic independence for $e^{\beta_1}, \ldots, e^{\beta_n}$, are known. They used to be produced by means of Siegel-Shidlovskiĭ's method involving E-functions. In 1994, M. Ably [94 Ab] improved such an estimate from Nesterenko (1977) by means of Gel'fond-Chudnovsky's method, using the criterion of Jabbouri-Philippon [92 Jb]. The result in [94 Ab] is "effective", but not all constants are explicitly computed. A completely explicit measure is given in A. Sert's thesis [97 Se].

Here is the elliptic analog of the Lindemann-Weierstraß Theorem:

- *Let $\wp$ be a Weierstraß elliptic function with algebraic invariants. Denote by k the field of endomorphisms of $\wp$ (which is either $\mathbb{Q}$ – non CM case – or else an imaginary quadratic field – CM case –). If $\beta_1, \ldots, \beta_n$ are k-linearly independent complex numbers, then the field $\mathbb{Q}(\wp(\beta_1), \ldots, \wp(\beta_n))$ has transcendence degree $\geq n/[k : \mathbb{Q}]$ over $\mathbb{Q}$.*

As pointed out in [84 W1], the case of complex multiplication had been settled in 1983 by P. Philippon and G. Wüstholz, independently. In fact, the non CM case also follows from the works of P. Philippon and G. Wüstholz in 1983. In [86 Jb], E.M. Jabbouri produces quantitative results, including measures of algebraic independence in the CM case. His result deals more generally with Abelian varieties. His proof uses Philippon's zero estimate as well as the elimination techniques from [86 P]. Extensions of Lindemann-Weierstraß Theorem to algebraic groups, including the quantitative aspect, have also been considered in [87 P]. Jabbouri's measure of algebraic independence [86 Jb] has been refined by M. Ably in [92 Ab1], by means of a larger number of redundant variables. In the case of elliptic functions, Yu.V. Nesterenko's measure of algebraic independence for $\wp(\beta_1), \ldots, \wp(\beta_n)$ in the CM case [92 Ne] is better (see also [91 Ne]). The sharpest known results in this context are due to Yu.V. Nesterenko [95 Ne].

General results on large transcendence degree are obtained in [84 W2] for the exponential function in several variables, in [84 W3] for one parameter subgroups of products of linear groups and elliptic curves (giving mixed exponential-elliptic algebraic independence results) and in [86 W] for several parameters analytic subgroups of commutative algebraic groups. Here is an example (Corollary 15.4 of [86 W]), which displays a new kind of technical hypothesis.

Theorem 3.5 *Let A be a simple abelian variety of dimension g defined over the field of algebraic numbers, V a vector subspace of dimension n of the tangent space at the origin of A, and $Y = \mathbb{Z}y_1 + \cdots + \mathbb{Z}y_m$ a finitely generated subgroup of V of rank m. Let K be a subfield of $\mathbb{C}$ of transcendence degree t over $\mathbb{Q}$, such that $\exp_A Y \subset A(K)$. Assume that for any $\epsilon > 0$ there exists $H_0 > 0$ such that, for any $H \geq H_0$, any $(h_1, \ldots, h_m) \in \mathbb{Z}^m$ with*

$$0 < \max\{|h_1|, \ldots, |h_m|\} \leq H,$$

and any $\omega \in \ker \exp_A$, the condition

$$|h_1 y_1 + \cdots + h_m y_m - \omega| < \exp(-H^\epsilon)$$

implies $h_1 y_1 + \cdots + h_m y_m = \omega$. Then

$$t \geq \frac{mg}{n(m+2g)} - 1$$

For one parameter subgroups (i.e. $n = 1$), M. Ably [89 Ab], [92 Ab2] and R. Tubbs [91 Tu] improved the conclusion and got a strict inequality, which means $t \geq [mg/(m+2g)]$. The method is inspired by Diaz [89 Di1]. The elliptic analog of Diaz's Theorem 1.3 has been obtained by M. Ably [92 Ab2] and S.O. Shestakov [91 She], [92 She]:

Theorem 3.6 *Let $\wp$ be a Weierstraß elliptic function with algebraic invariants g_2, g_3. Denote by F the field of endomorphisms of the associated elliptic curve, and by $\delta = [F : \mathbb{Q}]$ the degree of F (that means $\delta = 2$ in the CM case, $\delta = 1$ otherwise). Let β be an algebraic number of degree $d > 2/\delta$ and u a complex number such that none of the d numbers $u, \beta u, \ldots, \beta^{d-1}u$ is a pole of $\wp$. Then the transcendence degree over $\mathbb{Q}$ of the field $\mathbb{Q}(\wp(u), \wp(\beta u), \ldots, \wp(\beta^{d-1}u))$ is at least*

$$\begin{cases} \left[\frac{d+1}{2}\right] & \text{for } \delta = 2 \text{ (CM case)}, \\ \left[\frac{d+1}{3}\right] & \text{for } \delta = 1. \end{cases}$$

In [91 Ab], M. Ably improves the result (adding 1 to the transcendence degree) in the special case where one of the numbers $\wp(u), \wp(\beta u), \ldots, \wp(\beta^{d-1}u)$ has extremely good approximations by algebraic numbers.

In [92 Ab2], M. Ably gives quantitative refinements to the result of [86 W] in the one dimensional case, involving a weakened technical hypothesis. These far reaching estimates on algebraic independence for one parameter subgroups of commutative algebraic groups have been refined by D. Caveny [96 Ca], who splits the dependence in the degree and height, like in G. Diaz' Theorem 3.2.

4 Modular Functions

The first results on algebraic independence of values of modular functions are due to D. Bertrand. The proofs involved elliptic functions. Only in 1995 a group of mathematicians in St Étienne[8] succeeded to prove a transcendence result by means of modular functions: they solve a problem of Mahler in the complex case and of Manin in the p-adic case by proving the transcendence of $J(q)$ when q is an algebraic (complex or p-adic) number in the domain $0 < |q| < 1$, and J is the modular invariant. In the complex case, this result solves a special case of the "mixed four exponentials Conjecture". In fact deep connections with the classical four exponentials exist [97 Di1].

One year later Nesterenko [96 Ne1], [96 Ne2] obtained a remarkable result of algebraic independence:

Theorem 4.1 (Nesterenko). *Let q be a complex or p-adic number satisfying $0 < |q| < 1$. Then three at least of the four numbers*

$$q, P(q), Q(q), R(q)$$

are algebraically independent.

These functions P, Q and R are the Eisenstein series of weights 2, 4 and 6 respectively. In the complex case, an equivalent statement in terms of Weierstraß elliptic functions reads as follows:

Corollary 4.2 *Let g_2 and g_3 be two complex numbers satisfying $g_2^3 \neq 27g_3^2$. Denote by ω_1, ω_2 a pair of fundamental periods of the elliptic curve $y^2 = 4x^3 - g_2x - g_3$, and by η_1, η_2 the corresponding quasi-periods. Then the transcendence degree over $\mathbb{Q}$ of the field*

$$\mathbb{Q}(g_2, g_3, \omega_1/\pi, \eta_1/\pi, e^{2i\pi\omega_1/\omega_2})$$

is at least 3.

Notice that without $e^{2i\pi\omega_1/\omega_2}$, it was known by G.V. Chudnovsky (see §3.1 above, as well as [84 Chu] and [84 W1]) that the transcendence degree is at least 2.

One of the most impressive consequences of Corollary 4.2 is:

Corollary 4.3 *The three numbers π, e^{π} and $\Gamma(1/4)$ are algebraically independent.*

Further references on this topic include [97 Ber], [97 Ne], [97 W], [98 Gr], [98 La], [98 P1], [98 Sar] and [98 W1]. In [96 Ne1], [96 Ne2], [97 Ne] and [98 P1], Yu.V. Nesterenko and P. Philippon give sharp quantitative measures of algebraic independence for the numbers in Corollary 4.3. Philippon's works [98 P1] and [98 P3] also apply to Mahler's method, and more generally to a new class of functions which he calls K-functions, completing the classes of E- and G-functions introduced by C.L. Siegel in 1929.

[8] Barré-Sirieix, Katia; Diaz, Guy; Gramain, François; Philibert, Georges – Une preuve de la conjecture de Mahler-Manin. Invent. Math. **124** (1996), no. 1–3, 1–9.

5 Diophantine Rings

P. Philippon has introduced a general concept of "Diophantine ring" in [92 P] (see also [95 P]). This is a commutative Noetherian ring, with a valuation and a "size inequality". Special cases are $\mathbb{Z}$ and rings of integers of algebraic number fields on one hand, $\mathbb{F}_q[T]$ and rings of functions of an algebraic curve on a finite field on the other hand. Also, if A is a Diophantine ring, then so is any ring which is finitely generated over A. In [92 P], Philippon extends his criterion of algebraic independence [86 P] to Diophantine rings. In [97 P], he proposes an axiomatic method for transcendence and algebraic independence. One main axiom is a Schwarz' Lemma. This axiomatic point of view enables him to contain most transcendence and algebraic independence results involving functions of a single variable, including Gel'fond's method, Baker's Theorem, Lindemann-Weierstraß Theorem, Mahler's method as well as results related to Carlitz exponential and Drinfel'd modules. Section 8 of [97 P] provides a very interesting speculative suggestion for solving Conjecture 1.2.

References

Among the main contributions during the period before 1984 which was covered by our previous survey [84 W1] was the work of G.V. Chudnovsky. His book [84 Chu] was published just after [84 W1] and includes many further results. Another important reference on this topic is Nesterenko's invited talk [91 Ne] at the International Congress of Mathematicians of Kyoto in 1990. Two surveys dealing more generally with transcendental number theory are [87 B...] and [98 W3]. Further general references for algebraic independence, including surveys, are [87 Ne], [96 Ca] and [98 La]. The forthcoming book [98 NP] will also provide detailed proofs.

A first version of this survey was just completed when the book [98 FN] was published. The last chapter "Algebraic independence of the values of analytic functions that have an addition law" includes a compilation of related results and provides a lot of further relevant information on this subject.

References are given in chronological ordering. We do not repeat the references given in [84 W1].

1984

[**84** BV] P. Bundschuh and Väänänen, K., On lower bounds for polynomials in the values of E-functions. II. *Topics in classical number theory,* vol. I, II (Budapest, 1981), 229–255, Colloq. Math. Soc. János Bolyai, **34**, North-Holland, Amsterdam-New York, 1984.

[**84** Chu] Chudnovsky, Gregory V., *Contributions to the theory of transcendental numbers.* Mathematical Surveys and Monographs, **19**. American Mathematical Society, Providence, R.I., 1984. xi+450 pp.

[**84** E] Endell, Regina, Zur algebraischen Unabhängigkeit gewisser Werte der Exponentialfunktion. *Number theory, Noordwijkerhout 1983* (Noordwijkerhout, 1983), 80–85, Lecture Notes in Math., **1068**, Springer, Berlin-New York, 1984.

[**84** Ne1] Nesterenko, Yuri V., On algebraic independence of algebraic powers of algebraic numbers. (Russian) Mat. Sb. (N.S.) **123** (165) (1984), no. 4, 435–459; English translation in Math. USSR-Sb. **51** (1985), 429–454.

[**84** Ne2] Nesterenko, Yuri V., Measure of algebraic independence for almost all pairs of p-adic numbers. (Russian) Mat. Zametki **36** (1984), no. 3, 295–304; English translation in Math. Notes Acad. Sc. USSR, **36** (1984), no. 3, 642–647.

[**84** P] Philippon, Patrice, Critères pour l'indépendance algébrique de familles de nombres. C. R. Math. Rep. Acad. Sci. Canada **6** (1984), no. 5, 285–290.

[**84** Sal] Salikhov, V. Kh., On the algebraic independence of the values of hypergeometric E-functions. (Russian) Uspekhi Mat. Nauk **39** (1984), no. 2(236), 185–186; English translation in Russian Math. Surveys **39** (1984), no. 2, 191–192.

[**84** W1] Waldschmidt, Michel, Algebraic independence of transcendental numbers. Gel'fond's method and its developments. *Perspectives in mathematics,* 551–571, Birkhäuser, Basel-Boston, Mass., 1984.

[**84** W2] Waldschmidt, Michel, Indépendance algébrique et exponentielles en plusieurs variables. *Number theory, Noordwijkerhout 1983* (Noordwijkerhout, 1983), 268–279, Lecture Notes in Math., **1068**, Springer, Berlin-New York, 1984.

[**84** W3] Waldschmidt, Michel, Algebraic independence of values of exponential and elliptic functions. J. Indian Math. Soc. (N.S.) **48** (1984), no. 1–4, 215–228.

[**84** Wü] Wüstholz, Gisbert, Recent progress in transcendence theory. *Number theory, Noordwijkerhout 1983* (Noordwijkerhout, 1983), 280–296, Lecture Notes in Math., **1068**, Springer, Berlin, 1984.

[**84** Z] Zhu, Yao Chen, On the algebraic independence of certain power series of algebraic numbers. Chinese Ann. Math. Ser. B **5** (1984), no. 1, 109–117. A Chinese summary appears in Chinese Ann. Math. Ser. A **5** (1984), no. 1, 124.

1985

[**85** Ad] Adams, William W., The algebraic independence of certain Liouville continued fractions. Proc. Amer. Math. Soc. **95** (1985), no. 4, 512–516.

[**85** Am] Amou, Masaaki, On algebraic independence of special values of gap series. Tôhoku Math. J. (2) **37** (1985), no. 3, 385–393.

[**85** Br] Brownawell, W. Dale, Effectivity in independence measures for values of E-functions. J. Austral. Math. Soc. Ser. A **38** (1985), no. 2, 227–240.

[**85** Mc] McQuillan, Donald L., Some results on algebraic independence. Arch. Math. (Basel) **45** (1985), no. 4, 336–341.

[**85** Ne1] Nesterenko, Yuri V., On a measure of the algebraic independence of the values of certain functions. (Russian) Mat. Sb. (N.S.) **128** (170) (1985), no. 4, 545–568, 576; English translation in Math. USSR-Sb. **56** (1987), 545–567.

[**85** Ne2] Nesterenko, Yuri V., On the measure of algebraic independence of the values of an elliptic function at algebraic points. (Russian) Uspekhi Mat. Nauk **40** (1985), no. 4(244), 221–222; English translation in Russian Math. Surveys **40** (1985) no. 4, 237–238.

[**85** Ne3] Nesterenko, Yuri V., Estimation of the measure of algebraic independence of numbers. (Russian) Diophantine approximations, Part 1, 65–77, Moskov. Gos. Univ., Mekh.-Mat. Fak., Moscow, 1985.

[**85** P] Philippon, Patrice, Sur les mesures d'indépendance algébrique. *Séminaire de théorie des nombres,* Paris 1983–84, 219–233, Progr. Math., **59**, Birkhäuser Boston, Boston, MA, 1985.

[**85** Shm] Shmelev, A.A., Simultaneous approximation of the exponential by transcendental numbers of a special form. (Russian) Ukr. Mat. Zh. **37** (1985), 232–237; English translation in Ukr. Math. J. **37** (1985), 198–202.

[85 Tr] Trubnikov, S.A., Degree of transcendence of some number fields. (Russian) Vestnik Moskov. Univ. Ser. I Mat. Mekh. 1985, no. 3, 22–25; English translation in Moscow Univ. Math. Bull. **40** (1985), no. 3, 31–36.

[85 W] Waldschmidt, Michel, Petits degrés de transcendance par la méthode de Schneider en une variable. C. R. Math. Rep. Acad. Sci. Canada **7** (1985), no. 2, 143–148.

[85 Z1] Zhu, Yao Chen, Criteria for the algebraic independence of complex numbers. Kexue Tongbao (English Ed.) **30** (1985), no. 9, 1139–1142.

[85 Z2] Zhu, Yao Chen, Algebraic independence property of values of certain gap series. Kexue Tongbao (English Ed.) **30** (1985), no. 3, 293–297.

[85 Z3] Zhu, Yao Chen, A note on algebraic independence of numbers. Acta Math. Sinica (N.S.) **1** (1985), no. 3, 251–254.

1986

[86 Br] Brownawell, W. Dale, The effectivity of certain measures of algebraic independence. *Séminaire de théorie des nombres,* Paris 1984–85, 41–50, Progr. Math., **63**, Birkhäuser Boston, Boston, MA, 1986.

[86 Jb] Jabbouri, El Mostafa, Mesures d'indépendance algébrique de valeurs de fonctions elliptiques et abéliennes. C. R. Acad. Sci. Paris Sér. I Math. **303** (1986), no. 9, 375–378.

[86 MW] Masser, David W. and Wüstholz, Gisbert, Algebraic independence of values of elliptic functions. Math. Ann. **276** (1986), no. 1, 1–17.

[86 Ne] Nesterenko, Yuri V., On a measure of the algebraic independence of the values of the exponential function. (Russian) Dokl. Akad. Nauk SSSR **286** (1986), no. 4, 817–821; English translation in Soviet Math. Dokl., **33** (1986), no. 1, 200–203.

[86 Ni1] Nishioka, Kumiko, Algebraic independence of certain power series of algebraic numbers. J. Number Theory **23** (1986), no. 3, 354–364.

[86 Ni2] Nishioka, Kumiko, Proof of Masser's conjecture on the algebraic independence of values of Liouville series. Proc. Japan Acad. Ser. A Math. Sci. **62** (1986), no. 6, 219–222.

[86 Ni3] Nishioka, Kumiko, Algebraic independence of three Liouville series. Arch. Math. (Basel) **47** (1986), no. 2, 117–120.

[86 O] Oleinikov, V.A., Conditions of mutual transcendence for hypergeometric series. (Russian) Vestnik Moskov. Univ. Ser. I Mat. Mekh. 1986, no. 1, 20–25; English translation in Moscow Univ. Math. Bull. **41** (1986), no. 1, 26–32.

[86 P] Philippon, Patrice, Critères pour l'indépendance algébrique. Inst. Hautes Études Sci. Publ. Math. No. **64** (1986), 5–52.

[86 TY] Toyoda, Masanori and Yasuda, Takeshi, On the algebraic independence of certain numbers connected with the exponential and the elliptic functions. Tokyo J. Math. **9** (1986), no. 1, 29–40.

[86 Tu1] Tubbs, Robert, Indépendance algébrique et groupes algébriques. C. R. Acad. Sci. Paris Sér. I Math. **302** (1986), no. 3, 91–94.

[86 Tu2] Tubbs, Robert, On the measure of algebraic independence of certain values of elliptic functions. J. Number Theory **23** (1986), no. 1, 60–79.

[86 W] Waldschmidt, Michel, Groupes algébriques et grands degrés de transcendance. With an appendix by J. Fresnel. Acta Math. **156** (1986), no. 3-4, 253–302.

1987

[87 Amr] Amoroso, Francesco, On the transcendence type of the classical numbers. Boll. Un. Mat. Ital. B (7) **1** (1987), no. 4, 1077–1089.

[87 B...] Bertrand, D., Emsalem, M., Gramain, F., Huttner, M., Langevin, M., Laurent, M., Mignotte, M., Moreau, J.-C., Philippon, P., Reyssat, É. and Waldschmidt, M., *Les nombres transcendants.* Mém. Soc. Math. France (N.S.) No. **13** (1984), 60 pp.

[87 Br] Brownawell, W. Dale, Large transcendence degree revisited. I. Exponential and non-CM cases. *Diophantine approximation and transcendence theory* (Bonn, 1985), 149–173, Lecture Notes in Math., **1290**, Springer, Berlin-New York, 1987.

[87 BT] Brownawell, W. Dale, Tubbs, Robert, Large transcendence degree revisited. II. The CM case. *Diophantine approximation and transcendence theory* (Bonn, 1985), 175–188, Lecture Notes in Math., **1290**, Springer, Berlin-New York, 1987.

[87 Di] Diaz, Guy, Grands degrés de transcendance pour des familles d'exponentielles. C. R. Acad. Sci. Paris Sér. I Math. **305** (1987), no. 5, 159–162.

[87 K] Kulagin, V.A., On the algebraic independencc of values of certain subclasses of E-functions. (Russian) Uspekhi Mat. Nauk **42** (1987), no. 2(254), 233–234; English translation in Russian Math. Surveys **42** (1987), no. 2, 285–286.

[87 Ne] Nesterenko, Yuri V., Measures of algebraic independence of numbers and functions. *Journées arithmétiques de Besançon* (Besançon, 1985). Astérisque No. **147-148** (1987), 141–149, 344.

[87 Ni] Nishioka, Kumiko, Conditions for algebraic independence of certain power series of algebraic numbers. Compositio Math. **62** (1987), no. 1, 53–61.

[87 P] Philippon, Patrice, Indépendance et groupes algébriques. *Number theory* (Montréal, Qué., 1985), 279–284, CMS Conf. Proc., **7**, Amer. Math. Soc., Providence, RI, 1987.

[87 Shi1] Shidlovskiĭ, Andreĭ B., Algebraic independence of the values of E-functions at algebraic points. (Russian) Vestnik Moskov. Univ. Ser. I Mat. Mekh. 1987, no. 1, 30–33, 101; English translation in Moscow Univ. Math. Bull. **42** (1987), no. 1, 37–41.

[87 Shi2] Shidlovskiĭ, Andreĭ B., *Transcendental numbers* (Russian) "Nauka", Moscow, 1987, 448 pp.

[87 Tu1] Tubbs, Robert, Elliptic curves in two-dimensional abelian varieties and the algebraic independence of certain numbers. Michigan Math. J. **34** (1987), no. 2, 173–182.

[87 Tu2] Tubbs, Robert, Algebraic groups and small transcendence degree. I. J. Number Theory **25** (1987), no. 3, 279–307.

[87 Z] Zhu, Yao Chen, Transcendence and algebraic independence of certain p-adic numbers. I. (Chinese) Acta Math. Sinica **30** (1987), no. 6, 742–752.

1988

[88 Ab] Ably, Mohammed, Mesure d'approximation simultanée. C. R. Acad. Sci. Paris Sér. I Math. **307** (1988), no. 8, 371–374.

[88 AMP] Allouche, J.-P., Mendès France, M. and van der Poorten, A. J., Indépendance algébrique de certaines séries formelles. Bull. Soc. Math. France **116** (1988), no. 4, 449–454.

[88 Amr] Amoroso, Francesco, On the distribution of complex numbers according to their transcendence types. Ann. Mat. Pura Appl. (4) **151** (1988), 359–368.

[88 Bec] Becker-Landeck, Paul-Georg, Maße für algebraische Unabhängigkeit nach einer Methode von Mahler. Acta Arith. **50** (1988), no. 3, 279–293.

[88 Beu] Beukers, Frits, Some new results on algebraic independence of E-functions. *New advances in transcendence theory* (Durham, 1986), 56–67, Cambridge Univ. Press, Cambridge-New York, 1988.

[88 BeBrH] Beukers, Frits, Brownawell, W. Dale and Heckman, Gert, Siegel normality. Ann. of Math. (2) **127** (1988), no. 2, 279–308.

[88 Bu] Bundschuh, Peter, A criterion for algebraic independence with some applications. Osaka J. Math. **25** (1988), no. 4, 849–858.

[**88** Di] Diaz, Guy, Mesure d'indépendance algébrique de α^{β}, α^{β^2}. *Séminaire de Théorie des Nombres,* Paris 1986–87, 129–135, Progr. Math., **75**, Birkhäuser Boston, Boston, MA, 1988.

[**88** Ne] Nesterenko, Yuri V., Effective estimates of algebraic independency measure of values of E-functions. (Russian) Vestnik Moskov. Univ. Ser. I Mat. Mekh. 1988, no. 4, 85–88; English translation in Moscow Univ. Math. Bull. **43** (1988), no. 4, 84–87.

[**88** Ph] Philibert, Georges, Une mesure d'indépendance algébrique. Ann. Inst. Fourier (Grenoble) **38** (1988), no. 3, 85–103.

[**88** Tu] Tubbs, Robert, A Diophantine problem on elliptic curves. Trans. Amer. Math. Soc. **309** (1988), no. 1, 325–338.

[**88** Z] Zhu, Yao Chen, Transcendence and algebraic independence of certain p-adic numbers. II. (Chinese) Acta Math. Sinica **31** (1988), no. 1, 54–66.

1989

[**89** Ab] Ably, Mohammed, Mesure d'approximation simultanée. Ann. Fac. Sci. Toulouse Math. (5) **10** (1989), no. 2, 259–289.

[**89** Amr] Amoroso, Francesco, Some remarks about algebraic independence measures in high dimension. Bull. Soc. Math. France **117** (1989), no. 3, 285–295.

[**89** BN] Becker, Paul-Georg and Nishioka, Kumiko, Measures for the algebraic independence of the values of Mahler type functions. C. R. Math. Rep. Acad. Sci. Canada **11** (1989), no. 3, 89–93.

[**89** BR] Bézivin, Jean-Paul and Robba, Philippe. A new p-adic method for proving irrationality and transcendence results. Ann. of Math. (2) **129** (1989), no. 1, 151–160.

[**89** Di1] Diaz, Guy, Grands degrés de transcendance pour des familles d'exponentielles. J. Number Theory **31** (1989), no. 1, 1–23.

[**89** Di2] Diaz, Guy, Généralisation d'un résultat de W.D. Brownawell-M. Waldschmidt. Bull. Soc. Math. France **117** (1989), no. 3, 267–283.

[**89** Lo] Lossov, N. I., On the algebraic independence of values of some classes of hypergeometric E-functions. (Russian) Uspekhi Mat. Nauk **44** (1989), no. 1(265), 201–202; English translation in Russian Math. Surveys **44** (1989), no. 1, 245–246.

[**89** Ne] Nesterenko, Yuri, Transcendence degree of some fields generated by values of the exponential function. (Russian) Mat. Zametki **46** (1989), no. 3, 40–49, 127; English translation in Math. Notes, **46** (1989), no. 3–4, 706–712 (1990).

[**89** Ni] Nishioka, Kumiko, Evertse theorem in algebraic independence. Arch. Math. (Basel) **53** (1989), no. 2, 159–170.

[**89** Sal] Salikhov, V.Kh., On algebraic independence of the values of hypergeometric E-functions. (Russian) Dokl. Akad. Nauk SSSR **307** (1989), no. 2, 284–287; English translation in Soviet Math. Dokl. **40** (1990), no. 1, 71–74.

[**89** Shi1] Shidlovskiĭ, Andreĭ B., Algebraic independence of values of E-functions at algebraic points. (Russian) Mat. Zametki **45** (1989), no. 1, 115–122, 141; English translation in Math. Notes **45** (1989), no. 1–2, 78–83.

[**89** Shi2] Shidlovskiĭ, Andreĭ B., *Transcendental numbers.* Translated from the Russian (see [87 Shi2]) by Neal Koblitz. de Gruyter Studies in Mathematics, **12**. Walter de Gruyter & Co., Berlin-New York, 1989. xx+466 pp.

[**89** Shi3] Shidlovskiĭ, Andreĭ B., Criterion of algebraic independence of the values of an E-function at algebraic points. (Russian) Vestnik Moskov. Univ. Ser. I Mat. Mekh. 1989, no. 6, 17–21, 97; English translation in Moscow Univ. Math. Bull. **44** (1989), no. 6, 20–25.

[**89** Tu] Tubbs, Robert, The algebraic independence of three numbers: Schneider's method. *Théorie des nombres* (Québec, 1987), 942–967, de Gruyter, Berlin-New York, 1989.

[**89** WZ] Waldschmidt, Michel and Zhu, Yao Chen, Algebraic independence of certain numbers related to Liouville numbers. (Chinese) Scientia Sinica, Ser.A, **9** (1989) 897–904; English translation in Sci. China Ser. A **33** (1990), no. 3, 257–268.

[**89** Z1] Zhu, Yao Chen, Criteria of algebraic independence of complex numbers over a field of finite transcendence type. Chinese Sci. Bull. **34** (1989), no. 3, 185–189.

[**89** Z2] Zhu, Yao Chen, On a criterion of algebraic independence of numbers. Chinese Sci. Bull. **34** (1989), no. 7, 540–542.

1990

[**90** BBR] Beukers, Frits Bézivin, Jean-Paul and Robba, Philippe, An alternative proof of the Lindemann-Weierstrass theorem. Amer. Math. Monthly **97** (1990), no. 3, 193–197.

[**90** Bu] Bundschuh, Peter, Indépendance algébrique par des méthodes d'approximations. *Cinquante ans de polynômes* (Paris, 1988), 116–122, Lecture Notes in Math., **1415**, Springer, Berlin-New York, 1990.

[**90** Chr] Cherepnev, M.A., On the algebraic independence of the values of hypergeometric E-functions. (Russian) Uspekhi Mat. Nauk **45** (1990), no. 6 (276), 155–156; English translation in Russian Math. Surveys **45** (1990), no. 6, 145–146.

[**90** Di] Diaz, Guy, Une nouvelle mesure d'indépendance algébrique pour $(\alpha^\beta, \alpha^{\beta^2})$. Acta Arith. **56** (1990), no. 1, 25–32.

[**90** Ni] Nishioka, Kumiko, Algebraic independence of certain power series. *Séminaire de Théorie des Nombres,* Paris 1987–88, 201–212, Progr. Math., **81**, Birkhäuser Boston, Boston, MA, 1990.

[**90** NN] Nishioka, Keiji and Nishioka, Kumiko, Algebraic independence of elementary functions and its application to Masser's vanishing theorem. Aequationes Math. **40** (1990), no. 1, 67–77.

[**90** Sal1] Salikhov, V.Kh., A criterion for the algebraic independence of values of a class of hypergeometric E-functions. (Russian) Mat. Sb. **181** (1990), no. 2, 189–211; English translation in Math. USSR-Sb. **69** (1991), no. 1, 203–226.

[**90** Sal2] Salikhov, V.Kh., Irreducibility of hypergeometric equations, and algebraic independence of values of E-functions. (Russian) Acta Arith. **53** (1990), no. 5, 453–471.

[**90** She] Shestakova, N.V., Estimates for reciprocal transcedence of certain numbers. (Russian) Vestnik Moskov. Univ. Ser. I Mat. Mekh. 1990, no. 6, 14–18; English translation in Moscow Univ. Math. Bull. **45** (1990), no. 6, 15–18.

[**90** Si] Simpson, Carlos T., Transcendental aspects of the Riemann-Hilbert correspondence. Illinois J. Math. **34** (1990), no. 2, 368–391.

[**90** Tu1] Tubbs, Robert, A note on some elementary measures of algebraic independence. Proc. Amer. Math. Soc. **109** (1990), no. 2, 297–304.

[**90** Tu2] Tubbs, Robert, Algebraic groups and small transcendence degree. II. J. Number Theory **35** (1990), no. 2, 109–127.

[**90** W1] Waldschmidt, Michel, Indépendance algébrique de nombres de Liouville. *Cinquante ans de polynômes (Fifty years of polynomials)* (Paris, 1988), 225–235, Lecture Notes in Math., **1415**, Springer, Berlin-New York, 1990.

[**90** W2] Waldschmidt, Michel, Transcendence problems connected with Drinfel'd modules. *International Symposium on Algebra and Number Theory* (Silivri, 1990). Istanbul Üniv. Fen Fak. Mat. Derg. **49** (1990), 57–75 (1993).

[**90** Wa] Wass, N.Ch., Algebraic independence of the values at algebraic points of a class of functions considered by Mahler. Dissertationes Math. (Rozprawy Mat.) **303** (1990), 61 pp.

[**90** Z] Zhu, Yao Chen, Criteria of algebraic independence of complex numbers over a field of finite transcendence type. II. Acta Math. Sinica (N.S.) **6** (1990), no. 1, 24–34.

1991

[**91** Ab] Ably, Mohammed, Indépendance algébrique de certains nombres admettant une "bonne" approximation diophantienne. C. R. Acad. Sci. Paris Sér. I Math. **313** (1991), no. 10, 653–655.

[**91** Am] Amou, Masaaki, Algebraic independence of the values of certain functions at a transcendental number. Acta Arith. **59** (1991), no. 1, 71–82.

[**91** Bec] Becker, Paul-Georg, Effective measures for algebraic independence of the values of Mahler type functions. Acta Arith. **58** (1991), no. 3, 239–250.

[**91** Chr] Cherepnev, M.A., On algebraic independence of values of hypergeometric E-functions satisfying linear nonhomogeneous differential equations. (Russian) Vestnik Moskov. Univ. Ser. I Mat. Mekh. 1991, no. 1, 27–33, 104; English translation in Moscow Univ. Math. Bull. **46** (1991), no. 1, 17–21.

[**91** K] Kulagin, V.A., On the algebraic independence of the values of some subclasses of hypergeometric E-functions. (Russian) Uspekhi Mat. Nauk **46** (1991), no. 5(281), 167–168; English translation in Russian Math. Surveys **46** (1991), no. 5, 189–190.

[**91** Ne] Nesterenko, Yuri V., Algebraic independence of values of analytic functions. *Proceedings of the International Congress of Mathematicians,* Vol. I, II (Kyoto, 1990), 447–457, Math. Soc. Japan, Tokyo, 1991.

[**91** Ni] Nishioka, Kumiko, Algebraic independence measures of the values of Mahler functions. J. reine angew. Math. **420** (1991), 203–214.

[**91** She] Shestakov, S.O., On transcendence degree of some fields generated by values of elliptic functions. (Russian) Vestnik Moskov. Univ. Ser. I Mat. Mekh. 1991, no. 3, 30–36; English translation in Moscow Univ. Math. Bull. **46** (1991), no. 3, 29–33.

[**91** Shi1] Shidlovskiĭ, Andreĭ B., On the algebraic independence of values of E-functions. (Russian) Uspekhi Mat. Nauk **46** (1991), no. 4(280), 175–176; English translation in Russian Math. Surveys **46** (1991), no. 4, 201–202.

[**91** Shi2] Shidlovskiĭ, Andreĭ B., On the algebraic independence of the values of E-functions connected by algebraic equations over the field of rational functions. (Russian) *Number theory and analysis* (Moscow, 1991). Trudy Mat. Inst. Steklov. **207** (1994), no. 6, 369–376; English translation in Proceedings of the Steklov Institute of Mathematics, **207** (1995), no. 6, 339–345.

[**91** Tu] Tubbs, Robert, Large transcendence degree. J. Austral. Math. Soc. Ser. A **51** (1991), no. 3, 400–422.

[**91** Z1] Zhu, Yao Chen, Algebraic independence of certain numbers over a field of finite transcendence type. Chinese Ann. Math. Ser. B **12** (1991), no. 4, 400–406. A Chinese summary appears in Chinese Ann. Math. Ser. A **12** (1991), no. 5, 644.

[**91** Z2] Zhu, Yao Chen, The algebraic independence of certain transcendental continued fractions. Acta Math. Sinica (N.S.) **7** (1991), no. 2, 127–134.

[**91** Zh] Zhukov, A.I., Effective estimates of measures of algebraic independence of the values of E-functions connected by algebraic equations over $\mathbf{C}(z)$. (Russian) Uspekhi Mat. Nauk **46** (1991), no. 2(278), 215–216; English translation in Russian Math. Surveys **46** (1991), no. 2, 262–263.

1992

[**92** Ab1] Ably, Mohammed, Sur l'analogue abélien du théorème de Lindemann-Weierstrass. C. R. Acad. Sci. Paris Sér. I Math. **315** (1992), no. 13, 1337–1339.

[**92** Ab2] Ably, Mohammed, Résultats quantitatifs d'indépendance algébrique pour les groupes algébriques. J. Number Theory **42** (1992), no. 2, 194–231.

[**92** Bec] Becker, Paul-Georg, Algebraic independence of the values of certain series by Mahler's method. Monatsh. Math. **114** (1992), no. 3–4, 183–198.

[**92** Del] Delaunay, Sandra, Grands degrés de transcendance, les hypothèses techniques dans le cas exponentiel. I. C. R. Acad. Sci. Paris Sér. I Math. **315** (1992), no. 13, 1333–1336.

[**92** Di] Diaz, Guy, Degré de transcendance 2 pour les familles d'exponentielles par la méthode de Baker. Séminaire d'arithmétique de St Etienne 1990–91–92, 1–23.

[**92** Go] Gorelov, V.A., Estimates for measures of algebraic independence of values of E-functions. (Russian) Izv. Vyssh. Uchebn. Zaved. Mat. 1992, no. 10, 6–11 (1993); English translation in Russian Math. (Iz. VUZ) **36** (1992), no. 10, 4–9.

[**92** Jb] Jabbouri, El Mostafa, Sur un critère pour l'indépendance algébrique de P. Philippon. *Approximations diophantiennes et nombres transcendants* (Luminy, 1990), 195–202, de Gruyter, Berlin, 1992.

[**92** K] Kulagin, V.A., Algebraic independence of values of a subclass of entire functions. (Russian) Mat. Zametki **51** (1992), no. 6, 46–51, 157; English translation in Math. Notes **51** (1992), no. 5–6, 561–565.

[**92** Ne] Nesterenko, Yuri V., On a measure of algebraic independence of the values of elliptic functions. *Approximations diophantiennes et nombres transcendants* (Luminy, 1990), 239–248, de Gruyter, Berlin, 1992.

[**92** P] Philippon, Patrice, Critères pour l'indépendance algébrique dans les anneaux diophantiens. C. R. Acad. Sci. Paris Sér. I Math. **315** (1992), no. 5, 511–515.

[**92** R] Roy, Damien, Matrices whose coefficients are linear forms in logarithms. J. Number Theory **41** (1992), no. 1, 22–47.

[**92** She] Shestakov, S.O., On the measure of algebraic independence of certain numbers. (Russian) Vestnik Moskov. Univ. Ser. I Mat. Mekh. 1992, no. 2, 8–12, 111; English translation in Moscow Univ. Math. Bull. **47** (1992), no. 2, 8–11.

[**92** Th] Thiery, Alain, Indépendance algébrique des périodes et quasi-périodes d'un module de Drinfel'd. *The arithmetic of function fields* (Columbus, OH, 1991), 265–284, Ohio State Univ. Math. Res. Inst. Publ., 2, de Gruyter, Berlin, 1992.

[**92** Y] Yu, Jing, Transcendence in finite characteristic. *The arithmetic of function fields* (Columbus, OH, 1991), 253–264, Ohio State Univ. Math. Res. Inst. Publ., 2, de Gruyter, Berlin, 1992.

1993

[**93** Br] Brownawell, W. Dale, Algebraic independence of Drinfel'd exponential and quasi-periodic functions. *Advances in number theory* (Kingston, ON, 1991), 341–365, Oxford Sci. Publ., Oxford Univ. Press, New York, 1993.

[**93** Ca] Caveny, Deanna M., U-numbers and T-numbers: some elementary transcendence and algebraic independence results. *Number theory with an emphasis on the Markoff spectrum* (Provo, UT, 1991), 43–52, Lecture Notes in Pure and Appl. Math., **147**, Dekker, New York, 1993.

[**93** CT] Caveny, Deanna and Tubbs, Robert, The arithmetic of well-approximated numbers. *Number theory with an emphasis on the Markoff spectrum* (Provo, UT, 1991), 53–59, Lecture Notes in Pure and Appl. Math., **147**, Dekker, New York, 1993.

[**93** Chn] Chen, Gong Liang, *Méthode de Baker pour les grands degrés de transcendance et relations de dépendance linéaire pour des logarithmes de nombres algébriques.* Thèse, Univ. Jean Monnet (St Etienne), 3 décembre 1993, 79 pp.

[**93** Del] Delaunay, Sandra, Grands degrés de transcendance, les hypothèses techniques dans le cas exponentiel. II. C. R. Acad. Sci. Paris Sér. I Math. **316** (1993), no. 1, 7–10.

[**93** Den1] Denis, Laurent, Indépendance algébrique sur le module de Carlitz. C. R. Acad. Sci. Paris Sér. I Math. **317** (1993), no. 10, 913–915.

[**93** Den2] Denis, Laurent, Transcendance et dérivées de l'exponentielle de Carlitz. *Séminaire de Théorie des Nombres,* Paris, 1991-92, 1–21, Progr. Math., **116**, Birkhäuser Boston, Boston, MA, 1993.

[**93** Mü] Müller, Ralf, Algebraische Unabhängigkeit der Werte gewisser Lückenreihen in nicht-archimedisch bewerteten Körpern. Results Math. **24** (1993), no. 3–4, 288–297.

1994

[**94** Ab] Ably, Mohammed, Une version quantitative du théorème de Lindemann-Weierstrass. Acta Arith. **67** (1994), no. 1, 29–45.

[**94** BBT] Becker, Paul-Georg, Brownawell, W. Dale and Tubbs, Robert, Gel'fond's theorem for Drinfel'd modules. Michigan Math. J. **41** (1994), no. 2, 219–233.

[**94** Ca] Caveny, Deanna M., Quantitative transcendence results for numbers associated with Liouville numbers. Proc. Amer. Math. Soc. **120** (1994), no. 2, 349–357.

[**94** Chr] Cherepnev, M.A., On algebraic independence of some subclasses of hypergeometric functions. (Russian) Mat. Zametki **55** (1994), no. 1, 117–129, 158; English translation in Math. Notes **55** (1994), no. 1–2, 78–85.

[**94** Ni] Nishioka, Kumiko, Algebraic independence by Mahler's method and S-unit equations. Compositio Math. **92** (1994), no. 1, 87–110.

[**94** Tö1] Töpfer, Thomas, An axiomatization of Nesterenko's method and applications on Mahler functions. J. Number Theory **49** (1994), no. 1, 1–26.

[**94** Tö2] Töpfer, Thomas, On the transcendence and algebraic independence of certain continued fractions. Monatsh. Math. **117** (1994), no. 3–4, 255–262.

1995

[**95** Chn] Chen, Gong Liang, Application de la méthode de Baker à de grands degrés de transcendance. C. R. Acad. Sci. Paris Sér. I Math. **320** (1995), no. 5, 523–528.

[**95** Chr] Cherepnev, M.A., Algebraic independence of values of hypergeometric E-functions. (Russian) Mat. Zametki **57** (1995), no. 6, 896–912, 960; English translation in Math. Notes **57** (1995), no. 6, 630–642.

[**95** Den1] Denis, Laurent, Indépendance algébrique et exponentielle de Carlitz. Acta Arith. **69** (1995), no. 1, 75–89.

[**95** Den2] Denis, Laurent, Dérivées d'un module de Drinfel'd et transcendance. Duke Math. J. **80** (1995), no. 1, 1–13.

[**95** Ne] Nesterenko, Yuri V., On a measure of algebraic independence of values of an elliptic function. (Russian) Izv. Ross. Akad. Nauk Ser. Mat. **59** (1995), no. 4, 155–178; English translation in Izv. Math. **59** (1995), no. 4, 815–838.

[**95** P] Philippon, Patrice, Nouveaux aspects de la transcendance. *Les Dix-huitièmes Journées Arithmétiques* (Bordeaux, 1993). J. Théor. Nombres Bordeaux **7** (1995), no. 1, 191–218.

[**95** R] Roy, Damien, Points whose coordinates are logarithms of algebraic numbers on algebraic varieties. Acta Math. **175** (1995), no. 1, 49–73.

[**95** RW] Roy, Damien and Waldschmidt, Michel, Quadratic relations between logarithms of algebraic numbers. Proc. Japan Acad. Ser. A Math. Sci. **71** (1995), no. 7, 151–153.

[**95** Tö1] Töpfer, Thomas, Algebraic independence of the values of generalized Mahler functions. Acta Arith. **70** (1995), no. 2, 161–181.

[**95** Tö2] Töpfer, Thomas, An axiomatization of Nesterenko's method and applications on Mahler functions II. Compositio Math. **95** (1995), no. 3, 323–342.

1996

[**96** An] André, Yves, G-fonctions et transcendance. J. reine angew. Math. **476** (1996), 95–125.

[**96** Br] Brownawell, W. Dale, Submodules of products of quasi-periodic modules. *Symposium on Diophantine Problems* (Boulder, CO, 1994). Rocky Mountain J. Math. **26** (1996), no. 3, 847–873.

[**96** Ca] Caveny, Deanna M., Commutative algebraic groups and refinements of the Gel'fond-Fel'dman measure. *Symposium on Diophantine Problems* (Boulder, CO, 1994). Rocky Mountain J. Math. **26** (1996), no. 3, 889–935.

[**96** Del1] Delaunay, Sandra, Grands degrés de transcendance pour la fonction exponentielle; modification des hypothèses techniques dans la méthode de Diaz. Ann. Fac. Sci. Toulouse Math. (6) **5** (1996), no. 1, 57–68.

[**96** Del2] Delaunay, Sandra, Grands degrés de transcendance pour la fonction exponentielle; modification des hypothèses techniques dans la méthode de Brownawell. Ann. Fac. Sci. Toulouse Math. (6) **5** (1996), no. 1, 69–104.

[**96** Den] Denis, Laurent, Un critère de transcendance en caractéristique finie. J. Algebra **182** (1996), no. 2, 522–533.

[**96** Jd] Jadot, Christian, Critères pour l'indépendance algébrique et linéaire. Thèse, Univ. P. et M. Curie (Paris VI), 1996.

[**96** K] Kulagin, V.A., Algebraic independence of values of a certain class of hypergeometric functions. (Russian) Mat. Zametki **59** (1996), no. 3, 402–414, 479; English translation in Math. Notes **59** (1996), no. 3, 283–292.

[**96** Ne1] Nesterenko, Yuri V., Modular functions and transcendence problems — Un théorème de transcendance sur les fonctions modulaires. C. R. Acad. Sci. Paris Sér. I Math. **322** (1996), no. 10, 909–914.

[**96** Ne2] Nesterenko, Yuri V., Modular functions and transcendence questions. (Russian) Mat. Sb. **187** (1996), no. 9, 65–96; English translation in Math. USSR-Sb. **187** (1996), no. 9, 1319–1348.

[**96** Ne3] Nesterenko, Yuri V., On the Lindemann-Weierstrass theorem. (Russian) Vestnik Moskov. Univ. Ser. I Mat. Mekh. 1996, no. 3, 33–37, 92; English translation in Moscow Univ. Math. Bull., **51** (1996), no. 3, 27–30.

[**96** NS] Nesterenko, Yuri V. and Shidlovskiĭ, Andreĭ B., On the linear independence of values of E-functions. (Russian) Mat. Sb. **187** (1996), no. 8, 93–108; English translation in Math. USSR-Sb., **187** (1996), no. 8, 1197–1211.

[**96** Ni1] Nishioka, Kumiko, *Mahler functions and transcendence.* Lecture Notes in Mathematics, **1631**. Springer-Verlag, Berlin, 1996. viii+185 pp.

[**96** Ni2] Nishioka, Kumiko, Algebraic independence of Mahler functions and their values. Tôhoku Math. J. (2) **48** (1996), no. 1, 51–70.

[**96** Ni3] Nishioka, Kumiko, A new proof of Masser's vanishing theorem. Proc. Amer. Math. Soc. **124** (1996), no. 11, 3271–3274.

[**96** Ta] Tanaka, Taka-Aki, Algebraic independence of the values of power series generated by linear recurrences. Acta Arith. **74** (1996), no. 2, 177–190.

[**96** Tu] Tubbs, Robert, Algebraic independence and products of Drinfeld modules. *Symposium on Diophantine Problems* (Boulder, CO, 1994). Rocky Mountain J. Math. **26** (1996), no. 3, 1165–1181.

[**96** V] Vasilev, K.G., Algebraic independence of the periods of abelian integrals. (Russian) Mat. Zametki **60** (1996), no. 5, 681–691; English translation in Math. Notes **60** (1996), no. 5, 510–518.

[**96** W] Waldschmidt, Michel, Dependence of logarithms on commutative algebraic groups. *Symposium on Diophantine Problems* (Boulder, CO, 1994). Rocky Mountain J. Math. **26** (1996), no. 3, 1199–1223.

[**96** Zu] Zudilin, V.V., Lower bounds for polynomials in the values of some entire functions. (Russian) Mat. Sb. (N.S.) **187** (165) (1996), no. 12, 57–86; English translation in Math. USSR-Sb., **187** (1996), no. 12, 1791–1818.

1997

[**97** Ber] Bertrand, Daniel, Theta functions and transcendence. The Ramanujan Journal, **1** (1997), no. 4, 339–350.

[**97** Den] Denis, Laurent, Indépendance algébrique en caractéristique deux. J. Number Theory **66** (1997), no. 1, 183–200.

[**97** Di1] Diaz, Guy, La conjecture des quatre exponentielles et les conjectures de D. Bertrand sur la fonction modulaire. J. Théor. Nombres Bordeaux **9** (1997), 229–245.

[**97** Di2] Diaz, Guy, Une nouvelle propriété d'approximation diophantienne. C. R. Acad. Sci. Paris Sér. I Math. **324** (1997), no. 9, 969–972.

[**97** Ga] Galochkin, A.I., Estimates of the number of zeros of some functions with algebraic Taylor coefficients. (Russian) Mat. Zametki **61** (1997), no. 6, 817–824; English translation in Math. Notes Acad. Sc. USSR, **61** (1997), no. 6, 687–692.

[**97** Ne] Nesterenko, Yuri V., On the measure of algebraic independence of the values of Ramanujan functions. Trudy Mat. Inst. Steklov. **218** (1997), 299–334; English translation in Proceedings of the Steklov Institute of Mathematics, **218** (1997), 294–331.

[**97** Ni] Nishioka, Kumiko, Algebraic independence of reciprocal sums of binary recurrences. Monatsh. Math. **123** (1997), no. 2, 135–148.

[**97** P] Philippon, Patrice, Une approche méthodique pour la transcendance et l'indépendance algébrique de valeurs de fonctions analytiques. J. Number Theory **64** (1997), no. 2, 291–338.

[**97** Pr] Präve, Alexandra, Über den Transzendenzgrad von Periodenquotienten. Dissertation Dokt. Natur., J.W. Goethe-Universität, Frankfurt am Main, Dec. 1997, 52 pp.

[**97** RW1] Roy, Damien and Waldschmidt, Michel, Approximation diophantienne et indépendance algébrique de logarithmes. Ann. scient. Ec. Norm. Sup. **30** (1997), no. 6, 753–796.

[**97** RW2] Roy, Damien and Waldschmidt, Michel, Simultaneous approximation and algebraic independence. The Ramanujan Journal, **1** (1997), no. 4, 379–430.

[**97** Se] Sert, Alain, Théorème de Lindemann-Weierstraß et déterminants d'interpolation. Thèse Doctorat, Univ. Méditerranée Aix-Marseille II, Oct. 1997, 78 pp.

[**97** To] Toshimitsu, Takeshi, q-additive functions and algebraic independence. Arch. Math. (Basel) **69** (1997), no. 2, 112–119.

[**97** W] Waldschmidt, Michel, Sur la nature arithmétique des valeurs de fonctions modulaires. *Séminaire Bourbaki,* 49è année, 1996–97, No. 824; Soc. Math. France, Astérisque **245** (1997), 105–140.

[**97** Ya] Yanchenko, A.Ya., On the measure of algebraic independence of values of derivatives of a p-adic modular function. (Russian) Mat. Zametki **61** (1997), no. 3, 431–440; English translation in Math. Notes Acad. Sc. USSR, **61** (1997), no. 3, 352–358.

[**97** Y] Yu, Jing, Analytic homomorphisms into Drinfeld modules. Ann. of Math. (2) **145** (1997), no. 2, 215–233.

[**97** Zu] Zudilin, V.V., On the measure of linear and algebraic independence for values of entire hypergeometric functions. (Russian) Mat. Zametki **61** (1997), no. 2, 302–304; English translation in Math. Notes Acad. Sc. USSR, **61** (1997), no. 2, 246–248.

1998

[**98** An1] André, Yves, Quelques conjectures de transcendance issues de la géométrie algébrique. Manuscript. Institut de Mathématiques de Jussieu, Prépublication **121** (Mai 1997). Révisé en Janvier 1998.

[**98** An2] André, Yves, Séries Gevrey de type arithmétique. II: transcendance sans transcendance. Institut de Mathématiques de Jussieu, Prépublication **136** (Octobre 1997).

[**98** Br] Brownawell, W. Dale, Transcendence in positive characteristic. *Number Theory* (Tiruchirapalli, India, 1996), 317–332, Contemporary Math. **210**, Amer. Math. Soc., 1998.

[**98** FN] Fel'dman, N.I. and Nesterenko, Yuri V., *Number theory. IV. Transcendental Numbers.* Encyclopaedia of Mathematical Sciences, **44**. Springer-Verlag, Berlin, 1998. iii+345 pp. ISBN: 3-540-61467-2.

[**98** Gr] Gramain, François, Transcendance et fonctions modulaires. *Journées Arithmétiques 1997 (Limoges)*, J. Tn. Nombres Bordeaux, à paraître.

[**98** La] Laurent, Michel, New methods in algebraic independence. *Number Theory – Diophantine, Computational and Algebraic Aspects,* K. Györy, A. Pethö and V.T. Sòs, eds., Proc. Conf. Number Theory Eger 1996, W. de Gruyter, Berlin (1998), 311–330.

[**98** LR1] Laurent, Michel and Roy, Damien, Criteria of algebraic independence with multiplicities and interpolation determinants. Trans. Amer. Math. Soc. **351** (1999), no. 5, 1845–1870.

[**98** LR2] Laurent, Michel and Roy, Damien, Sur l'approximation algébrique en degré de transcendance un. Annales Instit. Fourier **49** (1999), 27–55.

[**98** Ne] Nesterenko, Yuri V., Algebraic independence of π and e^{π}. In *Number theory and its applications,* Cem Y. Yıldırım and Serguei Stepanov Ed., Lecture Notes in Pure and Applied Mathematics **204**, 121–149, Marcel Dekker Inc., 1999.

[**98** NP] Nesterenko, Yuri V. and Philippon, Patrice, *Introduction to algebraic independence theory.* Proc. Conf. CIRM (Luminy 1997). To appear.

[**98** P1] Philippon, Patrice, Indépendance algébrique et K-fonctions. J. reine angew. Math. **497** (1998), 1–15.

[**98** P2] Philippon, Patrice, Approximations algébriques des points dans les espaces projectifs. J. Number Theory, to appear.

[**98** P3] Philippon, Patrice, Mesures d'approximation de valeurs de fonctions analytiques. Acta Arith. **88** (1999), no. 2, 113–127.

[**98** Sar] Saradha, Natrajan, Transcendance measure of η/ω. Acta Arith., to appear.

[**98** W1] Waldschmidt, Michel, Transcendance et indépendance algébrique de valeurs de fonctions modulaires. Proceedings of the Fifth Conference of the Canadian Number Theory Association, éd.: Rajiv Gupta et Kenneth Williams, CRM Proceedings and Lecture Notes, Amer. Math. Soc. **19** (1999), 353–375.

[**98** W2] Waldschmidt, Michel, From simultaneous approximations to algebraic independence. In *Number Theory and its Applications*, Cem Y. Yıldırım and Serguei Stepanov Ed., Lecture Notes in Pure and Applied Mathematics **204**, 283–305, Marcel Dekker Inc., 1999.

[**98** W3] Waldschmidt, Michel, Un demi-siècle de transcendance. *Développement des mathématiques au cours de la seconde moitié du XXe siècle.* Ouvrage collectif, Birkhäuser Verlag, to appear.

Universite P. et M. Curie (Paris VI)
Institut Mathematique de Jussieu
Problemes Diophanties, Case 247
4 Place Jussieu
F-75252 Paris Cedex 05
E-Mail: miw@math.jussieu.fr

MIX
Papier aus verantwortungsvollen Quellen
Paper from responsible sources
FSC® C105338

If you have any concerns about our products,
you can contact us on
ProductSafety@springernature.com

In case Publisher is established outside the EU,
the EU authorized representative is:
Springer Nature Customer Service Center GmbH
Europaplatz 3, 69115 Heidelberg, Germany

Printed by Libri Plureos GmbH
in Hamburg, Germany